Mathematical Logic and Computation

This new book on mathematical logic by Jeremy Avigad gives a thorough introduction to the fundamental results and methods of the subject from the syntactic point of view, emphasizing logic as the study of formal languages and systems and their proper use. Topics include proof theory, model theory, the theory of computability, and axiomatic foundations, with special emphasis given to aspects of mathematical logic that are fundamental to computer science, including deductive systems, constructive logic, the simply typed lambda calculus, and type-theoretic foundations.

Clear and engaging, with plentiful examples and exercises, it is an excellent introduction to the subject for graduate students and advanced undergraduates who are interested in logic in mathematics, computer science, and philosophy, and an invaluable reference for any practicing logician's bookshelf.

JEREMY AVIGAD is Professor in the Department of Philosophy and the Department of Mathematical Sciences at Carnegie Mellon University. His research interests include mathematical logic, formal verification, automated reasoning, and the philosophy and history of mathematics. He is Director of the Charles C. Hoskinson Center for Formal Mathematics at Carnegie Mellon University.

Mathematical Logic and Computation

Jeremy Avigad
Carnegie Mellon University

Shaftesbury Road, Cambridge CB2 8EA, United Kingdom

One Liberty Plaza, 20th Floor, New York, NY 10006, USA

477 Williamstown Road, Port Melbourne, VIC 3207, Australia

314–321, 3rd Floor, Plot 3, Splendor Forum, Jasola District Centre, New Delhi – 110025, India

103 Penang Road, #05–06/07, Visioncrest Commercial, Singapore 238467

Cambridge University Press is part of Cambridge University Press & Assessment, a department of the University of Cambridge.

We share the University's mission to contribute to society through the pursuit of education, learning and research at the highest international levels of excellence.

www.cambridge.org
Information on this title: www.cambridge.org/9781108478755

DOI: 10.1017/9781108778756

First published 2023

A catalogue record for this publication is available from the British Library.

Library of Congress Cataloging-in-Publication Data
Names: Avigad, Jeremy, author.
Title: Mathematical logic and computation / Jeremy Avigad, Carnegie Mellon University, Pennsylvania.
Description: First edition. | Cambridge, United Kingdom ; Boca Raton : Cambridge University Press, 2023. | Includes bibliographical references and index.
Identifiers: LCCN 2022006053 | ISBN 9781108478755 (hbk) | ISBN 9781108778756 (ebook)
Subjects: LCSH: Logic, Symbolic and mathematical.
Classification: LCC QA9 .A92 2023 | DDC 511.3–dc23 / eng20220726
LC record available at https://lccn.loc.gov/2022006053

ISBN 978-1-108-47875-5 Hardback

Contents

Preface

In the phrase *mathematical logic*, the word "mathematical" is ambiguous. It can be taken to specify the methods used, so that the phrase refers to the mathematical study of the principles of reasoning. It can be taken to demarcate the type of reasoning considered, so that the phrase refers to the study of specifically mathematical reasoning. Or it can be taken to indicate the purpose, so that the phrase refers to the study of logic with an eye toward mathematical applications.

In the title of this book, the word "mathematical" is intended in the first two senses but not in the third. In other words, mathematical logic is viewed here as a mathematical study of the methods of mathematical reasoning. The subject is interesting in its own right, and it has mathematical applications. But it also has applications in computer science, for example, to the verification of hardware and software and to the mechanization of mathematical reasoning. It can inform the philosophy of mathematics as well, by providing idealized models of what it means to do mathematics.

One thing that distinguishes logic as a discipline is its focus on language. The subject starts with formal expressions that are supposed to model the informal language we use to define objects, state claims, and prove them. At that point, two distinct perspectives emerge. From a *semantic* perspective, the formal expressions are used to say things about abstract mathematical objects and structures. They can be used to characterize classes of structures like groups, rings, and fields; to characterize particular structures, like the Euclidean plane or the real numbers; or to describe relationships within a particular structure. From that point of view, mathematical logic is the science of reference, definability, and truth, clarifying the semantic notions that determine the relationship between mathematical language and the mathematical structures it describes.

This book adopts a more *syntactic* perspective, in which the primary objects of interest are the expressions themselves. From that point of view, formal languages are used to reason and calculate, and what we care about are the rules that govern their proper use. We will use formal systems to understand patterns of mathematical inference and the structure of mathematical definitions and proofs, and we will be interested in the things that we can do with these syntactic representations. We won't shy away from the use of semantic methods, but our goal is to use semantics to illuminate syntax rather than the other way around.

There are a number of reasons why a syntactic approach is valuable. The mathematical theory of syntax is independently interesting and informative. A focus on syntactic objects is also more closely aligned with computer science, since these objects can be represented as data and acted on by algorithms. Finally, there are philosophical benefits. Because a general theory of finite strings of symbols is all that is needed to work with expressions, a syntactic perspective provides a means of studying mathematical reasoning – including the use of infinitary objects and methods – without importing strong mathematical presuppositions at the outset.

Another notable feature of this book is its focus on computation. On the one hand, we expect mathematics to give us a broad conceptual understanding. At the empirical borders of the subject, this serves to organize and explain our scientific observations, but our desire for understanding is not limited to empirical phenomena. On the other hand, we also expect mathematics to tell us how to calculate trajectories and probabilities so that we can make better predictions and decisions, and act rationally toward securing our pragmatic goals. There is a tension between conceptual understanding and calculation: computation is important, but we often see further, and reason more efficiently, by suppressing computational detail.

The tension is partially captured by the logician's distinction between *classical* logic, on the one hand, and *intuitionistic* or *constructive* logic, on the other. Classical logic is meant to support a style of reasoning that supports abstraction and idealization, whereas intuitionistic logic is more directly suited to computational interpretation. Mathematics today is resolutely classical, but mathematics without calculation is almost a contradiction in terms. And while the end goal of computer science is practical computation, abstraction is essential to designing and reasoning about computational systems. Here we will study both classical and constructive logic, with an eye toward understanding the relationships between the two.

The connections between logic and computation run deep: we can compute with formal expressions, we can reason formally about computation, and we can extract computational content from formal derivations. The style of presentation I have adopted here runs the risk of being judged too mathematical by computer scientists and too computationally minded for pure mathematicians, but I have aimed to strike a balance and, hopefully, bring the two communities closer together.

Audience The material here should be accessible at an advanced undergraduate or introductory graduate level. I have tried to keep the presentation self-contained, but the emphasis is on proving things about logical systems rather than illustrating and motivating their use. Readers who have had a prior introduction to logic will be able to navigate the material here more quickly and comfortably.

Notation Most of the notation in this book is conventional, but challenges inevitably arise when juxtaposing material from fields where conventions differ. The most striking instance of such a challenge is the use of *binders* like quantifiers and lambda abstraction: mathematical logicians generally take such operations to bind tightly, while computer scientists typically give them the widest scope possible. Another mismatch arises with respect to function application, since theoretical computer scientists often write $f\,x$ instead of $f(x)$. So, where a mathematical logician would write

$$\exists x\, A(x) \to \exists x\, (A(x) \wedge \forall y\, (R(y, x) \to \neg A(y))),$$

a computer scientist might write

$$(\exists x.\, A\, x) \to \exists x.\, A\, x \wedge \forall y.\, R\, y\, x \to \neg A\, y.$$

As a compromise, this book uses the mathematician's notation for symbolic logic but adopts the computer scientist's conventions for computational systems like the simply typed lambda calculus. The differences are most pronounced in Chapters 14 and 17, where logic and type theory come together.

Teaching The material in this book can be used to teach a number of different courses, and the dependencies between topics are mild. Chapters 1–7 provide a thorough introduction to the syntax and semantics of first-order logic. Classically minded mathematicians can easily tune out anything to do with intuitionistic logic, while computer scientists and other interested parties can spend more time with it. Ignoring intuitionistic logic might leave time for the cut elimination theorem and its applications, while skipping cut elimination might leave time to dabble in algebraic semantics.

For students already familiar with propositional and first-order logic, Chapters 8–14 provide a fairly self-contained presentation of formal arithmetic, computability, and the incompleteness theorems. A course on computability and incompleteness can start with Chapter 8, skip Section 8.5, continue to Chapter 11, refer back to Sections 10.2 and 10.5 for background on arithmetic definability, skip Sections 11.8 and 11.9, and then move on to Chapter 12.

A course on proof theory can focus on the cut elimination theorem and its applications (Chapters 6 and 7) and the simply typed lambda calculus (Chapter 13). Chapter 14 can serve as the focus of a course on computational interpretations of arithmetic, building on material in Chapters 8–11 and 13; for that purpose, Sections 8.5, 9.5, 10.6–10.7, and 11.6–11.9 can be omitted. Similarly, Chapter 16 can serve as the focus for an introduction to subsystems of second-order arithmetic and reverse mathematics, supported by a quick tour of Chapters 8–11 and Sections 15.4 and 15.5. Sections 11.8 and 11.9, in particular, were written with Chapter 16 in mind.

Most of the material after Chapter 7 is not strongly tied to any particular deductive system. Natural deduction is sometimes used by default, but that choice is not essential to the exposition.

Acknowledgments

I have had the privilege of learning logic with colleagues, mentors, and countless students over the years. Among those to whom I am indebted for comments on drafts of this book are Seulkee Baek, Krishan Canzius, Mario Carneiro, Lucas Clark, Ulrich Kohlenbach, Gustavo Lacerda, Fernando Langlois, Steffen Lempp, Oualid Merzouga, Paulo Oliva, Jason Rute, Alex Smith, Henry Towsner, and Jin Wei. This list is woefully incomplete. I owe a special thanks to Jasmin Blanchette, Madeleine Harlow, and Stephen Mackereth for exceptionally close readings and detailed corrections. The derivations in this book have been typeset in LaTeX by Samuel Buss's *Bussproofs* package.

The bibliographical notes at the end of each chapter implicitly acknowledge another debt. I myself learned the topics presented here from a number of excellent sources, including Shoenfield's *Mathematical Logic*, *The Handbook of Mathematical Logic*, Chang and Keisler's *Model Theory*, Troelstra and Van Dalen's *Constructive Mathematics*, Troelstra and Schwichtenberg's *Basic Proof Theory*, Hájek and Pudlák's *Metamathematics of First-Order Arithmetic*, Soare's *Recursively Enumerable Sets and Degrees*, Girard's *Proofs and Types*, Hindley and Seldin's *Lambda-Calculus and Combinators*, Troelstra's epic *Metamathematical Investigation of Intuitionistic Arithmetic and Analysis*, and Simpson's *Subsystems of Second-Order Arithmetic*. I have drawn extensively on all of these sources and many others. I hope that my presentation does the material justice and helps keep it accessible to the next generation of logicians.

I am indebted to numerous colleagues, including Steve Awodey and Wilfried Sieg, for sharing their insights and expertise. I am especially indebted to Jack Silver and Solomon Feferman, who introduced me to logic and set me on my way. Finally, this book is dedicated to my wife, Elinor, and to my daughters, Rebecca, Jordana, and Ariella. They are the ones who make it all worthwhile.

1

Fundamentals

Every branch of mathematics has its subject matter, and one of the distinguishing features of logic is that so many of its fundamental objects of study are rooted in language. The subject deals with terms, expressions, formulas, theorems, and proofs. When we speak about these notions informally, we are talking about things that can be written down and communicated with symbols. One of the goals of mathematical logic is to introduce formal definitions that capture our intuitions about such objects and enable us to reason about them precisely.

At the most basic level, syntactic objects can be viewed as strings of symbols. For concreteness, we can identify symbols with particular set-theoretic objects, but, for most purposes, it does not matter what they are; all that is needed is that they are distinct from one another. A set of symbols is called an *alphabet*, and a *string* of symbols from the alphabet A is just a finite sequence of elements of A. Notions like the length of a string s and the concatenation $s^\frown t$ of two strings s and t are carried over from sequences. If $a_0, \ldots, a_{k-1}$ are symbols in some alphabet, '$a_0 \ldots a_{k-1}$' should be interpreted as the sequence $(a_0, \ldots, a_{k-1})$. These representations give much of logic a finitary, combinatorial, and computational flavor.

Nonetheless, abstraction can be helpful. What is essential about expressions like $((x + 7) \cdot (y + 9))$ is that they are built up from simple expressions – in this case, variables and numerical constants – using fixed operations in a systematic way. We ought to be able to prove things about such expressions inductively, and define operations on such expressions recursively, without descending to the level of symbols and strings. Functional programming languages often support recursive definitions on such inductively defined types.

The goal of this chapter is to develop a foundation for reasoning about syntax. While the definitions and theorems here underwrite many of the fundamental patterns of reasoning and inference in this book, most of those patterns are intuitively clear and natural when taken at face value. As a result, it would be reasonable to skim this chapter and refer back to it as necessary.

In logic, we state things about formal statements and prove things about formal proofs. This apparent circularity is sometimes confusing to novices. Philosophers and logicians often distinguish a *language* under study from the *metalanguage* used to study it, and a formal axiomatic *theory* from the *metatheory* that embodies the methods that are used to reason about it. Here our metatheory is simply everyday mathematics, as it is found in ordinary textbooks in algebra, analysis, or number theory. It is the subject matter, not the principles of reasoning, that sets mathematical logic apart.

1.1 Languages and Structures

Mathematics deals with structures. A *group* consists of a set of elements G together with a distinguished element 1 of G, a binary operation $\cdot$ on G, and an inverse function $x \mapsto x^{-1}$ from G to G, such that these data satisfy the group axioms. A *field* consists of a larger set of data, satisfying a different set of axioms. A *partial order* on a set A consists of a binary relation $\leq$ on A that is reflexive, antisymmetric, and transitive. An *equivalence relation* on a set A is a binary relation $\equiv$ on A that is reflexive, symmetric, and transitive.

Each of these can be viewed as a *structure* satisfying some *axioms*. We will later determine what sort of thing an axiom is and what it means to satisfy one. But first, we need to say what a structure is. In the examples above, each particular structure provides an interpretation of a certain set of symbols, such as $\{1, \cdot, \cdot^{-1}\}$ or $\{\equiv\}$, that are intended to denote functions or relations. Such a specification is known as a *language*.

Definition 1.1.1. A *language* is a triple (Γ, Δ, a), where Γ and Δ are disjoint sets of symbols and a is a function from $\Gamma \cup \Delta$ to $\mathbb{N}$. Γ is said to be the set of *function symbols* of the language, Δ is the set of *relation symbols*, and a assigns to each function and relation symbol its *arity*. If f is an element of Γ and $a(f) = k$, then f is said to be a *k-ary function symbol.* If R is an element of Δ and $a(R) = k$, then R is said to be a *k-ary relation symbol.*

Intuitively, a function is something that returns a value, whereas a relation is something that may or may not hold of its arguments. We can think of a 0-ary function as a constant value, that is, a function that returns a value without taking any arguments. Similarly, we can think of a 0-ary relation as a constant truth value. In the examples above, the language of groups has a 0-ary function symbol, 1, a binary function symbol, $\cdot$, and a unary function symbol, $\cdot^{-1}$. The language of equivalence relations has a single binary relation symbol, $\equiv$.

The word "language" is misleading since a language is really a specification of a basic vocabulary from which complex expressions can be built. Later on, we will also consider other kinds of specifications. The present notion is also called a *signature*, and sometimes a *first-order language* to distinguish it from other kinds of languages. First-order languages can be used to reason about algebraic structures like groups and fields; to reason about particular structures like the natural numbers and the real numbers; or to give foundational accounts of the entire universe of mathematical objects.

Definition 1.1.2. If $L = (\Gamma, \Delta, a)$ is a language, a *structure* for L (or an *L-structure*) consists of a set U and a function I that assigns to each k-ary function symbol in Γ a k-ary function from U to U and to each k-ary relation symbol in Δ a k-ary relation on U.

An L-structure is also often called a *model for* L, or simply a *model* when the language is understood. If $\mathfrak{A} = (U, I)$ is an L-structure, we typically write $|\mathfrak{A}|$ for the set U, called the *universe of* $\mathfrak{A}$, $f^{\mathfrak{A}}$ instead of $I(f)$ for the interpretation of the function symbol f in $\mathfrak{A}$, and $R^{\mathfrak{A}}$ instead of $I(R)$ for the interpretation of the relation symbol R. For example, if L is a language with one 0-ary function symbol c (i.e. a constant symbol), two binary function symbols f and g, and one binary relation symbol R, then we can interpret L in the structure with universe $\mathbb{N}$, the constant 0, functions $+$ and $\cdot$, and relation $\leq$. For convenience, we will typically refer to this as the structure $(\mathbb{N}, 0, +, \cdot, \leq)$, leaving the correspondence with the symbols c, f, g, and R to be inferred from context.

If G and H are groups, a *homomorphism* $\varphi\colon G \to H$ is a function that maps 1^G to 1^H and respects multiplication and inverses. Saying that φ respects multiplication means that $\varphi(g_1 \cdot^G g_2) = \varphi(g_1) \cdot^H \varphi(g_2)$ holds for every g_1 and g_2 in G, and saying that it respects inverses means that $\varphi(g^{-1}) = \varphi(g)^{-1}$ holds for every g in G, where the first inverse is computed in G and the second one is computed in H. (By the usual abuse of notation, we have written G for both the group structure and the underlying carrier set, $|G|$.) If $\equiv$ is an equivalence relation on a set A and $\sim$ is an equivalence relation on a set B, then a homomorphism from $(A, \equiv)$ to $(B, \sim)$ is a function $\varphi\colon A \to B$ with the property that whenever $a_1 \equiv a_2$, we have $\varphi(a_1) \sim \varphi(a_2)$. Both of these are instances of a general notion.

Definition 1.1.3. Let L be a language and let $\mathfrak{A}$ and $\mathfrak{B}$ be L-structures. Then a *homomorphism* φ from $\mathfrak{A}$ to $\mathfrak{B}$ is a function from $|\mathfrak{A}|$ to $|\mathfrak{B}|$ that satisfies the following two requirements:

- For every k-ary function symbol f of L and every tuple $a_0, \ldots, a_{k-1}$ of elements of $|\mathfrak{A}|$, $\varphi(f^{\mathfrak{A}}(a_0, \ldots, a_{k-1})) = f^{\mathfrak{B}}(\varphi(a_0), \ldots, \varphi(a_{k-1}))$.
- For every k-ary relation symbol R of L and every tuple $a_0, \ldots, a_{k-1}$ of elements of $|\mathfrak{A}|$, if $R^{\mathfrak{A}}(a_0, \ldots, a_{k-1})$, then $R^{\mathfrak{B}}(\varphi(a_0), \ldots, \varphi(a_{k-1}))$.

Notice that the implication in the second clause of the definition of a homomorphism does not necessarily reverse: a homomorphism is required to *preserve* each of the relations in the source structure but not *reflect* them. A homomorphism is called an *embedding* if it is injective and satisfies the following strengthening of the second clause:

- For every k-ary relation symbol R of L and every tuple $a_0, \ldots, a_{k-1}$ of elements of $|\mathfrak{A}|$, $R^{\mathfrak{A}}(a_0, \ldots, a_{k-1})$ if and only if $R^{\mathfrak{B}}(\varphi(a_0), \ldots, \varphi(a_{k-1}))$.

In other words, an embedding both preserves and reflects the relations.

A homomorphism $\varphi\colon \mathfrak{A} \to \mathfrak{B}$ is an *isomorphism* if there is a homomorphism $\psi\colon \mathfrak{B} \to \mathfrak{A}$ such that $\psi \circ \varphi$ is the identity on $\mathfrak{A}$ and $\varphi \circ \psi$ is the identity on $\mathfrak{B}$. Exercise 1.1.1 asks you to show that $\varphi\colon \mathfrak{A} \to \mathfrak{B}$ is an isomorphism if and only if it is a surjective embedding. Two structures for a language L are said to be *isomorphic* if there is an isomorphism between them. A homomorphism from a structure $\mathfrak{A}$ to itself is called an *endomorphism*, and an isomorphism from a structure $\mathfrak{A}$ to itself is called an *automorphism*.

Think of a homomorphism φ from $\mathfrak{A}$ to $\mathfrak{B}$ as a translation between the two structures. The first clause says that applying a function in $\mathfrak{A}$ to some elements in the universe of $\mathfrak{A}$ and then translating to $\mathfrak{B}$ gives the same result as translating the elements to $\mathfrak{B}$ and applying the corresponding function in $\mathfrak{B}$. The second clause says that relations in $\mathfrak{A}$ are preserved by the translation. For example, $\varphi(x) = 2x$ is an embedding of $(\mathbb{Z}, 0, +, \leq)$ in $(\mathbb{Z}, 0, +, \leq)$, and $\psi(x) = 0$ is a homomorphism between those two structures. The identity function on $\mathbb{Z}$ is an embedding of $(\mathbb{Z}, 0, 1, +, \cdot, \leq)$ in $(\mathbb{R}, 0, 1, +, \cdot, \leq)$, and the function $h(x) = e^x$ is an isomorphism of $(\mathbb{R}, 0, +, \leq)$ and $(\mathbb{R}^{>0}, 1, \cdot, \leq)$, where $\mathbb{R}^{>0}$ denotes the positive real numbers.

With the definitions above, saying that $\mathfrak{A}$ is an L-structure means that we have chosen specific symbols to represent the relevant data. Mathematically, groups are sometimes written with multiplicative notation $1, \cdot, \cdot^{-1}$, sometimes with additive notation 0, $+$, and $-$, and sometimes with neutral symbols, such as e, $\cdot$, and i. In the terminology introduced here, each language gives rise to a different kind of a structure, say, multiplicative group structures, additive group structures, and general group structures. Whether or not this is a good thing is subject to debate, but, in any case, any structure of one kind can be mapped to a structure of one of another kind in such a way that the identity function on the underlying set will be an isomorphism between them. There are other approaches to talking about languages and structures; one is to take a signature to be a specification of arities without a choice of function symbols, and a structure for that signature to be an ordered sequence of interpretations. The various approaches are generally intertranslatable.

In Chapter 5, we will make use of the notion of a *quotient* construction for structures. Let $\mathfrak{A}$ be a structure for a language L and $\sim$ be an equivalence relation on $|\mathfrak{A}|$. Suppose furthermore that all the functions and relations in $\mathfrak{A}$ respect the congruence, as described in Appendix A.2. We define $\mathfrak{A}/\!\sim$ to be the structure with universe $|\mathfrak{A}|/\sim$, where for every k-ary function symbol f of L, $f^{\mathfrak{A}/\sim}([a_0], \ldots, [a_{k-1}])$ is defined to be $[f^{\mathfrak{A}}(a_0, \ldots, a_{k-1})]$, and for every k-ary relation symbol R of L, $R^{\mathfrak{A}/\sim}([a_0], \ldots, [a_{k-1}])$ holds if and only if $R^{\mathfrak{A}}(a_0, \ldots, a_{k-1})$. The function $\varphi(a) = [a]$ is then a surjective homomorphism from $\mathfrak{A}$ to $\mathfrak{A}/\!\sim$, and $a \sim b$ holds of elements of $|\mathfrak{A}|$ if and only if $\varphi(a) = \varphi(b)$. Thus, the quotient construction turns the equivalence relation $\sim$ on $\mathfrak{A}$ into equality on $\mathfrak{A}/\!\sim$.

Exercises

1.1.1. Show that a function $\varphi\colon \mathfrak{A} \to \mathfrak{B}$ between two L-structures is an isomorphism if and only if it is a surjective embedding.

1.1.2. Show that the composition of two homomorphisms is a homomorphism, and similarly for embeddings and isomorphisms.

1.1.3. Show that isomorphism is an equivalence relation.

1.1.4. For each of the following pairs, show that the two structures are isomorphic:
   a. $((a, b), <)$ and $((c, d), <)$, where $a, b, c, d \in \mathbb{R}$, $a < b$, $c < d$, and (a, b) denotes the open interval $\{x \mid a < x < b\}$
   b. $((0, 2), 1, <)$ and $(\mathbb{R}, 0, <)$
   c. $(\mathbb{R}, 0, +, <)$ and $(\mathbb{R}^{>0}, 1, \cdot, <)$, where $\mathbb{R}^{>0}$ denotes the positive real numbers.

1.1.5. For each of the following pairs, show that the two structures are *not* isomorphic:
   a. $(\mathbb{N}, <)$ and $(\mathbb{N}, >)$
   b. $((0, 1), <)$ and $((0, 1], <)$, where $(0, 1]$ denotes the half-closed interval $\{x \mid 0 < x \leq 1\}$
   c. $((0, 1) \cup (1, 2), <)$ and $((0, 2), <)$.

1.1.6. Determine all the endomorphisms and automorphisms of each of the following structures: $(\mathbb{N}, <)$, $(\mathbb{N}, +)$, $(\mathbb{R}, +)$, $(\mathbb{R}, +, <)$, and $(\mathbb{R}, \cdot)$.

1.1.7. Verify the last claim in this section: if $\mathfrak{A}$ is any structure and $\sim$ is an equivalence relation on $|\mathfrak{A}|$, then $\varphi(a) = [a]$ is a surjective homomorphism from $\mathfrak{A}$ to $\mathfrak{A}/\!\sim$ that preserves the relations in $\mathfrak{A}$, and for every a and b in $|\mathfrak{A}|$, $a \sim b$ if and only if $\varphi(a) = \varphi(b)$.

1.2 Inductively Defined Sets

The natural numbers can be characterized inductively as a set that contains a distinguished element, 0, and is closed under an injective operation, $\mathrm{succ}(n)$, that returns the *successor* of n. The inductive character amounts to the fact that the sequence 0, $\mathrm{succ}(0)$, $\mathrm{succ}(\mathrm{succ}(0)), \ldots$ exhausts the set of natural numbers, in the following sense: if $A \subseteq \mathbb{N}$ contains 0 and is closed under succ, then $A = \mathbb{N}$. If we associate any property P of natural numbers with the set of numbers satisfying P, the preceding statement amounts to the principle of induction on $\mathbb{N}$.

It is often useful to characterize sets of expressions in a similar fashion. We can view arithmetic expressions like $((x+7) \cdot (y+9))$ as being built up from variables and numeric constants by the syntactic operations of forming sums and products. The utility of such a perspective is not limited to syntax, and we often come across sets and structures in mathematics that are generated in such a way. For example, if G is a group and S is a subset of G, the subgroup $\langle S \rangle$ generated by S is the smallest subset of G containing S and closed under the group operations. Similarly, the collection of Borel subsets of $\mathbb{R}$ is the smallest collection of subsets of $\mathbb{R}$ containing the open sets and closed under the operations of forming complements and countable unions.

In this section, we will develop a very general, abstract framework for describing sets of elements that are defined inductively, from the bottom up. Our high-level approach is somewhat heavy-handed, and we could certainly develop a theory of syntax in more concrete terms. But as the examples below indicate, the approach has applications beyond defining terms and expressions, and abstracting the common features provides a clear understanding of the essential features of the constructions.

We start with a set U of objects, a *universe*, within which the construction takes place. A *rule on* U is just a pair (S, a), where S is a subset of U and a is an element of U. We will think of a set of rules as a recipe for constructing a set of objects, where the rule (S, a) says "if the elements of S are in the set, then a must be in the set as well." Our goal is to construct a set that consists of only those elements that are *required* to be there by the rules. Notice that if S is the empty set, then the rule (S, a) asserts outright that a must be in the set we construct. Such rules are the starting point for the construction. If there are no such rules, the empty set itself satisfies all the requirements.

Let $\mathfrak{R}$ be a set of rules on U. Say that a set B is *inductive with respect to* $\mathfrak{R}$ or *closed under* $\mathfrak{R}$ if it meets the specification above, that is, for each rule $(S, a) \in \mathfrak{R}$, if $S \subseteq B$, then $a \in B$. Let

$$A = \bigcap \{B \subseteq U \mid B \text{ is inductive}\}.$$

In words, A is the intersection of all inductive subsets of U, so an element a of U is in A if and only if it is in *every* inductive subset of U. The next proposition shows that A is the set we are after.

Proposition 1.2.1. *The following hold:*

1. *A is inductive.*
2. *If $B \subseteq A$ is inductive, then $B = A$.*

Proof For the first claim, suppose $(S, a) \in \mathfrak{R}$ and $S \subseteq A$. By the definition of A, $S \subseteq B$ for every inductive subset B of U. But if B is inductive and $S \subseteq B$, then a is in B. So a is in every inductive set as well, and so $a \in A$.

For the second claim, if $B \subseteq U$ is inductive, then $A \subseteq B$, since every element of A is in every inductive set. Since we are assuming $B \subseteq A$, we have $B = A$. ☐

The first part of Proposition 1.2.1 says that A is closed under the rules, while the second part of the proposition says that A is the smallest such set. In practice, the relevant set of rules is often described with a list of conditions, as in the following example.

Example. Let G be any group and S be any nonempty subset of G. Then the *subgroup of G generated by S* is the smallest subset H of the carrier of G that satisfies the following:

- If g is any element of S, then g is in H.
- If g_1 and g_2 are in H, then so is $g_1 g_2$.
- If g is in H, then so is g^{-1}.

To express this in formal terms, take the universe U to be the carrier of G and take $\mathfrak{R}$ to be the union of the following three sets:

- $\{(\emptyset, g) \mid g \in S\}$
- $\{(\{g_1, g_2\}, g_1 g_2) \mid g_1, g_2 \in G\}$
- $\{(\{g\}, g^{-1}) \mid g \in G\}$.

Then H is the subset of G defined inductively by $\mathfrak{R}$.

Notice that there is a theorem implicit in our use of the phrase "the subgroup of G generated by S," namely, that the inductively defined set is in fact a subgroup. This follows easily from the closure under the rules. The inductive character implies that if K is any other subgroup of G containing S, then $H \subseteq K$.

I will leave it to you to carry out similar translations for the next three examples.

Example. The collection of *Borel subsets of* $\mathbb{R}$ is the smallest subset B of $\mathcal{P}(\mathbb{R})$ that satisfies the following:

- If a, b are in $\mathbb{R}$, then the open interval (a, b) is in B.
- If S is in B, so is $\overline{S}$, the complement of S.
- If $(S_i)_{i \in \mathbb{N}}$ is a countable sequence of subsets of $\mathbb{R}$ and each S_i is in B, then so is $\bigcup_i S_i$.

Example. Let A be any set and R be any binary relation on A. Then the *transitive closure* of R is the relation R' defined inductively as follows:

- For any $a, b \in A$, if $R(a, b)$, then $R'(a, b)$.
- For any $a, b, c \in A$, if $R'(a, b)$ and $R'(b, c)$, then $R'(a, c)$.

The closure and inductive properties imply that R' is the smallest transitive relation on A containing R.

Example 1.2.2. Let U be a collection of sets containing $\emptyset$ and closed under the function $\mathrm{succ}(a) = a \cup \{a\}$. (The set-theoretic *axiom of infinity* states precisely that there exists such a set U.) Then, in set-theoretic terms, the set of natural numbers, $\mathbb{N}$, can be defined as the smallest set containing $\emptyset$ and closed under succ. Similarly, let U be a collection of sets containing $\emptyset$ and closed under succ and countable unions. Then the smallest subset of U with these properties is exactly the set of countable ordinals.

Example 1.2.3. More generally, suppose U is any set of objects and $\mathfrak{F}$ is a set of functions from U to U of various arities, including constants. Then we can define the set A to be the smallest subset of U closed under all the elements of $\mathfrak{F}$. More precisely, A is the smallest subset of U containing all the constants in $\mathfrak{F}$, and closed under all the functions. This set has the following two properties:

- It contains all the constants in $\mathfrak{F}$ and is closed under each of the functions.
- If $B \subseteq A$ has these properties, then $B = A$.

Since this example will figure prominently in this book, I will describe the set of rules explicitly. For each k-ary function $f \in \mathfrak{F}$ and sequence $a_0, \ldots, a_{k-1}$, we add the rule $(\{a_0, \ldots, a_{k-1}\}, f(a_0, \ldots, a_{k-1}))$. Intuitively, this says that as soon as $a_0, \ldots, a_{k-1}$ are included in the set, we should include $f(a_0, \ldots, a_{k-1})$ as well. Each constant c in $\mathfrak{F}$ corresponds to the rule $(\emptyset, c)$.

The terminology commonly used to describe these constructions varies. In the general situation, we may say that A is the *smallest set closed under the rules in* $\mathfrak{R}$ or the set *inductively generated by* $\mathfrak{R}$. Similar language is used to describe the construction of Example 1.2.3. For example, we have already described the natural numbers as the *smallest set containing* 0 *and closed under* succ in Example 1.2.2, and we might also say that $\mathbb{N}$ is *inductively generated from* $\{0\}$ *by* succ.

The second claim of Proposition 1.2.1 is really an induction principle for A. It implies that in order to show that every element of A has some property P, it suffices to show that for each rule (S, a) in $\mathfrak{R}$, if $S \subseteq A$ and P holds of every element of S, then P holds of a. To see that this principle follows from Proposition 1.2.1, given P, let B be the set of elements of A satisfying P. The hypothesis of the principle says exactly that B is closed under the set of rules in $\mathfrak{R}$, and the second claim of Proposition 1.2.1 then implies that every element of A satisfies P.

The abstract characterization of A as the intersection of all inductive subsets of U is clever, but, from a foundational point of view, it is heavy-handed: we have defined A by reference to the collection of *all* the subsets of U, which may be very large. To make matters worse, it also contains the very object A that is being defined, making it a prototypical example of an *impredicative definition*. We can often provide a more explicit description of the set we are after. If the set of rules $\mathfrak{R}$ has the property that each rule $(S, a) \in \mathfrak{R}$ is finite, which is to say, the set S is finite, then we define a finite sequence $(a_0, \ldots, a_{n-1})$ of elements of U to be a *formation sequence* (again with respect to $\mathfrak{R}$) if, for each i, there is a rule (S, a_i) with $S \subseteq \{a_0, \ldots, a_{i-1}\}$. Intuitively, this says that every element of the formation sequence is justified by previous elements of the sequence.

Proposition 1.2.4. *With U and $\mathfrak{R}$ as above, let A be the subset of U defined inductively by $\mathfrak{R}$. Then A is equal to the set of elements a of U such that there is a formation sequence containing a.*

Proof Let B be the set of elements a of U such that there is a formation sequence containing a. To show that A is a subset of B, we use the induction principle for A. Suppose (S, a) is a rule in $\mathfrak{R}$ and there is a formation sequence for each element of S. Since S is finite, we can concatenate these and append a to obtain a formation sequence for a.

On the other hand, to show that B is a subset of A, let $(a_0, \ldots, a_{n-1})$ be any formation sequence. Then, using the definition of a formation sequence, it is easy to show by induction on the natural numbers that for each i, $a_i \in A$. □

The constructions described here can be described in other terms. Given a universe U and a set $\mathfrak{R}$ of rules, define the function $\Gamma\colon \mathcal{P}(U) \to \mathcal{P}(U)$ by

$$\Gamma(B) = \{a \in U \mid \text{for some } S \subseteq B, (S, a) \in \mathfrak{R}\}.$$

In other words, $\Gamma(B)$ consists of the elements that should be added to B in conformance with the rules. It is easy to check that Γ is *monotone*: whenever $B \subseteq C$, we have $\Gamma(B) \subseteq \Gamma(C)$. Exercise 1.2.2 asks you to show that any monotone operator from $\mathcal{P}(U)$ to $\mathcal{P}(U)$ has a *least fixed point*, which is to say, there is a set A such that $\Gamma(A) = A$ and whenever $\Gamma(B) = B$, $A \subseteq B$. The construction we have described here is the special case where Γ is defined from a set of rules, as above. Exercise 1.2.4 asks you to show that, conversely, given any monotone operator Γ, the least fixed point of Γ can be characterized as the smallest set closed under a suitable set of rules.

Exercises

1.2.1. For each of the examples of inductive definitions in this section, define a corresponding set of rules explicitly.

1.2.2. Let U be a set, let Γ be a monotone function from $\mathcal{P}(U)$ to $\mathcal{P}(U)$, and let

$$A = \bigcap \{B \in \mathcal{P}(U) \mid \Gamma(B) \subseteq B\}.$$

Note that $\Gamma(U) \subseteq U$, so A is the intersection of a nonempty set.

a. Show that $\Gamma(A) \subseteq A$. (Hint: show $\Gamma(A) \subseteq B$ whenever $\Gamma(B) \subseteq B$.)
b. Show that $A \subseteq \Gamma(A)$. So $A = \Gamma(A)$ is a fixed point of Γ.
c. Show that if B is any other fixed point, $A \subseteq B$.

So A is the least fixed point. The construction generalizes to arbitrary complete lattices. The statement that every monotone function on a complete lattice has a least fixed point is called the *Knaster–Tarski* theorem.

1.2.3. The following provides an alternative "bottom-up" definition of the least fixed point of Γ, using principles of transfinite recursion along suitable ordinals. (It requires some basic set theory.)

Let Γ be monotone. Define a sequence of subsets of U by transfinite recursion on the ordinals, as follows:

- $A_0 = \emptyset$
- $A_{\alpha+1} = \Gamma(A_\alpha)$
- $A_\lambda = \bigcup_{\alpha<\lambda} A_\alpha$, for limit ordinals λ.

Do the following:

a. Show that whenever $\alpha < \beta$, $A_\alpha \subseteq A_\beta$.
b. Show, using cardinality considerations, that for some α, $A_\alpha = A_{\alpha+1}$. (In fact, $\alpha < |U|^+$, the least cardinal larger than the cardinality of U.) After that, the process stabilizes, so A_α is a fixed point of Γ.
c. Show that if B is any fixed point of Γ, then for every β, $A_\beta \subseteq B$. In particular, A_α is the least fixed point.

1.2.4. Let Γ be a monotone function from $\mathcal{P}(U)$ to $\mathcal{P}(U)$ and define the set of rules $\mathfrak{R} = \{(S, a) \mid a \in \Gamma(S)\}$.

a. Show that for any $B \subseteq U$, $\Gamma(B) \subseteq B$ if and only if B is closed under $\mathfrak{R}$.
b. Show that the least fixed point of Γ is exactly the subset of U inductively generated by $\mathfrak{R}$.

1.2.5. Show that every monotone function from $\mathcal{P}(U)$ to $\mathcal{P}(U)$ also has a *greatest fixed point*.

1.3 Terms and Formulas

We will now apply the abstract machinery we have just developed to the relatively concrete tasks of defining syntactic objects like terms and formulas. Let L be a language without any relation symbols, or, if there are any, just ignore them for now. Fix a stock of variables $x_0, x_1, x_2, \ldots$ different from the symbols of L. (We will generally be interested in languages with countably many symbols, and we will generally need only countably many variables. In principle, however, nothing prevents us from using uncountably many symbols and variables, and this is often useful in model theory.) We want the *terms* of L to be expressions built up from the variables and constant symbols using the function symbols in the language. For example, if L has a constant symbol, c, a unary function symbol, f, and two binary function symbols, h and k, then the following are examples of terms in L:

$$c, \ x_0, \ x_1, \ f(c), \ f(x_0), \ h(f(x_0), x_1), \ k(f(c), h(f(x_0), x_1)).$$

Formally, we will take a term in this language to be a string of symbols in alphabet that includes the symbols in F, the variables, symbols for the open- and close-parentheses, and a comma. We assume that all these symbols are distinct from one another.

Definition 1.3.1. Let $L = (\Gamma, \Delta, a)$ be a language. The set of *terms* of L is the smallest set of strings over the alphabet above satisfying the following:

- If x is a variable, then x is a term.
- If c is a constant symbol in Γ (a 0-ary function symbol), then c is a term.
- If $f \in \Gamma$ has arity k and $t_0, \ldots, t_{k-1}$ are all terms, then so is $f(t_0, \ldots, t_{k-1})$.

I have taken some notational liberties in this definition. The first condition says, more precisely, that if x is any variable, then the string 'x' is a term. The conclusion of the third condition says, more properly, that the string '$f($'$^\frown t_0{}^\frown$','$^\frown \ldots ^\frown$','$^\frown t_k{}^\frown$')' is a term. I will generally rely on the more convenient manner of presentation above and leave these details implicit.

There are additional syntactic nuances. Assuming we have fixed a set of variables $\{x_0, x_1, \ldots\}$, the inscription "x" in the definition is a variable ranging over these symbols. Similarly, the inscriptions "t_0," ..., "t_{k-1}" in the third condition range over terms, which is to say, they are variables ranging over syntactic expressions. Bearing in mind the distinction between theory and metatheory, these are sometimes called *metavariables*, to distinguish them between the symbols for variables in the formal language we are constructing. Once again, I will avoid such ponderous language and trust you to be mindful of the difference.

Notice that this definition is an instance of Example 1.2.3, since the set of terms is generated by the following collection of functions on strings:

- for each variable x, the string 'x'
- for each 0-ary symbol c in Γ, the string 'c'
- for each k-ary symbol f in F with $k > 0$, the k-ary function $\bar{f}$ which takes as input arbitrary strings $t_0, \ldots, t_{k-1}$ and assembles the string $f(t_0, \ldots, t_{k-1})$.

We can use induction to prove some basic facts about terms.

Proposition 1.3.2. *Every term in a language L has the same number of left and right parentheses.*

Proof The claim is true of the base cases, namely, the variables and constants. And, assuming it is true for $t_0, \ldots, t_{k-1}$, it is also true of the string of symbols $f(t_0, \ldots, t_{k-1})$. □

The following is even easier to prove by induction:

Proposition 1.3.3. *Let t be any term of L. Then either t is a constant symbol, or t is a variable, or there are a k-ary function symbol f of L and a sequence of terms $s_0, \ldots, s_{k-1}$ such that t is the string $f(s_0, \ldots, s_{k-1})$.*

What is not nearly as obvious is that a given term falls into one of these three categories in a unique way. Establishing this stronger fact has to rely on the specific details of the definition. If we had allowed infix notation like $t_0 + t_1$ and $t_0 \cdot t_1$ and failed to include parentheses, then a term like $x_0 + x_1 \cdot x_2$ would be ambiguous; it could arise from applying the $+$ construction to x_0 and $x_1 \cdot x_2$, or the $\cdot$ construction to $x_0 + x_1$ and x_2. We therefore need to prove that our current definition avoids such ambiguity.

Theorem 1.3.4. *On the set of terms, the generating functions are injective and their images are disjoint from one another.*

Proof It is easy to see that the images of all the operations are disjoint, because each string begins with the corresponding symbol, and we are assuming that the variables, constant symbols, and function symbols are all distinct. So all that remains is to show that each term-forming operation is injective.

So suppose f is a k-ary function symbol with $k > 0$, $t_0, \ldots, t_{k-1}, t'_0, \ldots, t'_{k-1}$ are all terms, and $f(t_0, \ldots, t_{k-1})$ and $f(t'_0, \ldots, t'_{k-1})$ are equal, which is to say, they are the same string of symbols. Dropping the first two characters and the last character, we have that the two strings '$t_0, \ldots, t_{k-1}$' and '$t'_0, \ldots, t'_{k-1}$' are the same. We need to show that each t_j is equal to t'_j, which is to say, the two strings are the same. In other words, we need to show that a string of symbols cannot be parsed as the concatenation of k-terms, separated by commas, in two distinct ways.

To prove this, we first establish an auxiliary claim. If s and t are strings of symbols, say that s is a *proper initial segment of* t if s is strictly shorter than t, and the two sequences of characters agree up to the length of s. I claim that if t is any term and s is a proper initial segment of t, then s is *not* a term. We prove this by induction on terms. The only proper initial segment of a constant or variable is the empty string, which is not a term (prove this by induction on terms as well). In the induction step, suppose f is a k-ary function symbol with $k > 0$, and s is a proper initial segment of $f(t_0, \ldots, t_{k-1})$. Then it has one of the following forms:

- the empty string
- f
- $f($
- $f(t_0, \ldots, t_j$
- $f(t_0, \ldots, t_j, t'$

where, in the last case, t' is a proper initial segment of t_{j+1}. It is not hard to verify that the first three are not terms (again, using induction on terms). We have already established that every term has the same number of left and right parentheses, so the fourth case has more left parentheses than right parentheses, and hence is not a term. To handle the last case, we need to establish the slightly stronger assertion that any initial segment of a term has no more right parentheses than left parentheses, which is again easy to do, by induction on terms.

Returning to the main proof, since '$t_0, \ldots, t_{k-1}$' and '$t'_0, \ldots, t'_{k-1}$' are the same string, either t_0 is equal to t'_0, or one is a proper initial segment of the other. By the previous claim, the latter is impossible, so t_0 and t'_0 are equal. Dropping these and the subsequent comma from each string, we have that '$t_1, \ldots, t_{k-1}$' and '$t'_1, \ldots, t'_{k-1}$' are the same string. Proceeding iteratively in this way, we obtain that each t_j is equal to t'_j, as required. □

Theorem 1.3.4 is sometimes said to express the *unique readability of terms*, since it says that for any term t, there is only one way that t can be interpreted as having been generated by the rules. In computational terms, it says that any string denoting a term can be *parsed* in a unique way. Below we will use this fact to justify a principle of recursion on terms.

Let us consider another example of an inductively defined set of expressions. In Chapter 2, we will make use of the language of propositional logic. To define that language, we begin with the alphabet

$$\wedge, \vee, \rightarrow, \bot, (,),$$

as well as a stock of propositional variables $P_0, P_1, P_2, \ldots$.

Definition 1.3.5. The set of *propositional formulas* is the smallest set of strings over the alphabet above satisfying the following:

- If P is a propositional variable, then P is a propositional formula.
- $\bot$ is a propositional formula.
- If A and B are propositional formulas, then $(A \wedge B)$, $(A \vee B)$, and $(A \rightarrow B)$ are also propositional formulas.

In the intended interpretation, the variables $P_0, P_1, \ldots$ range over basic, atomic assertions, or *propositions*. The symbol $\bot$ stands for falsity, that is, any false proposition. The binary connectives are used to build up more complex propositions; $(A \wedge B)$, $(A \vee B)$, and $(A \rightarrow B)$ are intended to denote "A and B," "A or B," and "if A then B," respectively.

Once again, we have unique readability:

Theorem 1.3.6. *On the set of propositional formulas, the generating functions are injective and their images are disjoint from one another.*

The proof is left as an exercise.

Exercises

1.3.1. Show that unique readability of terms still holds even if we omit parentheses and commas.

1.3.2. Prove the unique readability property for propositional formulas.

1.4 Trees

There are several generally inequivalent mathematical notions of a *tree* in the literature. Rather than survey the range of possibilities, I will describe the notion that is used in this book, and then indicate some of the features that distinguish it from other definitions.

Definition 1.4.1. Let Σ be any set. A *tree on* Σ is a set of finite sequences of elements of Σ that is closed under initial segments.

The phrase "closed under initial segments" means that if τ is in the tree and $\sigma \subseteq \tau$ then σ is in the tree as well. For instance, the following set is a tree on $\{0, 1\}$:

$$\{(), (0), (1), (0, 0), (0, 1), (1, 1), (0, 1, 0)\}.$$

This tree can be depicted graphically as in Figure 1.1. Elements of the tree are called *nodes*.

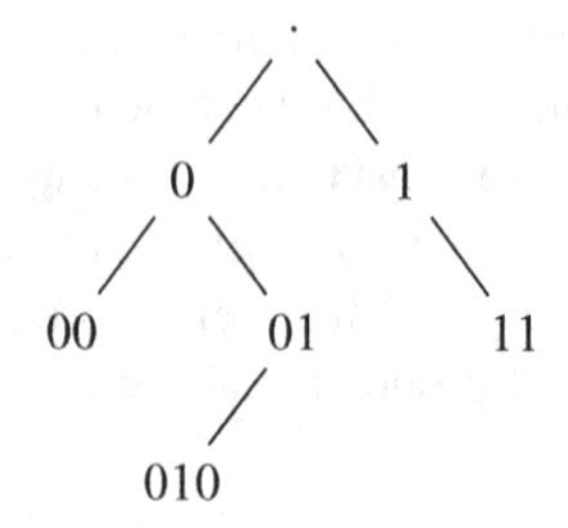

Figure 1.1 A tree on $\{0, 1\}$.

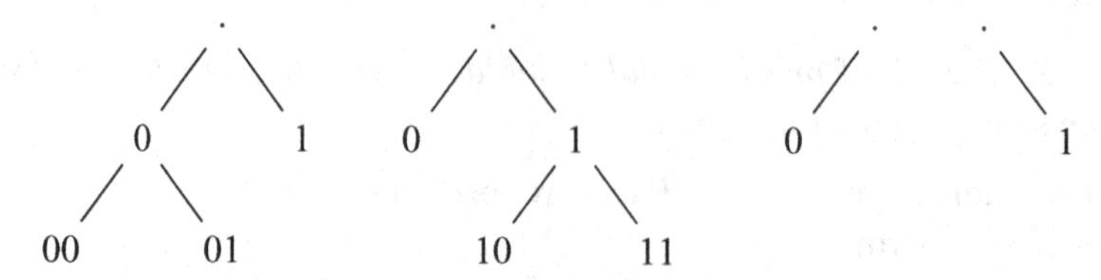

Figure 1.2 More trees on $\{0, 1\}$.

If σ is a node of the tree, the *children* of σ are the nodes of the form $\sigma^\frown(a)$ that are in the tree, the *descendants* of σ are the nodes of which σ is a proper initial segment, and the *ancestors* of σ are the proper initial segments of σ. I will generally write $\sigma^\frown a$ instead of $\sigma^\frown(a)$ when appending a single element, and, similarly, I will write $a^\frown\sigma$ when prepending a single element.

Our definition of a tree has the following properties:

- Every tree other than the empty tree has a single distinguished root, $()$.
- The names of the nodes matter. For example, the first two trees in Figure 1.2 are different from one another, and the last two are also different from one another.
- Every node has only finitely many ancestors.

If T is a tree on Σ and σ is a finite sequence of elements of Σ, then $T\restriction_\sigma$ denotes the subtree rooted at σ, that is, the set $\{\tau \mid \sigma^\frown\tau \in T\}$. If $(T_a)_{a\in\Sigma}$ is a sequence of trees (some of them possibly empty) indexed by Σ, we will write $\sup_{a\in\Sigma} T_a$ for the tree

$$\{()\} \cup \{a^\frown\sigma \mid a \in \Sigma, \sigma \in T_a\}.$$

Thus $\sup_{a\in\Sigma} T_a$ is built up from the subtrees T_a in such a way that for each $a \in \Sigma$, $(\sup_{a\in\Sigma} T_a)\restriction_{(a)} = T_a$. Conversely, if T is nonempty, $T = \sup_{a\in\Sigma} (T\restriction_{(a)})$.

Definition 1.4.2. Let Σ be any set. The set of *well-founded trees on* Σ is defined inductively as follows:

- The empty tree is a well-founded tree.
- If $(T_a)_{a\in\Sigma}$ is a sequence of well-founded trees, then $\sup_{a\in\Sigma} T_a$ is a well-founded tree.

From the very nature of the definition, we have a principle of induction: any property that holds of the empty tree and is maintained under taking supremums necessarily holds of all well-founded trees. The following provides a classically equivalent characterization of what

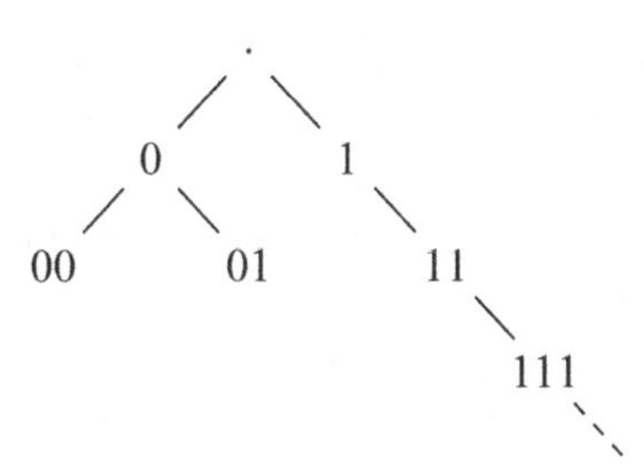

Figure 1.3 A tree on $\{0, 1\}$ with an infinite path.

it means to be well founded. If T is a tree, a *path* through T (or *branch* of T) is a set P of nodes such that

- every two nodes σ and τ are comparable, which is to say, either $\sigma \subseteq \tau$ or $\tau \subseteq \sigma$; and
- P is *maximal* with that property, which is to say, if ρ is comparable with every element of P, then ρ is in P.

Equivalently (up to change of representation), we can take a path through a nonempty tree to be either a finite or an infinite sequence $(a_0, a_1, a_2, \ldots)$ such that for each n, $(a_0, \ldots, a_{n-1})$ is in the tree, and if the path is finite there is no node extending it in the tree. As an example, the tree on $\{0, 1\}$ shown in Figure 1.3 has two finite paths and one infinite path:

Proposition 1.4.3. *A tree on Σ is well founded if and only if it has no infinite path.*

Proof Show by induction on the set of well-founded trees that no well-founded tree has an infinite path. Clearly the empty tree does not have an infinite path, and if T_a does not have an infinite path for each a in Σ, then neither $\sup_{a\in\Sigma} T_a$: if $(a_0, a_1, a_2, \ldots)$ were an infinite path through $\sup_{a\in\Sigma} T_a$, then $(a_1, a_2, a_3, \ldots)$ would be an infinite path through T_{a_0}.

Conversely, suppose T is not well founded. Then T is nonempty and for some a_0, $T\restriction_{(a_0)}$ is not well founded. This implies that $T\restriction_{(a_0)}$ is nonempty, and for some (a_1), $T\restriction_{(a_0,a_1)}$ is not well founded. Continuing in this way, we obtain an infinite path $a_0, a_1, \ldots$ through T. □

A tree is said to be *finitely branching* if every node has finitely many children. In particular, every tree on a finite set Σ is finitely branching. For finitely branching trees, it is easier to say what it means to be well founded:

Proposition 1.4.4 (Kőnig's lemma). *Let T be a finitely branching tree on Σ. The following are equivalent:*

1. *T is finite.*
2. *For some n, every path through T has length at most n.*
3. *T is well founded.*

Proof For any tree, clearly 1 implies 2 and 2 implies 3. To show that 3 implies 1, use induction on well-founded trees T to show that if T is finitely branching, it is finite. The claim is clearly true of the empty tree. Suppose, on the other hand, $T = \sup_{a\in\Sigma} T_a$. Since T is finitely branching, each T_a is finitely branching, and at most finitely many of them are nonempty. By the inductive hypothesis, each of these trees T_a is finite, so $T = \sup_{a\in\Sigma} T_a$ is finite as well. □

The implication from 3 to 1 can also be proved using the equivalent characterization of well-foundedness given by Proposition 1.4.3. Suppose T is finitely branching and infinite. Define a sequence of nodes as follows: let σ_0 be the root, and assuming σ_i has been chosen, let σ_{i+1} be any child of σ_i such that $T\restriction_{\sigma_{i+1}}$ is infinite. We can do this because the root has infinitely many descendants, and assuming σ_i has infinitely many descendants, the fact that σ_i has only finitely many children implies that at least one of these children has infinitely many descendants. Then the set $\{\sigma_0, \sigma_1, \sigma_2, \ldots\}$ is an infinite path through T, showing that T is not well founded.

Keep in mind that for arbitrary trees, the implications from 1 to 2 and from 2 to 3 do not reverse; the exercises below ask you to find counterexamples. In the proof above, the phrase "let σ_{i+1} be any child of σ_i with infinitely many descendants" used the axiom of choice implicitly. If Σ can be well ordered (see Appendix A.2) then this use can be avoided; for example, if T is a tree on $\mathbb{N}$ we can choose σ_{i+1} to be the *leftmost* child of σ_i with the desired property, that is, the one that is least in the lexicographic order. Exercise 1.4.2 below shows that Kőnig's lemma is really a kind of compactness principle.

It is often useful to consider trees whose nodes are labeled by elements from some set S. Formally, this is just a tree T paired with a function $\ell\colon T \to S$. Sometimes the underlying set Σ is unimportant; we only care that the children of a node have a distinguished order. If every node has only finitely or at most countably many children, we can take Σ to be $\mathbb{N}$, and restrict attention to trees in which the children of a node σ constitute a finite or infinite set of the form $\{\sigma^\frown 0, \sigma^\frown 1, \ldots\}$.

Exercises

1.4.1. Show that for trees that are not finitely branching, the easy implications in Kőnig's lemma do not reverse. In particular:

a. Describe a tree on $\mathbb{N}$ that does not have arbitrarily long branches but is infinite.

b. Describe a tree on $\mathbb{N}$ that is well founded but has arbitrarily long finite branches.

1.4.2. Kőnig's lemma is closely related to the topological statement that a countable product of finite discrete spaces is compact in the product topology. (See Appendix A.4.) In particular, Kőnig's lemma for trees on $\{0, 1\}$ is equivalent to the statement that $\{0, 1\}^{\mathbb{N}}$ is compact. We can think of $\{0, 1\}^{\mathbb{N}}$ as the set of infinite binary sequences. A basis for the product topology on this space is given by sets of the form $[\sigma]$, where σ is a finite binary sequence and $[\sigma]$ is the set of infinite binary sequences that extend σ. A space is *compact*, by definition, if every covering by open sets has a finite subcover. Equivalently, a space is compact if whenever S is a collection of closed subsets of the space and every finite subcollection of S has a nonempty intersection, then S has a nonempty intersection. Given a tree T on $\{0, 1\}$, identify infinite paths through T with elements of $\{0, 1\}^{\mathbb{N}}$.

a. For each natural number i, let C_i be the subset of $\{0, 1\}^{\mathbb{N}}$ whose initial segments of length i are in the tree. Show that C_i is clopen, that is, both closed and open.

b. Show that the intersection of the C_i's is nonempty if and only if there is an infinite path through T.

c. Show of Kőnig's lemma for trees on $\{0, 1\}$ follows from the assertion that $\{0, 1\}^{\mathbb{N}}$ is compact.

d. Show that, conversely, the compactness of $\{0, 1\}^{\mathbb{N}}$ follows from Kőnig's lemma. (Hint: Note that any open set can be written as a union of basic open sets. Given a collection S of basic open sets that covers $\{0, 1\}^{\mathbb{N}}$, consider the tree of sequences τ such that $\sigma \not\subseteq \tau$ for any $[\sigma]$ in S. Intuitively, T is the tree of sequences that have not yet entered $\bigcup S$.)

1.4.3. Set theorists sometimes define a tree to be a partial order $(A, \leq)$ such that for every x in A the set $\{y \mid y \leq x\}$ is a well-order, and such that A has a minimum element (the root of the tree). Show that trees, in our sense, correspond to partial orders $(A, \leq)$ where for every x in A the set $\{y \mid y \leq x\}$ is finite.

1.4.4. Let T be a tree on Σ, and suppose Σ is well ordered. Define the *Kleene–Brouwer order* on the nodes of T as follows: say $\sigma \preceq \tau$ if and only if $\sigma \supseteq \tau$ or for some i, $(\sigma)_i \neq (\tau)_i$, and $(\sigma)_i$ is less than $(\tau)_i$ in the order on Σ for the least such i. Visually, this means that $\sigma \preceq \tau$ if and only if σ extends τ or is to the left of it in the tree.

Show that the Kleene–Brouwer order on T is a linear order, and that it is a well-order if and only if T is well founded. (Hint: There are at least three ways to show that if T is well founded then the Kleene Brouwer order is a well-order. One option is to use induction on well-founded trees to show that every nonempty subset S of T has a least element in the Kleene–Brouwer order. Another is to suppose T is not well founded in the Kleene–Brouwer order, let $\sigma_0, \sigma_1, \sigma_2, \ldots$ be an infinite descending sequence, and then carefully choose a sequence $\tau_0 \subseteq \tau_1 \subseteq \ldots$ such that each τ_{i+1} is the leftmost child of τ_i with some σ_j below it. A variant of this second proof is to let T' be the tree of initial segments of the σ_is, and use the fact that T' cannot be finitely branching to produce an infinite descending sequence.)

1.5 Structural Recursion

We can express the principle of recursion on the natural numbers as follows:

Theorem 1.5.1. *Let B be any set, $a \in B$, and $f : \mathbb{N} \times B \to B$. Then there is a unique function $\varphi : \mathbb{N} \to B$ such that*

- *$\varphi(0) = a$, and*
- *$\varphi(n+1) = f(n, \varphi(n))$ for every $n \in \mathbb{N}$.*

This says that we can define a function φ from the natural numbers to any set B by specifying $\varphi(0)$ and then specifying $\varphi(n+1)$ in terms of n and $\varphi(n)$. For example, we can define the factorial function by specifying $\varphi(0) = 1$ and $\varphi(n+1) = n \cdot \varphi(n)$. Notice that the theorem states not only that there exists a function meeting that specification, but, moreover, that there is exactly one.

We have seen that the set of natural numbers is inductively generated from 0 by the function $\mathrm{succ}(n) = n + 1$. Intuitively, that is what makes the principle of recursion work; the specification of φ in Theorem 1.5.1 mirrors the inductive construction. We would like to have analogous principles of recursion on, say, the set of terms and the set of propositional formulas. That would allow us to define a notion of complexity for terms as follows:

$$\begin{aligned} \mathrm{complexity}(x) &= 1, \text{ for any variable } x \\ \mathrm{complexity}(c) &= 1, \text{ for any constant } c \\ \mathrm{complexity}(f(t_0, \ldots, t_{k-1})) &= \mathrm{complexity}(t_0) + \cdots + \mathrm{complexity}(t_{k-1}) + 1. \end{aligned}$$

We could similarly define a notion of depth:

$$\begin{aligned} \mathrm{depth}(x) &= 0, \text{ for any variable } x \\ \mathrm{depth}(c) &= 0, \text{ for any constant } c \\ \mathrm{depth}(f(t_0, \ldots, t_{k-1})) &= \max(\mathrm{depth}(t_0), \ldots, \mathrm{depth}(t_{k-1})) + 1. \end{aligned}$$

A moment's reflection shows that we have to be careful, however. We would like to specify the clauses of the recursive definition in terms of the rules that comprise the corresponding inductive definition of the set, but, a priori, there is nothing that guarantees the existence of a function meeting this specification. As an example, let A be the smallest set of natural numbers satisfying the following:

- 0 is in A.
- If n is in A, then so is $n+2$.
- If n is in A, then so is $n+4$.

This is just a roundabout way of defining the set of even numbers, and the third clause is redundant. But now suppose we try to use this description to define a function $\varphi \colon A \to \mathbb{N}$ recursively, using the following conditions:

- $\varphi(0)=0$
- $\varphi(n+2)=\varphi(n)+1$, for every n in A
- $\varphi(n+4)=\varphi(n)+3$, for every n in A

This specification is contradictory: any function φ satisfying these conditions must satisfy

$$\varphi(4)=\varphi((0+2)+2)=\varphi(0+2)+1=(\varphi(0)+1)+1=(0+1)+1=2,$$

as well as

$$\varphi(4)=\varphi(0+4)=\varphi(0)+3=0+3=3,$$

which is clearly impossible.

What went wrong is that in the definition of A there are two ways that 4 can enter the set, and each one gives rise to a different recipe for computation. When it comes to the set of terms, however, the unique readability theorem, Theorem 1.3.4, rules out that possibility. What we want is a general theorem that provides a principle of recursion in situations like this.

We will restrict our attention to the special case of Example 1.2.3, where A is generated by a set $\mathfrak{F}$ of functions from U to U of various arities.

Definition 1.5.2. Let A be inductively generated by $\mathfrak{F}$. A is *freely generated* by $\mathfrak{F}$ if the following hold:

- Each function $f \in \mathfrak{F}$ is injective on A.
- The images of the functions in $\mathfrak{F}$ on A are disjoint from one another.

The second condition of the definition means that if f and f' are distinct elements of $\mathfrak{F}$, f is k-ary, and f' is ℓ-ary, then whenever $a_0, \ldots, a_{k-1}, a'_0, \ldots, a'_{\ell-1}$ are in A, $f(a_0, \ldots, a_{k-1})$ is not equal to $f'(a'_0, \ldots, a'_{\ell-1})$. Theorem 1.3.4 shows that the set of terms is freely generated by the term forming operations from the variables and constant symbols, and Theorem 1.3.6 shows that the set of propositional formulas is freely generated from the set of propositional variables.

We can now generalize Theorem 1.5.1 to any freely generated inductive structure.

Theorem 1.5.3. *Suppose A is inductively and freely generated by $\mathfrak{F}$. Let B be any nonempty set, and suppose that with each f in $\mathfrak{F}$ we associate a function g_f from B to B of the same* *arity. Then there is a unique function* $\varphi\colon A \to B$ *such that for every k-ary function* f *and* $a_0, \ldots, a_{k-1}$ *in* A *we have*

$$\varphi(f(a_0, \ldots, a_{k-1})) = g_f(\varphi(a_0), \ldots, \varphi(a_{k-1})).$$

This is often called the *recursion theorem* for freely generated inductive structures. It asserts the existence of a function φ that translates A to B in such a way that applying f in A translates to applying g_f in B. Here k may be 0, so in the base case we have equations of the form $\varphi(c) = d$, where d is a constant of B associated with the constant c in A.

Proof For any partial function $h\colon A \to B$, say h is *good* if it satisfies the specification as far as it is defined; in other words, whenever $f(a_0, \ldots, a_{k-1})$ is in the domain of h, then so are $a_0, \ldots, a_{k-1}$, and

$$h(f(a_0, \ldots, a_{k-1})) = g_f(h(a_0), \ldots, h(a_{k-1})).$$

I claim that if h_1 and h_2 are both good and defined at $a \in A$, then $h_1(a) = h_2(a)$. Let us prove this by induction on A. Suppose $a = f(a_0, \ldots, a_{k-1})$, and, inductively, suppose that the claim holds of $a_0, \ldots, a_{k-1}$. By the definition of "good," each a_i is in the domain of both h_1 and h_2, and by the inductive hypothesis we have $h_1(a_i) = h_2(a_i)$ for each $i < k$. Again by the definition of "good," we now have

$$h_1(a) = g_f(h_1(a_0), \ldots, h_1(a_{k-1})) = g_f(h_2(a_0), \ldots, h_2(a_{k-1})) = h_2(a),$$

as required.

Let $\varphi = \bigcup\{g \mid g \text{ is good}\}$. Here we are relying on the set-theoretic definition of a function as a set of ordered pairs, so for every $a \in A$ and $b \in B$, $(a, b) \in \varphi$ if and only if there is some good function h such that $h(a) = b$. By the previous paragraph, φ is a partial function from A to B, which is to say, for every a there is at most one b such that $(a, b) \in \varphi$.

I now claim that φ is good. Suppose $a = f(a_0, \ldots, a_{k-1})$ is in the domain of φ. Then there is a partial function h such that h is good and a is in the domain of h. Since h is good, it is also defined on $a_0, \ldots, a_{k-1}$, and satisfies

$$h(a) = g_f(h(a_0), \ldots, h(a_{k-1})).$$

But then φ is also defined on $a_0, \ldots, a_{k-1}$ and $\varphi(a_i) = h(a_i)$ for each $i = 0, \ldots, k-1$. So

$$\varphi(a) = h(a) = g_f(h(a_0), \ldots, h(a_{k-1})) = g_f(\varphi(a_0), \ldots, \varphi(a_{k-1})),$$

as required.

Next, I claim that φ is in fact a total function from A to B, which is to say, for every a in A, a is in the domain of φ. Prove this by induction on A. Suppose $a = f(a_0, \ldots, a_{k-1})$, and, inductively, that the claim holds of $a_0, \ldots, a_{k-1}$. Then it is not hard to check that $\varphi \cup \{(a, g_f(\varphi(a_0), \ldots, \varphi(a_{k-1})))\}$ is a good function – that is, we can use the specification g_f to extend the definition of φ to a. By the definition of φ, this means that φ is already defined at a.

This shows the existence of a function f meeting the specification of the theorem. But we have already shown uniqueness: if φ_1 and φ_2 both meet the specification, then they are both good, which means that they agree at every element of their domain. □

As with our treatment of inductive definitions, you may consider the set-theoretic machinery to be heavy-handed. Exercise 1.5.3 asks you to provide an alternative proof of the theorem, using formation sequences, in the case where the generating rules are finite.

Since the set of terms is freely generated, we can evaluate them in any L-structure, modulo an assignment of elements of $|\mathfrak{A}|$ to the variables. Specifically, let $\mathfrak{A}$ be any L-structure and let σ be a function which assigns to each variable x an element of $|\mathfrak{A}|$. Then we can define a function $\text{eval}_{\mathfrak{A},\sigma}$ from the set of terms of L to $|\mathfrak{A}|$ as follows:

- $\text{eval}_{\mathfrak{A},\sigma}(x) = \sigma(x)$ for every variable x.
- $\text{eval}_{\mathfrak{A},\sigma}(c) = c^{\mathfrak{A}}$ for every constant symbol c.
- $\text{eval}_{\mathfrak{A},\sigma}(f(t_0, \ldots, t_{k-1})) = f^{\mathfrak{A}}(\text{eval}_{\mathfrak{A},\sigma}(t_0), \ldots, \text{eval}_{\mathfrak{A},\sigma}(t_{k-1}))$, where f is any k-ary function symbol of L and $t_0, \ldots, t_{k-1}$ is any sequence of terms.

We will often use notation like $[\![t]\!]^{\mathfrak{A},\sigma}$ for a function like $\text{eval}_{\mathfrak{A},\sigma}$ that assigns a semantic value, representing the denotation of an expression in a structure.

Let L be a purely functional language, which is to say, a language with no relation symbols. Let T be the set of terms with no variables. Then T itself becomes an L-structure when each function symbol f is interpreted as the corresponding term-forming operation. This structure is known as the *term algebra for L*. In that case, Theorem 1.3.4 says exactly that the term algebra is inductively and freely generated by the interpretations of L in that structure. The use of the word "the" in the phrase "the term algebra for L" is justified by the following:

Theorem 1.5.4. *Let L be a purely functional language, and let $\mathfrak{M}_1$ and $\mathfrak{M}_2$ be two L-structures that are inductively and freely generated by the interpretations of the function symbols of L. Then there is a unique isomorphism between $\mathfrak{M}_1$ and $\mathfrak{M}_2$.*

Proof By the recursion theorem there are unique homomorphisms $f: \mathfrak{M}_1 \to \mathfrak{M}_2$ and $g: \mathfrak{M}_2 \to \mathfrak{M}_1$. The composition $g \circ f: \mathfrak{M}_1 \to \mathfrak{M}_1$ is also a homomorphism, and by the uniqueness part of Theorem 1.5.3, it must be equal to the identity. Similarly, $f \circ g$ is the identity, so f is an isomorphism from $\mathfrak{M}_1$ to $\mathfrak{M}_2$. □

More generally, the set of terms involving a set of variables X together with the same term-forming operations is called the *term algebra for L over X*. There is not much of a principled distinction between variables and constants, so we can think of the term algebra for L over the set of variables X as the ordinary term algebra for a language L_X consisting of the symbols of L together with additional constant symbols for the elements of X. As a result, the term algebra T_X for L over the variables X can be characterized abstractly as a freely generated L_X structure. The set of strings defined in Section 1.3 meets that criterion, and any two such structures are isomorphic.

This shows that our particular choice of symbols and strings to represent terms does not really matter. We could equally well have defined terms to be trees with nodes labeled by the function symbols of L and variables. (See Exercise 1.5.6.) All that matters is that terms of L_X form an L_X structure and that we have the expected principles of induction and recursion. What then characterizes the function $\text{eval}_{\mathfrak{A},\sigma}$ is that it is the unique homomorphism from the term algebra T_X, viewed as an L-structure, to $\mathfrak{A}$ that maps each variable x to $s(x)$.

In any case, now that we have the principle of recursion, we can start defining functions on terms. As suggested above, we can define functions complexity(t) and depth(t) from the set of terms to the natural numbers. We can even give a recursive specification of the function length(t) that returns the number of symbols in the string representation of a term, and prove that it has that property. We can also use the following recursive clauses to define a function that takes any term and returns the set of variables it contains:

- $\text{variables}(x) = \{x\}$
- $\text{variables}(f(t_0, \ldots, t_{k-1})) = \text{variables}(t_0) \cup \ldots \cup \text{variables}(t_{k-1})$.

By induction on terms, we can show that variables(t) is always finite. A term t is said to be *closed* if $\text{variables}(t) = \emptyset$. More interestingly, we can show by induction that the function $[\![t]\!]^{\mathfrak{A},\sigma}$ defined above only depends on the values assigned to the variables of t:

Proposition 1.5.5. *Let* $\sigma, \sigma' : X \to |\mathfrak{A}|$ *and suppose* $\sigma(x) = \sigma'(x)$ *for every* $x \in \text{variables}(t)$. *Then* $[\![t]\!]^{\mathfrak{A},\sigma} = [\![t]\!]^{\mathfrak{A},\sigma'}$.

The function subterms(t) assigns to each term t the its set of terms that occur within it:

- $\text{subterms}(x) = \{x\}$
- $\text{subterms}(c) = \{c\}$
- $\text{subterms}(f_i(t_0, \ldots, t_{k-1})) = \{f_i(t_0, \ldots, t_{k-1})\} \cup \text{subterms}(t_0) \cup \ldots \cup \text{subterms}(t_{k-1})$.

A term s is said to be a *subterm of* t if $s \in \text{subterms}(t)$.

We can also use recursion to say what it means to substitute a term s for a variable x in another term t. This operation is fundamental in logic, and the various forms of notation used to denote it in the literature include $t[s/x]$, $t[x/s]$, $[s/x]t$, $t[x := s]$, $t[x \leftarrow s]$, $t[x \mapsto s]$, and t^x_s. We will adopt the first notation. Keeping s and x fixed, the definition is by recursion on t:

- $w[s/x] = \begin{cases} s & \text{if } w \text{ is the variable } x \\ w & \text{if } w \text{ is any other variable} \end{cases}$
- $c[s/x] = c$, for any constant c
- $(f(r_0, \ldots, r_{k-1}))[s/x] = f(r_0[s/x], \ldots, r_{k-1}[s/x])$.

For an example of a proof by induction, consider the following:

Proposition 1.5.6. *For every pair of terms t and s and variable x*, $\text{depth}(t[s/x]) \leq \text{depth}(t) + \text{depth}(s)$.

Proof Fix s and x and use induction on t. If t is a constant or a variable other than x, then $\text{depth}(t[s/x]) = \text{depth}(t) = 0 \leq \text{depth}(t) + \text{depth}(s)$. If t is the variable x, then $\text{depth}(t[s/x]) = \text{depth}(s) = \text{depth}(t) + \text{depth}(s)$.

In the inductive step, suppose t is of the form $f_i(t_0, \ldots, t_{k-1})$, and that the proposition is true of each t_i in place of t. Then

$$\begin{aligned} \text{depth}(f_i(t_0, \ldots, t_{k-1})[s/x]) &= \text{depth}(f_i(t_0[s/x], \ldots, t_{k-1}[s/x])) \\ &= \max(\text{depth}(t_0[s/x]), \ldots, \text{depth}(t_{k-1}[s/x])) + 1 \\ &\leq \max(\text{depth}(t_0) + \text{depth}(s), \ldots, \text{depth}(t_{k-1}) + \text{depth}(s)) + 1 \\ &= \max(\text{depth}(t_0), \ldots, \text{depth}(t_{k-1})) + \text{depth}(s) + 1 \\ &= \text{depth}(f_i(t_0, \ldots, t_{k-1})) + \text{depth}(s), \end{aligned}$$

as required. □

You can verify that the following properties also hold.

Proposition 1.5.7. *Suppose x and y are distinct variables, and r, s, and t are terms.*

1. *If $x \notin$ variables(t), then $t[s/x] = t$.*
2. *If $x \in$ variables(t), then* variables$(t[s/x]) = ($variables$(t) \setminus \{x\}) \cup$ variables(s).
3. *If x is not a variable of r, then $(t[s/x])[r/y] = (t[r/y])[(s[r/y])/x]$.*

The last claim shows that substitutions do not commute, since in the expression $(t[s/x])$ $[r/y]$, the term r is substituted not only for the occurrences of y in t but also for occurrences of y in occurrences of s in $t[s/x]$. Given a tuple of distinct variables $x_0, \ldots, x_{n-1}$, there is a natural notion of *simultaneous substitution* $t[s_0/x_0, \ldots, s_{n-1}/x_{n-1}]$ that can be defined by recursion on t. One need only replace the base case in the definition of single substitution by

$$w[s_0/x_0, \ldots, s_{n-1}/x_{n-1}] = \begin{cases} s_i & \text{if } w \text{ is the variable } x_i \\ w & \text{if } w \text{ is any other variable.} \end{cases}$$

Functional programming languages often provide built-in means of describing data types inductively and defining functions on such data recursively. The principle of induction on an inductively defined set is sometimes called *structural induction*, and the corresponding principle of recursion is sometimes called *structural recursion*. Taken together, these principles often provide a clean mechanism for defining operations on syntactic objects and reasoning about them, but they need not be adhered to slavishly. The existence of functions like length, complexity, and depth means that we can also use ordinary induction and recursion on the natural numbers. For example, we can define a function f on the set of terms by specifying the value of $f(t)$ in terms of the value of f on terms of smaller complexity, and we can prove that a property P holds of every term, by showing that $P(t)$ holds under the assumption that it holds for terms of smaller complexity.

Exercises

1.5.1. On the surface, the principle of recursion on the natural numbers, Theorem 1.5.1, is slightly more general than the principal of recursion offered by Theorem 1.5.3, in that the argument, n, appears on the right side of the second equation in Theorem 1.5.1. Strengthen Theorem 1.5.3 so that each g_f takes additional arguments from A, and φ is required to satisfy:

$$\varphi(f(a_0, \ldots, a_{k-1})) = g_f(a_0, \ldots, a_{k-1}, \varphi(a_0), \ldots, \varphi(a_{k-1})).$$

You can do this with straightforward modifications to the proof of Theorem 1.5.3.

1.5.2. Generalize the principal of recursion, Theorem 1.5.3, to obtain a principle of recursion on the set of well-founded trees, as defined by Definition 1.4.2.

1.5.3. Give an alternative, more explicit proof of the recursion theorem in the case where all the generating rules are finite, using the characterization of the inductively generated set in terms of formation sequences.

1.5.4. Let L be the language with one constant symbol c, two unary functions f and g, and one binary function h. Show that the free algebra on L is isomorphic to the set of finite trees on $\{0, 1\}$, where c is mapped to the tree with a single node, and f, g, and h build composite trees, where the root has only a left subtree, only a right subtree, or both, respectively.

1.5.5. On the set of finite trees on $\{0, 1\}$, define depth by

- depth$(\emptyset) = -1$
- depth$(T) = \max($depth$(T\restriction_{(0)})$, depth$(T\restriction_{(1)})) + 1$ for $T \neq \emptyset$,

and define size by

- $\text{size}(\emptyset) = 0$
- $\text{size}(T) = 1 + \text{size}(T\restriction_{(0)}) + \text{size}(T\restriction_{(1)})$ for $T \neq \emptyset$.

Equivalently, size(T) is just the cardinality of the set T.

a. Show by induction that for every tree a we have $\text{size}(a) \leq 2^{\text{depth}(a)+1} - 1$.
b. Define the function numleaves(a) that returns the number of leaves of a, and show that for every a we have $\text{size}(a) \geq 2 \cdot \text{numleaves}(a) - 1$.

1.5.6. Let T be a tree on Σ. Recall that a *labeling* of the nodes of T with elements of a set A is just a function $\ell \colon T \to A$. Show that given a language $L = (F, \emptyset, a)$ with no relation symbols, we could have taken the set of terms T_X with variables in X to be a set of well-founded trees on $\mathbb{N}$ with labels from $F \cup X$. For example, you might represent the term $f(g(a), x)$ by the tree $\{(), (0), (0, 0), (1)\}$ where $()$ has the label f, (0) has the label g, $(0, 0)$ has the label a, and (1) has the label x.

Specifically, you should define a set of labeled trees inductively, and then show that it is a freely-generated L_X structure and hence isomorphic to the term algebra.

1.6 Bound Variables

In Chapter 4, we will consider first-order logic, which introduces the universal quantifier, $\forall$, and the existential quantifier, $\exists$. The first is used to say that every object in a domain under consideration has a given property, while the second is used to say that at least one object has the property. For example, $\forall x\, (x = x)$ says, informally, that every object is equal to itself, and $\exists x\, (x > 0)$ says that some object is greater than 0. Quantifiers can be nested, so that $\forall x\, \exists y\, (y > x)$ says that for every object, there is one greater than it. The variables in these expressions are said to be *bound*, and the quantifiers are said to be *binding operations* or *binders*.

Adding binders to a language introduces two complications that we will address here. The details are fiddly, however, and you may wish to skip this section until you get to Chapter 4, and then refer back to it as necessary.

The first complication is that, intuitively, we want to think of bound variables as mere placeholders. The expression $\forall x\, (x = x)$ can be read "everything is equal to itself," an expression which does not mention x at all. We would like to think of this expression as being the same as $\forall y\, (y = y)$, much the way we think of $\sum_{i=1}^{n} i^2$ and $\sum_{j=1}^{n} j^2$ is being the same sum.

The second complication is that the notion of substitution becomes dicey. We would like to say that the expression $\forall x\, \exists y\, (y > x)$ is true of the natural numbers, and this should mean that the expression $\exists y\, (y > x)$ is true of any term we wish to substitute for x. But the assertion is patently false if we substitute $y + 1$ for x, since we end up with the expression $\exists y\, (y > y + 1)$. What happened is that the variable y in the expression $y + 1$ was *captured* by the existential binder $\exists y$, changing the meaning of the term.

There are two options for dealing with the first complication. The first is to stay close to syntax and deny that $\forall x\, (x = x)$ and $\forall y\, (y = y)$ are the same formula. We can say what it means for two formulas to be *equivalent up to renaming of bound variables*, or *α-equivalent*, and the things we say about formulas will generally respect this equivalence. For instance, if A is provable in some formal calculus and A' is α-equivalent to A, then A' will generally be provable as well. With the first option, however, A and A' are different formulas.

The second option is to quotient the set of expressions by α-equivalence, so that when we speak of a formula A, we are really talking about an equivalence class of syntactic expressions. We view operations on formulas as operations on these equivalence classes and we view assertions about formulas as assertions about these equivalence classes.

In some respects, there is not much difference between the two approaches. With the first option, we deal with expressions and prove that our operations and concepts respect α-equivalence after the fact. With the second option, we deal with abstract classes, and we have to prove that our operations and concepts respect α-equivalence of expressions in order to lift them to operations and concepts at the abstract level. But the second option does encapsulate a mathematical commitment. With the first option, we could, in principle, define operations and concepts that do not respect the equivalence; we could refer to "the variable bound by the universal quantifier in $\forall x\,(x=x)$," or say that "x occurs as a bound variable in $\forall x\,(x=x)$." These phrases do not make sense if we think of $\forall x\,(x=x)$ as denoting the same object as $\forall y\,(y=y)$.

With respect to the first option, there are various ways of dealing with the complication arising from capture in substitution. One possibility is simply to give a name to the problem, and rule it out: we say what it means for a term t to be *free for* a variable x in an expression A, and we qualify our definitions and assertions so that t is only substituted for x when this is the case. A second possibility is to maintain separate stocks of free and bound variables, so that the problem never arises. Yet a third possibility is to modify the definition of substitution so that bound variables get renamed as needed, so that the result of substituting $y+1$ for x in $\exists y\,(y>x)$ is something like $\exists z\,(z>y+1)$.

None of these variations is fully satisfactory. The first involves inserting qualifications everywhere and checking that they are maintained. The second requires us to rename a variable whenever we wish to apply a binder. With the third, even though we can fix a recipe for executing the renaming, the choice is somewhat arbitrary. Moreover, because of the renamings, statements we make about substitutions will generally hold only up to α-equivalence, cluttering up our statements.

With the second option – viewing formulas as equivalence classes – we have no choice but to define substitution in a way that respects α-equivalence. When we say what it means to substitute a term like $y+1$ for a variable like x in an expression like $\exists y\,(y>x)$, we cannot distinguish the latter from $\exists z\,(z>x)$. So we have to choose a definition that is independent of the syntactic representative.

In this book, we will identify formulas that differ only up to the names of their bound variables, and we will define substitution accordingly. This commits us to making sure that everything we say respects α-equivalence.

Recall that a language, L, consists of a set F of function symbols, a set R of relation symbols, and function a assigning an arity to each. To illustrate the handling of bound variables, we now define the set of *first-order formulas in L*. As suggested above, we will proceed in two stages. First, we will define a set of *syntactic formulas* inductively. Then we will say what it means for two syntactic formulas to be α-equivalent. Then we will define the set of formulas to be the quotient of the set of syntactic formulas by the equivalence relation. Finally, we will lift a number of functions defined on the set of syntactic formulas to functions on the set of formulas.

A syntactic formula (relative to L) will be a string of symbols over an alphabet that includes all the symbols in L, a stock of variables $x_0, x_1, x_2, \ldots$, parentheses, the comma symbol, $\bot$, $\wedge$, $\vee$, and $\to$, the equality symbol $=$, and the quantifiers $\forall$ and $\exists$. We have already defined the set of terms of L, and the next definition builds on that.

Definition 1.6.1. The set of syntactic formulas is defined inductively as follows:

- $\bot$ is a syntactic formula.
- If s and t are terms, $s = t$ is a syntactic formula.
- If $t_0, \ldots, t_{k-1}$ are terms and R is a k-ary relation symbol of L, then $R(t_0, \ldots, t_{k-1})$ is a syntactic formula.
- If A and B are syntactic formulas, then so are $(A \wedge B)$, $(A \vee B)$, and $(A \to B)$.
- If A is a syntactic formula and x is a variable, then $(\forall x\, A)$ and $(\exists x\, A)$ are syntactic formulas.

Syntactic formulas of the form $s = t$, $\bot$, and $R(t_0, \ldots, t_{k-1})$ are said to be *atomic*. They give rise to the base cases in proofs by induction. The same terminology will carry over to the set of formulas. (Relations are sometimes also called *predicates*, and so first-order logic is also known as *predicate logic*.)

As we did with the set of terms, we can show that this set is freely generated by the clauses, yielding the corresponding principles of induction and recursion. In particular, we can define notions of depth and complexity, as we did for the set of terms. We can also define the set of free variables $\mathrm{fv}(A)$ of a formula expression A recursively as follows:

- $\mathrm{fv}(\bot) = \emptyset$
- $\mathrm{fv}(s = t) = \mathrm{variables}(s) \cup \mathrm{variables}(t)$
- $\mathrm{fv}(R(t_0, \ldots, t_{k-1})) = \mathrm{fv}(t_0) \cup \ldots \cup \mathrm{fv}(t_{k-1})$
- $\mathrm{fv}(A \circ B) = \mathrm{fv}(A) \cup \mathrm{fv}(B)$, where $\circ$ is any of the propositional connectives.
- $\mathrm{fv}(\forall x\, A) = \mathrm{fv}(\exists x\, A) = \mathrm{fv}(A) \setminus \{x\}$.

The last clause reflects the fact that x becomes bound in $\forall x\, A$ and $\exists x\, A$, and hence is no longer a free variable. By induction on syntactic formulas, it is not hard to show that the set of free variables of a formula is always finite. Intuitively, the truth value of a formula will depend on the values that are assigned to the free variables. A formula with no free variables is called a *sentence*.

Let us now define a notion of substitution for syntactic formulas that renames bound variables dynamically. To do that, we need an auxiliary function, $\mathrm{variables}(A)$, that returns the set of variables occurring in the expression A, including the bound ones. The definition is straightforward, and the statement "z is a variable of A" in the definition below means that z is either a free or a bound variable in this sense. The notion $A[t/x]$ of substitution with renaming is defined with the following clauses:

- $(r = s)[t/x]$ is the formula $(r[t/x] = s[t/x])$.
- $(R(s_0, \ldots, s_{k-1}))[t/x] = R(s_0[t/x], \ldots, s_{k-1}[t/x])$.
- $(A \circ B)[t/x] = (A[t/x] \circ B[t/x])$.
- $(\forall x\, A)[t/x] = \forall x\, A$.
- $(\forall y\, A)[t/x] = \forall y\, (A[t/x])$ when $y \neq x$, and either y is not a variable of t or x is not a free variable of A.
- $(\forall y\, A)[t/x] = \forall z\, ((A[z/y])[t/x])$ when $y \neq x$, y is a variable of t, x is a free variable of A, and z is the first variable in the sequence $x_0, x_1, x_2, \ldots$ such that z is neither a variable of t nor of A.
- $(\exists x\, A)$ is defined similarly, replacing $\forall$ by $\exists$ in the preceding three clauses.

The second-to-last clause concerns the case where we have to rename the bound variable y because otherwise the corresponding variable in t would be captured by the substitution. The nested appearance of $A[z/y]$ on the right-hand side of that equation means that the definition is not a structural recursion on A, but because the substitution $A[z/y]$ does not increase the complexity of the formula, both recursive calls involve a formula that is less complex than the one on the left-hand side, which means that the entire definition can be justified by a suitable recursion on the natural numbers.

Define the notion of α-equivalence between syntactic formulas, $A \equiv_\alpha B$, to be the smallest satisfying the following:

- $\forall x\, A \equiv_\alpha \forall y\, (A[y/x])$ and $\exists x\, A \equiv_\alpha \exists y\, (A[y/x])$, for any formula A and variable y that is not a free variable of A.
- $\equiv_\alpha$ respects all the propositional connectives (so $A \wedge B \equiv_\alpha A' \wedge B'$ if $A \equiv_\alpha A$ and $B \equiv_\alpha B'$, and so on).

Syntactic formulas that are α-equivalent are also said to be *equivalent up to renaming of bound variables*.

We now define the set of *formulas* in L to be the set of syntactic formulas modulo α-equivalence. With some care, we can show that the definition $A[t/x]$ respects α-equivalence, which is to say, $A \equiv_\alpha A'$ implies $A[t/x] \equiv A'[t/x]$. We can also show that all the formula-building operations, namely, composition of formulas with $\wedge$, $\vee$, $\to$ and even the binding operations $A \mapsto \forall x\, A$ and $A \mapsto \exists x\, A$ respect α-equivalence. So do the functions complexity(A) and depth(A), so all these can be lifted to functions on formulas. When the dust settles, we have the following facts and principles of reasoning.

We start with lifted versions of all of the constructions that went into the definition of syntactic formulas, corresponding to the clauses of Definition 1.6.1: for each k-ary relation symbol R and sequence of terms $t_0, \ldots, t_{k-1}$, there is a formula $R(t_0, \ldots, t_{k-1})$; given two formulas A and B, there are formulas $A \wedge B$, $A \vee B$, and $A \to B$; and given a formula A and a variable x, there are formulas $\forall x\, A$ and $\exists x\, A$.

The set of formulas is generated inductively by these constructions. In particular, we can prove that every formula has a certain property by showing that the atomic formulas do and showing that the property is preserved by the constructions. The set of formulas is *not* freely generated by these constructions, however: we have $\forall x\, A = \forall y\, A'$ for distinct pairs (x, A) and (y, A'), and similarly for the existential quantifier. But this is the only thing that can go wrong: any formula falls uniquely under one of the cases listed in Definition 1.6.1, and if A is of the form $R(t_0, \ldots, t_{k-1})$, then R and $t_0, \ldots, t_{k-1}$ are uniquely determined; if A is of the form $C \wedge D$, then C and D are uniquely determined; and so on. Moreover, if A is of the form $\forall x\, C$, then for every variable y not among the free variables of A, there is a unique formula C such that $A = \forall y\, C$. We can think of C as the body of A relative to the choice of variable y, or the result of *opening* the binder in A with y. Moreover, if x and y are distinct, $\forall x\, C = \forall y\, C'$ if and only if y is not free in $C' = C[y/x]$, which, symmetrically, is the same as saying that x is not free in C' and $C = C'[x/y]$. Similar considerations hold for the existential formula.

As a result of all this, we can still define a function f by recursion of formulas, provided that in defining clause for $f(A)$ when A begins with a universal or existential quantifier, we fix some determinate method of choosing a variable not free in A and define $f(A)$ in terms of the result of opening A with that variable. Or, better: define the value of $f(\forall x\, C)$ in terms of x and C in such a way that we assign the same value for any y and C' satisfying $\forall x\, C =$

$\forall y\, C'$, and similarly for the existential quantifier. Of course, we can also define functions by recursions on complexity or depth, or any other well-founded measure we wish to assign.

Henceforth, we will work with formulas, rather than syntactic formulas.

Notice that although we identify the formulas $\exists y\,(y > x)$ and $\exists z\,(z > x)$, the formula $\exists y\,(y > w)$ is not the same as either of these. The first two formulas say that there is something greater than x, while the third says that there is something greater than w. We express this by saying that x is a free variable in the first two expressions, whereas w is a free variable in the third. Intuitively, a formula says something about its free variables.

It is sometimes important to know that we can represent formulas as finite objects, for example, by representing each formula as a natural number or string. To that end, thinking of a formula as an infinite set of equivalent expressions will not do. What we can do is, instead, choose a canonical representative of each class, such as the expression in the class that comes first in alphabetical order. This will not make our operations computationally efficient, but it is sufficient to show that they are computable in principle.

When it comes to implementing software that manipulates formulas, of course, efficient representation is a central concern. Since the names of the bound variables are generally irrelevant, one solution is to replace them with *de Bruijn indices*. Rather than associate a variable with each binder, each would-be occurrence of a bound variable is replaced by a natural number indicating how many binders one has to traverse before reaching the one associated with the variable. For example, the formula $\forall x\,(P(x) \wedge \exists y\, R(y, x))$ is represented by the expression $\forall\,(P(0) \wedge \exists\, R(0, 1))$. In the expression $P(0)$, the 0 points to the universal quantifier, which is the closest binder in scope. In $R(0, 1)$, the 0 points to the existential quantifier, and the 1 points to the universal quantifier. Operations on formulas then act on these representations.

We will not be concerned with efficient representations here. Rather, our approach to identifying formulas up to renaming of bound variables is intended to clarify and explain what the more efficient implementations are meant to achieve.

Exercises

1.6.1. Fix a first-order language L. Say a term t is *free for* x in a syntactic formula A if, recursively, one of the following cases holds:

- A is atomic.
- A is of the form $(\neg C)$, and t is free for x in C.
- A is of the form $(C \circ D)$, and t is free for x in C and D.
- A is of the form $\forall y\, C$ or $\exists y\, C$, $y \neq x$, and t is free for x in C.

Do the following:

a. Show that, assuming there are infinitely many variables, we can always find a formula A' that is α-equivalent to A such that t is free for x in A'. (Assuming that there are infinitely many variables means that we can always choose a "fresh" variable that does not occur anywhere in A.)
b. Give a simpler definition of the substitution operation on the assumption that t is free for x in A, and show that it agrees with the definition given above in that case.

1.6.2. Show that syntactic formulas A and A' are equivalent if and only if one of the following cases holds:

- A is atomic and $A = A'$.
- A is of the form $C \circ D$, A' is of the form $C' \circ D'$, $C \equiv_\alpha C'$, and $D \equiv_\alpha D'$.
- A is of the form $\exists y\, C$, A' is of the form $\exists z\, C'$, and there are $B \equiv_\alpha B'$ and a variable w such that $C = B[y/w]$ and $C' = B'[z/w]$; and similarly for $\forall$.

1.6.3. Show that substitution respects α-equivalence: if $A \equiv_\alpha A'$, then $A[t/x] \equiv_\alpha A'[t/x]$.

1.6.4. Show that if y is not free in A, then $A[y/x][x/y] = A$.

Bibliographical Notes

Languages and structures These form the basis for a model-theoretic approach to logic, described in Chapter 5; see the references there.

Inductive definitions In 1888, in an essay titled *Was sind und was sollen die Zahlen?* (roughly, "What are the numbers, and what should they be?"), Richard Dedekind gave an inductive characterization of the natural numbers similar to the one presented in this section. For a more on inductively defined sets and structures, see Aczel (1977), Section 16.5, and Chapter 17.

Recursive definitions and inductively defined data structures play an important role in computer science. In addition to this chapter, sections 8.4 and 17.6 especially hint at some of relevant theoretical considerations. See, for example, Harper (2016) and Pierce (2002) for the perspective from the semantics of programming languages.

Bound variables Barendregt (1981) discusses ways of identifying formulas up to renaming of bound variables in the context of the lambda calculus, and Charguéraud (2012) provides a discussion of implementation issues.

History of logic and foundations Kneale and Kneale (1962) is a history of logic through the early twentieth century and Mancosu et al. (2009) is a history of logic from 1900 to 1935. van Heijenoort (1967) is a sourcebook consisting of seminal papers in logic from the late nineteenth century to the middle of the twentieth century, and Ewald (1996) is a sourcebook with writings on the foundations of mathematics from the seventeenth century to the early nineteenth century. The bibliography and bibliographical notes to Kleene (1952) provide a sense of the state of mathematical logic by the early 1950s, and the *Handbook of Mathematical Logic*, Barwise (1977), provides a sense of the landscape of the subject by the 1970s.

The field is now far too vast to provide a synoptic overview. To readers interested in learning more about the recent history of the subject, I recommend following up on the references to individual topics that are found at the end of each chapter. Here I will highlight a few that have either helpful historical information or extensive bibliographies, or both. For proof theory, the *Handbook of Proof Theory* (Buss (1998a)), is thorough, and Troelstra and Schwichtenberg (2000) is another good reference. For model theory, Hodges (1993) is a good place to start. For computability theory, Soare (2016) has a chapter on the history of the subject as well as reading recommendations. For computability, there are the *Handbook of Computability Theory* (Griffor (1999)), Odifreddi (1989), and Odifreddi (1999). For intuitionistic logic and constructivism, Troelstra and van Dalen (1988), Beeson (1985), and Troelstra (1973) are good references. For set theory, Kanamori (2003) provides a wealth of information. Hájek and Pudlák (1998) provides extensive bibliographical information related to the metamathematics of classical first-order arithmetic. For logic in automated reasoning, *The Handbook of Automated Reasoning* (Robinson and Voronkov (2001)) is an extensive reference, and Harrison (2009) provides informative bibliographical notes and references. For philosophical topics, there is *The Oxford Handbook of Philosophy of Mathematics and Logic* (Shapiro (2005)). The essays in Mancosu (2008) survey some more recent topics of interest.

2

Propositional Logic

In Chapter 1, we briefly considered the language of propositional logic, which can be used to build compound expressions out of basic propositions, using connectives like *and*, *or*, and *implies*. In this chapter, we will consider formal deductive systems for propositional logic, including both *classical* and *intuitionistic* versions. Roughly speaking, intuitionistic logic is designed to reflect constructive or computational forms of reasoning, whereas classical logic incorporates nonconstructive methods that are commonly used in mathematics, like proof by contradiction. We will also consider a variant of intuitionistic logic known as *minimal* logic, which has a slightly better computational interpretation.

In Chapter 3, there is a precise semantic account of what a deductive system for classical logic is supposed to do. A propositional formula is *valid*, or a *tautology*, if it expresses a true proposition under every interpretation of the propositional variables. The role of a proof system is to derive tautologies, and an ideal proof system will enable us to derive all of them.

The familiar method of truth tables shows that there is a straightforward algorithm for determining whether a propositional formula is classically valid. Why then bother with a deductive system? For one thing, it is often more efficient to present a proof of a tautology than to write out its full truth table. More importantly, deductive systems provide a more faithful model of mathematical reasoning: when proving a theorem, we do not typically use truth tables to determine whether or not an inference is valid, but, rather, justify the inference using a sequence of smaller, more basic ones. Formal systems are designed to model that practice.

The semantics of intuitionistic and minimal propositional logic are not nearly as straightforward as that of classical logic. In those cases, the deductive systems themselves may provide the best characterization of what it means for a formula to be valid. Moreover, when it comes to first-order logic, the set of valid statements (in any of the classical, intuitionistic, or minimal variants) is not algorithmically decidable; the problem of deciding the validity of a sentence in first-order logic is algorithmically equivalent to the halting problem. Characterizing the valid sentences of first-order logic in terms of provability in a formal deductive system is an important part of establishing that equivalence.

2.1 The Language of Propositional Logic

Recall that in Section 1.3 we defined the set of *propositional formulas* as the smallest set of strings over a suitable alphabet satisfying the following:

- If P is a propositional variable, then P is a propositional formula.
- $\bot$ is a propositional formula.
- If A and B are propositional formulas, so are $(A \wedge B)$, $(A \vee B)$, and $(A \to B)$.

This set is freely generated, so we have principles of induction and recursion on formulas. The set of propositional formulas can be viewed as the term algebra over the set of propositional variables for a language with a constant, $\bot$, and binary functions for $\wedge$, $\vee$, and $\to$.

Remember also that $\bot$ is intended to denote any false statement, and $(A \wedge B)$, $(A \vee B)$, and $(A \to B)$ are intended to denote "A and B," "A or B," and "if A then B," respectively. Rather than add another symbol for negation, we will take $(\neg A)$, intended to denote "not A," to abbreviate $(A \to \bot)$. You can think of this as saying that A implies something absurd, like saying "if A is true, then pigs have wings." We will see below that this gives $\neg A$ the formal deductive properties it ought to have. We can take the symbol $\top$, for "true," to abbreviate $\bot \to \bot$, and we can take $(A \leftrightarrow B)$, for "A if and only if B," abbreviates $((A \to B) \wedge (B \to A))$. Note that expanding this abbreviation can be costly if we are concerned about the length of the underlying strings.

When writing propositional formulas informally, it is convenient to leave out parentheses, with the understanding that they are implicitly inserted according to the following order of operations:

1. $\neg$ binds most tightly.
2. $\wedge$ and $\vee$ come next, read from left to right.
3. $\to$ and $\leftrightarrow$ come last.

So, for example, $\neg P \wedge Q \to R \vee \neg S$ is really $(((\neg P) \wedge Q) \to (R \vee (\neg S)))$. Some logicians adopt the convention that $\wedge$ binds more tightly than $\vee$. Also, some logicians adopt the convention that multiple arrows are read from right to left, so that the formula $P \to Q \to R \to S$ can be read as the assertion that S follows from hypotheses P, Q, and R. In cases like these, I will use the extra parentheses for clarity.

We define the following function by recursion on formulas:

- $\text{complexity}(\bot) = \text{complexity}(P) = 1$
- $\text{complexity}(A \circ B) = \text{complexity}(A) + \text{complexity}(B) + 1$.

Here and in the definitions below, $P, Q, R, \ldots$ stand for propositional variables and $\circ$ stands for any of the binary connectives $\wedge$, $\vee$, or $\to$. The function $\text{depth}(A)$ computes the depth of a formula's parse tree:

- $\text{depth}(\bot) = \text{depth}(P) = 0$
- $\text{depth}(A \circ B) = \max(\text{depth}(A), \text{depth}(B)) + 1$.

The function $\text{variables}(A)$ maps each formula A to the set of propositional variables occurring in A:

- $\text{variables}(\bot) = \emptyset$
- $\text{variables}(P) = \{P\}$
- $\text{variables}(C \circ D) = \text{variables}(C) \cup \text{variables}(D)$.

The function $\text{subformulas}(A)$ assigns to each formula A its set of subformulas:

- $\text{subformulas}(\bot) = \{\bot\}$
- $\text{subformulas}(P) = \{P\}$
- $\text{subformulas}(C \circ D) = \text{subformulas}(C) \cup \text{subformulas}(D) \cup \{C \circ D\}$.

As with terms, we define $A[B/P]$ to be the result of substituting the formula B for the variable P in A. The notion has properties that are similar to the properties of substitution on terms.

Exercises

2.1.1. The set of propositional formulas in *prefix* form is defined inductively as follows:

- $\bot$ is a prefix formula.
- Any variable P is a prefix formula.
- If A and B are prefix formulas, so are $\wedge AB$, $\vee AB$, and $\to AB$.

This is just another notation for propositional formulas in which the connectives come in front of the arguments rather than between them. Note that we have dispensed with parentheses. Ordinary propositional formulas, as we have defined them, are said to be in *infix* form.

a. Convert $\wedge{\to}P_1P_2{\vee}P_3P_4$ to a regular propositional formula.
b. Convert $((P_1 \vee P_2) \to (P_3 \vee P_4))$ to a prefix formula.
c. Show that the set of prefix formulas is freely generated.
d. Use definition by recursion to define translation functions from prefix to infix, and vice versa.
e. Come up with the corresponding definition and translations for *postfix* formulas, in which the connectives come after the arguments.

2.1.2. Prove that for any formula A, $|\text{subformulas}(A)| \leq \text{complexity}(A)$.

2.1.3. Prove by an appropriate induction on formulas that if A is a subformula of B and B is a subformula of C, then A is a subformula of C.

2.2 Axiomatic Systems

Axiomatic deductive systems are neither pleasant to use nor convenient to study, but they have played an important historical role. The following set of axioms is essentially that used by the David Hilbert in the 1930s:

1. $A \to (B \to A)$
2. $(A \to (B \to C)) \to ((A \to B) \to (A \to C))$
3. $A \to (B \to A \wedge B)$
4. $A \wedge B \to A$
5. $A \wedge B \to B$
6. $A \to A \vee B$
7. $B \to A \vee B$
8. $(A \to C) \to ((B \to C) \to (A \vee B \to C))$.

These are more properly said to denote axiom *schemas*. For example, the first schema says that for any propositional formulas A and B, the formula $A \to (B \to A)$ is an axiom. The only rule is *modus ponens*: From $A \to B$ and A, we are allowed to conclude B. Axioms 1–8 with modus ponens constitute minimal logic. For intuitionistic logic, add the following axiom:

9. $\bot \to A$.

This is known as *ex falso sequitur quodlibet*, which means, "from falsity, anything follows." For classical logic, replace this with the following axiom:

9′. $\neg\neg A \to A$.

Schema 9′ is sometimes called *double-negation elimination*.

Were we to take $\top$, $\neg$ and $\leftrightarrow$ as basic connectives as well, we would add the following axioms.

- $\top$
- $(A \to \bot) \to \neg A$
- $\neg A \to (A \to \bot)$
- $(A \leftrightarrow B) \to (A \to B)$
- $(A \leftrightarrow B) \to (B \to A)$
- $(A \to B) \to ((B \to A) \to (A \leftrightarrow B))$.

These all follow from the definitions of $\top$, $\neg$, and $\leftrightarrow$ that we adopted above.

Working with axiomatic systems can be tedious. Even proving something as simple as $P \to P$ requires a fair amount of work:

1	$P \to (P \to P)$	an instance of 1
2	$P \to ((P \to P) \to P)$	an instance of 1
3	$(P \to ((P \to P) \to P)) \to$ $((P \to (P \to P)) \to (P \to P))$	an instance of 2
4	$((P \to (P \to P)) \to P \to P)$	from 2,3
5	$P \to P$	from 4,1.

We haven't yet said exactly what a proof is. The following definition tells us what it means for something to be a proof of a formula, A, from a set of formulas, Γ.

Definition 2.2.1. A *proof with hypotheses in* Γ is a sequence $(B_0, \ldots, B_{n-1})$ of propositional formulas such that for each $k < n$, either B_k is a formula in Γ, or B_k is an instance of an axiom schema, or there are $i, j < k$ such that B_k follows from B_i and B_j by modus ponens. If d is a proof with hypotheses in Γ and A is the last element of d, then A is said to be the *conclusion* of d, and d is said to be *a proof of* A *from* Γ.

The phrase "B_k follows from B_i and B_j by modus ponens" means simply that B_i is the formula $B_j \to B_k$. The formulas B_i are sometimes called the *lines* of the proof. Some authors use the term *derivation* to distinguish it from the informal notion of proof, so we prove things about derivations, instead of proving things about proofs. I will use the two words interchangeably, however, when talking about the formal proofs. Whether we are talking about classical, intuitionistic, or minimal logic, the following definition is fundamental:

Definition 2.2.2. Let Γ be any set of propositional formulas and let A be any propositional formula. Then A *is provable from* Γ, written $\Gamma \vdash A$, if there is a proof of A from Γ.

The notation $\Gamma \vdash A$ is often read Γ *proves* A. If A is provable without hypothesis, we write $\vdash A$ instead of $\emptyset \vdash A$. When it is necessary to specify whether classical, intuitionistic, or minimal provability is intended, I will use $\Gamma \vdash_{\mathrm{C}} A$, $\Gamma \vdash_{\mathrm{I}} A$, or $\Gamma \vdash_{\mathrm{M}} A$, respectively.

We will consider a number of equivalent characterizations of the provability relation. Even restricting attention to axiomatic proof systems, there are a number of variations. For example, in Definition 2.2.1, there can be ambiguity as to whether an element of the sequence is to be viewed as an axiom, a hypothesis, or a consequence of previous elements. (There are contrived proofs where a particular formula can be viewed as any of the three.) Since formulas can be repeated, when an element of the sequence is a consequence of prior ones, there can be ambiguity as to which ones form the hypotheses for the applied rule. It is sometimes convenient to augment the notion of a proof so that it carries additional information that resolves these ambiguities. If each conclusion of the modus ponens rule is equipped with pointers to the relevant hypotheses, the underlying structure of the proof is essentially a directed acyclic graph, and so such proofs are sometimes referred to as *DAG-like*.

Proofs in which each formula is used only once as the hypothesis of a rule are called *tree-like*. Equivalently, we can take tree-like proofs to be labeled well-founded trees, with each node labeled with a formula and a rule of inference.

Definition 2.2.3. The set of *tree-like proofs* for any of the variants of the axiomatic system described above is generated inductively according to the following clauses:

- For any formula A, the one-node tree with the label (A, hypothesis) is a tree-like proof.
- If A is an axiom, the one-node tree with the label (A, axiom) is a tree-like proof.
- If d and e are tree-like proofs, the first component of the label of the root of d is $A \to B$, and the first component of the label of the root of e is A, then the labeled tree consisting of a root labeled (B, modus ponens), with left subtree d and right subtree e, is a tree-like proof.

Were there any ambiguity, we could have used a distinct label for each axiom schema on our list. This type of definition can clearly be extended to any deductive system described in terms of axioms and rules. It is often convenient to view axioms as 0-ary rules, that is, rules with no hypotheses.

Because Definition 2.2.3 is an inductive definition, we have a principle of induction on tree-like proofs. The rules above correspond to only partial functions, because we cannot combine two arbitrary proofs using modus ponens. As a result, Theorem 1.5.3, the principle of recursive definition, does not apply exactly as it is stated. But the set of tree-like proofs is a subset of the set of *all* well-founded trees labeled with such data, and that set is freely generated using the obvious formation rules. Thus we can define functions by recursion on the set of all labeled trees and just ignore the ones that do not correspond to bona fide proofs.

If d is a tree-like proof, the first component A of the label of the root is called the *conclusion* of d, and d is said to be a proof of A (from hypotheses). The second component of the label of the root is called the *last rule of d*. As an example of a definition by recursion, the function hypotheses(d) is defined as follows:

- If the root of d is labeled (A, axiom), then hypotheses$(d) = \emptyset$.
- If the root of d is labeled (A, hypothesis), then hypotheses$(d) = \{A\}$.
- Otherwise, d has two immediate subtrees, e and f. In that case, hypotheses$(d) =$ hypotheses$(e) \cup$ hypotheses(f).

The following proposition shows that the tree-like notion of proof provides an alternative characterization of provability:

Proposition 2.2.4. *For any Γ and A, $\Gamma \vdash A$ if and only if there is a tree-like proof of A whose hypotheses are all in Γ.*

Proof Suppose d is a proof of A from Γ. By induction on the natural numbers, it is easy to show that there is a tree-like proof of each line of d, and hence a tree-like proof of A.

Conversely, by induction on the set of tree-like proofs, it is easy to show that whenever there is a tree-like proof with hypotheses in Γ, there is an ordinary proof of the same conclusion with hypotheses in Γ. This is obvious in the base cases. In the induction step, suppose the last rule of d is modus ponens applied to subproofs e and f. By the inductive hypothesis, there are ordinary proofs of the conclusions of e and f from hypotheses in Γ. Concatenating them and appending the conclusion of d yields an ordinary proof of the conclusion of d. □

It is an interesting fact that we can even characterize the provability relation without mentioning proofs at all. The following proposition provides one way of doing this.

Proposition 2.2.5. *Suppose Γ is a set of propositional formulas. Then the set of formulas provable from hypotheses in Γ is generated inductively by the following clauses:*

- *If $A \in \Gamma$, then A is provable from hypotheses in Γ.*
- *If A is an axiom, then A is provable from hypotheses in Γ.*
- *If $A \to B$ and A are provable from hypotheses in Γ, then so is B.*

Proof By induction on the length of any proof d of a with hypotheses in Γ, it is easy to see that each line of d is in the inductively defined set. Conversely, by induction on that set, it is easy to show that every formula in the set has a proof with hypotheses in Γ. □

In hindsight, we can see that our definition of an axiomatic proof is just the notion of a formation sequence for the corresponding inductive definition. Thus we have three equivalent formulations of the provability relation: A is provable from Γ if and only if there is a proof, if and only if there is a tree-like proof, and if and only if A is in the inductively defined set of consequences of Γ. In Section 2.4 and Chapter 6, we will consider other systems of formal deduction, giving rise to additional notions of formal proof. In Chapter 3, we will consider semantic and algebraic characterizations of the provability relation as well.

Which is the right characterization? From a proof-theoretic point of view, provability is paramount, but, for many purposes, the choice of a deductive system is largely irrelevant. All that we require of a notion of proof is that it provides concrete witnesses to the provability relation that we are interested in studying. Some branches of proof theory are more concerned with the specific properties of various deductive systems, as well as transformations between proofs, the possibility of specifying normal forms, and the design of automated search procedures. In proof complexity, one is interested in upper and lower bounds on the lengths of proof. Many of the results in this book involve implicit translations between proofs, and at times notions of length and complexity will come into play. But, for the most part, we will primarily be interested in the provability relation rather than particular proofs.

Exercises

2.2.1. Show that if $\Gamma \vdash A \to B$ and $\Gamma \vdash B \to C$ then $\Gamma \vdash A \to C$, whether we consider provability in minimal, intuitionistic, or classical logic.

2.2.2. Show that minimal logic proves $(A \to B) \to (\neg B \to \neg A)$.

2.2.3. Show that minimal logic proves $A \wedge B \to B \wedge A$.

2.2.4. Show that we get an equivalent proof system for classical logic by replacing $9'$ by the principle $(\neg A \to \neg B) \to (B \to A)$, which implies that every implication is equivalent to its contrapositive.

2.2.5. In *Principia Mathematica*, Russell and Whitehead presented a set of axioms for classical propositional logic, one of which was shown by Bernays to be redundant in 1919. The Russell–Whitehead system is based on the connectives $\vee$ and $\neg$, with $A \to B$ defined as $\neg A \vee B$. With Bernays's modification, the axioms are as follows:

1. $A \vee A \to A$
2. $A \to A \vee B$
3. $A \vee B \to B \vee A$
4. $(A \to B) \to (C \vee A \to C \vee B)$.

The only rule of inference is modus ponens. Note that axiom 4 yields the transitivity of implication, so that from $A \to B$ and $B \to C$ we can derive $A \to C$. The axiom schema that Bernays showed to be redundant in the Russell–Whitehead system was $A \vee (B \vee C) \to (A \vee B) \vee C$. Show that this can be derived from the other axioms. Here is one way to go about it:

a. Derive $A \vee (B \vee C) \to A \vee ((A \vee B) \vee C)$.
b. With this, show that it suffices to prove $A \to ((A \vee B) \vee C)$.
c. Derive $A \to ((A \vee B) \vee C)$.

2.3 The Provability Relation

The next three propositions hold for all the classical, intuitionistic, and minimal versions of the provability relation, and they carry over to first-order logic as well. Indeed, these properties characterize a broad class of notions of provability.

Proposition 2.3.1 (Weakening). *If* $\Gamma \vdash A$ *and* $\Gamma' \supseteq \Gamma$ *then* $\Gamma' \vdash A$

Proof Suppose $\Gamma' \supseteq \Gamma$. To show that every consequence of Γ is a consequence of Γ', we use induction on the set of consequences of Γ, as characterized by Proposition 2.2.5. If A is an axiom or an element of Γ, it is clear that $\Gamma' \vdash A$. Otherwise, for some formula B, we have $\Gamma \vdash B \to A$ and $\Gamma \vdash B$. By the inductive hypothesis, we have $\Gamma' \vdash B \to A$ and $\Gamma' \vdash B$. Hence $\Gamma' \vdash A$, as required. □

This property is also sometimes known as *monotonicity*. As an exercise, you should give proofs corresponding to the other characterizations of the provability relation that we have seen. In particular, if d is a proof (or tree-like proof) of a formula A from hypotheses in Γ, it is not hard to show that d is also a proof (respectively, a tree-like proof) of A from hypotheses in Γ'.

Proposition 2.3.2 (Finite character). *If* $\Gamma \vdash A$ *then for some finite subset* Γ' *of* Γ, $\Gamma' \vdash A$.

Proof Again, we use induction on the set of consequences of Γ. The proposition clearly holds if A is an axiom or an element of Γ, since in those cases we can take $\Gamma' = \emptyset$ or $\Gamma' = \{A\}$, respectively. For the induction step, suppose there are finite sets $\Gamma_0, \Gamma_1 \subseteq \Gamma$ such that $\Gamma_0 \vdash B \to A$ and $\Gamma_1 \vdash B$. Let $\Gamma' = \Gamma_0 \cup \Gamma_1$. By weakening, $\Gamma' \vdash B \to A$ and $\Gamma' \vdash B$, and hence $\Gamma' \vdash A$. □

Proposition 2.3.2 is the syntactic analogue of the semantic notion of *compactness*, discussed in Chapter 3. Once again, you should be able to fashion proofs that correspond to the other characterizations of the provability relation. In particular, if d is a tree-like proof of A from hypotheses in Γ, it is not hard to show that we can take $\Gamma' = \text{hypotheses}(d)$.

Theorem 2.3.3 (The deduction theorem). *If* $\Gamma \cup \{A\} \vdash B$ *then* $\Gamma \vdash A \to B$.

Proof Fixing Γ and A, we use induction on the set of consequences of $\Gamma \cup \{A\}$. If B is an axiom, since $B \to (A \to B)$ is an instance of axiom 1, we can prove $A \to B$ with no hypotheses, and so $\Gamma \vdash A \to B$. If B is an element of Γ, the same argument shows $\Gamma \vdash A \to B$. If B is A itself, the example in the last section shows how to derive $B \to B$, and so $\Gamma \vdash A \to B$.

We therefore only need to consider the case where for some formula C, $\Gamma \cup \{A\} \vdash C \to B$ and $\Gamma \cup \{A\} \vdash C$. By the inductive hypothesis, we have $\Gamma \vdash A \to (C \to B)$ and $\Gamma \vdash A \to C$. Using an instance of axiom 2, we have

$$\Gamma \vdash (A \to (C \to B)) \to ((A \to C) \to (A \to B)).$$

Using two applications of modus ponens we have $\Gamma \vdash A \to B$. □

Notice that we have really proved something stronger: the deduction theorem holds for any axiomatic system that includes axioms 1 and 2 and has modus ponens as the only rule of inference. As an example of how the deduction theorem can be used, let us show that for any formulas A, B, and C, minimal logic proves $(A \to B) \to ((B \to C) \to (A \to C))$. The following list of inferences is a *metaproof*, a proof that there is a formal derivation.

1	$\{A \to B, B \to C, A\} \vdash A$	hypothesis
2	$\{A \to B, B \to C, A\} \vdash A \to B$	hypothesis
3	$\{A \to B, B \to C, A\} \vdash B$	from 1,2
4	$\{A \to B, B \to C, A\} \vdash B \to C$	hypothesis
5	$\{A \to B, B \to C, A\} \vdash C$	from 3,4
6	$\{A \to B, B \to C\} \vdash A \to C$	by the deduction theorem
7	$\{A \to B\} \vdash (B \to C) \to (A \to C)$	by the deduction theorem
8	$\vdash (A \to B) \to ((B \to C) \to (A \to C))$	by the deduction theorem.

The following is an easy consequence of the deduction theorem.

Proposition 2.3.4 (Transitivity). *Suppose that we have $\Gamma \vdash A_i$ for every $i < k$, and $\Gamma \cup \{A_0, \ldots, A_{k-1}\} \vdash B$. Then $\Gamma \vdash B$.*

Proof If $\Gamma \cup \{A_0, \ldots, A_{k-1}\}$ proves B, then, by the deduction theorem, Γ proves $A_0 \to (A_1 \to \cdots \to (A_{k-1} \to B))$. Applying modus ponens k times we have that Γ proves B. □

Some presentations of propositional logic include a substitution rule: assuming we have derived a formula A without using any hypotheses, we are allowed to conclude $A[B/P]$ for any B. The following proposition shows that even though adding such a rule may make proofs shorter and more convenient, it does not change the notion of provability. In the statement of the proposition, $\Gamma[B/P]$ denotes $\{C[B/P] \mid C \in \Gamma\}$.

Proposition 2.3.5 (Substitution). *If $\Gamma \vdash A$ then for any formula B and any propositional variable P, $\Gamma[B/P] \vdash A[B/P]$.*

Proof By induction on the provability relation. □

We express the statement of Proposition 2.3.5 by saying that the substitution rule is *admissible*, which is to say, it can be added to the calculus without changing the provability relation.

We have seen that the deduction theorem is a convenient tool for showing that various formulas are derivable. When one is forced to work with an axiomatic system, one quickly learns that it is helpful to build up a battery of such rules. Here are some that are useful.

Proposition 2.3.6. *For any Γ, A, B, and C:*

- *If $\Gamma \vdash A$ and $\Gamma \vdash B$ then $\Gamma \vdash A \wedge B$.*
- *If $\Gamma \vdash A \wedge B$ then $\Gamma \vdash A$ and $\Gamma \vdash B$.*
- *If $\Gamma \vdash A$ or $\Gamma \vdash B$ then $\Gamma \vdash A \vee B$.*
- *If $\Gamma \vdash A \vee B$, $\Gamma \cup \{A\} \vdash C$ and $\Gamma \cup \{B\} \vdash C$, then $\Gamma \vdash C$.*

For intuitionistic logic:

- *If $\Gamma \vdash \bot$, then $\Gamma \vdash A$.*

For classical logic:

- *If $\Gamma \cup \{\neg A\} \vdash \bot$, then $\Gamma \vdash A$.*

Proof All of these inferences are justified using the corresponding axiom schema and modus ponens, as well as (in two cases) an appeal to the deduction theorem. □

Exercises

2.3.1. Give an alternative proof of Proposition 2.3.4 using induction on the set of consequences of $\Gamma \cup \{A_0, \ldots, A_{k-1}\} \vdash B$.

2.3.2. Prove Proposition 2.3.5:
  a. using induction on the set of consequences of Γ.
  b. using a recursion on proof trees to take a tree-like proof of A from Γ to a tree-like proof of $A[B/P]$ from $\Gamma[B/P]$.

2.3.3. Prove Proposition 2.3.6.

2.3.4. Redo problems 2.2.1 and 2.2.3 from the last set of exercises, now using the deduction theorem.

2.3.5. Consider the following two versions of a proof system for classical logic based on the connectives $\to$ and $\bot$:

- schemas 1, 2, and 9′ with modus ponens
- schemas 1, 2, and the schema $(\neg B \to \neg A) \to (A \to B)$ with modus ponens.

Show that the proof systems are equivalent, by showing that each one can prove the third schema of the other.

2.4 Natural Deduction

We have seen that, when reasoning about provability in an axiomatic proof system, it is helpful to derive rules that encapsulate common patterns of reasoning. Given that they are so common, it is reasonable to ask, why not take these rules as primitive? This is the motivation behind Gerhard Gentzen's systems of *natural deduction*.

The first guiding principle is foreshadowed by the list of axioms in Section 2.2: rules are associated with each connective to characterize its proper usage. In particular, for each logical connective we have *introduction rules*, which tell us what is needed to justify a statement that involves that connective, and *elimination rules*, which tell us what we may legitimately infer from such a statement. For example, here are the rules for conjunction:

$$\frac{A \quad B}{A \wedge B}\wedge\mathrm{I} \qquad \frac{A \wedge B}{A}\wedge\mathrm{E_l} \qquad \frac{A \wedge B}{B}\wedge\mathrm{E_r}$$

Here letters l and r are mnemonic for *left* and *right* respectively.

Gentzen's second innovation is more notable. In the last section, we saw that the deduction theorem is useful in that it lets us prove an implication by adding the antecedent to the set of hypotheses temporarily. Gentzen's idea was to model the deductive process in such a way that the notion of a proof from hypotheses is fundamental, and in such a way that hypotheses can not only be introduced in the course of a derivation, but also eliminated, or *canceled*. In an ordinary mathematical proof, we often introduce an assumption for the sake of showing that something follows. So, at any point in a proof, we are working in the *context*, or *scope*, of a list of assumptions. The same is true in natural deduction: whenever we claim to have proved something, the claim is relative to the assumptions that we have made. We should therefore read the introduction rule for $\wedge$ as follows: given a proof of A from some hypotheses and a proof of B from hypotheses, we obtain a proof of $A \wedge B$ from the union of the two sets of hypotheses. The first elimination rule for $\wedge$ says that given a proof of $A \wedge B$ from some hypotheses, we obtain a proof of A from the same set of hypotheses. These are the rules for implication:

$$\begin{array}{c}\overline{A}\\ \vdots\\ \dfrac{B}{A \to B}\to\mathrm{I}\end{array} \qquad \frac{A \to B \quad A}{B}\to\mathrm{E}$$

The introduction rule is the interesting one, since it involves canceling a hypothesis. Informally, it says that in order to prove $A \to B$, it suffices to assume A and conclude B. The three dots suggest a proof of B in which the assumption A can be used any number of times. In concluding $A \to B$, this assumption is made explicit. In the resulting proof, then, A is no longer an assumption; it has been *canceled*. More precisely, the introduction rule for $\to$ should be read as follows: given a proof of B from some hypotheses, which may include A, We obtain a proof of $A \to B$, from the same set of hypotheses except for the fact that A may be canceled. In essence, Theorem 2.3.3, the deduction theorem, has become one of the defining rules for implication.

The rules for disjunction are as follows:

$$\frac{A}{A \vee B}\vee\mathrm{I_l} \qquad \frac{B}{A \vee B}\vee\mathrm{I_r} \qquad \begin{array}{c} \quad\ \ \overline{A} \quad \overline{B}\\ \quad\ \ \vdots \quad \vdots\\ \dfrac{A \vee B \quad C \quad C}{C}\vee\mathrm{E}\end{array}$$

The elimination rule models the natural process of proving C from $A \vee B$ by branching on cases: "Suppose $A \vee B$. Case 1: A holds ... hence C. Case 2: B holds ... hence C. Either way, we have C." Notice that in the resulting inference, the hypotheses A and B are canceled.

All the rules we have seen so far are sound for minimal logic. For intuitionistic logic, add *ex falso sequitur quodlibet*:

$$\frac{\bot}{A}\ \bot\mathrm{E}$$

This can be viewed as an elimination rule for $\bot$. For classical logic, add *reductio ad absurdum*:

$$\begin{array}{c}\overline{\neg A}\\ \vdots\\ \dfrac{\bot}{A}\end{array}$$

Remembering that $\neg A$ abbreviates $A \to \bot$, the natural introduction and elimination rules for negation follow from the corresponding rules for implication:

$$\begin{array}{c}\overline{A}\\ \vdots\\ \dfrac{\bot}{\neg A}\ \neg\mathrm{I}\end{array} \qquad \frac{\neg A \quad A}{\bot}\ \neg\mathrm{E}$$

Similarly, $\top$ is provable outright, and we have the following derived rules for $\leftrightarrow$:

$$\begin{array}{c}\overline{A} \quad \overline{B}\\ \vdots \quad\ \ \vdots\\ \dfrac{B \quad A}{A \leftrightarrow B}\ \leftrightarrow\mathrm{I}\end{array} \qquad \frac{A \leftrightarrow B \quad A}{B}\ \leftrightarrow\mathrm{E}_l \qquad \frac{A \leftrightarrow B \quad B}{A}\ \leftrightarrow\mathrm{E}_r$$

Don't confuse *reductio ad absurdum* with negation introduction. The former is not valid in minimal logic, whereas the latter is.

All the rules we have listed so far can only be applied to existing proofs, so we need to have a rule with no hypotheses to get started. Here it is:

$$A$$

This is called the *assumption rule*, and has a straightforward interpretation: we can always prove A, assuming A as a hypothesis. The fact that this is an open hypothesis is diagrammatically clear from the fact that there is no line over it. If and when it is canceled, we put a line over it.

Reading a natural deduction proof can be difficult because hypotheses are introduced and canceled at various times. It helps to label each canceled hypothesis and mark the point at which it is canceled with the corresponding label. We will use numbers for this purpose, though any choice of symbols will do. For example, the following is a proof of $B \to (A \wedge B)$ from hypothesis A:

$$\dfrac{\dfrac{A \quad \overline{B}^{\,1}}{A \wedge B}}{B \to A \wedge B}\ 1$$

One more instance of $\to$I yields a proof of $A \to (B \to A \wedge B)$:

$$\dfrac{\dfrac{\dfrac{\dfrac{}{A}\,2 \qquad \dfrac{}{B}\,1}{A \wedge B}}{B \to A \wedge B}\,1}{A \to (B \to A \wedge B)}\,2$$

Note that this proof goes through in minimal logic. Further examples are provided in the next section.

There is some legalistic fine print associated with the implication introduction rule. (Similar considerations apply to disjunction elimination as well.) The wording above was carefully chosen so that we do not *need* the hypothesis A to conclude $A \to B$; if you know B outright, you know $A \to B$. We need this flexibility, for example, to derive the first schema in our axiomatic proof systems:

$$\dfrac{\dfrac{\dfrac{}{B}\,1}{A \to B}}{B \to (A \to B)}\,1$$

Furthermore, nothing is harmed if we leave some of the hypotheses open when we could have canceled them; again, this only weakens the proof. If we adopt similar conventions with respect to *reductio ad absurdum*, we can view it as a generalization of the *ex falso* rule.

I have presented an informal description of natural deduction. When it comes to presenting a formal definition, it is not immediately clear how we should handle hypotheses and their cancellation at various stages. To do so, let us modify our informal description so that each node of our proof tree has explicit information about the hypotheses that are open at that stage. A pair (Γ, A) in which Γ is a finite set of formulas and A is a propositional formula is called a *sequent*, and is written $\Gamma \vdash A$. Intuitively, this sequent should be read as the assertion that we have established A from hypotheses in Γ. If Γ is a set of formulas and A is a formula, it is convenient to write Γ, A for $\Gamma \cup \{A\}$. More generally, it is convenient to leave off curly braces when listing the elements of a finite set. With this new mode of presentation, the natural deduction rules are expressed as follows:

$$\dfrac{}{\Gamma, A \vdash A}$$

$$\dfrac{\Gamma \vdash A \qquad \Gamma \vdash B}{\Gamma \vdash A \wedge B}\,\wedge\text{I} \qquad\qquad \dfrac{\Gamma \vdash A_0 \wedge A_1}{\Gamma \vdash A_i}\,\wedge\text{E}$$

$$\dfrac{\Gamma \vdash A_i}{\Gamma \vdash A_0 \vee A_1}\,\vee\text{I} \qquad\qquad \dfrac{\Gamma \vdash A \vee B \qquad \Gamma, A \vdash C \qquad \Gamma, B \vdash C}{\Gamma \vdash C}\,\vee\text{E}$$

$$\dfrac{\Gamma, A \vdash B}{\Gamma \vdash A \to B}\,\to\text{I} \qquad\qquad \dfrac{\Gamma \vdash A \to B \qquad \Gamma \vdash A}{\Gamma \vdash B}\,\to\text{E}$$

For intuitionistic logic, add *ex falso sequitur quodlibet*:

$$\dfrac{\Gamma \vdash \bot}{\Gamma \vdash A}\,\bot\text{E}$$

For classical logic, add *reductio ad absurdum*:

$$\frac{\Gamma, \neg A \vdash \bot}{\Gamma \vdash A} \text{RAA}$$

For example, the following proof tree witnesses $\vdash A \to (B \to A \wedge B)$:

$$\cfrac{\cfrac{\cfrac{\cfrac{}{A, B \vdash A} \quad \cfrac{}{A, B \vdash B}}{A, B \vdash A \wedge B} \wedge\text{I}}{A \vdash B \to A \wedge B} \to\text{I}}{\vdash A \to (B \to A \wedge B)} \to\text{I}$$

As in Section 2.2, we now have at least three options at our disposal: we can define a (DAG-like) natural deduction proof to be a sequence of lines (sequents), in which each is justified from previous sequents by one of the rules above; we can define a (tree-like) natural deduction proof to be a tree labeled appropriately by sequents and rules; or we can define the set of provable sequents inductively. All of these will work and will yield equivalent notions of provability. For concreteness, let us choose the third option.

Definition 2.4.1. The set of sequents provable in natural deduction (for minimal, intuitionistic, or classical logic) is defined to be the smallest set of sequents closed under the natural deduction rules.

Proposition 2.4.2 (Weakening). *If $\Gamma \vdash A$ is provable in natural deduction and $\Gamma' \supseteq \Gamma$, then $\Gamma' \vdash A$ is provable as well.*

Proposition 2.4.2 is easily proved by induction on the set of provable sequents, since the claim is clearly true of the assumption rule and is maintained by the others. As with the substitution rule, the proposition asserts that weakening is an admissible rule in the calculus. It is not a *derived* rule: in our formulation, in general, there is no way of going from $\Gamma \vdash A$ to $\Gamma' \vdash A$. But admissibility means that the rule can be added *conservatively*, in other words, without changing the set of provable sequents.

Weakening can be composed with the other rules to obtain more liberal forms of these rules. For example, conjunction introduction can be expressed as follows:

$$\frac{\Gamma \vdash A \qquad \Delta \vdash B}{\Gamma, \Delta \vdash A \wedge B}$$

It is often convenient to use such modified forms of the rules. So, for example, the preceding proof could be written more concisely as follows:

$$\cfrac{\cfrac{\cfrac{\cfrac{}{A \vdash A} \quad \cfrac{}{B \vdash B}}{A, B \vdash A \wedge B}}{A \vdash B \to A \wedge B}}{\vdash A \to (B \to A \wedge B)}$$

Note that in a sequent $\Gamma \vdash A$, the set Γ is required to be finite. To connect natural deduction to our previous notion of provability, it is natural to guess that $\Gamma \vdash A$ will hold if and only if the sequent $\Gamma' \vdash A$ is provable in natural deduction for some finite $\Gamma' \subseteq \Gamma$. The following theorem shows that this is indeed the case, and holds whether we consider the minimal, intuitionistic, or classical version of provability.

Theorem 2.4.3. *For any set of formulas Γ and any formula A, Γ proves A if and only if, for some finite $\Gamma' \subseteq \Gamma$, the sequent $\Gamma' \vdash A$ is provable in natural deduction.*

Proof For the forward direction, use induction on the set of consequences of Γ. The desired conclusion is clearly true for each element A of Γ, and if it holds of A and $A \to B$, then using weakening and the implication elimination rule it holds of B. As a result, it suffices to show that each axiom of the corresponding Hilbert-style proof system is provable in natural deduction.

For the reverse direction, use induction on the set of provable sequents to show that whenever $\Gamma' \vdash A$ is a provable sequent, Γ' proves A. This is straightforward using the deduction theorem and Proposition 2.3.6. □

The right-hand side of the equivalence in Theorem 2.4.3 can be taken as an alternative definition of provability. The properties enumerated in Section 2.3 carry over, but finite character and the deduction theorem are built into the new characterization.

The formal definition of provability in natural deduction does not exactly mirror the informal diagrammatic presentation with which we began. In the rule for implication introduction in the formal definition, A may or may not be an element of Γ. The case where it is corresponds to an inference in which A is not canceled as a hypothesis. But our diagrammatic representation allows us to cancel different hypotheses in different places, whereas, in contrast, the formal definition keeps track of which *formulas* are still open hypotheses, rather than which *instances*. For example, there are three patterns of cancellation that are consistent with the proof at left, of which the proof at right is one:

$$\dfrac{\dfrac{\dfrac{A \vdash A \qquad A \vdash A}{A \vdash A \wedge A}}{A \vdash A \to A \wedge A}}{\vdash A \to (A \to A \wedge A)} \qquad\qquad \dfrac{\dfrac{\dfrac{\dfrac{}{A}{}^{2} \qquad \dfrac{}{A}{}^{1}}{A \wedge A}}{A \to A \wedge A}{}^{1}}{A \to (A \to A \wedge A)}{}^{2}$$

Moreover, the formal definition allows us to introduce extraneous hypotheses that are not used in the derivation, e.g.:

$$\dfrac{A, B \vdash B}{A \vdash B \to B}$$

In contrast, there is no way to indicate an extraneous hypothesis in this diagram:

$$\dfrac{\dfrac{}{B}{}^{1}}{B \to B}{}^{1}$$

In Section 13.8, we will see that the typed lambda calculus suggests a way of describing natural deduction proofs that is closer to the informal diagrammatic description. Whatever formal presentation we choose, however, what is important is that ultimately we end up with the same notion of provability. To see that this is the case here, note that every diagrammatic proof corresponds to a formal one in a natural way (in each sequent, first take Γ to be the set of uncanceled hypotheses, and then weaken as necessary). Conversely, given any formal proof there is at least one diagrammatic proof of the conclusion from a subset of the hypotheses found in the formal proof.

1	$(A \to C) \wedge (B \to C)$	
2	$A \to C$	$\wedge E_1$, 1
3	$B \to C$	$\wedge E_2$, 1
4	$A \vee B$	
5	A	
6	C	$\to$E, 2, 5
7	B	
8	C	$\to$E, 3, 7
9	C	$\vee$E, 4, 5–6, 7–8
10	$A \vee B \to C$	$\to$I, 4–9
11	$(A \to C) \wedge (B \to C) \to (A \vee B \to C)$	$\to$I, 1–10

Figure 2.1 A Fitch diagram.

There is another convenient representation of natural deduction proofs that is worth mentioning, known as a *Fitch diagram*. Fitch diagrams are essentially linear proofs, except that lines appear in a nested system of boxes. Within each box, the open hypotheses consist of any assumption in the box itself, as well as any assumption in any box that contains it. To introduce a new assumptions, we begin a new box, and canceling an assumption amounts to closing off a box. Figure 2.1 shows an example. Not every sequence of lines can be translated directly to a Fitch diagram, but you can verify (formally, by induction on proof trees) that any tree-like proof can be expressed as a Fitch diagram. So we can always obtain a Fitch diagram from a DAG-like proof by translating it to a tree-like proof first.

Exercises

2.4.1. Show by induction on natural deduction derivations that whenever $\Gamma \vdash B$ is derivable in intuitionistic logic, then for some finite set of formulas $\{A_0, \ldots, A_{k-1}\}$, the sequent

$$\Gamma, \bot \to A_0, \ldots, \bot \to A_{k-1} \vdash B$$

is derivable in natural deduction for minimal logic.

2.4.2. Give natural deduction proofs of $(A \to (B \to C)) \to (A \wedge B \to C)$ represented as

a. a tree labeled by formulas (and cancellations at the leaves),
b. a tree labeled by sequents, and
c. a Fitch diagram.

2.5 Some Propositional Validities

Of the three types of logic we have seen so far, minimal logic has the best computational interpretation, in which formulas can be seen as datatype specifications of their own proofs. For example, a proof of $A \wedge B$ can be viewed as an ordered pair consisting of a proof of A paired with a proof of B, a proof of $A \to B$ can be viewed as a procedure that transforms a proof of A to a proof of B, and so on. With this interpretation, the intuitionistic principle $\bot \to A$, *ex falso sequitur quodlibet*, assumes the existence of a procedure transforming a proof of falsity into the proof of any formula at all. Computationally, this is somewhat palatable because don't expect there to be any proofs of falsity. We will verify below that the passage to classical logic is equivalent to adding the law of the excluded middle, $A \vee \neg A$. With the constructive interpretation, we are justified in making this assertion only when we have either a proof of A or a proof of $\neg A$. The classical principle is also known as *tertium non datur*, "the third [case] does not hold." It is often associated with a Platonic or realist worldview, on which assertions are taken to describe states of affairs, independent of our knowledge of them. On that view, any meaningful assertion is either true or false, whether or not we know which is the case.

These characterizations are informal. In chapters to come, we will develop conceptual tools that will help us be much more precise about the senses in which intuitionistic and minimal logic are constructive and classical logic is not. In the meanwhile, the best way to get a sense of the different types of logic is to work with them formally.

Two propositions A and B are said to be *equivalent* (in minimal, intuitionistic, or classical logic, respectively) if $A \leftrightarrow B$ is provable, in which case we write $A \equiv B$. It is easy to see that this notion of equivalence is an equivalence relation. The following proposition provides a deductive schema that asserts that substituting equivalent formulas for a variable P in a formula A yields equivalent formulas.

Proposition 2.5.1. *For any A, B, C and propositional variable P, minimal logic proves* $(B \leftrightarrow C) \to (A[B/P] \leftrightarrow A[C/P])$.

The proof is left as an exercise.

The following proposition presents some important general laws that are common to all three versions of logic.

Proposition 2.5.2. *The following are provable in minimal propositional logic:*

1. *Commutativity of* $\wedge$: $A \wedge B \leftrightarrow B \wedge A$
2. *Commutativity of* $\vee$: $A \vee B \leftrightarrow B \vee A$
3. *Associativity of* $\wedge$: $(A \wedge B) \wedge C \leftrightarrow A \wedge (B \wedge C)$
4. *Associativity of* $\vee$: $(A \vee B) \vee C \leftrightarrow A \vee (B \vee C)$
5. *Distributivity of* $\wedge$ *over* $\vee$: $A \wedge (B \vee C) \leftrightarrow (A \wedge B) \vee (A \wedge C)$
6. *Distributivity of* $\vee$ *over* $\wedge$: $A \vee (B \wedge C) \leftrightarrow (A \vee B) \wedge (A \vee C)$
7. $(A \to (B \to C)) \leftrightarrow (A \wedge B \to C)$
8. $(A \to B) \to ((B \to C) \to (A \to C))$
9. $(A \vee B \to C) \leftrightarrow (A \to C) \wedge (B \to C)$
10. $\neg(A \wedge \neg A)$
11. $\neg(A \to B) \to \neg B$.

If you want to get a feel for natural deduction and minimal logic, you should work out proofs of as many of these as you can. When it comes to minimal logic, there is a simple heuristic for searching for proofs: work backward from the goal using introduction rules and work forward from the hypothesis using elimination rules until all the pieces come together. For example, here is a proof of the forward direction of claim 5:

$$\dfrac{\dfrac{\dfrac{}{A \wedge (B \vee C)}\,{}^{2}}{B \vee C} \qquad \dfrac{\dfrac{\dfrac{\dfrac{}{A \wedge (B \vee C)}\,{}^{2}}{A} \quad \dfrac{}{B}\,{}^{1}}{A \wedge B}}{(A \wedge B) \vee (A \wedge C)} \qquad \dfrac{\dfrac{\dfrac{\dfrac{}{A \wedge (B \vee C)}\,{}^{2}}{A} \quad \dfrac{}{C}\,{}^{1}}{A \wedge C}}{(A \wedge B) \vee (A \wedge C)}}{\dfrac{(A \wedge B) \vee (A \wedge C)}{(A \wedge (B \vee C)) \rightarrow (A \wedge B) \vee (A \wedge C)}\,{}^{2}}\,{}^{1}$$

Here is a proof of the forward direction of claim 7:

$$\dfrac{\dfrac{\dfrac{\dfrac{}{A \rightarrow (B \rightarrow C)}\,{}^{2} \quad \dfrac{\dfrac{}{A \wedge B}\,{}^{1}}{A}}{B \rightarrow C} \qquad \dfrac{\dfrac{}{A \wedge B}\,{}^{1}}{B}}{\dfrac{C}{A \wedge B \rightarrow C}\,{}^{1}}}{(A \rightarrow (B \rightarrow C)) \rightarrow (A \wedge B \rightarrow C)}\,{}^{2}$$

Many familiar classical equivalences are not equivalent in minimal logic. The next proposition gives some examples of the directions that *are* minimally valid. In the next few chapters, we will develop tools that allow us to show that minimal logic does not generally prove the reverse implications.

Proposition 2.5.3. *The following implications are provable in minimal logic:*

1. $(A \rightarrow B) \rightarrow (\neg B \rightarrow \neg A)$
2. $A \rightarrow \neg\neg A$
3. $A \vee B \rightarrow \neg(\neg A \wedge \neg B)$
4. $\neg A \vee \neg B \rightarrow \neg(A \wedge B)$
5. $\neg(A \vee B) \leftrightarrow \neg A \wedge \neg B$.

The fact that the third and fourth implications do not reverse supports the understanding that the intuitionistic and minimal versions of disjunction are stronger than the classical version, since in the former cases one is required to know which disjunct holds. Here is a proof of the forward direction of claim 5:

$$\dfrac{\dfrac{\dfrac{\dfrac{\dfrac{}{\neg(A \vee B)}\,{}^{3} \quad \dfrac{\dfrac{}{A}\,{}^{1}}{A \vee B}}{\bot}}{\neg A}\,{}^{1} \qquad \dfrac{\dfrac{\dfrac{}{\neg(A \vee B)}\,{}^{3} \quad \dfrac{\dfrac{}{B}\,{}^{2}}{A \vee B}}{\bot}}{\neg B}\,{}^{2}}{\dfrac{\neg A \wedge \neg B}{\neg(A \vee B) \rightarrow \neg A \wedge \neg B}\,{}^{3}}$$

Claim 1 is the special case of claim 8 of Proposition 2.5.2 in which C is taken to be $\bot$, and, similarly, claim 5 is a special case of claim 9 of Proposition 2.5.2.

The following proposition has to do with the nature of double negations in minimal logic. It will be useful in Section 2.7.

Proposition 2.5.4. *The following are provable in minimal logic:*

1. $\neg A \leftrightarrow \neg\neg\neg A$
2. $(A \to \neg B) \leftrightarrow (\neg\neg A \to \neg B)$
3. $\neg\neg(A \vee B) \leftrightarrow \neg(\neg A \wedge \neg B)$
4. $\neg\neg(A \to B) \to (\neg\neg A \to \neg\neg B)$
5. $\neg\neg(A \wedge B) \leftrightarrow (\neg\neg A \wedge \neg\neg B)$.

Here is a proof of the converse direction of claim 1:

$$\dfrac{\dfrac{\dfrac{\dfrac{}{\neg\neg\neg A}{\,}^{3} \qquad \dfrac{\dfrac{\dfrac{}{A}{\,}^{2} \quad \dfrac{}{\neg A}{\,}^{1}}{\bot}}{\neg\neg A}{\,}^{1}}{\bot}}{\neg A}{\,}^{2}}{\neg\neg\neg A \to \neg A}{\,}^{3}$$

The other claims are left as exercises.

In the next section, we will see that the differences between intuitionistic and minimal logic only show up with respect to falsity and negation. The following proposition provides some examples of intuitionistic validities that are not provable in minimal logic.

Proposition 2.5.5. *The following are provable in intuitionistic logic:*

1. $\bot \leftrightarrow A \wedge \neg A$
2. $\neg A \to (A \to B)$
3. $(\neg A \vee B) \to (A \to B)$
4. $A \vee B \to (\neg A \to B)$
5. $\neg(A \to B) \to \neg\neg A$
6. $\neg\neg(A \to B) \leftrightarrow (\neg\neg A \to \neg\neg B)$
7. $A \vee \bot \leftrightarrow A$
8. $A \wedge \bot \leftrightarrow \bot$.

Once again, the best way to get a feel for intuitionistic logic is to work through some proofs. For example, the following is a proof of claim 4:

$$\dfrac{\dfrac{\dfrac{\dfrac{}{A \vee B}{\,}^{3} \qquad \dfrac{\dfrac{\dfrac{}{A}{\,}^{1} \quad \dfrac{}{\neg A}{\,}^{2}}{\bot}}{B} \qquad \dfrac{}{B}{\,}^{1}}{B}{\,}^{1}}{\neg A \to B}{\,}^{2}}{A \vee B \to (\neg A \to B)}{\,}^{3}$$

The next proposition provides classically valid statements that are not intuitionistically valid in general. Claim 6 together with claim 5 of Proposition 2.5.3 are known as *De Morgan's laws*, and claim 10 is known as *Peirce's law*.

Proposition 2.5.6. *The following are all provable in classical logic:*

1. $A \vee \neg A$
2. $A \leftrightarrow \neg\neg A$

3. $A \vee B \leftrightarrow \neg(\neg A \wedge \neg B)$
4. $A \wedge B \leftrightarrow \neg(\neg A \vee \neg B)$
5. $(A \to B) \leftrightarrow (\neg A \vee B)$
6. $\neg(A \wedge B) \leftrightarrow \neg A \vee \neg B$
7. $\neg(A \to B) \leftrightarrow (A \wedge \neg B)$
8. $(A \to B) \leftrightarrow (\neg B \to \neg A)$
9. $(A \to C \vee D) \leftrightarrow ((A \to C) \vee (A \to D))$
10. $(((A \to B) \to A) \to A)$.

An extended heuristic is often needed to find natural deduction proofs in classical logic. As with minimal and intuitionistic logic, one should first work backward from the conclusion and forward from the hypothesis. But then, when all else fails, try a proof by contradiction: assume the negation of your goal and derive a contradiction. Here is a proof of the law of the excluded middle:

$$\dfrac{\dfrac{\dfrac{\dfrac{\dfrac{\overline{\neg(A \vee \neg A)}^{\,2} \qquad \dfrac{\overline{A}^{\,1}}{A \vee \neg A}}{\bot}}{\neg A}\,{\scriptstyle 1}}{A \vee \neg A} \qquad \overline{\neg(A \vee \neg A)}^{\,2}}{\bot}}{A \vee \neg A}\,{\scriptstyle 2}$$

Here is a proof of claim 2, double-negation elimination:

$$\dfrac{\dfrac{\dfrac{\overline{\neg\neg A}^{\,2} \qquad \overline{\neg A}^{\,1}}{\bot}}{A}\,{\scriptstyle 1}}{\neg\neg A \to A}\,{\scriptstyle 2}$$

Note that *reductio* is used when canceling hypothesis 1. Claim 8 says that every implication is equivalent to its contrapositive. Using claims 8 and 2, the conclusion of the following proof is equivalent to the forward direction of claim 7:

$$\dfrac{\dfrac{\dfrac{\dfrac{\overline{\neg(A \wedge \neg B)}^{\,3} \qquad \dfrac{\overline{A}^{\,2} \quad \overline{\neg B}^{\,1}}{A \wedge \neg B}}{\bot}}{B}\,{\scriptstyle 1}}{A \to B}\,{\scriptstyle 2}}{\neg(A \wedge \neg B) \to (A \to B)}\,{\scriptstyle 3}$$

Exercises

2.5.1. Prove Proposition 2.5.1.

2.5.2. Prove the following in natural deduction for minimal logic:
a. $(A \vee B) \vee C \to A \vee (B \vee C)$
b. $\neg\neg(A \vee B) \leftrightarrow \neg(\neg A \wedge \neg B)$
c. $\neg\neg(A \to B) \to (\neg\neg A \to \neg\neg B)$

d. $\neg\neg(A \wedge B) \leftrightarrow (\neg\neg A \wedge \neg\neg B)$
e. $\neg(A \to B) \to \neg B$.

2.5.3. Prove the following in natural deduction for intuitionistic logic:
a. $\neg(A \to B) \to \neg\neg A$
b. $(\neg\neg A \to \neg\neg B) \to \neg\neg(A \to B)$.

2.5.4. Prove the following in natural deduction for classical logic:
a. $(\neg A \vee B) \leftrightarrow (A \to B)$
b. $(A \to B) \vee (B \to A)$
c. $((A \to B) \to A) \to A)$.

2.5.5. Prove $\neg(A \leftrightarrow \neg A)$ in minimal logic.

2.5.6. Prove any of the remaining laws enumerated in this section.

2.6 Normal Forms for Classical Logic

Any propositional formula can be expressed in various classically equivalent ways. For example, with the equivalences $(A \to B) \equiv (\neg A \vee B)$ and $(A \wedge B) \equiv \neg(\neg A \vee \neg B)$, any formula can be expressed using only the connectives $\vee$ and $\neg$. A set of connectives is said to be *complete* for classical logic if any formula is equivalent to a formula using only the connectives in the set. This notion extends to any new connectives we may define. For example, *exclusive or* is defined by

$$(A \oplus B) \equiv (A \vee B) \wedge \neg(A \wedge B),$$

nand is defined by

$$(A \mid B) \equiv \neg(A \wedge B),$$

and *nor* is defined by

$$(A \downarrow B) \equiv \neg(A \vee B).$$

The symbol $\mid$ is sometimes called the *Sheffer stroke*. In the exercises, you are asked to show that each of $\{\to, \bot\}$, $\{\mid\}$, and $\{\downarrow\}$ is a complete set of connectives for classical logic, but neither $\{\to, \vee, \wedge\}$ nor $\{\bot, \leftrightarrow\}$ is complete. Definition 3.1.16 says what it means to be a complete set of connectives in semantic terms.

It is common to express classical formulas in terms of the connectives $\wedge$, $\vee$, and $\neg$. Propositional variables and $\bot$ are called *atomic* formulas, and an atomic formula or its negation is called a *literal*. A propositional formula is said to be *in conjunctive normal form*, or *CNF*, if it is of the form $\bigwedge_{i=0}^{n-1} \bigvee_{j=0}^{m_j} A_{i,j}$, where each $A_{i,j}$ is a literal. Here the expression $\bigwedge_{i=0}^{n-1} B_i$ abbreviates $B_0 \wedge \cdots \wedge B_{m-1}$, and similarly for disjunction. Dually, it is said to be in *disjunctive normal form*, or *DNF*, it is of the form $\bigvee_{i=0}^{n-1} \bigwedge_{j=0}^{m_j} A_{i,j}$, where each $A_{i,j}$ is a literal.

Proposition 2.6.1. *For every propositional formula A, there are a CNF formula A^{cnf} and a DNF formula A^{dnf} such that classical logic proves $A \leftrightarrow A^{\text{cnf}}$ and $A \leftrightarrow A^{\text{dnf}}$.*

Without loss of generality, we can assume that A involves only the connectives $\wedge$, $\vee$, and $\neg$, and we can then prove both claims simultaneously by induction on the set of such formulas. By De Morgan's laws, the negation of a CNF formula is equivalent to a DNF formula and vice versa. Clearly the conjunction of two CNF formulas is equivalent to a CNF formula and the disjunction of two DNF formulas is equivalent to a DNF formula. So it suffices to show that the disjunction of two CNF formulas is equivalent to a CNF formula, and dually for DNF formulas. The first claim is a consequence of the following lemma, which can be proved by induction:

Lemma 2.6.2. *For every n, m, $A_0, \ldots, A_{n-1}, B_0, \ldots, B_{m-1}$, we have*

$$\left(\bigwedge_{i<m} A_i\right) \vee \left(\bigwedge_{j<n} B_j\right) \equiv \bigwedge_{i<m, j<n} (A_i \vee B_j).$$

The dual claim with $\wedge$ and $\vee$ switched is proved the same way.

A propositional formula A is *satisfiable* if there is a truth assignment that makes it true, and it is *valid* if every truth assignment makes it true. Section 3.1 makes these notion precise, and shows that a formula is valid if and only if it is provable. We will see in Section 3.1 that a putting a formula in conjunctive normal form yields a quick test for its validity, and dually, putting a formula in disjunctive yields a quick test for its satisfiability. Unfortunately, putting a formula in either CNF or DNF can result in an exponential increase in length. (See Exercise 2.6.7.)

There is, however, an efficient way of putting formulas in *negation normal form*, or NNF. Such formulas are, by definition, are built up from literals using only $\wedge$ and $\vee$, although alternations between these connectives are allowed. To put a propositional formula in negation normal form, first get rid of implications by replacing each subformula $A \to B$ by $\neg A \vee B$. Then use De Morgan's laws to push negations inwards, canceling double negations along the way.

In more detail, the set of formulas in negation normal form is defined inductively as follows:

- $\bot$ and $\top$ (i.e. $\neg\bot$) are in negation normal form.
- For each propositional variable P, P and $\neg P$ are in negation normal form.
- If A and B are in negation normal form, so are $A \vee B$ and $A \wedge B$.

Consider the act of negating an NNF formula A and then using De Morgan's laws to push the negation inward. This gives rise to the following negation operation $\sim A$, for NNF formulas:

- $\sim\bot = \top$
- $\sim\top = \bot$
- $\sim P = \neg P$
- $\sim(\neg P) = P$
- $\sim(A \wedge B) = \sim A \vee \sim B$
- $\sim(A \vee B) = \sim A \wedge \sim B$.

Notice that $\sim$ is not a new logical symbol, but, rather, an operation defined in the metatheory that has the effect of switching atomic formulas P with their negations $\neg P$ and switching $\wedge$ with $\vee$ everywhere in a formula. It is clear that if A is in negation normal form then so is $\sim A$, and that $\sim A$ is equivalent to $\neg A$.

Given any formula A in the full language of propositional logic, define its negation normal form equivalent A^{nnf} inductively, as follows:

- $\bot^{\text{nnf}} = \bot$
- $P^{\text{nnf}} = P$
- $(A \to B)^{\text{nnf}} = {\sim}(A)^{\text{nnf}} \vee (B)^{\text{nnf}}$
- $(A \wedge B)^{\text{nnf}} = A^{\text{nnf}} \wedge B^{\text{nnf}}$
- $(A \vee B)^{\text{nnf}} = A^{\text{nnf}} \vee B^{\text{nnf}}$.

Then A^{nnf} is in negation normal form and is classically equivalent to A. The translation can be made even more efficient using the identities $A \vee \bot \leftrightarrow A$, $A \wedge \bot \leftrightarrow \bot$, $A \vee \top \leftrightarrow \top$, and $A \wedge \top \leftrightarrow A$, in which case $(\neg A)^{\text{nnf}}$ is exactly ${\sim}A$.

Exercises

2.6.1. Show that if S is a complete set of connectives and A has at least one variable, then A is equivalent to a formula A' with connectives in S in which the propositional variables occurring in A' are among those in A. If S includes either $\top$ or $\bot$, then we can drop the assumption that A has at least one variable.

2.6.2. Show that $\{\wedge, \neg\}$ and $\{\to, \bot\}$ are complete sets of connectives.

2.6.3. a. Show that $\{\to, \vee, \wedge\}$ is not a complete set of connectives. (You will have to assume that at least *some* propositional formulas are not provable, like $\bot$. Hint: show that if A is any formula built up from variables $P_0, \ldots, P_{n-1}$ using these connectives, then $P_0 \wedge \cdots \wedge P_{n-1} \to A$ is provable.)

b. Conclude that $\{\to, \vee, \wedge, \leftrightarrow, \top\}$ is not a complete set of connectives.

2.6.4. a. Show that $\{\bot, \leftrightarrow\}$ is not a complete set of connectives.

b. Conclude that $\{\bot, \top, \neg, \leftrightarrow, \oplus\}$ is not complete.

2.6.5. Using the property $A \vee (B \wedge C) \equiv (A \vee B) \wedge (A \vee C)$ and the dual statement with $\wedge$ and $\vee$ switched, put $(P_1 \wedge P_2) \vee (Q_1 \wedge Q_2) \vee (R_1 \wedge R_2)$ in conjunctive normal form.

2.6.6. Using the semantic notions defined in Section 3.1, show that there is an easy test to determine whether a formula in disjunctive normal form is satisfiable, and, dually, there is an easy test to determine whether a formula in conjunctive normal form is valid. Together with an algorithm for putting formulas into CNF or DNF, this provides an algorithm to determine whether or not a propositional formula is valid.

2.6.7. Let A_n be the formula $P_0 \oplus P_1 \oplus \cdots \oplus P_{n-1}$. This describes the *parity* function, which is to say, A_n is true if and only if an odd number of the inputs are true. Show that if B is a DNF equivalent of A_n, then each disjunct of B, if it is satisfiable at all, must mention all the variables. Show that, as a result, there must be at least 2^{n-1} disjuncts. (To do this, you can use the semantic notions from Section 3.1.)

Note that eliminating $\oplus$ in favor of $\wedge$, $\vee$, $\neg$, and $\bot$ using the definition of $\oplus$ naively can result in an exponential increase in the complexity of the formula. Nonetheless, show that for every A_n there is an equivalent formula C in those connectives whose size is bounded by a quadratic function in n. (Hint: first use a recursive construction to handle the case where $n = 2^m$.)

2.6.8. Show that every propositional formula is equivalent to either $\bot$, $\top$, or a formula in negation normal form in which $\bot$ and $\top$ do not occur.

2.7 Translations Between Logics

We will now explore some of the relationships between classical, intuitionistic, and minimal propositional logic. To facilitate comparison, we will use $\vdash_C$, $\vdash_I$, and $\vdash_M$ to denote the corresponding notions of provability.

First, let us consider some of the relationships between minimal and intuitionistic logic. Since there are no rules governing falsity in minimal logic, $\bot$ acts like a propositional variable. Writing $A[B/\bot]$ to denote the result of replacing $\bot$ by B in A and interpreting $\Gamma[B/\bot]$ in the obvious way, we have the following:

Proposition 2.7.1. *If $\Gamma \vdash_M A$, then $\Gamma[B/\bot] \vdash_M A[B/\bot]$.*

The proof is essentially the same as the proof of the substitution theorem, Proposition 2.3.5. The following is also often useful:

Proposition 2.7.2. *Consider minimal logic together with the rule* ex falso *restricted to propositional variables, so that from $\bot$ we can conclude P only when P is a variable. The resulting provability relation coincides with intuitionistic provability.*

Proof It suffices to show that in the restricted system can derive the *ex falso* rule for arbitrary formulas A. This is immediate in the base cases, and it is not hard to show that assuming we can derive A and B from $\bot$, we can also derive $A \wedge B$, $A \vee B$, and $A \to B$. □

As a corollary to these two propositions, we have the following:

Theorem 2.7.3. *Intuitionistic logic is conservative over minimal logic for formulas that do not involve $\bot$. In other words, if A is such a formula and $\vdash_I A$, then $\vdash_M A$.*

Proof Suppose $\vdash_I A$, where $\bot$ does not occur in A. By Proposition 2.7.2, we can assume that *ex falso* has been restricted to propositional variables, and clearly we can replace the rule by the axiom schema $\bot \to P$. So for some tuple of propositional variables $P_0, \ldots, P_{k-1}$, we have $\{\bot \to P_0, \ldots, \bot \to P_{k-1}\} \vdash_M A$. By Proposition 2.7.1 we can substitute $P_0 \wedge \cdots \wedge P_{k-1}$ for $\bot$, which leaves A unchanged and renders the left-hand side provable in minimal logic. □

Given a propositional formula A, let A^* denote the result of replacing each propositional variable P by $P \vee \bot$. Formally, the translation is defined by recursion on formulas:

- $\bot^* = \bot$
- $P^* = P \vee \bot$ for propositional variables P
- $(A \circ B)^* = A^* \circ B^*$ for any binary connective $\circ$.

Given a set of formulas Γ, let Γ^* be $\{A^* \mid A \in \Gamma\}$. The following theorem provides an easy translation from intuitionistic logic to minimal logic.

Theorem 2.7.4. *Let Γ be any set of propositional formulas, and A any formula.*

1. $\vdash_I A \leftrightarrow A^*$.
2. *If $\Gamma \vdash_I A$, then $\Gamma^* \vdash_M A^*$*

Proof From Proposition 2.5.5, we have $\vdash_I P \leftrightarrow P \vee \bot$, and the first claim follows easily. For the second claim, we use the characterization given by Proposition 2.7.2. We can assume that Γ is finite and that the sequent $\Gamma \vdash A$ is provable in natural deduction with the restricted *ex falso* rule. Since the $*$-translation commutes with the logical connectives, the only rule we have to think about is *ex falso* itself. So, suppose $\Gamma \vdash P$ is obtained from $\Gamma \vdash \bot$. By the inductive hypothesis, we know $\Gamma^* \vdash \bot$ is provable in minimal logic, and from that we can conclude $\Gamma^* \vdash P \vee \bot$. □

Let us now consider classical logic. It is easy to see that *reductio ad absurdum* is equivalent over minimal logic to the principle of double-negation elimination, $\neg\neg A \to A$. The following proposition shows that, over intuitionistic logic, double-negation elimination is equivalent to the law of the excluded middle.

Proposition 2.7.5. *In intuitionistic logic, the axiom schema $A \vee \neg A$ is equivalent to the axiom schema $\neg\neg A \to A$.*

Proof The proof of $A \vee \neg A$ using *reductio* in Section 2.5 can easily be turned into a proof of $A \vee \neg A$ from $\neg\neg(A \vee \neg A) \to A \vee \neg A$ in minimal logic. In the other direction, the following is a proof of $\neg\neg A \to A$ from $A \vee \neg A$:

$$\cfrac{\cfrac{A \vee \neg A \qquad \cfrac{}{A}\,{}^{1} \qquad \cfrac{\cfrac{\cfrac{}{\neg\neg A}\,{}^{2} \quad \cfrac{}{\neg A}\,{}^{1}}{\bot}}{A}}{A}\,{}^{1}}{\neg\neg A \to A}\,{}^{2}$$

□

Note that Proposition 2.7.5 does not assert that for every A, $A \vee \neg A$ is equivalent to $\neg\neg A \to A$ in intuitionistic logic. Rather, it asserts that they are equivalent schemas, so that any instance of one can be derived from an appropriate instance of the other.

The Gödel–Gentzen *double-negation translation* is a remarkably effective tool for interpreting classical theories in constructive ones. It provides a translation of classical logic to minimal logic, defined inductively as follows:

- $\bot^{\mathrm{N}} = \bot$
- $P^{\mathrm{N}} = \neg\neg P$, for any propositional variable P
- $(A \wedge B)^{\mathrm{N}} = A^{\mathrm{N}} \wedge B^{\mathrm{N}}$
- $(A \vee B)^{\mathrm{N}} = \neg(\neg A^{\mathrm{N}} \wedge \neg B^{\mathrm{N}})$
- $(A \to B)^{\mathrm{N}} = A^{\mathrm{N}} \to B^{\mathrm{N}}$.

The name of the translation is explained by the fact that in minimal logic $(A \vee B)^{\mathrm{N}}$ is equivalent to $\neg\neg(A^{\mathrm{N}} \vee B^{\mathrm{N}})$. If Γ is a set of sentences, let Γ^{N} denote $\{B^{\mathrm{N}} \mid B \in \Gamma\}$.

Theorem 2.7.6. *For any propositional formula A and any set of formulas Γ:*

1. $\vdash_{\mathrm{C}} A \leftrightarrow A^{\mathrm{N}}$.
2. *If $\Gamma \vdash_{\mathrm{C}} A$ then $\Gamma^{\mathrm{N}} \vdash_{\mathrm{M}} A^{\mathrm{N}}$.*

The first statement is easy to prove by induction on formulas, using the fact that the schemas $A \leftrightarrow \neg\neg A$ and $(A \vee B) \leftrightarrow \neg(\neg A \wedge \neg B)$ are classically provable. To prove the second part, we need a definition and a lemma.

Definition 2.7.7. A propositional formula is *negative* if it does not contain $\vee$ and every propositional variable is negated.

$$\dfrac{\neg(\neg C^{\mathrm{N}} \wedge \neg D^{\mathrm{N}}) \qquad \dfrac{\dfrac{\dfrac{\dfrac{}{\neg A^{\mathrm{N}}}\,3 \qquad \begin{array}{c}\dfrac{}{C^{\mathrm{N}}}\,1\\ \vdots \\ A^{\mathrm{N}}\end{array}}{\bot}}{\neg C^{\mathrm{N}}}\,1 \qquad \dfrac{\dfrac{\dfrac{}{\neg A^{\mathrm{N}}}\,3 \qquad \begin{array}{c}\dfrac{}{D^{\mathrm{N}}}\,2\\ \vdots \\ A^{\mathrm{N}}\end{array}}{\bot}}{\neg D^{\mathrm{N}}}\,1}{\neg C^{\mathrm{N}} \wedge \neg D^{\mathrm{N}}}}{\dfrac{\bot}{\neg\neg A^{\mathrm{N}}}\,3}$$

Figure 2.2 Translating the disjunction rule.

In other words, the set of negative propositional formulas is the smallest set containing $\bot$ and $\neg P$ for propositional variables P, and closed under conjunction and implication. For any formula A, the formula A^{N} is negative, so the following lemma implies that A^{N} is equivalent to $\neg\neg A^{\mathrm{N}}$.

Proposition 2.7.8. *If A is negative, $\vdash_{\mathrm{M}} \neg\neg A \leftrightarrow A$.*

Proof We already know that $A \to \neg\neg A$ is provable in minimal logic for any A. For the other direction, use induction on A and the fact that the formulas $\neg\neg\neg A \to \neg A$, $\neg\neg(A \wedge B) \to \neg\neg A \wedge \neg\neg B$, and $\neg\neg(A \to B) \to (A \to \neg\neg B)$ are all provable in minimal logic. □

Proof of Theorem 2.7.6 For the second part, use induction on classical natural deduction proofs of $\Gamma \vdash A$. The cases where the last inference is an assumption or one of the rules for $\wedge$ or $\to$ are straightforward. Suppose the last inference is the left introduction rule for $\vee$, which is to say, the proof ends by concluding $\Gamma \vdash A \vee B$ from $\Gamma \vdash A$. By the inductive hypothesis, we have a proof of $\Gamma^{\mathrm{N}} \vdash A^{\mathrm{N}}$, to which we append the following proof of $\neg(\neg A^{\mathrm{N}} \wedge \neg B^{\mathrm{N}})$ from A^{N}:

$$\dfrac{\dfrac{\dfrac{\dfrac{}{\neg A^{\mathrm{N}} \wedge \neg B^{\mathrm{N}}}\,1}{\neg A^{\mathrm{N}}} \qquad A^{\mathrm{N}}}{\bot}}{\neg(\neg A^{\mathrm{N}} \wedge \neg B^{\mathrm{N}})}\,1$$

The right introduction rule for $\vee$ is handled similarly. Suppose the last inference is an $\vee$-elimination rule:

$$\dfrac{\Gamma \vdash C \vee D \qquad \Gamma, C \vdash A \qquad \Gamma, D \vdash A}{\Gamma \vdash A}$$

By the inductive hypothesis, we have a proof of $\Gamma^{\mathrm{N}} \vdash \neg(\neg C^{\mathrm{N}} \wedge \neg D^{\mathrm{N}})$ and proofs of $\Gamma^{\mathrm{N}}, C^{\mathrm{N}} \vdash A^{\mathrm{N}}$ and $\Gamma, D^{\mathrm{N}} \vdash A^{\mathrm{N}}$. Figure 2.2 shows how to obtain a natural deduction proof of $\Gamma^{\mathrm{N}} \vdash \neg\neg A^{\mathrm{N}}$, leaving Γ^{N} in the background. Proposition 2.7.8 then finishes it off. The only remaining rule is *reductio*, which is easily handled using Proposition 2.7.8. □

The double-negation translation sprinkles negations throughout a formula. In the exercises, you are asked to consider what is known as the *Kolmogorov translation*, which is even more liberal in adding such negations. If we are interested in interpreting classical logic to intuitionistic logic instead of minimal logic, the following proposition and theorem show that, in fact, two negations suffice.

Theorem 2.7.9. *For every propositional formula A,* $\vdash_I \neg\neg A \leftrightarrow A^N$.

The proof, which proceeds by induction on A, is left as an exercise.

If Γ is a set of propositional formulas, let $\neg\neg\Gamma$ denote $\{\neg\neg A \mid A \in \Gamma\}$.

Corollary 2.7.10. *For every propositional formula A and set of propositional formulas Γ, if $\Gamma \vdash_C A$ then $\neg\neg\Gamma \vdash_I \neg\neg A$.*

Proof Suppose $\Gamma \vdash_C A$. By Theorem 2.7.6, $\Gamma^N \vdash_M A^N$. By Theorem 2.7.9, this implies $\neg\neg\Gamma \vdash_I \neg\neg A$. □

In the case where $\Gamma = \emptyset$, Corollary 2.7.10 is known as *Glivenko's theorem.*

Restricting attention to formulas in negation normal form is particularly convenient when dealing with classical logic. For such formulas, the following provides a variant of Glivenko's theorem.

Theorem 2.7.11. *Let A be in negation normal form. Then* $\vdash_M \neg({\sim}A) \leftrightarrow A^N$.

This can be proved using a straightforward induction on formulas. It implies that if A is in negation normal form and A is provable classically, then $\neg({\sim}A)$ is provable in minimal logic.

Corollary 2.7.12. *Let $A, B_0, \ldots, B_{n-1}$ be formulas in negation normal form. If $\{B_0, \ldots, B_{n-1}\} \vdash_C A$, then $\{B_0, \ldots, B_{n-1}\} \vdash_M \neg({\sim}A)$.*

Proof If $\{B_0, \ldots, B_{n-1}\} \vdash_C A$, then ${\sim}B_0 \vee \cdots \vee {\sim}B_{n-1} \vee A$ is provable classically. By Theorem 4.6.9, this means that $\neg(B_0 \wedge \cdots \wedge B_{n-1} \wedge {\sim}A)$ is provable in minimal logic, and hence $\{B_0, \ldots, B_{n-1}\} \vdash_M \neg({\sim}A)$. □

We will consider additional variations on the double-negation translation in Section 4.6, in the context of first-order logic.

Exercises

2.7.1. Show that if A is any propositional formula with variables $P_0, \ldots, P_{k-1}$, then $\{P_0 \vee \neg P_0, \ldots, P_{k-1} \vee \neg P_{k-1}\} \vdash_I A \vee \neg A$.

2.7.2. A double-negation translation due to Kolmogorov adds double negations in front of *every* subformula. Show that for every propositional formula A, the Kolmogorov translation of A is equivalent to the Gödel–Gentzen translation of A.

2.7.3. Prove Theorem 2.7.9.

2.7.4. Prove Theorem 2.7.11.

2.8 Other Deductive Systems

Other deductive systems for propositional logic are useful for various purposes. In Chapter 6, we will consider sequent calculi, which have nice metamathematical properties and are important for automated reasoning.

Categorical logic provides a deductive system for minimal logic that is aligned with the algebraic semantics presented in Chapter 3. Take sequents to be of the form $A \vdash B$, which expresses that the formula A entails the formula B. Start with the axiom schema $A \vdash A$ and add the rule "from $A \vdash B$ and $B \vdash C$ conclude $A \vdash C$," which expresses the transitivity of the entailment relation. The three logical rules can be used in either direction:

- $A \vdash B \wedge C$ if and only if $A \vdash B$ and $A \vdash C$.
- $A \vee B \vdash C$ if and only if $A \vdash C$ and $B \vdash C$.
- $A \vdash B \to C$ if and only if $A \wedge B \vdash C$.

One obtains a system for intuitionistic logic by adding $\bot \vdash A$ and $A \vdash \top$ as axioms. This system can be extended to first- and higher-order logic.

There are deductive systems for classical logic that use smaller sets of connectives. If we restrict the connectives to $\to$ and $\bot$, it suffices to take schemas 1 and 2 in the axiomatic system of Section 2.2 together with the schema $(\neg B \to \neg A) \to (A \to B)$.

The search for short axiomatizations for classical logic has given rise to some interesting curiosities. In 1917, Jean Nicod presented an axiomatic deductive system where the only connective is the Sheffer stroke. The system has only one axiom schema,

$$((A \mid (B \mid C)) \mid (D \mid (D \mid D))) \mid ((E \mid B) \mid ((A \mid E) \mid (A \mid E))),$$

and one rule:

$$\frac{A \mid (B \mid C) \qquad A}{C}$$

Together with modus ponens, Meredith's axiom is also a complete proof system for classical logic in which we take $\to$ and $\neg$ as the only primitives:

$$((((A \to B) \to (\neg C \to \neg D)) \to C) \to E) \to ((E \to A) \to (D \to A)).$$

One can also provide a proof system for propositional logic by writing down equational axioms for a Boolean algebra (see Section 3.2). Proving a theorem A is tantamount to proving $A = \top$ in a proof system for equational reasoning, as described in Section 4.4. The following is a single identity axiomatizing Boolean algebras in terms of the Sheffer stroke:

$$((A \mid B) \mid C) \mid (A \mid ((A \mid C) \mid A)) = C.$$

If we allow a *nand* with multiple arities, we can define $\bot$ as nand() and $\neg A$ as nand(A), and then define the rest of the logical connectives in the usual way. Propositions of the form

$$\mathrm{nand}(\vec{A}, \vec{B}, \mathrm{nand}(\vec{B}))$$

are tautologically true, since if none of the B_i are false, then nand($\vec{B}$) is. The following rule is sound:

$$\frac{\mathrm{nand}(\vec{A}, \vec{B}) \qquad \mathrm{nand}(\vec{A}, \mathrm{nand}(\vec{B}))}{\mathrm{nand}(\vec{A})}$$

To see this, reason as follows: either all the formulas B_i are true, in which case the first premise guarantees the conclusion, or at least one B_i is false, in which case nand($\vec{B}$) is true and the second premise guarantees the conclusion. This axiom and rule constitute a complete proof system, provided we think of the arguments to nand as a set of formulas or allow arbitrary permutations of the arguments in the rules.

The field of *proof complexity* is concerned with questions as to the lengths of proofs. Axiomatic systems like the ones we have considered here, based on finitely many axiom schemas and modus ponens, are known as *Frege systems*. There are polynomial-time translations between any two such systems, and Frege systems are also polynomial-time equivalent to natural deduction. We obtain *extended Frege systems* by adding the substitution rule: from A, derive $A[B/P]$ for any formula B and variable P. Up to polynomial-time equivalence, this is equivalent to adding definitions: at any point in a proof, one can introduce a new propositional constant P to stand for a propositional formula A together with the axiom $P \leftrightarrow A$. Given a proof of a formula B in an extended Frege system, where B does not contain any of the new constants, it is straightforward to obtain a proof of B in the associated Frege system by iteratively replacing each new constant symbol P by the formula A it abbreviates. This can make a proof much longer, and it is still an open question as to whether extended Frege systems are substantially more efficient than Frege systems in general.

Proof systems for classical propositional logic are especially important in automated reasoning, since combinatorial problems can often be encoded as propositional search problems. Remember that a propositional formula A is *satisfiable* if there is a truth assignment that makes it true. Section 3.1 shows that a formula A is classically provable if and only if $\neg A$ is unsatisfiable. This means that having an algorithm to determine whether or not a propositional formula is provable is the same as having an algorithm to decide whether or not a propositional formula is satisfiable, modulo switching the input from A to $\neg A$ and switching yes and no in the output. The literature on decision procedures for propositional logic is sometimes cast in terms of provability and sometimes in terms of satisfiability, so it is helpful to be comfortable translating between the two presentations.

Replacing A by $\neg A$ in the previous discussion, we have that a formula A is refutable (that is, $\neg A$ is provable) if and only if A is unsatisfiable. A number of algorithms are designed to determine the refutability or satisfiability of a formula in conjunctive normal form. Recall that a *literal* is an atomic formula or a negated atomic formula. A *clause* is a disjunction of literals, where we take the empty disjunction to be $\bot$. Since disjunction is associative, commutative, and idempotent, it is reasonable to identify a clause $A_0 \vee \cdots \vee A_{n-1}$ with the set $\{A_0, \ldots, A_{n-1}\}$. With this identification, we consider two clauses the same when they contain the same set of literals, and the empty clause corresponds to $\bot$. We can then identify a formula A in conjunctive normal form with a set Γ of clauses, where A corresponds to the conjunction of clauses in Γ. In that case, A is refutable, and hence unsatisfiable, if and only if $\Gamma \vdash \bot$. The goal is therefore to design an algorithm that either refutes a set of clauses Γ or finds a satisfying assignment. If one of the clauses in Γ contains a pair P and $\neg P$, then it is equivalent to $\top$, and it can be deleted from Γ without changing the outcome. So without loss of generality we can assume that there are no such clauses in Γ.

Exercise 2.6.7 should raise concerns about the efficiency of translating questions about satisfiability in general to questions about satisfiability of formulas in conjunctive normal form, since it shows that in general there is no efficient way of assigning to an arbitrary propositional formula an equivalent one in conjunctive normal form. But we can get by with something weaker: given a propositional formula A, we want a set Γ of clauses that is *equisatisfiable* with A, so that A is satisfiable (respectively, refutable) if and only if Γ is. Here the idea of extending the propositional variables in A with new definitions is useful. If

A and B are propositional variables, we can express definitions $P \leftrightarrow A \wedge B$, $P \leftrightarrow A \vee B$, $P \leftrightarrow (A \to B)$, and $P \leftrightarrow (A \leftrightarrow B)$ as sets of clauses. For example, $P \leftrightarrow A \wedge B$ is represented by the set $\{\neg P \vee A, \neg P \vee B, \neg A \vee \neg B \vee P\}$. Exercise 2.8.2 asks you to find similar definitions for the other connectives and to verify that this provides an efficient way of associating an equisatisfiable set of clauses Γ to any propositional formula A. This procedure is known as the *Tseitin transformation*.

Now suppose Γ is a set of clauses, C is a clause, and L is a literal. Since $C \vee L$ is a consequence of L, the set of clauses $\Gamma \cup \{L, C \vee L\}$ is refutable if and only if $\Gamma \cup \{L\}$ is refutable. Similarly, since C is a consequence of L and $C \vee {\sim}L$, the set of clauses $\Gamma \cup \{L, C \vee {\sim}L\}$ is refutable if and only if $\Gamma \cup \{L, C\}$ is refutable. So as soon as we have a singleton L among our clauses, we can delete any other clause that contains it, and we can delete ${\sim}L$ from any clause in which it occurs, all while maintaining refutability. Finally, if neither L nor ${\sim}L$ occurs in any clause in Γ, then $\Gamma \cup \{L\}$ is refutable if and only if Γ is, since in any proof of $\bot$ from $\Gamma \cup \{L\}$, we can substitute $\top$ for $\{L\}$. Hence we can drop L outright. In sum, if Γ contains a singleton clause L, then L can be eliminated entirely from Γ to obtain an equirefutable (and hence equisatisfiable) set Γ'.

For any propositional variable P, the fact that classical logic proves $P \vee \neg P$ means that Γ is refutable if and only if both $\Gamma \cup \{P\}$ and $\Gamma \cup \{\neg P\}$ are. This provides the basis for a decision procedure: pick a variable P, consider $\Gamma \cup \{P\}$ and $\Gamma \cup \{\neg P\}$, and then use the method described in the previous paragraph to eliminate P from each set of clauses. If either of the resulting sets contains an empty clause, it is refutable, and if a set itself is empty, it is satisfiable. Otherwise, pick another variable P' and continue in this way on each branch. Since the number of variables decreases at each step, we have the following:

Theorem 2.8.1. *Starting with a set of clauses Γ and iteratively splitting either results in an empty set of clauses in each branch of the search, in which case Γ is refutable, or reaches the empty set of clauses on some branch, in which case Γ satisfiable.*

In practice, it pays to eliminate singleton clauses eagerly. This is known as *unit propagation*. One can also show that if any variable P occurs only positively in the clauses or only negatively (that is, as $\neg P$), then deleting the clauses in which P occurs preserves refutability. This is known as *pure literal elimination*, and it is generally also applied eagerly. The resulting decision procedure, based on unit propagation, pure literal elimination, and the splitting rule, is known as the *Davis–Putnam–Logemann–Loveland* procedure or *DPLL*.

When implementing the DPLL procedure, there is an art to deciding which variable to split on. Fast decision procedures for propositional logic, known as *SAT solvers*, also employ a method known as *conflict driven clause learning* to narrow the search space by analyzing branches that lead to refutations and learning new clauses that will hopefully result in quicker refutations on other branches. See the bibliographical notes at the end of this chapter for more information.

Another decision procedure, known as the *Davis–Putnam* procedure, is based on the *resolution rule*:

$$\frac{C \vee P \qquad D \vee \neg P}{C \vee D}$$

Here C and D are clauses and P is any propositional variable. The clauses $C \vee P$ and $D \vee \neg P$ are called a *complementary pair* and $C \vee D$ is their *resolvent*. If, by applying the resolution rule repeatedly, we derive the empty clause, we have shown that Γ is refutable. We will now show that the converse is true, namely, that if Γ is refutable, then it can be refuted by systematically applying the resolution rule.

Lemma 2.8.2. *Let Γ be any set of clauses and let P be any propositional variable. Let $C_0, \ldots, C_{n-1}$ be all the clauses in Γ that contain P, and write $C_i = C_i' \vee P$. Similarly, let $D_0, \ldots, D_{m-1}$ be all the clauses in Γ that contain $\neg P$, with $D_j = D_j' \vee \neg P$. Let $E_0, \ldots, E_{k-1}$ be the clauses that include neither P nor $\neg P$. Then Γ is refutable if and only if the set of clauses*

$$\Gamma' = \{C_i \vee D_j\}_{i<n, j<m} \cup \{E_0, \ldots, E_{k-1}\}$$

is refutable.

Proof Writing $C = \bigwedge_i C_i'$, $D = \bigwedge_j D_j'$, and $E = \bigwedge_k E_k$, we have that $\bigwedge \Gamma$ is equivalent to $(C \vee P) \wedge (D \vee \neg P) \wedge E$ and $\bigwedge \Gamma'$ is equivalent to $(C \vee D) \wedge E$. As an exercise, you can show that classical logic proves $(C \vee P) \wedge (D \vee \neg P) \to C \vee D$. Hence, if Γ' is refutable, then so is Γ. Conversely, suppose there is a proof of $\bot$ from $(C \vee P) \wedge (D \vee \neg P) \wedge E$. Replacing P by $\neg C$, we obtain a proof of $\bot$ from $(C \vee D) \wedge E$. □

Theorem 2.8.3. *If Γ is refutable, it can be refuted using only the resolution rule.*

The algorithm for finding such a refutation is straightforward. Lemma 2.8.2 shows that given a set of clauses, we can pick any variable P, form the resolvent of all pairs of the form $C \vee P$ and $D \vee \neg P$, and then delete the clauses involving P or $\neg P$. The result is a set of clauses in one fewer variable. Unit propagation corresponds to the resolution rule on a variable P where one of the clauses is either P or $\neg P$. The Davis–Putnam procedure also applies unit propagation and the pure literal rule eagerly, before choosing the next variable to resolve on.

Our descriptions of DPLL and the Davis–Putnam procedure involve replacing a set of clauses by another set of clauses that is equirefutable. In the literature, it is more common to reason about these steps in semantic terms, showing, equivalently, that they are equisatisfiable. Notice that if you demonstrate equirefutability and equisatisfiability independently, then the fact that the either procedure terminates shows that a set of clauses is satisfiable if and only if it is not refutable. We will see in Section 3.1 that this is one formulation of the completeness theorem for propositional logic.

Exercises

2.8.1. Show that the proof system based on a multiple-arity nand is a complete proof system, as follows:

a. Show that weakening is an admissible rule: if $\text{nand}(\vec{A})$ is derivable, so is $\text{nand}(\vec{A}, \vec{B})$.

b. Show that these rules are derivable in the system with weakening:

$$\frac{\text{nand}(\vec{A}, \text{nand}(B_0)) \qquad \cdots \qquad \text{nand}(\vec{A}, \text{nand}(B_{n-1}))}{\text{nand}(\vec{A}, \text{nand}(B_0, \ldots, B_{n-1}))}$$

and

$$\frac{\text{nand}(\vec{A}, \vec{B})}{\text{nand}(\vec{A}, \text{nand}(\text{nand}(\vec{B})))}$$

c. Argue that given a valid conclusion, we can work backward using these two rules until we reach axioms.

A similar argument is used to establish the completeness of the cut-free sequent calculus for classical propositional logic in Section 6.3.

2.8.2. Show how each propositional connectives can be defined with a set of clauses, and verify that the Tseitin transformation behaves as advertised.

2.8.3. Verify that the DPLL steps described in this section preserve refutability, in the sense that the output of a step is refutable if and only if all the inputs are. Do the same for satisfiability.

2.8.4. Verify that the resolution rule preserves refutability and, dually, satisfiability.

Bibliographical Notes

Historical information. For the early history of deductive systems for propositional logic, see van Heijenoort (1967). See also the notes and historical references in Troelstra and Schwichtenberg (2000).

Other proof systems For more about the proof system inspired by categorical logic that is described in Section 2.8, see Lambek and Scott (1988). At the time of writing, the best references for short axiomatizations of propositional logic and Boolean algebras are on *Wikipedia*, in articles named "List of Hilbert systems" and "Minimal axioms for Boolean algebra." The axiomatizations by Nicod and Meredith can be found in Mendelson (2015), and the short equational axiomatization of Boolean algebras based on the Sheffer stroke is presented in McCune et al. (2002).

Decision procedures for propositional logic The standard reference for SAT solvers and decision procedures for propositional logic is Biere et al. (2021). For more on resolution, Davis–Putnam, and DPLL, see Bradley and Manna (2007), Harrison (2009), Kroening and Strichman (2016), Robinson and Voronkov (2001), and Troelstra and Schwichtenberg (2000).

Proof complexity For more on propositional proof complexity, see Krajíček (1995), Urquhart (1995), and Pudlák (1998).

3

Semantics of Propositional Logic

Let Γ be a set of propositional formulas and let A be a propositional formula. There are at least two ways to interpret the informal statement that A follows from Γ: that it is possible to prove A from Γ, and that A must be true in any circumstances in which the hypotheses in Γ are true. Formal notions of proof make precise the first sense in which A follows from Γ. Making the second sense precise requires developing an account of the meaning of a propositional formula and the circumstances under which it is true. This is what a *semantics* is supposed to do.

In Section 3.1, we will present the standard truth-table semantics for classical propositional logic. We will then be able to show that our deductive system for classical logic is *sound*, which is to say, if A is provable from Γ, then A follows from Γ in the semantic sense. We will also show that classical logic is *complete* for the semantics, which is the converse of soundness: if A is a semantic consequence of Γ, then A is in fact provable from Γ.

The semantics of intuitionistic logic is more involved. Intuitively, an intuitionistic proof of A from Γ is supposed to provide us with a construction or procedure that transforms evidence for the hypotheses in Γ to evidence for A. In Section 13.8 and Chapter 14, we will see some of the ways that this plays out formally. In this chapter, we will present another semantics, Kripke semantics, and show that our deductive systems are sound and complete with respect to that. This provides an alternative perspective on the meaning of intuitionistic logic.

There are good reasons to be interested in semantics even if you are primarily interested in provability. Semantics provides a useful tool for studying proof systems, shedding light on what they can and cannot do. In particular, the easiest way to show that a formula A is *not* provable from a set of hypotheses Γ is to give a countermodel. We will see in the chapters to come that model-theoretic arguments often provide perspicuous proofs of purely syntactic results. Different semantic perspectives are useful for different purposes, and we will consider a variety of approaches, including algebraic ones.

The influence runs both ways. Just as semantic notions can illuminate systems of logical deduction, deductive systems can often be used to illuminate objects and structures that have intrinsic mathematical interest. *Model theory* uses classical logic to study structures that arise in algebra, analysis, and topology, and *categorical logic* generally uses logics with a more intuitionistic flavor to that same end. Even though such applications are not a focus of this book, the presentation in this chapter and in Chapter 5 provides some indications in that direction.

3.1 Classical Logic

From the point of view of classical logic, given a formula such as $P \wedge (Q \vee R)$, we think of the variables P, Q, and R as standing for propositions that can be true or false. Whether the compound expression $P \wedge (Q \vee R)$ is true depends on the truth of the variables. More importantly, that is all it depends on: the truth value of $P \wedge (Q \vee R)$ is a function of the truth values of P, Q, and R.

Let $\mathbb{B}$ be the set $\{\top, \bot\}$, where we think of $\top$ as denoting the value *true*, and $\bot$ as denoting the value *false*. It does not matter whether we take these to be the symbols we used in our syntax or other mathematical objects; any two-element set will do.

Definition 3.1.1. A *truth assignment* is a function from the set of propositional variables to $\mathbb{B}$.

Let v be a truth assignment. Using recursion on the set of propositional formulas, we extend v to a function $\bar{v}$ as follows:

- $\bar{v}(P) = v(P)$ for every propositional variable P
- $\bar{v}(\bot) = \bot$
- $\bar{v}(A \wedge B) = \begin{cases} \top & \text{if } \bar{v}(A) = \top \text{ and } \bar{v}(B) = \top \\ \bot & \text{otherwise} \end{cases}$
- $\bar{v}(A \vee B) = \begin{cases} \top & \text{if } \bar{v}(A) = \top \text{ or } \bar{v}(B) = \top \\ \bot & \text{otherwise} \end{cases}$
- $\bar{v}(A \to B) = \begin{cases} \top & \text{if } \bar{v}(A) = \bot \text{ or } \bar{v}(B) = \top \\ \bot & \text{otherwise.} \end{cases}$

If we view the set of propositional formulas as a term algebra over the set of propositional variables, $\bar{v}(A)$ is just the evaluation of A in $\mathbb{B}$, now viewed as a structure with the corresponding interpretations of $\wedge$, $\vee$, and $\to$. As a result, we will write $[\![A]\!]^v$ instead of $\bar{v}(A)$ and read this as *the truth value of A under v*. As expected, we have the following:

Proposition 3.1.2. *Let v and v' be truth assignments such that $v(P) = v'(P)$ for every variable P occurring in A. Then $[\![A]\!]^v = [\![A]\!]^{v'}$.*

Definition 3.1.3. Let A be a propositional formula and let v be a truth assignment.

- We say that v *satisfies* A, written $v \models A$, if $[\![A]\!]^v = \top$. In that case, we also say that A is *true under* v.
- If Γ is a set of formulas, $v \models \Gamma$ means that $v \models B$ for every $B \in \Gamma$. In this case, we say that v *satisfies* Γ.
- A formula A or a set of formulas Γ is *satisfiable* if it is satisfied by some truth assignment.
- A formula A is *valid*, written $\models A$, if it is true under every truth assignment, that is, if $v \models A$ for every v. In that case, we also say that A is a *tautology*.
- A is a *logical consequence* of Γ, written $\Gamma \models A$, if every truth assignment v that satisfies Γ also satisfies A. In that case, we also say that A is a *semantic consequence* of Γ, that Γ *logically entails* A, or, more simply, that Γ *entails* A.

Notice that the symbol $\models$ means different things in the expressions $v \models A$ and $\Gamma \models A$. Proposition 3.1.2 provides an algorithm for deciding validity: to decide whether or not A is a tautology, simply evaluate A under all possible assignments to its variables.

Theorem 3.1.4 (soundness). *For every set Γ of propositional formulas and every propositional formula A, if $\Gamma \vdash A$ then $\Gamma \models A$.*

Proof We are free to choose any system we want to represent provability, and so we use induction on derivations in natural deduction to show that if a sequent $\Gamma \vdash A$ is derivable, then $\Gamma \models A$.

If $\Gamma \vdash A$ is an assumption, then A is in Γ, and clearly $\Gamma \models A$.

Suppose we derive $\Gamma \vdash A \wedge B$ from $\Gamma \vdash A$ and $\Gamma \vdash B$ using the $\wedge$-introduction rule. Let v be any truth assignment that satisfies Γ. By the inductive hypothesis, v satisfies A and B, and hence it satisfies $A \wedge B$.

Suppose we derive $\Gamma \vdash A$ from $\Gamma \vdash A \wedge B$. Let v be any assignment that satisfies Γ. By the inductive hypothesis, v satisfies $A \wedge B$, and hence it satisfies A. The case for the right $\wedge$-elimination rule is similar.

Suppose we derive $\Gamma \vdash A \to B$ from $\Gamma, A \vdash B$. Let v be any truth assignment that satisfies Γ. If $[\![A]\!]^v = \bot$, then $[\![A \to B]\!]^v = \top$, as required. Otherwise, $[\![A]\!]^v = \top$, and by the inductive hypothesis we have $[\![B]\!]^v = \top$, which implies $[\![A \to B]\!]^v = \top$.

Suppose we derive $\Gamma \vdash B$ from $\Gamma \vdash A \to B$ and $\Gamma \vdash A$. Let v be any truth assignment that satisfies Γ. Then, by the inductive hypothesis, v satisfies both $A \to B$ and A. By the definition of satisfaction for $A \to B$, we have that v satisfies B, as required.

The cases for the $\vee$ rules and *reductio* are similar. □

At the outset, we would not expect to be able to prove an arbitrary propositional variable P or an arbitrary implication $P \to Q$. But it is surprisingly hard to show that these are not provable without invoking the soundness theorem. In fact, without semantics, it is difficult to show that there is *anything* that propositional logic doesn't prove. In Chapter 6, however, we will develop syntactic means of doing so.

As far as proving soundness is concerned, there is nothing special about natural deduction. To show soundness with respect to a Hilbert-style calculus, for example, we simply show that all the axioms are true under any assignment to the variables, and that truth is preserved by every rule of inference.

The converse of soundness is *completeness*.

Theorem 3.1.5 (completeness). *For every set Γ of propositional formulas and every propositional formula A, if $\Gamma \models A$ then $\Gamma \vdash A$.*

As for soundness, our proof of completeness is not sensitive to the choice of deductive calculus. Rather, the properties that the deductive system has to satisfy will be made manifest by our proof. To start, it is helpful to have an equivalent formulation of the completeness theorem.

Definition 3.1.6. A set of propositional formulas is *inconsistent* if it proves $\bot$, and *consistent* otherwise.

Proposition 3.1.7. *Let Γ be any set of propositional formulas. The following are equivalent for classical and intuitionistic logic:*

- *Γ is inconsistent.*
- *For some formula A, Γ proves both A and $\neg A$.*
- *For every formula A, Γ proves A.*

For minimal logic, the first two are equivalent and are implied by the third.

The next proposition is also easy to prove, by unfolding the definitions of satisfiability and entailment.

Proposition 3.1.8. *Let Γ be any set of propositional formulas. The following are equivalent:*

- *Γ is unsatisfiable.*
- $\Gamma \models \bot$.

The following proposition expresses an important relationship between provability and consistency. It only holds for classical logic.

Proposition 3.1.9. *Let Γ be any set of propositional formulas, and let A be any formula. The following are equivalent:*

- $\Gamma \vdash A$.
- *$\Gamma \cup \{\neg A\}$ is inconsistent.*

So are the following:

- $\Gamma \models A$.
- *$\Gamma \cup \{\neg A\}$ is unsatisfiable.*

In minimal logic, we have $\Gamma \vdash \neg A$ if and only if $\Gamma \cup \{A\}$ is inconsistent. One direction of the first equivalence above relies on the fact that in classical logic, $\Gamma \vdash \neg\neg A$ if and only if $\Gamma \vdash A$. With these propositions in hand, we can show that the completeness theorem is equivalent to the following:

Theorem 3.1.10. *Every consistent set of propositional formulas is satisfiable.*

To see that Theorem 3.1.10 implies Theorem 3.1.5, suppose $\Gamma \nvdash A$. Then, by Proposition 3.1.9, $\Gamma \cup \{\neg A\}$ is consistent, and by Theorem 3.1.10, it is satisfiable, which shows $\Gamma \not\models A$. To see that Theorem 3.1.5 implies Theorem 3.1.10, suppose Γ is consistent. Then $\Gamma \nvdash \bot$, and so, by Theorem 3.1.5, $\Gamma \not\models \bot$. By Proposition 3.1.8, this means that Γ is satisfiable.

The completeness theorem has the following important corollary, known as the *compactness* theorem:

Theorem 3.1.11 (Compactness). *For any set Γ of propositional formulas and any propositional formula A, if $\Gamma \models A$, then for some finite subset $\Gamma' \subseteq \Gamma$, we have $\Gamma' \models A$. Equivalently, for any set of propositional formulas Γ, if every finite subset of Γ is satisfiable, then so is Γ.*

Proof We will prove the first statement here, and I will leave it to you to verify that it is equivalent to the second statement. Suppose $\Gamma \models A$. By Theorem 3.1.5, we have $\Gamma \vdash A$. If we let Γ' be the finite set of hypotheses used in a proof of A from Γ, then we have $\Gamma' \vdash A$. By soundness, $\Gamma' \models A$. □

The sets Γ that are mentioned in the formulations of the completeness theorem, expressed as either Theorem 3.1.5 or Theorem 3.1.10, may be infinite. Indeed, the statement of the compactness theorem is interesting only in that case. We will prove Theorem 3.1.10 directly, since it is this proof that extends most straightforwardly to the case of first-order logic in Chapter 5. But, in the exercises below, you will see that it is easier to prove Theorem 3.1.5 in the case where Γ is finite. This provides another strategy for proving the completeness theorem:

- Prove the weak form of the completeness theorem in which Γ is restricted to finite sets.
- Prove the compactness theorem independently.
- Show that these two facts imply the general statement of Theorem 3.1.5.

This approach is outlined in Exercises 3.1.6 and 3.1.7.

We now turn to the proof of Theorem 3.1.10. Given a consistent set of formulas, Γ, we need to find a satisfying truth assignment. We will adopt an interesting strategy: we will make the task seemingly harder by expanding Γ to a larger set, Γ'. The resulting set, Γ', will have so much information that it will be straightforward to read off a satisfying assignment.

Definition 3.1.12. A set Γ of propositional formulas is *maximally consistent* if

- it is consistent, and
- for every formula $B \notin \Gamma$, $\Gamma \cup \{B\}$ is inconsistent.

In other words, Γ is maximally consistent if it is consistent but adding any new formula makes it inconsistent. Exercise 3.1.4 below asks you to show that if v is a truth assignment, the set $\{A \mid [\![A]\!]^v = \top\}$ is maximally consistent. Lemma 3.1.15 below shows that, in a sense, this is the only example: any maximally consistent set arises from some truth assignment v in this way.

Lemma 3.1.13. *Let Γ be any consistent set of propositional formulas. Then there is a maximally consistent set $\Gamma' \supseteq \Gamma$.*

Proof We will assume that there are only countably many propositional variables, and hence countably many propositional formulas. This means that it is possible to enumerate them as a sequence $B_0, B_1, B_2, \ldots$. For the more general case, use Zorn's lemma, as described in Appendix A.1.

Define a sequence of sets Γ_i by

$$\Gamma_0 = \Gamma$$

$$\Gamma_{i+1} = \begin{cases} \Gamma_i \cup \{B_i\} & \text{if this is consistent} \\ \Gamma_i & \text{otherwise.} \end{cases}$$

Let $\Gamma' = \bigcup_i \Gamma_i$. By induction, each set Γ_i is consistent. This implies that Γ' is consistent, because given any proof of $\bot$ from Γ', the finite set of hypotheses used would be contained in some Γ_i. To see that Γ' is maximally consistent, suppose $A \notin \Gamma'$. For some i, $A = B_i$, and we have $A \notin \Gamma_{i+1}$ because $\Gamma_{i+1} \subseteq \Gamma'$. By construction, this means that $\Gamma_i \cup \{A\}$ is inconsistent, which implies that $\Gamma' \cup \{A\}$ is inconsistent. □

Lemma 3.1.14. *Let Γ be a maximally consistent set of propositional formulas. Then Γ has all the following properties:*

1. *Γ is deductively closed: if $\Gamma \vdash A$, then $A \in \Gamma$.*
2. *$\bot \notin \Gamma$.*

3. *$A \wedge B \in \Gamma$ if and only if $A \in \Gamma$ and $B \in \Gamma$.*
4. *$A \to B \in \Gamma$ if and only if $A \notin \Gamma$ or $B \in \Gamma$.*
5. *$A \vee B \in \Gamma$ if and only if $A \in \Gamma$ or $B \in \Gamma$.*

Proof For claim 1, suppose $\Gamma \vdash A$. Since Γ is consistent, $\Gamma \cup \{A\}$ is consistent. Since Γ is maximally consistent, $A \in \Gamma$.

Claim 2 follows from the fact that Γ is consistent. For claim 3, suppose $A \wedge B \in \Gamma$. Then $\Gamma \vdash A$ and $\Gamma \vdash B$, and hence, by claim 1, $A \in \Gamma$ and so $B \in \Gamma$. Conversely, suppose $A \in \Gamma$ and $B \in \Gamma$. Then $\Gamma \vdash A \wedge B$, and hence $A \wedge B$ is in Γ.

For claim 4, suppose $A \to B \in \Gamma$. If A is in Γ, then $\Gamma \vdash B$ and $B \in \Gamma$. So, either $A \notin \Gamma$ or $B \in \Gamma$. Conversely, suppose either $A \notin \Gamma$ or $B \in \Gamma$. In the first case, $\Gamma \cup \{A\}$ is inconsistent, which implies $\Gamma \vdash \neg A$. This implies $\Gamma \vdash A \to B$, and so $A \to B \in \Gamma$. In the second case, $\Gamma \vdash B$, and so $\Gamma \vdash A \to B$. This implies $A \to B \in \Gamma$. So, either way, $A \to B \in \Gamma$.

The proof of claim 5 is similar to the proof of claim 4. □

Lemma 3.1.15. *Suppose Γ is a maximally consistent set of formulas. Define a truth assignment v by setting $v(P) = \top$ if and only if P is in Γ. Then for every propositional formula A, we have $[\![A]\!]^v = \top$ if and only if A is in Γ.*

Proof Use induction on propositional formulas A. If A is a propositional variable, the claim is true by the definition of v, and in all the other cases, we can apply Lemma 3.1.14. □

This yields a proof of the completeness theorem.

Proof of Theorem 3.1.10 Suppose Γ is consistent. By Lemma 3.1.13, there is a maximally consistent set $\Gamma' \supseteq \Gamma$. By Lemma 3.1.15, there is a truth assignment v satisfying Γ' and hence Γ. □

Semantic notions also allow us to express what it means for a set of connectives to be complete. Any connective, taking any number of arguments, is defined by its truth table. In other words, we define a new connective by extending the clauses in the definition of $[\![A]\!]^v$ to accommodate the new symbol. By Proposition 3.1.2, any formula A with free variables $P_0, \ldots, P_{n-1}$ gives rise to an n-ary function f_A from $\mathbb{B}$ to $\mathbb{B}$ that maps the tuple of values $a_0, \ldots, a_{n-1}$ to $[\![A]\!]^v$, where v is any truth assignment such that $v(P_i) = a_i$ for each $i < n$.

Definition 3.1.16. A set of connectives together with their interpretations as functions on $\mathbb{B}$ is said to be *complete* if for every n-ary function f from $\mathbb{B}$ to $\mathbb{B}$ there is a formula A involving only those connectives and the variables $P_0, \ldots, P_{n-1}$, such that $f = f_A$.

Exercises

3.1.1. Prove the following from the definitions of satisfiability and entailment: if $A_0, \ldots, A_{k-1}$ and B are propositional formulas, then $\{A_0, \ldots, A_{k-1}\} \models B$ if and only if $\models A_0 \wedge \cdots \wedge A_{k-1} \to B$.

3.1.2. Prove or find a counterexample to each of the following:
   a. For every set of formulas Γ, every formula A, and every formula B, if $\Gamma \models A \wedge B$, then $\Gamma \models A$ and $\Gamma \models B$.
   b. For every set of formulas Γ, every formula A, and every formula B, if $\Gamma \models A \vee B$, then $\Gamma \models A$ or $\Gamma \models B$.

3.1.3. Let Γ be a set of propositional formulas. Show that the following are equivalent:
- Γ is maximally consistent.
- Γ is consistent, and for every formula A, either A or $\neg A$ is in Γ.

3.1.4. Let v be any truth assignment, and let $\Gamma = \{A \mid [\![A]\!]^v = \top\}$. Show that Γ is maximally consistent.

3.1.5. A formula A is said to be *independent* of a set of formulas Γ if $\Gamma \nvdash A$ and $\Gamma \nvdash \neg A$. Suppose Γ is a consistent set of formulas, A is independent of Γ, and B is independent of $\Gamma \cup \{A\}$. Show that there are at least three different maximally consistent sets containing Γ.

3.1.6. The following is a more direct proof of the compactness theorem for propositional logic. A set of formulas Γ is said to be *finitely satisfiable* if every finite subset is satisfiable. The compactness theorem states that every finitely satisfiable set of formulas is satisfiable.

a. Show that if Γ is finitely satisfiable and P is an propositional variable, then either Γ, P or $\Gamma, \neg P$ is finitely satisfiable.

b. Let $P_0, P_1, P_2, \ldots$ enumerate all the propositional variables. Iteratively define a sequence of formulas $B_0, B_1, B_2, \ldots$ by

$$B_i = \begin{cases} P_i & \text{if } \Gamma, B_0, \ldots, B_{i-1}, P_i \text{ is finitely satisfiable} \\ \neg P_i & \text{otherwise.} \end{cases}$$

Let $\Delta = \{B_0, B_1, B_2, \ldots\}$. Show that $\Gamma \cup \Delta$ is finitely satisfiable.

c. Observe that there is exactly one truth assignment v that satisfies Δ. Suppose A is any formula in Γ whose variables are among $P_0, \ldots, P_{k-1}$. Since $\Gamma \cup \Delta$ is finitely satisfiable, the set $\{A, B_0, \ldots, B_{k-1}\}$ is satisfiable. Show that it is satisfied by v.

Since v satisfies every formula in Γ, we are done.

3.1.7. The following gives a direct, finitary proof of the weak form of the completeness theorem.

a. If A is any propositional formula and v is any truth assignment, let $\Gamma_{A,v}$ denote the finite set that contains P for every variable P occurring in A such that $v(P) = \top$, and $\neg P$ for every variable P occurring in A such that $v(P) = \bot$. Show by induction on A that $\Gamma_{A,v} \vdash A$ if $[\![A]\!]^v = \top$ and $\Gamma_{A,v} \vdash \neg A$ if $[\![A]\!]^v = \bot$.

b. Show that for any finite set of formulas Γ, any formula A, and any propositional variable P, if $\Gamma, P \vdash A$ and $\Gamma, \neg P \vdash A$, then $\Gamma \vdash A$.

c. Conclude that for any propositional formula A, if $\models A$, then $\vdash A$.

3.2 Algebraic Semantics for Classical Logic

With truth-table semantics, formulas A take values $[\![A]\!]^v$ in the set $\mathbb{B} = \{\top, \bot\}$. The symbols for conjunction, disjunction, and implication are then interpreted by corresponding operations on $\mathbb{B}$. We can generalize this semantics by allowing formulas to take truth values in any set, as long as that set is equipped with a structure that can interpret the relevant operations. For classical logic, a *Boolean algebra* provides exactly the right structure. Evaluating formulas in an arbitrary Boolean algebra not only provides a more general semantics but also offers a new perspective on the original two-valued semantics.

Let P be a set and let $\leq$ be a binary relation on P. Recall that $(P, \leq)$ is a partial order if $\leq$ is reflexive, transitive, and antisymmetric.

Definition 3.2.1. A *lattice* is a partial order $(P, \leq)$ in which every pair of elements has a least upper bound and a greatest lower bound.

We will use $a \wedge b$ to denote the greatest lower bound of a and b; it is also sometimes called the *meet*. Saying it is the greatest lower bound of a and b means that the following hold:

- $a \wedge b \leq a$.
- $a \wedge b \leq b$.
- For every c, if $c \leq a$ and $c \leq b$, then $c \leq a \wedge b$.

Equivalently, we can say that for every c, $c \leq a \wedge b$ if and only if $c \leq a$ and $c \leq b$. It is easy to see that if a and b have a greatest lower bound then it is unique.

Similarly, the least upper bound of a and b, also called the *join* of a and b, is written $a \vee b$ and satisfies the following:

- $a \leq a \vee b$.
- $b \leq a \vee b$.
- For every c, if $a \leq c$ and $b \leq c$, then $a \vee b \leq c$.

Equivalently, for every c, $a \vee b \leq c$ if and only if $a \leq c$ and $b \leq c$.

The properties of meet and join imply that they are associative and commutative, and satisfy the following absorption laws:

- $a \vee (a \wedge b) = a$
- $a \wedge (a \vee b) = a$.

Exercise 3.2.1 asks you to show that $a \wedge a = a$ and $a \vee a = a$ follow from the absorption laws.

If we think of the elements of a lattice as being generalized truth values, $a \leq b$ says that b is more true than a, or that it is implied by, or entailed by, a. In any lattice, $a \leq b$ is equivalent to $a \wedge b = a$, and also equivalent to $a \vee b = b$. Conversely, given a structure $(P, \wedge, \vee)$ that is commutative, associative, and satisfies the absorption laws, the assertions $a \wedge b = a$ and $a \vee b = b$ are equivalent. Taking either of these as a definition of $a \leq b$ makes the structure into a lattice.

We need to impose additional conditions to ensure that we will have a sound interpretation of classical logic. A lattice is *bounded* if there are greatest and least elements, denoted $\top$ and $\bot$, respectively. These are usually called *top* and *bottom*, respectively. A lattice is *distributive* if meet and join satisfy the following identities for every a, b, and c:

- $a \wedge (b \vee c) = (a \wedge b) \vee (a \wedge c)$
- $a \vee (b \wedge c) = (a \vee b) \wedge (a \vee c)$.

Exercise 3.2.2 asks you to show that these two principles are equivalent: any lattice satisfying one for every a, b, and c automatically satisfies the other.

Finally, an element a of a bounded lattice is *complemented* if there exists a b such that $a \wedge b = \bot$ and $a \vee b = \top$. In a bounded distributive lattice, if a has a complement, then it is unique. We will write it as $\neg a$.

Definition 3.2.2. A *Boolean algebra* is a bounded, distributive lattice in which every element has a complement.

We have already seen one example: the set $\mathbb{B} = \{\top, \bot\}$ is a Boolean algebra with $\bot < \top$ and $\wedge$, $\vee$, and $\neg$ forming the operations of meet, join, and complement, respectively. For another example, if U is any set, the collection of subsets of U forms a Boolean algebra under set inclusion, $\subseteq$. Here, meet, join, and complement are interpreted by $\cap$, $\cup$, and the operation $X \mapsto U \setminus X$.

It is not hard to check that if $(P, \leq)$ is a Boolean algebra, then so is $(P, \geq)$. The least and greatest elements are swapped, as are meet and join, while the complement operation remains the same. $(P, \geq)$ is said to be the *dual* of $(P, \leq)$, and the fact that the dual of a Boolean algebra is again a Boolean algebra is known as *duality*. Duality implies that any statement that holds in every Boolean algebra also holds in the dual of every algebra. So any valid identity expressed in terms of meets, joins, complements, $\top$, and $\bot$ remains valid when we swap meets with joins and $\top$ with $\bot$.

Let $\mathfrak{B}$ be a Boolean algebra. Any assignment v mapping propositional variables to $\mathfrak{B}$ extends to a valuation $[\![A]\!]^v$ of every propositional formula, where we assign

$$[\![A \to B]\!]^v = \neg [\![A]\!]^v \vee [\![B]\!]^v,$$

and where we interpret the symbols $\wedge$, $\vee$, and $\bot$ by their counterparts in $\mathfrak{B}$. Semantic notions carry over to the new setting: we say a formula A is *valid* with respect to Boolean algebra semantics, written $\models A$, if $[\![A]\!]^v = \top$ for every assignment v taking values in any Boolean algebra. Similarly, we say that Γ entails A, and write $\Gamma \models A$, if $[\![A]\!]^v = \top$ for every assignment v such that every formula $B \in \Gamma$ satisfies $[\![B]\!]^v = \top$.

Since $\mathbb{B}$ is a Boolean algebra, the new semantics generalizes the two-valued semantics. Providing a broader notion of interpretation generally makes it harder to prove soundness, since there are more interpretations to consider, but easier to prove completeness, for exactly the same reason. In fact, the completeness of propositional logic with respect to Boolean-valued semantics follows immediately from completeness with respect to the two-valued semantics:

Theorem 3.2.3. *For any Γ and A, if $\Gamma \models A$ then $\Gamma \vdash A$.*

Proof If Γ entails A with respect to valuations in every Boolean algebra, then, in particular, it entails A with respect to valuations in $\mathbb{B}$. By Theorem 3.1.5, this implies $\Gamma \vdash A$. □

We will see below that it is also not hard to prove completeness directly. Remember that soundness is expressed as follows:

Theorem 3.2.4. *For any Γ and A, if $\Gamma \vdash A$, then $\Gamma \models A$.*

Proof Without loss of generality, we can assume Γ is finite. With that assumption, we will prove something stronger: given any Boolean algebra $\mathfrak{B}$ and any assignment v of an element of $\mathfrak{B}$ to each propositional variable, whenever $\Gamma \vdash A$, we have $\bigwedge\{[\![D]\!]^v \mid D \in \Gamma\} \leq [\![A]\!]^v$.

To that end, it suffices to show that the conclusion holds whenever a sequent $\Gamma \vdash A$ is derivable in natural deduction for classical logic. Fix $\mathfrak{B}$ and v, and use induction on derivations. The proof is similar to the proof of soundness for two-valued semantics, occasionally requiring a bit of calculation in a Boolean algebra. We will consider a few illustrative cases. In each case, we will use a to denote $[\![A]\!]^v$, b to denote $[\![B]\!]^v$, c to denote $[\![C]\!]^v$, and d to denote $\bigwedge\{[\![D]\!]^v \mid D \in \Gamma\}$.

Suppose the last rule of the proof yields $\Gamma \vdash B$ from $\Gamma \vdash A \to B$ and $\Gamma \vdash A$. From the inductive hypotheses, we have $d \leq \neg a \vee b$ and $d \leq a$. Thus we have

$$d \leq (\neg a \vee b) \wedge a = (\neg a \wedge a) \vee (b \wedge a) = \bot \vee (b \wedge a) = b \wedge a \leq b.$$

Suppose the last rule of the proof yields $\Gamma \vdash A \to B$ from $\Gamma, A \vdash B$. By the inductive hypothesis, we have $d \wedge a \leq b$. Then we have

$$\neg a \vee b \geq \neg a \vee (d \wedge a) = (\neg a \vee d) \wedge (\neg a \vee a) = \neg a \vee d \geq d.$$

The rules for $\wedge$ introduction and elimination and $\vee$ introduction follow straightforwardly from the definitions of meet and join. Suppose the last rule of the proof yields $\Gamma \vdash C$ from $\Gamma \vdash A \vee B$ together with $\Gamma, A \vdash C$ and $\Gamma, B \vdash C$. Then we have, by the inductive hypothesis, $d \leq a \vee b$, $d \wedge a \leq c$, and $d \wedge b \leq c$. This yields

$$c \geq (d \wedge a) \vee (d \wedge b) = d \wedge (a \vee b) \geq d \wedge d = d,$$

as required. □

To prove completeness directly without relying on the completeness theorem for truth-table semantics, define an order on the set of all propositional formulas whereby $A \leq B$ if $\{A\} \vdash B$. This is not a partial order because we may have $A \leq B$ and $B \leq A$ for distinct formulas A and B. But we get a partial order if we mod out by the associated equivalence relation, $A \equiv B$, which holds if and only if $\vdash A \leftrightarrow B$. Let $\mathfrak{B}$ be this quotient, so $\mathfrak{B}$ is the set of equivalence classes $[A] = \{B \mid \vdash A \leftrightarrow B\}$. The syntactic operations of forming conjunctions and disjunctions respect the equivalence relation, and hence extend to operations $\wedge$ and $\vee$ on the quotient structure. It is not hard to check that $\mathfrak{B}$ is a Boolean algebra with $\top = [\top]$ and $\bot = [\bot]$. If we let v be the assignment which maps each propositional variable P to $[P] \in \mathfrak{B}$, then for every formula A, we have $[\![A]\!]^v = [A]$. For any formula A, we have $\vdash A$ if and only if $\vdash A \leftrightarrow \top$, which happens if and only if $[\![A]\!]^v = \top$. This proves the following weak form of the completeness theorem: if A is valid with respect to assignments in any Boolean algebra, then $[\![A]\!]^v = \top$ for this particular assignment, and hence $\vdash A$.

A slight variation on this construction yields the full completeness theorem. Fix any set Γ of propositional formulas and now define $A \leq B$ to mean $\Gamma \cup \{A\} \vdash B$, and then proceed as before. Now two formulas A and B are equivalent if and only if $\Gamma \vdash A \leftrightarrow B$, and in the quotient structure $[A] = \top$ if and only if $\Gamma \vdash A$. This structure is sometimes called the *Tarski–Lindenbaum algebra for* Γ. This construction yields the completeness theorem: if A is true under any Boolean-algebra assignment v that assigns $\top$ to all the formulas in Γ, then $[\![A]\!]^v = \top$ under the particular assignment $v(P) = [P]$ in the Tarski–Lindenbaum algebra for Γ, which means $\Gamma \vdash A$.

Boolean-valued semantics also provides another way of formulating the proof of the completeness theorem for two-valued classical semantics. This is essentially just an algebraic reformulation of the proof in the last section, so we only sketch the details.

Definition 3.2.5. Let $(L, \leq)$ be a bounded lattice. A *filter* F on L is a subset of L satisfying the following three properties:

1. $\top$ is in F.
2. If x is in F and $x \leq y$, then y is in F.
3. If x and y are in F, then so is $x \wedge y$.

In a moment, we will see that given a filter F on a Boolean algebra $\mathfrak{B}$, it is possible to form a quotient $\mathfrak{B}/F$ with a homomorphism from $\mathfrak{B}$ to $\mathfrak{B}/F$ that sends every element of $\mathfrak{B}/F$ to $\top$. To that end, we want to think of F as a set of true elements of $\mathfrak{B}$. The conditions in the definition of a filter then say that if x is true and $x \leq y$ then y is true, and if x and y are both true then so is $x \wedge y$. A filter is *proper* if it is a proper subset of L.

Given a filter F on $\mathfrak{B}$, to define the quotient, first define the relation $x \preceq_F y$ on $\mathfrak{B}$ to mean $\neg x \vee y \in F$, and define $x \equiv_F y$ to mean that $x \preceq_F y$ and $y \preceq_F x$. Intuitively, $x \preceq_F y$ means that F thinks that x implies y, and $x \equiv_F y$ means that F thinks that they are equivalent. The exercises ask you to show that $\preceq_F$ is a preorder, which implies that $\equiv_F$ is an equivalence relation. The exercises also ask you to show that the operations $\wedge$, $\vee$, and $\neg$ on $\mathfrak{B}$ respect $\equiv_F$. So the quotient structure $\mathfrak{B}/\equiv_F$, which we will write $\mathfrak{B}/F$, is also a Boolean algebra, and the map $x \mapsto [x]$ from $\mathfrak{B}$ to $\mathfrak{B}/F$ is a homomorphism. Moreover, we have $[x] = \top$ if and only if $x \in F$, so in the new Boolean algebra $\mathfrak{B}/F$ every element of F has been collapsed to $\top$.

If S is any set of elements of $\mathfrak{B}$, there is a smallest filter F containing S, namely,

$$F = \{x \mid \text{for some finite } T \subseteq S, x \geq \bigwedge T\}.$$

It is not hard to show that the set F just defined is a filter, that is, closed upward and under $\wedge$. It is also not hard to see that if F' is any other filter that includes S, then $F' \subseteq F$, since F' also contains every finite intersection of elements of F and is closed upward. F is therefore called the *filter generated by* S.

Under what conditions will $\mathfrak{B}/F$ be (isomorphic to) the two-element Boolean algebra? A moment's reflection shows that that happens if and only if for every x in $\mathfrak{B}$, either $x \leftrightarrow \top \in F$ or $x \leftrightarrow \bot \in F$, which is to say, either x or $\neg x$ is in F. If for some x both occur then the formula $\top \leftrightarrow \bot$ is in F, which means that $\mathfrak{B}/F$ has been collapsed to a degenerate Boolean algebra in which $\top = \bot$. We can rule that out by requiring F to be proper as well. The following definition provides an alternative characterization of such filters.

Definition 3.2.6. Let $\mathfrak{B}$ be a Boolean algebra. An *ultrafilter* U on $\mathfrak{B}$ is a maximal proper filter, that is, a filter U with the property that whenever F is a filter on $\mathfrak{B}$ and $U \subsetneq F$, then $F = \mathfrak{B}$.

Lemma 3.2.7. *Let U be a filter on $\mathfrak{B}$. Then U is an ultrafilter if and only if for every x in $\mathfrak{B}$, x is in U if and only if $\neg x$ is not in U.*

Proof Suppose U is an ultrafilter, and suppose x is in U. Since U is proper, $\neg x$ is not in U. Conversely, suppose x is not in U, and consider the filter generated by $U \cup \{x\}$. Since this properly extends U and U is a maximal proper filter, the filter generated by $U \cup \{x\}$ is equal to $\mathfrak{B}$, and, in particular, contains $\bot$. By the observations above, this means that there is a finite subset $S \subseteq U$ such that $\bigwedge S \wedge x = \bot$, which implies $\bigwedge S \leq \neg x$. But since U is a filter, $\bigwedge S \in U$, and hence $\neg x \in U$, as required.

In the other direction, suppose U has the property given in the statement of the lemma. Then U is proper because $\bot = \neg\top$ is not in U. Suppose U is a strict subset of a filter F. Then for any $x \in F \setminus U$, $\neg x$ is in U, and hence in F as well. This means that $x \wedge \neg x$, which is equivalent to $\bot$, is in F. This implies $F = \mathfrak{B}$. □

In a manner analogous to the proof of Lemma 3.1.13, we can show the following:

Lemma 3.2.8. *Let F be any proper filter on $\mathfrak{B}$. Then there is an ultrafilter $U \supseteq F$.*

This provides the following alternative proof of the completeness theorem, Theorem 3.1.5. Suppose Γ is a consistent set of propositional formulas. Then $\mathfrak{B}$, the Tarski–Lindenbaum algebra relative to Γ, is a proper Boolean algebra, that is, one in which $\top \neq \bot$. Moreover, in $\mathfrak{B}$, every formula in Γ evaluates to $\top$. Let U be any ultrafilter in $\mathfrak{B}$, so that $\mathfrak{B}/U$ is isomorphic to $\mathbb{B}$. Then composing the evaluation of propositional variables in $\mathfrak{B}$ with the projection to $\mathfrak{B}$ to $\mathbb{B}$, we have a two-valued truth assignment that satisfies Γ.

Exercises

3.2.1. Show that in any partial order with meets and joins, the following two absorption laws hold:
- $a \vee (a \wedge b) = a$
- $a \wedge (a \vee b) = a$.

Show also that $a \wedge a = a$ and $a \vee a = a$ follow from the absorption laws. (Hint: for the first start with $a \wedge a$, and use an instance of absorption to replace the second a.)

3.2.2. Show that in any lattice, the following two principles are equivalent:
- $a \wedge (b \vee c) = (a \wedge b) \vee (a \wedge c)$
- $a \vee (b \wedge c) = (a \vee b) \wedge (a \vee c)$.

3.2.3. Show that if an element of a bounded distributive lattice has a complement, then it is unique.

3.2.4. Show that the following identities hold in any Boolean algebra:

a. $\neg a \vee (a \wedge b) = \neg a \vee b$
b. $(\neg a \vee \neg b) \wedge (a \wedge b) = \bot$
c. $(\neg a \vee \neg b) \vee (a \wedge b) = \top$.

By duality, the versions and $\wedge$ and $\vee$ swapped also hold. The last two identities establish establish De Morgan's laws, $\neg(a \wedge b) = (\neg a \vee \neg b)$ and $\neg(a \vee b) = (\neg a \wedge \neg b)$.

3.2.5. Let $\mathfrak{B}$ be a boolean algebra and let F be a filter on $\mathfrak{B}$. Given the definition of $\preceq_F$ and $\equiv_F$ after Definition 3.2.5, show that $\preceq_F$ is a preorder and that the operations $\wedge$, $\vee$, and $\neg$ on $\mathfrak{B}$ respect the relation $\equiv_F$.

3.2.6. Prove Lemma 3.2.8.

3.2.7. Let $\mathfrak{B}$ be the Tarski–Lindenbaum algebra relative to the empty set, let Γ be a set of formulas, and let $F = \{[A] \mid \Gamma \vdash A\}$. Show that F is a filter, and that $\mathfrak{B}/F$ is isomorphic to the Tarski–Lindenbaum algebra relative to Γ.

3.3 Intuitionistic Logic

Since intuitionistic and minimal logic are subsets of classical logic, deductive systems for intuitionistic and minimal logic are also sound with respect to truth-table semantics and with respect to interpretations in any Boolean algebra. But they are not complete for that semantics: the formula $P \vee \neg P$ evaluates to $\top$ in every Boolean algebra, but it is not intuitionistically provable. If we are looking for a semantics to characterize intuitionistic provability, we need a semantics in which such formulas are not valid.

With a *Kripke model* for propositional logic, instead of evaluating formulas with respect to a single truth assignment, we consider a set of states, each of which encodes partial information as to which propositional variables are true. We start with a partial order $(K, \leq)$. The elements $\alpha, \beta, \ldots$ of K, which are known as the *nodes* or *worlds* of the Kripke model, are supposed to represent partial descriptions of a state of affairs. At a node α, the Kripke model specifies a truth assignment, which indicates which propositional variables are known to be true at that node. You can think of $\beta \leq \alpha$ as saying that the information at β extends the information at α, and we therefore require that the set of propositional variables that are known to be true at a node $\beta \leq \alpha$ includes the set of variables that are known to be true at α. If you think of $\beta \leq \alpha$ as meaning that the information at β arrives later than the information at α, then, as time progresses, we can only learn that more of the basic variables are true. Formally, we have the following:

Definition 3.3.1. A *Kripke model* for intuitionistic logic consists of a partial order $(K, \leq)$ and a function that assigns to each $\alpha \in K$ a truth assignment v_α satisfying the following monotonicity condition: for any proposition P and pair of conditions α, β in K, if $v_\alpha(P) = \top$ and $\beta \leq \alpha$, then $v_\beta(P) = \top$.

We will write $\alpha \Vdash P$ and say *P is true at α* or *α forces P* when $v_\alpha(P) = \top$. The next definition extends the notion of being true at α to arbitrary formulas A in such a way as to preserve monotonicity.

Definition 3.3.2. Fix a Kripke model, $\mathfrak{K} = (K, \leq, v)$. The relation $\alpha \Vdash A$ is extended to arbitrary propositional formulas recursively, as follows.

- $\alpha \Vdash P$ if and only if $v_\alpha(P) = \top$.
- $\alpha \Vdash A \wedge B$ if and only if $\alpha \Vdash A$ and $\alpha \Vdash B$.
- $\alpha \Vdash A \vee B$ if and only if $\alpha \Vdash A$ or $\alpha \Vdash B$.
- $\alpha \Vdash A \to B$ if and only if for every $\beta \leq \alpha$, if $\beta \Vdash A$ then $\beta \Vdash B$.

For intuitionistic logic, we add the clause

- $\alpha \nVdash \bot$.

In minimal logic, $\bot$ is treated as a propositional variable instead.

Proposition 3.3.3. *For any pair of nodes, α and β, and any propositional formula A, if $\alpha \Vdash A$ and $\beta \leq \alpha$, then $\beta \Vdash A$.*

Proof Use induction on propositional formulas. The base case where A is a propositional variable follows from the definition of a Kripke model.

To handle the case for $\wedge$, suppose the claim holds for A and B, suppose $\alpha \Vdash A \wedge B$, and suppose $\beta \leq \alpha$. By definition, $\alpha \Vdash A$ and $\alpha \Vdash B$. By the inductive hypothesis, $\beta \Vdash A$ and $\beta \Vdash B$, and hence $\beta \Vdash A \wedge B$. The case for $\vee$ is handled similarly.

The only case that requires some thought is the clause for implication. Suppose $\alpha \Vdash A \to B$, and $\beta \leq \alpha$. We need to show $\beta \Vdash A \to B$. So suppose $\gamma \leq \beta$, and $\gamma \Vdash A$. Then $\gamma \leq \alpha$, and since $\alpha \Vdash A \to B$, $\gamma \Vdash B$, as required. □

Notice that the inductive hypothesis is not used in the case for implication. You should think about what would have gone wrong if we defined $\alpha \Vdash A \to B$ to mean that either $\alpha \nVdash A$ or $\alpha \Vdash B$. Imagine a Kripke model with two nodes, α and β, with $\beta \leq \alpha$. Suppose no propositional variable is true at α, and then P becomes true at β. With the alternative definition, we would have $\alpha \nVdash P$, and hence $\alpha \Vdash P \to Q$. But then at β, $\beta \Vdash P$ but $\beta \nVdash Q$, so $\beta \nVdash P \to Q$. This violates monotonicity. The problem is that, at α, we have to consider the possibility that P may become true in the future, and we have to insist that whenever that happens, Q becomes true as well. This is exactly what the clause for implication does. Once we settle

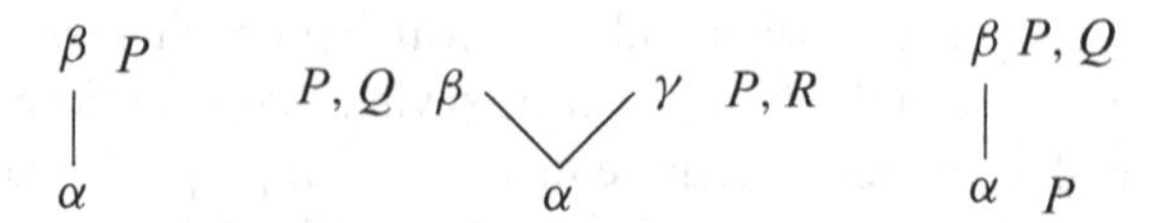

Figure 3.1 Some Kripke models.

on the idea of relativizing truth to each world and insisting that information increases over time, the definition virtually writes itself.

Since $\neg A$ is defined to be $A \to \bot$, we have $\alpha \Vdash \neg A$ if and only if there is no $\beta \leq \alpha$ such that $\beta \Vdash A$. So saying $\neg A$ is true at α is not just a matter of saying that A is not true at α. We can assert $\neg A$ at α if and only if we can see, at α, that no matter what information we later receive, A will never become true.

The convention of using $\beta \leq \alpha$ to denote that β is stronger than α may seem counterintuitive, but it is consistent with the use of $\beta \leq \alpha$ in a Boolean algebra to denote the fact that β entails α. Indeed, suppose β, α, and A are elements of the Tarski–Lindenbaum algebra. If $\alpha = [\alpha']$ and $\beta = [\beta']$, then whenever $\alpha' \models A$ and $\beta \leq \alpha$, then $\beta' \models A$ as well. We are now in a similar situation, where we think of $\alpha \Vdash A$ as saying that A is entailed by α.

In the context of Kripke semantics, say $\Gamma \models A$ if, for every Kripke model $\mathfrak{K}$ and every node α of $\mathfrak{K}$, if $\alpha \Vdash B$ for every $B \in \Gamma$, then $\alpha \Vdash A$. Soundness with respect to this semantics is then expressed as follows.

Theorem 3.3.4. *If $\Gamma \vdash A$ in intuitionistic or minimal logic, then $\Gamma \models A$ in the corresponding version of Kripke semantics.*

Proof Fix a Kripke model, $\mathfrak{K}$. It is straightforward to show by induction that whenever $\Gamma \vdash A$ is derivable in natural deduction, then for every node α, if $\alpha \Vdash B$ for every $B \in \Gamma$, then $\alpha \Vdash A$. For brevity, we will express the antecedent of this implication by writing $\alpha \Vdash \Gamma$. The steps are straightforward, and I will focus on the rules for implication to illustrate.

Suppose the last rule invoked obtains $\Gamma \vdash A \to B$ from $\Gamma, A \vdash B$. Suppose $\alpha \Vdash \Gamma$, $\beta \leq \alpha$, and $\beta \Vdash A$. By monotonicity, we have $\beta \Vdash \Gamma$, and so $\beta \Vdash \Gamma, A$. By the inductive hypothesis, we have $\beta \Vdash B$, as required.

Suppose the last rule invoked obtains $\Gamma \vdash B$ from $\Gamma \vdash A \to B$ and $\Gamma \vdash A$, and suppose $\alpha \Vdash \Gamma$. By the inductive hypothesis, we have $\alpha \Vdash A \to B$ and $\alpha \Vdash A$. Since $\alpha \leq \alpha$, from the clause for implication we have $\alpha \Vdash B$, as required. □

Since $\bot$ is false in any Kripke model for intuitionistic logic, soundness implies that $\bot$ is not derivable. For some more interesting examples, consider the Kripke models depicted in Figure 3.1. In each case, the labels next to each node indicate which propositional variables are true at a node, and nodes higher up in the diagram are less than the ones below them in the order. The first model is a counterexample to $P \vee \neg P$: the variable P is not true at α, but neither is $\neg P$, since P becomes true at β. That same model is also a counterexample to $\neg\neg P \to P$: we have $\alpha \Vdash \neg\neg P$ because no node forces $\neg P$, but $\alpha \nVdash P$.

The second model is a counterexample to $(P \to Q \vee R) \to (P \to Q) \vee (P \to R)$. At any node where P is forced, so is $Q \vee R$, so α forces the antecedent. But $\alpha \nVdash P \to Q$, since $\gamma \Vdash P$ and $\gamma \nVdash Q$. Similarly, $\alpha \nVdash P \to R$, since $\beta \Vdash P$ and $\beta \nVdash R$.

I will leave it to you to check that the third Kripke model is a counterexample to $\neg(P \wedge \neg Q) \to (P \to Q)$. Let us now show that Kripke models provide a complete semantics for intuitionistic and minimal logic.

Definition 3.3.5. A set of formulas Γ is *saturated* if the following hold:

- Γ is deductively closed, that is, if $\Gamma \vdash A$ then $A \in \Gamma$.
- If $A \vee B \in \Gamma$, then $A \in \Gamma$ or $B \in \Gamma$.

When dealing with intuitionistic logic rather than minimal logic, we also require that Γ is consistent.

The following lemma holds for both intuitionistic and minimal logic:

Lemma 3.3.6. *Suppose $\Gamma \nvdash A$. Then there is a saturated set $\Gamma' \supseteq \Gamma$ such that $\Gamma' \nvdash A$.*

Proof As in the proof of Lemma 3.1.13, we will focus on the case where the language is countable. In the general case, we can use Zorn's lemma instead. Let $(C_i \vee D_i)_{i \in \mathbb{N}}$ be an enumeration of all the disjunctions in the language such that each disjunction occurs infinitely often. Define a sequence $(\Gamma_i)_{i \in \mathbb{N}}$ recursively so that each Γ_i is deductively closed and does not prove A, as follows. Let Γ_0 be the deductive closure of Γ, that is, the set of formulas provable from Γ. Assuming Γ_i has been defined, if Γ_i does not prove $C_i \vee D_i$, set $\Gamma_{i+1} = \Gamma_i$. Otherwise, we have $\Gamma_i \vdash C_i \vee D_i$, but $\Gamma_i \nvdash A$. This means that either $\Gamma_i \cup \{C_i\} \nvdash A$ or $\Gamma_i \cup \{D_i\} \nvdash A$. In the first case, take Γ_{i+1} to be the set of deductive consequences of $\Gamma \cup \{C_i\}$, and otherwise take Γ_{i+1} to be the set of deductive consequences of $\Gamma \cup \{D_i\}$. Let Γ' be $\bigcup_i \Gamma_i$. Then Γ' is saturated and does not prove A. □

Notice that, in the case of intuitionistic logic, the fact that Γ' doesn't prove A implies that Γ' is consistent.

The following is a strong form of the completeness theorem. With Lemma 3.3.6, it shows that there is a single Kripke model with the following property: if $\Gamma \nvdash A$, there is some node that forces all the formulas in Γ but does not force A.

Theorem 3.3.7. *Let $\mathfrak{K}$ be the Kripke model whose nodes are all the consistent saturated sets, where $\alpha \leq \beta$ is defined to be $\alpha \supseteq \beta$ and for every variable P, $\alpha \Vdash P$ is defined to hold if and only if $P \in \alpha$. Then for every propositional formula A, $\alpha \Vdash A$ if and only if $\alpha \vdash_{\mathrm{I}} A$.*

Proof It is not hard to check that whenever α is a saturated set and A and B are any formulas, the following hold:

- $A \wedge B \in \alpha$ if and only if $A \in \alpha$ and $B \in \alpha$.
- $A \vee B \in \alpha$ if and only if $A \in \alpha$ or $B \in \alpha$.
- $\bot$ is not in α.

We prove Theorem 3.3.7 by induction on A. When A is a propositional variable, the claim follows from the definition of a Kripke model. The cases for $\wedge$, $\vee$, and $\bot$ follow immediately from the observations in the previous paragraph.

To handle the case for implication, suppose $\alpha \Vdash A \to B$. We need to show $\alpha \vdash A \to B$. Suppose otherwise; then $\alpha \cup \{A\} \nvdash B$ and, by Lemma 3.3.6, there is a saturated set $\beta \supseteq \alpha \cup \{A\}$ such that $\beta \nvdash B$. But then $\beta \leq \alpha$, and by the inductive hypothesis we have $\beta \Vdash A$ and $\beta \nVdash B$, contrary to the fact that $\alpha \Vdash A \to B$.

Conversely, suppose $\alpha \vdash A \to B$, $\beta \leq \alpha$, and $\beta \Vdash A$. By the inductive hypothesis, we have $\beta \vdash A$, and hence $\beta \vdash B$. By the inductive hypothesis again, this means $\beta \Vdash B$. □

I will leave it to you to check that Theorem 3.3.7 holds for provability in minimal logic instead of intuitionistic logic if we do not require the nodes to be consistent and we say $\alpha \Vdash \bot$ if and only if $\bot \in \alpha$.

We close this section with some observations about Kripke models. First, in the definition of a Kripke model, it is enough to require (P, α) to be a preorder rather than a partial order. The definition of a Kripke model then implies that if α is equivalent to β – that is, if $\alpha \leq \beta$ and $\beta \leq \alpha$ – then α and β force exactly the same formulas.

Next, given any Kripke model $\mathfrak{K}$ and node α, we obtain another Kripke model $\mathfrak{K}'$ by restricting to nodes $\beta \leq \alpha$. It is not hard to check that for any $\beta \leq \alpha$ and any formula A, $\beta \Vdash A$ in $\mathfrak{K}'$ if and only if $\beta \Vdash A$ in $\mathfrak{K}$. So, from the point of view of the completeness theorem, it is enough to consider *pointed* Kripke models, that is, Kripke models in which the underlying partial set has a greatest element. In fact, Exercise 3.3.1 shows that we can even assume that the underlying partial order is a tree.

For intuitionistic logic, we can loosen the restriction that no node of the Kripke model forces $\bot$, and instead require only that whenever a node forces $\bot$, it forces every propositional variable. At such nodes, every propositional formula becomes true. Such models are sometimes called *exploding* Kripke models.

Finally, if Γ is finite and doesn't prove some formula A, we can extract from our completeness proof a *finite* Kripke model that forces Γ but not A. The details are spelled out in Exercise 3.3.3. This provides decision procedures for intuitionistic and minimal logic: to decide whether a formula A is provable, simultaneously search for a proof and a countermodel.

Exercises

3.3.1. Show that without loss of generality, we can restrict our attention to Kripke models in which the underlying partial order is a tree, as follows. Let $\mathfrak{K}$ be any Kripke model. Without loss of generality, we can assume $\mathfrak{K}$ has greatest element, $\top$. Define a new Kripke model $\mathfrak{K}'$ by taking the nodes of $\mathfrak{K}'$ to be the set of finite strictly decreasing sequences $\sigma = (\alpha_0, \ldots, \alpha_{k-1})$ of nodes of $\mathfrak{K}$, with $\tau \leq' \sigma$ if and only if $\tau \supseteq \sigma$. For $k \geq 1$, define $v_\sigma^{\mathfrak{K}'}$ to be equal to $v_{\alpha_{k-1}}^{\mathfrak{K}}$, and take $v_{()}^{\mathfrak{K}'} = v_\top^{\mathfrak{K}}$. Show that for every such σ and formula A, $\sigma \Vdash A$ in $\mathfrak{K}'$ if and only if $\alpha_{k-1} \Vdash A$ in $\mathfrak{K}$.

3.3.2. Provide Kripke models that show that the following schemas are not intuitionistically valid:

a. $\neg A \vee \neg\neg A$

b. $\neg(A \wedge B) \to \neg A \vee \neg B$

c. $\neg\neg(A \vee B) \to \neg\neg A \vee \neg\neg B$

d. $((A \to B) \to A) \to A$

e. $(\neg A \to B \vee C) \to (\neg A \to B) \vee (\neg A \to C)$.

3.3.3. This problem shows that a propositional formula A is provable if and only if it is true in every finite Kripke model.

Let $\mathfrak{K}$ be any Kripke model for propositional logic, and let Γ be any finite set of propositional formulas that is closed under subformulas. Define a new Kripke model $\mathfrak{K}'$ as follows. For each α in $\mathfrak{K}$, define the node $[\alpha]$ of $\mathfrak{K}'$ to be the set $\{A \in \Gamma \mid \alpha \Vdash A\}$. Since each set $[\alpha]$ is a subset of Γ, there are only finitely many such sets. Order these by reverse inclusion, so that $[\alpha] \leq [\beta]$ if and only if $[\alpha] \supseteq [\beta]$. For any propositional variable P, define $[\alpha] \Vdash P$ in $\mathfrak{K}'$ to hold if and only if $P \in [\alpha]$. Show that for every propositional formula A in Γ, $[\alpha] \Vdash A$ in $\mathfrak{K}'$ if and only if $\alpha \Vdash A$ in $\mathfrak{K}$.

3.3.4. a. Suppose α is a minimal node in a Kripke model, that is, for every $\beta \leq \alpha$, $\beta = \alpha$. (So, for example, α is a leaf if the Kripke model is a tree.) Show that for every formula A, $\alpha \Vdash A$ if and only if $v_\alpha \models A$, where v_α is the truth assignment defined by $v(P) = \top$ if and only if $\alpha \Vdash P$. In other words, the forcing relation at α coincides with the classical valuation.

b. Recall that Glivenko's theorem states that a propositional formula A is provable classically if and only if $\neg\neg A$ is provable intuitionistically. Use the fact above to give a semantic proof of this theorem. (Hint: Let $\mathfrak{K}$ be a finite Kripke model such that $\neg\neg A$ is not forced at any node. Show there is a node α such that $\alpha \Vdash \neg A$.)

3.3.5. a. Show $\wedge$ is not definable in intuitionistic logic from $\to$, $\vee$, and $\bot$, by considering a Kripke model with three distinct elements, α, β, γ, where $\gamma < \alpha$, $\gamma < \beta$, $\alpha \Vdash P$, $\beta \Vdash Q$, $\gamma \Vdash P$, and $\gamma \Vdash Q$. Show that for every formula A built up from P, Q, $\to$, $\bot$, and $\vee$, if A is forced at γ, then it is forced at either α or β.

b. Similarly, show $\to$ is not definable from $\wedge$, $\vee$, $\neg$, and $\bot$. Use a Kripke model similar to the one above, except that no propositional variables are forced at β. Then show that for every formula A built up from P, Q, and the other connectives, if $\beta \Vdash A$, then $\alpha \Vdash A$.

3.4 Algebraic Semantics for Intuitionistic Logic

Remember that a Boolean algebra is a bounded distributive lattice with complements. For intuitionistic logic, the demand for complements is too strong. An element c is said to be a *pseudocomplement* of a if c is a greatest element satisfying $c \wedge a = \bot$. This is equivalent to saying that, for every element d, we have $d \wedge a = \bot$ if and only if $d \leq c$. It is not hard to show that if there is any element c with this property, it is unique. Such an element c can play the role of $\neg a$: defining $\neg a$ to be c, we have $\neg a \wedge a = \bot$, and if $d \wedge a = \bot$, then $d \leq \neg a$.

To interpret implication, we need a bit more. An element c is said to be a *pseudocomplement of a relative to b* if and only for every d we have $d \wedge a \leq b$ if and only if $d \leq c$. If we write such a c as $a \to b$, then we have a natural characterization of implication: for every d,

$$d \wedge a \leq b \quad \Leftrightarrow \quad d \leq (a \to b).$$

In analogy to the definition of meets and joins in a lattice, we can characterize $a \to b$ as follows:

- $(a \to b) \wedge a \leq b$.
- For every d, if $d \wedge a \leq b$, then $d \leq (a \to b)$.

It is not hard to check that if c is any other element satisfying these two conditions in place of $(a \to b)$, then we have $c \leq (a \to b)$ and $(a \to b) \leq c$, and hence $c = (a \to b)$. In other words, when they exist, relative pseudocomplements are unique. In categorical logic, the relative pseudocomplement $a \to b$ is also called an *exponential* and sometimes written b^a.

A bounded, distributive lattice in which relative pseudocomplements exist is called a *Heyting algebra*. In a Heyting algebra, we have $a \leq b$ if and only if $\top \wedge a \leq b$, which, in turn, happens if and only if $(a \to b) = \top$. So, in a sense, $a \to b$ is an internalization of the $\leq$ relation. Exercise 3.4.3 shows that we can equivalently define a Heyting algebra to be a bounded lattice with relative pseudocomplements, since distributivity follows from that.

Given a Heyting algebra $\mathfrak{H}$, we can now consider assignments v to propositional variables that take values in $\mathfrak{H}$. As with classical logic, this extends to a valuation $[\![A]\!]^v$ of all propositional formulas. If $\Gamma \cup \{A\}$ is any set of propositional formulas, we will say that Γ entails A in Heyting-valued semantics if the following holds: for every assignment of v mapping propositional variables to elements of a Heyting algebra $\mathfrak{H}$, if $[\![B]\!]^v = \top$ for every B in Γ, then $[\![A]\!]^v = \top$.

Theorem 3.4.1. *Intuitionistic logic is sound and complete with respect to Heyting-valued semantics. In other words, for every* Γ *and* A, $\Gamma \vdash A$ *if and only if* $\Gamma \models A$.

Proof To prove soundness, it suffices to show that every rule of natural deduction preserves entailment, in the sense that whenever a sequent $\Gamma \vdash A$ is derivable, we have $\bigwedge\{[\![D]\!]^v \mid D \in \Gamma\} \leq [\![A]\!]^v$. Compared to Theorem 3.2.4, only the interpretation of implication has changed, so we focus on the corresponding rules.

Consider the case of $\to$ introduction, whereby $\Gamma \vdash A \to B$ is obtained from $\Gamma, A \vdash B$. As in the proof of Theorem 3.2.4, we use d to denote $\bigwedge\{[\![D]\!]^v \mid D \in \Gamma\}$, a to denote $[\![A]\!]^v$, and b to denote $[\![B]\!]^v$. Then, from the inductive hypothesis, we have $d \wedge a \leq b$. By the definition of the relative pseudocomplement, this implies $d \leq a \to b$, as required.

For $\to$ elimination, suppose $\Gamma \vdash B$ is derived from $\Gamma \vdash A \to B$ and $\Gamma \vdash A$. By the inductive hypotheses, we have $d \leq a \to b$ and $d \leq a$. By the definition of the relative pseudocomplement, we have $d \wedge a \leq b$, and since $d = d \wedge a$, we have $d \leq b$.

Just as we could prove the completeness of classical logic with respect to truth-table semantics using an algebraic construction from the Tarski–Lindenbaum algebra, we can now prove the completeness of intuitionistic logic with respect to Kripke models using the Tarski–Lindenbaum construction with respect to intuitionistic provability. A filter F in a lattice is said to be *prime* if it has the property that whenever $x \vee y \in F$, then either $x \in F$ or $y \in F$. In particular, the equivalence classes of elements of a saturated set form a prime filter in the intuitionistic Tarski–Lindenbaum algebra. Given a Heyting algebra and an assignment v of elements of $\mathfrak{H}$ to propositional variables, consider the Kripke model whose nodes are the prime filters of $\mathfrak{H}$, with $\alpha \Vdash P$ if and only if $v(P) \in \alpha$. We can then show that for every formula A, $\alpha \Vdash A$ if and only if $[\![A]\!]^v \in \alpha$. If $\Gamma \nvdash A$, we can take $\mathfrak{H}$ to be the Tarski–Lindenbaum algebra relative to Γ. Since $\{[\top]\}$ is a filter in $\mathfrak{H}$ that does not include $[\![A]\!]^v$, we can find a prime filter α of $\mathfrak{H}$ that does not contain $[\![A]\!]^v$. Hence α is a node of the Kripke model that forces Γ but not A. □

We will see below that the completeness of intuitionitic logic with respect to Heyting-valued semantics also follows from completeness with respect to Kripke models.

We will now interpolate two important classes of models in between Kripke models and assignments in a Heyting algebra. *Topological semantics* arises from the observation that the open sets of any topological space form a Heyting algebra. Specifically, let $(X, \mathcal{T})$ be a topological space, where $\mathcal{T}$ is the collection of open subsets of X. (For a brief overview of topology, see Appendix A.4.) Since $\mathcal{T}$ is closed under binary intersections and arbitrary unions, $\mathcal{T}$ is a bounded distributive lattice with intersection as meet, union as join, the empty set as the bottom element, and X as the top element. Define $U \to V$ by

$$U \to V = \mathrm{int}(\overline{U} \cup V),$$

where $\mathrm{int}(W)$ is the *interior* of W, the union of all the open sets contained in W, and $\overline{U} = X \setminus U$ denotes the complement of U. To show that $U \to V$ is the pseudocomplement of U relative to V, we need to show

$$W \cap U \subseteq V \quad \Leftrightarrow \quad W \subseteq \mathrm{int}(\overline{U} \cup V).$$

To do this, first note that $W \cap U \subseteq V$ is equivalent to $W \subseteq \overline{U} \cup V$. But since W is open, this happens if and only if $W \subseteq \mathrm{int}(\overline{U} \cup V)$, as required.

In fact, the collection of open subsets of a topological space is more than just a Heyting algebra; it is *complete*.

Definition 3.4.2. A lattice $(L, \leq)$ is *complete* if it has arbitrary meets and joins, that is, any set S of elements has a greatest lower bound and a least upper bound.

Thus we can talk about complete Heyting algebras, complete Boolean algebras, and so on. Saying that an element x is the greatest lower bound of S (that is, the meet of S) means two things:

- It is a lower bound, which is to say, $x \leq y$ for every y in S.
- It is the greatest element with this property, which is to say, if $z \leq y$ for every y in S, then $z \leq x$.

The least upper bound, or join, of a set S is defined dually. The meet of the empty set is always $\top$, since every element is a lower bound of the empty set. The join of the empty set is always $\bot$.

In fact, having arbitrary meets implies having arbitrary joins, and vice versa:

Proposition 3.4.3. *Let $(L, \leq)$ be any partial order. If L has arbitrary meets or arbitrary joins, then it is a complete bounded lattice.*

Suppose, for example, L has arbitrary joins, which is to say, least upper bounds. Then if S is any set, we can define the greatest lower bound of S to be the least upper bound of the set of all lower bounds to S. Exercise 3.4.2 below asks you to show that this works. Dually, once we have arbitrary meets, we have arbitrary joins, and once we have both, we have $\top$ and $\bot$.

Exercise 3.4.3 asks you to show that any lattice with relative pseudocomplements is distributive. Similarly, we can show that any complete lattice with relative pseudocomplements satisfies an infinitary version of the distributivity law, namely, that for any set S, we have

$$a \wedge \bigvee S = \bigvee \{a \wedge b \mid b \in S\}. \tag{3.1}$$

Thus we can characterize a complete Heyting algebra as a complete lattice with relative pseudocomplements. Conversely, if L is a complete lattice satisfying the infinitary distributive law, we can define $a \to b$ to be the join of those elements c such that $a \wedge c \leq b$,

$$a \to b = \bigvee \{c \mid a \wedge c \leq b\}. \tag{3.2}$$

Exercise 3.4.4 asks you to work out the details.

In the Heyting algebra of open sets of a topological space, the join of a collection S of open sets is clearly $\bigcup S$. Exercise 3.4.5 asks you to show that the meet is $\mathrm{int}(\bigcap S)$. The following theorem says that all of the semantics just discussed are sound and complete for intuitionistic logic.

Theorem 3.4.4. *For any set of propositional formulas $\Gamma \cup \{A\}$, the following are equivalent:*

1. $\Gamma \vdash A$ *intuitionistically.*
2. $\Gamma \models A$ *in Kripke semantics.*
3. $\Gamma \models A$ *in topological semantics.*
4. $\Gamma \models A$ *in complete Heyting algebras.*
5. $\Gamma \models A$ *in Heyting algebras.*

Proof We have already shown that interpretations in Heyting algebras are sound for intuitionistic logic, which means that 1 implies 5. Since every complete Heyting algebra is a Heyting algebra and every topological model is a complete Heyting algebra, we have that 5 implies 4 and 4 implies 3. The fact that Kripke semantics is complete means that 2 implies 1. So, to close the circle, we only need to show that 3 implies 2.

We can do this by showing that every Kripke model gives rise to a topological model that validates the same entailments. Let $\mathfrak{K} = (K, \leq, D)$ be any propositional Kripke model. Define a topology on K by saying that a set U is open if and only if it is downward closed, meaning that whenever $\alpha \in U$ and $\beta \leq \alpha$ then $\beta \in U$. This is equivalent to saying that whenever α is in U then the set $\downarrow(\alpha) = \{\beta \mid \beta \leq \alpha\}$ is a subset of U. Then for any open set U, we have $\alpha \in U$ if and only if $\downarrow(\alpha) \subseteq U$. Since each set $\downarrow(\alpha)$ is open, these sets form a subbasis for the topology. Also, for any set V, α is in the interior of V if and only if $\downarrow(\alpha) \subseteq V$.

Define a valuation v by $v(P) = \{\alpha \mid \alpha \Vdash P\}$. By monotonicity, $v(P)$ is open. Let us show by induction on formulas that for every A, we have $[\![A]\!]^v = \{\alpha \mid \alpha \Vdash P\}$. From the definition of the forcing relation, it is easy to check that $[\![\bot]\!]^v = \{\alpha \mid \alpha \Vdash \bot\}$, $[\![A \wedge B]\!]^v = \{\alpha \mid \alpha \Vdash A \wedge B\}$, and $[\![A \vee B]\!]^v = \{\alpha \mid \alpha \Vdash A \vee B\}$, assuming the inductive hypothesis holds for A and B. Again assuming the inductive hypothesis for A and B, we have

$$\begin{aligned} \alpha \in [\![A \to B]\!]^v &\Leftrightarrow \alpha \in \mathrm{int}(\overline{[\![A]\!]^v} \cup [\![B]\!]^v) \\ &\Leftrightarrow \downarrow(\alpha) \subseteq \overline{[\![A]\!]^v} \cup [\![B]\!]^v \\ &\Leftrightarrow \forall \beta \leq \alpha\, ((\beta \nVdash A) \vee (\beta \Vdash B))) \\ &\Leftrightarrow \beta \Vdash A \to B. \end{aligned}$$

□

Exercises

3.4.1. Let $\mathfrak{P} = (P, \leq)$ be a partial order with meets. Suppose that S is downward closed and T is any set. Show that the following two statements are equivalent:

a. For every $\beta \leq \alpha$, if β is in S, then β is in T.
b. For every β, if $\beta \in S$, then $\alpha \wedge \beta \in T$.

This provides an equivalent way of stating the clause for implication in a Kripke model when the underlying partial order has meets.

3.4.2. Show that a lattice with arbitrary joins also has arbitrary meets, by defining $\bigwedge S$ to be the join of the set of all lower bounds to S and showing that this has the desired properties.

3.4.3. Let $(L, \leq)$ be any lattice with relative pseudocomplements.

a. Show that for every a, b, c, and d in L, $(a \vee b) \wedge c \leq d$ if and only if $(a \wedge c) \vee (b \wedge c) \leq d$.
b. Instantiating d to $(a \vee b) \wedge c$ and $(a \wedge c) \vee (b \wedge c)$ respectively, show that L is a distributive lattice.

3.4.4. Show that a complete lattice satisfies the infinitary distributivity property (3.1) if and only if it has relative pseudocomplements as defined by (3.2).

3.4.5. Show that if S is a set of elements in the Heyting algebra of open sets of a topological space, $\bigwedge S = \mathrm{int}(\bigcap S)$.

3.4.6. Prove the Knaster–Tarski theorem: if L is a complete lattice, every monotone function $f : L \to L$ has a least fixed point. In other words, if f is monotone, there is an a in L such that $f(a) = a$ and for every b satisfying $f(b) = b$, $a \leq b$. (See Exercise 1.2.2.)

3.5 Variations

Abstracting the properties of topological semantics suggests a generalization of Kripke semantics that I will call *generalized Beth semantics*. Suppose $(X, \mathfrak{T})$ is a topological space and $\mathfrak{B}$ is a basis for that space. The latter means that $\mathfrak{B}$ is a collection of open sets with the property that every open set is a union of elements of $\mathfrak{B}$. Fix an assignment v of open sets to propositional variables, and, for $\alpha \in \mathfrak{B}$, write $\alpha \Vdash A$ to mean that $[\![A]\!]^v \subseteq \alpha$. If we think of topological semantics as assigning to every formula A the set of points where A holds, $\alpha \Vdash A$ means that A is true on the basis set α.

This forcing notion is similar to that given by a Kripke model, in that the assignment v determines which propositional variables are forced by α. As with Kripke models, we have the following monotonicity condition: if $\alpha \Vdash A$ and $\beta \subseteq \alpha$, then $\beta \Vdash A$. But forcing now satisfies an additional condition that runs in the opposite direction, showing that $\alpha \Vdash A$ when a sufficient collection of subsets of α forces A. Specifically, if C is a collection of subsets of α, say *C covers α* if $\bigcup C = \alpha$. Then we have the following: if C covers α and $\beta \Vdash A$ for every β in C, then $\alpha \Vdash A$.

In fact, the similarities to forcing in Kripke models run deeper: the specification of $\alpha \Vdash P$ for propositional variables P determines the forcing relation for all formulas, according to the following clauses:

- $\alpha \nVdash \bot$.
- $\alpha \Vdash A \wedge B$ if and only if $\alpha \Vdash A$ and $\alpha \Vdash B$.
- $\alpha \Vdash A \to B$ if and only if for every $\beta \leq \alpha$, if $\beta \Vdash A$ then $\beta \Vdash B$.
- $\alpha \Vdash A \vee B$ if and only if there is a covering C of α such that for every $\beta \in C$, $\beta \Vdash A$ or $\beta \Vdash B$.

The only condition that has changed is the clause for disjunction: rather than require α to force A or B, it suffices to find a covering of α such that each element forces one or the other.

Abstracting the features of this forcing relation yields a more general semantics for intuitionistic propositional logic.

Definition 3.5.1. Let $(X, \leq)$ be any partial order. A *covering notion* for X is a relation between conditions $\alpha \in X$ and sets S of conditions less than or equal to α, read *S covers α*, satisfying the following three conditions:

- *trivial cover:* $\{\alpha\}$ covers α
- *transitivity:* If A covers α, and for each β in A the set B_β covers β, then $\bigcup_{\beta \in A} B$ covers α.
- *stability:* If A covers α and $\beta \leq \alpha$, then there is a set B covering β such that for each γ in B there is a δ in A such that $\gamma \leq \delta$.

Definition 3.5.2. A *generalized Beth model* for intuitionistic logic consists of a partial order, $(X, \leq)$, a covering notion for X, and a function that assigns to each $\alpha \in X$ a truth assignment v_α, satisfying the following two conditions:

- monotonicity, as for Kripke models; and
- covering: for every condition α, propositional variable P, and set C covering α, if $v_\beta(P) = \top$ for every $\beta \in C$, then $v_\alpha(P) = \top$.

The relation $\alpha \Vdash A$ is defined recursively for arbitrary propositional formulas A, according to the clauses above.

As for Kripke models, the requirements on the forcing relation for propositional variables then carry over to the full forcing relation.

Proposition 3.5.3. *The monotonicity and covering properties extend to all propositional formulas: if* $\alpha \Vdash A$ *and* $\beta \leq \alpha$, *then* $\beta \Vdash A$, *and if* $\beta \Vdash A$ *for every element* β *of some covering of* α, *then* $\alpha \Vdash A$.

Notice that if $(X, \leq)$ is any partial order, then we obtain a trivial covering relation by declaring that the only cover of α is $\{\alpha\}$ itself. For this covering relation, generalized Beth semantics collapses to Kripke semantics.

Theorem 3.5.4. *Generalized Beth semantics is sound and complete for intuitionistic propositional logic.*

Proof For soundness, the only thing we need to change in the proof of Theorem 3.3.4 is the handling of the rules for disjunction.

Suppose we derive $\Gamma \vdash A \vee B$ from $\Gamma \vdash A$, and suppose $\alpha \Vdash \Gamma$. By the inductive hypothesis, $\alpha \Vdash A$, and since $\{\alpha\}$ covers α, we have $\alpha \Vdash A \vee B$, as required. The case where we derive $\Gamma \vdash A \vee B$ from $\Gamma \vdash B$ is analogous.

Suppose we derive $\Gamma \vdash C$ from $\Gamma \vdash A \vee B$ together with $\Gamma, A \vdash C$ and $\Gamma, B \vdash C$. Suppose $\alpha \Vdash \Gamma$. By the inductive hypotheses, $\alpha \Vdash A \vee B$, which means that there is a covering D of α such that for every $\beta \in D$, either $\beta \Vdash A$ or $\beta \Vdash B$. By the covering property, to show $\alpha \Vdash C$, it suffices to show $\beta \Vdash C$ for every $\beta \in D$. If $\beta \Vdash A$, then $\beta \Vdash \Gamma, A$, and hence, by the inductive hypothesis, $\beta \Vdash C$. The case where $\beta \Vdash B$ is analogous.

Completeness is immediate: since Kripke models are special cases of generalized Beth models, any entailment that holds in every generalized Beth model also holds in every Kripke model, so completeness with respect to Kripke semantics implies completeness with respect to generalized Beth semantics. □

We can prove the weak form of the completeness theorem more directly by constructing a generalized Beth model with a node $\top$ such that for any propositional formula A, $\vdash A$ if and only if $\top \Vdash A$. When proving completeness with respect to propositional logic, we needed Lemma 3.3.6 to construct saturated sets, which were, in turn, needed to handle the clause for disjunction. Now that we have weakened the requirements for forcing a disjunction, we no longer need each condition α to decide which disjunct is forced, and the following construction does the trick.

Define the nodes of a generalized Beth model $\mathfrak{B}$ to be propositional formulas, and say $\alpha \leq \beta$ if $\{\alpha\} \vdash \beta$. Write the latter more conveniently as $\alpha \vdash \beta$. Assuming $\beta_0 \leq \alpha, \ldots, \beta_{n-1} \leq \alpha$, say that $\{\beta_0, \ldots, \beta_{n-1}\}$ covers α if and only if $\alpha \vdash \beta_0 \vee \cdots \vee \beta_{n-1}$. For a propositional variable P, say $\alpha \Vdash P$ if and only if $\alpha \vdash P$. Exercise 3.5.2 asks you to show that the covering notion just defined is indeed a suitable notion of covering, and that the model just defined has the property that for every propositional formula A, $\alpha \Vdash A$ if and only if $\alpha \vdash A$.

We have seen that topological models and Kripke models give rise to instances of generalized Beth models. The original notion of a Beth model is a special case. Let Σ be any set of symbols. A tree on Σ is said to be *pruned* if every node has at least one successor. This ensures that all the paths through the tree are infinite and every node is an element of an infinite path. Beth took conditions to be nodes of a pruned tree T on some set Σ, where $\beta \leq \alpha$ means that β extends α. In a Beth model, a set of conditions C covers a condition α

if for every $\beta \leq \alpha$, there is some $\gamma \leq \beta$ such that β is in C; this is equivalent to saying that every infinite path through α passes through some element of C. You can think of the paths through T as possible evolutions of our state of knowledge; saying that C covers α means that, at state α, we can be sure that at some point later on our state of knowledge will land somewhere in C, although we are not sure where.

By Exercise 3.3.1, any Kripke model can be converted to a Kripke model on a tree. Exercise 3.5.3 asks you to show that any such model can be turned into a Beth model, simply by repeating any maximal elements. Theorem 3.5.4 and the completeness of Kripke semantics then implies that intuitionistic logic is sound and complete with respect to Beth semantics as well.

We have invested a good deal of effort in developing very general semantics for intuitionistic logic, but some of the constructions pull back to classical logic via the double-negation translation. Under that translation, every formula is classically equivalent to one in which the disjunction symbol $\vee$ does not occur, so we would not expect Beth semantics to bring anything new to the table. Suppose we are given a Kripke model which further satisfies the condition that for every α, $\Vdash \neg\neg P$ if and only if $\Vdash P$. Say that such a Kripke model is *stable under double negation*. Define weak forcing, $\alpha \Vdash_w A$, to mean $\alpha \Vdash A^{\mathrm{N}}$ in the usual sense of a Kripke model. Then the soundness of Kripke model semantics and Theorem 2.7.6 imply:

Theorem 3.5.5. *A is provable classically if and only if $\Vdash_w A$ in every Kripke model that is stable under double negation.*

Moreover, the following proposition characterizes weak forcing:

Proposition 3.5.6. *Let $\mathfrak{K}$ be a Kripke model that is stable under double negation. Then for every condition α and every propositional formula A, $\alpha \Vdash_w \neg\neg A$ if and only if $\alpha \Vdash_w A$. Moreover, for any propositional formulas A and B:*

- *$\alpha \Vdash_w A \wedge B$ if and only if $\alpha \Vdash_w A$ and $\alpha \Vdash_w B$.*
- *$\alpha \Vdash_w A \to B$ if and only if for every $\beta \leq \alpha$, if $\beta \Vdash_w A$, then $\beta \Vdash_w B$.*
- *$\alpha \Vdash_w A \vee B$ if and only if for every $\beta \leq \alpha$, there is a $\gamma \leq \beta$ such that $\gamma \Vdash_w A$ or $\gamma \Vdash_w B$.*

Exercise 3.5.4 shows that we can interpret weak forcing in a Kripke model that is stable under double-negation as an instance of a generalized Beth model.

We will see in Chapter 5 that the notion of forcing extends to first-order logic. This, in turn, is the basis for the notion of forcing that has been so fruitful in set theory. Given a Kripke model that is not stable under double negation, we can always obtain one that is by defining $\alpha \Vdash' P$ to mean $\alpha \Vdash \neg\neg P$. In fact, Glivenko's theorem yields something stronger. Say a proposition formula A is *generically valid* in a Kripke model if for every α, there is a $\beta \leq \alpha$ such that $\beta \Vdash A$.

Theorem 3.5.7. *A is provable classically if and only if A is generically valid in every Kripke model.*

Exercises

3.5.1. Show that with the definition of forcing in a generalized Beth model, monotonicity and the covering property extend to all formulas.
   a. *Monotonicity:* If $\alpha \Vdash A$ and $\beta \leq \alpha$, then $\beta \Vdash A$.
   b. *Covering property:* If C covers α and $\beta \Vdash A$ for every $\beta \in C$, then $\alpha \Vdash A$.

3.5.2. Show that the generalized Beth model $\mathfrak{B}$ described in this section has the properties claimed. Specifically, take the underlying partial order to be the set of propositional formulas with $\alpha \leq \beta$ defined as $\alpha \vdash \beta$. Say $\{\beta_0, \ldots, \beta_{n-1}\}$ covers α if and only if each $\beta_i \leq \alpha$ and $\alpha \vdash \beta_0 \vee \cdots \vee \beta_{n-1}$. For a propositional variable P, say $\alpha \Vdash P$ if and only if $\alpha \vdash P$.

a. Show that "C covers α" is a covering notion, that is, it satisfies the trivial cover, transitivity, and stability properties.
b. Show that the definition of $\alpha \Vdash p$ gives rise to a generalized Beth model, i.e. that monotonicity and the covering property hold.
c. Show by induction on formulas A that for every α, $\alpha \Vdash A$ if and only if $\alpha \vdash A$.
d. Conclude that if A is not provable, then $\top \nVdash A$ in this model.

3.5.3. Show that any Kripke model whose nodes are the nodes of a tree can be turned into a Beth model, whose root forces exactly the same propositional formulas.

3.5.4. Let $(P, \leq)$ be any partial order. Given $\alpha \in P$ and any set C of elements less than or equal to α, say C covers α if for every $\beta \leq \alpha$, there is a $\gamma \leq \beta$ such that γ is less than or equal to some element of C. Show that C is a covering notion. (Hint: for stability, given A covering α and $\beta \leq \alpha$, let B be the set of elements that are less than or equal to β and also less than or equal to some element of A.)

Note that with this notion of covering, the condition $\alpha \Vdash_w A \vee B$ in the characterization of the weak forcing relation is equivalent to saying that there is a set C covering α such that for every β in C, $\beta \Vdash A$ or $\beta \Vdash B$. In particular, if $\alpha \Vdash_w A \vee B$, we can take C to be the set of β such that $\beta \Vdash_w A$ or $\beta \Vdash_w B$.

Bibliographical Notes

Algebraic semantics Bell and Slomson (1969) is a good reference for Boolean algebras and classical model-theoretic constructions.

Semantics of intuitionistic logic For more on the semantics of intuitionistic logic, see Troelstra and van Dalen (1988) and Mints (2000b).

Generalized Beth models The covering notion given by Definition 3.5.2 is a special case of the notion of a *basis for a Grothendieck topology*; see Mac Lane and Moerdijk (1992).

4

First-Order Logic

The goal of the *logicist* program, initiated by Gottlob Frege in the second half of the nineteenth century and pursued by Bertrand Russell and Alfred North Whitehead in the early twentieth century, was to reduce all or at least substantial parts of mathematics to pure logic. The common view today is that the program has failed, since all the foundations for mathematics that have been proposed require assumptions that do not seem to have a purely logical character. Thus the common view is that mathematics is best understood as the result of logical reasoning from properly mathematical axioms.

There is not, however, full agreement as to where to draw the line between logic and mathematics. Philosophers since Gottfried Leibniz have characterized logic as a means of determining *necessary* truths, that is, those assertions that are true in all possible worlds. George Boole and Frege characterized logic as a codification of the fundamental laws of thought. In the early twentieth century, the logical positivist movement tried to characterize the principles of logic as a matter of linguistic convention. Others, like Russell, have characterized logic as the general principles of reasoning that can be applied to any domain of inquiry. With this in mind, some have maintained that logic should be ontologically neutral: if its laws are supposed to hold in any possible domain, they should hold in domains that are empty, and so no logical principle should presuppose or imply the existence of objects (other than, perhaps, purely logical ones).

Whatever the extent of logical reasoning may be, first-order logic fares well on most accounts, since it describes constructs like "and," "every," "some," and "if . . . then," which play a fundamental role in our language and thought. Our deductive systems for first-order logic will assume that the intended domain is nonempty, but they are nonetheless sound for domains with just a single element, showing that the logic itself does not presuppose a strong ontology.

The differences between classical and intuitionistic logic reflect different understandings of the nature of logic reasoning. What these differences come to in practical terms will be clarified as we continue to study these two versions of logic and their formal properties.

4.1 The Language of First-Order Logic

Remember that a first-order language, L, is a tuple (Γ, Δ, a) in which Γ is the set of function symbols, Δ is the set of relation symbols, and a is a function that assigns an arity to each element of $\Gamma \cup \Delta$. In contrast to propositional logic, the expressions of first-order logic are defined in two stages: first we define the set of terms, and then we define the set of formulas. The set of terms of L has already been defined in Section 1.3 inductively as follows:

- Each variable x is a term.
- If $c \in \Gamma$ is a constant symbol (a 0-ary function symbol), then c is a term.
- If $f \in \Gamma$ is a k-ary function symbol with $k > 0$ and $t_0, \ldots, t_{k-1}$ are all terms, then so is $f(t_0, \ldots, t_{k-1})$.

The set of terms is freely generated, so we have principles of induction and recursion on terms. In particular, we can define functions like variables(t), depth(t), and the substitution operation $t[s/x]$ as in Section 1.3, and we can prove that they have the expected properties.

We define the set of first-order formulas of L as described in Section 1.6, identifying formulas up to renaming of their free variables. Thus the set of formulas is generated inductively by the following clauses:

- $\bot$ is a formula.
- If s and t are terms, $s = t$ is a formula.
- If $t_0, \ldots, t_{k-1}$ are terms and R is a k-ary relation symbol of L, then $R(t_0, \ldots, t_{k-1})$ is a formula.
- If A and B are formulas, then so are $(A \wedge B)$, $(A \vee B)$, and $(A \to B)$.
- If A is a formula and x is a variable, then $(\forall x\, A)$ and $(\exists x\, A)$ are formulas.

The set of formulas is not freely generated by these conditions: if y is not free in A, then $\forall x\, A$ and $\forall y\, A[y/x]$ are considered the same. We can nonetheless define functions by recursion on the complexity of A, as long as we specify the value at $\forall x\, A$ in terms of the value at A in such a way that the same value is assigned if we consider $A[y/x]$ instead, for any y not free in A.

In the definition above, the equality symbol is used with infix notation, but otherwise it behaves just like a binary relation symbol in the language. We will, however, continue to give it special status as one of the basic logical symbols. It is sometimes useful to consider first-order logic without the equality symbol, in which case I will refer to *first-order logic without equality*. By default, *first-order logic* is first-order logic with equality.

Formulas of the form $s = t$, $\bot$, and $R(t_0, \ldots, t_{k-1})$ are said to be *atomic*. These constitute the base cases in proofs that proceed by induction on formulas. As was the case with terms, a formula with free variables is only meaningful relative to an assignment to these free variables. A formula with no free variables is called a *sentence*.

In the metatheory, we have already begun to use $x, y, z, \ldots$ to range over variables, $r, s, t, \ldots$ to range over terms, and $A, B, C, \ldots$ to range over formulas. The conventions for dropping parentheses carry over from propositional logic, with the added stipulation that $\forall$ and $\exists$ bind tightest of all, and are read from right to left. So, for example, the formula

$$\forall x\, A \wedge \exists y\, B \to C$$

is understood as

$$(((\forall x\, A) \wedge (\exists y\, B)) \to C).$$

Be warned that conventions vary. I have already noted in the preface that in computer science the convention is to give universal and existential quantifiers the widest scope possible, so that $\forall x\, A \wedge B$ abbreviates $\forall x\, (A \wedge B)$.

Since 0-ary relation symbols can be viewed as propositional variables, we can think of the syntax of first-order logic as including that of propositional logic. It is often clearer to write binary relations like $<$ using infix notation, but that should be viewed as a convention taking place in the metatheory.

Define the set of *free variables* of a formula recursively, as follows:

- $\text{variables}(\bot) = \emptyset$.
- $\text{variables}(s = t) = \text{variables}(s) \cup \text{variables}(t)$.
- $\text{variables}(R(t_0, \ldots, t_{k-1})) = \text{variables}(t_0) \cup \ldots \cup \text{variables}(t_{k-1})$.
- $\text{variables}(A \circ B) = \text{variables}(A) \cup \text{variables}(B)$, where $\circ$ is any of the propositional connectives.
- $\text{variables}(\forall x\, A) = \text{variables}(\exists x\, A) = \text{variables}(A) \setminus \{x\}$.

The last clause reflects the fact that x becomes bound in $\forall x\, A$ and $\exists x\, A$, and hence is no longer a free variable. Indeed, the definition would be ill-posed if we did not eliminate x, since it would depend on the choice of a representative of $\forall x\, A$ among expressions that are considered equivalent up to renaming of their bound variables.

There is a useful convention that allows us to treat formulas with free variables as though they are functions that return propositions. Saying "let $A(x)$ be a formula" specifies that A is a formula and, at the same time, highlights a particular variable x that may or may not occur in A. Later references to $A(t)$ are then interpreted as $A[t/x]$. A similar understanding applies to terms, which is to say, if t has been introduced as $t(x)$, the expression $t(s)$ subsequently denotes $t[s/x]$. These conventions should bring to mind the standard mathematical practice of defining a function $f(x)$ with an expression like $x^2 + bx + c$, after which $f(u)$ denotes the result of substituting u for x in the expression. Multiple substitutions in formulas $A(x_0, \ldots, x_{k-1})$ and terms $t(x_0, \ldots, x_{k-1})$ are dealt with in a similar way, using the notion of simultaneous substitution, as described in Section 1.5. I will often write $\vec{x}$ to denote a tuple of variables $x_0, \ldots, x_{k-1}$ and $\vec{t}$ to denote a tuple of terms $t_0, \ldots, t_{k-1}$.

4.2 Quantifiers

When it comes to first-order logic, most of the deductive apparatus from propositional logic carries over with little change. But how shall we interpret a proof of a formula A when A has free variable? We will think of a free variable x of A as representing an arbitrary, unspecified object, and therefore understand the proof as showing that A is true for any value of x. On that view, we are allowed to conclude $\forall x\, A$ from a proof of A.

To obtain axiomatic systems for minimal, intuitionistic, and classical logic, we now read the axioms of Section 2.2 as schemas involving first-order formulas. We add the following axioms:

- $\forall x\, A \to A[t/x]$
- $A[t/x] \to \exists x\, A$.

In addition to modus ponens, we now have the following rules:

- From $A \to B$ conclude $A \to \forall x\, B$, provided x is not free in A.
- From $A \to B$ conclude $\exists x\, A \to B$, provided x is not free in B.

The first of these two rules is a slight generalization of the rule "from B, conclude $\forall x\, B$." The difference is that we are allowed to do this modulo a hypothesis, A, that does not depend on x.

The rule for the existential quantifier is dual; if B follows from the assumption that A holds of an arbitrary x, then B follows from $\exists x\, B$. In these rules, the side conditions are sometimes called *eigenvariable conditions*.

Presentations that do not identify formulas up to renaming of bound variables typically allow variables to be renamed when applying the quantifier rules. In such a presentation, the rules read as follows:

- From $A \to B$ conclude $A \to \forall y\, B[y/x]$, provided x is not free in A, and y is not free in B unless $y = x$.
- From $A \to B$ conclude $\exists y\, A[y/x] \to B$, provided x is not free in B, and y is not free in A unless $y = x$.

The restrictions ensure that the new bound variable y is *fresh*. With our conventions, however, $\forall y\, B[y/x]$ and $\forall x\, B$ are exactly the same formula, so we can adopt the simpler formulations.

We have to be careful in defining the notion of provability from hypotheses: what are we to make of a hypothesis, A, that has a free variable, x? One option is to view x as, essentially, a constant whose value is fixed throughout the proof. On this view, the conclusion $\forall x\, A$ is not warranted. Another option is to view A as asserting its *universal closure*, i.e. the formula $\forall\, \vec{z} A$ where $\vec{z}$ is a list of all the free variables of A in some fixed order. In that case $\forall x\, A$ *does* follow. For the time being we will avoid ambiguity and confusion by restricting the notion of provability to hypotheses that are *sentences*. Then, as before, we will say $\Gamma \vdash A$ if and only if there is a proof of A with hypotheses in Γ, where a proof is a sequence of formulas in which each formula either is a hypothesis, is an axiom, or follows from previous formulas by one of the rules of inference.

To extend our systems of natural deduction to first-order logic, add the following rules:

$$\frac{A}{\forall x\, A}\ \forall\mathrm{I} \qquad \frac{\forall x\, A}{A[t/x]}\ \forall\mathrm{E}$$

In the introduction rule, we place the restriction that x is not free in any open hypothesis, which corresponds to the eigenvalue condition above. Similarly, we have the rules for the existential quantifier:

$$\frac{A[t/x]}{\exists x\, A}\ \exists\mathrm{I} \qquad \frac{\exists x\, A \quad \begin{matrix}\overline{A}\\ \vdots \\ B\end{matrix}}{B}\ \exists\mathrm{E}$$

Once again, in the elimination rule, we assume that x is not free in B or any hypothesis other than A. The elimination rule exhibits a pattern of reasoning that is similar to that of disjunction elimination: to show that B holds on assumption $\exists x\, A$, let x be an arbitrary object satisfying A and show that B follows.

In sequent form, the natural deduction rules are expressed as follows.

$$\frac{\Gamma \vdash A}{\Gamma \vdash \forall x\, A}\ \forall\mathrm{R} \qquad \frac{\Gamma \vdash \forall x\, A}{\Gamma \vdash A[t/x]}\ \forall\mathrm{E}$$

$$\frac{\Gamma \vdash A[t/x]}{\Gamma \vdash \exists x\, A}\ \exists\mathrm{I} \qquad \frac{\Gamma \vdash \exists x\, A \qquad \Gamma, A \vdash B}{\Gamma \vdash B}\ \exists\mathrm{E}$$

The set of sequents provable in natural deduction is defined, as before, as the set that is generated inductively by these rules. If Γ is a set of sentences and A is any formula, we can show as usual that Γ proves A if and only if there is a natural deduction proof of $\Gamma' \vdash A$ for some finite $\Gamma' \subseteq \Gamma$.

If $\Gamma \vdash A(x)$, then $\Gamma \vdash \forall x\, A(x)$, and hence $\Gamma \vdash A(c)$ for any constant symbol c. Conversely, if c does not occur in Γ or $A(x)$ and $\Gamma \vdash A(c)$, we can think of c as being arbitrary. As a result, in first-order logic, there isn't much of a difference between a variable and a fresh constant. The following proposition makes this precise.

Proposition 4.2.1. *Let Γ be a set of sentences, let $A(x)$ be a formula, and let c be a constant that does not occur in Γ or A. If $\Gamma \vdash A(c)$, then $\Gamma \vdash A(x)$.*

The idea is that given a proof of $A(c)$ from some finite subset of Γ, we can replace c by x everywhere in the proof. This is not quite correct: if x occurs somewhere in the original proof and is eliminated by a quantifier rule, then replacing c by x might violate an eigenvariable condition. A solution is to first replace c by a variable y which does not occur anywhere in the proof, thereby avoiding such conflicts. This results in a proof of $A(y)$ from Γ. From that, we can derive $\forall y\, A(y)$ and $A(x)$.

Carrying out the proof rigorously requires defining what it means for a variable to occur in a proof. The following lemma provides another approach which renames eigenvariables as needed. It is tailored specifically to natural deduction, but it has Proposition 4.2.1 as an immediate corollary. In the statement of the lemma, if $\vec{x}$ is any tuple of variables, $\vec{c}$ is any tuple of constants, and $\vec{t}$ and $\vec{s}$ are tuples of terms the same length as $\vec{x}$ and $\vec{c}$ respectively, we write $A[\vec{t}/\vec{x}, \vec{s}/\vec{c}]$ for the result of simultaneously substituting those terms for the corresponding variables and constants. Similarly, we write $\Gamma[\vec{t}/\vec{x}, \vec{s}/\vec{c}]$ for $\{B[\vec{t}/\vec{x}, \vec{s}/\vec{c}] \mid B \in \Gamma\}$.

Lemma 4.2.2. *Suppose $\Gamma \vdash A$ is derivable in natural deduction. Then for any $\vec{x}$, $\vec{c}$, $\vec{s}$, and $\vec{t}$, $\Gamma[\vec{t}/\vec{x}, \vec{s}/\vec{c}] \vdash A[\vec{t}/\vec{x}, \vec{s}/\vec{c}]$.*

Proof The lemma is clearly true of any sequent derived by the assumption rule, and in most cases it is straightforward to show that each rule commutes with the substitution. The only interesting cases are $\forall$ introduction and $\exists$ elimination. For the first of these, suppose the last rule obtains $\Gamma \vdash \forall y\, B$ from $\Gamma \vdash B$. By the eigenvariable condition, y does not occur in Γ. Let z be a variable that does not occur in Γ, B, $\vec{x}$, $\vec{t}$, or $\vec{s}$. By the inductive hypothesis, there is a derivation of $\Gamma[\vec{t}/\vec{x}, z/y, \vec{s}/\vec{c}] \vdash B[\vec{t}/\vec{x}, z/y, \vec{s}/\vec{c}]$, and since y does not occur in Γ, the antecedent is $\Gamma[\vec{t}/\vec{x}, \vec{s}/\vec{c}]$. Using the $\forall$ introduction rule, we have a proof of $\Gamma[\vec{t}/\vec{x}, \vec{s}/\vec{c}] \vdash \forall z\, (B[\vec{t}/\vec{x}, z/y, \vec{s}/\vec{c}])$. The assumptions on z are then enough to ensure that we have

$$
\begin{aligned}
(\forall y\, B)[\vec{t}/\vec{x}, \vec{s}/\vec{c}] &= (\forall z\, B[z/y])[\vec{t}/\vec{x}, \vec{s}/\vec{c}] \\
&= \forall z\, (B[z/y][\vec{t}/\vec{x}, \vec{s}/\vec{c}]) \\
&= \forall z\, (B[\vec{t}/\vec{x}, z/y, \vec{s}/\vec{c}]),
\end{aligned}
$$

which is what we need. The $\exists$ elimination rule is handled similarly. □

The next two propositions present key properties of the first-order quantifiers. In two instances, where only one direction of a classical equivalence holds in minimal logic, we write $A \leftarrow B$ for $B \rightarrow A$.

Proposition 4.2.3. *The following are derivable in minimal logic:*

1. $\forall x\, A \leftrightarrow A$ *if x is not free in A*
2. $\exists x\, A \leftrightarrow A$ *if x is not free in A*
3. $\forall x\, (A \wedge B) \leftrightarrow \forall x\, A \wedge \forall x\, B$
4. $\exists x\, (A \wedge B) \leftrightarrow \exists x\, A \wedge B$ *if x is not free in B*
5. $\forall x\, (A \vee B) \leftarrow \forall x\, A \vee B$ *if x is not free in B*
6. $\exists x\, (A \vee B) \leftrightarrow \exists x\, A \vee \exists x\, B$
7. $\forall x\, (A \to B) \leftrightarrow (\exists x\, A \to B)$ *if x is not free in B*
8. $\exists x\, (A \to B) \to (\forall x\, A \to B)$ *if x is not free in B*
9. $\forall x\, (A \to B) \leftrightarrow (A \to \forall x\, B)$ *if x is not free in A*
10. $\exists x\, (A \to B) \to (A \to \exists x\, B)$ *if x is not free in A*
11. $\neg \exists x\, A \leftrightarrow \forall x\, \neg A$
12. $\neg \forall x\, A \leftarrow \exists x\, \neg A$.

For example, here is a proof of the forward direction of claim 4:

$$\dfrac{\dfrac{\dfrac{}{\exists x\,(A(x)\wedge B)}\,2 \qquad \dfrac{\dfrac{\dfrac{\dfrac{}{A(x)\wedge B}\,1}{A(x)}}{\exists x\,A(x)} \qquad \dfrac{\dfrac{}{A(x)\wedge B}\,1}{B}}{\exists x\,A(x)\wedge B}}{\exists x\,A(x)\wedge B}\,1}{\exists x\,(A(x)\wedge B)\to\exists x\,A(x)\wedge B}\,2$$

Here is proof of the converse direction:

$$\dfrac{\dfrac{\dfrac{\dfrac{}{\exists x\,A(x)\wedge B}\,2}{\exists x\,A(x)} \qquad \dfrac{\dfrac{\dfrac{}{A(x)}\,1 \qquad \dfrac{\dfrac{}{\exists x\,A(x)\wedge B}\,2}{B}}{A(x)\wedge B}}{\exists x\,(A(x)\wedge B)}}{\exists x\,(A(x)\wedge B)}\,1}{\exists x\,A(x)\wedge B\to\exists x\,(A(x)\wedge B)}\,2$$

I have used the notation $A(x)$ instead of A as a heuristic, to emphasize that A may depend on x whereas B may not. The fact that x is not free in B is only required in the forward direction, to satisfy the eigenvariable condition for the exists elimination rule.

Proposition 4.2.4. *The following are derivable in classical logic:*

1. $\exists x\, A \leftrightarrow \neg \forall x\, \neg A$
2. $\forall x\, A \leftrightarrow \neg \exists x\, \neg A$
3. $\forall x\, (A \vee B) \leftrightarrow \forall x\, A \vee B$ *if x is not free in B*
4. $\exists x\, (A \to B) \leftrightarrow (\forall x\, A \to B)$ *if x is not free in B*
5. $\exists x\, (A \to B) \leftrightarrow (A \to \exists x\, B)$ *if x is not free in A.*

Here is a proof of the right-to-left direction of claim 1:

$$\dfrac{\dfrac{\neg\forall x\,\neg A \qquad \dfrac{\dfrac{\dfrac{\dfrac{}{\neg\exists x\,A}\,2 \qquad \dfrac{\dfrac{}{A}\,1}{\exists x\,A}}{\bot}}{\neg A}\,1}{\forall x\,\neg A}}{\bot}}{\exists x\,A}\,2$$

Proposition 4.2.4 shows that including both $\forall$ and $\exists$ among the basic connectives of classical logic is redundant. If we express all our formulas in terms of $\forall$, the following provides what is perhaps the quickest deductive characterization of classical provability in first-order logic: as axioms, take

1. any substitution instance of a propositional tautology, and
2. $\forall x\,A \to A[t/x]$

and for rules, allow

1. modus ponens: from $A \to B$ and A conclude B
2. generalization: from $A \to B$ conclude $A \to \forall x\,B$, as long as x is not free in A.

In the exercises below you are asked to show that if we modify the axioms, we can get by with only one rule, modus ponens.

That our axioms and rules presuppose that the domain of discourse is nonempty can be seen from the fact that $\forall x\,A(x) \to \exists x\,A(x)$ is derivable for any formula A, as well as $\exists x\,(A(x) \to A(x))$. (When we have the rules for equality, we will see that $\exists x\,(x = x)$ is also derivable.) In Section 4.8, we will consider variants of first-order logic that allow for the possibility of an empty domain.

First-order logic is often used to model the notion of logical consequence in the context of scientific reasoning. From a logical perspective, a first-order *theory* is any set of sentences that is closed under the provability relation. In other words, a theory is a set T such that whenever T proves A, A is already in T. As a consequence of the transitivity of the provability relation, it is easy to check that if Γ is any set of sentences, then the set $T = \{A \mid \Gamma \vdash A\}$ is a theory. This set T is called the *theory axiomatized by* Γ.

It is often useful to classify first-order formulas in terms of the structure of the patterns of quantifiers they use. A first-order formula is said to be *universal* if it consists of a block of universal quantifiers (possibly empty) followed by a quantifier-free formula. Replacing "universal" by "existential" yields the notion of an *existential* formula. A theory that is axiomatized by a set of universal formulas is said to be *universally axiomatized.* We will have more to say about such theories in Section 7.1.

Exercises

4.2.1. Give examples that show what can go wrong when eigenvariable conditions in the quantifier rules are violated.

4.2.2. Consider the following equivalence:

$$\exists x\,(A \to B) \leftrightarrow (\forall x\,A \to B),$$

where x is not free in B. Using natural deduction, show that the forward direction is provable in minimal logic and the other is provable classically.

4.2.3. Prove the following in the natural deduction proof system for minimal logic:
a. $\forall x\,(A \to B) \to (\forall x\, A \to \forall x\, B)$
b. $\forall x\, A \to \exists x\, A$
c. $\neg\neg\exists x\, A \leftrightarrow \neg\forall x\, \neg A$
d. $\exists x\, A \to \neg\forall x\, \neg A$
e. $\neg\neg\forall x\, A \to \forall x\, \neg\neg A$.

4.2.4. Prove the following equivalences in the natural deduction proof system for classical logic:
a. $\neg\forall x\, \neg A \leftrightarrow \exists x\, A$
b. $(A \to \exists x\, B) \leftrightarrow \exists x\,(A \to B)$, if x is not free in A (independence of premise).

4.2.5. Suppose I tell you that, in a town, there is a (male) barber that shaves all and only the men who do not shave themselves. Formalize this claim, and show, in minimal logic, that it implies a contradiction.

4.2.6. Justify the remaining claims in Proposition 4.2.3.

4.2.7. Justify the remaining claims in Proposition 4.2.4. The "if" direction of 3 is especially tricky. One strategy is to rely on the classical fact $\exists x\, \neg A \vee \neg\exists x\, \neg A$, and reason by cases. For a more direct proof, show that from $\neg\exists x\,(A \to B)$ we can derive $\forall x\, A$.

4.2.8. Confirm that the theory axiomatized by a set of sentences, Γ, is indeed a theory in the sense defined above.

4.2.9. Consider classical first-order logic expressed in a language with basic connectives $\vee$, $\neg$, and $\exists$. Take $A \to B$ to be an abbreviation for $\neg A \vee B$, and adopt similar conventions for $A \wedge B$ and $\forall x A$. Now consider the formal system with the following axioms:
a. $A \vee \neg A$
b. $A[t/x] \to \exists x\, A$.

Allow the following rules of inference:
a. From A conclude $B \vee A$.
b. From $A \vee A$ conclude A.
c. From $A \vee (B \vee C)$ conclude $(A \vee B) \vee C$.
d. From $A \vee B$ and $\neg A \vee C$ conclude $B \vee C$.
e. If x is not free in B, from $A \to B$ conclude $\exists x\, A \to B$.

Show that this provides another formulation of classical first-order logic.

4.2.10. This exercise asks you to show that we can omit the generalization rule and obtain axiomatic proof systems for first-order logic that have require only one rule, modus ponens. First, for each axiom of first-order logic, A, also allow all generalizations, $\forall \vec{x}\, A$, for any sequence of variable $\vec{x}$. Also, add (all generalizations of) the following axiom schemas:
a. $\forall x\,(A \to B) \to (\forall x\, A \to \forall x\, B)$.
b. $A \to \forall x\, A$ where x is not free in A.

Do the following:
a. Show that, in this new axiomatic system, whenever $\Gamma \vdash A$, we have $\Gamma \vdash \forall \vec{x}\, A$, for any sequence of variables A.
b. Show that, whenever $\Gamma \vdash A \to B$ and x is not free in A, we have $\Gamma \vdash A \to \forall x\, B$.

4.3 Equality

In first-order logic, the expression $s = t$ is used to say that s and t denote the same object, or, at least, entities that are indistinguishable given the linguistic resources at hand. In other words, we assume that equality is an equivalence relation and that it is a congruence with respect to all the function and relation symbols in the language. These properties are expressed by the universal closures of the following axioms:

1. $x = x$
2. $x = y \to y = x$
3. $x = y \wedge y = z \to x = z$
4. $\vec{x} = \vec{y} \to f(\vec{x}) = f(\vec{y})$ for each function symbol f
5. $\vec{x} = \vec{y} \to (R(\vec{x}) \to R(\vec{y}))$ for each relation symbol R.

Here $\vec{x}$ and $\vec{y}$ denote tuples of variables of the appropriate lengths, and I have written $\vec{x} = \vec{y}$ for k-tuples $\vec{x}$ and $\vec{y}$ instead of $x_0 = y_0 \wedge \cdots \wedge x_{k-1} = y_{k-1}$. We can assume R is not the equality symbol in 5; conversely, Exercise 4.3.2 shows that if we allow R to be equality then we can dispense with axioms 2 and 3. Exercise 4.3.3 shows that it suffices instead to use the following axiom and axiom schema:

1. $x = x$
2. $x = y \to (A(x) \to A(y))$ for any atomic formula A.

The important observation is that we do not need to state equality axioms explicitly for arbitrary terms and formulas:

Proposition 4.3.1. *The following are provable (in minimal, intuitionistic, and classical logic) for any term $t(x)$ and formula $A(x)$:*

1. $x = y \to t(x) = t(y)$
2. $x = y \to (A(x) \to A(y))$.

Proof The first claim is proved by induction on terms and the second is proved by induction on formulas. □

In systems of natural deduction, it is more natural to express the axioms above as rules, allowing arbitrary instances of terms in place of the variables:

$$\frac{}{t = t} \qquad \frac{s = t}{t = s} \qquad \frac{s = t \quad t = u}{s = u}$$

$$\frac{\vec{s} = \vec{t}}{f(\vec{s}) = f(\vec{t})} \qquad \frac{\vec{s} = \vec{t} \quad R(\vec{s})}{R(\vec{t})}$$

Here $\vec{s}$ and $\vec{t}$ denote arbitrary tuples of terms of appropriate length, and the bar over the reflexivity rule indicates that the rule has no premises. As above, we can use the following rules instead, where $A(x)$ is any atomic formula:

$$\frac{}{t = t} \qquad \frac{s = t \quad A(s)}{A(t)}$$

Proposition 4.3.1 tells us that the following are derived rules for any term $r(x)$ and any formula $A(x)$:

$$\frac{s = t}{r(s) = r(t)} \qquad \frac{s = t \quad A(s)}{A(t)}$$

The initial five rules provide a useful characterization of equality, but when it comes to *using* a deductive system for equality, it is nice to have the last two substitution rules, and, perhaps, symmetric versions with $s = t$ replaced by $t = s$.

We can study the rules of equality in isolation. Consider a system of natural deduction in which every sequent is of the form $u_0 = v_0, \ldots, u_{n-1} = v_{n-1} \vdash s = t$. We allow only the first four of the basic equality rules listed above, that is, reflexivity, symmetry, transitivity, and the congruence rules for the function symbols. We also include the assumption rule, as usual. I will refer to this as the *equational fragment of first-order logic*. In a sequent formulation, we can require the same hypotheses in both premises of a binary rule, so, for example, transitivity is written as follows:

$$\frac{\Gamma \vdash r = s \qquad \Gamma \vdash s = t}{\Gamma \vdash r = t}$$

Alternatively, we can keep track of the hypotheses that are actually used, using $s = t \vdash s = t$ for the assumption rule and writing transitivity as follows:

$$\frac{\Gamma \vdash r = s \qquad \Delta \vdash s = t}{\Gamma, \Delta \vdash r = t}$$

For consistency with our presentation of natural deduction, we will follow the first approach.

We will see in Chapter 5 that this proof system is complete, in the sense that if $\Gamma \cup \{A\}$ consists of equations between closed terms and A holds in every structure that satisfies Γ, the A is derivable in the system above. This implies that if a sequent $\Gamma \vdash A$ is provable with all of first-order logic, using either the classical, intuitionistic, or minimal version, it is derivable from the equality rules alone. In Chapter 6, we will have a purely syntactic proof of this fact.

Even without this strong result, however, we can derive an important property of the system.

Theorem 4.3.2. *If a sequent $\Gamma \vdash A$ is provable in the equational fragment of first-order logic, there is a proof in which every term that appears occurs as a subterm of some formula in $\Gamma \cup \{A\}$.*

Proof First, notice that in any proof of an equation $s = t$, we can permute the symmetry rule upward so that it is only applied to hypotheses. For example, the pattern on the left, consisting of a transitivity inference followed by symmetry, can be replaced by the pattern on the right:

$$\frac{\dfrac{s = r \qquad r = t}{s = t}}{t = s} \qquad \frac{\dfrac{r = t}{t = r} \qquad \dfrac{s = r}{r = s}}{t = s}$$

A similar transformation works for instances of the congruence rule followed by symmetry, and iterating these transformations has the desired effect. Henceforth, we therefore restrict attention to proofs in which symmetry is only applied at hypotheses.

The main difficulty in proving Theorem 4.3.2 involves the instances of transitivity:

$$\frac{\Gamma \vdash s = r \qquad \Gamma \vdash r = t}{\Gamma \vdash s = t}$$

The problem is that the middle term, r, vanishes from the proof, and if r is not a subterm of a formula in Γ, there is nothing that guarantees that it will be a subterm of the conclusion. Call a proof *good* if, for every instance of such a transitivity rule, r is indeed a subterm of a formula in Γ. It is not hard to show, by induction, that a good proof of a sequent $\Gamma \vdash A$ has the property stated in the theorem.

To prove the theorem, we first prove the following auxiliary claim: given good proofs of $\Gamma \vdash s = r$ and $\Gamma \vdash r = t$, there is a good proof of $\Gamma \vdash s = t$. To do so, we use induction on the sum of the number of nodes in the two proofs.

If the proof of either $\Gamma \vdash s = r$ or $\Gamma \vdash r = t$ is an instance of reflexivity, then r is identical with either s or t and the other proof is the one we want. Similarly, if either is a hypothesis or symmetry applied to a hypothesis, then r is a subterm of a formula in Γ, and we can apply transitivity to obtain a good proof of $\Gamma \vdash s = t$.

Suppose one of the subproofs – say, the left one – is again an instance of transitivity, so that applying transitivity would result in a proof that ends like so:

$$\dfrac{\dfrac{s = u \qquad u = r}{s = r} \qquad r = t}{s = t}$$

We can permute the transitivity rules like so:

$$\dfrac{s = u \qquad \dfrac{u = r \qquad r = t}{u = t}}{s = t}$$

By the hypothesis that the original proof of $s = r$ is good, u is a subterm of a formula in Γ. Since the proof of $u = r$ is smaller than the proof of $s = r$, we can apply the inductive hypothesis to the pair on the right and obtain a good proof of $u = t$. By hypothesis, the proof of $s = u$ is also good, so we can apply transitivity to obtain a good proof.

The only case remaining is where both the final inferences are congruences, so that applying transitivity would result in a proof that ends like so:

$$\dfrac{\dfrac{s_0 = r_0 \quad \cdots \quad s_{n-1} = r_{n-1}}{f(\vec{s}) = f(\vec{r})} \qquad \dfrac{r_0 = t_0 \quad \cdots \quad r_{n-1} = t_{n-1}}{f(\vec{r}) = f(\vec{t})}}{f(\vec{s}) = f(\vec{t})}$$

In this case, we can again permute the transitivity rule upward:

$$\dfrac{\dfrac{s_0 = r_0 \qquad r_0 = t_0}{s_0 = t_0} \qquad \cdots \qquad \dfrac{s_{n-1} = r_{n-1} \qquad r_{n-1} = t_{n-1}}{s_{n-1} = t_{n-1}}}{f(\vec{s}) = f(\vec{t})}$$

The fact that every provable sequent $\Gamma \vdash A$ has a good proof now follows by an easy induction on the derivation of $\Gamma \vdash A$. Reflexivity, a hypothesis, or a hypothesis followed by symmetry are all good proofs, and applying a congruence rule preserves goodness. The only other case is the transitivity rule, which is handled by the claim that we just proved. □

This provides us with an algorithm to determine whether or not the sequent $\Gamma \vdash A$ is provable in the equational fragment of first-order logic: make a list of all the subterms occurring in Γ and A, start with the hypotheses in Γ, and then iteratively apply rules of inference to derive new equalities between subterms, until A has been proved or there are no more equalities to be had. In the latter case, Theorem 4.3.2 guarantees that the sequent is not provable.

Exercises

4.3.1. An old jazz standard says "everybody loves my baby, but my baby don't love nobody but me." Formalize this, and show that it follows that I am my baby.

4.3.2. Show that the equality axioms 1–3 follow from the following two:

a. $x = x$

b. $x = y \wedge x = z \rightarrow y = z$.

Hence axioms 1 and 4 above can be taken to subsume axioms 2 and 3.

4.3.3. Show that all the equality axioms follow from the following axiom and schema:

a. $x = x$

b. $x = y \rightarrow (A(x) \rightarrow A(y))$ for any atomic formula A.

4.3.4. Prove Proposition 4.3.1.

4.4 Equational and Quantifier-Free Logic

It is sometimes useful to consider an extension of the equational fragment of first-order logic where the hypotheses are taken to be implicitly universally quantified. For example, a *group* is a structure for a language with a constant symbol, 1, a binary operation, $\cdot$, and a unary inverse, satisfying the following axioms:

- $\forall x, y, z\, ((x \cdot y) \cdot z = x \cdot (y \cdot z))$
- $\forall x\, (x \cdot 1 = x)$
- $\forall x\, (1 \cdot x = x)$
- $\forall x\, (x \cdot x^{-1} = 1)$
- $\forall x\, (x^{-1} \cdot x = 1)$.

We may want to have a proof system for deriving equations from these axioms. We can leave off the quantifiers but still have the effect of treating formulas with variables as implicitly universally quantified if we add a substitution rule,

$$\frac{A(x)}{A(t)}$$

for any term t. The result of adding the substitution rule to the equational fragment of first-order logic is known as *equational logic*. We have the following:

Theorem 4.4.1. *Let Γ be a set of equations and let $s = t$ be an equation. Then the following are equivalent:*

1. *$s = t$ is provable from substitution instances of Γ in the equational fragment of first-order logic.*
2. *$s = t$ is provable from Γ in equational logic.*
3. *$s = t$ is provable from the universal closures of formulas in Γ in classical first-order logic.*

The implication from 1 to 2 is immediate, since we can use the substitution rule to obtain substitution instances of any formula in Γ. The implication from 2 to 3 is also immediate, since we can use the $\forall$ elimination rule to instantiate universal quantifiers. It is not hard to prove 1 from 2 by induction on derivations. The implication from 3 to 1 and 2 is the most interesting, since it shows that even adding full classical logic with quantifiers does not change the provable entailments between equations. We will see a model-theoretic proof of this in Section 5.2 and a proof-theoretic proof in Section 6.6.

Although the terminological distinction between equational logic and the equational fragment of first-order logic is subtle, the difference is substantial. We saw in the last section that the question as to whether something is provable in the equational fragment of first-order logic is algorithmically decidable. This is no longer true when we add the substitution rule. Roughly speaking, the problem is that we cannot predict in advance which substitution instances of the hypotheses might be needed to prove the conclusion. In computer science and universal algebra, the problem of determining whether two first-order terms are equal modulo a set of identities is known as the *word problem*. For example, the word problem for groups is as follows: given a finite set S of identities between closed terms in the language of groups and an equation $s = t$, does $s = t$ hold in the free group modulo S, and hence in all groups satisfying S? It is known that the word problem for groups is undecidable, which is to say, there is no algorithm that can reliably answer that question in general. For more information, see the discussion at the end of Section 12.5 and the notes to Chapter 12.

Just as it is sometimes useful to isolate the equational fragment of first-order logic, it is also sometimes useful to isolate the quantifier-free fragment of first-order logic, that is, the rules for the propositional connectives and equality. In analogy with *equational logic*, I will use the phrase *quantifier-free logic* to refer to the quantifier-free fragment of first-order logic together with the substitution rule. In analogy to Theorem 4.4.1, we have the following relationship between quantifier-free logic and the quantifier-free fragment of first-order logic:

Theorem 4.4.2. *Let Γ be a set of quantifier-free formulas and let A be a quantifier-free formula. Then the following are equivalent:*

1. *A is provable from substitution instances of Γ in the quantifier-free fragment of first-order logic.*
2. *A is provable from Γ in quantifier-free logic.*
3. *A is provable from the universal closures of formulas in Γ in full first-order logic.*

In contrast to Theorem 4.4.1, here we need to specify whether we are talking about classical, intuitionistic, or minimal logic, since "quantifier-free logic" means something different in each case. For classical logic, Section 5.2 provides a model-theoretic proof, and Section 7.1 provides a proof-theoretic proof. The latter can be straightforwardly adapted to intuitionistic and minimal logic as well.

Recall from Section 4.2 that a formula is *universal* if it consists of a block of universal quantifiers followed by a quantifier-free formula. When we prove a formula A with free variables, the intended interpretation is that the conclusion holds for any value that can be assigned to the variables. So we can think of quantifier-free logic as a means of deriving (implicitly) universal conclusions from (implicitly) universal hypotheses.

We observed above that the word problem for groups is undecidable. Since the theory of groups can be axiomatized by a finite set of identities, we have the undecidability of this more general version of the word problem:

Theorem 4.4.3. *The question as to whether an equation follows from a finite set of identities in first-order logic is undecidable.*

Since equality itself can be axiomatized by a finite set of universal formulas, this implies that the question as to whether a quantifier-free formula can be derived from a finite set of universal axioms in first-order logic without equality is also undecidable.

But the question as to whether a quantifier-free formula, A, is derivable in pure first-order logic, without any additional axioms, *is* decidable. By Theorem 4.4.2, the question is equivalent to asking whether A is derivable in the quantifier-free fragment of first-order logic. In fact, Section 6.6 shows that it suffices to consider instances of the equality axioms for subterms of terms occurring in A, of which there are only finitely many. (Exercise 5.2.2 suggests a model-theoretic argument for the classical case.) This reduces the problem to the decidability of propositional logic. Thus we have:

Theorem 4.4.4. *The universal fragment of first-order logic with equality is decidable.*

In other words, there is an algorithm for determining whether a universal formula is valid. In the field of automated reasoning, practical implementations are usually based on an algorithm known as *congruence closure*.

Exercise

4.4.1. Prove Theorem 4.4.4. (See also Sections 5.2, 7.1, and 12.5.)

4.5 Normal Forms for Classical Logic

As with propositional logic, when dealing with classical first-order logic it is often useful to express formulas in various normal forms. For example, we can always eliminate $\exists$ in favor of $\forall$ or vice versa, and we can restrict the propositional connectives to any complete set. The definition of negation normal form carries over from Section 2.6; in the first-order setting the set of formulas in negation normal form is defined to be the smallest set of formulas that contains all the atomic formulas and their negations and is closed under conjunction, disjunction, and application of the universal and existential quantifiers. We extend the definition of the negation operator with the clauses $\sim(\forall x\, A) \equiv \exists x\, (\sim A)$ and $\sim(\exists x\, A) \equiv \forall x\, (\sim A)$.

A formula is said to be in *prenex* form if it consists of a string of quantifiers followed by a quantifier-free formula. More formally, the set of prenex formulas is defined to be the smallest set of formulas containing all the quantifier-free formulas and closed under universal and existential quantification. Classically, we can always put a formula in prenex form by putting it in negation normal form and then using these rules, which hold when x is not free in B:

$$\begin{aligned}
\forall x\, A \vee B &\equiv \forall x\, (A \vee B) \\
\forall x\, A \wedge B &\equiv \forall x\, (A \wedge B) \\
\exists x\, A \vee B &\equiv \exists x\, (A \vee B) \\
\exists x\, A \wedge B &\equiv \exists x\, (A \wedge B),
\end{aligned}$$

We can always rename bound variables to ensure that these rules apply. Of course, it is not necessary to put a formula in negation normal form in order to bring the quantifiers to the front; for example, we can use the equivalences

$$\begin{aligned}
\forall x\, A \to B &\equiv \exists x\, (A \to B) \\
\exists x\, A \to B &\equiv \forall x\, (A \to B) \\
A \to \forall x\, B &\equiv \forall x\, (A \to B) \\
A \to \exists x\, B &\equiv \exists x\, (A \to B)
\end{aligned}$$

to bring quantifiers to the front of an implication.

Though it is often convenient, a formula's prenex form is only a crude representation of its logical structure. In general, the prenex form is not unique, since one has to make choices as to which quantifiers to bring to the front first. One typically minimizes the number of alternations between blocks of universal and existential quantifiers, but even that restriction leaves room for variation. For example, the formula $\exists x\, A(x) \to \exists y\, B(y)$ is equivalent to both $\forall x\, \exists y\, (A(x) \to B(y))$ and $\exists y\, \forall x\, (A(x) \to B(y))$.

Exercise

4.5.1. Which of the rules above for bringing quantifiers to the front of a formula are valid in minimal or intuitionistic logic?

4.6 Translations Between Logics

In Section 2.7, we considered translations between the classical, intuitionistic, and minimal versions of propositional logic. We now extend these to first-order logic. The interpretation of intuitionistic logic in minimal logic goes through with few changes. If we let A^* denote the result of replacing each atomic subformula C of A (other than $\bot$) by $C \vee \bot$, we still have:

Theorem 4.6.1. *Let Γ be any set of sentences and let A be any formula.*

1. $\vdash_{\mathrm{I}} A \leftrightarrow A^*$.
2. *If $\Gamma \vdash_{\mathrm{I}} A$, then $\Gamma^* \vdash_{\mathrm{M}} A^*$.*

We can extend the Gödel–Gentzen double-negation translation to first-order logic with the following clauses:

- $(\forall x\, A)^{\mathrm{N}} = \forall x\, A^{\mathrm{N}}$
- $(\exists x\, A)^{\mathrm{N}} = \neg \forall x\, \neg A^{\mathrm{N}}$.

In minimal logic, $\neg \forall x\, \neg A$ is equivalent to $\neg\neg \exists x\, A$. But with the choice of $(\exists x\, A)^{\mathrm{N}}$ above, A^{N} is always *negative*, in the following sense.

Definition 4.6.2. A formula is *negative* if $\vee$ and $\exists$ do not occur in it and every atomic formula other than $\bot$ is negated.

Equivalently, the set of negative formulas is the smallest set containing $\bot$ and negated atomic formulas, and closed under $\wedge$, $\to$, and $\forall$. As in Section 2.7, we can show that, for any negative formula A, the formula $\neg\neg A \to A$ is provable in minimal logic. The only new case is handled using the fact that $\neg\neg \forall x\, A \to \forall x\, \neg\neg A$ is provable there. We will see in Section 5.3 that the converse direction, $\forall x\, \neg\neg A \to \neg\neg \forall x\, A$, is not generally derivable in minimal or even intuitionistic logic. This implication is known as the *double-negation shift*. As in Section 2.7, we have the following theorem and corollary.

Theorem 4.6.3. *Let Γ be any set of sentences and let A be any formula.*

1. $\vdash_{\mathrm{C}} A \leftrightarrow A^{\mathrm{N}}$.
2. *If $\Gamma \vdash_{\mathrm{C}} A$ then $\Gamma^{\mathrm{N}} \vdash_{\mathrm{M}} A^{\mathrm{N}}$.*

Corollary 4.6.4. *Classical logic is conservative over minimal logic for negative formulas: if* $\vdash_C A$ *and* A *is negative, then* $\vdash_M A$.

There are many variations on the double-negation translation. Glivenko's theorem does not hold for first-order logic, but we can still reduce the number of negations that are added. Define a translation inductively as follows:

$$\begin{aligned} A_K &= A \\ (A \circ B)_K &= A_K \circ B_K \\ (\exists x\, A)_K &= \exists x\, A_K \\ (\forall x\, A)_K &= \forall x\, \neg\neg A_K. \end{aligned}$$

In the first clause, A is assumed to be atomic, and in the second clause, $\circ$ stands for any of the binary propositional connectives. Define the *Kuroda translation of* A, denoted A^K, to be $\neg\neg A_K$. So the Kuroda translation inserts a double-negation at the beginning of the formula and after each universal quantifier.

Lemma 4.6.5. *For every formula* A, $\vdash_I A^K \leftrightarrow A^N$.

The proof proceeds by induction on A. If Γ is a set sentences, as usual, let Γ^K denote $\{A^K \mid A \in \Gamma\}$. Together with Theorem 4.6.3, this yields:

Theorem 4.6.6 (Kuroda's theorem). *For every formula* A *and set of sentences* Γ, *if* $\Gamma \vdash_C A$ *then* $\Gamma^K \vdash_I A^K$.

Note that, as with Glivenko's theorem, intuitionistic logic rather than minimal is required in the conclusion of Kuroda's theorem.

The following translation works on formulas expressed in terms of the connectives $\{\neg, \wedge, \vee, \forall, \exists\}$. It maps each formula A to the formula $A^S = \neg A_S$, so A_S is supposed to represent an intuitionistic version of the *negation* of A. The map from A to A_S is defined recursively as follows, where in the first clause A is assumed to be atomic:

$$\begin{aligned} A_S &= \neg A \\ (\neg A)_S &= \neg A_S \\ (A \wedge B)_S &= A_S \vee B_S \\ (A \vee B)_S &= A_S \wedge B_S \\ (\forall x\, A)_S &= \exists x\, A_S \\ (\exists x\, A)_S &= \neg\exists x\, \neg A_S \end{aligned}$$

The translation $A \mapsto A^S$ is parsimonious about adding negations: it adds one at the beginning of a formula, one in front of every atomic subformula, and two for every existential quantifier.

Theorem 4.6.7. *Minimal logic proves* $A^S \leftrightarrow A^N$.

Proof Both translations are unchanged if we replace every existential quantifier by the classical equivalent in terms of $\neg$ and $\forall$, so we can assume that there are no existential quantifiers in A. The cases where A is atomic or a negation are immediate. For $\vee$, we have

$$(A \vee B)^S = \neg(A_S \wedge B_S) \equiv \neg(\neg\neg A_S \wedge \neg\neg B_S) \equiv \neg(\neg A^N \wedge \neg B^N) = (A \vee B)^N,$$

using the inductive hypothesis for A and B. For $\wedge$, we have

$$(A \wedge B)^{\mathrm{S}} = \neg(A_{\mathrm{S}} \vee B_{\mathrm{S}}) \equiv \neg A_{\mathrm{S}} \wedge \neg B_{\mathrm{S}} \equiv A^{\mathrm{N}} \wedge B^{\mathrm{N}} = (A \wedge B)^{\mathrm{N}}.$$

For $\forall$, we have

$$(\forall x\, A)^{\mathrm{S}} = \neg\exists x\, A_{\mathrm{S}} \equiv \forall x\, \neg A_{\mathrm{S}} \equiv \forall x\, A^{\mathrm{N}} = (\forall x\, A)^{\mathrm{N}}.$$ □

If we put A in negation normal form first, we can be even more parsimonious with negations. Define $A^{\mathrm{M}} = \neg A_{\mathrm{M}}$, where A_{M} is defined recursively as follows, with A in the first two clauses atomic:

$$\begin{aligned}
A_{\mathrm{M}} &= \neg A \\
(\neg A)_{\mathrm{M}} &= A \\
(A \wedge B)_{\mathrm{M}} &= A_{\mathrm{M}} \vee B_{\mathrm{M}} \\
(A \vee B)_{\mathrm{M}} &= A_{\mathrm{M}} \wedge B_{\mathrm{M}} \\
(\forall x\, A)_{\mathrm{M}} &= \exists x\, A_{\mathrm{M}} \\
(\exists x\, A)_{\mathrm{M}} &= \neg\exists x\, ({\sim}A)_{\mathrm{M}}.
\end{aligned}$$

Theorem 4.6.8. *Minimal logic proves $A^{\mathrm{M}} \leftrightarrow A^{\mathrm{N}}$.*

The proof is left as an exercise. Of course, Theorems 4.6.7 and 4.6.8 imply that Theorem 4.6.3 holds with the N-translation replaced by the S-translation or M-translation.

For propositional logic, Theorem 2.7.11 shows that for any propositional formula A in negation normal form, A^{N} is equivalent to $\neg({\sim}A)$ in minimal logic. In first-order logic, the first implies the second, but in general the second is weaker. Nonetheless, we can still prove the following first-order variant of Corollary 2.7.12.

Theorem 4.6.9. *Let $A, B_0, \ldots, B_{n-1}$ be in negation normal form. If $\{B_0, \ldots, B_{n-1}\} \vdash_{\mathrm{C}} A$ then $\{B_0, \ldots, B_{n-1}\} \vdash_{\mathrm{M}} \neg({\sim}A)$.*

To prove this, we need the following lemma, which can be proved using a straightforward induction on formulas:

Lemma 4.6.10. *If A is in negation normal form, then $\vdash_{\mathrm{M}} A \to A^{\mathrm{N}}$.*

If we replace by A by ${\sim}A$ in the statement of the lemma and take the contrapositive of the formula in the conclusion, we have that $\neg({\sim}A)^{\mathrm{N}} \to \neg({\sim}A)$ is provable in minimal logic. The fact that ${\sim}A$ is equivalent to $\neg A$ in classical logic implies that $\neg({\sim}A)^{\mathrm{N}}$ is equivalent to A^{N} in minimal logic, which shows that A^{N} implies $\neg({\sim}A)$ in minimal logic. We can now prove Theorem 4.6.9.

Proof If $\{B_0, \ldots, B_{n-1}\} \vdash_{\mathrm{C}} A$ then $\{B_0, \ldots, B_{n-1}, {\sim}A\} \vdash_{\mathrm{C}} \bot$. By Theorem 2.7.6, we have $\{B_0^{\mathrm{N}}, \ldots, B_{n-1}^{\mathrm{N}}, ({\sim}A)^{\mathrm{N}}\} \vdash_{\mathrm{M}} \bot$. We then have $\{B_0, \ldots, B_{n-1}, {\sim}A\} \vdash_{\mathrm{M}} \bot$ by Lemma 4.6.10, and hence $\{B_0, \ldots, B_{n-1}\} \vdash_{\mathrm{M}} \neg({\sim}A)$. □

If we take A to be $\exists x\, \neg P(x)$ for a predicate symbol P, then $\neg({\sim}A) \to A^{\mathrm{N}}$ is $\neg\forall x\, P(x) \to \neg\forall x\, \neg\neg P(x)$. This is, in turn, equivalent to $\forall x\, \neg\neg P(x) \to \neg\neg\forall x\, P(x)$ in minimal logic. This is the principle of double-negation shift alluded to above, which shows that $\neg({\sim}A)$ is strictly weaker than A^{N} in intuitionistic logic. As a result, the translation $A \mapsto \neg({\sim}A)$ from classical to minimal logic is not modular: Exercise 4.6.6 asks you to show that from $\neg(\sim(A \to B)^{\mathrm{nnf}})$ and $\neg(\sim A^{\mathrm{nnf}})$ it is, in general, not possible to conclude $\neg(\sim B^{\mathrm{nnf}})$ intuitionistically.

Exercises

4.6.1. Prove the following analogue of Glivenko's theorem: a formula A is provable in classical logic if and only if $\neg\neg A$ is provable in minimal logic together with the schema of double-negation shift: $\forall x\, \neg\neg B(x) \to \neg\neg\forall x\, B(x)$.

4.6.2. Show that over minimal logic double-negation shift is equivalent to the schema $\neg\neg\forall x\, (A(x) \vee \neg A(x))$:

a. Show that $\forall x\, (A(x) \vee \neg A(x))$ implies $\forall x\, \neg\neg A(x) \to \neg\neg\forall x\, A(x)$ for any formula $A(x)$.
b. Show that if $B(x)$ is $A(x) \vee \neg A(x)$, then double-negation shift for B implies $\neg\neg\forall x\, (A(x) \vee \neg A(x))$.

4.6.3. Prove Lemma 4.6.5.

4.6.4. Prove Theorem 4.6.8.

4.6.5. Show that the formula $\neg\forall x\, P(x) \to \neg\forall x\, \neg\neg P(x)$ is equivalent to the formula $\forall x\, \neg\neg P(x) \to \neg\neg\forall x\, P(x)$ in minimal logic.

4.6.6. Show that from $\neg(\sim(A \to B)^{\mathrm{nnf}})$ and $\neg(\sim A^{\mathrm{nnf}})$ it is, in general, not possible to conclude $\neg(\sim B^{\mathrm{nnf}})$ in intuitionistic logic. (Hint: Take A to be $\forall x\, P(x)$ for some predicate symbol $P(x)$ and take B to be $\bot$.)

4.6.7. Use the classical cut elimination theorem in Section 6.4 to provide another proof of Theorem 4.6.9. (Hint: Show that the rules of the one-sided cut-free calculus correspond to left rules of the two-sided calculus for minimal logic. So if Γ is a set of formulas in negation normal form, $\Gamma \vdash \bot$ is derivable in classical logic if and only if it is derivable in minimal logic.)

4.6.8. Let $Q_1x_1\, Q_2x_2\, \ldots Q_kx_k\, A$ be any prenex formula. Show that minimal logic proves

$$(Q_1x_1\, Q_2x_2\, \ldots Q_kx_k\, A)^{\mathrm{N}} \to \neg Q'_1x_1\, Q'_2x_2\, \ldots Q'_kx_k\, \neg A^{\mathrm{N}},$$

where Q'_i is $\forall$ if Q_i is $\exists$ and vice versa.

4.6.9. The next sequence of problems foreshadows developments from Chapter 10. *Peano arithmetic* (PA) is the first-order theory in a language with nonlogical symbols 0, succ, +, and $\cdot$, axiomatized by the universal closures of the following:

- $\mathrm{succ}(x) \neq 0$
- $\mathrm{succ}(x) = \mathrm{succ}(y) \to x = y$
- $x + 0 = x$
- $x + \mathrm{succ}(y) = \mathrm{succ}(x + y)$
- $x \cdot 0 = x$
- $x \cdot \mathrm{succ}(y) = (x \cdot y) + x$,

and the axiom of induction for each formula A:

$$A(0) \wedge \forall x\, (A(x) \to A(\mathrm{succ}(x))) \to \forall x\, A(x).$$

Heyting arithmetic (HA) is given by the same axioms, but taken over *intuitionistic* logic.

Show that for every formula A in the language of arithmetic, if A is provable in PA, then A^{N} is provable in HA. (Note that you need to show that the double-negation translations of the axioms are provable in HA.)

4.6.10. Note that the only atomic formulas in the language of arithmetic are of the form $s = t$.

a. Show that HA proves $\forall x, y\, (x = y \vee x \neq y)$.
b. Show that for any quantifier-free formula A, HA proves the following are equivalent: A, $\neg\neg A$, and A^{N}. (In Chapter 10, we will extend this to a wider class of formulas.)
c. Conclude that whenever PA proves $\forall x\, \exists y\, A(x, y)$ with A quantifier-free, HA proves $\forall x\, \neg\neg\exists y\, A(x, y)$.

4.6.11. This problem and the next one strengthen the previous result to show that whenever PA proves $\forall x\, \exists y\, A(x, y)$ with A quantifier-free, then HA proves $\forall x\, \exists y\, A(x, y)$ as well. First, remember the *-translation from intuitionistic logic to minimal logic, in which atomic formulas C are replaced by $C \vee \bot$.

a. Show that if HA proves A, then the axioms of arithmetic imply A^* in *minimal logic*.
b. Show that if A is quantifier-free, there is a formula A' in negation normal form such that HA proves $A \leftrightarrow A'$. (Use the fact that this is true of classical logic, the double-negation translation, and the previous problem.)
c. Show that if A is in negation normal form, then $\vdash_{\mathrm{M}} A^* \leftrightarrow A \vee \bot$.

4.6.12. Now suppose PA proves $\forall x\, \exists y\, A(x, y)$, with A quantifier-free. By the above, we know that HA proves $\forall x\, \neg\neg\exists y\, A(x, y)$, and we can assume without loss of generality that A is in negation normal form. If we expand the language of HA to include a new constant symbol, c, we have that HA proves $\neg\neg\exists y\, A(c, y)$.

a. Use the *-translation to show that HA over minimal logic proves

$$(\exists y\, (A(c, y) \vee \bot) \to \bot) \to \bot.$$

b. Substitute $\exists y\, A(c, y)$ for $\bot$, and show that HA proves $\exists y\, A(c, y)$.

Replacing c by a variable, x, we have the desired conclusion. This trick, known as the *Friedman–Dragalin translation*, is discussed in further detail in Section 10.4.

4.7 Definite Descriptions

Suppose that, working with a formal language designed to reason about the natural numbers, we say what it means for z to be a greatest common divisor of x and y. Suppose we then prove that every pair of natural numbers x and y has a unique greatest common divisor. (Let us suppose that our definition specifies a value even in the case where x and y are both 0.) From that moment on we would be justified in referring to *the* greatest common divisor of x and y. Such a specification of the greatest common divisor is an example of what is known as a *definite description*, and the goal of this section is to explain how to use first-order logic to interpret the corresponding use of the word "the."

The phrase "the x such that $A(x)$" presupposes that there is a unique x satisfying A. The following definition formalizes the notion of unique existence:

Definition 4.7.1. If $A(x)$ is any formula, the expression $\exists! x\, A(x)$, read *there is a unique x such that $A(x)$*, is the formula $\exists x\, (A(x) \wedge \forall y\, (A(y) \to y = x))$.

The exercises provide some equivalent formulations. From the assumption $\exists! x\, A$ we can show that, for any formula B, the statements $\exists x\, (A \wedge B)$ and $\forall x\, (A \to B)$ are equivalent. Assuming $\exists! x\, A$, either can be taken to represent the statement that B holds of *the* x satisfying A.

In general, A can have variables other than x, in which case, the phrase "the x such that A" describes a function of these parameters. Assuming we have shown $\forall \vec{x}\, \exists! y\, A(\vec{x}, y)$, we should be justified in introducing a new function symbol $f(\vec{x})$ to denote the unique y satisfying A.

The next theorem and its proof show how we can interpret the use of such a function in the original language. The process is delicate because the function f can be composed with other function symbols, but the method is a straightforward extension of our analysis of the word "the."

The idea of expanding a theory to one in a larger language without altering the consequences in the original language is so common that it warrants some additional terminology.

Definition 4.7.2. Let Γ be a set of sentences in a language, L, and let $\Gamma' \supseteq \Gamma$ be a set of sentences in a larger language, L'. Suppose that whenever B is a formula in L and $\Gamma' \vdash B$ in first-order logic for L', then $\Gamma \vdash B$ in first-order logic for L. Then Γ' is said to be a *conservative extension* of Γ.

Let L be a first-order language, $f(\vec{x})$ be a new function symbol, and L_f be the language L extended by f. Let $A(\vec{x}, y)$ be a formula in L with at most the free variables shown. Think of $A(\vec{x}, y)$ as a way of expressing $f(\vec{x}) = y$ without mentioning f explicitly. Using the terminology just introduced, the theorem we are after is the following:

Theorem 4.7.3. *If* $\Gamma \vdash \forall \vec{x}\, \exists! y\, A(\vec{x}, y)$*, then* $\Gamma \cup \{\forall \vec{x}\, A(\vec{x}, f(\vec{x}))\}$ *is a conservative extension of* Γ.

The theorem holds for classical, intuitionistic, and minimal logic.

We will prove Theorem 4.7.3 by defining a translation that maps each formula B of L_f to a formula B^* of L so that the following hold:

1. If B is in L, then $B^* \leftrightarrow B$ is provable in first-order logic for L_f.
2. From hypothesis $\forall \vec{x}, y\, (A(\vec{x}, y) \leftrightarrow y = f(\vec{x}))$, we can prove $B^* \leftrightarrow B$ for every B in L_f.
3. If Γ and B are formula of L_f and $\Gamma \cup \{\forall \vec{x}\, A(\vec{x}, f(\vec{x}))\}$ proves B in first-order logic for L_f, then $\Gamma^* \cup \{\forall \vec{x}\, \exists! y\, A(\vec{x}, y)\}$ proves B^* in first-order logic for L.

In other words, the map $B \mapsto B^*$ translates statements in the language with f to statements in which references to f are replaced by the description $A(\vec{x}, y)$. All three claims will hold whether provability is interpreted in terms of classical, intuitionistic, or minimal logic. The first claim asserts that the translation leaves formulas in L essentially unchanged. The second guarantees that the translation means what we think it means: assuming A describes f, B^* and B say the same thing. The third claim is that the translation respects provability. To see that this suffices to prove Theorem 4.7.3, suppose that Γ proves $\forall \vec{x}\, \exists! y\, A(\vec{x}, y)$, and suppose $\Gamma \cup \{\forall \vec{x}\, A(\vec{x}, f(\vec{x}))\}$ proves some formula B in L. By the first and third properties, we have that Γ together with $\forall \vec{x}\, \exists! y\, A(\vec{x}, y)$ proves B. But since we are assuming that Γ proves $\forall \vec{x}\, \exists! y\, A(\vec{x}, y)$, we have that Γ proves B, as required.

Before defining the translation of formulas, we will define a translation of terms. Specifically, we will assign to each term $t(\vec{x})$ in the language L_f with the free variables shown a formula $C_t(\vec{x}, y)$ in the language of L that represents the statement $y = t(\vec{x})$ without mentioning f. (Throughout, I will assume y is a variable different from all the xs.) The assignment is defined by induction on terms of L_f:

- For each variable x, $C_x(x, y)$ is the formula $y = x$.
- If t is a term of the form $f(t_0(\vec{x}), \ldots, t_{k-1}(\vec{x}))$, where f is the new function symbol, then $C_t(\vec{x}, y)$ is the formula

$$\exists z_0, \ldots, z_{k-1}\, (C_{t_0}(\vec{x}, z_0) \wedge \cdots \wedge C_{t_{k-1}}(\vec{x}, z_{k-1}) \wedge A(\vec{z}, y)),$$

where $z_0, \ldots, z_{k-1}$ are variables different from $\vec{x}$ and y. (I have assumed that each t_i has the same parameters $\vec{x}$ only for notational simplicity; otherwise, only the variables among $\vec{x}$ occurring in t_i will occur in C_{t_i}.)

- If t is a term of the form $g(t_0(\vec{x}), \ldots, t_{m-1}(\vec{x}))$ for some function symbol g of L, then $C_t(\vec{w}, y)$ is the formula

$$\exists z_0, \ldots, z_{m-1} \, (C_{t_0}(\vec{x}, z_0) \wedge \cdots \wedge C_{t_{k-1}}(\vec{x}, z_{m-1}) \wedge y = g(\vec{z})).$$

By induction on terms, it is straightforward to show the following:

1. For every term $t(\vec{x})$ of L, minimal logic proves $\forall \vec{x} \, (C_t(\vec{x}, y) \leftrightarrow y = t(\vec{x}))$.
2. For every term $t(\vec{x})$ of L_f, minimal logic with $\forall \vec{x}, y \, (A(\vec{x}, y) \leftrightarrow y = f(\vec{x}))$ proves $\forall \vec{x} \, (C_t(\vec{x}, y) \leftrightarrow y = t(\vec{x}))$.
3. For every term $t(\vec{x})$ of L_f, minimal logic together with $\forall \vec{x} \, \exists! y \, A(\vec{x}, y)$ proves $\forall \vec{x} \, \exists! y \, C_t(\vec{x}, y)$.

We also need to know that the translation respects substitution. Suppose $s(\vec{x})$ and $t(\vec{x}, w)$ are terms with the free variables shown. Then assuming $w = s(\vec{x})$, saying $y = t(\vec{x}, w)$ is equivalent to saying $y = t[s/w](\vec{x})$. Modulo our translation, this amounts to the following:

$$C_s(\vec{x}, w) \to (C_t(\vec{x}, w, y) \leftrightarrow C_{t[s/w]}(\vec{x}, y)). \tag{4.1}$$

We can show that this is provable in minimal logic from $\forall \vec{x} \, \exists! y \, A(\vec{x}, y)$ using a straightforward induction on t. In the base case where t is a variable other than w, there is nothing to do. If t is the variable w, we need to show

$$C_s(\vec{x}, w) \to (y = w \leftrightarrow C_s(\vec{x}, y)).$$

Assuming $C_s(\vec{x}, w)$, the forward implication is immediate, and the reverse implication follows from the third claim listed above. In the induction step, where t is a composite term, it suffices to unwrap the definition of C_t and the definition of $t[s/x]$ and make use of the inductive hypothesis and claim 3.

Now that we can describe terms of L_f in the language of L, we can define a translation mapping formulas B of L_f to formulas B^* of L, recursively, as follows:

- If s and t are terms of L_f, $(s(\vec{x}) = t(\vec{x}))^*$ is the formula $\exists y \, (C_s(\vec{x}, y) \wedge C_t(\vec{x}, y))$.
- If R is an ℓ-ary relation symbol of L_f and $t_0(\vec{x}), \ldots, t_{\ell-1}(\vec{x})$ are terms of L_f, then $(R(t_0, \ldots, t_{\ell-1}))^*$ is the formula

$$\exists z_0, \ldots, z_{\ell-1} \, (C_{t_0}(\vec{x}, z_0) \wedge \cdots \wedge C_{t_{\ell-1}}(\vec{x}, z_{\ell-1}) \wedge R(z_0, \ldots, z_{\ell-1})).$$

- $(B \circ C)^* = B^* \circ C^*$ for any propositional connective $\circ$, and $\bot^* = \bot$.
- $(\forall x \, B)^* = \forall x \, B^*$ and $(\exists x \, B)^* = \exists x \, B^*$.

It is then straightforward to show the desired properties 1 and 2 of the translation of formulas from the corresponding properties of the translation of terms. Once again, we need to know that the translation behaves well with respect to substitution, namely, that for any formula $B(\vec{x})$ and term $t(\vec{x})$, from $\forall \vec{x} \, \exists! y \, A(\vec{x}, y)$ minimal logic can prove

$$C_t(\vec{x}, w) \to (B^*(\vec{x}, w) \leftrightarrow B(\vec{x}, t(\vec{x}))^*). \tag{4.2}$$

This is proved by induction on B, using (4.1) in the base cases.

All we have left to do is prove property 3, namely, that if Γ and B are in the language L_f and $\Gamma \cup \{\forall \vec{x}\, A(\vec{x}, f(\vec{x}))\}$ proves B, then $\Gamma^* \cup \{\forall \vec{x}\, \exists! y\, A(\vec{x}, y)\}$ proves B^* in L. First, we prove the claim in the case where $\forall \vec{x}\, A(\vec{x}, f(\vec{x}))$ is omitted from the hypotheses in the assumption. In that case, we proceed by induction on derivations in natural deduction. The case for the assumption rule is immediate, and because the translation commutes with propositional connectives, the propositional rules are also easily handled. Handling the reflexivity, symmetry, and transitivity of equality is straightforward. We can take the remaining equality rule to be "from $s(\vec{x}) = t(\vec{x})$ and $B(\vec{x}, s(\vec{x}))$ conclude $B(\vec{x}, t(\vec{x}))$." The hypothesis translates to $\exists w\, (C_s(\vec{x}, w) \wedge C_t(\vec{x}, w))$ and, for that w, (4.2) implies that $B(\vec{x}, s(\vec{x}))^*$ and $B(\vec{x}, t(\vec{x}))^*$ are both equivalent to $B^*(\vec{x}, w)$.

The introduction rule for the universal quantifier and the elimination rule for the existential quantifier are easily handled. Suppose the last inference is an instance of the elimination rule for the universal quantifier, yielding $B(\vec{x}, t(\vec{x}))$ from $\forall w\, B(\vec{x}, w)$. The hypothesis translates to $\forall w\, B^*(\vec{x}, w)$. We can also prove $\exists! w\, C_t(\vec{x}, w)$, and (4.2) implies that, for that w, the conclusion is equivalent to $B^*(\vec{x}, w)$. The elimination rule for the existential quantifier is handled similarly.

To handle the hypothesis $\forall \vec{x}\, A(\vec{x}, f(\vec{x}))$, first replace this by the equivalent formula $\forall \vec{x}, y\, (f(\vec{x}) = y \rightarrow A(\vec{x}, y))$. By the previous argument, B^* is provable from Γ^* together with the translation of this hypothesis. But since A is in L, the latter is equivalent to $\forall \vec{x}, y\, (A(\vec{x}, y) \rightarrow A(\vec{x}, y))$, which is provable outright.

To sum up, we have the following theorem:

Theorem 4.7.4. *Let L be a first-order language, let $A(\vec{x}, y)$ be a formula of L with the free variables shown, let Γ, B be formulas of L, let $f(\vec{x})$ be a new function symbol, and let L_f denote the first-order language obtained by adding f to L. Suppose $\Gamma \cup \{\forall \vec{x}\, A(\vec{x}, f(\vec{x}))\}$ proves B in classical, intuitionistic, or minimal logic for L_f. Then $\Gamma \cup \{\forall \vec{x}\, \exists! y\, A(\vec{x}, y)\}$ proves B in the same system for L.*

In particular, if Γ proves $\forall \vec{x}\, \exists! y\, A(\vec{x}, y)$ then Γ proves B, and Theorem 4.7.3 follows.

I will refer to Theorems 4.7.3 and 4.7.4 as the *elimination of definite descriptions*. The process can be repeated, allowing us to conservatively extend our axiomatic theory with any number of function definitions. Similarly, we can conservatively extend a theory with a new relation symbol, $R(x_0, \ldots, x_{k-1})$, with defining axiom $\forall \vec{x}\, (R(\vec{x}) \leftrightarrow A(\vec{x}))$. Such definitions are more readily eliminated by replacing R by A everywhere. If T is any theory and T' is obtained from T by any number of extensions of these two kinds, T' is said to be a *definitional extension* of T.

When defining an axiomatic theory, it is often preferable to keep the number of primitives to a minimum, making it easier to study the theory and its metamathematical properties. When reasoning about what such a theory can do, however, it is useful to have a more expressive language. The use of definitional extensions is a way of having one's cake and eating it too.

The methods we have just developed also give us a way of eliminating function symbols in favor of relation symbols in a first-order language. Let L' be a language with function symbols $f_0, \ldots, f_{n-1}$. For each k-ary function symbol $f_i(\vec{x})$, choose a $(k+1)$-ary relation symbol $F_i(\vec{x}, y)$ and let L be the language with each f_i replaced by F_i. Let Π be the set of axioms $\forall \vec{x}\, \exists! y\, F_i(\vec{x}, y)$. Then, for each i, f_i can be viewed as the function defined by F_i. Our translation takes any formula A in the language of L' to a formula $\hat{A}$ in the language of L.

Proposition 4.7.5. *Let L, L', Π and the map $A \mapsto \hat{A}$ be as above. Then A is provable in L' if and only if $\hat{A}$ is provable from Π in L.*

Proof Suppose A is provable in L'. Then we also have, trivially, that A is provable from the set of hypotheses $\{\ldots, \forall \vec{y}\, F_i(\vec{y}, f_i(\vec{y})), \ldots\}$ in $L \cup L'$. Viewing the functions f_i as witnessing the definite descriptions afforded by F_i and applying Theorem 4.7.4, we have that $\hat{A}$ is provable from Π in L.

Conversely, translate formulas B of L' to formulas B^* of L by replacing every atomic formula $F_i(\vec{s}, t)$ by $f_i(\vec{s}) = t$. Under this translation, every sentence in Π translates to an assertion of the form $\forall \vec{x}\, \exists! y\, (f_i(\vec{x}) = y)$, which is provable in L'. As a result, if B is provable from Π in L, then B^* is provable in L'. In particular, if $\hat{A}$ is provable from Π in L, then $(\hat{A})^*$ is provable in L'. But it is not hard to show that for every formula C in L', $(\hat{C})^* \leftrightarrow C$ is provable in L'. So A is provable in L'. □

What happens if, in Theorems 4.7.3 and 4.7.4, we drop uniqueness by replacing $\forall \vec{x}\, \exists! y\, A(\vec{x}, y)$ by $\forall \vec{x}\, \exists y\, A(\vec{x}, y)$? Assuming the weaker hypothesis, we can think of $A(\vec{x}, y)$ as providing an *indefinite description* of an element y depending on $\vec{x}$. A function f satisfying $\forall \vec{x}\, A(\vec{x}, f(\vec{x}))$ is guaranteed to return *some* y satisfying $A(\vec{x}, y)$, but in some cases more than one value of y may work. Such a function may therefore have to choose among the possible values of y.

We will see Chapter 7 that, somewhat surprisingly, adding such choice functions also yields a conservative extension of classical first-order logic. It does not yield a conservative extension of intuitionistic or minimal logic, but in those cases, we will find a simple axiomatization of the new consequences that are obtained.

Exercises

4.7.1. Show that the following are pairwise equivalent in minimal logic.

a. $\exists! x\, A(x)$
b. $\exists x\, \forall y\, (A(y) \leftrightarrow x = y)$
c. $\exists x\, A(x) \wedge \forall y, y'\, (A(y) \wedge A(y') \to y = y')$.

This provides us with various ways of expressing the notion "there exists a unique x satisfying A." Notice that in the third expression, the existence and uniqueness claims have been separated.

4.7.2. Show that from the assumption $\exists! x\, A$ we can prove

$$\forall x\, (A \to B) \leftrightarrow \exists x\, (A \wedge B).$$

4.7.3. Fill in the details left out of the proof of Theorem 4.7.4.

4.8 Sorts and Undefined Terms

In this section we will consider two variations on first-order logic. Both increase the expressiveness of the language, but they can be interpreted straightforwardly using ordinary first-order logic.

As we have described first-order logic, there is only one universe of objects, and variables and quantifiers range over the entire universe. For many purposes, it is useful to have different *sorts* of objects in the universe, with each variable and quantifier ranging over objects of a particular sort. For Euclidean geometry, we might want to have a universe of points,

$p, q, r, \ldots$ and a universe of lines, $l, m, n, \ldots$, with a relation $\mathrm{on}(p, l)$ that relates points and lines. The statement "two distinct points determine a line" would then be represented as

$$\forall p, q\, (p \neq q \to \exists! l\, (\mathrm{on}(p, l) \wedge \mathrm{on}(q, l))).$$

Here the quantifiers and variables are assumed to range over the appropriate sorts.

It is straightforward to modify the description of first-order logic to accommodate this. A many-sorted language comes with a set of sorts, and it specifies the sorts of the arguments to each function and relation symbol and the sort of each function symbol's output. In particular, there is a different equality symbol for each sort. Terms are defined as before, except that there is a stock of variables for each sort, and the term-forming operations are required to conform to the sort specifications in the language. Thus, each term inherits a sort, and the terms in an atomic formula $R(t_0, \ldots, t_{n-1})$ are required to conform to specification of R's expected arguments. The quantifier axioms and rules are essentially the same, except that the sorts of the relevant variables and terms have to match. For example, in the axiom $A(t) \to \exists x\, A(x)$, the variable x is assumed to have the same sort as the term t.

The resulting logic is easily translated into ordinary first-order logic by adding a new predicate $S(x)$ for each basic sort S of the many-sorted language and assigning a distinct variable x^S of the first-order language to each sorted variable x^S of the many-sorted language. In ordinary first-order logic, we can define the *relativized quantifiers* $\forall^S$ and $\exists^S$ by setting

$$\forall^S x\, A(x) \equiv \forall x\, (S(x) \to A(x))$$

and

$$\exists^S x\, A(x) \equiv \exists x\, (S(x) \wedge A(x)).$$

For each function symbol $f(x_0^{S_0}, \ldots, x_{k-1}^{S_{k-1}})$ of the many-sorted language expecting arguments of the sort shown and returning an argument of sort S_k, add an axiom to ordinary first-order logic,

$$S_0(x^{S_0}) \wedge \cdots \wedge S_{k-1}(x^{S_{k-1}}) \to S_k(f(x_0^{S_0}, \ldots, x_{k-1}^{S_{k-1}})).$$

Then translate each formula $A(x_0^{S_0}, \ldots, x_{k-1}^{S_{k-1}})$ of the many-sorted language with the free variables shown to the first-order formula $S_0(x^{S_0}) \wedge \cdots \wedge S_{k-1}(x^{S_{k-1}}) \to \hat{A}(x_0^{S_0}, \ldots, x_{k-1}^{S_{k-1}})$, where $\hat{A}$ is defined inductively as follows:

- $\widehat{\forall x^S A} = \forall^S x \hat{A}$.
- $\widehat{\exists x^S A} = \exists^S x \hat{A}$.
- The $\hat{\cdot}$-translation fixes atomic statements and commutes with the other logical connectives.

Proposition 4.8.1. *Let Γ be the additional axioms described above. Then for any formula A in the language of many-sorted first-order logic, the latter proves the formula $A(x_0^{S_0}, \ldots, x_{k-1}^{S_{k-1}})$ if and only if ordinary first-order logic, together with Γ, proves*

$$S_0(x_0^{S_0}) \wedge \cdots \wedge S_{k-1}(x_{k-1}^{S_{k-1}}) \to \hat{A}(x_0^{S_0}, \ldots, x_{k-1}^{S_{k-1}}).$$

Proof (sketch) The forward direction is obtained by a straightforward induction on proofs. The other direction is more subtle, because, a priori, an ordinary first-order proof in the extended language may not respect any of the sort conventions specified by the many-sorted language. One strategy is to restrict attention to formulas that are *well sorted* in the sense that all the quantifiers are relativized to sorts and it is possible to assign sorts to the variables in a manner that is consistent with those conventions. We can then appeal to the cut elimination theorem, discussed in Chapter 6, to argue that we can restrict attention to proofs in which there is an assignment of sorts to variables which makes all the formulas in the proof well-sorted. To each well-sorted formula A, we can then assign a formula $\bar{A}$ which replaces each relativized quantifier with the corresponding many-sorted quantifier, and each atomic formula $S(t)$ with an equality $t = t$ of the appropriate sort. The first-order proof then translates back to a many-sorted proof of the $\bar{\cdot}$-translation of the conclusion. Finally, by induction on formulas, we show that many-sorted logic can prove that this is equivalent to the original formula, A.

Exercise 4.8.3 below challenges you to find a proof that avoids the use of cut elimination. Alternatively, we can use a model-theoretic argument, using the methods developed in Chapter 7. □

The other variation on first-order logic we will consider is designed to model logics where function symbols may be only partially defined. For example, in a theory of the real numbers it is natural to use a division symbol but take $1/0$ to be an undefined term. In such a framework, variables are still assumed to range over elements of the universe, but since the function symbols denote partial functions, a term of the form $f(t_0, \ldots, t_{k-1})$ may be undefined. To extend first-order logic to handle this, add a new predicate symbol, E, where $E(t)$ is intended to express the fact that *t exists*, or *t is defined*, or *t denotes*. It is conventional to write $t\downarrow$ for $E(t)$, and $t\uparrow$ for $\neg E(t)$.

In a natural deduction formulation, the propositional axioms and rules remain the same, as do the introduction rule for the universal quantifier and elimination rule for the existential quantifier. But the elimination rule for the universal quantifier and the introduction rule for the existential quantifier are now expressed as follows, to reflect the fact that some terms may be undefined:

$$\frac{\forall x\, A(x) \qquad t\downarrow}{A(t)} \qquad \frac{A(t) \qquad t\downarrow}{\exists x\, A(x)}$$

The following axioms and rules reflect the fact that variables are assumed to range over (existing) elements of the domain, but functions may be partial:

$$\frac{}{x\downarrow} \qquad \frac{}{c\downarrow} \qquad \frac{f(t_0, \ldots, t_{n-1})\downarrow}{t_i\downarrow} \qquad \frac{R(t_0, \ldots, t_{n-1})}{t_i\downarrow}$$

The third rule implies that if any of $t_0, \ldots, t_{n-1}$ are undefined, then $f(t_0, \ldots, t_{n-1})$ is undefined. The fourth rule says that any relation, including equality, can only hold of terms that are defined. If s and t are terms, the relation $s \simeq t$ of *Kleene equality* is defined to be $s\downarrow \vee\, t\downarrow \to s = t$. With the rule above, if $s \simeq t$ and either s or t is defined, then they are both defined and equal.

The equality axioms for reflexivity, symmetry, and transitivity are as before, as well as the congruence property for relations. For functions symbols, however, we replace the congruence property by the following:

$$\frac{\vec{s} = \vec{t}}{f(\vec{s}) \simeq f(\vec{t})}$$

This reflects the fact that $f(\vec{s})$ and $f(\vec{t})$ may be undefined.

This extension of first-order logic is known as the *logic of partial terms*. It is not hard to interpret this system in ordinary first-order logic with a new predicate E. Because functions in ordinary first-order logic are total, we have to allow the first-order domain to contain elements that don't exist and use the predicate E to specify the ones that do. We can explicitly restrict variables and quantifiers to range over elements that exist by translating $A(x_0, \ldots, x_{n-1})$ to the formula $E(x_0) \wedge E(x_1) \wedge \cdots \wedge E(x_{n-1}) \rightarrow A^*(x_0, \ldots, x_{n-1})$, where A^* is defined inductively as follows:

- $(R(t_0, \ldots, t_{n-1}))^* = E(t_0) \wedge \cdots \wedge E(t_{n-1}) \wedge R(t_0, \ldots, t_{n-1})$.
- $(\forall x\, A(x))^* = \forall x\, (E(x) \rightarrow A(x))$
- $(\exists x\, A(x))^* = \exists x\, (E(x) \wedge A(x))$.
- Otherwise, the $*$-translation commutes with logical connectives.

In addition, let Γ be the set of axioms

$$\{\forall x_0, \ldots, x_{n-1}\, (E(f(x_0, \ldots, x_{n-1})) \rightarrow E(x_0) \wedge \cdots \wedge E(x_{n-1}))\}.$$

Proposition 4.8.2. *Logic of partial terms proves* $A(x_0, \ldots, x_{n-1})$ *if and only if* $\Gamma \vdash E(x_0) \wedge E(x_1) \wedge \cdots \wedge E(x_{n-1}) \rightarrow A^*(x_0, \ldots, x_{n-1})$ *in ordinary first-order logic.*

Proof (sketch) Once again, it is straightforward to prove the forward direction by induction on derivations, while the backward direction is more subtle. As for Proposition 4.8.1, we can use a model-theoretic argument or appeal to the cut elimination theorem to restrict attention to proofs where all variables are assumed to satisfy the predicate E. □

The logic of partial terms provides a natural way to introduce definite and indefinite descriptions axiomatically. It allows us to safely introduce a partial function symbol $f(\vec{x})$ to denote the unique value of y satisfying $A(\vec{x}, y)$ assuming there is one, with axioms $\forall \vec{x}\, (f(\vec{x}) \downarrow \leftrightarrow \exists! y\, A(\vec{x}, y))$ and $\forall \vec{x}\, (f(\vec{x}) \downarrow \rightarrow A(\vec{x}, f(\vec{x})))$. Writing $f(\vec{t})$ then presupposes that there is a unique y such that $A(\vec{t}, y)$ holds; if there isn't, the expression is simply undefined.

Exercises

4.8.1. Prove the forward direction of Proposition 4.8.1.

4.8.2. Prove the backward direction using a model-theoretic argument or cut elimination.

4.8.3. Prove the backward direction more directly. Note that for any variable x occurring in formula A in the first-order language, if x appears as an argument to a function or relation other than equality, there is only one sort that can be consistently assigned to x. Conversely, on the many-sorted side, we can take relations to be false when they are applied to arguments of the wrong sort. Argue that we can translate proofs in first-order logic to proofs in many-sorted logic by keeping track of all possible sort assignments to the variables, at once.

4.8.4. Prove the forward direction of Proposition 4.8.2.

4.8.5. Prove the backward direction.

Bibliographical Notes

Axiomatizations of first-order logic The axiomatization of first-order logic given in Exercise 4.2.9 can be found in Shoenfield (2001). First-order logic as presented here presupposes a nonempty universe; for a variant of first-order logic that is also valid for the empty universe, see Lambek and Scott (1988).

Equational logic and universal algebra A structure for a language with no relation symbols is known as an *algebra*, and a class of algebras axiomatized by identities is known as a *variety*. Equational logic provides a complete proof system for establishing the identities that hold in a given variety. *Birkhoff's theorem* shows that a class of algebras is a variety if and only if it is closed under homomorphic images, subalgebras, and direct products; see Bergman (2015).

Automated reasoning Practical decision procedures for the quantifier-free fragment of classical logic usually rely on the *congruence closure* algorithm to handle equational reasoning. *SMT solvers* provide decision procedures for the quantifier-free fragment of classical first-order logic extended by combinations of theories that also have a decidable quantifier-free fragment, such as the ones discussion in Section 11.2. Here *SMT* stands for *satisfiability modulo theories*; see Harrison (2009), Bradley and Manna (2007), and Kroening and Strichman (2016). Since proving a formula $A(\vec{x})$ with free variables $\vec{x}$ is the same as proving $\forall \vec{x}\, A(\vec{x})$, a decision procedure for the quantifier-free consequences of a theory T is the same as a decision procedure for the universal consequences of T, also known as the *universal fragment* of T.

As noted in Section 4.4, the question as to whether an entailment holds in quantifier-free logic with the substitution rule, or, equivalently, the question as to whether a universal formula is a first-order consequence of some universal axioms, is generally undecidable. We will see in Section 7.5 that questions about provability in full classical logic can straightforwardly be reduced to this case. SMT solvers often allow universal axioms and use heuristics to instantiate the quantifiers, but they generally do not provide complete search procedures. That belongs to the realm of first-order theorem provers, for which see Robinson and Voronkov (2001) and the notes at the end of Chapters 6 and 7.

Double-negation translations Original papers on the double-negation translation can be found in van Heijenoort (1967) and in Gödel (1986–2003). Ferreira and Oliva (2012) provides a thorough analysis of the variations; see also Troelstra and Schwichtenberg (2000). Theorem 4.6.7 is implicit in the presentation of the Dialectica interpretation in Shoenfield (2001).

Definition descriptions The analysis of definite descriptions is found in the landmark paper, Russell (1905). Russell and Whitehead used the notation $\iota x.A$ to denote "the x satisfying A" in *Principia Mathematica*.

Logics for partial terms Feferman (1995) provides an overview of logics for partial terms. Beeson (1985), Troelstra and Schwichtenberg (2000), and Troelstra and van Dalen (1988) are also good references.

5

Semantics of First-Order Logic

In this chapter, we extend semantic notions that were introduced in Chapter 3 to first-order logic. Because the language of first-order logic is more complex than that of propositional logic, the semantics is more complex as well. Whereas propositional variables were interpreted as ranging over truth values, we now have to interpret first-order variables as ranging over objects in some domain of interpretation, and we have to explain the meaning of terms and formulas that contain such variables. But although there are more details to contend with, the ideas are quite similar.

5.1 Classical Logic

Fix a first-order language, L. Definition 1.1.2 says that an *L-structure* $\mathfrak{M}$ consists of a nonempty set, $|\mathfrak{M}|$, called the *universe of* $\mathfrak{M}$, an interpretation of each function symbol f by a function $f^{\mathfrak{M}}$ on $|\mathfrak{M}|$ of the right arity, and an interpretation of each relation symbol R by a relation $R^{\mathfrak{M}}$ on $|\mathfrak{M}|$ of the right arity. Our goal here is to specify what it means for an L-structure to *satisfy* a sentence or set of sentences of L, in analogy to the sense in which a truth assignment can be said to satisfy a propositional formula or set of propositional formulas. When we think of the sentence or sentences as axioms, we generally to refer to an L-structure as a *model*.

Let $\mathfrak{M}$ be an L-structure, let t be any term in L, and let σ be an assignment of elements of $|\mathfrak{M}|$ to variables. Section 1.5 provides a recursive specification that assigns an element $[\![t]\!]^{\mathfrak{M},\sigma}$ of the universe of $\mathfrak{M}$. We interpret this as the *denotation* or *evaluation* of t in $\mathfrak{M}$ under the assignment σ. In Section 1.5, we saw that the denotation only depends on the values that σ assigns to the variables of t. As a result, we could take σ to be an assignment to just those variables instead. Given a term $t(x_0, \ldots, x_{n-1})$ with the free variables shown, we can use $[\![t]\!]^{\mathfrak{M}}$ to denote the n-ary function defined by setting $[\![t]\!]^{\mathfrak{M}}(a_0, \ldots, a_{n-1})$ equal the denotation of t under any assignment that maps each variable x_i to a_i.

In the classical, model-theoretic interpretation of first-order logic, a term therefore names an element of a structure, modulo an assignment of values to the free variables. In contrast, a formula is supposed to assert something that is either true or false, again modulo such an assignment. Our goal is therefore to specify a value $[\![A]\!]^{\mathfrak{M},\sigma}$ that is equal to either $\top$ or $\bot$. Adapting the semantic notation introduced in Chapter 3, will write $\mathfrak{M} \models_\sigma A$ for the statement $[\![A]\!]^{\mathfrak{M},\sigma} = \top$. Read this as saying that *A is true in* $\mathfrak{M}$ under assignment σ, that *A holds in* $\mathfrak{M}$ under σ, or that $\mathfrak{M}$ and σ *satisfy* A. If A has free variables $x_0, \ldots, x_{n-1}$, we may write, suggestively, $\mathfrak{M} \models A(a_0, \ldots, a_{n-1})$ to say that A is true in $\mathfrak{M}$ under any assignment that maps $x_0, \ldots, x_{n-1}$ to $a_0, \ldots, a_{n-1}$.

Fixing a model $\mathfrak{M}$, the value $[\![A]\!]^{\mathfrak{M},\sigma}$ is defined by recursion on formulas A. To handle the quantifiers, it helps to adopt the following notation: if σ is an assignment of elements of $|\mathfrak{M}|$ to variables, x is a variable, and a is an element of $|\mathfrak{M}|$, then $\sigma[x \mapsto a]$ denotes the assignment that maps x to a and otherwise agrees with σ.

- $\mathfrak{M} \models_\sigma t = t'$ if and only if $[\![t]\!]^{\mathfrak{M},\sigma} = [\![t']\!]^{\mathfrak{M},\sigma}$.
- $\mathfrak{M} \models_\sigma R(t_0, \dots, t_{n-1})$ if and only if $R^{\mathfrak{M}}([\![t_0]\!]^{\mathfrak{M},\sigma}, \dots, [\![t_{n-1}]\!]^{\mathfrak{M},\sigma})$.
- $\mathfrak{M} \models_\sigma \bot$ is always false.
- $\mathfrak{M} \models_\sigma A \wedge B$ if and only if $\mathfrak{M} \models_\sigma A$ and $\mathfrak{M} \models_\sigma B$.
- $\mathfrak{M} \models_\sigma A \vee B$ if and only if $\mathfrak{M} \models_\sigma A$ or $\mathfrak{M} \models_\sigma B$.
- $\mathfrak{M} \models_\sigma A \to B$ if and only if $\mathfrak{M} \not\models_\sigma A$ or $\mathfrak{M} \models_\sigma B$.
- $\mathfrak{M} \models_\sigma \forall x\, A$ if and only if for every $a \in |\mathfrak{M}|$, $\mathfrak{M} \models_{\sigma[x \mapsto a]} A$.
- $\mathfrak{M} \models_\sigma \exists x\, A$ if and only if for some $a \in |\mathfrak{M}|$, $\mathfrak{M} \models_{\sigma[x \mapsto a]} A$.

In words, $t = t'$ holds in $\mathfrak{M}$ under σ if and only if t and t' denote the same element of the universe of $\mathfrak{M}$ under σ. Similarly, $R(t_0, \dots, t_{n-1})$ holds in $\mathfrak{M}$ under σ if and only if the interpretation of R holds of the elements denoted by $t_0, \dots, t_{n-1}$ under σ. The clauses describing the evaluation of formulas under the propositional connectives are carried over from the propositional setting. A formula $\forall x\, A$ holds in $\mathfrak{M}$ under σ if and only if A holds no matter what element of the universe of $\mathfrak{M}$ is assigned to x, and a formula $\exists x\, A$ holds if and only if there is some assignment to x for which A holds in $\mathfrak{M}$. According to the dictates of Section 1.6, we need to specify a choice of variable with which to open the bound quantifiers in the recursion scheme, but we can show after the fact that the result does not depend on the particular choice. We can think of a formula $A(x_0, \dots, x_{n-1})$ with the free variables shown as defining an n-ary relation $[\![A]\!]^{\mathfrak{M}}$ on the universe of $|\mathfrak{M}|$, whereby $[\![A]\!]^{\mathfrak{M}}(a_0, \dots, a_{n-1})$ means $\mathfrak{M} \models A(a_0, \dots, a_{n-1})$.

There is an alternative, equivalent approach to defining satisfiability of formulas in a model that avoids the need to carry around the assignment σ. If L is a language and $\mathfrak{M}$ is a structure for L, let $L(\mathfrak{M})$ denote the language obtained by adding a new constant $\bar{a}$ for each element $a \in \mathfrak{M}$. We now restrict the evaluation to closed terms and sentences in the expanded language, and the clauses for the universal and existential quantifiers now read as follows:

- $\mathfrak{M} \models \forall x\, A(x)$ if and only if for every $a \in |\mathfrak{M}|$, $\mathfrak{M} \models A(\bar{a})$.
- $\mathfrak{M} \models \exists x\, A(x)$ if and only if for some $a \in |\mathfrak{M}|$, $\mathfrak{M} \models A(\bar{a})$.

Exercise 5.1.1 asks you to show that for any formula $A(x_0, \dots, x_{n-1})$ with the free variables shown, $\mathfrak{M} \models A(\bar{a}_0, \dots, \bar{a}_{n-1})$ in the new sense if and only if $\mathfrak{M} \models_\sigma A$ in the old sense for any σ such that $\sigma(x_i) = a_i$ for each $i < n$.

We will later have occasion to speak of a formula $A(\vec{x}, \vec{a})$ with *parameters* from a model $\mathfrak{M}$. This can be understood as a formula of $L(\mathfrak{M})$ in which each new constant $\bar{a}$ is interpreted as the corresponding element $a \in |\mathfrak{M}|$, or, equivalently, a formula $A(\vec{x}, \vec{y})$ with a distinguished tuple of variables $\vec{y}$ and a specification $\vec{a}$ of the elements of $|\mathfrak{M}|$ that they are assigned to them.

Now that we have defined the notion of satisfiability, the remaining semantic notions carry over from Chapter 3.

Definition 5.1.1. Let L be a language, let A be a sentence of L, let Γ be a set of sentences of L, and let $\mathfrak{M}$ be an L-structure.

- $\mathfrak{M} \models \Gamma$ means that $\mathfrak{M} \models B$ for every $B \in \Gamma$. In this case, we say that $\mathfrak{M}$ *satisfies* Γ, or that $\mathfrak{M}$ is a *model* of Γ.
- A is *satisfiable* if it is true in some model, and similarly for Γ.
- A is *valid*, written $\models A$, if it is true in every model.
- A is a *logical consequence* of Γ, written $\Gamma \models A$, if every model of Γ is also a model of A. In that case, we also say that A is a *semantic consequence* of Γ, that Γ *logically entails* A, or, more simply, that Γ *entails* A.

As shorthand, one often describes a structure by simply listing its components, starting with the universe. For example, $(\mathbb{N}, 0, 1, +, \cdot, <)$ is a structure for a language with two constant symbols, two binary function symbols, and one binary relation symbol, with the corresponding interpretations.

We will need to know that our semantic notions play well with substitution, in the following sense.

Proposition 5.1.2. *Let s and t be terms of a language L, let A be a formula of L, let x be a variable, and let $\mathfrak{M}$ be an L-structure. Let σ be an assignment of elements of $|\mathfrak{M}|$ to variables and let $a = [\![s]\!]^{\mathfrak{M},\sigma}$.*

1. $[\![t[s/x]]\!]^{\mathfrak{M},\sigma} = [\![t]\!]^{\mathfrak{M},\sigma[x \mapsto a]}$
2. $\mathfrak{M} \models_\sigma A[s/x]$ *if and only if* $\mathfrak{M} \models_{\sigma[x \mapsto a]} A$.

In words, the first claim says that substituting s for x in a term t and then evaluating it in $\mathfrak{M}$ has the same effect as evaluating t with x assigned to result of evaluating s. The second claim is the analogous statement for a formula A. The two claims can be proved by induction on terms and formulas, respectively.

We now turn to the task of proving the soundness and completeness of classical first-order logic with respect to the semantics just described. For the rest of this section, let us assume we have fixed a language, L.

Theorem 5.1.3 (soundness). *For every set of sentences Γ and every sentence A, if $\Gamma \vdash A$, then $\Gamma \models A$.*

Proof By induction on natural deduction derivations, we can show a more general claim: if Γ is a finite set of formulas and A is any formula, then for every assignment σ, if $\Gamma \vdash A$ is derivable and $\mathfrak{M} \models_\sigma \Gamma$, then $\mathfrak{M} \models_\sigma A$. The propositional rules are handled just as for propositional logic. For the $\forall$ introduction rule, suppose we derive $\Gamma \vdash \forall x\, A$ from $\Gamma \vdash A$, and suppose $\mathfrak{M} \models_\sigma \Gamma$. To show $\mathfrak{M} \models_\sigma \forall x\, A$, let a be any element of $|\mathfrak{M}|$; we need to show $\mathfrak{M} \models_{\sigma[x \mapsto a]} A$. By the eigenvariable condition, x does not occur free in Γ, and since s and $\sigma[x \mapsto a]$ agree on all the other variables, $\mathfrak{M} \models_{\sigma[x \mapsto a]} \Gamma$. By the inductive hypothesis, we have $\mathfrak{M} \models_{\sigma[x \mapsto a]} A$.

For the $\forall$ elimination rule, suppose we derive $\Gamma \vdash A[t/x]$ from $\Gamma \vdash \forall x\, A$, and suppose $\mathfrak{M} \models_\sigma \Gamma$. By the inductive hypothesis, we have $\mathfrak{M} \models_\sigma \forall x\, A$. From the definition of the satisfaction relation, we have $\mathfrak{M} \models_{\sigma[x \mapsto [\![t]\!]^{\mathfrak{M},\sigma}]} A$. By Proposition 5.1.2, this is equivalent to $\mathfrak{M} \models_\sigma A[t/x]$, as required.

The introduction and elimination rules for the existential quantifier, as well as the rules for equality, are left as an exercise. □

Exercise 5.1.3 suggests an alternative approach to proving Theorem 5.1.3 that avoids the appeal to Proposition 5.1.2 by adopting a more general inductive claim: if $\Gamma \vdash A$, then for any tuple of variables $\vec{x}$, any tuple of terms $\vec{t}$, any model $\mathfrak{M}$, and any assignment σ, if $\mathfrak{M} \models_\sigma \Gamma[\vec{t}/\vec{a}]$, then $\mathfrak{M} \models_\sigma A[\vec{t}/\vec{a}]$.

Theorem 5.1.4 (completeness). *For every set of sentences Γ and every sentence A, if $\Gamma \models A$, then $\Gamma \vdash A$.*

First we will prove completeness for first-order logic without equality, and then we will extend the proof to handle equality as well. As for propositional logic, we will say that a set of sentences Γ is *consistent* if $\Gamma \nvdash \bot$. And, as was the case with propositional logic, completeness is equivalent to the following:

Theorem 5.1.5. *For every set of sentences Γ, if Γ is consistent, then it has a model.*

In the case of propositional logic, we were required to come up with a truth assignment satisfying Γ. Now we need a model, that is, a universe and functions and relations defined on that universe. Since all we have to work with are the sentences in Γ, the only option available is to build the universe out of syntax. For example, we might take the universe to be the set of closed terms in the language and interpret each function symbol as the corresponding term-forming operation. One complication is that if there are no constant symbols in L, there won't be any terms. So our first step is to expand Γ to a set of sentences in a larger language, L', to ensure that there are enough constants.

Definition 5.1.6. Let Γ be a set of sentences in a language L. Γ is a *Henkin set* if the following holds: for every sentence $\exists x\, A(x)$ in L, there is a constant c such that $\Gamma \vdash \exists x\, A(x) \to A(c)$.

A constant c associated with the sentence $\exists x\, A(x)$ in this way is called a *Henkin constant.*

Lemma 5.1.7. *Let Γ be a set of sentences in a language L. Then there is a language L' extending L and a set of sentences $\Gamma' \supseteq \Gamma$ of L' such that*

- *Γ' is a Henkin set, and*
- *Γ' is a conservative extension of Γ.*

Recall from Definition 4.7.2 that saying that Γ' is a conservative extension of Γ means that if $\Gamma' \vdash B$ and B is a formula of L, then $\Gamma \vdash B$. To prove Lemma 5.1.7, we need the following lemma, which shows that we can conservatively add a single Henkin constant.

Lemma 5.1.8. *Let Γ be a set of sentences in a language L, let $\exists x\, A(x)$ be any sentence, and let c be a new constant. Suppose $\Gamma \cup \{\exists x\, A(x) \to A(c)\} \vdash B$, where B is in L. Then $\Gamma \vdash B$.*

Proof Suppose $\Gamma \cup \{\exists x\, A(x) \to A(c)\} \vdash B$. By the deduction theorem,

$$\Gamma \vdash (\exists x\, A(x) \to A(c)) \to B.$$

By Proposition 4.2.1, we can replace c by a fresh variable, so we have

$$\Gamma \vdash \forall y\, ((\exists x\, A(x) \to A(y)) \to B).$$

This last sentence is equivalent to $\exists y\, (\exists x\, A(x) \to A(y)) \to B$, and the antecedent is provable in classical logic, so $\Gamma \vdash B$. □

Proof of Lemma 5.1.7 Let us focus on the case where L is a countable language. The argument can be adapted to uncountable languages using Zorn's lemma. (See Section A.3.)

Let L' be L together with countably many new constants. Enumerate all sentences C_0, C_1, $C_2, \ldots$ of L' that are of the form $\exists x\, A(x)$. Inductively define a sequence of formulas D_0, D_1, $D_2, \ldots$ so that each D_i is a Henkin axiom for C_i using a constant that does not occur in $D_0, \ldots, D_{i-1}$. Let $\Gamma' = \Gamma \cup \{D_0, D_1, D_2, \ldots\}$. Then every sentence $\exists x\, A(x)$ in L' has a Henkin axiom in Γ'. By Lemma 5.1.8, for each i, $\Gamma \cup \{D_0, \ldots, D_{i+1}\}$ is conservative over $\Gamma \cup \{D_0, \ldots, D_i\}$, and hence, by induction, conservative over Γ. Since anything provable from Γ' is provable from $\Gamma \cup \{D_0, \ldots, D_i\}$ for some i, this means that Γ' is conservative over Γ. □

The exercises below provide an alternative proof of this lemma. We can now prove Theorem 5.1.4, which asserts the completeness of our deductive systems for classical first-order logic, in the case where we do not include equality as a special relation symbol.

Proof of Theorem 5.1.4 for first-order logic without equality Let Γ be a consistent set of sentences in L. By Lemma 5.1.7, let Γ' be a Henkin set in a larger language, L', such that Γ' is conservative over Γ. In particular, this means that Γ' is consistent, since conservativity implies that if Γ' proves $\bot$ then so does Γ. Any model $\mathfrak{M}$ of Γ' becomes a model of Γ is we ignore the interpretations of the extra constants, so it suffices to show that Γ' has a model.

As in the proof of Lemma 3.1.13, extend Γ' to a maximally consistent set Δ in the same language. Notice that Δ is still a Henkin set because the language has not changed. I claim that Δ has all the following properties:

1. Δ is deductively closed: if $\Delta \vdash A$, then $A \in \Delta$.
2. $\bot \notin \Delta$.
3. $A \wedge B \in \Delta$ if and only if $A \in \Delta$ and $B \in \Delta$.
4. $A \to B \in \Delta$ if and only if $A \notin \Delta$ or $B \in \Delta$.
5. $A \vee B \in \Delta$ if and only if $A \in \Delta$ or $B \in \Delta$.
6. $\exists x\, A(x)$ is in Δ if and only if for some closed term t, $A(t) \in \Delta$.
7. $\forall x\, A(x)$ is in Δ if and only if for every closed term t, $A(t) \in \Delta$.

Claims 1–5 are handled just as in the propositional case. For claim 6, suppose $\exists x\, A(x)$ is in Δ. Since Δ is a Henkin set, the sentence $\exists x\, A(x) \to A(c)$ is in Δ for some constant c. So $\Delta \vdash A(c)$, and hence $A(c) \in \Delta$. Conversely, suppose for some term t, $A(t) \in \Delta$. Then $\Delta \vdash \exists x\, A(x)$, and hence $\exists x\, A(x) \in \Delta$.

The proof of claim 7 is similar. The forward direction is immediate, and since Δ is a Henkin set, for some constant c, $\exists x\, \neg A(x) \to \neg A(c)$ is in Δ. If $A(t)$ is in Δ for every closed term t, then $A(c)$ is in Δ. Together these imply that $\neg \exists x\, \neg A(x)$, which is equivalent to $\forall x\, A(x)$, is in Δ.

Just as a maximally consistent set of propositional formulas enabled us to read off a satisfying truth assignment in Section 3.1, a maximally consistent Henkin set Δ enables us to read off a satisfying model, $\mathfrak{M}$, as follows:

- $|\mathfrak{M}| = \{t \mid t \text{ is a closed term of } L'\}$.
- For each k-ary function symbol f, if $t_0, \ldots, t_{k-1}$ are elements of $|\mathfrak{M}|$, we set $f^{\mathfrak{M}}(t_0, \ldots, t_{k-1})$ equal to the term $f(t_0, \ldots, t_{k-1})$.
- For each k-ary relation symbol R, if $t_0, \ldots, t_{k-1}$ are elements of $|\mathfrak{M}|$, we specify that $R^{\mathfrak{M}}(t_0, \ldots, t_{k-1})$ holds if and only if $R(t_0, \ldots, t_{k-1})$ is in Δ.

The construction is entirely syntactic: the universe of the model consists of terms and the interpretation $f^{\mathfrak{M}}$ of each function symbol f is an operation on terms.

By induction on terms, we have that if $t(x_0, \ldots, x_{k-1})$ is any term in L' with the variables shown and σ is any assignment, then

$$[\![t]\!]^{\mathfrak{M},\sigma} = t(\sigma(x_0), \ldots, \sigma(x_{k-1})).$$

Similarly, using the properties of Δ enumerated above, we have that if $A(x_0, \ldots, x_{k-1})$ is any formula of L' and σ is any assignment,

$$\mathfrak{M} \models_\sigma A(\vec{x}) \quad \Leftrightarrow \quad A(\sigma(x_0), \ldots, \sigma(x_{k-1})) \in \Delta.$$

In particular, since Δ includes Γ, if we restrict $\mathfrak{M}$ to interpret the symbols of L, we have a model of Γ. □

Let $\mathfrak{M}$ be any L-structure and let $L(\mathfrak{M})$ be the expanded language with a new constant, $\bar{a}$, for every element a of $\mathfrak{M}$. If A is a sentence in this expanded language, we can abuse notation slightly and write $\mathfrak{M} \models A$ when the expansion of $\mathfrak{M}$ to L' satisfies A. It is not hard to check that the set $\{A \mid \mathfrak{M} \models A\}$ is a maximally consistent Henkin set. Our proof of the completeness theorem shows that every maximally consistent Henkin set arises in this way.

An advantage to the approach we have followed is that it is modular: first, we add Henkin axioms, and then we extend to a maximally consistent set. An alternative approach is to add build a maximally consistent set and add suitable constants at the same time. In the countable case, let L', as above, be a language with countably many new constant symbols. Let $B_0, B_1, B_2, \ldots$ enumerate all sentences in the language, and let $\exists x\, C_0(x), \exists x\, C_1(x), \ldots$ enumerate all sentences of the form $\exists x\, C(x)$. Then modify the construction in the proof of Lemma 3.1.13 to define a sequence $\Gamma = \Gamma_0, \Gamma_1, \Gamma_2, \ldots$ so that at stage $i+1$, we

- add B_i to Γ_i, assuming the result is consistent, and
- if the resulting set proves $\exists x\, C_i(x)$, add $C_i(c)$, where c does not occur in any B_j or C_j for $j < i$.

The exercises below ask you to show that the set $\Gamma_\omega = \bigcup_i \Gamma_i$ satisfies the properties required in the previous proof.

All we have left to do is extend the proof of the completeness theorem to first-order logic with equality. Suppose a set of sentences Γ proves a formula A in that setting. Let Δ be the set of axioms for equality described in Section 4.3. Then, treating the equality symbol as an ordinary binary relation symbol, $\Gamma \cup \Delta$ proves A in first-order logic *without* equality. In particular, if Γ is consistent with respect to first-order logic with equality, then $\Gamma \cup \Delta$ is consistent with respect to first-order logic without equality. We have already shown that $\Gamma \cup \Delta$ has a model, $\mathfrak{M}$, in which the equality symbol is interpreted as an ordinary binary relation symbol. Let $\sim$ denote the interpretation of equality in $\mathfrak{M}$. The fact that $\mathfrak{M}$ satisfies Δ means that $\sim$ is reflexive, symmetric, and transitive. It is also a congruence with respect to every function and relation symbol. This means that if f is an n-ary function symbol, $a_0, \ldots, a_{n-1}, b_0, \ldots, b_{n-1}$ are elements of U, and $a_i \sim b_i$ for every i, then $f^{\mathfrak{M}}(a_0, \ldots, a_{n-1}) \sim f^{\mathfrak{M}}(b_0, \ldots, b_{n-1})$, and similarly for relation symbols.

We could stop here: what we now know that if we allow equality to be interpreted by any such relation, we have a complete semantics for classical first-order logic with equality. But it would be nice to know that we can rely on the narrower semantics where the equality symbol is interpreted by the equality relation itself on the elements of the universe. For that purpose, quotient structure $\mathfrak{M}/{\sim}$, defined in Section 1.1, has the properties we need. Recall that it is defined as follows:

- The universe of $\mathfrak{M}/{\sim}$ is the set of equivalence classes of elements of $|\mathfrak{M}|$, i.e. the set $\{[a] \mid a \in |\mathfrak{M}|\}$, where $[a]$ denotes the equivalence class $\{b \in |\mathfrak{M}| \mid b \sim a\}$.
- For each function symbol f of L, the interpretation $f^{\mathfrak{M}/\sim}$ is defined by

$$f^{\mathfrak{M}/\sim}([a_0], \ldots, [a_{n-1}]) = [f^{\mathfrak{M}}(a_0, \ldots, a_{n-1})].$$

- For each relation symbol R of L other than equality, the interpretation $R^{\mathfrak{M}/\sim}$ is defined by

$$R^{\mathfrak{M}/\sim}([a_0], \ldots, [a_{n-1}]) \quad \Leftrightarrow \quad R^{\mathfrak{M}}(a_0, \ldots, a_{n-1}).$$

Equality in L is interpreted as equality on the universe of $\mathfrak{M}/{\sim}$. The fact that $\sim$ is an equivalence relation is needed to ensure that this model is well defined, which is to say, the definitions of the functions and relations on $\mathfrak{M}/{\sim}$ do not depend on the choice of representatives of the equivalence classes. We can show inductively that if $t(x_0, \ldots, x_{n-1})$ is any term in the language of L with the variables shown and $a_0, \ldots, a_{n-1}$ are elements of $|\mathfrak{M}|$, then

$$[\![t([a_0], \ldots, [a_{n-1}])]\!]^{\mathfrak{M}/\sim} = [[\![t(a_0, \ldots, a_{n-1})]\!]^{\mathfrak{M}}]. \tag{5.1}$$

Furthermore, if $A(x_0, \ldots, x_{n-1})$ is any formula of L with the free variables shown and $a_0, \ldots, a_{n-1}$ are elements of $|\mathfrak{M}|$, we have

$$\mathfrak{M}/{\sim} \models A([a_0], \ldots, [a_{n-1}]) \quad \Leftrightarrow \quad \mathfrak{M} \models A(a_0, \ldots, a_{n-1}).$$

Remember that, on the left, we are interpreting the equality symbol by equality in the model, whereas on the right, it is interpreted by the relation $\sim$. All the work in proving this equivalence is in the base case, for atomic formulas of the form $R(t_0, \ldots, t_{n-1})$ and $t_1 = t_2$. In these cases, the equivalence follows straightforwardly from the definition of $\mathfrak{M}$ and equation (5.1).

In particular, $\mathfrak{M}$ and $\mathfrak{M}/{\sim}$ agree on sentences of L, and so $\mathfrak{M}/{\sim}$ is the desired model of Γ, in which the equality symbol is interpreted as equality.

Exercises

5.1.1. Show that the two definitions of $\mathfrak{M} \models_\sigma A$ given above – the first by direct recursion, the second by defining satisfiability first for closed formulas in the language $L(\mathfrak{M})$ – coincide.

5.1.2. Carry out the cases for the $\exists$ introduction and elimination rules, and the rules for equality, in the soundness theorem for first-order logic, Theorem 5.1.3.

5.1.3. Spell out the details of the alternative approach to proving soundness that was suggested right after the proof of Theorem 5.1.3.

5.1.4. Consider the following alternative proof of Lemma 5.1.7. Given a set of sentences Γ in L, set $L_0 = L$ and $\Gamma_0 = \Gamma$. Inductively, for each natural number i define L_{i+1} to be L_i with new Henkin constants for each formula $\exists x\, A(x)$ in L_i, and let Γ_{i+1} be Γ_i together with the new

Henkin axioms. Let L_ω be the union of all the languages and let $\Gamma_\omega = \bigcup_i \Gamma_i$. Show that Γ_ω has the desired properties.

5.1.5. Show that if $\mathfrak{M}$ is any L-structure then the set $\{A \in L(\mathfrak{M}) \mid \mathfrak{M} \models A\}$ is a maximally consistent Henkin set.

5.2 Equational and Quantifier-Free Logic

We can get more mileage out of the quotient construction. We will now use it to fulfill a promise made in Section 4.3, namely, to show that equational logic is complete with respect to equational entailments. Fix a language L without relation symbols and assume that L contains at least one constant. The following completes the proof of Theorem 4.4.1.

Proposition 5.2.1. *Let Γ be a set of universally quantified equations $\forall \vec{x}\,(u = v)$. If s and t are closed terms and $\Gamma \models s = t$, then $s = t$ is provable from substitution instances of formulas in Γ in the equational fragment of first-order logic.*

Proof Let L be the language of Γ and let $\mathfrak{M}$ be the *term model* for L, that is, the model whose universe is the set of closed terms in L, interpreting each n-ary function symbol f by the function which maps terms $r_0, \ldots, r_{n-1}$ to the term $f(r_0, \ldots, r_{n-1})$. For any two closed terms q and r, define $q \sim r$ to mean that $q = r$ is provable from substitution instances of Γ in the equational fragment of first-order logic. It follows directly from the rules for equality that $\sim$ is an equivalence relation and that it respects the function symbols in the language, in the sense that if f is an n-ary function symbol and $q_i \sim r_i$ for each $i < n$, then $f(q_0, \ldots, q_{n-1}) \sim f(r_0, \ldots, r_{n-1})$.

I claim that $\mathfrak{M}/\!\sim$ is a model of Γ. The construction of $\mathfrak{M}/\!\sim$ was designed to make this true, but let us spell out the details. Suppose $\forall x_0, \ldots, x_{n-1}\,(u(\vec{x}) = v(\vec{x}))$ is in Γ and let $[r_0], \ldots, [r_{n-1}]$ be any tuple of elements of the universe of $\mathfrak{M}/\!\sim$. Then under the assignment which maps each x_i to r_i, $u(\vec{x})$ denotes $[u(\vec{r})]$ and $v(\vec{x})$ denotes $[v(\vec{r})]$. Since the equation $u(\vec{r}) = v(\vec{r})$ is a substitution instance of a formula in Γ, we have $u(\vec{r}) \sim v(\vec{r})$. This implies $[u(\vec{r})] = [v(\vec{r})]$, as required.

Since $\Gamma \models s = t$, we have that $s = t$ is true in $\mathfrak{M}/\!\sim$. But in that model, s denotes $[s]$ and t denotes $[t]$, so we have $[s] = [t]$. This implies $s \sim t$, and, by definition, this means that $s = t$ is provable from substitution instances of formulas in Γ in the equational fragment of first-order logic. □

The restriction to closed terms s and t in the previous proposition is only for notational convenience, since, by Proposition 4.2.1, we can replace variables by fresh constants. We can also fulfill another promise from Section 4.4 and show that classical first-order logic is conservative over the deductive system for quantifier-free classical first-order logic.

Proposition 5.2.2. *Let Γ be a set of universal sentences and let A be a quantifier-free formula. If $\Gamma \models A$, then A is provable from substitution instances of Γ in the quantifier-free fragment of first-order logic.*

Proof Once again, replacing variables in A by fresh constants, it does not hurt to assume that A is a sentence. Suppose there is no proof of A from substitution instances of Γ in quantifier-free first-order logic. Let Δ be the set of such instances together with the formula $\neg A$ and all instances of the axioms for equality. Treating every atomic sentence in the language as a propositional formula, we have that Δ is propositionally consistent. By the completeness theorem for propositional logic, there is a truth assignment v that satisfies all the atomic formulas in Δ.

Define the relation $s \sim t$ on closed formulas to hold if $v \models s = t$. Because Δ includes all instances of the axioms for equality, we have that $\sim$ is an equivalence relation and a congruence with respect to all the function and relation symbols. Define a model $\mathfrak{M}$ whose universe consists of the equivalence classes of closed terms under $\sim$ by setting $f^{\mathfrak{M}}([t_0], \ldots, [t_{n-1}]) = [f(t_0, \ldots, t_{n-1})]$ as before, and declaring that $R^{\mathfrak{M}}([t_0], \ldots, [t_{n-1}])$ holds if and only if $v \models R(t_0, \ldots, t_{n-1})$. By induction on quantifier-free formulas, we have that for every quantifier-free formula $A(\vec{x})$ and any tuple of terms $t_0, \ldots, t_{n-1}$, $\mathfrak{M} \models A([t_0], \ldots, [t_{n-1}])$ if and only if $v \models A(t_0, \ldots, t_{n-1})$. In particular, every quantifier-free substitution instance of a formula in Γ is true in $\mathfrak{M}$, and hence every sentence in Γ is true in $\mathfrak{M}$. But also, since $\neg A$ is in Δ, A is false in $\mathfrak{M}$. So there is no first-order proof of A from Γ. □

Exercise 5.2.2 asks you to modify this construction to show that if a quantifier-free formula A is provable in classical first-order logic, then in fact there is a propositional proof of A from instances of equality axioms in which all the terms that appear are subterms of terms that occur in A. Since there are only finitely many such axioms, this provides a decision procedure for the quantifier-free fragment of first-order logic, and hence the universal fragment as well. (See Theorem 4.4.4.)

Exercises

5.2.1. Show that the relation $\sim$ in the statement of Theorem 5.2.1 is the smallest equivalence relation such that $u \sim v$ holds whenever $u = v$ is in Γ'.

5.2.2. Use a model-theoretic argument to prove the last claim in this section. (Hint: assuming there is a truth assignment that makes all such instances of equality axioms true but A false, construct a countermodel whose universe consists of equivalence classes of closed terms plus one additional default value.)

5.3 Intuitionistic Logic

In Chapter 3, we saw that Kripke models provide a natural generalization of classical truth-table semantics to intuitionistic and minimal logic. In this section, we will see that classical model-theoretic semantics carries over in a similar way. Once again, we will start with first-order logic without equality. A Kripke structure $\mathfrak{K}$ for a language L consists of a partial order $(K, \leq)$ together with an assignment of a structure, $\mathfrak{K}_\alpha$, to each node $\alpha \in K$, in such a way that the following conditions hold for every α and every $\beta \leq \alpha$:

- $|\mathfrak{K}_\beta| \supseteq |\mathfrak{K}_\alpha|$.
- If $\vec{a} \in |\mathfrak{K}_\alpha|$, $f^{\mathfrak{K}_\alpha}(\vec{a}) = f^{\mathfrak{K}_\beta}(\vec{a})$.
- If $\vec{a} \in |\mathfrak{K}_\alpha|$ and $R^{\mathfrak{K}_\alpha}(\vec{a})$, then $R^{\mathfrak{K}_\beta}(\vec{a})$.

As with Kripke semantics for propositional logic, think of $\beta \leq \alpha$ as asserting that β is stronger than α, in the sense that it provides more information or corresponds to a later point in time. The monotonicity clauses then assert that when we pass to a stronger condition, more elements of the domain become visible, and more atomic facts are seen to be true.

The second clause guarantees that the interpretation of the function symbols on old elements remains stable as new elements appear.

Let $\mathfrak{K}$ be a Kripke structure for L. As for classical logic, given any term of L and any assignment σ of elements of $|\mathfrak{K}_\alpha|$ to variables, each term t of L has a denotation $[\![t]\!]^{\alpha,\sigma}$. The forcing relation $\alpha \Vdash_\sigma A$ is now defined by recursion on formulas A, where in each case σ is assumed to be an assignment of variables to elements of $|\mathfrak{K}_\alpha|$. In the base case, where A is atomic, we use the classical notion of truth in $\mathfrak{K}_\alpha$, and the clauses for the propositional connectives carry over from Section 3.3. Only the clauses for the quantifiers are new.

- $\alpha \Vdash_\sigma A$ if and only if $\mathfrak{K}_\alpha \models_\sigma A$, when A is an atomic formula.
- $\alpha \Vdash_\sigma A \wedge B$ if and only if $\alpha \Vdash A$ and $\alpha \Vdash B$.
- $\alpha \Vdash_\sigma A \vee B$ if and only if $\alpha \Vdash A$ or $\alpha \Vdash B$.
- $\alpha \Vdash_\sigma A \to B$ if and only if $\forall \beta \leq \alpha\ (\beta \Vdash A \to \beta \Vdash B)$.
- $\alpha \Vdash_\sigma \exists x\, A(x)$ if and only if $\exists a \in |\mathfrak{K}_\alpha|\ (\alpha \Vdash_{\sigma[x \mapsto a]} A)$
- $\alpha \Vdash_\sigma \forall x\, A(x)$ if and only if $\forall \beta \leq \alpha\ \forall a \in |\mathfrak{K}_\beta|\ (\beta \Vdash_{\sigma[x \mapsto a]} A)$

In words, α forces $\exists x\, A(x)$ if there is an element a in the universe at α such that α forces $A(a)$. The clause for $\forall x\, A(x)$ is more interesting: to say that α forces $\forall x\, A(x)$, we need to know not only that A is forced to hold of every element of the current domain, but, moreover, if any elements appear at a later node, β, A is forced to hold of those elements too. As with propositional logic, to extend the interpretation from minimal to intuitionistic logic, we add:

- $\alpha \nVdash_\sigma \bot$

Once again, a straightforward induction on formulas yields monotonicity:

Lemma 5.3.1. *For every α, β, and σ, if $\alpha \Vdash_\sigma A$ and $\beta \leq \alpha$, then $\beta \Vdash_\sigma A$.*

Proof The most interesting case involves the universal quantifier. Indeed, in this case the definition of forcing is designed just for this purpose. Suppose $\alpha \Vdash_\sigma \forall x\, A$ and $\beta \leq \alpha$. By the clause for forcing for the universal quantifier, we know that for every $\gamma \leq \alpha$ and every $a \in |\mathfrak{K}_\alpha|$, $\gamma \Vdash_{\sigma[x \mapsto a]} A$. Since any γ less than or equal to β is also less than or equal to α, the previous statement holds with β in place of α.

The case of the existential quantifier is handled by the fact that for every $\beta \leq \alpha$, we have $|\mathfrak{K}_\beta| \supseteq |\mathfrak{K}_\alpha|$. □

It is not hard to verify that the analogue of Proposition 5.1.2, the substitution lemma, holds for Kripke models as well. In other words, if $a = [\![t]\!]^{\mathfrak{K}_\alpha,\sigma}$, then $\alpha \Vdash_\sigma A[t/x]$ if and only if $\alpha \Vdash_{\sigma[x \mapsto a]} A$. In particular, if x does not occur in A and $\alpha \Vdash_\sigma A$, then $\alpha \Vdash_{\sigma[x \mapsto a]} A$ for every a.

We now establish the soundness of first-order minimal and intuitionistic logic with respect the Kripke semantics, still restricting our attention to first-order logic without equality. If $\mathfrak{K}$ is a Kripke model, α is any node, σ is any assignment of elements of $|\mathfrak{K}_\alpha|$ to variables, and Γ is a set of sentences, say $\alpha \Vdash_\sigma \Gamma$ if and only if $\alpha \Vdash_\sigma B$ for every B in Γ. If $\Gamma \cup \{A\}$ is a set of sentences, say $\Gamma \Vdash A$ if, for every Kripke model $\mathfrak{K}$ and node α such that $\alpha \Vdash \Gamma$, we have $\alpha \Vdash A$.

Theorem 5.3.2. *For every set of sentences Γ and every sentence A, if $\Gamma \vdash A$, then $\Gamma \Vdash A$.*

Proof We fix a Kripke model $\mathfrak{K}$ and show the following by induction on derivations: for every derivable sequent $\Gamma \vdash A$, every node α, and every assignment σ, if $\alpha \Vdash_\sigma \Gamma$, then $\alpha \Vdash_\sigma A$. The proof is similar to the proof of soundness for propositional logic, Theorem 3.3.7. To handle the introduction rule for the universal quantifier, suppose the last line of the derivation obtains $\Gamma \vdash \forall x A$ from $\Gamma \vdash A$, and suppose $\alpha \Vdash_\sigma \Gamma$. We need to show $\alpha \Vdash_\sigma \forall x A$, so suppose $\beta \leq \alpha$ and $a \in |\mathfrak{K}_\beta|$. By monotonicity and the fact that x does not occur in Γ, we have $\beta \Vdash_{\sigma[x \mapsto a]} \Gamma$. By the inductive hypothesis, this yields $\beta \Vdash_{\sigma[x \mapsto a]} A$, as required.

For the $\forall$ elimination rule, suppose the last line of the derivation obtains $\Gamma \vdash A[t/x]$ from $\Gamma \vdash \forall x A$ and suppose $\alpha \Vdash_\sigma \Gamma$. By the inductive hypothesis, we have $\alpha \Vdash_\sigma \forall x A$. Let $a = [\![t]\!]^{\mathfrak{K}_\alpha,\sigma}$. Since $\alpha \leq \alpha$, we have $\alpha \Vdash_{\sigma[x \mapsto a]} A$, and hence $\alpha \Vdash_\sigma A[t/x]$.

The rules for the existential quantifier are left to you. □

The proof of the completeness theorem for intuitionistic and minimal first-order logic without equality is similar to the proof of completeness for intuitionistic and minimal propositional logic. Given a set of sentences Γ and a sentence A, if Γ does not prove A, we want to build a Kripke model with a node that forces Γ but not A. Section 5.1 used classical logic to extend Γ to a Henkin set, enabling us to build a syntactic model. We do not have that option here, so we need use another strategy to add constants to the language. To that end, we generalize the notion of a *saturated* set from Definition 3.3.5.

Definition 5.3.3. A set of first-order sentences Γ is *saturated* if the following hold:

- Γ is deductively closed.
- If $A \vee B \in \Gamma$, then $A \in \Gamma$ or $B \in \Gamma$.
- If $\exists x A(x) \in \Gamma$, then $A(c) \in \Gamma$ for some c.

When dealing with intuitionistic logic instead of minimal logic, we also require that Γ is consistent.

In analogy to Lemma 3.3.6 for propositional logic, we have:

Lemma 5.3.4. *Let Γ be a set of sentences in a language L and A be a sentence in L, and suppose $\Gamma \nvdash A$. Then there is a saturated set $\Gamma' \supseteq \Gamma$ in a larger language, L', such that $\Gamma' \nvdash A$.*

Proof As usual, we will focus on the case where L is countable. Let L' be a language with countably many new constants $c_0, c_1, c_2, \ldots$. The proof is similar to the proof of Lemma 3.3.6 for the propositional case, except we now divide the construction into even and odd steps. Start with Γ_0 equal to the deductive closure of Γ. In step $2i$, we consider a formula $C_i \vee D_i$ in Γ_{2i}, and add one or the other to Γ_{2i+1}, as in the proof of Lemma 3.3.6. At stage $2i+1$ we take the ith formula $\exists x\, C_i(x)$ in an enumeration of L'. If it is in Γ_{2i+1}, we set Γ_{2i+2} to be the set of all deductive consequences of Γ_{2i+1} together with $C_i(c)$, for a constant c that has not been used yet. Finally, we set $\Gamma' = \bigcup_i \Gamma_i$. □

Theorem 5.3.5. *For every set of sentences Γ and every sentence A, if $\Gamma \Vdash A$ then $\Gamma \vdash A$.*

As in Section 3.3, we will construct a single Kripke model $\mathfrak{K}$ that is rich enough to serve as a counterexample for any entailment in L that is not provable. First, construct a sequence of languages $L = L_0, L_1, L_2, \ldots$ where each augments the previous one with countably many new constants. A node of $\mathfrak{K}$ is any consistent, saturated set of sentences in L_i, for some i. If α and β are nodes of $\mathfrak{K}$, then $\beta \leq \alpha$ if and only if $\beta \supseteq \alpha$.

If α is a saturated subset of L_i, the universe of $\mathfrak{K}_\alpha$ consists of the closed terms of L_i. We interpret the function and relation symbols of L as in Section 5.1 so that for any atomic sentence A of L_i, $\mathfrak{K}_\alpha \models A$ if and only if $A \in \Gamma$. The fact that each node α is contained in one of the languages L_i enables us to use Lemma 5.3.4 to find extensions of α that are saturated in L_{i+1}, which has countably many new constants. The completeness theorem for first-order logic without equality is therefore a consequence of the following.

Theorem 5.3.6. *In the Kripke model $\mathfrak{K}$ just described, given any node α in the language L_i, any formula $A(x_0, \ldots, x_{n-1})$ in L_i with the free variables shown, and any assignment σ of variables to closed terms, $\alpha \Vdash_\sigma A$ if and only if $A(\sigma(x_0), \ldots, \sigma(x_{n-1})) \in \alpha$.*

Proof The proof is by induction on A. In the base case where A is atomic, the claim follows immediately from the construction. The propositional connectives are handled just as in the proof of completeness for propositional logic, Theorem 3.3.7, so we only have to consider the cases for the quantifiers.

Suppose $\forall y\, A(y, \sigma(x_0), \ldots, \sigma(x_{n-1})) \in \alpha$, and suppose $\beta \leq \alpha$. We need to show that for every element $a \in |M_\beta|$, $\beta \Vdash_{\sigma[y \mapsto a]} A(y, x_0, \ldots, x_{n-1})$. By the inductive hypothesis, this happens if and only if $A(a, \sigma(x_0), \ldots, \sigma(x_{n-1})) \in \beta$, and the latter follows from $\beta \supseteq \alpha$ and the fact that α is closed under provability.

Conversely, suppose $\forall y\, A(y, \sigma(x_0), \ldots, \sigma(x_{n-1})) \notin \alpha$. Then, choosing a constant c in L_{i+1}, we have $\alpha \nvdash A(c, \sigma(x_0), \ldots, \sigma(x_{n-1}))$. By Lemma 5.3.4, there is a saturated set β in L_{i+2} such that $A(c, \sigma(x_0), \ldots, \sigma(x_{n-1})) \notin \beta$. By the inductive hypothesis, we have $\beta \nVdash_{\sigma[y \mapsto c]} A(\sigma(y), \sigma(x_0), \ldots, \sigma(x_{n-1}))$. Since $\beta \leq \alpha$, this means $\alpha \nVdash_\sigma \forall y\, A(y, \sigma(x_0), \ldots, \sigma(x_{n-1}))$.

The case for the existential quantifier is straightforward, given that a sentence $\exists y\, A(y)$ is in a saturated set α if and only if $A(c)$ is in α for some constant c. $\square$

We now define a Kripke model for first-order logic with equality to be a Kripke model equipped with an equivalence relation at each node that respects the interpretations of the functions and relations and is preserved when we pass to a stronger node. Equivalently, a Kripke model for first-order logic with equality is just a Kripke model in which equality is interpreted as a relation satisfying the equality axioms. In that case, there is nothing more to say: a formula A is provable from Γ in first-order logic with equality just in case it is provable from Γ and the equality axioms in first-order logic without special treatment of equality, and the completeness of natural deduction with respect to the former the former follows its completeness with respect to the latter.

In general, we cannot apply a quotient construction to obtain a Kripke model where the equality symbol is interpreted by equality at every node. For example, to show that $\forall x, y\, (x = y \lor x \neq y)$ is not intuitionistically valid, we need a Kripke model and a node with elements a, b such that a and b are not equal at that node but become equal later on. Our definition of a Kripke model does not accommodate collapses of this sort. However, Exercise 5.3.1 asks you to show that for intuitionistic logic with the additional axiom $\forall x, y\, (x = y \lor x \neq y)$, we can restrict attention to Kripke models in which the equality symbol is interpreted by equality. Such models are called *normal*.

As for propositional logic, we can use Kripke models to show unprovability.

Proposition 5.3.7. *None of these three schemas are intuitionistically valid:*

- $\forall x\, (A \lor B(x)) \to A \lor \forall x\, B(x)$, *where x is not free in A*
- $(A \to \exists x\, B(x)) \to \exists x\, (A \to B(x))$
- $\neg\neg\forall x\, (B(x) \lor \neg B(x))$.

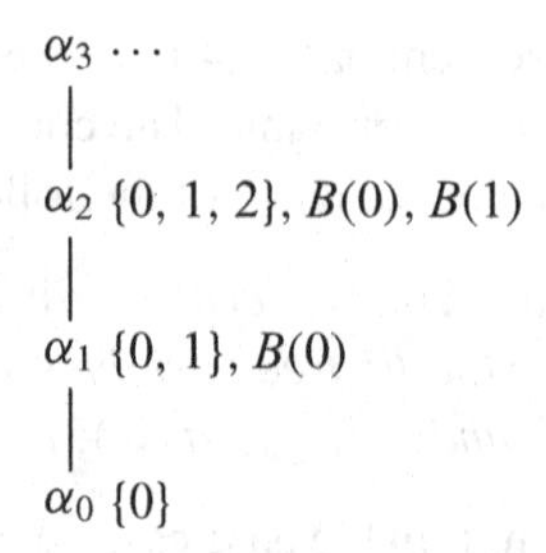

Figure 5.1 A Kripke model.

Proof We will give countermodels in which A is a propositional variable (a 0-ary predicate symbol) and $B(x)$ is a unary predicate symbol. Consider these two models:

β $\{u, v\}, A, B(u)$ | β $\{u, v\}, A, B(v)$

α $\{u\}, B(u)$ | α $\{u\}$

At each node, we indicate the universe at that node and the atomic facts that hold at that node. In the model on the left, $\forall x\,(A \vee B(x))$ is forced at α, since $A \vee B(u)$ is forced at α and β and $A \vee B(v)$ is forced at β. But neither A nor $\forall x\, B(x)$ is forced at α, the latter since $B(v)$ is not forced at β.

In the model on the right, α forces $A \rightarrow \exists x\, B(x)$, because at every node where A holds, there is an element where B holds as well. But at α there is no element a satisfying $A \rightarrow B(a)$.

For the third schema, consider the Kripke model in Figure 5.1. The nodes $\alpha_0, \alpha_1, \alpha_2, \ldots$ extend indefinitely, the universe at each α_i is $\{0, \ldots, i\}$, and at node α_{i+1}, $B(0), \ldots, B(i)$ are specified to hold. Then at every node α_i, α_i does not force either $B(i)$ or $\neg B(i)$. Unwrapping the forcing definition, this shows that α_0 does not force $\neg\neg\forall x\,(B(x) \vee \neg B(x))$. □

In the exercises, you are asked to show that the first schema, $\forall x\,(A \vee B(x)) \rightarrow A \vee \forall x\, B(x)$, holds in any Kripke model in which every node has the same universe. Intuitionistic logic with that schema is also complete with respect that semantics, and the schema is therefore known as *constant domain*, or (CD). Note that (CD) together with $\forall x\,(B(x) \vee \neg B(x))$ implies $\forall x\, B(x) \vee \exists x\, \neg B(x)$.

The second schema is known as *independence of premise*, or (IP), and is discussed in Chapter 14. The third schema is equivalent to the schema of *double-negation shift*, or (DNS), and has already been discussed in Section 4.6 and Exercise 4.6.2.

The variations on propositional Kripke models discussed at the end of Section 3.3 carry over to first-order logic. In particular, we can consider generalizations in which the underlying order is a preorder rather than a partial order, and we can consider exploding Kripke models for intuitionistic logic, in which some nodes force every formula, including $\bot$. Also, we can once again extend the forcing relation to classical logic. Restricting the forcing definition to Kripke models with constant domain, the clauses for the quantifiers are now as follows:

- $\alpha \Vdash_{w,\sigma} \forall x\, A$ if and only if for every $a \in |\mathfrak{K}_\alpha|$, $\alpha \Vdash_{w,\sigma[x\mapsto a]} A$.
- $\alpha \Vdash_{w,\sigma} A$ if and only if for every $\beta \leq \alpha$, there is a $\gamma \leq \beta$ and an $a \in |M_\gamma|$ such that $\gamma \Vdash_{w,\sigma[x\mapsto a]} A$.

This forcing relation is important in set theory.

In Section 3.5, we considered generalizations of Kripke models for propositional logic, namely, Beth models and a generalization of Beth models based on the notion of a covering relation. We also saw that it is possible to interpret classical forcing in terms of the latter. These ideas carry over to first-order logic and even higher-order logic; see the notes at the end of this chapter and at the end of Chapter 15.

Exercises

5.3.1. Use a quotient construction to show that first-order intuitionistic logic with decidable equality (i.e. the axiom $\forall x, y\,(x = y \vee x \neq y)$) is complete with respect to Kripke semantics in which the equality symbol is interpreted by equality at each node.

5.3.2. Give Kripke-model counterexamples to the following formulas:
a. $\neg\forall x\, A(x) \to \exists x\, \neg A(x)$
b. $(A \to \exists x\, B(x)) \to \exists x\, (A \to B(x))$
c. $\forall x\, \neg\neg A(x) \to \neg\neg\forall x\, A(x)$.

5.3.3. Consider the schema $\forall x\, (A(x) \vee B) \to \forall x\, A(x) \vee B$, where x is not free in B. Show that this schema holds in any Kripke model with constant domain.

5.4 Algebraic Semantics

Definition 3.4.2 tells us that a Boolean algebra or Heyting algebra is *complete* if it is closed under arbitrary meets and joins. We can generalize classical model-theoretic semantics so that formulas take values in an arbitrary complete Boolean algebra $\mathfrak{B}$ using these meets and joins to interpret the quantifiers. Such a model $\mathfrak{M}$ consists of a nonempty set, $|\mathfrak{M}|$, an interpretation of each n-ary function symbol f as an n-ary function $f^{\mathfrak{M}}$ on $|\mathfrak{M}|$, and an interpretation of each n-ary relation symbol R as an n-ary function from $|\mathfrak{M}|^n \to \mathfrak{B}$, providing $R^{\mathfrak{M}}(a_0, \ldots, a_{n-1})$ with a truth value in $\mathfrak{B}$ for each n-tuple $\vec{a}$ in $|\mathfrak{M}|$. The resulting structure is known as a *Boolean-valued model*. As before, every term t evaluates to an element $t^{\mathfrak{M},\sigma}$ in $|\mathfrak{M}|$ given an assignment σ to the free variables. We can then evaluate atomic formulas $[\![R(t_0, \ldots, t_{n-1})]\!]^{\mathfrak{M},\sigma} = R^{\mathfrak{M}}(t_0^{\mathfrak{M},\sigma}, \ldots, t_{n-1}^{\mathfrak{M},\sigma})$ and extend the means of evaluating the propositional connectives described in Section 3.2 with the following clauses for the quantifiers:

- $[\![\forall x\, A(x)]\!]^{\mathfrak{M},\sigma} = \bigwedge_{a \in |\mathfrak{M}|} [\![A]\!]^{\mathfrak{M},\sigma[x \mapsto a]}$
- $[\![\exists x\, A(x)]\!]^{\mathfrak{M},\sigma} = \bigvee_{a \in |\mathfrak{M}|} [\![A]\!]^{\mathfrak{M},\sigma[x \mapsto a]}$.

The usual model-theoretic semantics is the special case where $\mathfrak{B}$ is the two-element complete Boolean algebra $\{\top, \bot\}$.

It is straightforward to establish soundness. As in the proof of soundness for propositional logic with respect to Boolean-valued semantics, we want to show that if a sequent $\Gamma \vdash A$ is provable, then, for any σ, $\bigwedge_{B \in \Gamma} [\![B]\!]^{\mathfrak{M},\sigma} \leq [\![A]\!]^{\mathfrak{M},\sigma}$. For convenience, I will abbreviate the left-hand side of that inequality as $[\![\Gamma]\!]^{\mathfrak{M},\sigma}$. Let us consider the natural deduction rules for the existential quantifier. For the introduction rule, suppose for some t we have $[\![\Gamma]\!]^{\mathfrak{M},\sigma} \leq [\![A(t)]\!]^{\mathfrak{M},\sigma}$. Since

$$[\![A(t)]\!]^{\mathfrak{M},\sigma} = [\![A(x)]\!]^{\mathfrak{M},\sigma[x \mapsto [\![t]\!]^{\mathfrak{M},\sigma}]} \leq [\![\exists x\, A(x)]\!]^{\mathfrak{M},\sigma},$$

we have $[\![\Gamma]\!]^{\mathfrak{M},\sigma} \leq [\![A(t)]\!]^{\mathfrak{M},\sigma}$, as required. For the elimination rule, suppose we have $[\![\Gamma]\!]^{\mathfrak{M},\sigma} \leq [\![\exists x\, A(x)]\!]^{\mathfrak{M},\sigma}$ and $[\![\Gamma, A(x)]\!]^{\mathfrak{M},\sigma} \leq [\![B]\!]^{\mathfrak{M},\sigma}$ for every σ, where x is not free in Γ or B. The latter implies $[\![A(x)]\!]^{\mathfrak{M},\sigma} \leq [\![\bigwedge \Gamma \to B]\!]^{\mathfrak{M},\sigma}$ for every σ, and since the right-hand side doesn't change when we replace σ by $\sigma[x \mapsto a]$ for any $a \in |\mathfrak{M}|$, we have $[\![A(x)]\!]^{\mathfrak{M},\sigma[x\mapsto a]} \leq [\![\bigwedge \Gamma \to B]\!]^{\mathfrak{M},\sigma}$ for every $a \in |\mathfrak{M}|$. By the definition of $[\![\exists x\, A(x)]\!]^{\mathfrak{M},\sigma}$, this means $[\![\exists x\, A(x)]\!]^{\mathfrak{M},\sigma} \leq [\![\bigwedge \Gamma \to B]\!]^{\mathfrak{M},\sigma}$, and since $[\![\Gamma]\!]^{\mathfrak{M},\sigma} \leq [\![\exists x\, A(x)]\!]^{\mathfrak{M},\sigma}$, we have $[\![\Gamma]\!]^{\mathfrak{M},\sigma} \leq [\![B]\!]^{\mathfrak{M},\sigma}$, as required.

Since we have completeness with respect to two-valued semantics, we have completeness with respect to complete Boolean algebras as well.

Theorem 5.4.1. *Classical logic is sound and complete with respect to Boolean-valued models.*

What happens if, instead, we evaluate formulas with respect to a complete Heyting algebra instead of a complete Boolean algebra? Establishing the soundness of intuitionistic logic with respect to this semantics is straightforward. Let us show that intuitionistic logic is also complete with respect to this semantics. We will focus on first-order logic without equality, since we can handle equality by interpreting it as a relation symbol satisfying the equality axioms, as we did in the last section.

The construction we will use is similar to the proof of completeness for classical propositional logic via the Tarski–Lindenbaum algebra that was presented in Section 3.2. Let Γ be a set of sentences in some language, L. We want to show that for any formula A, if $\Gamma \models A$, then $\Gamma \vdash A$. We will take the universe of $\mathfrak{M}$ to consist of all terms in L, including terms with free variables. Let S be the set of all formulas in L, including formulas with free variables, and say two such formulas A and B are equivalent, written $A \equiv B$, if $\Gamma \vdash A \leftrightarrow B$ in intuitionistic logic. Let $\mathfrak{H} = S/{\equiv}$.

It is straightforward to show that $\mathfrak{H}$ has the structure of a Heyting algebra when we define $A \leq B$ by $\Gamma \vdash A \to B$. The meets, joins, and pseudocomplements are induced on the quotient by the syntactic operations of forming conjunctions, disjunctions, and implications.

$\mathfrak{H}$ is not a complete algebra; we will deal with that in a moment. But, in $\mathfrak{H}$, we have that $[\exists x\, A(x)]$ is the least upper bound for the elements $[\exists x\, A(t)]$ as t ranges over terms in the language, and $\forall x\, A(x)$ is the greatest lower bound for those elements. For the first claim, it is straightforward that $[A(t)] \leq [\exists x\, A(x)]$ for every term t, so $[\exists x\, A(x)]$ is an upper bound on the set of elements $[A(t)]$. To show that it is the least such, suppose that B is another upper bound, so that we have $[A(t)] \leq [B]$ for every term t. This means that Γ proves $A(t) \to B$ for every term t. Taking t to be a variable x that does not occur in B, we have that $\Gamma \vdash \exists x\, A(x) \to B$, and so $[\exists x\, A(x)] \leq [B]$, as required.

As was the case with the syntactic construction of a two-valued model in Section 5.1, we have $[\![t(x_0, \ldots, x_{n-1})]\!]^{\mathfrak{M},\sigma} = t(\sigma(x_0), \ldots, \sigma(x_{n-1}))$ for every term t. Similarly, we have $[\![A(x_0, \ldots, x_{n-1})]\!]^{\mathfrak{M},\sigma} = [A(\sigma(x_0), \ldots, \sigma(x_{n-1})]$. So, for any sentence A, we have $[\![A]\!]^{\mathfrak{M},\sigma} = [\top]$ if and only if $[A] = [\top]$, which happens if and only if $\Gamma \vdash A$.

Our model takes values in $\mathfrak{H}$, which we only know to be a Heyting algebra, rather than a complete Heyting algebra. We have established that $\mathfrak{H}$ has enough meets and joins to interpret the quantifiers. The good news is that lattice theory provides various ways of embedding lattices into larger ones with better closure properties. This is the result we need:

Theorem 5.4.2. *Every Heyting algebra can be embedded in a complete Heyting algebra in a way that preserves existing meets and joins.*

The construction we will use to prove Theorem 5.4.2 is known as the MacNeille completion, and it will take us on a detour through some lattice theory.

Let $(P, \leq)$ be any partial order. An *order ideal* is any subset S of P that is closed downward, that is, such that $x \in S$ and $y \leq x$ imply $y \in S$ for any $x, y \in P$. An *order filter* is, dually, a subset of S that is closed upward. Given $x \in P$, the set $\downarrow(x) = \{y \mid y \leq x\}$ is clearly an order ideal, and the set $\uparrow(x) = \{y \mid y \leq x\}$ is clearly an order filter. These are called the *principal order ideal* and *principal order filter* generated by x, respectively.

Given a subset $S \subseteq P$, let $u(S)$ denote the set of all *upper bounds on* S, that is, the set of element y such that $y \geq x$ for every x in S. Dually, let $l(S)$ be the set of lower bounds on S. Define a *cut* in P to be a pair (L, U) of subsets of P such that $L = l(U)$ and $U = u(L)$. It is easy to check that if (L_0, U_0) and (L_1, U_1) are cuts, then $L_0 \subseteq L_1$ if and only if $U_0 \subseteq U_1$. When this happens, we say $(L_0, U_0) \leq (L_1, U_1)$. Clearly this defines a partial order on the set of all cuts.

The *MacNeille completion* of $(P, \leq)$ is defined to be the set of all cuts in P with the order just defined. If x is any element of P, then $(\downarrow(x), \uparrow(x))$ is a cut. We define a function φ from $(P, \leq)$ to its MacNeille completion by $\varphi(x) = (\downarrow(x), \uparrow(x))$. For every x and y in P have $x \leq y$ if and only if $\varphi(x) \leq \varphi(y)$, so φ is an order-preserving embedding. (An exercise below asks you to check that for any subset S of a partial order, $S \subseteq u(l(S))$. The conditions $L = l(U)$ and $U = u(L)$ on a cut imply $L = l(u(L))$, and given any set L satisfying this condition, we can recover U as $U = u(L)$. Thus we can equivalently take the MacNeille completion to be the collection of sets L satisfying $L = l(u(L))$, ordered by inclusion.)

In Section 3.2, we came across the notion of a *filter* in a bounded lattice. The dual notion is that of an *ideal*: a subset I of a lattice $(L, \leq)$ is an ideal if it is an order ideal such that $\bot \in I$ and $x \vee y$ is in I whenever x and y are in I. I leave it as an exercise to check that for any subset S of a bounded lattice, $u(S)$ is always a filter and $l(S)$ is always an ideal. The next two propositions tell us that the MacNeille completion of a Heyting algebra satisfies the conclusion of Theorem 5.4.2.

Proposition 5.4.3. *For any partial order $(P, \leq)$, the MacNeille completion of P is a bounded complete lattice, and the embedding φ preserves all meets and joins that exist in P.*

Proof Clearly $(\emptyset, P)$ is the least element of the MacNeille completion and $(P, \emptyset)$ is the greatest element. To show the existence of arbitrary meets, let S be a subset of the MacNeille completion and let $\hat{L} = \bigcap\{L \mid (L, U) \in S\}$. Then for any cut (L, U) in S, $\hat{L} \subseteq L$, and hence $u(\hat{L}) \supseteq u(L) = U$ and $l(u(\hat{L})) \subseteq l(U) = L$. Hence $l(u(\hat{L})) \subseteq \bigcap\{L \mid (L, U) \in S\} = \hat{L}$, so $(\hat{L}, u(\hat{L}))$ is a cut. It is straightforward to check that this is the greatest lower bound on S. A dual argument gives the existence of arbitrary joins.

Saying that φ preserves all meets that exist in P means that if a set S has a greatest lower bound $\bigwedge S$ in P, then $\varphi(\bigwedge S) = \bigwedge_{x \in S} \varphi(x)$. To show that, it is enough to check that the first element of each cut is the same, that is, $\downarrow(\bigwedge S) = \bigcap_{x \in S} \downarrow(x)$. But this follows from the chain

$$\begin{aligned} y \in \downarrow\left(\bigwedge S\right) &\Leftrightarrow y \leq \bigwedge S \\ &\Leftrightarrow y \leq x \text{ for every } x \text{ in } S \\ &\Leftrightarrow y \in \bigcap_{x \in S} \downarrow(x). \end{aligned}$$

The dual argument shows preservation of joins. □

Remember from Section 3.4 that in any Heyting algebra, the relative pseudocomplement $x \to y$ can be characterized as $\bigvee\{z \mid x \wedge z \leq y\}$. So if $(P, \leq)$ is a Heyting algebra in Proposition 5.4.3, that part of the structure is preserved by the embedding. But that doesn't guarantee that the MacNeille completion itself is a Heyting algebra. By Exercise 3.4.3, we would have that conclusion if we could show that the MacNeille completion is a distributive lattice, but, unfortunately, this is not always the case. Instead, we prove the result directly.

Proposition 5.4.4. *If $\mathfrak{H}$ is a Heyting algebra, then its MacNeille completion is a Heyting algebra as well.*

Proof Given cuts (L, U) and (L', U'), define

$$\hat{L} = \{x \in H \mid \forall y \in L\, (x \wedge y \in L')\},$$

and let $\hat{U} = u(\hat{L})$. I claim

- $(\hat{L}, \hat{U})$ is a cut, that is, $\hat{L} = l(\hat{U}) = l(u(\hat{L}))$, and
- $(\hat{L}, \hat{U})$ is the relative pseudocomplement of (L, U) and (L', U').

To prove the first claim, since the forward inclusion always holds, let x be an element of $l(u(\hat{L}))$ and let us show that x is in $\hat{L}$. To that end, let y be any element of L; it suffices to show $x \wedge y$ is in L'. Since $L' = l(U')$, it suffices to show $x \wedge y \leq z$ for any z in U'. So let z be any element of U'. Since $\mathfrak{H}$ is a Heyting algebra, $x \wedge y \leq z$ is equivalent to $x \leq y \to z$. Since x is in $l(u(\hat{L}))$, it suffices to show $y \to z$ is in $u(\hat{L})$. Let w be any element of $\hat{L}$; we need to show $w \leq y \to z$. Since y is in L, the definition of $\hat{L}$ gives $w \wedge y \in L'$, and since z is in U', we have $w \wedge y \leq z$. Since $\mathfrak{H}$ is a Heyting algebra, this is equivalent to $w \leq y \to z$, as required.

To prove the second claim, in order to show that $(\hat{L}, \hat{U})$ is the relative pseudocomplement $(L, U) \to (L', U')$, we need to show that for any cut $(\bar{L}, \bar{U})$, we have $(\bar{L}, \bar{U}) \leq (\hat{L}, \hat{U})$ if and only if $(\bar{L}, \bar{U}) \wedge (L, U) \leq (L', U')$. By the definition of wedge and the order in the MacNeille completion, this is equivalent to showing that for any cut $(\bar{L}, \bar{U})$, $\bar{L} \cap L \subseteq L'$ if and only if $\bar{L} \subseteq \hat{L}$.

Suppose the left-hand side holds and $x \in \bar{L}$. To show $x \in \hat{L}$, suppose $y \in L$. Since $\bar{L}$ and L are both closed downward, $x \wedge y$ is in $\bar{L} \cap L$, and hence in L', as required. Conversely, suppose the right-hand side holds and x is in $L \cap L$. Then x is in $\hat{L}$, and so $x \wedge x = x \in L'$, as required. □

Thus, we have shown:

Theorem 5.4.5. *Intuitionistic logic is sound and complete with respect to Heyting-valued semantics.*

In the exercises, you are asked to show that the MacNeille completion of a Boolean algebra is again a Boolean algebra. Exercise 5.4.5 provides an alternative proof of Theorem 5.4.2.

Exercises

5.4.1. Show that for any subset S of a partial order, $S \subseteq u(l(S))$.

5.4.2. Show that for any partial order P, $(\downarrow(a), \uparrow(a))$ is a cut.

5.4.3. Show that for any subset S of a bounded lattice, $u(S)$ is always a filter and $l(S)$ is always an ideal.

5.4.4. Show that the MacNeille completion of a Boolean algebra is again a Boolean algebra.

5.4.5. This exercise provides an alternative proof of Theorem 5.4.2. Let $(H, \leq)$ be a Heyting algebra. Say a *complete ideal* $I \subseteq H$ is an order ideal closed under all existing joins in H; in other words, if $S \subseteq I$ has a least upper bound $\bigvee S$ in H, then $\bigvee S$ is in I. Consider the set $\hat{H}$ of all complete ideals ordered by the inclusion relation. Show that this structure meets the requirements of Theorem 5.4.2 as follows.

a. Show that if $S \subseteq \hat{H}$, the meet $\bigwedge S$ of S is given by $\bigcap S$.
b. Show that if $S \subseteq \hat{H}$, the join $\bigvee S$ of S is given by

$$\bigvee S = \{U \mid U = \bigvee W \text{ for some } W \subseteq \bigcup S\}.$$

c. Show that $\hat{H}$ satisfies the infinitary distributive law (3.1) of Section 3.4.
d. Show that the map $\varphi(x) \colon x \mapsto \downarrow(x)$ is an embedding of H into $\hat{H}$ that preserves existing meets and joins.

5.5 Definability

In this section and the next, we consider some results of classical *model theory*. Even though the focus of this book is on provability rather than semantics, here we will see that semantic notions provide a helpful perspective on the meaning of a formal system. Of course, they are also often interesting in their own right.

Given a formal language and a notion of satisfaction $\mathfrak{M} \models A(\vec{x})$, we are often interested in the extent to which expressions in the language can be used to describe relations and structures of interest. In this section we will consider two notions of definability:

- A set of sentences Γ is said to *define* the class of structures $\{\mathfrak{M} \mid \mathfrak{M} \models \Gamma\}$. Similarly, a single sentence A is said to define the class $\{\mathfrak{M} \mid \mathfrak{M} \models A\}$. (Since any set can serve as the universe of a structure, the class of structures defined by a set of sentences or a single sentence is generally too large to be a set.)
- Fixing a model $\mathfrak{M}$, a formula $A(x_0, \ldots, x_{n-1})$ is said to *define* the n-ary relation R on $|\mathfrak{M}|$ with the property that, for any tuple $\vec{a}$, $R(\vec{a})$ holds if and only if $\mathfrak{M} \models A(\vec{a})$. Similarly, a formula $A(\vec{x}, y)$ is said to define the function $f(\vec{x})$ on $\mathfrak{M}$ if it defines the graph of f, that is, the relation $f(\vec{x}) = y$. More precisely, if $A(\vec{x})$ is a formula in the language of the model, $A(\vec{x})$ is said to define the corresponding relation *without parameters*. Recall from Section 5.1 that, given a tuple of parameters $\vec{b}$ from the universe of $\mathfrak{M}$, we can talk about a formula $A(\vec{x}, \vec{b})$ that uses those parameters. A function or relation that is defined by such a formula is said to be *definable from the parameters* $\vec{b}$.

Here are some examples of definability in the first sense:

- A *partial order* is a structure with a binary relation satisfying the following:
  - $\forall x\, (x \leq x)$
  - $\forall x, y, z\, (x \leq y \wedge y \leq z \to x \leq z)$
  - $\forall x, y\, (x \leq y \wedge y \leq x \to x = y)$.

  In other words, the set with these three axioms defines the class of partial orders. Adding the axiom $\forall x, y\, (x \leq y \vee y \leq x)$ yields the class of *total*, or *linear* orders. If we define $x < y$ to mean $x \leq y \wedge x \neq y$, adding the axiom $\forall x, y\, (x < y \to \exists z\, (x < z \wedge z < y))$ to that yields the class of *dense linear orders*.

- In the first-order language with no function or relation symbols other than equality, a structure is a set. For every n, the sentence

$$A_n \equiv \exists x_0, \ldots, x_{n-1} \bigwedge_{i \neq j} x_i \neq x_j$$

defines the class of structures with at least n distinct elements. The set of sentences $\{A_n\}_{n \in \mathbb{N}}$ defines the class of infinite structures.
- A *directed graph* is just a structure with a binary relation E. Here $E(a, b)$ is interpreted as the statement that there is an *edge* from a to b. It is sometimes useful to rule out self-loops by adding the axiom $\exists x \neg E(x, x)$. Edges in a directed graph have a direction, so $E(a, b)$ and $E(b, a)$ are independent of one another. We can define the class of *undirected graphs* to be graphs where the edge relation is symmetric, that is, the directed graphs satisfying $\forall x, y \,(E(x, y) \to E(y, x))$.
- The usual axioms for groups, rings, and fields define the class of groups, rings, and fields, respectively.

For examples of definitions in the second sense, fix the structure $\mathfrak{N}$ to be $(\mathbb{N}, +)$. Then the predicate "is equal to 0" on the universe of that structure is defined by the formula $A_0(x) \equiv x + x = x$, the relation "is less than or equal to" is defined by the formula $A_{\leq}(x, y) \equiv \exists z \,(x + z = y)$, and the predicate "is even" is defined by the formula $A_{\text{even}}(x) \equiv \exists y \,(y + y = x)$. None of these definitions require parameters from the model, which is to say, the definitions don't mention any elements of the universe of the particular model in which the definition is interpreted. For an example that does, consider the set of nonnegative real numbers in the structure $(\mathbb{R}, \leq)$. We will see below that this set is not definable in the language with $\leq$ alone, but it clearly is definable by the formula $0 \leq x$ if we are allowed to refer to 0.

Even when we deal with formal derivations, we generally take ourselves to be proving things about objects, functions, and relations in structures that are of interest to us. As a result, it is important to know what sorts of things can be expressed in the languages we have at our disposal. This is one way in which the theory of first-order definability is helpful.

Showing that a class of structures is definable is usually straightforward: one describes a sentence, or a set of sentences, defining it. Showing that a class of structures is *not* definable generally requires more ingenuity. Recall the notion of isomorphism from Section 1.1. An important observation is that first-order logic cannot distinguish isomorphic structures.

Theorem 5.5.1. *Suppose φ is an isomorphism between two structures $\mathfrak{A}$ and $\mathfrak{B}$, $A(x_0, \ldots, x_{n-1})$ is a formula with the free variables shown, and $a_0, \ldots, a_{n-1}$ are elements of $|\mathfrak{A}|$. Then*

$$\mathfrak{A} \models A(a_0, \ldots, a_{n-1}) \quad \textit{if and only if} \quad \mathfrak{B} \models A(\varphi(a_0), \ldots, \varphi(a_{n-1})).$$

In particular, isomorphic structures satisfy exactly the same sentences. The proof is a straightforward induction on formulas, using the defining clauses for the satisfaction relation.

The compactness theorem is a useful tool. As was the case with propositional logic, it can be stated in two forms.

Theorem 5.5.2. *Let Γ be any set of sentences.*

- *If every finite subset of Γ has a model, then Γ has a model.*
- *If $\Gamma \models A$ for some sentence A, then for some finite $\Gamma' \subseteq \Gamma$, $\Gamma' \models A$.*

As for propositional logic, these are easily shown to be equivalent, and both follow immediately from soundness and completeness.

Theorem 5.5.3. *Let Γ be any set of first-order sentences with arbitrarily large finite models. Then Γ has an infinite model.*

Proof Let A_n be a first-order sentence that holds in a model if and only if the model has at least n elements. The hypothesis implies that every finite subset of $\Gamma \cup \{A_0, A_1, A_2, \ldots\}$ has a model. By compactness, the whole set has a model, □

Corollary 5.5.4. *The class of finite sets cannot be defined by any set Γ of first-order sentences.*

Proof Suppose the class of finite sets is defined by the set Γ. For each n, let A_n be the sentence that says that there are at least n elements of the universe. Then every finite subset of $\Gamma \cup \{A_0, A_1, A_2, \ldots\}$ has a model but the full set does not, contrary to the compactness theorem. □

Corollary 5.5.5. *The class of infinite sets cannot be defined by any finite set of first-order formulas.*

Proof Suppose the class were defined by the finite set Γ. Then it would be defined by the single sentence $A \equiv \bigwedge \Gamma$, in which case $\neg A$ would define the class of finite sets, contrary to Corollary 5.5.4. □

A binary relation R on a set A is said to be *well founded* if every nonempty subset $S \subseteq A$ has an R-minimal element; in other words, if for every nonempty S there is an x in S such that for no z in S do we have $R(z, x)$. Exercise 5.5.2 below asks you to show that this is equivalent to saying that there is no infinite sequence of elements $x_0, x_1, x_2, \ldots$ such that for every i, $R(x_{i+1}, x_i)$. (See also Appendix A.2.)

Theorem 5.5.6. *The class of well-founded relations is not first-order definable.*

Proof Suppose Γ defines the class of well-founded relations in a language with a single binary relation symbol, R. Add countably many constants $c_0, c_1, c_2, \ldots$ and consider the set of sentences $\Delta = \Gamma \cup \{R(c_{i+1}, c_i)\}$. Let Δ' be any finite subset of Δ, and suppose n is the largest number such that the sentence $R(c_{n+1}, c_n)$ appears in Δ. Then we obtain a model of Δ' with universe $\{0, 1, \ldots, n+1\}$ by taking R to be $<$ and interpreting c_0 to be $n+1$, c_1 to be n, and so on. (It doesn't matter what we assign to c_j when $j > n+1$.) By compactness, Δ has a model, $\mathfrak{M}$. But in that model, we have $R^{\mathfrak{M}}(c_{i+1}^{\mathfrak{M}}, c_i^{\mathfrak{M}})$ for every i, contrary to our initial assumption and the fact that $\mathfrak{M}$ is a model of Γ. □

As far as the second sort of definability, one way to show that a relation is not definable in a structure (without parameters) is to show that there is an automorphism of the structure that is not compatible with the relation. For example:

Theorem 5.5.7. *The relation $<$ is not definable in $\mathfrak{R} = (\mathbb{R}, 0, +)$.*

Proof The function $f(x) = -x$ is an automorphism of the structure. Suppose $A(x, y)$ defines the less-than relation. Then we have $\mathfrak{R} \models A(1, 2)$. By Theorem 5.5.1, we then have $\mathfrak{R} \models A(-1, -2)$, contradicting the fact that $-1 \not< -2$. □

Proving interesting undefinability results generally requires more work than that. Exercise 5.5.10 offers an example where compactness together with an automorphism does the trick. Another powerful body of methods are based on what are called *Ehrenfeucht–Fraissé games*. We can sometimes obtain undefinability results by bringing computability issues into play. For example, it is impossible define multiplication in $(\mathbb{N}, +)$, because the theory of this structure is decidable whereas the theory of $(\mathbb{N}, +, \cdot)$ is not. (See Section 12.5.)

One can ask whether it is possible to define a class consisting of a single structure using a set of sentences in first-order logic. Given a set of sentences, Γ, let $\mathrm{Mod}(\Gamma)$ denote the class of its models. In the other direction, given a structure, $\mathfrak{M}$, let $\mathrm{Th}(\mathfrak{M})$ be the *theory of* $\mathfrak{M}$, that is, the set of all sentences true of $\mathfrak{M}$. What we are asking is whether there is a Γ such that $\mathrm{Mod}(\Gamma)$ consists of a single element, or, equivalently, whether there is a model $\mathfrak{M}$ such that $\mathrm{Th}(\mathfrak{M})$ has no other models.

A moment's reflection shows that this is too much to ask for: by Theorem 5.5.1, the best we can do is define a structure up to isomorphism, i.e. define the class of structures isomorphic to the one we are after. A set of sentences Γ such that all the elements of $\mathrm{Mod}(\Gamma)$ are isomorphic is said to be *categorical.* We will see in the next section that even that is too ambitious: a theory is categorical if and only if it is the theory of a *finite* structure. If λ is a cardinal number, a set of sentences Γ is said to be λ*-categorical* if all its models of cardinality λ are isomorphic. Indeed, some interesting structures have λ-categorical theories. We will show, however, that the theory of the natural numbers is not $\aleph_0$-categorical, which is to say, there are countable models of the theory of the natural numbers that are not isomorphic to the standard interpretation. (In fact, the theory is not λ-categorical for any λ.)

Two structures $\mathfrak{M}$ and $\mathfrak{N}$ for a language L are said to be *elementarily equivalent* if $\mathrm{Th}(\mathfrak{M}) = \mathrm{Th}(\mathfrak{N})$. This means that $\mathfrak{M}$ and $\mathfrak{N}$ satisfy exactly the same first-order sentences, and hence are indistinguishable, as far as first-order logic is concerned. By Theorem 5.5.1, if $\mathfrak{M}$ and $\mathfrak{N}$ are isomorphic then they are elementarily equivalent, but the converse does not hold. For example, we will see in Section 11.2 that the structure $(\mathbb{Q}, <)$ is elementarily equivalent to $(\mathbb{R}, <)$, but the two structures are not isomorphic, since $\mathbb{Q}$ is countable but $\mathbb{R}$ isn't.

Theorem 5.5.8. *Let* $\mathfrak{N}$ *be the structure* $(\mathbb{N}, 0, \mathrm{succ}, +, \cdot, <)$. *There is a countable structure* $\mathfrak{M}$ *such that*

- $\mathfrak{M}$ *is elementarily equivalent to* $\mathfrak{N}$, *and*
- $\mathfrak{M}$ *is not isomorphic to* $\mathfrak{N}$.

Proof Let L be the language of $\mathfrak{N}$, and L' be L together with a new constant c. Let Γ be the following set of sentences in L':

$$Th(\mathfrak{N}) \cup \{0 < c, \mathrm{succ}(0) < c, \mathrm{succ}(\mathrm{succ}(0)) < c, \ldots\}.$$

Every finite subset Γ' of Γ has a model of the form $(\mathbb{N}, 0, S, +, \cdot, <, m)$, where m is a natural number that is large enough to satisfy the finitely many sentences involving c in Γ'. By compactness, Γ has a model. Let $\mathfrak{M}$ be the result of restricting this structure to the language L. We have $\mathfrak{M} \models \mathrm{Th}(\mathfrak{N})$ since $\mathrm{Th}(\mathfrak{N})$ is a subset of Γ'., and so $\mathfrak{M}$ is elementarily equivalent to $\mathfrak{N}$. On the other hand, by construction, there is an element a in the universe of $\mathfrak{M}$ that is bigger than $0^{\mathfrak{M}}$, $\mathrm{succ}(0)^{\mathfrak{M}}$, $\mathrm{succ}(\mathrm{succ}(0))^{\mathfrak{M}}, \ldots$. This implies that $\mathfrak{M}$ is not isomorphic to $\mathfrak{N}$. □

The elements $0^{\mathfrak{M}}$, $\text{succ}(0)^{\mathfrak{M}}$, $\text{succ}(\text{succ}(0))^{\mathfrak{M}}, \ldots$ of such a nonstandard model $\mathfrak{M}$ of arithmetic are called the *standard* elements of $\mathfrak{M}$, and the remaining elements of the universe are called the *nonstandard* elements. In a sense, a nonstandard number is infinitely large. But only in a sense; as far as $\mathfrak{M}$ is concerned, such an element is a natural number just like any other.

What does a nonstandard model of arithmetic look like? All of the following sentences are true of the natural numbers, and hence true in any such model.

1. $<$ is a linear order.
2. Every element has a unique successor.
3. Every element other than 0 has a unique predecessor.
4. Every element is even or odd.
5. For every x and y, if $x > y$ then $x + x > x + y$.
6. If x is even, then there is a number y such that $y + y = x$.
7. There are infinitely many primes.
8. Fermat's last theorem holds.

If we draw a picture, we see that every nonstandard number is a member of a countable chain that is order-isomorphic to the integers, since it has infinitely many predecessors and infinitely many successors. Statement 5 can be used to show that there are infinitely many such chains, and statement 6 can be used to show that between any two such chains, there is another one. (See Exercise 5.5.13.)

The same trick can be used to construct nonstandard models of any theory of the real numbers, in which there are positive real numbers that are smaller than $1, 1/2, 1/3, 1/4, \ldots$. Such numbers can be considered infinitely small, and, interestingly enough, can be used to make some sense of the informal understanding of infinitesimals in the historical development of calculus. Using nonstandard models to justify reasoning about infinitesimals forms the basis of a field known as *nonstandard analysis*.

Exercises

5.5.1. Let $\mathfrak{A} = (A, R)$ be a graph. A *path* from a to b is a finite sequence of vertices $a_0, \ldots, a_n$ such that $a_0 = a$, $a_n = b$, and for each $i < n$, $R(a_i, a_{i+1})$. A graph is said to be *connected* if there is a path between every two vertices. Use the compactness theorem to show that the class of connected graphs is not definable in first-order logic.

5.5.2. Let R be a binary relation on a set A. Show that the following are pairwise equivalent:

a. R is well founded.

b. Suppose $S \subseteq A$ has the property that whenever y is in S for every y such that $R(y, x)$, then x is in S. Then $S = A$.

c. There is no infinite $\prec$-decreasing sequence $x_0 \succ x_1 \succ x_2 \succ \ldots$.

5.5.3. Show that the class of torsion groups is not definable.

5.5.4. Show that there is no first-order formula that defines the positive real numbers in $(\mathbb{R}, <)$.

5.5.5. Show that addition (i.e. the relation $x + y = z$) is not definable in $(\mathbb{R}, 0, <)$.

5.5.6. Show that $\leq$ is not definable in $(\mathbb{Z}, 0, +)$ or $(\mathbb{R}, 0, +)$.

5.5.7. Show that $\leq$ is definable in $(\mathbb{R}, +, \cdot)$.

5.5.8. Show that $\leq$ is definable in $(\mathbb{Z}, +, \cdot)$. (Hint: look up Lagrange's theorem on sums of four squares.)

5.5.9. Show that multiplication is not definable in $(\mathbb{R}, 0, 1, +, \leq)$ or $(\mathbb{Z}, 0, +, \leq)$. (See Section 12.5.)

5.5.10. Show that $<$ is not definable in $(\mathbb{N}, 0, \text{succ})$. (Hint: Use compactness to obtain an elementarily equivalent structure $\mathfrak{A}$ with nonstandard elements a and b such that neither one can be obtained from the other by finitely many applications of the successor function. Then show there is an automorphism of this structure that switches a and b.)

5.5.11. Show that for any structure $\mathfrak{A}$, the theory of $\mathfrak{A}$, $\text{Th}(\mathfrak{A})$, is maximally consistent.

5.5.12. Show that if two sets Γ_1 and Γ_2 have exactly the same class of models, then the theories axiomatized by Γ_1 and Γ_2 are the same.

5.5.13. Let $\mathfrak{M}$ be a nonstandard model of arithmetic, and let a and b be nonstandard elements such that $\mathfrak{M} \models a + n \leq b$ for every standard number n in $\mathfrak{M}$. Show that there is an element c such that $\mathfrak{M} \models a + n \leq c$ and $\mathfrak{M} \models c + n \leq b$ for every standard number n.

5.5.14. Show that there are uncountably many non-isomorphic countable structures elementarily equivalent to $(\mathbb{N}, 0, \text{succ}, +, \cdot, <)$. (Hint: Let $p_0, p_1, \ldots$ enumerate the prime numbers. Show that for every subset $S \subseteq \mathbb{N}$, there is a countable model of arithmetic with an element a such that for every standard natural number n, p_n divides a in the model if and only if $n \in S$.)

5.6 Some Model Theory

This section introduces some additional concepts and results that are fundamental to the model theory of classical first-order logic.

Let L_1 and L_2 be languages and suppose L_1 is a subset of L_2, in the sense that every function, relation, and constant symbol of L_1 is also a symbol of L_2. Given an L_1-structure $\mathfrak{A}$, it is often useful to consider an *expansion* of $\mathfrak{A}$ to an L_2-structure $\mathfrak{B}$. This means that $\mathfrak{A}$ and $\mathfrak{B}$ have the same universe and interpret the symbols of L_1 the same way, so $\mathfrak{B}$ only adds interpretations of the additional symbols of L_2. If $\mathfrak{B}$ is an expansion of $\mathfrak{A}$, then $\mathfrak{A}$ is said to be a *reduct* of $\mathfrak{B}$. For example, $(\mathbb{N}, 0, <)$ is a reduct of $(\mathbb{N}, 0, 1, +, \cdot, <)$ for suitable L_1 and L_2.

If $\mathfrak{A}$ and $\mathfrak{B}$ are structures for the same language, $\mathfrak{A}$ is a *substructure* of $\mathfrak{B}$, written $\mathfrak{A} \subseteq \mathfrak{B}$, if the universe of $\mathfrak{A}$ is a subset of the universe of $\mathfrak{B}$, the functions of $\mathfrak{A}$ are the restrictions of the functions of $\mathfrak{B}$ to the universe of $\mathfrak{A}$, and the relations of $\mathfrak{A}$ are just the restrictions of the relations of $\mathfrak{B}$ to the universe of $\mathfrak{A}$. For example, $(\mathbb{N}, 0, +, \cdot)$ is a substructure of $(\mathbb{R}, 0, +, \cdot)$, but $(\mathbb{Q}, <)$ is not a substructure of $(\mathbb{R}, >)$.

$\mathfrak{A}$ is an *elementary substructure* of $\mathfrak{B}$ if it is a substructure of $\mathfrak{B}$ and the following holds: whenever $a_0, \ldots, a_{k-1}$ are elements of the universe of $\mathfrak{A}$ and $A(x_0, \ldots, x_{k-1})$ is a formula, then $\mathfrak{A} \models A(a_0, \ldots, a_{k-1})$ if and only if $\mathfrak{B} \models A(a_0, \ldots, a_{k-1})$. This says more than that $\mathfrak{A}$ is a substructure of $\mathfrak{B}$ and that $\mathfrak{A}$ is elementarily equivalent to $\mathfrak{B}$, since $\mathfrak{A}$ and $\mathfrak{B}$ have to agree not just on the truth of sentences in the underlying language, but on sentences with parameters from the universe of $\mathfrak{A}$ as well. For example, $(\mathbb{N}^{>0}, <)$ is a substructure of $(\mathbb{N}, <)$, and since the two structures are isomorphic, they are elementarily equivalent. But in the first structure, the sentence $\exists x\, (x < y)$ is false when y is 1, whereas it is true in the second. The following, known as the Tarski–Vaught test, gives a necessary and sufficient condition for one structure to be an elementary substructure of another.

Proposition 5.6.1. *Suppose $\mathfrak{A}$ and $\mathfrak{B}$ are structures for a language L, and $\mathfrak{A}$ is a substructure of $\mathfrak{B}$. Then $\mathfrak{A}$ is an elementary substructure of $\mathfrak{B}$ if and only if the following holds: whenever $A(\vec{x}, y)$ is any formula in L, $\vec{c}$ is any tuple of elements from $|\mathfrak{A}|$, and b is an element of $|\mathfrak{B}|$ such that $\mathfrak{B} \models A(\vec{c}, b)$, then there is an element a of $|\mathfrak{A}|$ such that $\mathfrak{B} \models A(\vec{c}, a)$.*

The fact that the condition holds when $\mathfrak{A}$ is an elementary substructure of $\mathfrak{B}$ is immediate. The other direction can be proved using induction on formulas, and the details are left as an exercise.

If φ is an embedding of $\mathfrak{A}$ into $\mathfrak{B}$, then φ is also an isomorphism of $\mathfrak{A}$ with its image in $\mathfrak{B}$, interpreted as a substructure of $\mathfrak{B}$. It is called an *elementary embedding* if the image of $\mathfrak{A}$ is an elementary substructure of $\mathfrak{B}$.

It is not hard to check that the nonstandard model of arithmetic asserted to exist by Theorem 5.5.8 is an elementary extension of a structure that is isomorphic to the standard model, $\mathfrak{N}$. This is because the language of arithmetic has names 0, succ(0), succ(succ(0)), ... for each natural number and the theory specifies the behavior of the function and relation symbols on those terms. In general, it is not the case that every element of a structure is denoted by some closed term in the underlying language. But if $\mathfrak{A}$ is an L-structure, we can always add a new constant $\bar{a}$ for each element a of $\mathfrak{A}$ to obtain the language $L(\mathfrak{A})$. The *diagram* of $\mathfrak{A}$ is defined to be the set of atomic and negated atomic sentences of $L(\mathfrak{A})$ that are true in $\mathfrak{A}$. The *complete diagram* of $\mathfrak{A}$ is defined to be the set of all the first-order sentences of $L(\mathfrak{A})$ that are true in $\mathfrak{A}$.

Proposition 5.6.2. *$\mathfrak{A}$ is isomorphic to a substructure of $\mathfrak{B}$ if and only if $\mathfrak{B}$ can be expanded to a model of the diagram of $\mathfrak{A}$. $\mathfrak{A}$ is isomorphic to an elementary substructure of $\mathfrak{B}$ if and only if $\mathfrak{B}$ can be expanded to a model of the complete diagram of $\mathfrak{A}$.*

Proof If $\varphi\colon |\mathfrak{A}| \to |\mathfrak{B}|$ is an isomorphism, then $\mathfrak{B}$ can be expanded to a model of the diagram of $\mathfrak{A}$ by interpreting each constant $\bar{a}$ by $\varphi(a)$. Conversely, if $\mathfrak{B}$ is a model of the diagram of $\mathfrak{A}$, the map $\varphi\colon a \mapsto \bar{a}^{\mathfrak{B}}$ is an isomorphism. The proof of the second part of the proposition is similar. □

Proposition 5.6.2 can be used is to prove the *Löwenheim–Skolem theorems*. Say the *cardinality* of a language, *L*, is the cardinality of the set of all constant, function, and relation symbols. The following is a refined statement of Theorem 5.1.5.

Theorem 5.6.3. *Let Γ be a consistent set of sentences in a language, L, and let λ be the maximum of the cardinality of L and the smallest infinite cardinal, $\aleph_0$. Then Γ has a model of cardinality at most λ.*

Proof The set of terms used in the syntactic construction in Section 5.1 has cardinality at most λ. □

Theorem 5.6.4. (Upward Löwenheim–Skolem Theorem). *Let $\mathfrak{A}$ be any infinite structure for a language L. Then for any infinite cardinal λ greater than or equal to the cardinality of L and the cardinality of $|\mathfrak{A}|$, there is a $\mathfrak{B}$ such that $|\mathfrak{B}| = \lambda$ and $\mathfrak{A}$ is an elementary substructure of $\mathfrak{B}$.*

Proof (sketch) Let C be a set of λ-many new constants, and let Γ be the complete diagram of $\mathfrak{A}$ together with the set of atomic sentences $\{c \neq d \mid c, d \in C\}$. Every finite subset of Γ is consistent, since it can be interpreted in $\mathfrak{A}$, assigning the constants in C to any distinct elements. By Theorem 5.6.3, Γ has a model $\mathfrak{B}$ of cardinality at most λ, and hence of cardinality exactly λ. Since $\mathfrak{B}$ is a model of the complete diagram of $\mathfrak{A}$, $\mathfrak{A}$ is isomorphic to an elementary substructure of $\mathfrak{B}$. With some set-theoretic surgery – replacing the relevant elements of $\mathfrak{B}$ with the corresponding elements of $\mathfrak{A}$ – we can arrange it so that $\mathfrak{A}$ is an elementary substructure of $\mathfrak{B}$. □

Theorem 5.6.5 (Downward Löwenheim–Skolem Theorem). *Let $\mathfrak{A}$ be any structure for a language L. Then for any infinite cardinal λ greater than or equal to the cardinality of L and less than or equal to the cardinality of $|\mathfrak{A}|$, there is a $\mathfrak{B}$ such that $|\mathfrak{B}| = \lambda$ and $\mathfrak{B}$ is an elementary substructure of $\mathfrak{A}$.*

First we need a lemma.

Lemma 5.6.6. *Let L be a language, and $\mathfrak{M} \subseteq \mathfrak{N}$ structures for L. Let A, B, and C be sentences of $L(\mathfrak{M})$ such that A is quantifier-free, C is existential, and D is universal. Then*

1. *$\mathfrak{M} \models A$ if and only if $\mathfrak{N} \models A$.*
2. *If $\mathfrak{M} \models C$, then $\mathfrak{N} \models C$.*
3. *If $\mathfrak{N} \models D$, then $\mathfrak{M} \models D$.*

Proof The first claim is proved with a straightforward induction on A, and the other claims follow easily. □

One sometimes expresses the conclusion of the lemma by saying that quantifier-free formulas are *absolute* between submodels, existential formulas are *upward persistent*, and universal formulas are *downward persistent*.

Proof of Theorem 5.6.5 (sketch) Let λ be the cardinality of L, or the least uncountable cardinal, $\aleph_0$, if L is finite. We need to use the fact from Section 7.5 that $\mathfrak{A}$ can be expanded with λ-many Skolem functions such that, under the assumption that these are Skolem functions, every formula in the expanded language is equivalent to one that is quantifier-free. Now start with any subset $S \subseteq |\mathfrak{A}|$ of cardinality λ and close it under these Skolem functions. The result is still a set of cardinality λ. Let $\mathfrak{B}$ be the substructure of $\mathfrak{A}$ with this universe. By Lemma 5.6.6 the same quantifier-free formulas with parameters from $|\mathfrak{B}|$ are true in $\mathfrak{A}$ and $\mathfrak{B}$. Since the Skolem functions in $\mathfrak{A}$ are still Skolem functions in $\mathfrak{B}$, $\mathfrak{B}$ is an elementary substructure of $\mathfrak{A}$. □

Consider Zermelo–Fraenkel set theory, ZF, the first-order theory defined in Chapter 17. In ZF, we can formally define the real numbers and prove that they are uncountable. However, the downward Löwenheim–Skolem theorem (in fact, even Theorem 5.6.3) implies that ZF, if consistent, has a countable model. The existence of a countable model in which the statement "there are uncountably many real numbers" holds is known as *Skolem's paradox*.

Skolem's paradox isn't really a paradox. In any model of ZF, there are objects in the model that serve as numbers, functions, and so on. Because it is a model of ZF there is no object in the model that satisfies the statement that it is a bijection from the natural numbers to the reals. But the model is just an ordinary mathematical object, and our mathematics tells us that there is a bijection between the natural numbers and those objects in the model that satisfy the property of being a real number. This shows that, when dealing with models of foundational theories, it is important to distinguish between ordinary mathematical statements and formal statements that may or may not hold in such a model. Skolem's paradox is analogous to the fact that we can discuss models of the theory of the natural numbers in which we can see, from the outside, that there are elements of the universe with infinitely many elements below them in the sense of the model.

Let Γ be a set of universal axioms and let T be the theory axiomatized by Γ. Lemma 5.6.6 implies that the class of models of T is closed under substructures, which is to say, whenever $\mathfrak{B}$ is in the class and $\mathfrak{A} \subseteq \mathfrak{B}$, then $\mathfrak{A}$ is in the class also. Interestingly, the converse also holds:

Theorem 5.6.7. *If T is any theory, then T is universally axiomatizable if and only if the class of models of T is closed under substructures.*

This provides an interesting correspondence between a syntactic property and a semantic one.

Proof We already have the forward direction. For the reverse direction, suppose the class of models of T is closed under substructures, and let

$$\Gamma = \{A \mid A \text{ is universal sentence and } T \vdash A\}.$$

I claim that any model of Γ is also a model of T, so that Γ provides the necessary axiomatization. It suffices to show the contrapositive, which states that if some structure is not a model of T, then it is not a model of Γ either.

To show this, suppose $\mathfrak{A}$ is a structure that does not satisfy T. Since the class of models of T is closed under substructures, $\mathfrak{A}$ is not a substructure of any model of T. Let Δ be the diagram of $\mathfrak{A}$. By Proposition 5.6.2, $\Delta \cup T$ is inconsistent. By compactness, some finite subset of this set is inconsistent. Hence there is a finite set $\{A_0, \ldots, A_{n-1}\}$ of literals in Δ such that T proves $\neg A_0(\vec{c}) \vee \cdots \vee \neg A_{n-1}(\vec{c})$, where $\vec{c}$ enumerates all the names for elements of $\mathfrak{A}$ that occur in these sentences. Since these constants do not occur in T, Proposition 4.2.1 implies that T proves the sentence $\forall \vec{x}\, (\neg A_0(\vec{x}) \vee \cdots \vee \neg A_{n-1}(\vec{x}))$. But this a universal sentence, and hence it is in Γ. It is false in $\mathfrak{A}$, since, if $\vec{a}$ are the elements named by $\vec{c}$, we have $\mathfrak{A} \models A_i(\vec{a})$ for each i. So $\mathfrak{A}$ is not a model of Γ. □

A sentence A is said to be $\forall\exists$ if it consists of a (possibly empty) block of universal quantifiers followed by a (possibly empty) block of existential quantifiers, followed by a quantifier-free formula. In other words, A is $\forall\exists$ if it is of the form $\forall \vec{x}\, \exists \vec{y}\, B(\vec{x}, \vec{y})$, where B is quantifier-free.

Fix a language L. A theory T in L is said to be *preserved under increasing chains* if whenever $\mathfrak{A}_0 \subseteq \mathfrak{A}_1 \subseteq \mathfrak{A}_2 \subseteq \mathfrak{A}_3 \subseteq \ldots$ is a sequence of models of T, each a substructure of the next, then $\bigcup_i \mathfrak{A}_i$ is a model of T as well. (This last structure denotes the structure whose universe is $\bigcup_i |\mathfrak{A}_i|$, with the functions and relations derived from the corresponding ones on the $\mathfrak{A}_i$'s.) In the exercises, you are asked to prove the following:

Theorem 5.6.8. *A theory T is axiomatized by a set of $\forall\exists$ sentences if and only it is preserved under increasing chains.*

I will close this section with one last bit of storytelling. Our construction of a nonstandard model of arithmetic involved building a model satisfying the set of sentences $\{c > 0, c > \mathrm{succ}(0), c > \mathrm{succ}(\mathrm{succ}(0)), \ldots\}$, where c is a constant outside the language initially under consideration. Replacing c by a variable, x, results in what is known as a *1-type*. More generally, a set Γ of formulas in a language L whose variables are among a finite tuple $x_0, \ldots, x_{n-1}$ is said to be an *n-type*, or just a *type*. It is a *complete type* if for every formula

$A(\vec{x})$ in those variables, either $A(\vec{x})$ or $\neg A(\vec{x})$ is in Γ. An n-type Γ is said to be *realized* in a model $\mathfrak{A}$ if and only if there is a tuple $a_0, \ldots, a_{n-1}$ such that $\mathfrak{A} \models \Gamma(a_0, \ldots, a_{n-1})$.

Let $\mathfrak{A}$ be a structure in a language L, let X be a subset of $|\mathfrak{A}|$, and let $\Gamma(\vec{x})$ be a type in $L(X)$. In the exercises you are asked to show that the following are equivalent:

1. $\Gamma(\vec{c})$ is consistent with the complete diagram of $\mathfrak{A}$.
2. For every finite subset $\{A_0, \ldots, A_{n-1}\}$ of Γ, $\mathfrak{A} \models \exists \vec{x}\, (A_0 \wedge \cdots \wedge A_{n-1})$.
3. Γ is realized in some elementary extension of $\mathfrak{A}$.

Such a type Γ is said to be a *type of* $\mathfrak{A}$ *over* X, or simply a *type over* X when $\mathfrak{A}$ is understood. Intuitively, a type over X is a description of a finite tuple of objects in terms of X, whose existence is consistent with $\mathfrak{A}$.

Just as we were able to realize a single 1-type in constructing our nonstandard model of arithmetic, we can iterate the process and realize lots of types. A structure $\mathfrak{A}$ is said to be λ-*saturated* if, for every $X \subseteq |\mathfrak{A}|$ with $|X| < \lambda$, every type over X is realized in $\mathfrak{A}$. A structure $\mathfrak{A}$ is *saturated* if it is λ-saturated for λ equal to the cardinality of the universe of $\mathfrak{A}$.

Theorem 5.6.9. *Let T be a consistent theory with an infinite model. Then, for every λ, T has a λ-saturated model. Assuming the generalized continuum hypothesis, T has a saturated model.*

In a sense, a saturated model is one in which anything that can possibly happen does happen. For that reason, they have interesting model-theoretic properties, and provide fruitful terrain for model-theoretic constructions. For more information, see the notes at the end of this chapter.

Exercises

5.6.1. Prove the correctness of the Tarski–Vaught test, Proposition 5.6.1.

5.6.2. A theory T is said to be *complete* if for every sentence A, either $T \vdash A$ or $T \vdash \neg A$. The *Łoś–Vaught test* says that if T is a theory with only infinite models and T is categorical in some infinite cardinal, then T is complete. Prove this.

5.6.3. Consider the theory in the language with 0 and succ axiomatized by the following sentences:

- $\forall x\, (\mathrm{succ}(x) \neq 0)$
- $\forall x, y\, (\mathrm{succ}(x) = \mathrm{succ}(y) \rightarrow x = y)$
- $\forall x\, (x \neq 0 \rightarrow \exists y\, (\mathrm{succ}(y) = x))$
- For each i, the sentence $\forall x\, \mathrm{succ}^i(x) \neq x$, where $\mathrm{succ}^i(x)$ abbreviates the expression succ(succ $(\cdots(\mathrm{succ}(x))\cdots))$ with i occurrences of succ.

a. Show that this theory is not categorical for countable structures.
b. Show that this theory *is* categorical for uncountable structures. Hence, by the Łoś–Vaught test, it is the complete theory of $(\mathbb{N}, 0, \mathrm{succ})$.

5.6.4. Prove Theorem 5.6.8 as follows:

a. Show the forward direction: if a theory T is axiomatized by a set of $\forall\exists$ sentences, then it is preserved under increasing chains.
b. Suppose $\mathfrak{A}$ and $\mathfrak{B}$ are structures for L with $\mathfrak{A} \subseteq \mathfrak{B}$. Say that $\mathfrak{A}$ is a 1-elementary substructure of $\mathfrak{B}$, written $\mathfrak{A} \preceq_1 \mathfrak{B}$, if for every universal formula $\forall \vec{x}\, A(\vec{x}, \vec{y})$ and every sequence $\vec{a}$ of elements of $|\mathfrak{A}|$,

$$\mathfrak{A} \models \forall \vec{x}\, A(\vec{x}, \vec{a}) \quad \Leftrightarrow \quad \mathfrak{B} \models \forall \vec{x}\, \varphi(\vec{x}, \vec{a}).$$

Show that if $\mathfrak{A} \preceq_1 \mathfrak{B}$, there is a model $\mathfrak{C}$ such that $\mathfrak{B} \subseteq \mathfrak{C}$ and $\mathfrak{A} \preceq \mathfrak{C}$ (in other words, $\mathfrak{A}$ is an elementary substructure of $\mathfrak{M}$). (Hint: It suffices to show that the diagram of $\mathfrak{B}$ together with the complete diagram of $\mathfrak{A}$ is consistent. Suppose otherwise, and use compactness to get a contradiction.)

c. Suppose T is preserved under unions of chains, and let

$$\Gamma = \{A \mid A \text{ is a } \forall\exists \text{ sentence and } T \models A\}.$$

Show that if $\mathfrak{A}$ is any model of Γ, then there is a model $\mathfrak{B} \succeq_1 \mathfrak{A}$ such that $\mathfrak{B}$ is a model of T.

d. Use Proposition 5.6.1 to show that given a sequence of models

$$\mathfrak{A}_1 \preceq \mathfrak{A}_2 \preceq \mathfrak{A}_3 \preceq \ldots,$$

where each $\mathfrak{A}_i$ is an elementary substructure of $\mathfrak{A}_{i+1}$, we have that each $\mathfrak{A}_i$ is a substructure of $\bigcup_i \mathfrak{A}_i$.

e. Show that Γ axiomatizes T. Clearly everything in Γ is a consequence of T. Conversely, it suffices to show that any model $\mathfrak{A}$ of Γ is a model of T. Build a chain of models

$$\mathfrak{A} \subseteq \mathfrak{A}_0 \subseteq \mathfrak{A}_1 \subseteq \mathfrak{A}_2 \subseteq \ldots$$

such that for each i, $\mathfrak{A}_{2i} \preceq_1 \mathfrak{A}_{2i+1}$, $\mathfrak{A}_{2i} \preceq \mathfrak{A}_{2i+2}$, and $\mathfrak{A}_{2i+1}$ is a model of T.

5.6.5. Show that the three characterizations of a type over a set $X \subseteq |\mathfrak{A}|$ presented before Theorem 5.6.9 are equivalent.

Bibliographical Notes

Classical model theory There are a number of good introductions to the model theory of classical structures, including Chang and Keisler (1990), Hodges (1993), and Marker (2002). These cover all the topics discussed in Sections 5.5 and 5.6. Ehrenfeucht-Fraissé games are described in Hodges's book, and also in numerous textbooks on finite model theory, such as Ebbinghaus and Flum (2006) and Libkin (2004).

Intuitionistic model theory Troelstra and van Dalen (1988) provides a thorough introduction to the semantics of intuitionistic logic. Fitting (1969) and Mints (2000b) are also good references. For Kripke models in particular see also Moschovakis (2018) and Smoryński (1973). Beth models are discussed in Troelstra and van Dalen (1988) and Smoryński (1973). For sheaf-theoretic generalizations of Beth models for first-order logic, see Palmgren (1997) and Butz and Johnstone (1998).

For the completeness of intuitionistic logic with the schema (CD) discussed in Section 5.3 with respect to Kripke models with constant domain, see Görnemann (1971). For a computational interpretation of (CD) see Aschieri (2018), and for a computational interpretation of (DNS), see Escardó and Oliva (2017).

Lattices Balbes and Dwinger (1974), Davey and Priestley (2002), and Johnstone (1986) are good references for the lattice-theoretic constructions described in Section 5.4. Our construction of the MacNeille completion follows Johnstone. The construction presented in Exercise 5.4.5 is described in Troelstra and van Dalen (1988).

6

Cut Elimination

In Chapters 2 and 4, we considered two kinds of deductive systems, namely, axiomatic systems and natural deduction. In this chapter, we will consider yet a third kind of deductive system, sequent calculi. Whereas systems of natural deduction are pleasant to reason *with*, an important feature of a sequent calculus is that it is pleasant to reason *about*.

In natural deduction there are introduction and elimination rules for each of the connectives. In contrast, a sequent calculus has only introduction rules, but now connectives can be introduced on both the left and right sides of the sequent. Reading the rules upward, an introduction rule on the left side tells us how to work forward from a hypothesis, and an introduction rule on the right side tells us how to work backward from the goal. The one exception is the *cut rule*, essentially a form of modus ponens, which is necessary for an efficient translation of proofs from natural deduction. In this chapter we will see that the sequent calculus is complete even without the cut rule, and we will consider explicit procedures that show how, effectively, to eliminate cuts from proofs.

It is the cut rule that allows us to prove a formula B by first proving a lemma $A \to B$ and then establishing A, or to prove a formula $C(t)$ by proving the more general formula $\forall x\, C(x)$ and then instantiating the universal quantifier. The cut elimination theorem tells us that such detours can always be avoided, resulting in a more direct proof of the conclusion. Every formula that appears in a cut-free proof is a subformula of a formula that appears in the final sequent, a useful fact that we will exploit in the next chapter.

6.1 An Intuitionistic Sequent Calculus

To start, we will focus on first-order logic without equality. As was the case with natural deduction, sequent calculi for intuitionistic and minimal logic are based on the notion of a sequent, $\Gamma \vdash A$, where Γ is a finite set of formulas and A is any formula. For minimal logic, the rules for the propositional connectives are listed in Figure 6.1. The first rule is called the

$$\frac{}{\Gamma, A \vdash A}$$

$$\frac{\Gamma, A_i \vdash B}{\Gamma, A_0 \wedge A_1 \vdash B}\ \wedge\mathrm{L} \qquad \frac{\Gamma \vdash A \qquad \Gamma \vdash B}{\Gamma \vdash A \wedge B}\ \wedge\mathrm{R}$$

$$\frac{\Gamma, A \vdash C \qquad \Gamma, B \vdash C}{\Gamma, A \vee B \vdash C}\ \vee\mathrm{L} \qquad \frac{\Gamma \vdash A_i}{\Gamma \vdash A_0 \vee A_1}\ \vee\mathrm{R}$$

$$\frac{\Gamma \vdash A \qquad \Gamma, B \vdash C}{\Gamma, A \to B \vdash C}\ \to\mathrm{L} \qquad \frac{\Gamma, A \vdash B}{\Gamma \vdash A \to B}\ \to\mathrm{R}$$

Figure 6.1 The intuitionistic sequent calculus.

assumption rule. Proposition 6.1.1 shows that we can restrict that rule to atomic formulas and then derive it for all the others, so we adopt that restriction. In the $\vee R$ rule, i can be either 0 or 1. The quantifier rules are as follows:

$$\frac{\Gamma, A[t/x] \vdash B}{\Gamma, \forall x\, A \vdash B}\ \forall\mathrm{L} \qquad \frac{\Gamma \vdash B}{\Gamma \vdash \forall x\, B}\ \forall\mathrm{R}$$

$$\frac{\Gamma, A \vdash B}{\Gamma, \exists x\, A \vdash B}\ \exists\mathrm{L} \qquad \frac{\Gamma \vdash B[t/x]}{\Gamma \vdash \exists x\, B}\ \exists\mathrm{R}$$

The usual eigenvalue restrictions hold, which is to say, in the right $\forall$ rule we assume that x is not free in Γ, and in the left $\exists$ rule we assume that x is not free in Γ, B. For intuitionistic logic, add the following:

$$\frac{}{\Gamma, \bot \vdash A}\ \bot\mathrm{L}$$

As in Proposition 2.7.2, we can restrict A to atomic formulas here as well.

The final rule, for both minimal and intuitionistic logic, is the *cut rule*:

$$\frac{\Gamma \vdash A \qquad \Gamma, A \vdash B}{\Gamma \vdash B}\ \mathrm{Cut}$$

Proofs that do not use the cut rule are said to be cut-free. The set of provable sequents can now be defined to be the inductive closure of these rules, and similarly for the set of sequents that are provable without cut.

Proposition 6.1.1. *Provability in the sequent calculus has the following properties:*

1. *If $\Gamma \vdash A$ is provable and $\Gamma' \supseteq \Gamma$, then $\Gamma' \vdash A$ is also provable.*
2. *$\Gamma, A \vdash A$ is provable for any A.*

The same holds for provability without cut.

Proof The first claim, known as *weakening*, follows from an easy induction on the set of provable sequents. The second claim involves a straightforward induction on A. For example, assuming $\Gamma, A \vdash A$ and $\Gamma, B \vdash B$ are provable, the following takes care of $A \to B$:

$$\frac{\dfrac{\Gamma, A \vdash A \qquad \Gamma, B \vdash B}{\Gamma, A, A \to B \vdash B}}{\Gamma, A \to B \vdash A \to B}$$

Here there is an implicit appeal to weakening in the first inference. The other connectives are handled similarly. □

Proposition 6.1.1 justifies the use of more liberal forms of the rules in which the sets of hypotheses in the premises need not be the same. For example, we could equally well have adopted the following version of the $\wedge$-introduction rule:

$$\frac{\Gamma \vdash A \qquad \Delta \vdash B}{\Gamma, \Delta \vdash A \wedge B}$$

The original form of the rule makes sense when reading upward from the conclusion, since it provides us with a recipe for establishing $\Gamma \vdash A \wedge B$. The more liberal form is more natural when reasoning forward from the premises to the conclusion. We will henceforth use the more liberal versions of the rules without further comment.

When looking for proofs in natural deduction, it is heuristically useful to use elimination rules to work forward from hypotheses and introduction rules to work backward from the goal. When read from the bottom up, the rules of the sequent calculus model this process: right rules correspond to working backward from the goal and left rules correspond to working forward from hypotheses. The natural deduction proof of $A \wedge B \rightarrow B \wedge A$ in Section 2.4 can be interpreted as a translation of the following sequent proof:

$$\cfrac{\cfrac{\cfrac{\cfrac{}{B \vdash B}}{A \wedge B \vdash B} \qquad \cfrac{\cfrac{}{A \vdash A}}{A \wedge B \vdash A}}{A \wedge B \vdash B \wedge A}}{\vdash A \wedge B \rightarrow B \wedge A}$$

But it can also be seen as arising from the following proof:

$$\cfrac{\cfrac{\cfrac{\cfrac{\cfrac{}{B \vdash B} \qquad \cfrac{}{A \vdash A}}{A, B \vdash B \wedge A}}{A \wedge B, B \vdash B \wedge A}}{A \wedge B \vdash B \wedge A}}{\vdash A \wedge B \rightarrow B \wedge A}$$

Note that both of these proofs are cut-free. We will see in Section 13.8 that cut-free proofs in intuitionistic and minimal logic generally correspond to *normal* proofs in natural deduction, though this correspondence is a many-one map.

Theorem 6.1.2. *A sequent is provable in natural deduction for minimal or intuitionistic logic if and only if it if is provable in the corresponding version of the sequent calculus.*

Proof In the forward direction, we show by induction that if a sequent is derivable in natural deduction, it is derivable in the sequent calculus. The introduction rules in natural deduction correspond to right introduction rules in the sequent calculus. The elimination rules, however, require some work. For example,

$$\cfrac{\vdots \\ \Gamma \vdash A \wedge B}{\Gamma \vdash A}$$

in natural deduction translates to

$$\dfrac{\begin{array}{c}\vdots \\ \Gamma \vdash A \wedge B\end{array} \qquad \dfrac{\overline{A \vdash A}}{A \wedge B \vdash A}}{\Gamma \vdash A}$$

in the sequent calculus. The $\rightarrow$-elimination rule

$$\dfrac{\begin{array}{c}\vdots \\ \Gamma \vdash A \rightarrow B\end{array} \qquad \begin{array}{c}\vdots \\ \Gamma \vdash A\end{array}}{\Gamma \vdash B}$$

in natural deduction translates to

$$\dfrac{\begin{array}{c}\vdots \\ \Gamma \vdash A \rightarrow B\end{array} \qquad \dfrac{\begin{array}{c}\vdots \\ \Gamma \vdash A\end{array} \qquad \dfrac{\overline{A \vdash A} \qquad \overline{B \vdash B}}{A, A \rightarrow B \vdash B}}{\Gamma, A \rightarrow B \vdash B}}{\Gamma \vdash B}$$

Note that in both instances we make use of the cut rule.

In the other direction, we need to prove by induction that any sequent derivable in the sequent calculus is also derivable in natural deduction. In this direction, the right introduction rules pose no problem, while the left introduction rules require some work. For example, a left $\wedge$ rule in the sequent calculus,

$$\dfrac{\begin{array}{c}\vdots \\ \Gamma, A \vdash C\end{array}}{\Gamma, A \wedge B \vdash C}$$

translates to

$$\dfrac{\dfrac{\begin{array}{c}\vdots \\ \Gamma, A \vdash C\end{array}}{\Gamma \vdash A \rightarrow C} \qquad \dfrac{\overline{A \wedge B \vdash A \wedge B}}{A \wedge B \vdash A}}{\Gamma, A \wedge B \vdash C}$$

In other words, we use modus ponens to simulate the inference. The other cases are similar and left for you to check. □

In order to describe what is perhaps the most useful feature of the sequent calculus, we first need a suitably general notion of a *subformula*.

Definition 6.1.3. The set of *(generalized) subformulas of* A, denoted $\mathrm{subform}(A)$, is defined by recursion on the depth of A, as follows:

- If C is atomic, $\mathrm{subform}(C) = \{C\}$.
- $\mathrm{subform}(C \circ D) = \{C \circ D\} \cup \mathrm{subform}(C) \cup \mathrm{subform}(D)$ for any propositional connective $\circ$.
- $\mathrm{subform}(\forall x\, C) = \{\forall x\, C\} \cup \bigcup\{\mathrm{subform}(C[t/x]) \mid t \text{ is a term}\}$.
- $\mathrm{subform}(\exists x\, C) = \{\exists x\, C\} \cup \bigcup\{\mathrm{subform}(C[t/x]) \mid t \text{ is a term}\}$.

A central property of cut-free proofs is this:

Proposition 6.1.4. *Every formula occurring in a cut-free proof of $\Gamma \vdash A$ is a generalized subformula of a formula in $\Gamma \cup \{A\}$.*

This is known as the subformula property, and is proved by an easy induction on cut-free proofs. Proposition 6.1.4 is markedly false in the presence of the cut rule, which gives a hint as to why this rule is difficult to use in automated proof search.

You may have noticed that Proposition 6.1.4 is not a property of the provability relation per se, but, rather, a property of formal derivations. In principle, we could avoid mention of these by defining, inductively, the relation "$\Gamma \vdash A$ is cut-free provable using formulas in Δ." But as we begin to discuss the cut elimination theorem, it will become increasingly difficult to avoid talking about formal derivations. So it is best to think about provability in terms of the existence of tree-like proofs.

In the absence of cut, we have to be careful as to how we define provability from hypotheses. As with natural deduction, we define $\Gamma \vdash A$ to mean that for some finite subset $\{B_0, \ldots, B_{k-1}\}$ of Γ there is a proof of the sequent $B_0, \ldots, B_{k-1} \vdash A$. In the exercises, you are asked to show that, in the system with cut, this is equivalent to saying that there is a proof of the sequent $\vdash A$ from weakenings of the sequents $\vdash B_0, \ldots, \vdash B_{k-1}$, treated as axioms. Without the cut rule, however, the second formulation is not equivalent to the first. For example, there is a cut-free proof of $P \wedge Q \vdash P$:

$$\dfrac{\overline{P \vdash P}}{P \wedge Q \vdash P}$$

But since, in a cut-free proof, connectives are only *introduced* into formulas, there is clearly no cut-free proof of $\vdash P$ from any weakening of $\vdash P \wedge Q$.

One also has to be careful with the treatment of bound variables when dealing with a cut-free calculus. If we neither identify formulas up to free variables nor allow renaming in the quantifier rules, then the cut-free calculus is not complete. For example, consider the sentence $\forall y\, (\forall x\, \forall y\, (A(x) \wedge B(y)) \to A(y))$ in a language with the two unary relation symbols shown. Notice that each of the quantifiers $\forall y$ binds a different occurrence of y. Working backward, we see that a cut-free proof has to begin like this:

$$\dfrac{\dfrac{\forall x\, \forall y\, (A(x) \wedge B(y)) \vdash A(y)}{\vdash \forall x\, \forall y\, (A(x) \wedge B(y)) \to A(y)}}{\vdash \forall y\, (\forall x\, \forall y\, (A(x) \wedge B(y)) \to A(y))}$$

But now we are stuck: the variable x should be instantiated to y, but y is already being used by the universal quantifier. The solution is to allow renaming in the quantifier rules so that in the last step we can replace y by a new variable:

$$\dfrac{\vdash \forall x\, \forall y\, (A(x) \wedge B(y)) \to A(z)}{\vdash \forall y\, (\forall x\, \forall y\, (A(x) \wedge B(y)) \to A(y))}$$

Identifying formulas up to renaming of bound variables avoids this issue altogether.

If the cut rule can be eliminated from proofs, why use it at all? The cut rule makes it easy to translate proofs from axiomatic systems and natural deduction, and we will see that in extreme cases avoiding the use of cut requires an iterated exponential increase in length. By allowing ourselves the use of the cut rule but guaranteeing that it can be eliminated, we can use the full calculus to establish provability and the cut-free calculus to establish important properties of the provability relation.

Exercises

6.1.1. Give a proof of the sequent $(A \to B) \wedge (B \to C) \vdash A \to C$ in the sequent calculus for minimal logic.

6.1.2. Complete the proof of Proposition 6.1.1.

6.1.3. Pick some other examples of minimal and intuitionistic laws and find proofs in the sequent calculus.

6.1.4. Fill in the details of the proof of Theorem 6.1.2.

6.1.5. Fill in the details of the proof of Proposition 6.1.4.

6.1.6. Show that for any sentences $B_0, \ldots, B_{k-1}$ and any formula A, there is a derivation of $B_0, \ldots, B_{k-1} \vdash A$ in the sequent calculus if and only if there is a derivation of $\vdash A$ from weakenings of the sequents $\vdash B_0, \ldots, \vdash B_{k-1}$.

6.2 Classical Sequent Calculi

To extend the sequent calculus to classical logic, one idea is simply to add a sequent form of *reductio ad absurdum*:

$$\frac{\Gamma, \neg A \vdash \bot}{\Gamma \vdash A}$$

But since $\neg A$ is not a subformula of A, this would mean giving up the subformula property, which is something we do not want to do. Fortunately, there is an alternative: if we modify the notion of a sequent to allow more than one formula on the right, we can capture classical logic while maintaining the subformula property.

To that end, we now take sequents to be of the form $\Gamma \vdash \Delta$, where Γ and Δ are finite sets of formulas. The informal interpretation of this sequent is that the conjunction of the formulas in Γ implies the *disjunction* for the formulas in Δ; in other words, if everything on the left holds, then something on the right does. The rules for the intuitionistic calculus are now modified to allow additional formulas on the right. The resulting rules are listed in Figure 6.2. As before, weakening is easily shown to be admissible, now in the sense that if $\Gamma' \supseteq \Gamma$, $\Delta' \supseteq \Delta$, and $\Gamma \vdash \Delta$ is provable, then so is $\Gamma' \vdash \Delta'$.

The following sequent proof of the law of the excluded middle explains how allowing extra formulas on the right results in classical logic:

$$\cfrac{\cfrac{\cfrac{\cfrac{}{A \vdash A, \bot}}{\vdash A, \neg A}}{\vdash A \vee \neg A, \neg A}}{\vdash A \vee \neg A}$$

The next theorem gives the precise sense in which natural deduction and the sequent calculus for classical logic coincide:

Theorem 6.2.1. $\Gamma \vdash A$ *is provable in natural deduction for classical logic if and only if it is provable in the sequent calculus for classical logic.*

This follows from the next three lemmas. The first lemma gives the forward direction.

$$\frac{}{\Gamma, A \vdash \Delta, A} \qquad \frac{}{\Gamma, \bot \vdash \Delta}\ \bot L$$

$$\frac{\Gamma, A_i \vdash \Delta}{\Gamma, A_0 \wedge A_1 \vdash \Delta}\ \wedge\mathrm{L} \qquad \frac{\Gamma \vdash \Delta, A \qquad \Gamma \vdash \Delta, B}{\Gamma \vdash \Delta, A \wedge B}\ \wedge\mathrm{R}$$

$$\frac{\Gamma, A \vdash \Delta \qquad \Gamma, B \vdash \Delta}{\Gamma, A \vee B \vdash \Delta}\ \vee\mathrm{L} \qquad \frac{\Gamma \vdash \Delta, A_i}{\Gamma \vdash \Delta, A_0 \vee A_1}\ \vee\mathrm{R}$$

$$\frac{\Gamma \vdash \Delta, A \qquad \Gamma, B \vdash \Delta}{\Gamma, A \to B \vdash \Delta}\ \to\mathrm{L} \qquad \frac{\Gamma, A \vdash \Delta, B}{\Gamma \vdash \Delta, A \to B}\ \to\mathrm{R}$$

$$\frac{\Gamma, A[t/x] \vdash \Delta}{\Gamma, \forall x\, A \vdash \Delta}\ \forall\mathrm{L} \qquad \frac{\Gamma \vdash A}{\Gamma \vdash \forall x\, A}\ \forall\mathrm{R}$$

$$\frac{\Gamma, A \vdash \Delta}{\Gamma, \exists x\, A \vdash \Delta}\ \exists\mathrm{L} \qquad \frac{\Gamma \vdash \Delta, A[t/x]}{\Gamma \vdash \Delta, \exists x\, A}\ \exists\mathrm{R}$$

$$\frac{\Gamma \vdash \Delta, A \qquad \Gamma, A \vdash \Delta}{\Gamma \vdash \Delta}\ \mathrm{Cut}$$

Figure 6.2 The classical sequent calculus.

Lemma 6.2.2. *If $\Gamma \vdash A$ is provable in classical natural deduction, then $\Gamma \vdash A$ is provable in the classical sequent calculus.*

Proof All the rules other than *reductio ad absurdum* are handled as before. To handle *reductio*, suppose that, inductively, we have a classical sequent proof of $\Gamma, \neg A \vdash \bot$. The following shows how to get a proof of $\Gamma \vdash A$:

$$\dfrac{\dfrac{\dfrac{}{A \vdash A, \bot}}{\vdash A, \neg A} \qquad \dfrac{\dfrac{\vdots}{\Gamma, \neg A \vdash \bot} \qquad \dfrac{}{\bot \vdash A}}{\Gamma, \neg A \vdash A}}{\Gamma \vdash A}$$

□

Translating the classical sequent calculus to natural deduction is more difficult. The trick is to move all the formulas into the set of hypotheses. First we need a lemma about classical natural deduction, whose proof is easy:

Lemma 6.2.3. *$\Gamma \vdash A$ is provable in classical natural deduction if and only if $\Gamma, \neg A \vdash \bot$ is.*

Remember that $\neg\Delta$ is defined to be $\{\neg A \mid A \in \Delta\}$.

Lemma 6.2.4. *If $\Gamma \vdash \Delta$ is provable in the classical sequent calculus, then $\Gamma, \neg\Delta \vdash \bot$ is provable in classical natural deduction.*

Proof By induction on derivations. The left introduction rules in the sequent calculus are handled as before. For the right introduction rules, we use Lemma 6.2.3 to move the relevant formulas back and forth. For example, to handle the right $\wedge$ rule, suppose, by the inductive hypothesis, we have proofs of $\Gamma, \neg\Delta, \neg A \vdash \bot$ and $\Gamma, \neg\Delta, \neg B \vdash \bot$ in natural deduction. By Lemma 6.2.3, we have proofs of $\Gamma, \neg\Delta \vdash A$ and $\Gamma, \neg\Delta \vdash B$ in natural deduction. By the $\wedge$-introduction rule, we have a proof of $\Gamma, \neg\Delta \vdash A \wedge B$. By Lemma 6.2.3, we have a proof of $\Gamma, \neg\Delta, \neg(A \wedge B) \vdash \bot$. The other right rules are handled similarly. □

The remaining direction of Theorem 6.2.1 now follows, since if $\Gamma \vdash A$ is provable in the sequent calculus, we have that $\Gamma, \neg A \vdash \bot$ is provable in natural deduction, and hence so is $\Gamma \vdash A$.

An advantage to the sequent calculus described above is that it is similar to the ones for intuitionistic and minimal logic. However, in the case of classical logic, the calculus can be simplified considerably. First, it is enough to restrict our attention to formulas in negation normal form, which means that we only have to deal with the connectives $\wedge$, $\vee$, $\forall$, and $\exists$. Second, as Lemma 6.2.4 suggests, two-sided sequents are unnecessary. If we move all the formulas to the right, we can take a sequent to be a finite set Γ of formulas, read disjunctively, corresponding to the sequent $\vdash \Gamma$ in the two-sided calculus. The rules of the one-sided sequent calculus for first-order logic without equality are then as follows:

$$\frac{}{\Gamma, \neg A, A}$$

$$\frac{\Gamma, A \qquad \Gamma, B}{\Gamma, A \wedge B}\ \wedge \qquad \frac{\Gamma, A_i}{\Gamma, A_0 \vee A_1}\ \vee$$

$$\frac{\Gamma, A}{\Gamma, \forall x\, A}\ \forall \qquad \frac{\Gamma, A[t/x]}{\Gamma, \exists x\, A}\ \exists$$

$$\frac{\Gamma, A \qquad \Gamma, \sim A}{\Gamma}$$

In the first rule, which expresses the law of the excluded middle, A is any atomic formula. In the $\forall$ rule, we assume x is not free in any formula in Γ. The last rule is the cut rule. remember that $\sim A$ is only a notation in the metatheory the result that denotes the result of switching $\wedge$ and $\vee$, $\forall$ and $\exists$, and atomic formulas and their negations in A. As usual, an easy induction on proofs shows that weakening is admissible. An advantage to this calculus is that we have cut down on the number of rules dramatically, which amounts to fewer cases when it comes to proving things by induction on derivations.

To link this calculus up with our prior notions of classical provability, the following theorem and its proof provide direct translations between the two-sided sequent calculus and the one-sided classical sequent calculus. Remember that, with care, we can define the translation of formulas to negation normal form so that every formula translates to $\bot$, $\top$, or a formula in which neither propositional constant appears. Similarly, we can contract a sequent $\Gamma, \bot$ to Γ, so in particular $\bot$ is represented by the empty sequent. We can also contract $\Gamma, \top$ to B for some fixed tautology B. These moves make it possible for us to dispense with $\top$ and $\bot$ in the language of the one-sided calculus.

Theorem 6.2.5. *$\Gamma \vdash \Delta$ is derivable in the two-sided calculus if and only if $\sim\Gamma^{\mathrm{nnf}}, \Delta^{\mathrm{nnf}}$ is derivable in the one-sided calculus.*

Proof The "only if" direction involves showing that proofs in the classical two-sided sequent calculus can be translated to proofs in the one-sided calculus. Let us consider, for example, the cases in which the last rules of inference are the left and right $\wedge$ rules, respectively.

Suppose the last inference in a sequent proof is

$$\frac{\Gamma \vdash \Delta, A \qquad \Gamma \vdash \Delta, B}{\Gamma \vdash \Delta, A \wedge B}$$

By the inductive hypothesis, we have a proof of $\sim\Gamma^{\mathrm{nnf}}$, Δ^{nnf}, A^{nnf} as well as a proof of $\sim\Gamma^{\mathrm{nnf}}$, Δ^{nnf}, B^{nnf} in the one-sided sequent calculus. By the $\wedge$ rule, we have a proof of $\sim\Gamma^{\mathrm{nnf}}$, Δ^{nnf}, $A^{\mathrm{nnf}} \wedge B^{\mathrm{nnf}}$, and since $(A \wedge B)^{\mathrm{nnf}}$ is $A^{\mathrm{nnf}} \wedge B^{\mathrm{nnf}}$, we are done.

Suppose the last inference in the sequent proof is

$$\frac{\Gamma, A \vdash \Delta}{\Gamma, A \wedge B \vdash \Delta}$$

By the inductive hypothesis, we have a sequent proof of $\sim\Gamma^{\mathrm{nnf}}$, $\sim A$, Δ. By the $\vee$ rule, we obtain a proof of $\sim\Gamma^{\mathrm{nnf}}$, $\sim A^{\mathrm{nnf}} \vee \sim B^{\mathrm{nnf}}$, Δ. As above, since $\sim(A \wedge B)^{\mathrm{nnf}}$ is $\sim A^{\mathrm{nnf}} \vee \sim B^{\mathrm{nnf}}$, we are done. You can check that these cases still go through when one of the premises is $\bot$ or $\top$. The other rules are handled similarly. Note, incidentally, that this translation takes cut-free proofs to cut-free proofs.

The "if" direction is trickier. In general, the mapping from arbitrary formulas to formulas in negation normal form is not one-one, since, for example, $\neg P \vee Q$ and $P \to Q$ both have the same translation. One option is to show, by induction, that whenever a sequent Π is derivable in the one-sided calculus, then $\Gamma \vdash \Delta$ is derivable in the two-sided calculus for *any* Γ and Δ such that $\sim\Gamma^{\mathrm{nnf}}$, Δ^{nnf} is equal to Π. This involves a lot of cases but has the advantage that cut-free proofs translate to cut-free proofs. An alternative is simply to show that $\vdash \Pi$ is derivable in the two-sided calculus, and then show that we can derive $\Gamma \vdash \Delta$ from Π. □

We have shown that a formula A is provable from a finite set of hypotheses Γ in an ordinary deductive system for classical logic if and only if $\sim\Gamma^{\mathrm{nnf}}$, A^{nnf} is derivable in the one-sided sequent calculus. Since the one-sided calculus provides a clear and concise characterization of classical provability, we will often use it to establish properties of classical first-order logic.

Exercises

6.2.1. a. Show by induction on formulas that it is possible to derive A, $\sim A$ in the one-sided sequent calculus for classical logic.

b. Conclude that the one-sided sequent calculus proves the law of the excluded middle, $A \vee \sim A$, for every formula A in negation normal form.

6.2.2. Prove Peirce's law, $((A \to B) \to A) \to A$, in the two-sided sequent calculus for classical logic. Then put the statement in negation normal form and give a proof in the one-sided sequent calculus.

6.2.3. Assuming x is not free in A, prove the sequent $A \to \exists x\, B \vdash \exists x\, (A \to B)$ in the classical two-sided sequent calculus.

6.2.4. Consider the following two possible definitions of $\Gamma \vdash A$ for the one-sided calculus:

For some finite $\{B_0, \ldots, B_{m-1}\} \subseteq \Gamma$, there is a proof of $\sim B_0, \ldots, \sim B_{n-1}, A$.

and

There is a proof of A from additional axioms Δ, B for each B in Γ.

Show that in the one-sided sequent calculus with the cut rule, the two are equivalent, but without the cut rule, $\{P, P \to Q\} \vdash Q$ holds in the first sense but not the second.

6.3 Cut-Free Completeness of Classical Logic

In this section, we will show that the one-sided sequent calculus without the cut rule is complete for classical first-order logic without equality. This gives us, indirectly, a cut elimination theorem: since the sequent calculus with cut is sound for our semantics, any sequent that is provable is valid, and hence, by completeness, provable without cut. Curiously, this does not provide us with an explicit means of translating a proof with cut to a proof without, although it guarantees that a blind (but systematic) search will eventually find one. In Section 6.4, we will consider an alternative, syntactic proof that provides an explicit translation.

Let us start with propositional logic, considered in terms of the one-sided sequent calculus. Remembering that sequents are read disjunctively, say that a sequent $A_0, \ldots, A_{k-1}$ is *valid* if and only if the disjunction $\bigvee A_i$ is a tautology. We will first prove the completeness theorem in the following restricted form:

Theorem 6.3.1. *If Δ is a valid propositional sequent, then Δ has a cut-free proof.*

Proof First, replace the $\vee$ rules with the following rule:

$$\frac{\Gamma, A, B}{\Gamma, A \vee B}$$

This rule is derivable in the previous system using two instances of the $\vee$ rule, and, conversely, the previous rules can be obtained from this one using weakening. With this change, all the rules of the system are *invertible*, which is to say, the conclusion of any rule is valid if and only if the premises are. Define the *size* of a formula to be the number of connectives it contains, not counting $\bot$ and negations (so the sizes of P and $\neg P$ are both 0). Define the size of a sequent to be the sum of the sizes of the formulas it contains.

To prove the theorem, use induction on the size of the sequent Δ. If the size is 0, every formula is either atomic or negated atomic. Since the sequent is valid, one of the variables has to occur with its negation, so the sequent is an axiom.

Otherwise, for some formulas C and B, the sequent is either of the form $\Delta', C \vee B$ or $\Delta', C \wedge B$. In the first case, the sequent is valid if and only if Δ', C, B is. By the inductive hypothesis, there is a proof of this latter sequent, which gives us a proof of $\Delta', C \vee B$. In the second case, the sequent is valid if and only if Δ', C and Δ', B both are. By the inductive hypothesis, there are proofs of both of these, and hence a proof of $\Delta', C \wedge B$. □

There is a picture associated with this proof: to find a formal proof of Δ, start building a tree with Δ at the bottom, and work upward, applying the rules backward. The procedure must stop after finitely many steps. If the last sequent on every path is an axiom, there is a proof of the sequent. Otherwise, at least one path ends with a sequent that is *not* an axiom. In that case, we can find a propositional assignment that makes this sequent, and hence the original one, false.

For example, suppose we want to decide whether the following sequent is valid:

$$P \wedge (\neg Q \vee \neg R), \neg Q \vee S, \neg P \vee (Q \wedge \neg S).$$

Working backward, we construct the tree in Figure 6.3. Since every leaf is an axiom, we have the desired proof. You should check that if we replace Q by another propositional variable U in the first formula of the original sequent, we can find a truth assignment that makes the resulting sequent false.

$$
\dfrac{\dfrac{\dfrac{\dfrac{}{P, \neg Q, S, \neg P, Q \wedge \neg S} \quad \dfrac{\dfrac{\dfrac{}{\neg R, \neg Q, S, \neg P, Q} \quad \dfrac{}{\neg R, \neg S, S, \neg P}}{\neg R, \neg Q, S, \neg P, Q \wedge \neg S}}{\neg Q \vee \neg R, \neg Q, S, \neg P, Q \wedge \neg S}}{P \wedge (\neg Q \vee \neg R), \neg Q, S, \neg P, Q \wedge \neg S}}{P \wedge (\neg Q \vee \neg R), \neg Q, S, \neg P \vee (Q \wedge \neg S)}}{P \wedge (\neg Q \vee \neg R), \neg Q \vee S, \neg P \vee (Q \wedge \neg S)}
$$

Figure 6.3 A cut-free proof.

The full completeness theorem says that for every set of formulas Γ and every formula A, if $\Gamma \models A$ then $\Gamma \vdash A$. As a corollary of the previous theorem, we have completeness in the following sense:

Theorem 6.3.2. *If $\Gamma \models A$ then for some finite subset $\Delta \subseteq \Gamma$ there is a cut-free proof of $\sim\Delta^{\mathrm{nnf}}, A^{\mathrm{nnf}}$.*

Proof By compactness, if $\Gamma \models A$, then for some finite subset $\Delta \subseteq \Gamma$, $\Delta \models A$. This means that, as a sequent, $\sim\Delta^{\mathrm{nnf}}, A^{\mathrm{nnf}}$ is valid, and hence has a cut-free proof. □

Can we find a more direct proof of completeness, without relying on the compactness theorem? In fact, we can, assuming that there are only countably many propositional variables. It should not be surprising that the proof uses a compactness principle of sorts, in the form of Kőnig's lemma. Remember that $\Gamma \models A$ if and only if $\Gamma \cup \{\neg A\}$ is unsatisfiable. So if we replace Γ by $\Gamma^{\mathrm{nnf}}, \sim A^{\mathrm{nnf}}$ in the following theorem, we obtain Theorem 6.3.2 as a consequence.

Theorem 6.3.3. *Suppose Γ is a set of sentences in negation normal form, and for every finite subset Δ of Γ, there is no cut-free proof of $\sim\Delta$. Then Γ is satisfiable.*

Proof (sketch) Assume the hypothesis and enumerate the formulas in Γ, so that

$$\Gamma = \{A_0, A_1, A_2, \ldots\}.$$

Build an infinite finitely branching tree in stages. At stage 0, unwind $\sim A_0$, searching for a proof. By the hypothesis, at least one leaf is not an axiom. Now, to each such leaf, append the formula $\sim A_1$, and continue unwinding. Again, at least one leaf of the resulting tree cannot be an axiom, because otherwise there would be a proof of $\sim A_0, \sim A_1$, obtained by adding copies of $\sim A_1$ to every node of the first tree. Repeat the process with $\sim A_2$, and so on.

Even though the labels of the nodes keep changing with the addition of new formulas, each stage only adds levels to the same underlying tree. By Kőnig's lemma (Proposition 1.4.4), there is a path P through this tree. Let

$$\Pi = \{\sim B \mid B \text{ appears in a sequent on } P\}.$$

Π has the following properties: first of all, no propositional formula and its negation appears in Π. Second, if $B \wedge C$ appears in Π, so do B and C, reading the $\vee$ rule backward. Finally, if $B \vee C$ appears in Π, so does either B or C.

Define a truth assignment v by setting

$$v(P) = \top \Leftrightarrow P \text{ is in } \Pi.$$

By induction on formulas A, the properties of Π ensure that $v(A) = \top$ for each A in Π. Since Γ is a subset of Π, v satisfies Γ. □

Let us now turn to first-order logic without equality. If $\Delta = \{A_0, \ldots, A_{k-1}\}$ is a finite set of first-order formulas, say Δ is *valid* if the universal closure of $\bigvee \Delta$ is true in every model.

Theorem 6.3.4. *If Δ is a valid first-order sequent, it has a cut-free proof.*

Proof We would like to use a proof that is similar to the one for propositional logic. Working backward, we will unwind Δ to obtain either a countermodel or a cut-free proof. But what does it mean to apply the quantifier rules backward? Remember that they are the following:

$$\frac{\Gamma, B}{\Gamma, \forall x\, B} \qquad \frac{\Gamma, B[t/x]}{\Gamma, \exists x\, B}$$

To handle the $\forall$ rule, let us add a new set of constants $c_0, c_1, c_2, \ldots$. Alternatively, you can think of these as variables that we agree to keep separate from the bound ones. When we get to a sequent of the form $\Gamma, \forall x\, B(x)$ in our unwinding procedure, we write above it $\Gamma, B(c)$ for some constant c that we have not used in the proof tree so far. If you think of the tree we are building as an attempted proof, then c acts as a free variable. If, on the other hand, you think of the tree as an attempt to build a counterexample, our actions embody the following reasoning: if $\Gamma, \forall x\, B$ is not valid, then we can build a model and find an element c such that every formula in $\Gamma, B[c/x]$ comes out false.

Dealing with the $\exists$ rule is trickier. In analogy to our treatment of the $\vee$ rule in the propositional case, we would like to replace the sequent $\Gamma, \exists x\, B$ with a sequent

$$\Gamma, B[t_0/x], B[t_1/x], B[t_2/x], \ldots,$$

where the sequence (t_i) enumerates all closed terms in the language. But infinite sequents are not allowed in our proof system, so we have to be more subtle, and add formulas of the form $B[t/x]$ one step at a time. This requires some bookkeeping. First, replace the finite sets of formulas by finite sequences, so that the formulas are listed in some order. When you get to $\Gamma, \exists x\, B$, replace it with

$$\exists x\, B, B[t_i/x], \Gamma,$$

where t_i is the first term such that $B[t_i/x]$ does not appear in Γ. In other words, process $\exists x\, B$ by instantiating the sentence at one term, and then move it back to the end of the queue. We are making a point of rotating sentences back to the queue after we have processed them, so that each sentence $\exists x\, B$ is processed infinitely often along any infinite path, if there is one.

If, at some finite stage, there is an axiom on each branch, we can read off a proof of Δ. Otherwise, either the process stops at some finite stage because on some branch branch which is not an axiom there is nothing left to do (which may happen if there are no existential quantifiers in the sequent), or the process goes on forever. In the first case, let P be the corresponding branch; in the second case, by Kőnig's lemma, let P be an infinite path through the tree. Let Π be the set $\{B \mid {\sim}B$ appears in some sequent of $P\}$. If we have done things right, Π will be a *Hintikka set*, which is to say it will be a set of sentences in the expanded language with the constants c_i with all the following properties:

1. If B is atomic, then B and $\neg B$ are not both in Π.
2. If $B \wedge C$ is in Π, so are B and C.
3. If $B \vee C$ is in Π, so is either B or C.
4. If $\forall x\, B$ is in Π, so is $B[t/x]$ for every closed term in the language.
5. If $\exists x\, B$ is in Π, then $B[c/x]$ is in Π, where c is one of the new constants.

If P is an infinite path then it does not contain an axiom, which guarantees claim 1. Claims 2 and 3 are established as in the propositional case. Our handling of the $\exists$ rule yields 4, our handling of the $\forall$ rule yields 5.

From Π we can then read off the desired model $\mathfrak{A}$. Let the universe of $\mathfrak{A}$ consist of the set of all closed terms in the language. For each function symbol f, define $f^{\mathfrak{A}}(t_0, \ldots, t_{k-1}) = f(t_0, \ldots, t_{k-1})$, and for each relation symbol R, let $R^{\mathfrak{A}}(t_0, \ldots, t_{k-1})$ hold if and only if $R(t_0, \ldots, t_{k-1})$ is in Π. An easy induction shows that for every $A \in \Pi$, we have $\mathfrak{A} \models A$. Since $\sim\Delta \subseteq \Pi$, $\mathfrak{A}$ is a model of $\sim\Delta$. □

By compactness, we therefore have the following:

Theorem 6.3.5. *Let Γ be any countable set of sentences, and let Δ be any sequent. If $\Gamma \models \Delta$, then for some finite subset $\Gamma' \subseteq \Delta$, there is a cut-free proof of $\sim\Gamma', \Delta$.*

Assuming the language is countable, we can alternatively modify the construction to avoid appealing to the compactness theorem, as we did for propositional logic.

In the next two sections, we will consider explicit cut elimination procedures for classical, intuitionistic, and minimal logic.

Exercises

6.3.1. Use tableau methods (that is, our completeness proof for the one-sided classical calculus) to find either a proof of the negation normal form the following propositional formula or a truth assignment that falsifies it:

$$((P \vee Q) \to R) \vee ((R \to P) \wedge (R \to Q)).$$

6.3.2. Show that every maximally consistent Henkin set of sentences is a Hintikka set, but the converse need not hold.

6.3.3. Fill in some of the details of the proof of Theorem 6.3.4 to show that every Hintikka set is satisfiable. In other words, show that the model constructed is, indeed, a model of Π. (Hint: it is technically smoother to show that for every formula $A(x_0, \ldots, x_{n-1})$ with the free variables shown and every assignment σ, if $A(\sigma(x_0), \ldots, \sigma(x_{n-1}))$ is in Π, then $\mathfrak{M} \models_\sigma A$.)

6.3.4. Consider Proposition 4.6.9 from Section 4.6. Show by induction on proofs that if $A_0, \ldots, A_{k-1}$ is provable in the classical one-sided calculus, then the sequent $\sim A_0, \ldots, \sim A_{k-1} \vdash \bot$ is provable in natural deduction for minimal logic. Show that this yields another proof of Proposition 4.6.9.

6.3.5. Recall from Section 2.8 the proof system for propositional logic based on nand() connectives with arbitrary arity, with axioms nand($\vec{A}$, $\vec{B}$, nand($\vec{B}$)) and the following rules:

$$\frac{\text{nand}(\vec{A}, \vec{B}) \qquad \text{nand}(\vec{A}, \text{nand}(\vec{B}))}{\text{nand}(\vec{A})}$$

Show that this system is complete, in the sense that if A is a tautology, it is derivable. (Hint: derive some additional invertible rules, and then use a completeness proof like the one for propositional logic, above.)

6.4 Cut Elimination for Classical Logic

Having shown nonconstructively that every sequent provable in the one-sided sequent calculus for classical first-order logic has a cut-free proof, we will now consider a proof that comes with an explicit procedure for eliminating cuts. We define the *depth* of a proof in the one-sided calculus to be the depth of the underlying tree, taking the depth of an axiom to be 0. Recall that, similarly, the depth of a formula in negation normal form is defined to be the depth of its parse tree, taking the depth of a literal (that is, an atomic formula or the negation of an atomic formula) to be 0. Given a proof, we define its cut rank to be one more than the maximum depth of any cut formula, so every cut formula in a proof with cut rank $r + 1$ has depth at most r. We will adopt the convention that a proof with cut rank 0 is one that is cut-free.

Let us write $\vdash^n_r \Gamma$ to assert that there is a proof of Γ with cut rank at most r and depth at most n. We continue to restrict our attention to first-order logic without equality for the time being. In the rules for the one-sided sequent calculus, we call the formulas in Γ the *side formulas* of the inference, the remaining formula in the conclusion the *principal formula* of the inference, and the remaining formulas in the premises the *main premises* of the inference. In an axiom $\Gamma, A, \neg A$, we consider both A and $\neg A$ to be principal formulas. Define iterated base k exponentiation by $k^r_0 = r$, $k^r_{n+1} = k^{k^r_n}$.

Theorem 6.4.1 (cut elimination theorem). *If* $\vdash^n_r \Gamma$ *in the classical one-sided sequent calculus for first-order logic, then* $\vdash^{2^n_r}_0 \Gamma$.

The iterated exponential may seem like a terrible upper bound on the increase in the depth of the proof. But it is unavoidable: see Exercise 6.4.4 and the notes at the end of this chapter.

The cut elimination theorem can be proved by induction, using the following lemma:

Lemma 6.4.2. *If* $\vdash^n_{r+1} \Gamma$*, then* $\vdash^{2^n}_r \Gamma$.

This lemma says that, given a proof of cut rank $r + 1$ (so that the maximum depth of a cut-formula is r), we can eliminate all the cuts on formulas of rank r with at most an exponential increase in depth. Applying Lemma 6.4.2 r-many times yields Theorem 6.4.1.

Toward proving Lemma 6.4.2, the following lemma shows that we can eliminate a single cut of depth r:

Lemma 6.4.3 (reduction lemma). *Suppose* $\vdash^m_r \Gamma, A$ *and* $\vdash^n_r \Gamma, {\sim}A$*, where* A *has depth* r*. Then we have* $\vdash^{m+n}_r \Gamma$.

Assuming the hypotheses to Lemma 6.4.3, it is easy to see that we can obtain a proof of depth at most $\max(m, n) + 1$ with cut rank $r + 1$: just apply the cut rule. The point of the lemma is that we can avoid this last cut, at the expense of adding the depths of the associated proof trees.

Our proof will give us an explicit recipe for doing this. This will, in turn, provide us with an explicit recipe for eliminating all the cuts of rank r: recursively eliminate all cuts on formulas of rank r below the final inference; and then, if the last inference is a cut on a formula of rank r, eliminate that too. The following shows that this suffices to give us the bound of Lemma 6.4.2.

Proof of Lemma 6.4.2 from Lemma 6.4.3 Use induction on proof trees. In the base case, the proof is an axiom, and there is nothing to do. In the inductive step, suppose $\vdash^{n+1}_{r+1} \Gamma$ is witnessed by a proof tree of depth at most $n+1$. If the last inference is not a cut on a formula of rank r, the conclusion follows easily by applying the inductive hypothesis to the proofs of the premises to the last rule, and then applying the last rule. Otherwise, the last inference is of the form

$$\dfrac{\begin{array}{cc} \vdots d_0 & \vdots d_1 \\ \Gamma, A & \Gamma, {\sim}A \end{array}}{\Gamma}$$

where d_0 and d_1 have cut rank at most $r+1$ and depth at most n. By the inductive hypothesis, there are proofs of Γ, A and $\Gamma, {\sim}A$ with cut rank at most r and depth at most 2^n. Now, by Lemma 6.4.3, there is a proof of Γ with cut rank r and depth at most $2^n + 2^n = 2^{n+1}$, as required. □

We have reduced our task to proving Lemma 6.4.3. The first strategy we will consider relies on the following two lemmas:

Lemma 6.4.4 (substitution lemma). *If* $\vdash^n_r \Gamma$ *then* $\vdash^n_r \Gamma[t/x]$.

Proof Use induction on n. If some variable of t is an eigenvariable in a $\forall$ rule, inductively substitute a fresh variable for that as well. □

Lemma 6.4.5 (inversion lemma). *We have the following:*

- *If* $\vdash^n_r \Gamma, A \wedge B$, *then* $\vdash^n_r \Gamma, A$ *and* $\vdash^n_r \Gamma, B$.
- *If* $\vdash^n_r \Gamma, \forall x\, A$ *then* $\vdash^n_r \Gamma, A[t/x]$.

Proof Each of these can be proved by a straightforward induction on proofs. By way of illustration, we will do the first. The only interesting case is where $A \wedge B$ is principal in the last inference. There is one caveat: we have to allow for the possibility that $A \wedge B$ is a side formula in the hypotheses but not a formula of Γ. In that case, the proof is of the following form:

$$\dfrac{\begin{array}{cc} \vdots d_0 & \vdots d_1 \\ \Gamma, A \wedge B, A & \Gamma, A \wedge B, B \end{array}}{\Gamma, A \wedge B}$$

Here d_0 and d_1 have depth at most $n-1$. This is, in fact, the most general case, since we can always weaken the hypotheses to add $A \wedge B$. Thus, here and below, we will always assume that the principal formula of an inference is a side formula as well.

Having dispensed with the caveat, we now simply apply the inductive hypothesis to either d_0 or d_1 depending on whether we are looking for a proof of Γ, A or Γ, B, and then we omit the last inference.

The second claim of the lemma is proved similarly, using Lemma 6.4.4. Once again, in the induction step, we assume without loss of generality that $\forall x\, A$ is a side formula of the last inference. □

We now turn to the proof of the reduction lemma.

Proof of Lemma 6.4.3 We use induction on $m+n$. By symmetry, we can assume that the formula A is either atomic, of the form $B \vee C$, or of the form $\exists x\, B$, in which case $\sim A$ is, respectively, the negation of an atomic formula, $\sim B \wedge \sim C$, or $\forall x \sim B$. Let d_0 be a proof of Γ, A of depth at most n and cut rank at most r, and let d_1 be a proof of $\Gamma, \sim A$ of depth at most m and cut rank at most r.

First, we show that if A is not principal in the last inference of d_0, the conclusion follows from the inductive hypothesis. This is the result of the fact that the cut rule can be permuted through any other rule. For example, suppose the last inference of d_0 is an instance of the $\wedge$-introduction rule,

$$\frac{\begin{matrix} \vdots d_{00} \\ \Gamma', D, A \end{matrix} \qquad \begin{matrix} \vdots d_{01} \\ \Gamma', E, A \end{matrix}}{\Gamma', D \wedge E, A}$$

where $\Gamma = \Gamma', D \wedge E$ and d_{00} and d_{01} have depth at most $m-1$ and cut rank at most r. By the inductive hypotheses, there are proofs of $\Gamma', D \wedge E, D$ and $\Gamma', D \wedge E, E$ with depth at most $m+n-1$ and cut rank at most r. Applying $\wedge$ introduction yields a proof of $\Gamma', D \wedge E = \Gamma$ with depth at most $m+n$ and cut rank at most r, as required. The other possibilities for the last inference of d_0 are handled similarly; note that if Γ, A is an axiom and A is not principal then Γ is an axiom as well.

So we are reduced to the case where A is principal in the last inference. By the assumption above, we have three cases to deal with. First, suppose A is atomic. Then Γ, A is an axiom and $\neg A$ is an element of Γ. In this case, d_1 is already a proof of Γ.

Second, suppose A is of the form $B \vee C$. Then d_1 is a proof of $\Gamma, \sim B \wedge \sim C$, and d_0 is either of the form

$$\frac{\begin{matrix} \vdots d_{00} \\ \Gamma, B \vee C, B \end{matrix}}{\Gamma, B \vee C}$$

where d_{00} has depth at most $m-1$ and cut rank r, or the analogous form in which C instead of B is the main hypothesis of the last inference. Note that we have assumed, without loss of generality, that $B \vee C$ is a side formula in the last inference.

We will treat only the first case, since the second is similar. Applying the inductive hypothesis to d_{00} and d_1, we obtain a proof d'_{00} of Γ, B with depth at most $m+n-1$ and cut rank at most r. Applying Lemma 6.4.5 to d_1, we also obtain a proof of $\Gamma, \sim B$ with cut rank r and depth at most n. Now applying the cut rule we obtain a proof of Γ with depth at most $m+n$ and cut rank at most r, since the depth of B is one less than the depth of A, i.e. $r-1$.

Finally, suppose A is of the form $\exists x\, B$. Then d_1 is a proof of $\Gamma, \forall x \sim B$, and d_0 is of the form

$$\frac{\begin{matrix} \vdots d_{00} \\ \Gamma, \exists x\, B, B[t/x] \end{matrix}}{\Gamma, \exists x\, B}$$

where d_{00} has depth at most $m-1$ and cut rank r. Applying the inductive hypothesis to d_{00} and d_1, we obtain a proof d'_{00} of $\Gamma, B[t/x]$ with depth at most $m+n-1$ and cut rank at most r. Applying Lemma 6.4.5 to d_1, we obtain a proof of $\Gamma, {\sim}B[t/x]$ with cut rank r and depth at most n. Once again we can now simply apply the cut rule, since the depth of A is $r-1$. □

It is worth thinking about the algorithmic procedure that is implicit in the argument. Given a proof of cut rank $r+1$, we iteratively decrease the cut rank by 1. At the first stage, we pick a topmost cut of rank $r+1$, that is, a cut of rank $r+1$ with no other such cuts above it. Then we find all the places in which the cut formula of the form $B \vee C$ or $\exists x\, B$ was introduced in the subproof with that formula, and use inversion on the other subproof to replace it with a smaller cut.

We will see that this strategy does not quite work when it comes to intuitionistic logic. Roughly, the problem is that one form of the inversion lemma fails. The following proof offers an alternative strategy that works equally well in the classical and intuitionistic settings. The idea is to permute cuts upward on both sides until the cut formula is principal in the last inference of both derivations, and then use a more symmetric procedure to eliminate the cuts.

Alternative proof of Lemma 6.4.3 Again we use induction on $m+n$, but now when the cut formula is not principal in the last inference of either proof, we apply a permutative conversion as above. Thus we only have to deal with the case where the cut formula is principal in the last inference of both proofs. As before, the case where A is atomic is straightforward.

Suppose A is of the form $B \vee C$, in which case ${\sim}A$ is ${\sim}B \wedge {\sim}C$. Then the proof d_0 is of the form

$$\dfrac{\begin{matrix}\vdots d_{00} \\ \Gamma, B \vee C, B\end{matrix}}{\Gamma, B \vee C}$$

and the proof d_1 is of the form

$$\dfrac{\begin{matrix}\vdots d_{10} \\ \Gamma, {\sim}B \wedge {\sim}C, {\sim}B\end{matrix} \qquad \begin{matrix}\vdots d_{11} \\ \Gamma, {\sim}B \wedge {\sim}C, {\sim}C\end{matrix}}{\Gamma, {\sim}B \wedge {\sim}C}$$

By the inductive hypothesis applied to d_0 and d_{10} we obtain a proof of $\Gamma, {\sim}B$ with depth at most $m+n-1$ and cut rank at most r. On the other hand, applying the inductive hypothesis to d_{00} and d_1 we obtain a proof of Γ, B with depth at most $m+n-1$ and cut rank at most r. The result is then obtained by applying the cut rule to these two proofs, since the depth of B is less than r.

Suppose A is instead of the form $\exists x\, B$, in which case ${\sim}A$ is $\forall x\, {\sim}B$. Then the proof d_0 is of the form

$$\dfrac{\begin{matrix}\vdots d_{00} \\ \Gamma, \exists x\, B, B[t/x]\end{matrix}}{\Gamma, \exists x\, B}$$

and the proof d_1 is of the form

$$\dfrac{\begin{matrix}\vdots d_{10} \\ \Gamma, \forall x\, {\sim}B, {\sim}B[y/x]\end{matrix}}{\Gamma, \forall x\, {\sim}B}$$

Applying the inductive hypothesis to d_0 and d_{10} we obtain a proof of Γ, $\sim B[y/x]$ with depth at most $m+n-1$ and cut rank at most r. Using the substitution lemma, we obtain a proof of Γ, $\sim B[t/x]$. Applying the inductive hypothesis to d_{00} and d_1 we obtain a proof of Γ, $B[t/x]$ with depth at most $m+n-1$ and cut rank at most r. Once again, applying the cut rule finishes it off. □

The modifications needed to adapt this proof to a two-sided calculus are discussed in the next section.

Exercises

6.4.1. Prove the following inversion lemmas:

a. If the sequent $\Gamma, A \to B \vdash \Delta$ has a proof in the classical two-sided sequent calculus, then so do the sequents $\Gamma, B \vdash \Delta$ and $\Gamma \vdash A, \Delta$, with proofs whose depth and cut rank are less than or equal to the original one.

b. If the sequent $\Gamma, A \vee B$ has a proof in the classical one-sided sequent calculus, then so does the sequent Γ, A, B, again with depth and cut rank less than or equal to the original proof.

6.4.2. Consider the one-sided sequent calculus for propositional rather than first-order logic. In this exercise you will find a better bound on the increase in the depth of proofs when eliminating cuts. You will also need the $\vee$ inversion rule from preceding problem. Do the following:

a. Prove the following version of the reduction lemma: Suppose $r \geq 1$, d_0 is a proof of Γ, A with cut rank r and depth m, d_1 is a proof of $\Gamma, \sim A$ with cut rank r and depth n, and A has depth r. (So, A is neither atomic nor negation atomic, and putting these two proofs together with a cut would result in a proof of cut rank $r+1$.) Show that there is a proof d of Γ with cut rank r, and depth at most $\max(m, n)+2$. (Hint: use the inversion lemma from the previous problem.)

b. Show that if $r \geq 1$ and d is a proof of Γ with cut rank $r+1$ and depth n, then there is a proof d' of Γ with cut rank r and depth at most $2n$.

c. Show that if d is a proof of Γ with cut rank $r \geq 1$ and depth n, there is a proof d' of Γ with cut rank 1, and depth $2^{r-1}n$.

d. Apply the usual reduction lemma to show that whenever d is a proof of Γ with cut rank $r \geq 1$ and depth n, there is a cut-free proof d' of depth at most $2^{2^{r-1}n}$.

6.4.3. The following is a rather contrived example to illustrate the cut-elimination procedure.

a. Derive the sequent $P \wedge \neg R, Q \wedge \neg R, \neg P \wedge \neg Q, R$ in the one-sided sequent calculus, corresponding to $P \to R, Q \to R, P \vee Q \vdash R$ in the two-sided calculus.

b. Derive the sequent $\neg Q \wedge \neg P, P \vee Q$, corresponding to $Q \vee P \vdash P \vee Q$.

c. Apply the cut rule to get a derivation of $P \wedge \neg R, Q \wedge \neg R, \neg Q \wedge \neg P, R$.

d. Eliminate the cuts by hand.

6.4.4. This problem shows that there is necessarily an iterated exponential increase in the length of proof when using a cut-free calculus.

Consider a language with a unary predicate A, constants 0 and 1, and symbols $+$ and $x \mapsto 2^x$. Define 2_0 to be the term 0 and 2_{n+1} to be 2^{2_n}. Let Δ be the following set of axioms:

- $2^0 = 1$
- $\forall x\, (2^x + 2^x = 2^{x+1})$
- $\forall x\, (0 + x = x)$
- $\forall x, y, z\, (x + (y + z) = x + (y + z))$

- $A(0)$
- $\forall x\,(A(x) \to A(x+1))$.

I will conflate these formulas and their negation normal form equivalents. We will show that, with cut, there are proofs of $\sim\Delta, A(2_n)$ with size that is linear in n, whereas without cut, any proof of this sequent has to have at least 2_n inferences.

Define a sequence of formulas $B_n(x)$ by defining $B_0(x)$ to be $A(x)$ and defining $B_{n+1}(x)$ to be $\forall y\,(B_n(y) \to B_n(y+2^x))$.

a. Show that, for every n, there is a proof of $\forall x\,(B_n(x) \to B_n(x+1))$ from the axioms above, with a constant number of steps. (You can describe the proof informally and use the substitution rule for equality as a single step.)
b. Show that, for every n, there is a proof of $B_n(0)$, with a number of steps bounded by a constant.
c. Show that, for every term t, there is a proof of $B_{n+1}(t) \to B_n(2^t)$ with a constant number of steps.
d. Conclude that, for every n, there is a proof of $A(2_n)$ with the number of steps linear in n.
e. Now suppose we have a cut-free proof of $\sim\Delta, A(2_n)$. Note that the elements of $\sim\Delta$ are existential sentences. Section 7.1 shows that deleting the $\exists$ rules in this proof yields a proof of a sequent $\sim\Delta', A(2_n)$ where Δ' consists of closed quantifier-free instances of Δ. Consider the instances $\neg A(t) \vee A(t+1)$ occurring in Δ'. Argue that there must be at least $2_n - 1$ such instances, because otherwise there is a model that falsifies $\sim\Delta', A(2_n)$, in which the universe is $\mathbb{N}$ and all the arithmetic functions have the usual interpretations and A is carefully chosen to make all the elements of Δ' true but $A(2_n)$ false.

Now for the fine print. Notice that the size of the formulas B_n grows exponentially in n, and that if we restrict the substitution rule for equality to atomic formulas, then a single substitution can require exponentially many steps. This is still meager compared to the iterated exponential increase. Nonetheless, we can avoid the expensive substitution by defining $B_{n+1}(x)$ instead to be $\forall y, z\,(B_n(y) \wedge z = y + 2^x \to B_n(z))$. In fact, we can avoid the use of equality in this example using instead a ternary relation $R(x, y, z)$ which is interpreted, as $z = x + 2^y \wedge A(x)$. Moreover, there are tricks that one can use to compress the size of the defined formulas. For more information, see the notes at the end of this chapter.

6.5 Cut Elimination for Intuitionistic Logic

Using the one-sided sequent calculus for classical logic has been convenient, cutting down the twelve rules of the two-sided calculus to only six. But it is straightforward to transfer the arguments to the two-sided calculus as well. For example, the case of the reduction lemma for the two-sided calculus where d_0 is of the form

$$\dfrac{\vdots d_{00} \atop \Gamma, B \wedge C, B \vdash \Delta}{\Gamma, B \wedge C \vdash \Delta}$$

and d_1 is a proof of $\Gamma \vdash \Delta, B \wedge C$ is structurally identical to the case in the proof of Lemma 6.4.3 where $\sim A$ is of the form $B \wedge C$. As before, we can apply the inductive hypothesis to d_{00} and d_1 to obtain a proof of $\Gamma, B \vdash \Delta$, apply the inversion lemma to d_1 to obtain a proof of $\Gamma \vdash \Delta, B$, and then apply a cut on B.

Most of the argument then carries over to minimal and intuitionistic logic without changes. For intuitionistic logic, we need to handle cuts of the following form:

$$\frac{\Gamma \vdash \bot \qquad \dfrac{}{\Gamma, \bot \vdash A}}{\Gamma \vdash A}$$

But it is easy to check that any proof of $\Gamma \vdash \bot$ can be transformed to a proof of $\Gamma \vdash A$ without increasing the depth or cut rank. The biggest differences arises when there is a cut on an implication, at which point the inversion strategy falls apart. In that case, we have to use a strategy where we permute cuts until the cut formula is principal in both inferences. In that case, the proof d_0 is of the form

$$\frac{\begin{array}{c}\vdots d_{00} \\ \Gamma, B \vdash C\end{array}}{\Gamma \vdash B \to C}$$

and d_1 is of the following form:

$$\frac{\begin{array}{c}\vdots d_{10} \\ \Gamma, B \to C \vdash B\end{array} \qquad \begin{array}{c}\vdots d_{11} \\ \Gamma, B \to C, C \vdash D\end{array}}{\Gamma, B \to C \vdash D}$$

Applying the inductive hypothesis to d_0 and d_{10} we obtain a proof e_0 of $\Gamma \vdash B$, and applying the inductive hypothesis to d_0 and d_{11} we obtain a proof e_1 of $\Gamma, C \vdash D$. The pieces can then be reassembled as follows:

$$\frac{\dfrac{\begin{array}{c}\vdots e_0 \\ \Gamma \vdash B\end{array} \qquad \begin{array}{c}\vdots d_{00} \\ \Gamma, B \vdash C\end{array}}{\Gamma \vdash C} \qquad \begin{array}{c}\vdots e_1 \\ \Gamma, C \vdash D\end{array}}{\Gamma \vdash D}$$

But now note that the depth of the resulting proof is *two* more than the depth of e_0. This results in slightly larger bounds in the reduction lemma and final theorem.

Lemma 6.5.1. *Suppose* $\vdash^m_r \Gamma \vdash A$ *and* $\vdash^n_r \Gamma, A \vdash D$ *in the minimal (respectively intuitionistic) sequent calculus, where* A *has depth* r*. Then we have* $\vdash^{2(m+n)}_r \Gamma \vdash D$.

Proof As in the alternative proof of Lemma 6.4.3. □

Lemma 6.5.2. *If* $\vdash^n_{r+1} \Gamma \vdash D$*, then* $\vdash^{4^n}_r \Gamma \vdash D$.

Theorem 6.5.3. *(cut elimination theorem) If* $\vdash^n_r \Gamma \vdash D$ *in the minimal or intuitionistic sequent calculus for first-order logic, then* $\vdash^{4^n_r}_0 \Gamma$.

Note that the various proofs of cut elimination go through virtually unchanged if we allow arbitrary formulas A in the axioms $\Gamma, A, {\sim}A$ in the one-sided sequent calculus, and, similarly, in the axioms $\Gamma, A \vdash \Delta, A$ in the two-sided calculus. Moreover, in most cases we can permute the order in which rules are applied. (The only time this can cause problems is when an application of the $\exists$ rule makes possible a subsequent $\forall$ rule by eliminating a free variable.) So cut elimination is nondeterministic in the sense that different strategies can lead to different cut-free proofs.

Exercises

6.5.1. Fill in some of the details of the proof of Lemma 6.5.1.

6.5.2. In Section 6.3, we saw that there are complete search procedures for classical propositional and first-order logic. The fact that there is no invertible version of the left $\to$ rule in intuitionistic and minimal logic makes it harder to describe complete search procedures in those settings. But there is a trick for propositional logic whereby we replace the left $\to$ rule with the following four rules, in the first of which we assume A is atomic:

$$\frac{\Gamma, A, B \vdash C}{\Gamma, A, A \to B \vdash C} \qquad \frac{\Gamma, B \to C \vdash A \to B \qquad \Gamma, C \vdash D}{\Gamma, (A \to B) \to C \vdash D}$$

$$\frac{\Gamma, A \to (B \to C) \vdash D}{\Gamma, A \wedge B \to C \vdash D} \qquad \frac{\Gamma, A \to C, B \to C \vdash D}{\Gamma, A \vee B \to C \vdash D}$$

Show that these rules are valid for minimal and intuitionistic logic, and see the bibliographical notes at the end of this chapter for a proof that the cut-free systems are complete.

6.5.3. Show that the cut elimination theorem for the classical one-sided sequent calculus follows from the cut elimination theorem for minimal logic, as follows:

a. Suppose $\{A_0, \ldots, A_{n-1}\}$ is provable in the classical one-sided sequent calculus, where each A_i is in negation normal form. By Theorem 4.6.9, conclude that $\sim A_0, \ldots, \sim A_{n-1} \vdash \bot$ is provable in minimal logic.

b. By the cut elimination theorem, this last sequent has a cut-free proof. Show that this gives rise to a cut-free proof of $\{A_0, \ldots, A_{n-1}\}$ in the classical sequent calculus. (Hint: compare the left rules in the sequent calculus for minimal logic to the corresponding rules in the classical calculus.)

6.6 Equality

Section 4.3 presented a few ways of incorporating equality in deductive systems for classical and intuitionistic logic. Theorem 4.3.2 provides a sense in which entailments between equalities can be proved from the bottom up. Here we will consider two approaches to adding equality to the sequent calculus: adding axioms, that is, initial sequents, and adding new rules. In each case, we will try to understand what can be said from the point of view of cut elimination.

Corresponding to an axiomatization of equality described in Section 4.3, we can take as axioms all weakenings of the following in any of our two-sided sequent calculi:

- $\vdash t = t$
- $s = t \vdash t = s$
- $r = s, s = t \vdash r = t$
- $\vec{s} = \vec{t} \vdash f(\vec{s}) = f(\vec{t})$
- $\vec{s} = \vec{t}, R(\vec{s}) \vdash R(\vec{t})$.

Call this set of sequents E. Adding them to the calculus as axioms means that they can be asserted without preconditions, just like the assumption axioms. We will consider all the displayed formulas to be principal in the corresponding inference.

Theorem 6.6.1. *Suppose a sequent is provable in the two-sided sequent calculus for classical, intuitionistic, or minimal logic with additional axioms E. Then there is a proof in which every cut has an equation $s = t$ as a principal formula.*

Proof The strategy extends the one described in Section 6.5: we move cuts upward until the cut formula is principal in both preceding inferences, and we ignore cuts whose principal formula is an equation. The only new case that needs to be handled occurs when the cut formula is atomic but not an equation. If one of the preceding inferences is an assumption axiom, as before, the cut can be eliminated entirely. Otherwise, the subproof in question has the following form:

$$\frac{\Gamma, \vec{s}=\vec{t}, R(\vec{s}) \vdash R(\vec{t}), \Delta \qquad \Gamma, \vec{t}=\vec{u}, R(\vec{t}) \vdash R(\vec{u}), \Delta}{\Gamma, \vec{s}=\vec{t}, \vec{t}=\vec{u}, R(\vec{s}) \vdash R(\vec{u}), \Delta}$$

Here we can assume without loss of generality that $\vec{s}=\vec{t}$, $\vec{t}=\vec{u}$, and $R(\vec{s})$ are all contained in Γ; they are displayed explicitly only to make it clear that the assumptions are equality axioms of the given form. In the case of intuitionistic or minimal logic, Δ is empty. We can avoid the cut by using transitivity to establish $\vec{s}=\vec{t}, \vec{t}=\vec{u} \vdash s_i = u_i$ for each i, followed by cuts with these equalities starting with the axiom $\Gamma, \vec{s}=\vec{u}, R(\vec{s}) \vdash R(\vec{u})$. □

For some purposes, it is more convenient to close the set of axioms under cuts. Let E' be the set of sequents that can be obtained using any number of cuts on elements of E or instances of the assumption axiom. The following theorem is an easy variation on the previous one.

Theorem 6.6.2. *Suppose a sequent is provable in the two-sided sequent calculus for classical, intuitionistic, or minimal logic with additional axioms E'. Then there is a cut-free proof.*

Corollary 6.6.3. *Suppose Γ, Δ contains only atomic formulas and $\Gamma \vdash \Delta$ is derivable in first-order logic with equality. Then $\Gamma \vdash \Delta$ is an element of E'.*

Proof By Theorem 6.6.2, if there is a proof of $\Gamma \vdash \Delta$ in classical logic, then there is a cut-free proof from E'. Since Γ, Δ contains no logical connectives, it must be itself be an element of E'. □

Adding the elements of E' as axioms is not an attractive option unless there is an algorithm to determine whether or not a given sequent is in E'. Fortunately, there is: in Section 4.4 we saw that if Γ, A consists only of atomic formulas, $\Gamma \vdash A$ is valid if and only if it is provable in the equational fragment of first-order logic, and, moreover, there is an algorithm to determine whether that is the case. This provides us with a decision procedure for E':

Theorem 6.6.4. *A sequent is an element of E' if and only if it is a weakening of a valid sequent $\Gamma \vdash A$ where Γ, A consists of atomic formulas.*

So we can determine whether a sequent $\Gamma \vdash \Delta$ is valid by checking whether $\Gamma' \vdash A$ is valid for some atomic formula A in Δ, where Γ' consists of the atomic formulas in Γ. To prove the theorem, note that if $\Gamma \vdash A$ is valid, then by Corollary 6.6.3, it is an element of E'. Conversely, each element of E is a weakening of a valid sequent consisting of atomic formulas, and it is not hard to check that the result of applying the cut rule to two such weakenings is again a sequent of that form. (See also Exercise 5.2.2, which provides a model-theoretic proof of this claim.)

An alternative strategy for incorporating equality in the two-sided sequent calculus involves adding the following two rules:

$$\frac{\Gamma, t=t \vdash \Delta}{\Gamma \vdash \Delta} \qquad \frac{\Gamma, s=t, A(s), A(t) \vdash \Delta}{\Gamma, s=t, A(s) \vdash \Delta}$$

We can restrict A to be atomic. These rules allow the following derivations:

$$\frac{\overline{\Gamma, t=t \vdash \Delta, t=t}}{\Gamma \vdash \Delta, t=t} \qquad \frac{\overline{\Gamma, s=t, A(s), A(t) \vdash \Delta, A(t)}}{\Gamma, s=t, A(s) \vdash \Delta, A(t)}$$

These two, in turn, suffice to axiomatize first-order logic with equality. The point is that adding these two rules restores cut elimination.

Theorem 6.6.5. *Suppose a sequent $\Gamma \vdash \Delta$ is derivable in the two-sided sequent calculus for classical, intuitionistic, or minimal logic with the additional two rules above. Then there is a cut-free proof.*

Proof The symmetric cut elimination procedures goes through virtually unchanged. The principal formulas are on the left of the sequent arrow. The only inferences in which an atomic formula is principal on the right are assumption axioms, and cuts on those axioms can be eliminated. □

The subformula property does not hold for proofs that are cut-free in the sense of Theorem 6.6.5, but it fails only mildly: as one travels upward through a proof, the additional formulas that appear are only trivial identities and variants of other atomic formulas in the sequent up to identities that are also found in the sequent. In such a proof the new rules can be permuted upward, so that they are used to close off a branch of the proof by making an equational entailment explicit.

All of the approaches described here carry over straightforwardly to the one-sided classical sequent calculus: we can add one-sided sequent versions of the basic equality axioms, or close them under cut, or add new rules. Exercise 6.6.2 shows that the semantic proof of cut elimination described in Section 6.3 can be extended to obtain the completeness of the calculus with additional rules.

Exercises

6.6.1. Prove the forward direction of Theorem 6.6.4 as suggested there, namely, by showing that if $\Pi \vdash B, \Delta$ and $\Pi', B \vdash \Delta'$ are both weakenings of valid atomic sequents of the form $\Gamma \vdash A$, then so is $\Pi, \Pi' \vdash \Delta, \Delta'$.

6.6.2. Suppose we extend the one-sided cut-free calculus with the following two rules to handle equality:

$$\frac{\Gamma, t \neq t}{\Gamma} \qquad \frac{\Gamma, s \neq t, {\sim}A(s), {\sim}A(t)}{\Gamma, s \neq t, {\sim}A(s)}$$

where t is any term and $A(s)$ is any atomic formula, possibly an equality. Extend the completeness proof from Section 6.3 to provide a model-theoretic proof that these rules suffice. That is, show that if Γ is a sequent of formulas in first-order logic with equality that does not have a cut-free proof, there is a model in which every formula in Γ is false.

6.7 Variations on Cut Elimination

In chapters to come, we will use the cut elimination theorem to study not just provability in pure logic but provability in axiomatic theories as well. Suppose a formula A is provable from a set of axioms Γ. Then for some finite subset $\Gamma' \subseteq \Gamma$, there is a cut-free proof of the sequent $\Gamma' \vdash A$ in the relevant two-sided calculus. By the subformula property, every formula occurring in the proof is a subformula of a formula in Γ' or a subformula of A. If we use the one-sided calculus for classical logic instead and assume everything is in negation normal form, we have that there is a cut-free proof of $\sim\Gamma', A$, and every formula occurring in such a proof is again a subformula of one occurring in the final sequent.

Often these facts are sufficient, but sometimes it is more useful to adopt a sequent calculus that is better adapted to the theory at hand, just as we did in Section 6.6. For example, we can add arbitrary axioms and rules, and, under the right circumstances, eliminate all cuts except for ones that trace their way back to principal formulas of those axioms and rules. The bibliographical notes indicate generalizations of the cut elimination theorem along these lines. Rather than develop the terminology needed to describe the most general scenarios, here we will discuss versions that are sufficient for the applications in this book. For example, our treatment of the equality axioms immediately generalizes as follows:

Theorem 6.7.1. *Let E be any set of two-sided sequents containing only atomic formulas. Let E' be the closure of E together with the assumption axioms under weakening, substitution, and cut. Then any sequent that is provable from E' in classical, intuitionistic, or minimal logic has a cut-free proof from E' in the same system. The analogous result holds for the one-side classical sequent calculus, where the sequents in E consist of literals.*

This formulation is useful for theories with axioms that are universally quantified equations or universally quantified implications between equations. If we do not close E' under cut, we instead obtain a proof that is cut-free except for cuts on atomic formulas.

In Section 7.4, we will consider an extension of the two-sided calculus with a rule of the following form, for a certain formula $A(\vec{x})$, with the free variables shown:

$$\frac{\Gamma, A(\vec{t}) \vdash \Delta}{\Gamma \vdash \Delta}$$

Here $\vec{t}$ is an arbitrary sequence of terms. In this extended system, we can derive $\vdash \forall\vec{x}\, A(\vec{x})$ as follows:

$$\cfrac{\cfrac{A(\vec{x}) \vdash A(\vec{x})}{\vdash A(\vec{x})}}{\vdash \forall\vec{x}\, A(\vec{x})}$$

So adding the rule is equivalent to adding the axiom $\forall\vec{x}\, A(\vec{x})$. But expressing the axiom as a rule means that the cut elimination theorem still holds, and the subformula property implies that every formula occurring in the proof is either a generalized subformula of a formula in the final sequent or a generalized subformula of $A(\vec{t})$, for some $\vec{t}$. A similar trick was used in the last section to introduce rules for equality.

Finally, in Chapter 10, we will study subsystems of *first-order arithmetic*, an axiomatic theory for reasoning about the natural numbers. In this context, the induction principle for the natural numbers can be expressed as a rule:

$$\frac{\Gamma \vdash \Delta, A(0) \qquad \Gamma, A(x) \vdash \Delta, A(x+1)}{\Gamma \vdash \Delta, A(t)}$$

Here the term t in the conclusion is arbitrary, and we assume that x is not free in any formula in Γ or Δ. The rule says that $A(t)$ follows from $A(0)$ and $A(x) \to A(x+1)$ for arbitrary x. The other axioms of first-order arithmetic are universal axioms of the form described in Theorem 6.7.1. In this system, we can eliminate all the cuts except for cuts that trace their way back to one of the formulas $A(t)$ occurring in the conclusion of the induction axiom.

Theorem 6.7.2. *Suppose a sequent $\Gamma \vdash \Delta$ is provable in a sequent calculus for first-order arithmetic. Then there is a proof of $\Gamma \vdash \Delta$ such that every cut formula is principal in an instance of the induction rule or an axiom other than the assumption axiom. Every formula occurring in such proof is either a generalized subformula of a formula in Γ, Δ or a generalized subformula of a formula $A(t)$ occurring in the conclusion of an instance of the induction rule.*

Proof (sketch) The second claim is proved by a straightforward induction on derivations with the stated property. To prove the first claim, we only need small modifications to the second (symmetric) proof of the cut elimination theorem that was presented in Section 6.4. When computing the cut-rank of a proof, simply do not count cuts on formulas that are principal in an induction rule anywhere above the cut. We have to check that with the addition of non-logical axioms and the induction rule, the substitution lemma still holds, and that a cut on a side formula of an induction rule can be permuted past that rule. We then only need to apply the reduction lemma in the case where the relevant formulas are not principal in a non-logical axiom or an instance of the induction rule in either subproof. □

6.8 Cut-Free Completeness of Intuitionistic Logic

It is possible to provide a semantic proof of the completeness of the cut-free sequent calculus for intuitionistic or minimal logic similar to the one for classical logic presented in Section 6.3. For example, we can describe explicit proof search procedures that either terminate with a proof or allow us to read off a Kripke model that provides a counterexample to the claim. But such arguments are considerably more delicate. Here, instead, I will present a proof that is inspired by the generalizations of Kripke models that were alluded to at the end of Section 5.3. The strategy will be to define a forcing relation in which conditions Γ, Δ, Π are finite sets of formulas, and to establish the following two claims:

- If $\Gamma \vdash A$, then $\Gamma \Vdash A$.
- If $\Gamma \Vdash A$, there is a cut-free proof of the sequent $\Gamma \vdash A$.

The first claim says that the forcing relation is sound for intuitionistic (or minimal) provability, whereas the second claim establishes that the cut-free calculus is complete for that semantics.

Take conditions $\Gamma, \Delta, \Pi, \ldots$ to be finite sets of formulas, and say Γ is *stronger than* Δ, written $\Gamma \preceq \Delta$, if $\Gamma \supseteq \Delta$. The clauses below provide an inductive definition of a relation S *covers* Γ between conditions Γ and finite sets S of conditions stronger than Γ:

1. $\{\Gamma\}$ covers Γ.
2. If $\{\Delta_0, \ldots, \Delta_{k-1}\}$ covers Γ, and for each i, S_i covers Δ_i, then $\bigcup_{i<k} S_i$ covers Γ.
3. If $A \vee B$ is in Γ, then $\{\Gamma \cup \{A\}, \Gamma \cup \{B\}\}$ covers Γ.
4. If $A \wedge B$ is in Γ, then $\{\Gamma \cup \{A\}\}$ covers Γ, and $\{\Gamma \cup \{B\}\}$ covers Γ.
5. If $A \to B$ is in Γ and there is a cut-free proof of $\Gamma \vdash A$, then $\{\Gamma \cup \{B\}\}$ covers Γ.
6. If $\forall x\, A$ is in Γ and t is any term, then $\{\Gamma \cup \{A[t/x]\}\}$ covers Γ.
7. If $\exists x\, A$ is in Γ and x is not free in Γ, then $\{\Gamma \cup \{A\}\}$ covers Γ.

Below I will drop the extra set brackets and say "$\Delta_0, \ldots, \Delta_{k-1}$ cover Γ" instead of "$\{\Delta_0, \ldots, \Delta_{k-1}\}$ covers Γ."

Lemma 6.8.1. *Suppose $\Delta_0, \ldots, \Delta_{k-1}$ cover Γ, and suppose there are cut-free proofs of $\Delta_0 \vdash A, \ldots, \Delta_{k-1} \vdash A$. Then there is a cut-free proof of $\Gamma \vdash A$.*

Proof Use induction on the covering relation. In fact, the clauses are equivalent to saying that there is a cut-free proof of $\Gamma \vdash A$ starting from the sequents $\Delta_0 \vdash A, \ldots, \Delta_{k-1} \vdash A$. □

Now define the notion of $\Gamma \Vdash A$ inductively, as follows:

1. If A is atomic, $\Gamma \Vdash A$ if and only if there is a cut-free proof of $\Gamma \vdash A$.
2. $\Gamma \Vdash C \wedge D$ if and only if $\Gamma \Vdash C$ and $\Gamma \Vdash D$.
3. $\Gamma \Vdash C \vee D$ if and only if there is a covering $\Delta_0, \ldots, \Delta_{k-1}$ of Γ such that for each i, $\Delta_i \Vdash C$ or $\Delta_i \Vdash D$.
4. $\Gamma \Vdash C \to D$ if and only if for every $\Delta \preceq \Gamma$, if $\Delta \Vdash C$ then $\Delta \Vdash D$.
5. $\Gamma \Vdash \forall x\, C(x)$ if and only if for every term t, $\Gamma \Vdash C(t)$.
6. $\Gamma \Vdash \exists x\, C(x)$ if and only if there is a covering $\Delta_0, \ldots, \Delta_{k-1}$ of Γ and a sequence of terms $t_0, \ldots, t_{k-1}$ such that for each i, $\Delta_i \Vdash C(t_i)$.

In the base case, we take cut-free provability to refer to intuitionistic or minimal logic, depending on which system we are concerned with.

In the next two lemmas, a *renaming* of variables is just an injective map from the set of variables to the set of variables. If σ is a renaming, then A^σ denotes the result of replacing each free variable x of A by $\sigma(x)$. In other words, if A has free variables $x_0, \ldots, x_{n-1}$, A^σ denotes the simultaneous substitution $A[\sigma(x_0)/x_0, \ldots, \sigma(x_{n-1})/x_{n-1}]$. Similarly, if Γ is a condition and σ is a renaming, then Γ^σ denotes $\{A^\sigma \mid A \in \Gamma\}$.

Lemma 6.8.2. *Suppose Γ is any condition, $\Delta_0, \ldots, \Delta_{k-1}$ are conditions covering Γ, and Π is a condition stronger than Γ. Then there is a renaming σ such that $\Delta_0^\sigma \cup \Pi, \ldots, \Delta_{k-1}^\sigma \cup \Pi$ covers Π.*

Proof A straightforward induction on the covering relation. The renamings are needed to handle the variable restriction in clause 7 of the definition of a covering relation. □

Lemma 6.8.3. *The forcing relation defined above satisfies the following:*

1. *Stability under renaming: if σ is any renaming of variables and $\Gamma \Vdash A$, then $\Gamma^\sigma \Vdash A^\sigma$.*
2. *Monotonicity: if $\Gamma \Vdash A$ and $\Delta \preceq \Gamma$ then $\Delta \Vdash A$.*
3. *The covering property: if $\Delta_0, \ldots, \Delta_{k-1}$ cover Γ and for each i, $\Delta_i \Vdash A$, then $\Gamma \Vdash A$.*

Proof Each claim can be proved using a straightforward induction on the formula A. Stability under renaming and Lemma 6.8.2 are needed to handle the claims for $\vee$ and $\exists$ when proving monotonicity. □

For intuitionistic logic, an easy induction on derivations shows that if there is a cut-free proof of $\Gamma \vdash \bot$ and A is any formula, then there is a cut-free proof of $\Gamma \vdash A$. This takes care of $\bot$ rule in the proof of the following lemma, which otherwise is the same for intuitionistic and minimal logic:

Lemma 6.8.4. *If A is provable, then A is forced.*

Proof A straightforward induction on derivations. □

Lemma 6.8.5. *Let A be any formula. Then*

1. $\{A\} \Vdash A$.
2. *If $\Gamma \Vdash A$, then there is a cut-free proof of $\Gamma \vdash A$.*

Proof We can prove both claims simultaneously by induction on A. I will focus on two illustrative cases.

For the first sample case, suppose A is a formula of the form $B \vee C$. Using the inductive hypothesis and monotonicity, we have $\{B \vee C, B\} \Vdash B$ and $\{B \vee C, C\} \Vdash C$. Hence, both these conditions force $B \vee C$, and since they cover $\{B \vee C\}$, we have $\{B \vee C\} \Vdash B \vee C$.

For the second claim, suppose $\Gamma \Vdash B \vee C$. By the definition of forcing and the inductive hypothesis, there are conditions $\Delta_0, \ldots, \Delta_{k-1}$ covering Γ such that for each i, there is a cut-free proof of $\Delta_i \vdash B$ or $\Delta_i \vdash C$. In particular, for each i, there is a cut-free proof of $\Delta_i \vdash B \vee C$. By Lemma 6.8.1, there is a cut-free proof of $\Gamma \vdash B \vee C$.

As a second example, suppose A is of the form $B \to C$. For the first claim, we need to show that if Γ is any condition and $\Gamma \Vdash B$, then $\Gamma \cup \{B \to C\} \Vdash C$. So suppose $\Gamma \Vdash B$. By the inductive hypothesis, there is a cut-free proof of $\Gamma \vdash B$. By the definition of covering, $\Gamma \cup \{C\}$ covers $\Gamma \cup \{B \to C\}$. Again by the inductive hypothesis, we have $\Gamma \cup \{C\} \Vdash C$. By Lemma 6.8.1, we have $\Gamma \cup \{B \to C\} \Vdash C$.

For the second claim, suppose $\Gamma \Vdash B \to C$. By the inductive hypothesis, we have $\{B\} \Vdash B$, and so $\Gamma, B \Vdash C$. Again by the inductive hypothesis, there is a cut-free proof of $\Gamma, B \vdash C$. This yields a cut-free proof of $\Gamma \vdash B \to C$. □

Theorem 6.8.6. *Any sequent provable in the intuitionistic sequent calculus has a cut-free proof.*

Proof Suppose there is a proof of $B_0, \ldots, B_{n-1} \vdash A$ in the sequent calculus with cut. Then $B_0 \wedge \cdots \wedge B_{n-1} \to A$ is provable intuitionistically, and so it is forced. By Lemma 6.8.5, for each i we have $\{B_i\} \Vdash B_i$. By monotonicity, we have $\{B_0, \ldots, B_{n-1}\} \Vdash B_0 \wedge \cdots \wedge B_{n-1}$, and hence $\{B_0, \ldots, B_{n-1}\} \Vdash A$. By Lemma 6.8.5 again, there is a cut-free proof of $B_0, \ldots, B_{n-1} \vdash A$. □

Exercise

6.8.1. Fill in some of the details elided in proofs in this section.

Bibliographical Notes

General references Gentzen's original papers on the sequent calculus, along with his proofs of the cut elimination theorems, can be found in his collected works, Gentzen (1969). Other general references include Kleene (1952), Schwichtenberg (1977), Takeuti (1987), Troelstra and Schwichtenberg (2000), Negri and von Plato (2001), Negri and von Plato (2011), and the *Handbook of Proof Theory* (Buss (1998a)), which includes Buss (1998c).

Proof complexity For more information on lower bounds on the increase in length of proof when avoiding the use of cut, see Pudlák (1998), Theorem 16.1.6 below, and the notes at the end of Chapters 10 and 16.

Proof search and automated reasoning In automated reasoning, the general search strategy described in Section 6.3 is known as *tableau search*. See Hähnle (2001) or Harrison (2009) for an overview. For a search strategy that is complete for intuitionistic logic, see Mints (2000b), and for the completeness of the systems of propositional logic described in Exercise 6.5.2, see Dyckhoff (1992, 2018).

Variations on cut elimination For more information on the kinds of variations on the cut elimination theorem discussed in Section 6.7, see Takeuti (1987), Buss (1998c), Troelstra and Schwichtenberg (2000), and Negri and von Plato (2001).

Algebraic proofs of cut elimination For algebraic approaches to cut elimination that generalize the ideas in Section 6.8, see Okada (2002) and the references there.

7

Properties of First-Order Logic

We have defined classical, intuitionistic, and minimal versions of provability in first-order logic, and we have begun to explore some of their basic properties. We have also developed syntactic and semantic tools that can help us in our analysis. In this chapter, we will put these tools to use toward obtaining a deeper understanding of first-order logic.

A number of important results show that we can find additional information implicit or hidden in mathematical proofs. For example, Herbrand's theorem says that given a classical proof of a purely existential statement from purely universal axioms, we can extract a finite set of terms explicitly witnessing the conclusion. The constructive nature of intuitionistic logic is illuminated by the fact that it satisfies even stronger properties, namely, the explicit definability and disjunction properties. The interpolation theorem says that any provable implication between sentences in different languages is mediated by a sentence in the language they have in common. Our proof will show how such an interpolant can be extracted from a derivation.

Other results in logic show that various styles of reasoning can be extended conservatively, or, put the other way around, that more elementary proofs can be extracted from arguments that, on the surface, make use of more powerful methods. Along these lines, we will see how classical and intuitionistic theories can be augmented by Skolem functions, and how and in what sense such uses can be eliminated. Such functions play an important role in automated reasoning.

7.1 Herbrand's Theorem

This section deals with provability in classical first-order logic with equality.

Theorem 7.1.1 (Herbrand's theorem). *Suppose $\exists \vec{y}\, A(\vec{y})$ is provable in classical first-order logic with equality, where $A(\vec{y})$ is quantifier-free, possibly with free variables other than the ones shown. Then there is a finite set $\{\vec{t}_0, \ldots, \vec{t}_{k-1}\}$ of tuples of terms in the same language such that $A(\vec{t}_0) \vee \cdots \vee A(\vec{t}_{k-1})$ is also provable in the quantifier-free fragment of first-order logic.*

If B is a universal or existential formula, a *quantifier-free instance* of B is any formula that is obtained by deleting the quantifiers and substituting terms for the quantified variables. The phrase *quantifier-free fragment of first-order logic* was introduced in Section 4.4 to refer to the rules for the propositional connectives and the equality relation. The notation in Theorem 7.1.1 is confusing: $\vec{y}$ denotes a tuple of variables, and each $\vec{t}_i$ is a tuple of terms of the same length as $\vec{y}$. If the language in question has at least one constant, we can assume without loss of generality that all the variables occurring in $\vec{t}_0, \ldots, \vec{t}_{k-1}$ are also free in

$\exists \vec{y}\, A(\vec{y})$, because otherwise we can substitute any constant for the extra variables occurring in $\vec{t}_0, \ldots, \vec{t}_{k-1}$ and the conclusion still holds. In particular, if $\exists \vec{y}\, A(\vec{y})$ is a sentence, then there is a finite set of tuples of closed terms witnessing the conclusion of the theorem. Since proving a formula with free variables is equivalent to proving its universal closure, we have the following:

Corollary 7.1.2. *Suppose $\forall \vec{x}\, \exists \vec{y}\, A(\vec{x}, \vec{y})$ is provable, where A has at most the free variables shown. Then, assuming the tuple $\vec{x}$ has length at least one or there is at least one constant in the language, there is a finite set of tuples of terms $\vec{t}_0, \ldots, \vec{t}_{k-1}$ with at most the variables $\vec{x}$ such that $A(\vec{t}_0) \vee \cdots \vee A(\vec{t}_{k-1})$ is provable in the quantifier-free fragment of first-order logic.*

Notice that we need to assume that either the tuple $\vec{x}$ has length at least one or there is at least one constant in the language to guarantee that there are any terms at all whose variables are among $\vec{x}$. We can obtain a further strengthening of this corollary.

Corollary 7.1.3. *Suppose Γ is a set of universal sentences that proves $\forall \vec{x}\, \exists \vec{y}\, A(\vec{x}, \vec{y})$, where A is quantifier-free with at most the free variables shown. Then, assuming the tuple $\vec{x}$ is nonempty or there is at least one constant in the language, there is a finite set $\{\vec{t}_0, \ldots, \vec{t}_{k-1}\}$ of tuples of terms with at most the variables $\vec{x}$ such that $A(\vec{x}, \vec{t}_0) \vee \cdots \vee A(\vec{x}, \vec{t}_{k-1})$ is provable in the quantifier-free fragment of first-order logic from finitely many quantifier-free instances of sentences in Γ.*

Proof If Γ proves $\forall \vec{x}\, \exists \vec{y}\, A(\vec{x}, \vec{y})$, then, by the deduction theorem, there is a finite subset $\{\forall \vec{z}_0\, B_0, \ldots, \forall \vec{z}_{\ell-1}\, B_{\ell-1}\}$ of Γ such that the formula

$$\forall \vec{z}_0\, B_0 \wedge \cdots \wedge \forall \vec{z}_{\ell-1}\, B_{\ell-1} \to \exists \vec{y}\, A(\vec{x}, \vec{y})$$

is provable in pure first-order logic with equality. Without loss of generality, we can assume that all the variables $\vec{x}, \vec{z}_0, \ldots, \vec{z}_{\ell-1}, \vec{y}$ are distinct. The last sentence is equivalent to

$$\exists \vec{y}, \vec{z}_0, \ldots, \vec{z}_{\ell-1}\, (B_0 \wedge \cdots \wedge B_{\ell-1} \to A(\vec{x}, \vec{y})).$$

Applying Herbrand's theorem, we obtain a finite set of tuples of terms witnessing the existentially quantified variables. In particular, there are tuples of terms $\vec{t}_0, \ldots, \vec{t}_{k-1}$ such that $A(\vec{t}_0) \vee \cdots \vee A(\vec{t}_{k-1})$ follows from instances of the equality axioms and the formulas $\forall \vec{z}_0\, B_0, \ldots, \forall \vec{z}_{\ell-1}\, B_{\ell-1}$, as required. □

This result easily extends to hypotheses of the form $\exists \vec{w}\, \forall \vec{z}\, B(\vec{w}, \vec{z})$, since we can replace the tuple $\vec{w}$ by a tuple $\vec{c}$ of fresh constants.

I will provide two proofs of Herbrand's theorem, the first syntactic and the second model-theoretic. The syntactic proof relies on the following lemma, which is just a slight generalization of the theorem itself.

Lemma 7.1.4. *Suppose Γ is a finite set of existential formulas and Γ is provable in the one-sided classical sequent calculus. Then there is a set Γ' of quantifier-free instances of formulas in Γ such that Γ' has a cut-free proof.*

Proof If Γ is provable, by the cut elimination theorem, it has a cut-free proof. The conclusion is obtained by simply deleting every instance of an $\exists$ rule.

More precisely, we Lemma 7.1.4 by induction on the cut-free proof d of Γ. If d is an axiom, then Γ is of the form $\Delta, A, \neg A$ where Δ is a set of existential formulas and A is atomic. In that case, any terms at all can be chosen as witnesses to the existential quantifiers in Γ. Suppose the last inference of d is an instance of the $\wedge$ rule:

$$\frac{\begin{matrix}\vdots & \quad & \vdots \\ \Delta, A & & \Delta, B\end{matrix}}{\Delta, A \wedge B}$$

Since $A \wedge B$ is existential and the outermost connective is not $\exists$, it must be quantifier-free, and the conclusion follows by applying the inductive hypothesis to the subproofs of Δ, A and Δ, B, and then applying the $\wedge$ rule one more time. The case where the last inference is an instance of the $\vee$ rule is similar.

Since all the formulas in Γ are existential, the last inference cannot be an instance of the $\forall$ rule. So the only remaining possibility is the rule for $\exists$. Consider a proof whose last inference is of the following form:

$$\frac{\begin{matrix}\vdots \\ \Delta, A(t)\end{matrix}}{\Delta, \exists x\, A(x)}$$

Since $A(t)$ is again existential, we can apply the inductive hypothesis to the subproof of $\Delta, A(t)$. This yields the desired conclusion, because any quantifier-free instance of $A(t)$ is a quantifier-free instance of $\exists x\, A(x)$. □

First proof of Herbrand's theorem Suppose that $\exists \vec{y}\, A(\vec{y})$ is provable, where A is quantifier-free. Without loss of generality, we can assume that A is in negation normal form, so there is a proof of $\{\exists \vec{y}\, A(\vec{y})\}$ in the one-sided sequent calculus. Apply the previous lemma to find a finite set $\{\vec{t}_0, \ldots, \vec{t}_{k-1}\}$ of terms and a cut-free proof of $\{A(\vec{t}_0), \ldots, A(\vec{t}_{k-1})\}$ from instances of the equality axioms. Since the proof is cut-free, it uses only the propositional rules. Applying the $\vee$ rule repeatedly yields a proof of $\bigvee_{i<k} A(\vec{t}_i)$. □

Second proof of Herbrand's theorem There is a short model-theoretic proof of Herbrand's theorem if we relax the requirement that the disjunction in the conclusion is provable in the quantifier-free fragment of first-order logic, and instead require only that it is provable in first-order logic. First, assume there is at least one constant in the language and that $\exists \vec{y}\, A(\vec{y})$ is a sentence. If the conclusion of Herbrand's theorem fails, then there is no finite set $\{\vec{t}_0, \ldots, \vec{t}_{k-1}\}$ of tuples of closed terms such that $A(\vec{t}_0) \vee \cdots \vee A(\vec{t}_{k-1})$ is provable. This means that the set

$$\{\neg A(\vec{t}) \mid \vec{t} \text{ is a tuple of closed terms in the language}\}$$

is consistent, and hence has a model $\mathfrak{M}$. Let $\mathfrak{M}' \subseteq \mathfrak{M}$ be the submodel whose universe consists of the denotations of the closed terms of $\mathfrak{M}$, that is,

$$|\mathfrak{M}'| = \{t^{\mathfrak{M}} \mid t \text{ is a closed term in the language}\}.$$

By Lemma 5.6.6, for each tuple $\vec{t}$, $\mathfrak{M}'$ satisfies $\neg A(\vec{t})$. But then $\mathfrak{M}'$ also satisfies $\forall \vec{x}\, \neg A(\vec{x})$, which shows that $\exists \vec{x}\, A(\vec{x})$ is not valid. By soundness, $\exists \vec{x}\, A(\vec{x})$ is not provable.

To handle the case where there are no constants in the language, just add a new constant, c. The previous proof shows that there are tuples of terms $\vec{t}_0, \ldots, \vec{t}_{k-1}$ in the larger language satisfying the conclusion of the theorem, and we can then replace c by a variable. To handle the case where $\exists \vec{y}\, A(\vec{y})$ has free variables $\vec{x}$, replace these by new constants, $\vec{d}$, apply the previous version, and then replace $\vec{d}$ by $\vec{x}$.

In fact, we have already seen a model-theoretic proof of the stronger version of Herbrand's theorem. Proposition 5.2.2 says that whenever Γ is a set of universal formulas, B is quantifier-free, and $\Gamma \models B$, then there is a quantifier-free proof of B from instances of the sentences in Γ. To see that this implies Herbrand's theorem, observe that $\exists \vec{x}\, A(\vec{x})$ is provable in classical logic if and only if $\forall \vec{x}\, \neg A(\vec{x}) \models \bot$, and the conclusion of Herbrand's theorem is equivalent to saying that $\bot$ is provable from instances of $\forall \vec{x}\, \neg A(\vec{x})$ in the quantifier-free fragment of first-order logic. □

The restriction to sets of universal sentences Γ in Corollary 7.1.3 is essential. For example, $\{\exists x\, A(x)\} \vdash \exists x\, A(x)$, but clearly there is no finite set $\{t_0, \ldots, t_{k-1}\}$ of terms such that $\{\exists x\, A(x)\} \vdash A(t_0) \vee \cdots \vee A(t_{k-1})$, since we can always construct a model in which A fails to hold of the elements denoted by $t_0, \ldots, t_{k-1}$ but holds of some other element. It is instructive to see what goes wrong with the inductive procedure described in the syntactic proof. The culprit is the eigenvariable condition on the $\forall$ rule. Consider, for example, the following derivation:

$$\dfrac{\dfrac{\neg P(x), P(x)}{\neg P(x), \exists x\, P(x)}}{\forall x\, \neg P(x), \exists x\, P(x)}$$

If we try to apply the procedure to this proof, inductively we have to extract a finite set of terms from the immediate subproof:

$$\dfrac{\neg P(x), P(x)}{\neg P(x), \exists x\, P(x)}$$

Taking $k = 1$ and $t_0 = x$ does the trick, with the axiom $\neg P(x), P(x)$ as the required proof. But then we cannot apply the $\forall$ rule, since x is free in the side formula $P(x)$.

The following closely related example shows the necessity of the restriction to existential formulas in the conclusion of Herbrand's theorem. Note that the formula $\exists x\, \forall y\, (P(y) \to P(x))$ is classically valid, so there is a cut-free derivation of the sequent $\{\exists x\, \forall y\, (\neg P(y) \vee P(x))\}$. But you should convince yourself that no matter what the underlying language is, there are no terms $t_0, \ldots, t_{k-1}$ such that the sequent

$$\forall y\, (\neg P(y) \vee P(t_0)), \ldots, \forall y\, (\neg P(y) \vee P(t_{k-1}))$$

is provable. (See Exercise 7.1.1.)

Finally, to see that multiple terms are necessary, we have $\{P(a) \vee P(b)\} \vdash \exists x\, P(x)$, but no single term can witness the conclusion. In general, there is no computable way of bounding the number of terms that may be to needed to witness an existential sentence, if we do not know a priori that the sentence is provable.

The following is an important generalization of Herbrand's theorem.

Theorem 7.1.5 (middle sequent theorem). *Suppose Γ is a provable sequent consisting of prenex formulas. Then there is a cut-free proof of Γ in which all the quantifier rules come last. In other words, there is a quantifier-free sequent Γ' such that Γ' is provable and Γ can be proved from Γ' using only the $\forall$ and $\exists$ rules.*

Proof Formally, this requires a couple of carefully chosen inductions, but the idea is that propositional inferences can be permuted above quantifier inferences. For example,

$$\dfrac{\dfrac{\Gamma, B, C}{\Gamma, \forall x\, B, C} \qquad \Gamma, \forall x\, B, D}{\Gamma, \forall x\, B, C \wedge D}$$

can be converted to

$$\dfrac{\dfrac{\Gamma, B, C \qquad \Gamma, \forall x\, B, D}{\Gamma, B, \forall x\, B, C \wedge D}}{\Gamma, \forall x\, B, C \wedge D}$$

The details are left as an exercise. □

In fact, most rules can be permuted freely; issues only arise with the quantifier rules. For example, in a sequence of inferences of the form

$$\dfrac{\dfrac{\Gamma, A(x), B(t(x))}{\Gamma, A(x), \exists y\, B(y)}}{\Gamma, \forall x\, A(x), \exists y\, B(y)}$$

the existential quantifier has to be applied first, to satisfy the eigenvariable condition for the universal quantifier. These constraints provide additional information, as we now show.

Theorem 7.1.6. *Suppose a prenex formula B of the form*

$$\forall x_0\, \exists y_0\, \forall x_1\, \exists y_1\, \ldots A(x_0, y_0, x_1, y_1, \ldots)$$

is provable, with A quantifier-free. Then there is a finite set Γ of the form

$$\{A(x_{0,0}, t_{0,0}, x_{0,1}, t_{0,1}, \ldots), \ldots, A(x_{k-1,0}, t_{k-1,0}, x_{k-1,1}, t_{k-1,1}, \ldots)\}$$

such that Γ is provable using only the propositional rules and equality axioms, and B can be proved from Γ using quantifier-rules only.

Proof This is a straightforward consequence of Theorem 7.1.5. □

It is easier to make sense of the previous theorem by focusing on a single quantifier alternation.

Corollary 7.1.7. *Suppose a sentence $\exists y\, \forall x\, A(y, x)$ is provable in a language with at least one constant symbol, with A quantifier-free. Then there is a finite sequence of terms $t_0, t_1(x_0), \ldots, t_{k-1}(x_0, x_1, \ldots, x_{k-2})$ with at most the free variables shown, such that*

$$A(t_0, x_0) \vee A(t_1(x_0), x_1) \vee \cdots \vee A(t_{k-1}(x_0, x_1, \ldots, x_{k-2}), x_{k-1})$$

is also provable.

There is a natural interpretation of this last formula. It says, roughly, "try t_0; if that doesn't work, try t_1; if *that* doesn't work, try t_2..."

Proof Suppose $\exists y\, \forall x\, A(y, x)$ is provable. By Theorem 7.1.6, there are terms $t_0, \ldots, t_{k-1}$ and variables $x_0, \ldots, x_{k-1}$ such that

$$A(t_0, x_0), A(t_1, x_1), \ldots, A(t_{k-1}, x_{k-1})$$

is also provable, and such that $\exists y\, \forall x\, A(x, y)$ is provable from this using the quantifier rules. We can assume that no variables other than $x_0, \ldots, x_{k-1}$ are free in the terms t_i, because otherwise we could replace these variables with a constant, without harming the rest of the proof. But, a priori, any of the variables x_i can occur in any of the terms t_i. So we need to show that on the assumption that $\exists y\, \forall x\, A(y, x)$ is provable from the sequent above, then the formulas can be renumbered so that in each t_i, all the variables are among $x_0, \ldots, x_{i-1}$.

More generally, use induction on k to show that if $\exists y\, \forall x\, A(y, x)$ is provable from

$$\exists y\, \forall x\, A(y, x), A(t_0, x_0), A(t_1, x_1), \ldots, A(t_{k-1}, x_{k-1})$$

then the formulas can be renumbered in this way. Since we can weaken the sequent above, this suffices for the conclusion. In the base case, $k = 0$, there is nothing to do. Now suppose the claim is true for k, and suppose $\exists y\, \forall x\, A(y, x)$ is provable from a sequent of the form

$$\exists y\, \forall x\, A(y, x), A(t_0, x_0), A(t_1, x_1), \ldots, A(t_{k-1}, x_{k-1}).$$

Clearly the next rule to be applied must be a $\forall$ rule. Without loss of generality, we can assume that it is applied to $A(t_{k-1}, x_{k-1})$; otherwise, just reindex. But this means that x_{k-1} is not a variable of $t_0, \ldots, t_{k-2}$, because this would violate the variable condition on the forall rule. It also cannot be a variable of t_{k-1}, since later we want to apply the $\exists$ rule to this formula to obtain $\exists y\, \forall x\, A(y, x)$, and we cannot do this unless $\forall x\, A(t_{k-1}, x)$ is of the form $(\forall x\, A(y, x))[t_{k-1}/y]$.

The term t_{k-1} may depend on all the other variables. But applying the $\forall$ rule followed immediately by the $\exists$ rule to the last formula, we get a proof of

$$\exists y\, \forall x\, A(y, x), A(t_0, x_0), A(t_1, x_1), \ldots, A(t_{k-2}, x_{k-2}).$$

Applying the inductive hypothesis, we can assume that the variables of each t_i are among $x_0, \ldots, x_{i-1}$. Together with our observations about t_{k-1}, this gives us the desired conclusion. □

Exercises

7.1.1. Prove that in any first-order language with a unary predicate $P(x)$, for any set $\{t_0, \ldots, t_{k-1}\}$ of terms, the following sequent is not provable:

$$\forall y\, (\neg P(y) \vee P(t_0)), \ldots, \forall y\, (\neg P(y) \vee P(t_{k-1})).$$

7.1.2. In a language with unary predicate symbols P and Q, we know that the sentence $\exists x\, P(x) \wedge \exists y\, Q(y) \to \exists z\, (P(z) \wedge Q(z))$ is not provable, because we can construct a countermodel. Use Herbrand's theorem, however, to give a purely syntactic proof that it is not provable. (Hint: If it is provable, there is a sequent proof of $\neg P(x), \neg Q(y), \exists z\, (P(z) \wedge Q(z))$. Apply Herbrand's theorem, and then $\wedge$-inversion to reduce the proof to a series of axioms; and then argue that they can't all be axioms.)

7.1.3. Consider the formula $\exists x\, (f(f(x) + 1) \neq x)$ in a language with symbols 0, 1, and $+$. Show that this is provable from the axiom $\forall x\, (x + 1 \neq 0)$ by exhibiting an explicit Herbrand disjunction. (This problem is by Ulrich Berger. Hint: suppose $f(f(0) + 1) = 0$ and $f(f(f(0)) + 1) = f(0)$ and show that $f(f(f(f(0)) + 1) + 1) \neq f(f(0)) + 1$.)

7.1.4. Fill in the details of the middle sequent theorem, Theorem 7.1.5.

7.1.5. Give a purely model-theoretic proof of Theorem 7.1.7. (Hint: Add new constants to the language and construct a set of sentences Γ such that for each closed term t in the language, there is a constant c such that $\neg A(t, c)$ is in Γ. Show that if the conclusion of the theorem does not hold, then Γ is consistent, and use that to obtain a model in which $\exists x \forall y A(x, y)$ is false.)

7.1.6. Consider a first-order language with a unary predicate P. Since $\exists x P(x) \to \exists y P(y)$ is classically equivalent to $\exists y \forall x (\neg P(x) \vee P(y))$, the latter is classically valid.
  a. Find a cut-free proof of $\exists y \forall x (\neg P(x) \vee P(y))$ in the one-sided sequent calculus.
  b. Show that, no matter what the underlying language is, there is no finite set of closed terms t_i such that $\bigvee_i \forall x (\neg P(x) \vee P(t_i))$ is provable.
  c. Find the terms guaranteed to exist by Theorem 7.1.7.
  d. Show that there is no cut-free proof of $\exists y \forall x (\neg P(x) \vee P(y))$ in which every universal quantifier rule precedes the existential quantifier rules.

7.2 Explicit Definability and the Disjunction Property

From a constructive point of view, a proof of a statement of the form $A \vee B$ should consist of either a proof of A or a proof of B, and a proof of a statement of the form $\exists x A(x)$ should yield an explicit description, t, of an object, and a proof of $A(t)$. The following two theorems show that intuitionistic logic is constructive in this sense.

Theorem 7.2.1 (disjunction property). *Suppose $A \vee B$ is provable in intuitionistic or minimal logic. Then so is either A or B.*

Theorem 7.2.2 (explicit definability property). *Suppose $\exists x A(x)$ is provable in intuitionistic or minimal logic. Then there is a term t such that $A(t)$ is provable as well.*

The second theorem is stronger than Herbrand's theorem in two ways: first, it applies to arbitrary formulas A, and, second, there is only a single term, t. The examples in Section 7.1 show that classical logic does not have the explicit definability property, and since $A \vee \neg A$ is always classically valid, it does not have the disjunction property either.

In Theorems 7.2.1 and 7.2.2, and in the rest of this section, the symbol $\vdash$ can be taken to represent provability in either intuitionistic or minimal logic. The results above follow immediately from the cut elimination theorem. For example, if $A \vee B$ is provable in intuitionistic or minimal logic, then there is a cut-free proof of $\vdash A \vee B$. But this is not an axiom, so the last inference in the proof must be the $\vee$ rule, applied to a proof of either $\vdash A$ or $\vdash B$. The following definition will allow us to state even stronger versions of the disjunction and explicit definability properties.

Definition 7.2.3. The set of *strictly positive subformulas* of a formula A, $\mathrm{spos}(A)$, is defined inductively, as follows:

- If A is atomic, $\mathrm{spos}(A) = \{A\}$.
- $\mathrm{spos}(A \wedge B) = \{A \wedge B\} \cup \mathrm{spos}(A) \cup \mathrm{spos}(B)$.
- $\mathrm{spos}(A \vee B) = \{A \vee B\} \cup \mathrm{spos}(A) \cup \mathrm{spos}(B)$.
- $\mathrm{spos}(A \to B) = \{A \to B\} \cup \mathrm{spos}(B)$.
- $\mathrm{spos}(\forall x A) = \{\forall x A\} \cup \mathrm{spos}(A)$.
- $\mathrm{spos}(\exists x A) = \{\exists x A\} \cup \mathrm{spos}(A)$.

Theorem 7.2.4. *Suppose Γ is a set of sentences and no strictly positive subformula of anything in Γ is of the form $C \vee D$. If $\Gamma \vdash A \vee B$, then $\Gamma \vdash A$ or $\Gamma \vdash B$.*

Proof Without loss of generality we can assume Γ is finite, and we can therefore use induction on cut-free proofs of sequents of the form $\Gamma \vdash A \vee B$. The last inference can't be an axiom, unless we are dealing with the ex falso rule in intuitionistic logic, in which case Γ contains $\bot$ and the conclusion is immediate.

If the last inference is a right inference rule, it must be the $\vee$ rule, in which case it is applied to a subproof of $\Gamma \vdash A$ or $\Gamma \vdash B$, and we are done.

Suppose the last inference is a left inference rule. The hypothesis on Γ implies that it can't be the $\vee$ rule. If it is a $\wedge$ rule, the proof has the following form:

$$\dfrac{\begin{array}{c}\vdots\\ \Gamma', C \vdash A \vee B\end{array}}{\Gamma', C \wedge D \vdash A \vee B}$$

In that case, applying the inductive hypothesis we obtain a proof of $\Gamma', C \vdash A$ or a proof of $\Gamma', C \vdash B$, to which we again apply the $\wedge$ rule on the left. The cases where the last inference is either a $\forall$ or an $\exists$ inference on the left are handled similarly.

Suppose the last inference is the left rule for $\to$. Then the proof has the following form:

$$\dfrac{\begin{array}{c}\vdots\\ \Gamma' \vdash C\end{array} \qquad \begin{array}{c}\vdots\\ \Gamma', D \vdash A \vee B\end{array}}{\Gamma', C \to D \vdash A \vee B}$$

The assumption that $\Gamma', C \to D$ has no strictly positive subformula that is a disjunction implies that the same holds for Γ', D. Applying the inductive hypothesis, we get a proof of $\Gamma', D \vdash A$ or $\Gamma', D \vdash B$, and we can again use the left $\to$ rule. □

Theorem 7.2.5. *Suppose Γ is a set of sentences such that no strictly positive subformula of a sentence in Γ is of the form $\exists x\, C$. Then if $\Gamma \vdash \exists x\, A(x)$, there is a tuple of terms $t_0, \ldots, t_{k-1}$ such that $\Gamma \vdash A(t_0) \vee \cdots \vee A(t_{k-1})$.*

Proof The proof is similar to the preceding one. If the last inference is a right $\exists$ rule, it is applied to a subproof of a sequent of the form $\Gamma \vdash A(t)$. The left rules are handled similarly, except that now we have to allow the possibility that the last inference is a left $\vee$ rule. In that case, the proof has the following form:

$$\dfrac{\begin{array}{c}\vdots\\ \Gamma', C \vdash \exists x\, A(x)\end{array} \qquad \begin{array}{c}\vdots\\ \Gamma', D \vdash \exists x\, A(x)\end{array}}{\Gamma', C \vee D \vdash \exists x\, A(x)}$$

Applying the inductive hypothesis to two subproofs, we obtain proofs of the form

$$\Gamma', C \vdash A(t_0) \vee \cdots \vee A(t_{k-1}) \quad \text{and} \quad \Gamma', D \vdash A(s_0) \vee \cdots \vee A(s_{\ell-1}).$$

Applying right $\vee$ rules, using the associativity of $\vee$, and applying a right $\vee$ rule we obtain a proof of

$$\Gamma', C \vee D \vdash A(t_0) \vee \cdots \vee A(t_{k-1}) \vee A(s_0) \vee \cdots \vee A(s_{\ell-1}),$$

as required. □

Definition 7.2.6. A formula A is said to be *Harrop* if no strictly positive subformula of A is of the form $C \vee D$ or $\exists x\, C$.

Notice that the set of Harrop formulas can be defined inductively, as follows:

- Any atomic formula is Harrop.
- If C and D are Harrop, so are $C \wedge D$ and $\forall x\, C$.
- If C is any formula and D is Harrop, then $C \to D$ is Harrop.

Intuitively, a formula is Harrop if no disjunction or existential assertion can be extracted from it using elimination rules. The sets of formulas described in the hypotheses of Theorems 7.2.4 and 7.2.5 have similar inductive characterizations, and are included in the set of Harrop formulas. So, a fortiori, these apply to Harrop formulas as well. As a corollary of Theorem 7.2.4 and 7.2.5, we therefore have:

Corollary 7.2.7. *Suppose* Γ *is a set of Harrop sentences and* Γ *proves* $\exists x\, A(x)$. *Then there is a term* t *such that* Γ *proves* $A(t)$.

Exercise

7.2.1. Show that if A is Harrop then $\neg\neg A \leftrightarrow A$ is provable in minimal logic from the assumptions $\neg\neg B \leftrightarrow B$ for each atomic subformula B of A. (Since any negative formula can be obtained from a Harrop formula by replacing atomic formulas by negated atomic formulas, this strengthens Proposition 2.7.8 and the generalization to first-order logic in Section 4.6.)

7.3 Interpolation Theorems

The Craig interpolation theorem is another notable application of the cut elimination theorem. It holds for the classical, intuitionistic, and minimal versions of the provability relation.

Theorem 7.3.1. *Let* A *and* B *be formulas such that* $A \to B$ *is provable. Then there is a sentence* C *containing only function symbols, relation symbols, constant symbols, and free variables that are found in both* A *and* B, *such that* $A \to C$ *and* $C \to B$ *are both provable.*

In the statement of the theorem, we count equality as a special logical symbol rather than a relation symbol. The formula C is called an *interpolant* for A and B. The philosophical interpretation is that if a formula A is to imply another formula B, it can only do it via something in their common language. For example, if A is a statement in the language of physics and C is a statement in the language of chemistry, an interpolant has to use terms common to the two subjects, such as mathematical terms or the concepts *mass* and *volume*.

We will prove the interpolation theorem in three stages. First, we will prove it for first-order logic without equality, without the requirement that the function symbols in the interpolant occur in both A and B. Then we will show that the conclusion still holds for first-order logic with equality. Then, finally, we will show how to obtain the additional requirement on function symbols.

For classical first-order logic, it is convenient to use the one-sided calculus. Theorem 7.3.1 requires that there is a sentence equivalent to $\bot$ without any function, relation, or constant symbols at all, and another such sentence that is equivalent to $\top$. If we include equality in the language of the one-sided calculus, we can use $\neg\forall x\,(x = x)$ and $\forall x\,(x = x)$, but without equality we need to include $\bot$ and $\top$ as formulas in negation normal form and add sequents $\Gamma, \top$ as axioms. Our first, restricted version of Theorem 7.3.1 is a consequence of the following lemma.

Lemma 7.3.2. *Suppose there is a cut-free proof of* Γ, Γ' *in the one-sided sequent calculus for classical first-order logic without equality. Then there is a formula* C *containing only the relation symbols, constant symbols, and free variables common to* Γ *and* Γ'*, such that* Γ, C *and* $\Gamma', {\sim}C$ *are both provable.*

Proof Use induction on the cut-free proof of Γ, Γ'.

Suppose Γ, Γ' is an axiom, $\Pi, B, \neg B$. There are four cases: both B and $\neg B$ are in Γ, in which case $\bot$ is an interpolant; B is in Γ and $\neg B$ is in Γ', in which case $\neg B$ is an interpolant; $\neg B$ is in Γ and B is in Γ', in which case B is an interpolant; or both are in Γ', in which case $\top$ is an interpolant.

For the other rules, without loss of generality, we can assume that the principal formula of the last inference is in Γ'. We don't require that Γ and Γ' are disjoint, so we have to allow for the possibility that the principal formula is in both sets, but the argument works just as well in that case.

Suppose the last inference is the $\wedge$ rule,

$$\frac{\Gamma, \Gamma'', A \qquad \Gamma, \Gamma'', B}{\Gamma, \Gamma'', A \wedge B}$$

where $\Gamma' = \Gamma'' \cup \{A \wedge B\}$. We have to allow for the possibility that $A \wedge B$ is in Γ'' but again the argument still works if this is the case. By the inductive hypothesis, there are interpolants C_1 and C_2, and proofs of the following:

- Γ, C_1
- $\Gamma'', A, {\sim}C_1$
- Γ, C_2
- $\Gamma'', B, {\sim}C_2$.

We can use these to obtain the following proofs:

$$\frac{\Gamma, C_1 \qquad \Gamma, C_2}{\Gamma, C_1 \wedge C_2} \qquad \frac{\dfrac{\Gamma'', A, {\sim}C_1}{\Gamma'', A, {\sim}C_1 \vee {\sim}C_2} \qquad \dfrac{\Gamma'', B, {\sim}C_2}{\Gamma'', B, {\sim}C_1 \vee {\sim}C_2}}{\Gamma'', A \wedge B, {\sim}C_1 \vee {\sim}C_2}$$

Taking C to be $C_1 \wedge C_2$ yields the desired interpolant. It is not hard to check that C satisfies the conditions of the theorem, assuming C_1 and C_2 do.

Suppose the last inference is of the form

$$\frac{\Gamma, \Gamma'', A}{\Gamma, \Gamma'', A \vee B}$$

where $\Gamma' = \Gamma'', A \vee B$. By the inductive hypothesis, there is an interpolant C such that Γ, C and $\Gamma'', A, {\sim}C$ are provable. But then $\Gamma'', A \vee B, {\sim}C$ is also provable, so C is the interpolant we want.

Suppose the last inference is of the form

$$\frac{\Gamma,\ \Gamma'',\ A(y)}{\Gamma,\ \Gamma'',\ \forall x\, A(x)}$$

By the inductive hypothesis, there is an interpolant C such that both $\Gamma,\ C$ and $\Gamma'',\ A(y),\ {\sim}C$ are provable. The eigenvariable condition and our requirement on the free variables of C implies that y is not free in C, so $\Gamma,\ \forall x\, A(x),\ {\sim}C$ is provable, as required.

Suppose the last inference is of the form

$$\frac{\Gamma,\ \Gamma'',\ A(t)}{\Gamma,\ \Gamma'',\ \exists x\, A(x)}$$

Again by the inductive hypothesis, there is an interpolant C with the property that $\Gamma,\ C$ and $\Gamma'',\ A(t),\ {\sim}C$ are provable. A subtlety arises in that t may have variables $\vec{u}$ and constants $\vec{d}$ that occur in Γ but not Γ''. In that case, C may contain these variables and constants even though they do not appear in $\Gamma' = \Gamma'',\ \exists x\, A(x)$. But in that case it is easy to check that if we choose fresh variables $\vec{v}$ to replace $\vec{d}$, the formula $\exists \vec{u},\ \vec{v}\ C[\vec{v}/\vec{d}]$ satisfies the requirements of the theorem. □

For intuitionistic and minimal logic, we have to use a two-sided calculus. The corresponding version of the interpolation theorem for first-order logic without equality is a consequence of the following.

Lemma 7.3.3. *Suppose* $\Gamma,\ \Gamma' \vdash D$ *is provable in the sequent calculus for intuitionistic or minimal logic. Then there is a formula* C *containing only relation symbols that are common to both* Γ *and* $\Gamma' \cup \{D\}$*, such that* $\Gamma \vdash C$ *and* $\Gamma',\ C \vdash D$ *are both provable.*

Proof Once again, we use induction on cut-free proofs. If $\Gamma,\ \Gamma' \vdash D$ is an axiom (other than *ex falso*), we have to consider two possibilities: either D is in Γ or it is in Γ'. In the first case, we can take $C = D$, and in the second case, we can take $C = \top$. If $\Gamma,\ \Gamma' \vdash D$ is an *ex falso* axiom, we take $C = \bot$ if $\bot$ is in Γ, and $C = \top$ if $\bot$ is in Γ'.

Dealing with the right rules is straightforward. For example, suppose the last rule is the right rule for implication, so the proof has the following form:

$$\frac{\begin{array}{c}\vdots\\ \Gamma,\ \Gamma',\ A \vdash B\end{array}}{\Gamma,\ \Gamma' \vdash A \to B}$$

By the inductive hypothesis, we have an interpolant C such that the following are provable:

- $\Gamma \vdash C$
- $\Gamma',\ C,\ A \vdash B.$

Then $\Gamma' \vdash A \to B$ is provable, and C is the desired interpolant. The right rules for conjunction, disjunction, and the quantifiers are handled similarly.

As far as the left rules, the most difficult cases to handle are the ones where there is more than one hypothesis. We leave the case where the last inference is the left rule for $\vee$ as an exercise. Suppose the last inference is the left rule for implication, with the principal formula in Γ. Then the proof has the following form:

$$\frac{\begin{array}{cc} \vdots & \vdots \\ \Gamma_1, \Gamma_2 \vdash A & \Gamma_1, \Gamma_2, B \vdash D \end{array}}{\Gamma_1, \Gamma_2, A \to B \vdash D}$$

with $\Gamma = \Gamma_1, A \to B$ and $\Gamma' = \Gamma_2$. In this case, the solution is tricky. First we apply the inductive hypothesis, switching the roles of Γ_1 and Γ_2 in the first instance, to get proofs of the following:

- $\Gamma_2 \vdash C_1$
- $\Gamma_1, C_1 \vdash A$
- $\Gamma_1, B \vdash C_2$
- $\Gamma_2, C_2 \vdash D$.

From the second and third proofs, we get a proof of $\Gamma_1, A \to B \vdash C_1 \to C_2$, as follows:

$$\frac{\dfrac{\begin{array}{cc} \vdots & \vdots \\ \Gamma_1, C_1 \vdash A & \Gamma_1, B \vdash C_2 \end{array}}{\Gamma_1, A \to B, C_1 \vdash C_2}}{\Gamma_1, A \to B \vdash C_1 \to C_2}$$

From the first and fourth proofs we get a proof of $\Gamma_2, C_1 \to C_2 \vdash D$. So $C_1 \to C_2$ is our interpolant. In the case where $\Gamma = \Gamma_1$ and $\Gamma' = \Gamma_2, A \to B$, we apply the inductive hypothesis to get interpolants C_1 and C_2 with proofs of these:

- $\Gamma_1 \vdash C_1$
- $\Gamma_2, C_1 \vdash A$
- $\Gamma_1 \vdash C_2$
- $\Gamma_2, C_2, B \vdash D$.

In that case, an exercise below asks you to verify that $C_1 \wedge C_2$ serves as an interpolant. Handling the other left rules is more straightforward. □

We now extend the result to first-order logic with equality, still without imposing any requirements on the function symbols that occur in the interpolant.

Proof of the second restricted version of Theorem 7.3.1. Suppose the formula $A \to B$ is provable in first-order logic with equality. Let E_0 be the conjunction of the reflexivity, symmetry, and transitivity axioms. Let E_1 be the conjunction of the equality axioms $\forall \vec{x}, \vec{y}\ (\vec{x} = \vec{y} \wedge R(\vec{x}) \to R(\vec{y}))$ for the relation symbols R occurring in A, and, similarly, let E_2 be the conjunction of the equality axioms for the relation symbols occurring in B. By the deduction theorem, there is a proof of the formula $E_0 \wedge E_1 \wedge E_2 \to (A \to B)$. But this is equivalent to $E_0 \wedge E_1 \wedge A \to (E_2 \to B)$. By the interpolation theorem for first-order logic without equality, there is an interpolant C such that $E_0 \wedge E_1 \wedge A \to C$ and $C \wedge E_2 \to B$ are both provable, and C has only relation symbols common to $E_0 \wedge E_1 \wedge A$ and $E_2 \to B$. But then $A \to C$ and $C \to B$ are both provable in first-order logic with equality since E_0, E_1, and E_2 are derivable there, and C has only relation symbols common to both A and B. □

The reason that extending the interpolation theorem to function symbols is a challenge is that they can be eliminated by the $\exists$ rule in the one-sided calculus and by the left $\forall$ and right $\exists$ rules in the two-sided calculus. But Proposition 4.7.5 gives us a way of translating any formula A with function symbols into a formula $\hat{A}$ where the function symbols are replaced by relation symbols. Suppose there is a proof of $A \to B$. Choose relation symbols F_i corresponding to the function symbols f_i of the language, let Π_0 be the set of axioms $\forall \vec{y}\, \exists! x\, F_i(\vec{y}, x)$ corresponding to function symbols occurring in A, and similarly, let Π_1 be the set of axioms for function symbols occurring in B. By Proposition 4.7.5, there is a proof of $\Pi_0 \wedge \hat{A} \to (\Pi_1 \to \hat{B})$. By the interpolation theorem for relational languages, there is an interpolant C involving only the relation symbols common to both $\hat{A}$ and $\hat{B}$, such that $\Pi_0 \wedge \hat{A} \to C$ and $\Pi_1 \wedge (C \to \hat{B})$ are both provable. Now use the other direction of Proposition 4.7.5 to replace the relation symbols by function symbols again. That is, replace each atomic formula $F_i(\vec{s}, t)$ by $f_i(\vec{s}) = t$ to obtain formulas $\hat{A}^*$, $\hat{B}^*$, and C^* in L, such that $\hat{A}^* \leftrightarrow A$ and $\hat{B}^* \leftrightarrow B$ are provable and C^* has only function and relation symbols common to both A and B. Then C^* is the desired interpolant. Thus we have obtained the complete statement of Theorem 7.3.1.

There are a number of variations on the interpolation theorem in the literature. To state a strengthening due to Roger Lyndon, we need the following definition.

Definition 7.3.4. The set of *positive subformulas* of a formula A, $\mathrm{pos}(A)$, and the set of *negative subformulas* of a formula A, $\mathrm{neg}(A)$, are defined by simultaneous recursion on A as follows:

- $\mathrm{pos}(A) = \{A\}$ for A atomic
- $\mathrm{neg}(A) = \emptyset$ for A atomic
- $\mathrm{pos}(A \wedge B) = \{A \wedge B\} \cup \mathrm{pos}(A) \cup \mathrm{pos}(B)$
- $\mathrm{neg}(A \wedge B) = \mathrm{neg}(A) \cup \mathrm{neg}(B)$
- $\mathrm{pos}(A \vee B) = \{A \vee B\} \cup \mathrm{pos}(A) \cup \mathrm{pos}(B)$
- $\mathrm{neg}(A \vee B) = \mathrm{neg}(A) \cup \mathrm{neg}(B)$
- $\mathrm{pos}(A \to B) = \{A \to B\} \cup \mathrm{neg}(A) \cup \mathrm{pos}(B)$
- $\mathrm{neg}(A \to B) = \mathrm{pos}(A) \cup \mathrm{neg}(B)$
- $\mathrm{pos}(\forall x\, A) = \{\forall x\, A\} \cup \mathrm{pos}(A)$
- $\mathrm{neg}(\forall x\, A) = \mathrm{neg}(A)$
- $\mathrm{pos}(\exists x\, A) = \{\exists x\, A\} \cup \mathrm{pos}(A)$
- $\mathrm{neg}(\exists x\, A) = \mathrm{neg}(A)$.

A relation symbol R is said to *occur positively in A* if an atomic formula of the form $R(t_0, \ldots, t_{k-1})$ occurs positively in A, and similarly for negative occurrences. You can check that the atomic formulas that occur positively in a formula A are precisely the ones that are not negated when the formula is put in negation normal form and the ones that occur negatively are the ones that are negated. Note that a subformula can occur both positively and negatively in a formula A.

Lyndon's strengthening of the interpolation theorem says that we can moreover require the interpolant C to have the property that every relation symbol occurring positively occurs positively in both A and B, and every relation symbol occurring negatively occurs negatively in both A and B. Exercise 7.3.6 asks you to prove this.

Exercises

7.3.1. Carry out the inductive step in the proof of Lemma 7.3.3 above, where the last step is a right $\wedge$ rule.

7.3.2. Carry out the inductive step in the proof of Lemma 7.3.3 above, where the last step is a left $\vee$ rule.

7.3.3. In the proof of the interpolation theorem for intuitionistic and minimal logic, in the case of a left $\to$ rule, verify the claim above that $C_1 \wedge C_2$ is the desired interpolant.

7.3.4. If we tried to prove the interpolation theorem for classical logic using a two-sided calculus, our main lemma would be as follows:

Lemma 7.3.5. *Suppose* $\Gamma, \Gamma' \vdash \Delta, \Delta'$ *is provable in the sequent calculus. Then there is a formula* C *containing only function, relation, and constant symbols that are common to both* Γ, Δ *and* Γ', Δ', *such that* $\Gamma \vdash C, \Delta$ *and* $\Gamma', C \vdash \Delta'$ *are both provable.*

Carry out some of the cases involved in proving the lemma in this form.

7.3.5. Show that if A^{nnf} is the negation normal form translation of A, then the same relation symbols occur positively and negatively in A and A^{nnf}.

7.3.6. Prove the Lyndon's strengthening of the Craig interpolation theorem: with the hypotheses of Theorem 7.3.1, we can find an interpolant C with the additional property that every relation symbol occurring positively occurs positively in both A and B, and every relation symbol occurring negatively occurs negatively in both A and B.

7.3.7. To show that we cannot extend Lyndon's theorem to say something about positive and negative occurrences of constants and function symbols, consider the sentence $\exists x\,(x = a \wedge \neg R(x)) \to \neg R(a)$, and note that a occurs positively in the antecedent of the implication and negatively in the conclusion. Show that a must occur in any interpolant.

7.3.8. In the context of classical *propositional* logic, the interpolation theorem states the following. Let $A(\vec{p}, \vec{q})$ and $B(\vec{p}, \vec{r})$ be formulas involving only the propositional variables shown, and suppose $\vdash A \to B$. Then there is a formula $C(\vec{p})$ involving at most the propositional variables shown, such that $\vdash A \to C$ and $\vdash C \to B$.

Come up with a semantic proof of this theorem, by showing that if $\models A \to B$, then there is a formula $C(\vec{p})$ with at most the free variables shown, such that $\models A \to C$ and $\models C \to B$. (Hint: If $A(p)$ is a propositional formula, let $\exists p\, A(p)$ abbreviate the formula $A(\top) \vee A(\bot)$, and let $\forall p\, A(p)$ abbreviate $A(\top) \wedge A(\bot)$. Show that if $A(\vec{p}, \vec{q}) \to B(\vec{p}, \vec{r})$ is provable, then $\exists \vec{q}\, A(\vec{p}, \vec{q})$ entails $\forall \vec{r}\, B(\vec{p}, \vec{r})$, and either one of these two formulas can serve as an interpolant.)

7.3.9. This exercise provides a model-theoretic proof of the interpolation theorem for classical logic. For simplicity, let us assume we are dealing with a first-order logic without equality and that the languages in question have no function symbols. It is not hard to extend the construction to work more generally.

Let L_1 be the language of A, L_2 be the language of B, and L_0 be the common language. Suppose there is no C in L_0 such that $A \to C$ and $C \to B$ are both provable. We want to show that $A \to B$ is not provable, which is to say, the set $\{A, \neg B\}$ is consistent.

Let $C = \{c_0, c_1, c_2, \ldots\}$ be a countable set of new constants. Let L_0' be L_0 together with C, and similarly for L_1' and L_2'. If Γ is a set of sentences in L_1' and Δ is a set of sentences in L_2', a sentence C in L_0' is said to *separate* Γ from Δ if $\Gamma \vdash C$ and $\Delta \vdash \neg C$. Γ and Δ are said to be *inseparable* if no such sentence separates them. Notice that, by hypothesis, $\{A\}$ and $\{\neg B\}$ are inseparable: if $C(\vec{c})$ were to separate them, where $\vec{c}$ is the tuple of new constants that appear in the formula, then $\forall \vec{x}\, C(\vec{x})$ would be an interpolant.

The following construction is similar to the construction of a maximally consistent Henkin set, but, instead, we construct maximal inseparable sets. Enumerate all the sentences $A_0, A_1, A_2, \ldots$ in L_1' and all the sentences $B_0, B_1, B_2, \ldots$ in L_2'. Set $\Gamma_0 = \{A\}$ and $\Delta_0 = \{\neg B\}$.

For each n, construct a finite set Γ_{n+1} of sentences in L_1' and a finite set Δ_{n+1} of sentences in L_2' as follows:

- Start by including everything Γ_m in Γ_{m+1}. If $\Gamma_m \cup \{A_m\}$ is not separable from Δ_m, add A_m to Γ_{m+1}.
- Moreover, if A_m is of the form $\exists x\, D(x)$ and A_m has been added to Γ_{m+1}, also add $D(c)$, for the first constant C that does not appear in $\Gamma_m \cup \Delta_m \cup \{A_m\}$.
- Similarly, let $\Delta_{m+1} \supseteq \Delta_m$. If $\Delta_m \cup \{B_m\}$ is not separable from Γ_{m+1}, add B_m to Δ_{m+1}.
- Moreover, if B_m is of the form $\exists x\, E(x)$ and B_m has been added to Δ_{m+1}, then also add $E(c)$, for the first constant c that does not appear in $\Gamma_{m+1} \cup \Delta_m \cup \{B_m\}$.

Finally, let $\Gamma_\omega = \bigcup_i \Gamma_i$ and $\Delta_\omega = \bigcup_i \Gamma_i$. Do the following:

a. Show that for every i, Γ_i and Δ_i are inseparable, and so Γ_ω and Δ_ω are inseparable.
b. Show that Γ_ω is a maximally consistent theory in L_1', and Δ_ω is a maximally consistent theory in L_2'. (Remember that you can show that a set Γ is maximally consistent by showing that for every formula φ, either φ or $\neg\varphi$ is in Γ but not both.)
c. Show that $\Gamma_\omega \cap \Delta_\omega$ is a maximally consistent theory in L_0'.

The usual model-theoretic construction then works: let $\mathfrak{M}$ be the model whose universe consists of the closed terms in $L_1' \cup L_2'$, in this case, all of which are constants. For a relation symbol R in L_1, say $R^{\mathfrak{M}}(\vec{a})$ holds if and only if $R(\vec{a}) \in \Gamma_\omega$, and similarly for the relation symbols in L_2. Then a routine induction on formulas shows that for every sentence D in L_1', $\mathfrak{M} \models D$ if and only if $D \in \Gamma_\omega$, and for every sentence E in L_2', $\mathfrak{M} \models E$ if and only if $E \in \Delta_\omega$. In particular, $\mathfrak{M} \models A \wedge \neg B$, as required.

7.3.10. Let Γ be a set of first-order sentences with a k-ary relation symbol P. Let P' be a new k-ary relation symbol, and let $\Gamma[P'/P]$ denote the result of replacing P by P' in Γ. One says that Γ *explicitly defines* P if there is a formula $B(y_1, \ldots, y_{k-1})$ that does not involve P such that

$$\Gamma \vdash \forall \vec{y}\, (P(\vec{y}) \leftrightarrow B(\vec{y})).$$

Here, B is the explicit definition of P. One says that Γ *implicitly defines* P if

$$\Gamma \cup \Gamma[P'/P] \vdash \forall \vec{y}\, (P(\vec{y}) \leftrightarrow P'(\vec{y})).$$

This implies that in any model of Γ, there is only one interpretation of P that satisfies all the sentences in Γ.

Beth's definability theorem says that for any Γ, Γ implicitly defines P if and only if it explicitly defines P. Prove this. (Hint: One direction is immediate. For the other direction, suppose Γ implicitly defines P. By the finite character of the provability relation, some finite subset of Γ implicitly defines P, so without loss of generality, we can assume that P is implicitly defined by a single formula D, involving the predicate P. Add k new constants $\vec{c}$ to the language. Since D implicitly defines P, we have

$$\vdash D \wedge P(\vec{c}) \to (D[P'/P] \to P'(\vec{c})).$$

Apply the interpolation theorem.)

7.4 Indefinite Descriptions

Section 4.7 described a way of eliminating definite descriptions from a proof, showing that if $\Gamma \cup \{\forall \vec{x}\, A(\vec{x}, f(\vec{x}))\}$ proves B and f does not occur in Γ, $A(\vec{x}, y)$, or B, then $\Gamma \cup \{\forall \vec{x}\, \exists! y\, A(\vec{x}, y)\}$ proves B. In particular, if Γ proves $\forall \vec{x}\, \exists! y\, A(\vec{x}, y)$, introducing a function symbol $f(\vec{x})$ to name these values of y results in a conservative extension of Γ.

The conservation result remains true even if we drop the uniqueness assumption, as long as we are dealing with classical first-order logic, or intuitionistic first-order logic without equality. It also holds for intuitionistic logic with decidable equality, that is, the axiom $\forall x, y\, (x = y \vee x \neq y)$. In this section, we will prove these results, and obtain precise information as to how the conclusion can fail for intuitionistic logic with equality.

Theorem 7.4.1. *Suppose* $\Gamma \cup \{\forall \vec{x}\, A(\vec{x}, f(\vec{x}))\} \vdash B$ *in classical logic with or without equality or in intuitionistic logic with decidable equality, where* f *does not occur in* $\Gamma \cup \{A(\vec{x}, y), B\}$. *Then* $\Gamma \cup \{\forall \vec{x}\, \exists y\, A(\vec{x}, y)\} \vdash B$ *as well.*

For classical logic with or without equality, there is an easy model-theoretic proof.

Proof of Theorem 7.4.1 for classical logic By soundness and completenesss, it suffices to assume $\Gamma \cup \{\forall \vec{x}\, A(\vec{x}, f(\vec{x}))\} \models B$ and show $\Gamma \cup \{\forall \vec{x}\, \exists y\, A(\vec{x}, y)\} \models B$. Let $\mathfrak{M}$ by any model of the sentences $\Gamma \cup \{\forall \vec{x}\, \exists y\, A(\vec{x}, y)\}$. Let $\hat{f}$ be any function which, for every tuple $\vec{a}$ in $|\mathfrak{M}|$, chooses a value b such that $\mathfrak{M} \models A(\vec{a}, b)$. Let $\mathfrak{M}'$ be the expansion of $\mathfrak{M}$ that interprets the symbol f as $\hat{f}$. Then $\mathfrak{M}' \models \Gamma \cup \{\forall \vec{x}\, A(\vec{x}, f(\vec{x}))\}$, and the hypothesis implies $\mathfrak{M}' \models B$. Since $\mathfrak{M}$ and $\mathfrak{M}'$ agree on sentences in the language of $\mathfrak{M}$, we have $\mathfrak{M} \models B$. □

This proof uses the axiom of choice, but without loss of generality we can restrict attention to the case where L is finite (since we can take Γ to be finite), and hence the case where the model in question is also countable. We can then define $\hat{f}$ by taking the first suitable element in a fixed enumeration of the universe.

For intuitionistic and minimal logic, model-theoretic proofs are possible, but they are more subtle. (See the bibliographical notes at the end of this chapter.) We will use the cut elimination theorem instead to establish those results. First, we will consider classical, intuitionistic, and minimal logic without equality, where the proof is somewhat simpler. We will treat the case of classical logic using a two-sided calculus with a set Δ of formulas on the right hand side. The same proof also covers intuitionistic and minimal logic if we view them as the special case where Δ consists of a single formula.

Fix the formula $A(\vec{x}, y)$ in the statement of Theorem 7.4.1. Without loss of generality, we can assume that the set Γ mentioned there is finite. Augment the two-sided sequent calculus with the following rule:

$$\frac{\Gamma, A(\vec{t}, f(\vec{t})) \vdash \Delta}{\Gamma \vdash \Delta}$$

Using this rule, we can derive $\vdash \forall \vec{x}\, A(\vec{x}, f(\vec{x}))$ as described in Section 6.7, so, for the set Γ in the statement of Theorem 7.4.1, we can derive $\Gamma \vdash B$ in the extended calculus. By the discussion in Section 6.7, the extended calculus still has cut elimination. Thus, it suffices to show that if $\Gamma \vdash B$ has a cut-free proof in the extended calculus and f does not occur in Γ, B, then $\Gamma \vdash B$ is provable in ordinary first-order logic.

The details of the proof are finicky, but the general idea is not difficult to understand. The final sequent, $\Gamma \vdash B$, does not contain f. Other formulas C in the proof may contain the symbol f, since terms involving f can be eliminated by the right $\exists$ rule, the left $\forall$ rule, or the additional rule of the extended calculus. But this means that none of the variables occurring in any of the arguments to f are bound in C. For example, if C were a formula like $\forall \vec{z}\, R(f(\vec{z}))$, f could not be eliminated in that way. So C must be of the form $C'[f(\vec{t}_0)/y_0, \ldots, f(\vec{t}_{n-1})/y_{n-1}]$ for some formula C' in the original language, L. The extended calculus has nothing specific to say about f, except that it has to satisfy the additional rule. So most of the proof will still go through if we replace each term $f(\vec{t}_i)$ by the variable y_i. We will still have to handle the additional rule in the extended calculus, which is replaced by a broken inference of the following form:

$$\frac{\Gamma, A(\vec{t}, y) \vdash \Delta}{\Gamma \vdash \Delta}$$

Here, the variable y replaces a term $f(\vec{t})$ that was there before. We will cope with that by leaving such formulas $A(\vec{t}, y)$ in the hypotheses until the last instance of $f(\vec{t})$ is eliminated in the original proof, at which point y is not free in Γ, Δ, or any other formula. At that point, we can eliminate $A(\vec{t}, y)$, possibly at the expense of introducing $\forall \vec{x}\, \exists y\, A(\vec{x}, y)$ into the hypotheses, as follows:

$$\cfrac{\cfrac{\Gamma, A(\vec{t}, y) \vdash \Delta}{\Gamma, \exists y\, A(\vec{t}, y) \vdash \Delta}}{\Gamma, \forall \vec{x}\, \exists y\, A(\vec{x}, y) \vdash \Delta}$$

The next lemma encapsulates a precise formulation of the argument just sketched. We will write $\ldots f(\vec{t}_i)/y_i \ldots$ instead of $f(\vec{t}_0)/y_0, \ldots, f(\vec{t}_{n-1})/y_{n-1}$ for the sake of brevity. As usual, $\Gamma[\ldots f(\vec{t}_i)/y_i \ldots]$ denotes $\{C[\ldots f(\vec{t}_i)/y_i \ldots] \mid C \in \Gamma\}$ and we'll adopt the corresponding notation $(\Gamma \vdash \Delta)[\ldots f(\vec{t}_i)/y_i \ldots]$ for sequents. When using this notation, we will always assume that the terms $f(\vec{t}_i)$ are distinct, and that each of the variables $\vec{y}$ occurs at least once in Γ or Δ.

Lemma 7.4.2. *Let Γ and Δ be finite sets of formulas in L. Suppose $y_0, \ldots, y_{n-1}$ is a tuple of variables occurring in $\Gamma \cup \Delta$, suppose L' consists of L together with a new function symbol, f, and suppose $f(\vec{t}_0), \ldots, f(\vec{t}_{n-1})$ are terms of L' that do not contain any of $y_0, \ldots, y_{n-1}$. If*

$$(\Gamma \vdash \Delta)[\ldots f(\vec{t}_i)/y_i \ldots]$$

is derivable in the extended calculus for L', then

$$\Gamma, \forall \vec{x}\, \exists y\, A(\vec{x}, y), \ldots, A(\vec{t}_i, y_i), \ldots \vdash \Delta$$

is derivable in the ordinary sequent calculus for L.

The particular case where $n = 0$ and Δ is the singleton $\{B\}$, together with the fact that $\forall \vec{x}\, A(\vec{x}, f(\vec{x}))$ is derivable in the extended calculus, yields Theorem 7.4.1.

Proof Use induction on cut-free derivations. Without loss of generality we can assume that the terms $f(\vec{t}_i)$ in the final sequent are distinct, because otherwise we can combine some of the y_is.

If the derivation is an instance of an assumption axiom, then there is an atomic formula C common to $\Gamma[\ldots f(\vec{t}_i)/y_i \ldots]$ and $\Delta[\ldots f(\vec{t}_i)/y_i \ldots]$. The assumption that the terms $f(\vec{t}_i)$ are distinct implies that C of the form $C'[\ldots f(\vec{t}_i)/y_i \ldots]$ for some C' common to both Γ and Δ, so $\Gamma \vdash \Delta$ is also an assumption axiom.

The cases where the last inference is a propositional inference or the *ex falso* axiom are straightforward, since we can apply the inductive hypothesis to all the subproofs and then use the same inference. The cases where the last inference is a right $\forall$ rule or left $\exists$ rule are similarly straightforward. For example, suppose the last inference in the original proof was a right $\forall$ rule with main premise C and principal formula $\forall z\, C$. The hypotheses imply that

$\forall z\, C$ is of the form $(\forall z\, C')[\dots f(\vec{t}_i)/y_i \dots]$ for some C', where z does not occur in any tuple $\vec{t}_i$. So the last inference can be written as follows:

$$\frac{(\Gamma \vdash \Delta, C')[\dots f(\vec{t}_i)/y_i \dots]}{(\Gamma \vdash \Delta, \forall z\, C')[\dots f(\vec{t}_i)/y_i \dots]}$$

Applying the inductive hypothesis to the subproof and then applying the right $\forall$ rule again yields

$$\frac{\Gamma, \forall \vec{x}\, \exists y\, A(\vec{x}, y), \dots, A(\vec{t}_i, y_i), \dots \vdash \Delta, C'}{\Gamma, \forall \vec{x}\, \exists y\, A(\vec{x}, y), \dots, A(\vec{t}_i, y_i), \dots \vdash \Delta, \forall z\, C'}$$

The case of a left $\exists$ rule is similar.

All that remains are the right $\exists$ and left $\forall$ rules and the additional rule of the extended calculus. These cases are more interesting because subterms of the form $f(\vec{t})$ may be eliminated by these rules. Suppose the last inference of the original proof is a right $\exists$ rule with principal formula $\exists z\, C(z)$ and main premise $C(s)$, where $\exists z\, C$ is of the form $(\exists z\, C')[\dots f(\vec{t}_i)/y_i \dots]$ for some C'. Write the term s as $s'[\dots f(\vec{t}_i)/y_i \dots f(\vec{t}'_j)/y'_j \dots]$ where s' is in L and the sequence $\dots f(\vec{t}'_j)/y'_j \dots$ includes the subterms of the form $f(t)$ that are not included among the terms $f(\vec{t}_i)$. Applying the inductive hypothesis and a right $\exists$ rule with major premise $C'(s')$ yields the following:

$$\frac{\Gamma, \forall \vec{x}\, \exists y\, A(\vec{x}, y), \dots, A(\vec{t}_i, y_i), \dots, A(\vec{t}'_j, y'_j), \dots \vdash \Delta, C'(s')}{\Gamma, \forall \vec{x}\, \exists y\, A(\vec{x}, y), \dots, A(\vec{t}_i, y_i), \dots, A(\vec{t}'_j, y'_j), \dots \vdash \Delta, \exists z\, C'(z)}$$

The only problem is the presence of the extra formulas $A(\vec{t}'_j, y'_j)$, since the variables y'_j no longer occur in $\Gamma \cup \Delta$. But we can then apply a left $\exists$ rule to transform this formula to $\exists y\, A(\vec{t}'_j, y)$, and then we can apply right $\forall$ rules to transform it to $\forall \vec{x}\, \exists y\, A(\vec{x}, y)$. Doing this for each j yields a proof of

$$\Gamma, \forall \vec{x}\, \exists y\, A(\vec{x}, y), \dots, A(\vec{t}_i, y_i), \dots \vdash \Delta, \exists z\, C'(z),$$

as required. The case of a right $\forall$ rule is similar.

Suppose the last rule is the additional rule of the extended calculus. Then the inference must have the following form:

$$\frac{(\Gamma, A(\vec{t}', y') \vdash \Delta)[\dots f(\vec{t}_i)/y_i \dots f(\vec{t})/y']}{(\Gamma \vdash \Delta)[\dots f(\vec{t}_i)/y_i \dots]}$$

where $\vec{t}$ can be written as $\vec{t}'[\dots f(\vec{t}_i)/y_i \dots]$ and $f(\vec{t})/y'$ may or may not already be included in the sequence $\dots f(\vec{t}_i)/y_i \dots$. By the inductive hypothesis, we get a proof of

$$\Gamma, \forall \vec{x}\, \exists y\, A(\vec{x}, y), \dots, A(\vec{t}_i, y_i), \dots, A(\vec{t}', y') \vdash \Delta.$$

If $f(\vec{t})/y'$ is already included among the substitutions $f(\vec{t}_i)/y_i$, we are done. Otherwise, y' occurs only in $A(\vec{t}', y')$, and we can apply an $\exists$ rule followed by $\forall$ rules to transform it to $\forall \vec{x}\, \exists y\, A(\vec{x}, y)$, as before. □

To appreciate the difficulties that arise when we turn to first-order logic with equality, consider a language L with a single binary relation symbol R and take $A(x, y)$ to be $R(x, y)$. Then from $\forall x\, R(x, f(x))$ we can prove

$$R(x_0, f(x_0)) \wedge R(x_1, f(x_1)) \wedge (x_0 = x_1 \rightarrow f(x_0) = f(x_1)),$$

and hence

$$\forall x_0\, \exists y_0\, \forall x_1\, \exists y_1\, (R(x_0, y_0) \wedge R(x_1, y_1) \wedge (x_0 = x_1 \to y_0 = y_1)).$$

Exercise 7.4.2 asks you to show that this sentence is provable classically from the assumption $\forall x\, \exists y\, R(x, y)$. In fact, it follows intuitionistically assuming only the decidability of equality, which is expressed by the sentence $\forall x, y\, (x = y \vee x \neq y)$. But Exercise 7.4.4 asks you to show that the sentence is not provable in intuitionistic logic alone. We can generalize this example by allowing any number of quantifier alternations, replacing each x_i by a tuple of variables $\vec{x}_i$, and replacing $R(x, y)$ by the formula $A(\vec{x}, y)$. When we do that, it turns out that the resulting formulas axiomatize exactly the consequences of $\forall \vec{x}\, \exists y\, A(\vec{x}, f(\vec{x}))$ in the original language L.

More precisely, given $A(\vec{x}, y)$ with at most the free variables shown, let D_n be the sentence

$$\forall \vec{x}_0\, \exists y_0 \ldots \forall \vec{x}_{n-1}\, \exists y_{n-1} \left(\bigwedge A(\vec{x}_i, y_i) \wedge \bigwedge_{i<j<n} (x_i = x_j \to y_i = y_j) \right).$$

Proposition 7.4.3. *The following hold:*

- *For each n, $\forall \vec{x}\, A(\vec{x}, f(\vec{x}))$ proves D_n in minimal logic.*
- *For each n, D_n is provable from $\forall \vec{x}\, \exists y\, A(\vec{x}, y)$ in intuitionistic logic with decidable equality, and hence in classical logic.*

Proposition 7.4.3 is left as an exercise. Our main theorem is as follows:

Theorem 7.4.4. *Suppose $\Gamma \cup \{\forall \vec{x}\, A(\vec{x}, f(\vec{x}))\} \vdash B$ in classical, intuitionistic, or minimal first-order logic with equality, where f does not occur in $\Gamma \cup \{A(\vec{x}, y), B\}$. Then for some n, $\Gamma \cup \{D_n\} \vdash B$ as well.*

Together with Proposition 7.4.3, this yields the rest of Theorem 7.4.1. The proof of Theorem 7.4.4 is similar to the proof of Theorem 7.4.1 for logic without equality, with some additional considerations. Letting L be the original language and L' be the language that adds f, we will consider an extended sequent calculus for L' with the same additional rule as before and equality expressed by the following rules:

$$\frac{\Gamma, t = t \vdash \Delta}{\Gamma \vdash \Delta} \qquad \frac{\Gamma, s = t, t = s \vdash \Delta}{\Gamma, s = t \vdash \Delta} \qquad \frac{\Gamma, r = s, s = t, r = t \vdash \Delta}{\Gamma, r = s, s = t \vdash \Delta}$$

$$\frac{\Gamma, R(\vec{s}), \vec{s} = \vec{t}, R(\vec{t}) \vdash \Delta}{\Gamma, R(\vec{s}), \vec{s} = \vec{t} \vdash \Delta} \qquad \frac{\Gamma, \vec{s} = \vec{t}, g(\vec{s}) = g(\vec{t}) \vdash \Delta}{\Gamma, \vec{s} = \vec{t} \vdash \Delta}$$

As in Section 6.6, the extended calculus still has cut elimination. Once again, the main challenge is to come up with the right inductive hypothesis.

Lemma 7.4.5. *Let Γ and Δ be finite sets of formulas in L. Suppose $y_0, \ldots, y_{m-1}$ is a tuple of variables occurring in $\Gamma \cup \Delta$ and $f(t_0), \ldots, f(t_{m-1})$ are terms of L' that do not contain any of $y_0, \ldots, y_{m-1}$, and suppose that*

$$(\Gamma \vdash \Delta)[\ldots f(\vec{t}_i)/y_i \ldots]$$

is derivable in the extended calculus for L'. For any $n > m$, let $\hat{D}_{n,\ldots,A(\vec{x}_i,t_i),\ldots}$ be the formula

$$\forall \vec{x}_m \, \exists y_m \cdots \forall \vec{x}_{n-1} \, \exists y_{n-1} \left(\bigwedge_{i<n} A(\vec{u}_i, y_i) \wedge \bigwedge_{i<j<n} (\vec{u}_i = \vec{u}_j \to y_i = y_j) \right),$$

where $\vec{u}_i$ denotes $\vec{t}_i$ for $i < m$ and $\vec{x}_i$ for $i \geq m$. Then for some $n > m$, the sequent $\Gamma, \hat{D}_{m,\ldots,A(\vec{t}_i,y_i),\ldots} \vdash \Delta$ is derivable in first-order logic with equality.

Taking $m = 0$ yields Theorem 7.4.4. The formula $\hat{D}_{n,\ldots,A(\vec{t}_i,y_i),\ldots}$ should be seen as an expression on its way to becoming D_n. Remember that whenever a variable x is not free in a formula E, minimal logic proves $\exists x\,(E \wedge F)$ equivalent to $\exists x\, E \wedge F$ and it proves $\forall x\,(E \wedge F)$ equivalent to $\forall x\, E \wedge F$. The formula $\hat{D}_{n,\ldots,A(\vec{t}_i,y_i),\ldots}$ therefore implies each of the conjuncts $A(\vec{t}_i, y_i)$ for $i < m$, so the formulas in the antecedent of the sequent found in the conclusion of Lemma 7.4.2 have been rolled into the one displayed here.

Proof The proof proceeds by induction on cut-free derivations, much as before. The cases for the assumption axioms, the axioms and rules for the propositional connectives, the left $\exists$ rule, and the right $\forall$ rule are straightforward: one applies the inductive hypothesis and uses the same rule in the target sequent calculus for L.

When a rule eliminates terms $f(\vec{t}_i)$ from the conclusion, we can once again apply left $\exists$ and $\forall$ rules to eliminate the corresponding y_i's from the hypothesis. For example, suppose the last inference of the original proof is a right $\exists$ rule with principal formula $\exists z\, C(z)$ and main premise $C(s)$. Then, as in the proof of Lemma 7.4.2, we obtain s' and C', apply the inductive hypothesis, and re-apply the $\exists$ rule as follows:

$$\frac{\Gamma, \hat{D}_{n,\ldots,A(\vec{t}_i,y_i),\ldots} \vdash \Delta, C'(s')}{\Gamma, \hat{D}_{n,\ldots,A(\vec{t}_i,y_i),\ldots} \vdash \Delta, \exists z\, C'(z)}$$

If we reindex the variables y_i so that the ones that in s' but not in the conclusion come last, we can then apply left $\exists$ and $\forall$ rules to replace $\hat{D}_{n,\ldots,A(\vec{t}_i,y_i),\ldots}$ in the antecedent of the conclusion by $\hat{D}_{n,\ldots,A(\vec{t}_j,y_j),\ldots}$, where the sequence $\ldots, A(\vec{t}_j, y_j), \ldots$ corresponds to the variables y_j that remain. The left $\forall$ rule, reflexivity, symmetry, transitivity, the congruence rules for the relation symbols, and the congruence rules for function symbols other than f are handled in a similar manner: apply the inductive hypothesis, apply the same rule, and then use left $\exists$ and $\forall$ rules to take care of any terms $f(\vec{t})$ that have vanished from the proof in the extended calculus.

The reason that the congruence rule for f is a special case is that any equation $f(\vec{u}) = f(\vec{v})$ is replaced by an equation $y_j = y_k$ in the original calculus. We will deal with that case below. But first let us deal with the only remaining rule, namely, the additional rule of the extended calculus. Applying the inductive hypothesis, we obtain a proof of

$$\Gamma, \hat{D}_{n,\ldots,A(\vec{t}_i,y_i),\ldots,A(\vec{t}',y'),} A(\vec{t}', y') \vdash \Delta.$$

Since $A(\vec{t}', y)$ is implied by the other displayed formula, it is redundant, and we can delete it. Then, if y' does not occur elsewhere in Γ or Δ, we can use left $\exists$ and $\forall$ rules to replace that formula by $\hat{D}_{n,\ldots,A(\vec{t}_i,y_i),\ldots}$, as required.

Finally, consider the congruence rule for f, which is ordinarily of the form:

$$\frac{\Pi, \vec{u} = \vec{v},\ f(\vec{u}) = f(\vec{v}) \vdash \Theta}{\Pi, \vec{u} = \vec{v} \vdash \Theta}$$

In the case at hand, this means that the premise must be of the form

$$(\Gamma, \vec{u}' = \vec{v}', y_j = y_k, \hat{D}_{n,\ldots,A(\vec{t}_i, y_i),\ldots,A(\vec{u}', y_j)}, A(\vec{v}', y_k) \vdash \Delta)[\ldots f(\vec{t}_i)/y_i \ldots],$$

where $\vec{t}_j$ is $\vec{u}$, $\vec{t}_k$ is $\vec{v}$, and $\vec{u}'$ and $\vec{v}'$ are versions of $\vec{u}$ and $\vec{v}$ with terms $f(\vec{t}_l)$ replaced by variables y_l. By the inductive hypothesis, we obtain a proof of

$$\Gamma, \vec{u}' = \vec{v}', y_j = y_k, \hat{D}_{n,\ldots,A(\vec{t}_i, y_i),\ldots,A(\vec{u}', y_j)}, A(\vec{v}', y_k) \vdash \Delta.$$

The formula $\hat{D}$ there implies $\vec{u}' = \vec{v}' \to y_j = y_k$ because that is included in the inner conjunct. So we obtain a proof of

$$\Gamma, \vec{u}' = \vec{v}', \hat{D}_{n,\ldots,A(\vec{t}_i, y_i),\ldots,A(\vec{u}', y_j)}, A(\vec{v}', y_k) \vdash \Delta.$$

Finally, if either of y_j or y_k does not occur in Γ or Δ, as usual we use left $\exists$ and $\forall$ rules to quantify them out of $\hat{D}$. □

Exercises

7.4.1. Use the double-negation translation to prove Theorem 7.4.1 for classical logic with equality from Theorem 7.4.4 for intuitionistic logic with equality.

7.4.2. Show that $\forall x_0\, \exists y_0\, \forall x_1\, \exists y_1\, (R(x_0, y_0) \wedge R(x_1, y_1) \wedge (x_0 = x_1 \to y_0 = y_1))$ is provable from $\forall x\, \exists y\, R(x, y)$ in intuitionistic logic with decidable equality.

7.4.3. More generally, prove Proposition 7.4.3.

7.4.4. Do the following.

a. Show that $\forall x_0\, \exists y_0\, \forall x_1\, \exists y_1\, (R(x_0, y_0) \wedge R(x_1, y_1) \wedge (x_0 = x_1 \to y_0 = y_1))$ is not provable from $\forall x\, \exists y\, R(x, y)$ in intuitionistic logic, by exhibiting a Kripke model where the former holds but the latter doesn't.

b. Do the same using a proof-theoretic argument, for example, using the middle-sequent theorem.

7.5 Skolemization in Classical Theories

To use an indefinite description in the context of classical logic, we can eliminate the need for a hypothesis of the form $\forall \vec{x}\, \exists y\, A(\vec{x}, y)$. For any formula $A(\vec{x}, y)$, we can classically prove the sentence

$$\forall \vec{x}\, \exists y\, (\exists z\, A(\vec{x}, z) \to A(\vec{x}, y)),$$

which asserts that for any $\vec{x}$ there is a value of y that satisfies A if anything does. The assumption that f witnesses this indefinite description is given by

$$\forall \vec{x}\, (\exists z\, A(\vec{x}, z) \to A(\vec{x}, f(\vec{x}))),$$

which is equivalent to

$$\forall \vec{x}, z\, (A(\vec{x}, z) \to A(\vec{x}, f(\vec{x}))).$$

Such an f is called a *Skolem function* for $\exists y\, A(\vec{x}, y)$, and either sentence above can be taken to be the corresponding *Skolem axiom*. The axiom says that, for any $\vec{x}$, if there is anything satisfying $A(\vec{x}, y)$, then $f(\vec{x})$ does. The results of Section 7.4 imply that we can extend classical first-order logic conservatively by adding new function symbols and their Skolem axioms. Without any assumptions on f, $A(\vec{x}, f(\vec{x}))$ implies $\exists y\, A(\vec{x}, y)$ in first-order logic. The Skolem axiom simply says that the converse statement also holds.

We can now go through a formula and iteratively eliminate existential quantifiers in favor of Skolem functions. I will assume that we are working with a language L' with the property that to every formula of the form $\exists x\, A(\vec{x}, y)$ in L' we have assigned a symbol $f_{\exists y A}(\vec{x})$ to serve as its Skolem function. Starting with a language L, we can obtain such a language L' as follows: let $L_0 = L$, for each natural number i let L_{i+1} be a language in which we have added Skolem functions for formulas in L_i, and let $L' = \bigcup_i L_i$.

Remember that an existential quantifier under a negation acts like a universal quantifier and vice versa. To avoid having to worry about the polarity of each quantifier, it is convenient to restrict attention to formulas in negation normal form. By recursion on depth, we assign to each such formula A its *Skolem normal form*A^S as follows:

- $A^S = A$ if A is atomic
- $(A \wedge B)^S = A^S \wedge B^S$
- $(A \vee B)^S = A^S \vee B^S$
- $(\forall x\, A)^S = \forall x\, A^S$
- $(\exists x\, A(x, \vec{y}))^S = (A(f(\vec{y}), \vec{y}))^S$, where A has the free variables shown and f is the Skolem function for $\exists x\, A(x, \vec{y})$.

Induction on the depth of A immediately yields the following:

1. A^S implies A in first-order logic.
2. Together with the axioms for the Skolem functions occurring in A^S, A implies A^S.
3. A^S has only universal quantifiers.

For example, assuming C and D are quantifier-free, to Skolemize

$$\forall x\, \exists y\, (C(x, y, u) \wedge \forall z\, \exists w\, D(x, y, z, w)),$$

we calculate as follows:

$$\begin{aligned}
&(\forall x\, \exists y\, (C(x, y, u) \wedge \forall z\, \exists w\, D(x, y, z, w)))^S \\
&\quad = \forall x\, (\exists y\, (C(x, y, u) \wedge \forall z\, \exists w\, D(x, y, z, w)))^S \\
&\quad = \forall x\, (C(x,\ f(x, u), y) \wedge \forall z\, \exists w\, D(x,\ f(x, u), z, w))^S \\
&\quad = \forall x\, (C(x,\ f(x, u), y)^S \wedge (\forall z\, \exists w\, D(x,\ f(x, u), z, w))^S) \\
&\quad = \forall x\, (C(x,\ f(x, u), y) \wedge \forall z\, (\exists w\, D(x,\ f(x, u), z, w))^S) \\
&\quad = \forall x\, (C(x,\ f(x, u), y) \wedge \forall z\, (D(x,\ f(x, u), z, g(x, u, x))^S) \\
&\quad = \forall x\, (C(x,\ f(x, u), y) \wedge \forall z\, D(x,\ f(x, u), z, g(x, u, z))).
\end{aligned}$$

Here f is the Skolem function for $\exists y\, (C(x, y, u) \wedge \forall z\, \exists w\, D(x, y, z, w)))$ and g is the Skolem function for $\exists w\, D(x,\ f(x, u), z, w)$.

Some authors define the Skolem normal form to be the equivalent formula obtained by bringing the universal quantifiers to the front. For the example above, that would be

$$\forall x, z\, (C(x, f(x, u), y) \wedge D(x, f(x, u), z, g(x, u, z))).$$

Below, I will blur the distinction and say that A^S is a universal formula, by which I mean that its prenexation is.

Some authors define Skolemization to be the variant $A^{S'}$ in which the last clause reads as follows:

- $(\exists x\, A(x, \vec{y}))^{S'} = (A(x, \vec{y}))^{S'}[f(\vec{y})/x]$.

Whereas before we substituted $f(\vec{y})$ first and the continued Skolemizing, here we Skolemize the inner formula first and then substitute $f(\vec{y})$. This avoids the need for iteration when we add Skolem function symbols to the language. You can check that in the example above the result would be

$$\forall x\, (C(x, f(x, u), y) \wedge \forall z\, D(x, f(x, u), z, g'(x, f(x, u), z))).$$

Here f is as before but now $g'(x, y, z)$ is the Skolem function for $\exists w\, D(x, y, z, w)$. So, in this second translation, $g'(x, f(x, u), z)$ plays the role of the Skolem function $g(x, u, z)$ in the first translation. You can check that $A^{S'}$ also has the three key properties enumerated above.

Informally, the idea is that A is equivalent to $\exists f_0, \ldots, f_{k-1}\, A^S$, where $f_0, \ldots, f_{k-1}$ are the Skolem functions that occur in A^S. This equivalence can be formalized in second-order logic with function symbols and an axiom of choice; see Chapter 15. In terms of first-order logic, the key properties of the translation are summarized in the next theorem.

Theorem 7.5.1. *Let A be any formula in a language L, let A^S be its Skolem normal form, and let $f_0, \ldots, f_{k-1}$ be the Skolem functions that occur in A^S. Then A^S is a universal formula and the following hold:*

1. $\vdash A^S \to A$.
2. *If Γ is the set of Skolem axioms for $f_0, \ldots, f_{k-1}$, then $\Gamma \vdash A \to A^S$.*
3. *If B is any formula of L then $\{A^S\} \vdash B$ if and only if $\{A\} \vdash B$.*

Proof We have already noted that the first two claims can be proved by an easy induction on the depth of A. The right-to-left direction of the third claim follows immediately from 1. For the converse, suppose $\{A^S\} \vdash B$. Let Γ be the set of Skolem axioms for the Skolem function symbols occurring in A. By claim 2, $\Gamma \cup \{A\} \vdash A^S$, and hence $\Gamma \cup \{A\} \vdash B$. By Theorem 7.4.1, $\{A\} \vdash B$. □

Although A^S and A are generally not equivalent, the left-to-right direction of the third claim says that they both have the same consequences in the original language. In particular, taking B to be $\bot$, we have that $\{A\} \vdash \bot$ if and only if $\{A^S\} \vdash \bot$. Together with the completeness theorem, this means that A^S is satisfiable if and only if A is, which is to say, A^S has a model if and only if A does.

The notion of *Herbrand normal form* is dual to the notion of Skolem normal form. If A is in negation normal form, we define the Herbrand normal form of A, denoted A^H, to be $\sim(\sim A)^S$. Returning to the example above, if A is the formula

$$\forall x\, \exists y\, (C(x, y, u) \wedge \forall z\, \exists w\, D(x, y, z, w)),$$

where C and D are quantifier-free, then A^H is the formula

$$\exists y\,(C(k(u), y, u) \wedge \exists w\, D(k(u), y, h(x, y, u), w)),$$

where $k(u)$ is a Skolem function for

$$\exists x\, \forall y\, ({\sim}C(x, y, u) \vee \exists z\, \forall w\, {\sim}D(x, y, z, w))$$

and $h(x, y, u)$ is a Skolem function for

$$\exists z\, \forall w\, {\sim}D(k(u), y, z, w).$$

Notice that any formula $\forall x\, A(x, \vec{y})$ is equivalent to $A(f(\vec{y}), \vec{y})$ under the assumption that $f(\vec{y})$ returns a counterexample if there is one. So whereas Skolemization replaces existential quantifiers with witnesses whenever they exist, Herbrandization replaces universal quantifiers with counterexamples whenever they exist. The following is dual to Theorem 7.5.1:

Theorem 7.5.2. *Let A be any formula in a language L, let A^H be its Herbrand normal form, and let $f_0, \ldots, f_{k-1}$ be the Skolem functions occurring in A^H. Then A^H is an existential formula and the following hold:*

1. $\vdash A \to A^H$.
2. *If Γ is the set of Skolem axioms for $f_0, \ldots, f_{k-1}$, then $\Gamma \vdash A^H \to A$.*
3. *If Δ is a set of sentences in L, then $\Delta \vdash A^H$ if and only if $\Delta \vdash A$.*

In model-theoretic terms, when Δ is empty, the last claim says that A^H is valid if and only A is.

Claim 3 is especially interesting because A^H is an existential formula. Applying Herbrand's theorem, we have the following:

Corollary 7.5.3. *Let A be any formula. Then A is provable in classical first-order logic if and only if some disjunction of instances of A^H is provable in the quantifier-free fragment of first-order logic.*

We can combine Skolem normal form and Herbrand normal form to generalize Theorems 7.5.1 and 7.5.2 as follows:

Theorem 7.5.4. *For any set of sentences Γ and any formula A, the following are equivalent:*

1. $\Gamma \vdash A$
2. $\Gamma^S \vdash A$
3. $\Gamma \vdash A^H$
4. $\Gamma^S \vdash A^H$.

Proof By claim 1 of Theorem 7.5.1 and claim 1 of 7.5.2, respectively, claim 1 of Theorem 7.5.4 implies 2 and 3, and each of these implies claim 4. So we only need to show that claim 4 implies claim 1.

The proof is just a slight generalization of the proofs of Theorems 7.5.1 and 7.5.2. Suppose $\Gamma^S \vdash A^H$. Then for an appropriate set Δ of Skolem axioms, using claim 2 of Theorem 7.5.1 and claim 2 of Theorem 7.5.2, $\Gamma \cup \Delta \vdash A$. Since the Skolem functions do not occur in Γ, we have $\Gamma \vdash A$ by Theorem 7.4.1.

Alternatively, we can use the deduction theorem to reduce the implication between clause 4 and clause 1 to claim 3 of Theorem 7.5.2. Suppose $\Gamma^S \vdash A^H$. Then for some finite subset

$\{B_0, \ldots, B_{k-1}\}$ of Γ, we have $\{B_0^S, \ldots, B_{k-1}^S\} \vdash A^H$. By the deduction theorem, we have $\vdash B_0^S \wedge \cdots \wedge B_{k-1}^S \to A^H$, which is equivalent to $(B_0 \wedge \cdots \wedge B_{k-1} \to A)^H$. By claim 3 of Theorem 7.5.2, $\vdash B_0 \wedge \cdots \wedge B_{k-1} \to A$ and hence $\Gamma \vdash A$. □

Since Γ^S is universal and A^H is existential, we can apply Herbrand's theorem.

Corollary 7.5.5. *Let Γ be a set of sentences and let A be any formula. Then $\Gamma \vdash A$ if and only if there is a proof of some disjunction of quantifier-free instances of A^H from quantifier-free instances of Γ^S in the quantifier-free fragment of first-order logic.*

This means that Skolem functions reduce theorem proving in classical first-order logic to theorem proving in quantifier-free logic as defined in Section 4.4. This gives rise to the method of *resolution theorem proving* in automated reasoning. When A is a sentence, we have that $\Gamma \vdash A$ if and only if $\Gamma, \neg A \vdash \bot$, so by Corollary 7.5.5, determining the provability of A from Γ is reduced to determining the refutability of a set Δ of universal sentences. Putting the quantifier-free part of each sentence in conjunctive normal form and using the identity $\forall \vec{x}\,(B \wedge C) \leftrightarrow \forall \vec{x}\, B \wedge \forall \vec{x}\, C$, we can assume that every sentence in Δ is of the form $\forall \vec{x}\, C$ where C is a clause. By Theorem 4.4.2, Δ is refutable if and only if some finite set of instances of these clauses is refutable, and by Theorem 2.8.3, that happens if and only if they can be refuted using the resolution rule. A resolution proof search therefore proceeds by looking for instantiations of elements $\forall \vec{x}\, C$ and $\forall \vec{y}\, D$ of Δ that permit a resolution step. Carrying out that step and generalizing over remaining variables yields a new universally quantified clause $\forall \vec{z}\, E$, and iterating this process until the empty clause is obtained shows that $\Gamma \vdash A$. With care, we can ensure that the search method is complete, so that if $\Gamma \vdash A$ then the empty clause is eventually obtained.

The reduction of classical first-order logic to quantifier-free logic with Skolem functions is also mathematically and philosophically interesting. Hilbert viewed such functions (more precisely, *epsilon terms*, which are closely related) as representing the *ideal* elements that are added to finitistic reasoning to allow reasoning over infinite domains.

Exercises

7.5.1. a. Find the Skolem normal form of (a negation normal form equivalent of) the following formula:

$$\forall x\, \exists y\, (A(x) \wedge B(y) \to \exists w\, \forall z\, (C(u, z) \to D(w))).$$

b. Find the Herbrand normal form of the same formula.

c. Find the Skolem normal form of an induction axiom,

$$\forall \vec{u}\, (A(0) \wedge \forall x\, (A(x) \to A(x+1)) \to \forall x\, A(x))$$

where $A(x)$ is quantifier-free with additional free variables $\vec{u}$.

d. Find the Skolem normal form of the same induction axiom, where $A(x)$ is an existential formula, $\exists y\, C(x, y)$. Once again, assume C is quantifier-free but has additional parameters $\vec{u}$.

7.5.2. Find the Herbrand normal form of $\exists y\, \forall x\, (\neg A(y) \vee A(x))$. Find a cut-free proof of this Herbrand normal form, as well as the terms guaranteed to exist by Herbrand's theorem.

Bibliographical Notes

Herbrand's theorem and generalizations For proof-theoretic proofs similar to the one presented here, see Troelstra and Schwichtenberg (2000), Schwichtenberg (1977), Buss (1998c), and Takeuti (1987). For model-theoretic proofs, see Chang and Keisler (1990), Hodges (1993) and Harrison (2009).

Explicit definability and the disjunction property The proof-theoretic proofs here are similar to the ones in Buss (1998c) and Troelstra and Schwichtenberg (2000). See also Troelstra and van Dalen (1988), Beeson (1985), Troelstra (1973), and see Chapter 14 for the explicit definability and disjunction properties for first-order arithmetic and other theories.

Interpolation theorems Proof-theoretic proofs of the interpolation theorem similar to the ones presented here can be found in Buss (1998a), Troelstra and Schwichtenberg (2000), and Takeuti (1987). Model-theoretic proofs for the classical case can be found in Chang and Keisler (1990), Hodges (1993), and Smullyan (1995). For more information and references see Buss (1998a), Troelstra and Schwichtenberg (2000), Harrison (2009), and Feferman (2008).

The fact that we use the cut elimination theorem leaves open the possibility that interpolants can be long. In the worst case, this is unavoidable; see Pudlák (1998).

Indefinite descriptions and Skolem functions Skolem functions are closely related to the *epsilon terms* occurring in Hilbert's epsilon calculus, and the *epsilon elimination theorems* essentially provide a syntactic procedure for eliminating indefinite descriptions. See Avigad and Zach (2002) for references. The proof presented in Section 7.4 draws on Mints (2000a). For a model-theoretic proof of Theorem 7.4.4, see Smoryński (1978). In the computer science literature, the process of eliminating Skolem functions is known as *deskolemization*, and there are still open questions as to the extent to which it can be done efficiently; see Baaz et al. (2012). For more on Skolemization and resolution theorem proving, see Bachmair and Ganzinger (2001), Baaz et al. (2001), and Harrison (2009).

8

Primitive Recursion

We now turn from pure logic to theories that describe portions of the mathematical universe. In Chapter 10, we will consider first-order arithmetic, a theory that describes the natural numbers. This may seem like a small start, since even parts of mathematics like elementary number theory and combinatorics deal with lots of finite objects other than numbers, such as finite sets, finite sequences, finite graphs, and finite groups. But we will see that, in the spirit of foundational reduction, objects like those can be represented as natural numbers and, with those representations, first-order arithmetic turns out to be a surprisingly robust theory.

In order to represent mathematical objects in this way, we need to make use of common operations, properties, and relations on the natural numbers. These include operations like addition and multiplication, relations like the order and divisibility relations, the property of being prime, the function f that on input n returns the nth prime, and so on. To that end, we will define a collection of functions called the *primitive recursive functions*, which are obtained from a set of very basic initial functions using only two operations, *composition* and *definition by primitive recursion*. We will see that many familiar functions on the natural numbers can be defined in this way.

The set of primitive recursive functions has some important virtues. It is easy to describe, and, at least from an intuitive standpoint, each primitive recursive function is clearly computable. At the same time, it is large enough for most purposes, and it includes most of the functions on finite objects that come up in practice.

We will see in Chapter 11 that the defining the primitive recursive functions is also helpful toward developing a theory of *arbitrary* computable functions and relations. But keep in mind that, when we talk about computability, we are concerned about computability in principle rather than practice. The defining schemas for the primitive recursive functions do not constitute a good programming language, and we will not worry about whether the implicit algorithms for evaluating the functions we define are efficient. In fact, many will be ridiculously inefficient. When it comes to practical computation, computer science offers much better languages for defining computable functions, and the field of computational complexity offers informative ways of measuring their efficiency. In that respect, you can think of the primitive recursive functions as a simple prototype, providing a first step toward reasoning about computable functions and their properties.

8.1 The Primitive Recursive Functions

To define the primitive recursive functions, we adopt the usual convention of interpreting constants as 0-ary functions. To start with, the constant 0 is primitive recursive, as is the successor function, $\mathrm{succ}(x) = x + 1$. We will also include among the primitive recursive functions the projection functions p_i^n, for each n and $i < n$, defined by

$$p_i^n(x_0, \ldots, x_{n-1}) = x_i.$$

To complete the definition of the set of primitive recursive functions, we specify that it is closed under the operations of composition and primitive recursion, which are defined as follows. If g is a k-ary function and $h_0, \ldots, h_{k-1}$ are ℓ-ary functions on the natural numbers, the *composition of g with $h_0, \ldots, h_{k-1}$* is the ℓ-ary function f defined by

$$f(x_0, \ldots, x_{\ell-1}) = g(h_0(x_0, \ldots, x_{\ell-1}), \ldots, h_{k-1}(x_0, \ldots, x_{\ell-1})).$$

Finally, if $g(z_0, \ldots, z_{k-1})$ is a k-ary function and $h(x, y, z_0, \ldots, z_{k-1})$ is a $k+2$-ary function, then the function *defined by primitive recursion from g and h* is the $k+1$-ary function f defined by the equations

$$f(0, z_0, \ldots, z_{k-1}) = g(z_0, \ldots, z_{k-1})$$
$$f(x+1, z_0, \ldots, z_{k-1}) = h(x, f(x, z_0, \ldots, z_{k-1}), z_0, \ldots, z_{k-1})$$

To summarize, we have the following:

Definition 8.1.1. The set of primitive recursive functions is the set of functions from the natural numbers to the natural numbers, of various arities, defined inductively by the following clauses:

- The constant 0 is primitive recursive.
- The successor function, succ, is primitive recursive.
- Each projection function p_i^n is primitive recursive.
- If f is a k-ary primitive recursive function and $g_0, \ldots, g_{k-1}$ are ℓ-ary primitive recursive functions, then the composition of f with $g_0, \ldots, g_{k-1}$ is primitive recursive.
- If f is a k-ary primitive recursive function and g is a $k+2$-ary primitive recursive function, then the function defined by primitive recursion from f and g is primitive recursive.

More concisely, the set of primitive recursive functions is the smallest set containing the constant zero, the successor function, and projection functions, and closed under composition and primitive recursion.

Since we can think of the number 1 as obtained by composition of the unary function succ(x) with the single 0-ary function 0, and similarly for each natural number, we have that the set of primitive recursive functions includes the set $\mathbb{N}$. The set of primitive recursive functions also contains the constant functions of arbitrary arity. For example, with a careful reading of the description of composition presented above, we can construe the function $z(x) = 0$ as the result of composing the constant 0 with zero-many unary functions.

The definition of composition above may seem too rigid since $g_0, \ldots, g_{k-1}$ are all required to have the same arity. But more general forms of composition can be obtained using the projection functions. For example, suppose f and g are ternary functions and h is the binary function defined by

$$h(x, y) = f(x, g(x, x, y), y).$$

Then the definition of h can be rewritten with the projection functions as

$$h(x, y) = f(p_0^2(x, y), g(p_0^2(x, y), p_0^2(x, y), p_1^2(x, y)), p_1^2(x, y)).$$

Then h is the composition of g with p_0^2, ℓ, and p_1^2, where ℓ is the function defined by the equation

$$\ell(x, y) = g(p_0^2(x, y), p_0^2(x, y), p_1^2(x, y)).$$

In other words, ℓ is the composition of g with p_0^2, p_0^2, and p_1^2.

Once we have chosen function symbols to represent the primitive recursive functions, we can build terms from these function symbols and variables in the usual way. A consequence of the fact that the set of primitive recursive functions has projections and is closed under composition is that if $t(x_0, \ldots, x_{n-1})$ is any term with at most the variables shown, there is an n-ary primitive recursive function f such that

$$f(x_0, \ldots, x_{n-1}) = t(x_0, \ldots, x_{n-1})$$

holds in the intended interpretation for every $x_0, \ldots, x_{n-1}$. In this case, f is said to be *explicitly defined* by t. This gives us an even more concise way to describe the set of primitive recursive functions, namely, as the smallest set containing the constant 0 and the successor function $\mathrm{succ}(x)$ and closed under explicit definition and primitive recursion.

As an example, let us show that addition is primitive recursive. We can define it informally with the following equations:

$$x + 0 = x$$
$$x + (y + 1) = \mathrm{succ}(x + y).$$

To show that this function can be obtained using the formal closure conditions above, note that the function $g(y, x) = x + y$ can be defined by the equations

$$g(0, x) = x$$
$$g(y + 1, x) = \mathrm{succ}(g(y, x)).$$

But even this is not a strict primitive recursive definition; we need to put it in the form

$$g(0, x) = k(x)$$
$$g(y + 1, x) = h(y, g(y, x), x)$$

for some unary primitive recursive function k and some ternary primitive recursive function h. We can take k to be p_0^1. On the other hand, h need only return the successor of its middle argument, so we can define h using composition, by

$$h(y, w, x) = \mathrm{succ}(p_1^3(y, w, x)).$$

Note that we have defined the function $g(y, x)$ meeting the recursive specification of $x + y$ with the order of the variables reversed. Luckily, addition is commutative, so here the difference is not important. But otherwise we could reverse the order of the variables by composing with projections, that is, define a function g' by $g'(x, y) = g(p_1^1(y, x), p_0^1(y, x)) = g(y, x)$.

In the exercises below, you are asked to try your hand at providing completely formal definitions of some basic mathematical functions. From now on, however, we will be less formal in our definitions and appeal often to closure under explicit definability. What is important is that when we define a function $f(x, \vec{z})$ by primitive recursion on x, we express $f(0, \vec{z})$ in terms of $\vec{z}$ and we express $f(x+1, \vec{z})$ in terms of x, $\vec{z}$, and $f(x, \vec{z})$. In particular, in the recursion step, we are only allowed to use the previous value of f and we are not allowed to change the parameters $\vec{z}$. In Section 8.4, we will justify the use of more flexible forms of primitive recursion.

Exercises

8.1.1. Show that multiplication is primitive recursive, strictly according to the definition.

8.1.2. Do the same for exponentiation, $f(x, y) = x^y$, and the factorial function.

8.2 Some Primitive Recursive Functions and Relations

We can now show that a number of common functions are primitive recursive.

- constants: for each natural number n, n is the 0-ary primitive recursive function $\mathrm{succ}(\mathrm{succ}(\cdots(\mathrm{succ}(0))\cdots))$.
- the identity function: $\mathrm{id}(x) = x$, i.e. p_0^1
- addition, $x + y$
- multiplication, $x \cdot y$, defined by

$$x \cdot 0 = 0, \quad x \cdot (y+1) = x \cdot y + x$$

- exponentiation, x^y (with 0^0 defined to be 1), defined by

$$x^0 = 1, \quad x^{y+1} = x^y \cdot x$$

- the factorial function, $x!$, defined by

$$0! = 1, \quad (x+1)! = (x+1) \cdot x!$$

- the predecessor function, $\mathrm{pred}(x)$, defined by

$$\mathrm{pred}(0) = 0, \quad \mathrm{pred}(x+1) = x$$

- truncated subtraction, $x \dot{-} y$, defined by

$$x \dot{-} 0 = x, \quad x \dot{-} (y+1) = \mathrm{pred}(x \dot{-} y)$$

- the maximum $\max(x, y)$ of two numbers x and y, defined by

$$\max(x, y) = x + (y \dot{-} x)$$

- the minimum function, $\min(x, y)$, whose definition is left as an exercise
- the distance between x and y, $\mathrm{dist}(x, y)$, which is equal to $(x \dot{-} y) + (y \dot{-} x)$.

Note that $x \dot{-} y$ is equal to 0 if y is greater than x.

The set of primitive recursive functions is further closed under the following two operations:

- finite sums: if $f(x, \vec{z})$ is primitive recursive, then so is the function
$$g(y, \vec{z}) \equiv \Sigma_{x<y} f(x, \vec{z}).$$
- finite products: if $f(x, \vec{z})$ is primitive recursive, then so is the function
$$h(y, \vec{z}) \equiv \Pi_{x<y} f(x, \vec{z}).$$

Finite sums are defined recursively by the equations

$$g(0, \vec{z}) = 0, \quad g(y+1, \vec{z}) = g(y, \vec{z}) + f(y, \vec{z})$$

and finite products are defined in a similar way. We can also define boolean operations, where 1 stands for true, and 0 for false:

- negation, defined by
$$\text{not}(0) = 1, \quad \text{not}(x+1) = 0,$$
or, alternatively, $\text{not}(x) = 1 \dot{-} x$
- conjunction, defined by
$$\text{and}(0, y) = 0, \quad \text{and}(x+1, y) = \text{tv}(y)$$
where $\text{tv}(y)$ is defined by $\text{tv}(0) = 0$, $\text{tv}(y+1) = 1$.

The function $\text{tv}(y)$ casts every number to a truth value by mapping anything nonzero to true. As inputs to $\text{and}(x, y)$ and $\text{not}(x)$, any nonzero value is interpreted as true, so the function $\text{tv}(y)$ only serves to normalize the output. It could have been defined just as well by $\text{tv}(y) = \min(y, 1)$ or $\text{tv}(y) = \text{not}(\text{not}(y))$. Other boolean operations like $\text{or}(x, y)$ and $\text{implies}(x, y)$ can be defined from $\text{and}(x, y)$ and $\text{not}(x, y)$ in the usual ways.

A relation $R(\vec{x})$ is said to be primitive recursive if its characteristic function,

$$\chi_R(\vec{x}) = \begin{cases} 1 & \text{if } R(\vec{x}) \\ 0 & \text{otherwise} \end{cases}$$

is primitive recursive. In other words, when we speak of a primitive recursive relation $R(\vec{x})$, we are referring to a relation of the form $\chi_R(\vec{x}) = 1$, where χ_R is a primitive recursive function which, on any input, returns either 1 or 0. For example, the relation

- $\text{Zero}(x)$, which holds if and only if $x = 0$,

corresponds to the function χ_{Zero}, defined using primitive recursion by

$$\chi_{\text{Zero}}(0) = 1, \quad \chi_{\text{Zero}}(x+1) = 0.$$

In fact, $\chi_{\text{Zero}}(x)$ is equal to $\text{not}(x)$.

We can compose relations with primitive recursive functions to define new relations. So the following are also primitive recursive:

- the equality relation, $x = y$, defined by $\text{Zero}(\text{dist}(x, y))$
- the less-than relation, $x \leq y$, defined by $\text{Zero}(x \dot{-} y)$.

Furthermore, using the boolean operations defined above, we have that the set of primitive recursive relations is closed under boolean operations:

- negation, $\neg P$
- conjunction, $P \wedge Q$
- disjunction, $P \vee Q$
- implication, $P \rightarrow Q$.

We can also define relations using bounded quantifiers:

- bounded universal quantification: if $R(x, \vec{z})$ is a primitive recursive relation, then so is the relation $\forall x < y\, R(x, \vec{z})$, which holds if and only if $R(x, \vec{z})$ holds for every x less than y.
- bounded existential quantification: if $R(x, \vec{z})$ is a primitive recursive relation, then so is $\exists x < y\, R(x, \vec{z})$.

We take $\forall x < 0\, R(x, \vec{z})$ to be true for the trivial reason that there are no values of x less than 0, and we take $\exists x < 0\, R(x, \vec{z})$ to be false. The characteristic function $g(y, \vec{z})$ of $\forall x < y\, R(x, \vec{z})$ can be defined using a finite product, but we can also define it directly by

$$g(0, \vec{z}) = 1, \quad g(y+1, \vec{z}) = \text{and}(g(y, \vec{z}),\, \chi_R(y, \vec{z})).$$

Bounded existential quantification can similarly be defined using $\text{or}(x, y)$ or by the equivalence $\exists x < y\, A \equiv \neg \forall x < y\, \neg A$. We interpret $\forall x \leq y$ as $\forall x < y+1$ and $\exists x \leq y$ as $\exists x < y+1$.

Here is another useful primitive recursive function:

- the conditional function, $\text{cond}(x, y, z)$, defined by

$$\text{cond}(x, y, z) = \begin{cases} y & \text{if } x \neq 0 \\ z & \text{otherwise.} \end{cases}$$

This is defined recursively by

$$\text{cond}(0, y, z) = z, \quad \text{cond}(x+1, y, z) = y.$$

This justifies the following:

- definition by cases: if $g_0(\vec{x}), \ldots, g_m(\vec{x})$ are primitive recursive functions and $R_0(\vec{x}), \ldots, R_{m-1}(\vec{x})$ are primitive recursive relations, then the function f defined by

$$f(\vec{x}) = \begin{cases} g_0(\vec{x}) & \text{if } R_0(\vec{x}) \\ g_1(\vec{x}) & \text{if } R_1(\vec{x}) \text{ and not } R_0(\vec{x}) \\ \vdots & \\ g_{m-1}(\vec{x}) & \text{if } R_{m-1}(\vec{x}) \text{ and none of the previous hold} \\ g_m(\vec{x}) & \text{otherwise} \end{cases}$$

  is also primitive recursive.

When $m = 1$, this is the function defined by

$$f(\vec{x}) = \text{cond}(\chi_{R_0}(\vec{x}), g_0(\vec{x}), g_1(\vec{x})).$$

For m greater than 1, just compose definitions of this form. We will also make good use of the following:

- bounded minimization: if $R(x, \vec{z})$ is primitive recursive, then so is the function $f(y, \vec{z}) = \min_{x<y} R(x, \vec{z})$, which returns the least x less than y such that $R(x, \vec{z})$ holds, if there is one, and y otherwise. It is defined recursively as follows:

$$f(0, \vec{z}) = 0$$

$$f(y+1, \vec{z}) = \begin{cases} f(y, \vec{z}) & \text{if } R(f(y, \vec{z}), \vec{z}) \\ y+1 & \text{otherwise.} \end{cases}$$

As with bounded quantification, $\min_{x\leq y} R(x, \vec{z})$ is interpreted as $\min_{x<y+1} R(x, \vec{z})$. We can similarly define a function $\max_{x<y} R(x, \vec{z})$ which returns the greatest x less than y such that $R(x, \vec{z})$ holds, if there is one, and the default value 0 otherwise.

In Section 8.3 we will make use of the operation $\max_{x<y} f(x, \vec{z})$ which returns the maximum value of the function $f(x, \vec{z})$ as x ranges over values less than y, defined by

$$\max_{x<0} f(x, \vec{z}) = 0, \quad \max_{x<y+1} f(x, \vec{z}) = \max(\max_{x<y} f(x, \vec{z}),\ f(y, \vec{z})).$$

Be careful: the notation is similar to $\max_{x<y} R(x, \vec{z})$, but when applied to a function instead of a relation, the meaning is different. We can similarly define $\min_{x<y} f(x, \vec{z})$, although there is no good choice of a value to assign when y is 0. In that case, we assign 0 by default.

All this provides us with a good deal of machinery to show that many common functions and relations are primitive recursive, including the following:

- the relation $\text{Divides}(x, y)$, written $x \mid y$, defined by

$$x \mid y \equiv \exists z \leq y\ (x \cdot z = y)$$

- the predicate $\text{Prime}(x)$, which asserts that x is prime, defined by

$$\text{Prime}(x) \equiv x \geq 2 \wedge \forall y \leq x\ (y \mid x \to y = 1 \vee y = x)$$

- the function nextp(x), which returns the first prime number larger than x, defined by

$$\text{nextp(x)} = \min_{y\leq x!+1} (y > x \wedge \text{Prime}(y))$$

Here we are relying on the fact that there is always a prime number between x and $x!+1$.

- the function $p(x)$, returning the xth prime, defined by

$$p(0) = 2, \quad p(x+1) = \text{nextp}(p(x)).$$

For convenience, we will write often write p_x instead of $p(x)$.

Exercises

8.2.1. Show that the function $\min(x, y)$ is primitive recursive.

8.2.2. Show that the following relations and the following function are primitive recursive.
a. $P(x)$, which holds if and only if x is divisible by at least three different prime numbers
b. $Q(x)$, which holds if and only if x is not a power of three (i.e. not of the form 3^y)
c. $f(x)$, which returns the sum of the first x nonzero squares, $1^2 + \ldots + x^2$.

8.2.3. Show that if $R(x, \vec{y})$ is a primitive recursive relation, then the function $\max_{x<z} R(x, \vec{y})$ is primitive recursive, where this function returns the largest value of x less than z satisfying $R(x, \vec{y})$, or 0 if there is no such x.

8.2.4. Show that the function $\text{div}(x, y)$ is primitive recursive, where $\text{div}(x, y)$ returns the whole number part of x divided by y if $y \neq 0$, and 0 if $y = 0$. Do the same for $\text{mod}(x, y)$, which returns the remainder of x divided by y if $y \neq 0$ and returns x if $y = 0$.

8.3 Finite Sets and Sequences

Up to this point, we have considered functions and relations on the natural numbers. In order to speak of primitive recursive functions and relations on things like finite groups and finite graphs, we need to represent these as natural numbers as well. Logicians often refer to this practice as *coding*, but the word *representation* is more accurate. By way of analogy, axiomatic set theory is a theory of sets, and to describe mathematics in set-theoretic terms, we have to represent objects like numbers and functions as sets. In a computer, data is stored as sequences of bits in memory, and programs that manipulate integers and arrays act on these representations. When we are dealing with primitive recursive functions, the fundamental objects are natural numbers, so if we want to view other mathematical objects through the lens of primitive recursion, we have to represent them as natural numbers.

It can be disconcerting that there are often multiple representations available with no strong reason to favor one over another. Such concerns are somewhat addressed by abstracting the properties of the representation, thereby providing a modular interface that specifies the design requirements and encapsulates the details of implementation. For example, to represent pairs of natural numbers as a single natural number, what we really want are a primitive recursive pairing operation (x, y) and primitive recursive projections $(z)_0$ and $(z)_1$ such that the projections of the pair (x, y) are x and y, respectively. We may also ask that the projections uniquely determine the pair, so that if z and w have the same projections, then $z = w$. Once we have specified the properties we care about and show that the specification can be met, we can forget the implementation. Later, when we work with the axiomatic theory PRA, we will also require that the relevant properties are formally derivable.

In this section, we will develop our representations of mathematical objects as follows:

- First, we will represent pairs (x, y) and define a primitive recursive pairing operation and projections.
- Then we will represent finite sets s and define a primitive recursive membership relation $x \in s$ as well as primitive recursive operations to form sets.
- Then we will represent finite sequences $(s_0, \ldots, s_{n-1})$ and define the relevant functions on those.

This approach is reasonably easy to implement in the formal axiomatic theory PRA, and it foreshadows the representations that we will need in Chapter 10, when we carry out a similar procedure in the context of first-order arithmetic. Once we have finite sets and sequences, set-theoretic language kicks in, and we can use familiar representations from axiomatic set theory. For example, a (finite) function between two finite sets is a finite set of ordered pairs, a finite group is a tuple consisting of a finite set and the relevant operations on that set, and so on.

To represent a pair (x, y) of numbers as a single natural number, we simply enumerate the pairs systematically as follows:

$$(0, 0),\ (0, 1),\ (1, 0),\ (0, 2),\ (1, 1),\ (2, 0),\ (0, 3),\ (1, 2),\ (2, 1),\ (3, 0),\ \ldots .$$

In other words, first we list the pairs whose elements sum to 0, then the pairs whose elements sum to 1, then the pairs whose elements sum to 2, and so on. Within each group, we first list the pair with first component 0, then 1, and so on. If you lay out the pairs on a two dimensional grid and then trace a path through the elements in this order, the line will form a zigzag. Logicians sometimes refer to an enumeration like this as *dovetailing* since it resembles a carpenter's dovetail joint. Under this scheme, 0 represents the pair (0, 0), 1 represents (0, 1), 2 represents (1, 0), etc.

More precisely, let $J(x, y) = (x + y)(x + y + 1)/2 + x$. To see that this corresponds to the enumeration above, notice that there are $1 + 2 + \ldots + n = n(n + 1)/2$ pairs (x, y) such that $x + y < n$, and these are numbered $0, \ldots, n(n + 1)/2 - 1$. The first pair (x, y) among those with $x + y = n$ is therefore $(x + y)(x + y + 1)/2$, and adding x then indexes the specific pair with that first component.

The function J is known as *Cantor's pairing function.* It is primitive recursive because addition, multiplication, and integer division are all primitive recursive (see Exercise 8.2.4). From $J(x, y)$, we can recover x and y by letting

$$K(z) = \min_{x \leq z} \exists y \leq z\, (J(x, y) = z)$$

and

$$L(z) = \min_{y \leq z} \exists x \leq z\, (J(x, y) = z).$$

This next lemma is left as an exercise:

Lemma 8.3.1. *For every z there exists a unique pair of x and y such that $J(x, y) = z$.*

Proposition 8.3.2. *There are primitive recursive functions $J(x, y)$, $K(z)$, and $L(z)$, such that for every x, y, and z:*

1. $K(J(x, y)) = x$.
2. $L(J(x, y)) = y$.
3. $J(K(z), L(z)) = z$.

Proof The first two claims follow from the uniqueness part of Lemma 8.3.1. For example, $K(J(x, y))$ returns an x' such that for some y', $J(x, y) = J(x', y')$, and the lemma implies $x' = x$. The third claim follows from these together with the existence part of Lemma 8.3.1. Specifically, given z, if we let x and y be such that $z = J(x, y)$, then from the first two claims we have $J(K(z), L(z)) = J(x, y) = z$. □

The third claim of Proposition 8.3.2 implies that every pair has a unique representation: if $K(z) = K(z')$ and $L(z) = L(z')$, then

$$z = J(K(z), L(z)) = J(K(z'), L(z')) = z'.$$

In other words, saying that z and z' represent the same pair is equivalent to saying $z = z'$. This is a nice feature of our encoding but not an essential one. Given *any* representation of pairs, we could write down a primitive recursive relation $P(z)$ that asserts that a number z is the smallest one that represents its pair. In a primitive recursive way, we could then define a new encoding such that each w represents the wth element z satisfying $P(z)$. That would turn the original encoding into one in which every pair is represented by one and only one natural number. Alternatively, we could keep the original encoding and restrict attention to representations satisfying P. Yet a third option is to live with multiple representations and avoid talking about equality of pairs directly. In other words, we could define $w \equiv z$ to mean $(w)_0 = (z)_0$ and $(w)_1 = (z)_1$, and we could then use that equivalence relation to express equality of pairs. These considerations are not restricted to pairs; similar options are available when representing other mathematical objects as well.

There are other reasonable schemes for representing pairs. One possibility is to define $J(x, y) = 2^x 3^y$, define $K(z)$ to be the largest x less than z such that 2^x divides z, and define $L(z)$ analogously. This would sacrifice the uniqueness of representations, since, for example, $2^x \cdot 3^y \cdot 5$ would also represent the pair (x, y).

We will use the following simple device to represent finite sets. Let $p_0, p_1, p_2, \ldots$ enumerate the prime numbers and define $x \in s$ to mean $p_x \mid s$. This way, the finite set $\{x_0, \ldots, x_{n-1}\}$ is represented by the number $p_{x_0} \cdots p_{x_{n-1}}$. With this definition of membership, representations are not unique, since the same set is also represented by the product of any nontrivial powers of those primes. We will adopt the convention that 0, like 1, represents the empty set.

Given any primitive recursive predicate $P(x, \vec{z})$ and any number y, define $g(x, \vec{z})$ to be equal to p_x if $P(x, \vec{z})$ holds and 1 otherwise. Then the set $\{x < y \mid P(x, \vec{z})\}$ is represented by the number $\prod_{x<y} g(x, \vec{z})$, which is a primitive recursive function of y and $\vec{z}$. Our representation of sets has the property that for every x and s, if $x \in s$, then $x < s$. In other words, s, when considered as a number, is a bound on its elements. If we define $\mathrm{bound}(s) = s$ to abstract this feature of the representation, we have the following:

Proposition 8.3.3. *There are a primitive recursive relation $x \in s$, a primitive recursive function $y, \vec{z} \mapsto \{x < y \mid P(x, \vec{z})\}$ for each primitive recursive relation P, and a primitive recursive function* $\mathrm{bound}(s)$ *such that the following hold:*

- *For every x, y, and $\vec{z}$, $x \in \{x < y \mid P(x, \vec{z})\}$ if and only if $x < y$ and $P(x, \vec{z})$.*
- *For every x and s, if $x \in s$, then $x < \mathrm{bound}(s)$.*

We can then define $\{x \in s \mid P(x, \vec{z})\}$ to be $\{x < \mathrm{bound}(s) \mid x \in s \wedge P(x, \vec{z})\}$. The exercises below ask you to show that the properties enumerated in Proposition 8.3.3 yield the following.

Proposition 8.3.4. *These are primitive recursive:*

- *a constant $\emptyset$ with the property that for every x, $x \notin \emptyset$*
- *a function $x \mapsto \{x\}$ with the property that for every y, $y \in \{x\}$ if and only if $y = x$*
- *the relation $s \subseteq t$ which holds if and only if for every x, if $x \in s$, then $x \in t$*
- *the relation $s \equiv t$ which holds if and only if $s \subseteq t$ and $t \subseteq s$*
- *a function $s \cap t$ with the property that for every x, $x \in s \cap t$ if and only if $x \in s$ and $x \in t$*
- *a function $s \cup t$ with the property that for every x, $x \in s \cup t$ if and only if $x \in s$ or $x \in t$*
- *a function $s \setminus t$ with the property that for every x, $x \in s \setminus t$ if and only if $x \in s$ and $x \notin t$.*

We can now define bounded quantification over finite sets:

$$\exists x \in s\, R(x, \vec{z}) \equiv \exists x < \mathrm{bound}(s)\ (x \in s \wedge R(x, \vec{z}))$$
$$\forall x \in s\, R(x, \vec{z}) \equiv \forall x < \mathrm{bound}(s)\ (x \in s \to R(x, \vec{z})).$$

We can also define products over finite sets,

$$\prod_{x \in s} f(x, \vec{z}) = \prod_{x < \mathrm{bound}(s)} g(x, \vec{z}),$$

where $g(x, \vec{z})$ is defined to be equal to $f(x, \vec{z})$ if $x \in s$ and 1 otherwise. Sums $\sum_{x \in s} f(x, \vec{z})$ can be defined similarly, as well as the maximum value $\max_{x \in s} f(x)$ of a function f on a set x, with the understanding that this value is equal to 0 if s is empty.

Using a bounded sum we can define the cardinality $\text{card}(s)$ of a finite set s by $\text{card}(s) = \sum_{x \in s} 1$. This satisfies the following:

- $\text{card}(\emptyset) = 0$
- $\text{card}(s \cup \{x\}) = \text{card}(s) + 1$, if $x \notin s$
- $\text{card}(s) = \text{card}(t)$ if $s \equiv t$.

If we think of $f(x, \vec{z})$ as a function that returns sets, then $\max_{x \in s} \text{bound}(f(x, \vec{z}))$ bounds all elements of the sets $f(x)$ for $x \in s$. So we can define a bounded union by

$$\bigcup_{x \in s} f(x, \vec{z}) = \{y < \max_{x \in s} \text{bound}(f(x, \vec{z})) \mid \exists x \in s\, (y \in f(x, \vec{z}))\}.$$

Then, for every y, we have $y \in \bigcup_{x \in s} f(x)$ if and only if $y \in f(x)$ for some x in s. In particular, if s is a set and f is any primitive recursive function, we can define the image of f on s by

$$f[s] = \bigcup_{x \in s} \{f(x)\}.$$

For every y, we have $y \in f[s]$ if and only if there is some $x \in s$ such that $f(x) = y$. The union of a finite set of finite sets $\bigcup s$ is defined to be $\bigcup_{x \in s} x$.

Remember that our representation of sets is not unique; for example, $\{0\}$ is represented by both 2 and 4. The property $s \equiv t$ defined in Proposition 8.3.4 says that s and t are *extensionally equal*, that is, they have the same elements. We can choose as canonical representatives of sets those numbers s that are nonzero and not divisible by the square of any prime, i.e. those numbers satisfying

$$\text{set}(s) \equiv s \neq 0 \wedge \forall x < s\,, p_x^2 \nmid s.$$

Note that if s is any finite set, if we let $s' = \{x \in s \mid \top\}$ then we have $s \equiv s'$ and $\text{set}(s')$.

Our representation of sets also has the property that any subset of the set represented by the number s has a representation that is again less than or equal to s. Hence we can define the power set operation by

$$\mathcal{P}(s) = \{t \leq s \mid \text{set(t)} \wedge t \subseteq s\}.$$

We can also define $\forall t \subseteq s\, R(t, \vec{z})$ to be $\forall t \in \mathcal{P}(s)\, R(t, \vec{z})$. In situations like this, we generally want to know that R respects extensional equality of sets, that is, if $t \equiv t'$ then $R(t, \vec{z})$ holds if and only if $R(t', \vec{z})$ holds. We can similarly define bounded existential quantification over subsets and bounded operations like min, max, sums, products, unions, and intersections.

We can now represent a finite function α by a set of ordered pairs, following the usual set-theoretic approach. It will be more convenient to represent the finite sequence $(s_0, \ldots, s_{n-1})$ by the pair (n, α), where α is the set of ordered pairs (i, s_i) with $i < n$. In other words, we define

- $\text{length}(\sigma) = (\sigma)_0$
- $\text{elt}(\sigma, i) = (\min_{p \in (\sigma)_1} (p)_0 = i)_1$.

I will henceforth write $(\sigma)_i$ for $\text{elt}(s, i)$. Note that if f is any primitive recursive function, the sequence $(f(0), \ldots, f(n-1))$ is a primitive recursive function of n: we get it by pairing the length, n, with the image of the map $i \to (i, f(i))$ on the set $\{0, \ldots, n-1\}$. Using this fact, it is not hard to show that the usual operations on finite sequences, such as concatenation, are primitive recursive.

In short, we have built up enough of a theory of finite sets to make it plausible that ordinary mathematical constructions of finite objects can be carried out in a primitive recursive way. In the next section, we will see that the primitive recursion supports more flexible forms of recursive definition, which adds further support to that claim.

From the point of view of computer science, finite sequences (which can also be viewed as lists, tuples, or arrays) are more natural than finite sets. After all, data is always stored in some order, and an algorithm that operates on a finite set has to traverse the data in some order, even to compute results that do not depend on the order chosen. As a result, another common strategy for representing mathematical objects as natural numbers is to start with the notion of a finite sequence and then define pairs, finite sets, and so on in terms of those. For example, we could represent the sequence $(a_0, \ldots, a_{n-2}, a_{n-1})$ with the number $p_0^{a_0} \cdots p_{n-2}^{a_{n-2}} p_{n-1}^{a_{n-1}+1}$. Adding one to the last exponent ensures that the length of the sequence is determined uniquely; otherwise, adding zeroes to the end of a sequence would not change the representation. Using the unique factorization theorem, there is a one-to-one correspondence between finite sequences and numbers other than 0. (We can take 0 to be an alternative representation of the empty sequence, or, to get uniqueness on the nose, modify the previous representation by subtracting one.) Sets can then be represented as lists of numbers, or lists of numbers in an increasing order, or finite binary sequences. The exercises ask you to show that the usual operations on these representations are primitive recursive.

Exercises

8.3.1. Prove Lemma 8.3.1.

8.3.2. Use pairing and projections to show that the Fibonacci sequence, defined by $a_0 = a_1 = 1$ and $a_{n+2} = a_{n+1} + a_n$, is primitive recursive.

8.3.3. Prove Proposition 8.3.4.

8.3.4. Show that if $R(x, \vec{z})$ is primitive recursive, then so are $\exists x \subseteq s\, R(x, \vec{z})$ and $\forall x \subseteq s\, R(x, \vec{z})$.

8.3.5. Show that the following are primitive recursive, where s and t denote sequences:

a. the relation "s is an initial segment of t"

b. the relation $R(s, t)$ which holds if and only if s is a subsequence of t; in other words, for some pair of sequences u and v, we have $t = u ^\frown s ^\frown v$

c. the function reverse(s) that reverses a finite sequence.

8.3.6. Show that the function sort(s) that returns a sequence with the same elements as s in nondecreasing order is primitive recursive. This requires some creativity. One idea is to define a function insert(a, s) which will insert an element a into a sorted list s. Then define a function helper(s, n) which returns a sorted list of the first n elements of s.

8.3.7. We did not use the pairing function J to develop the representation of finite sets. Show that once we have finite sets, we can instead use the Kuratowski representation of ordered pairs, $(x, y) = \{\{x\}, \{x, y\}\}$.

8.3.8. Show that the power set operation $\mathcal{P}(s)$ is primitive recursive without using the assumption that if t is a subset of s then $t \leq s$. (Hint: one way to do it is to use primitive recursion to define a function $f(x)$ that returns the power set of $\{0, \ldots, x-1\}$, and then use that to define the power set of s.)

8.3.9. Using the alternative representation of sequences described in the last paragraph of this section, define the following functions:
a. length(s)
b. append(s, a), the result of adding a to the end of the sequence s
c. elt(s, i)
d. concat(s, t), the result of concatenating s and t.

8.3.10. Write the pairing operation $J(x, y)$ as (x, y). For each n, we can define a representation of n-tuples via iterated pairing:

$$(a_0, a_1, \ldots, a_{n-1}) = (a_0, (a_1, \ldots a_{n-1})).$$

We can then represent any sequence s of length n by the pair (n, t), where t is the representation of s as an n-tuple. Carry out the previous exercise for this representation.

8.4 Other Recursion Principles

The goal of this section is to consider forms of recursive definition that do not fit the rigid schema of primitive recursion but nonetheless keep us within the set of primitive recursive functions. For example, it is often useful to define two functions simultaneously, as in the following schema:

$$\begin{aligned} f_0(0, \vec{z}) &= k_0(\vec{z}) \\ f_1(0, \vec{z}) &= k_1(\vec{z}) \\ f_0(x+1, \vec{z}) &= h_0(x, f_0(x, \vec{z}), f_1(x, \vec{z}), \vec{z}) \\ f_1(x+1, \vec{z}) &= h_1(x, f_0(x, \vec{z}), f_1(x, \vec{z}), \vec{z}). \end{aligned}$$

This augmented schema is called *simultaneous recursion* or *mutual recursion* because the recursive definition of f_0 relies on previous values of f_0 and f_1, and similarly for f_1. To see that the functions f_0 and f_1 are nonetheless primitive recursive whenever k_0, k_1, h_0, and h_1 are, first define an auxiliary function $F(x, \vec{z})$ that computes f_0 and f_1 in pairs as follows:

$$\begin{aligned} F(0, \vec{z}) &= (k_0(\vec{z}), k_1(\vec{z})) \\ F(x+1, \vec{z}) &= (h_0(x, (F(x, \vec{z}))_0, (F(x, \vec{z}))_1, \vec{z}), h_1(x, (F(x, \vec{z}))_0, (F(x, \vec{z}))_1, \vec{z})). \end{aligned}$$

This is an ordinary instance of primitive recursion, using k_0, k_1, h_0, h_1, and a primitive recursive pairing function. Now it is not hard to show by induction that for every x, we have $f_0(x, \vec{z}) = (F(x, \vec{z}))_0$ and $f_1(x, \vec{z}) = (F(x, \vec{z}))_1$, which shows that f_0 and f_1 are also primitive recursive.

Another useful way of defining functions is to define the value of $f(x+1, \vec{z})$ in terms of all the previous values $f(0, \vec{z}), \ldots, f(x, \vec{z})$, as in the following schema:

$$\begin{aligned} f(0, \vec{z}) &= g(\vec{z}) \\ f(x+1, \vec{z}) &= h(x, (f(0, \vec{z}), \ldots, f(x, \vec{z})), \vec{z}). \end{aligned}$$

Here the second argument to h is a number representing the sequence of values. The following schema captures this idea more succinctly:

$$f(x, \vec{z}) = h(x, (f(0, \vec{z}), \ldots, f(x-1, \vec{z})), \vec{z}).$$

When x is 0, the second argument to h is the empty sequence, $()$, in which case we can define h so that it returns $g(x)$. This is known as a *course-of-values* recursion. To show that the function f in the second schema is primitive recursive, define the auxiliary function F by

$$F(0, \vec{z}) = (h(0, (), \vec{z}))$$
$$F(x+1, \vec{z}) = F(x, \vec{z})^\frown h(x+1, F(x, \vec{z}), \vec{z}).$$

By induction we can show that for every x, $F(x, \vec{z}) = (f(0, \vec{z}), \ldots, f(x, \vec{z}))$, and then writing $f(x, \vec{z}) = (F(x, \vec{z}))_x$ shows that f is primitive recursive.

As a sample application, course-of-values recursion can be used to justify the following type of definition:

$$f(x, \vec{z}) = \begin{cases} h(x, f(k(x, \vec{z}), \vec{z}), \vec{z}) & \text{if } k(x, \vec{z}) < x \\ g(x, \vec{z}) & \text{otherwise.} \end{cases}$$

In other words, the value of f at x can be computed in terms of the value of f at any previous value, chosen by k.

Another more flexible variation on primitive recursion allows us to *change parameters* along the way. With ordinary primitive recursion, the value $f(x+1, \vec{z})$ is computed recursively in terms of $f(x, \vec{z})$, with the same values of $\vec{z}$ in the recursive call. We can, more generally, allow other primitive recursive functions to modify these. To simplify notation, I will assume that there is only one such parameter; this is not a serious restriction, since we can represent a tuple of parameters as a single one. In that case, the defining schema looks as follows:

$$f(0, z) = g(z)$$
$$f(x+1, z) = h(x, f(x, k(x, z)), z).$$

This, too, can be simulated with ordinary primitive recursion. Consider what happens when we compute the value of $f(3, z)$ by hand:

$$\begin{aligned} f(3, z) &= h(2, f(2, k(2, z)), z) \\ &= h(2, h(1, f(1, k(1, k(2, z))), k(2, z)), z) \\ &= h(2, h(1, h(0, f(0, k(0, k(1, k(2, z)))), k(1, k(2, z))), k(2, z)), z) \\ &= h(2, h(1, h(0, g(k(0, k(1, k(2, z)))), k(1, k(2, z))), k(2, z)), z). \end{aligned}$$

We can simulate this by defining an auxiliary function that computes the parameters

$$z, \quad k(2, z), \quad k(1, k(2, z)), \quad k(0, k(1, k(2, z)))$$

that are needed in the computation. Using ordinary primitive recursion, define the function $l(x, \vec{z}, u)$:

$$l(x, z, 0) = z$$
$$l(x, z, u+1) = k(x \dot{-} (u+1), l(x, z, u)).$$

In the example above, $l(3, z, 0)$, $l(3, z, 1)$, $l(3, z, 2)$, and $l(3, z, 3)$ are the parameters we need. Now define a function $m(x, z, u)$ using ordinary primitive recursion so as to satisfy the equation

$$m(x, z, u) = f(u, l(x, z, x \dot{-} u))$$

for every $u \leq x$. With the following definition of m, it is not hard to prove that equation by induction on u:

$$m(x, z, 0) = g(l(x, z, x))$$
$$m(x, z, u + 1) = h(u, m(x, z, u), l(x, z, x \dot{-} (u + 1))).$$

In particular, we have $m(x, z, x) = f(x, z)$. Turning this equation around, we have a definition of f in terms of m, showing that f is primitive recursive.

The form of recursion expressed by the next theorem is even more flexible.

Theorem 8.4.1. *Let $g(x)$, $h(x, y)$, $m(x)$, and $k(x)$ be primitive recursive. Then so is the function $f(x)$ satisfying the following recursive defining equation:*

$$f(x) = \begin{cases} h(x, f(k(x))) & \text{if } m(k(x)) < m(x) \\ g(x) & \text{otherwise.} \end{cases}$$

There is no harm in restricting to a single input, x, because we can always use our sequencing operations to combine data into tuples. Think of m as being some *measure* of the size of x. On input x, $f(x)$ is defined recursively in terms of $f(k(x))$, where $k(x)$ is smaller than x according to the measure we have chosen. To see that f is primitive recursive when g, h, k, and m are, use recursion with changeable parameters to define

$$F(0, x) = 0$$
$$F(y + 1, x) = \begin{cases} h(x, F(y, k(x))) & \text{if } m(k(x)) < m(x) \\ g(x) & \text{otherwise.} \end{cases}$$

By induction on y, we can then show that whenever $m(x) < y$, $F(y, x) = f(x)$, so $f(x) = F(m(x) + 1, x)$ is a primitive recursive definition of f.

I will close the section with a recursion principle that is powerful and easy to use yet can still be reduced to primitive recursion. If $f(x)$ is a primitive recursive function and s is a (number representing a) finite set, then I will write $f \upharpoonright s$ for the (number representing the) restriction of f to s, a finite function. The following theorem then says that we can define a function $f(x)$ recursively in terms of its values on an arbitrary finite set $u(x)$, as long as the elements $x' \in u(x)$ are smaller than x with respect to some measure, $m(x)$.

Theorem 8.4.2. *Let $g(x)$, $h(x, y)$, $m(x)$, and $u(x)$ be primitive recursive. Then so is the function $f(x)$ satisfying the following recursive defining equation:*

$$f(x) = \begin{cases} h(x, f \upharpoonright u(x)) & \text{if } m(x') < m(x) \text{ for every } x' \in u(x) \\ g(x) & \text{otherwise.} \end{cases}$$

This schema is reminiscent of the principle of recursion in Theorem 1.5.3 and its proof. Once again, the focus on unary functions f is not a strong restriction because we can take x to represent a tuple of values if necessary.

Proof For notational convenience, if v is a number representing a finite function, I will write $v(x)$ to denote the value of v on x, or 0 if x is not in the domain of v. Similarly, if s is a subset of the domain of v, I will write $v\upharpoonright s$ to denote the restriction of v to s. To define f, it suffices to define an auxiliary function $F(s)$ that returns the finite function $f\upharpoonright s$, because we can then define $f(x)$ to be $F(\{x\})(x)$.

We can reduce this task to an application of Theorem 8.4.1 as follows. $F(s)$ is supposed to compute the values of $f(x)$ for $x \in s$. To compute them, we need the values of $f(x')$ for $x' \in \bigcup_{x \in s} u(x)$, or, more precisely,

$$\left\{x' \in \bigcup_{x \in s} u(x) \;\middle|\; m(x') < m(x)\right\}.$$

Let us call this set $k(s)$ and use e to denote the finite function with empty domain. Then the function we want is

$$F(s) = \begin{cases} w(s, F(k(s))) & \text{if } k(s) \neq \emptyset \\ e & \text{otherwise,} \end{cases}$$

where $w(s, v)$ returns the finite function α that maps any $x \in s$ to

$$\alpha(x) = \begin{cases} h(x, v\upharpoonright u(x)) & \text{if } m(x') < m(x) \text{ for every } x' \in u(x) \\ g(x) & \text{otherwise.} \end{cases}$$

To see that the definition of F is of the form required by Theorem 8.4.1, define

$$\bar{m}(s) = \begin{cases} \max_{x \in s} m(x) & \text{if } k(s) \neq \emptyset \\ 0 & \text{otherwise.} \end{cases}$$

Then if $k(s) \neq \emptyset$, $\bar{m}(k(s)) < \bar{m}(s)$, since every element of $k(s)$ is strictly smaller than some element of s according to the measure m. Hence $k(s) \neq \emptyset$ if and only if $\bar{m}(k(s)) < \bar{m}(s)$, and so the definition of F has the form

$$F(s) = \begin{cases} w(s, F(k(s)) & \text{if } \bar{m}(k(s)) < \bar{m}(s) \\ e & \text{otherwise,} \end{cases}$$

as required. □

In Chapter 10, we will see that objects like terms and formulas can be represented in such a way that the usual operations and relations are primitive recursive way. Theorem 8.4.2 will be especially useful for defining functions by recursion on syntax.

Exercises

8.4.1. In the justification of recursion by parameters in this section, show that the equation $m(x, z, u) = f(u, l(x, z, x \dot{-} u))$ follows from the definition of m.

8.4.2. Fill in more of the details in the proof of Theorem 8.4.2 above.

8.5 Recursion along Well-Founded Relations

Recall the recursion schema from Theorem 8.4.1:

$$f(x) = \begin{cases} g(x, f(k(x))) & \text{if } m(k(x)) < m(x) \\ h(x) & \text{otherwise.} \end{cases}$$

This actually specifies a function because the measure $m(x)$ decreases in the recursive function call, and the usual order on the natural numbers is well founded, which means that these recursive calls cannot continue indefinitely.

A binary relation $\prec$ on a set A is said to be well founded if every nonempty subset of A has a $\prec$-least element, or, equivalently, if there is no infinite descending sequence $a_0 \succ a_1 \succ \dots$ of elements of A. (See Appendix A.2.) For example, define the lexicographic order $\prec$ by $(a_0, a_1) \prec (b_0, b_1)$ if and only if $a_0 < b_0$ or $a_0 = b_0$ and $a_1 < b_1$. This relation is well founded: given any nonempty set S of ordered pairs (a, b), if we let a be the least natural number appearing as the first element of such a pair and we let b be the least natural number that appears as the second element of pair whose first element is a, then (a, b) will be the lexicographically least element of S. In principle, replacing $<$ with any well-founded relation in the schema from Theorem 8.4.1 defines a total function, but the resulting function may not be primitive recursive, even if the functions g, h, k, and m, as well as the relation $\prec$, are all primitive recursive. In this section, we will identify a condition under which recursive definition using $\prec$ does give rise to a primitive recursive function, and show that, for each n, the lexicographic order on n-tuples of natural numbers meets the criterion.

To illustrate the central idea, I will provide a more direct primitive recursive definition of the function f meeting the specification in Theorem 8.4.1. Define the nth iterate of k, $k^n(x)$, by $k^0(x) = x$ and $k^{n+1}(x) = k(k^n(x))$. By induction we have that for every i, either $m(k^i(x)) \leq m(x) - i$ or there is a $j \leq i$ such that $m(k^{j+1}(x)) \geq m(k^j(x))$. In other words, either the measure of each subsequent element in the sequence $x, k(x), k(k(x)) \dots, k^i(x)$ decreases or there is at least one place where it fails to decrease. Since the sequence cannot decrease more than $m(x)$ times, there is some $j \leq m(x) + 1$ such that the second case holds. Consider the least such j. Then we know that the measure m of each element of the sequence $x, k(x), k^2(x), \dots, k^j(x)$ decreases, while $m(k^{j+1}(x)) \geq m(k^j(x))$. Now define a function $u(i, x)$ so that for every $i \leq j$, $u(i, x)$ is equal to $f(k^{j-i}(x))$. Specifically, writing $j = j(x)$ as a function of x, we can define u with the following equations:

$$u(0, x) = g(k^{j(x)}(x))$$
$$u(i+1, x) = h(k^{j(x)-(i+1)}(x), u(i, x)).$$

If we set $f(x) = u(j(x), x)$, we can show that f satisfies the specification in Theorem 8.4.1. If $m(k(x)) \geq m(x)$ then $j(x) = 0$ and $f(x) = g(x)$. Otherwise $j(x) > 0$ and the description of j implies that $j(k(x)) = j(x) - 1$. In that case, we have

$$f(x) = u(j(x), x) = u((j(x) - 1) + 1, x) = h(k(x), u(j(x) - 1, x)) = h(k(x), f(k(x))),$$

as required.

To modify the argument above to handle $\prec$ in place of $<$, it suffices to know that for any primitive recursive function $v(x, i)$ there is a primitive recursive function $w(x)$ such that for some $j \leq w(x)$, we have $v(x, j+1) \not\prec v(x, j)$. Fixing x, think of $v(x, j)$ as defining a sequence $v_0, v_1, v_2, \ldots$. We are asking that $w(x)$ returns a bound on how far we have to look to see that the sequence fails to decrease indefinitely. (In the construction for $<$, the sequence in question was $v(x, i) = m(k^i(x))$, in which case $w(x) = m(x) + 1$ was sufficient.) When such a function $w(x)$ exists, let us say that $\prec$ is *primitive-recursively well ordered.*

We now show that the lexicographic order $\prec$ on pairs has this property. Given the sequence $v_0, v_1, v_2, \ldots$, reason as follows. Let i_0 be the first component of v_0. Then, assuming $v_0, v_1, \ldots, v_{i_0+1}$ is decreasing, the second component can only decrease i_0 times before the first component decreases, so we know that the first component of v_{i_0+1} must be strictly smaller than the first component of v_0. Let i_1 be the second component of v_{i_0+1}. By the same reasoning, either the sequence fails to decrease before it reaches $v_{i_0+1+i_1+1}$, or the first component of this element is smaller than the first component of v_{i_0+1}. If n is the first component of v_0, it suffices to iterate this $n + 1$ times, because by that point, the first component can no longer decrease.

More formally, let $v(x, n)$ be any primitive recursive function. Define

$$z(x, 0) = (v(x, 0))_1 + 1$$
$$z(x, m + 1) = z(x, m) + (v(x, z(x, m)))_1 + 1,$$

and let $w(x) = z(x, (v(x, 0))_0 + 1)$. Then, following the argument above, assuming that the sequence $v(x, 0), v(x, 1), \ldots, v(x, w(x))$ is decreasing, we have that the first components of $v(x, z(x, 0)), v(x, z(x, 1)), \ldots, v(x, z(x, (v(x, 0))_0 + 1))$ are decreasing below the first component of $v(z, 0)$, a contradiction. So for some $j \le w(x)$, we have $v(x, j + 1) \not\prec v(x, j)$, as required.

We can extend the lexicographic order to triples, quadruples, and so on. For each fixed n, let $\prec_n$ be the lexicographic order in n-tuples. We therefore have the following:

Theorem 8.5.1. *Let $g(x)$, $h(x, y)$, $m(x)$, and $k(x)$ be primitive recursive. Then so is the function $f(x)$ satisfying the following recursive defining equation:*

$$f(x) = \begin{cases} h(x, f(k(x))) & \text{if } m(k(x)) \prec_n m(x) \\ g(x) & \text{otherwise.} \end{cases}$$

Finding a point at which a sequence fails to decrease with respect to $\prec_{n+1}$ requires a primitive recursive definition that makes use of a function that finds a point at which a related sequence fails to decrease with respect to $\prec_n$. In other words, modeling recursion for longer tuples requires more and more nested instances of primitive recursion. It is possible to extend the lexicographic order to arbitrary finite sequences by stipulating $\sigma \prec \tau$ whenever σ is shorter than τ. Call this relation $\prec_\omega$. Our discussion suggests that primitive recursion is not strong enough to replace $\prec_n$ by $\prec_\omega$ in Theorem 8.5.1. This is in fact the case, so Theorem 8.5.1 is sharp in this sense.

As in the last section, we can use Theorem 8.5.1 to justify the following strengthening of Theorem 8.4.2:

Theorem 8.5.2. *Let $g(x)$, $h(x, y)$, $m(x)$, and $u(x)$ be primitive recursive. Then so is the function $f(x)$ satisfying the following recursive defining equation:*

$$f(x) = \begin{cases} h(x, f\lceil u(x)) & \text{if } m(x') \prec_n m(x) \text{ for every } x' \in u(x) \\ g(x) & \text{otherwise.} \end{cases}$$

Exercise

8.5.1. Spell out a proof, more carefully, that for each n, $\prec_n$ is primitive-recursively well ordered.

8.6 Diagonalization and Reflection

We have seen that lots of functions from the natural numbers to the natural numbers are primitive recursive. Are there any functions that are not?

Cardinality considerations give an immediate answer. Each primitive recursive function is given by an explicit sequence of definitions, starting with the initial functions and using only composition and primitive recursion. It should be intuitively clear that we can systematically enumerate all these definition sequences, and hence obtain an enumeration of all the primitive recursive functions. (Any primitive recursive function will appear infinitely many times in the enumeration, since we can compose any definition with the identity function to obtain a new definition, but that does not cause problems in the arguments that follow.) As a result, there are only countably many primitive recursive functions. On the other hand, there are uncountably many functions from the natural numbers to the natural numbers, even if we restrict attention to unary functions. So there are functions that are not primitive recursive.

We can avoid reference to cardinality and present a direct argument. Using the methods of Section 8.3, we can represent the definition of any primitive recursive function as a natural number. Define a sequence of functions $f_0, f_1, f_2, \ldots$ by letting $f_x(y)$ be the unary function with definition represented by x, if x represents the definition of a unary function, or the constant 0 function otherwise. Define the function g by $g(x) = f_x(x) + 1$. Clearly g is not primitive recursive: if it were, it would have a definition represented by some number x, and then we would have both $g(x) = f_x(x)$ and $g(x) = f_x(x) + 1$, a contradiction.

Given that the collection of primitive recursive functions does not include all the functions from $\mathbb{N}$ to $\mathbb{N}$, one might ask whether it includes all the *computable* ones. We cannot answer this question definitively without a precise definition of computability, such as the one presented in Chapter 11. But the argument just presented should lead us to suspect that the answer will be no. After all, a representation x of the function f_x should enable us to systematically compute the value of $f_x(y)$ for each y. Assuming we can compute $f_x(y)$ uniformly from x and y, the function $g(x) = f_x(x) + 1$ will be an example of a function that is computable but not primitive recursive.

Let us come at this from a slightly different perspective. Given a reasonable representation of primitive recursive functions of arbitrary arity, let the function $\mathrm{eval}(x, s)$ denote the value of $f_x(\vec{y})$, where f_x is the function whose definition is represented by x and $\vec{y}$ is the tuple represented by s. (If x does not represent a function definition or s does not represent a tuple of the right length, we can set $\mathrm{eval}(x, s)$ equal to 0.) We have argued that, given a proper analysis of the notion of computability, we should have the following:

Theorem 8.6.1. *The function* $\mathrm{eval}(x, s)$ *is computable but not primitive recursive.*

After all, if $\mathrm{eval}(x, s)$ were primitive recursive, then $g(x) = \mathrm{eval}(x, (x)) + 1$ would be primitive recursive as well.

Theorem 8.6.1 states a positive result, namely, that eval is computable, and a negative result, namely, that it is not primitive recursive. We can provide additional information on the positive side. For each n, define $\text{eval}_n(x, s)$ to return $f_x(\vec{y})$ when s codes $\vec{y}$ and the definition x has at most n instances of primitive recursion, and zero otherwise. In other words, we restrict eval_n to functions of limited complexity, where the complexity is measured in terms of the number of instances of primitive recursion. Then we have:

Theorem 8.6.2. *For each n, $\text{eval}_n(x, s)$ is primitive recursive.*

This says that there are partial evaluation functions for the primitive recursive functions that are again primitive recursive. We will not take the time out to prove Theorem 8.6.2 here. Doing so requires presenting careful encodings of primitive recursive function definitions and writing down a number of explicit definitions, using the more liberal recursion schemas presented in Section 8.4. We will do something similar in Chapter 10, when we describe primitive recursive evaluation functions for terms and formulas in the language of first-order arithmetic. The exercises below encourage you to think about proving Theorem 8.6.2, possibly after reading the results in Chapter 10.

The ideas we have just explored illustrate recurring themes in logic. Languages and systems can often be stratified into hierarchies, and diagonalization is often used to produce something that lies outside some hierarchy. Tarski's theorem on the undefinability of truth and Gödel's incompleteness theorems, both discussed in Chapter 12, are examples.

It is possible to provide other examples of computable functions that are not primitive recursive. For example, define a hierarchy of functions $f_0, f_1, \ldots$ as follows:

$$f_0(x) = x + 1$$

$$f_{n+1}(x) = f_n^{x+2}(x + 1).$$

For each primitive recursive function f, the yth iterate $f^y(x)$ is a primitive recursive function of x and y. We can therefore use induction on n to show that each f_n is primitive recursive. The functions f_n are designed to grow very quickly, as can be seen by computing the first few values of f_3, f_4, and f_5.

If f and g are functions, f is said to *dominate* g (or *majorize* g) if for every x, $f(x) \geq g(x)$. The function f is said to *strictly dominate* g if for every x, $f(x) > g(x)$. f is said to *eventually dominate* g is there is some y such that $f(x) \geq g(x)$ for every $x \geq y$; in other words, f is greater than or equal to g for all but finitely many inputs. In the exercises below, you are asked to show that every unary primitive recursive function is dominated by one of the functions f_n defined above. As a result, if we define

$$f_\omega(x) = f_x(x),$$

then f_ω grows too fast to be primitive recursive. The sequence $f_0, f_1, f_2, \ldots$ in this example is an example of what logicians refer to as a *fast-growing hierarchy*.

Consider the following schema of definition by double recursion:

$$f(0, y) = g(y)$$

$$f(x + 1, 0) = h(x, f(x, k(x)))$$

$$f(x + 1, y + 1) = \ell(x, y, f(x + 1, y), f(x, m(x, y, f(x + 1, y)))).$$

In other words, to define $f(x+1, y+1)$, we are allowed to refer to $f(x+1, y)$, as well as $f(x, z)$ for *any* value of z computed from x, y, and $f(x+1, y)$. In the exercises below, you are asked to show that these equations define a unique function f from g, h, k, ℓ, and m. You are also asked to consider a fast-growing hierarchy g_n similar to f_n, and show that the function $A(n, x) = g_n(x)$ can be defined with that schema. The function $A(n, x)$ is one of a family of functions referred to as *the Ackermann function*, considered as variations of a function defined by Wilhelm Ackermann in 1928 as an early example of a computable function that is not primitive recursive.

Exercises

8.6.1. Define a schema for representing each definition of a primitive recursive function as a natural number so that the claims in this section hold.

8.6.2. Consider the hierarchy of functions $(f_n)_{n \in \mathbb{N}}$ defined above.

a. Show that each f_n is strictly increasing.
b. Show that for each n, f_{n+1} strictly dominates f_n.
c. Show that for each n and x, $f_{n+1}(x) \geq f_n(f_n(x))$.
d. Show by induction on n that if $h(\vec{x})$ is a primitive recursive function with a notation of depth n, then $h(\vec{x}) \leq f_n(\max(\vec{x}))$. (For 0-ary functions, take $\max(\vec{x})$ to be 0.) Remember that the functions with notations of depth 0 are the constant zero, the successor function, and the projection functions; functions with notations of depth $n+1$ are defined using composition and primitive recursion on functions on depth at most n.
e. The third part shows that, in particular, every unary primitive recursive function is dominated by one of the functions f_n. Show that f_ω does not have this property, so it is not primitive recursive.

8.6.3. Give explicit expressions for $f_0(x)$, $f_1(x)$, $f_2(x)$, $f_3(0)$, and $f_3(1)$.

8.6.4. In the definition of the f hierarchy, the expressions $n+2$ and $x+1$ were chosen to make it easier to prove the claims above. But slight variations typically do not have a strong effect on the rate of growth. For example, consider the following hierarchy introduced by Rózsa Péter:

$$g_0(x) = x + 1$$
$$g_{n+1}(x) = g_n^{x+1}(1).$$

The g hierarchy grows a little bit slower that the f hierarchy: you can check, for example, that $g_1(x) = x + 2$, $g_2(x) = 2x + 3$, $g_3(x) = 2^{x+3} - 3$. Also, in the definition of g_{n+1}, the iteration starts with 1 instead of x. In this exercise, you will show that for each n, f_n has the roughly same rate of growth as g_{n+1}, so that the g hierarchy exhausts the primitive recursive functions as well.

a. Verify the claims above about g_1, g_2, and g_3.
b. Show that each g_n is strictly increasing, and for each n, g_{n+1} strictly dominates g_n.
c. Show that for each n, there are natural numbers a and b such that $f_n(x) \leq g_{n+1}(ax+b)$. (Hints: Do this for $n = 0, 1, 2$ directly, and then use induction on n. Use the following facts: part (a) implies that for any natural numbers a and b, $ax + b \leq g_3(x) \leq g_n(x)$ when x is large enough and $n \geq 3$, and part (b) implies $g_n^x(1) \geq x + 1$ for every n. In the induction step, show that when x is large enough, $f_{n+1}(x) \leq g_{n+1}^{3x+4}(1) = g_{n+2}(3x+4)$. If "large enough" means $x \geq k$, we have $f_{n+1}(x) \leq f_{n+1}(x+k) \leq g_{n+2}(3x+3k+4)$ for every x.)
d. When $n \geq 1$, show $f_{n+1}(x) \leq g_{n+2}(g_{n+2}(x)) \leq g_{n+3}(x)$ for x big enough. So every primitive recursive function is eventually dominated by one of the g_n's.

8.6.5. Consider the schema of definition by double recursion above.

a. Show that the equations above define a unique function f from g, h, k, ℓ, and m. (Hint: show by induction that for every x, the function f_x which maps y to $f(x, y)$ is uniquely defined.)

b. Show that the function $w(x, y) = g_x(y)$, where g_x is the hierarchy defined in the last problem, can be defined by double recursion from primitive recursive functions. Hence double recursion cannot be justified using primitive recursion.

8.6.6. Sketch a proof of Theorem 8.6.2. As a warmup, first prove it in the case where n counts the number of instances of primitive recursion and composition in the definition.

Bibliographical Notes

General references A nice exposition of the primitive recursive functions can be found in Kleene (1952). Goodstein (1957) is a thorough early study of the primitive recursive functions, and Goodstein (1961) explores ways in which analysis can be carried out in a primitive recursive way.

General forms of recursion Rose (1984) and Odifreddi (1999) provide extensive information on various recursion schemas, some of which are reducible to primitive recursion and some of which properly extend it.

9

Primitive Recursive Arithmetic

Having established that many familiar functions and relations on the natural numbers are primitive recursive, we now present an axiomatic theory for reasoning about them. *Primitive recursive arithmetic*, or PRA, is naturally presented as a quantifier-free theory, though we will also consider a version in which it is presented as a first-order theory, and a minimal version based on equational logic. We will see that although PRA is fairly simple, it provides a robust framework for reasoning about finite objects and operations on them.

9.1 A Quantifier-Free Axiomatization

The first version of PRA that we will consider starts with quantifier-free first-order logic, as described in Section 4.4. The language has a function symbol for each primitive recursive function. More precisely, we assign a function symbol to each *description* of a primitive recursive function, which is to say, we start with a constant symbol 0, a symbol succ for the successor function, and symbols p_i^n for the projections, and iteratively add symbols for functions defined by composition and primitive recursion. Formulas in the deductive system are quantifier-free first-order formulas in this language. We interpret a formula with free variables as asserting its universal closure, and allow the substitution rule described in Section 4.4. (We will in fact state all the axioms and rules so that they are closed under substitution, so that substitution is an admissible rule and we could therefore do without it.)

The axioms governing the initial function symbols are the following:

- $\mathrm{succ}(x) \neq 0$
- for each n and $i < n$, the axiom $p_i^n(x_0, \ldots, x_{n-1}) = x_i$.

Here $\mathrm{succ}(x) \neq 0$ abbreviates $\neg\mathrm{succ}(x) = 0$. We also have defining equations for functions obtained by composition and primitive recursion:

- For each f defined by composition of g with $h_0, \ldots, h_{k-1}$,

$$f(x_0, \ldots, x_{\ell-1}) = g(h_0(x_0, \ldots, x_{\ell-1}), \ldots, h_{k-1}(x_0, \ldots, x_{\ell-1})).$$

- For each f defined by primitive recursion from g and h,

$$f(0, z_0, \ldots, z_{k-1}) = g(z_0, \ldots, z_{k-1})$$

and

$$f(\mathrm{succ}(x), z_0, \ldots, z_{k-1}) = h(x, f(x, z_0, \ldots, z_{k-1}), z_0, \ldots, z_{k-1}).$$

Finally, we add the quantifier-free induction rule:

- From $A(0)$ and $A(x) \to A(\mathrm{succ}(x))$ conclude $A(t)$ for any term t.

Section 9.4 below shows that the induction rule can be replaced by suitable axioms instead.

In the context of natural deduction, it is more natural to express induction as follows:

$$\dfrac{A(0) \qquad \begin{array}{c} \overline{A(x)} \\ \vdots \\ A(\mathrm{succ}(x)) \end{array}}{A(t)}$$

Here t is any term and x is not free in any hypothesis other than $A(x)$. It is not obvious that expressing the rule in natural deduction in this way does not change the notion of provability, since, in the original formulation, $A(0)$ and $A(x) \to A(\mathrm{succ}(x))$ had to be provable outright, and the natural deduction formulation allows them, more generally, to be provable from hypotheses. But if we denote the conjunction of all the hypotheses used in the two subproofs as C, the premises imply that $C \to A(0)$ and $C \wedge A(x) \to A(\mathrm{succ}(x))$ are derivable outright. From this we can derive

$$(C \to A(x)) \to (C \to A(\mathrm{succ}(x))),$$

and we can use the original formulation of the rule to derive $C \to A(t)$.

For each natural number n, it is common to let $\overline{n}$ denote the *numeral* representing n, defined by $\overline{0} = 0$ and $\overline{n+1} = \mathrm{succ}(\overline{n})$. So, if n is a natural number, $\overline{n}$ is a term in the language of arithmetic. But to avoid putting bars everywhere, I will instead adopt the convention that whenever n is a natural number and I treat n as a term in the language of arithmetic, then I mean $\overline{n}$. In particular, the symbol 1 in a formula abbreviates $\mathrm{succ}(0)$.

Formally, the fact that the set of primitive recursive functions is closed under explicit definition can be expressed as follows:

Proposition 9.1.1. *For each term $t(x_0, \ldots, x_{n-1})$ in the language of* PRA *with at most the free variables shown, there is a function symbol f such that* PRA *proves*

$$f(x_0, \ldots, x_{n-1}) = t(x_0, \ldots, x_{n-1}).$$

Remember that the predecessor function has defining equations $\mathrm{pred}(0) = 0$ and $\mathrm{pred}(\mathrm{succ}(x)) = x$. The following natural deduction derivation shows that the successor function is injective:

$$\dfrac{\dfrac{\dfrac{\dfrac{\overline{\mathrm{pred}(\mathrm{succ}(x)) = x}}{x = \mathrm{pred}(\mathrm{succ}(x))} \qquad \dfrac{\overline{\mathrm{succ}(x) = \mathrm{succ}(y)}^{\,1}}{\mathrm{pred}(\mathrm{succ}(x)) = \mathrm{pred}(\mathrm{succ}(y))}}{x = \mathrm{pred}(\mathrm{succ}(y))} \qquad \overline{\mathrm{pred}(\mathrm{succ}(y)) = y}}{x = y}}{\mathrm{succ}(x) = \mathrm{succ}(y) \to x = y}\,1$$

Addition and multiplication are defined as in Section 8.2. The following natural deduction proof shows that $x + 1$ is equal to $\mathrm{succ}(x)$:

$$\dfrac{\dfrac{}{x+\mathrm{succ}(0)=\mathrm{succ}(x+0)} \qquad \dfrac{\dfrac{}{x+0=x}}{\mathrm{succ}(x+0)=\mathrm{succ}(x)}}{x+\mathrm{succ}(0)=\mathrm{succ}(x)}$$

We can therefore use $x+1$ and $\mathrm{succ}(x)$ is interchangeably.

If we do not make use of the hypothesis in the induction rule, we have

$$\frac{A(0) \qquad A(\mathrm{succ}(x))}{A(t)}$$

That is, we can carry out proof by cases on whether a variable x is 0 or a successor. Here is an example.

$$\dfrac{\dfrac{\dfrac{\dfrac{\dfrac{}{0 \neq 0}\,1 \qquad \dfrac{}{0=0}}{\bot}}{0=\mathrm{succ}(\mathrm{pred}(0))}}{0 \neq 0 \to 0=\mathrm{succ}(\mathrm{pred}(0))}\,1 \qquad \dfrac{\dfrac{\dfrac{\dfrac{}{\mathrm{pred}(\mathrm{succ}(x))=x}}{x=\mathrm{pred}(\mathrm{succ}(x))}}{\mathrm{succ}(x)=\mathrm{succ}(\mathrm{pred}(\mathrm{succ}(x)))}}{\mathrm{succ}(x) \neq 0 \to \mathrm{succ}(x)=\mathrm{succ}(\mathrm{pred}(\mathrm{succ}(x)))}\,1}{x \neq 0 \to x=\mathrm{succ}(\mathrm{pred}(x))}$$

As our proofs get more elaborate, we will resort to proof sketches rather than detailed diagrams. In other words, we will often prove informally that there is a formal derivation without exhibiting one explicitly.

Proposition 9.1.3 below provides a more flexible form of induction. To justify it, we need a few facts about truncated subtraction, $x \dot{-} y$, which is defined by the equations $x \dot{-} 0 = x$ and $x \dot{-} \mathrm{succ}(y) = \mathrm{pred}(x \dot{-} y)$.

Proposition 9.1.2. PRA *proves the following:*

1. $0 \dot{-} x = 0$
2. $\mathrm{succ}(x) \dot{-} \mathrm{succ}(y) = x \dot{-} y$
3. $x \dot{-} x = 0$.

Proof The first claim is easy to prove by induction on x. For the second claim, use induction on y. When $y=0$, we have

$$\mathrm{succ}(x) \dot{-} \mathrm{succ}(0) = \mathrm{pred}(\mathrm{succ}(x) \dot{-} 0) = \mathrm{pred}(\mathrm{succ}(x)) = x = x \dot{-} 0.$$

In the inductive step, we have

$$\begin{aligned} \mathrm{succ}(x) \dot{-} \mathrm{succ}(\mathrm{succ}(y)) &= \mathrm{pred}(\mathrm{succ}(x) \dot{-} \mathrm{succ}(y)) \\ &= \mathrm{pred}(x \dot{-} y) \\ &= x \dot{-} \mathrm{succ}(y). \end{aligned}$$

Here we have used the inductive hypothesis in the second step. The third claim follows easily by induction. □

In Section 8.4, we considered recursion with changeable parameters. The rule given in the next proposition provides an analogous version of induction.

Proposition 9.1.3. *Let s, t, and $k(x, y)$ be any terms, possibly with free variables other than the ones shown. The following is a derived rule of primitive recursive arithmetic:*

$$\frac{A(0, y) \qquad A(x, k(x, y)) \to A(\mathrm{succ}(x), y)}{A(s, t)}$$

Proof By renaming variables, we can assume that x and y are not free in s and t. Let $\vec{z}$ denote the other free variables in A, k, s, and t. The idea is, as in Section 8.4, to define a function $\ell(u, \vec{z})$ that computes the relevant parameters backward, and then run the induction forward. Define $\ell(u, \vec{z})$ by

$$\ell(0, \vec{z}) = t$$
$$\ell(u + 1, \vec{z}) = k(\mathrm{pred}(s \dot{-} u), \ell(u, \vec{z})).$$

We want to show that for every $u \le s$, $A(u, \ell(s \dot{-} u, \vec{z}))$ holds, because then substituting s for u yields the desired conclusion. But at this stage it is tricky to restrict the induction to $u \le s$. One strategy, which you are encouraged to try as an exercise, is to use induction on u to show that $A(s \dot{-} (s \dot{-} u), \ell(s \dot{-} u, \vec{z}))$ holds for every u. We will pursue another strategy, namely, to show by induction that

$$A(s \dot{-} u, \ell(u, \vec{z})) \to A(s, t)$$

holds for every u. This suffices, since when we set $u = s$, we have $s \dot{-} u = 0$, and we can obtain $A(0, \ell(s, \vec{z}))$ by substituting $\ell(s, \vec{z})$ for y in the first hypothesis.

When $u = 0$, we have $\ell(u, \vec{z}) = t$ by definition, and the implication is trivial. For the induction step, it suffices to show

$$A(s \dot{-} \mathrm{succ}(u), \ell(\mathrm{succ}(u), \vec{z})) \to A(s \dot{-} u, \ell(u, \vec{z})),$$

which is equivalent to

$$A(\mathrm{pred}(s \dot{-} u), k(\mathrm{pred}(s \dot{-} u), \ell(u, \vec{z}))) \to A(s \dot{-} u, \ell(u, \vec{z})).$$

Generalizing $s \dot{-} u$ to an arbitrary value v, is suffices to prove

$$A(\mathrm{pred}(v), k(\mathrm{pred}(v), \ell(u, \vec{z}))) \to A(v, \ell(u, \vec{z})).$$

We can do this by cases on v: when v is 0, the conclusion follows from the first premise, and when v is replaced by $\mathrm{succ}(v)$, the conclusion is exactly the second premise. □

The following shows that we can restrict PRA to intuitionistic logic without changing the provability relation.

Proposition 9.1.4. PRA *over intuitionistic logic proves the law of the excluded middle.*

Proof By Exercise 2.7.1, it is sufficient to show that PRA proves the law of the excluded middle for atomic formulas $t_1 = t_2$, and for that it suffices to show that it proves $x = y \vee x \neq y$. By the previous proposition, it suffices to show that PRA proves $0 = y \vee 0 \neq y$ and

$$x = \mathrm{pred}(y) \vee x \neq \mathrm{pred}(y) \to \mathrm{succ}(x) = y \vee \mathrm{succ}(x) \neq y.$$

The first is proved easily by cases on y. For the second, uses cases on y as well. When $y = 0$, we have $\mathrm{succ}(x) \neq 0$. Replacing y by $\mathrm{succ}(y)$, from the assumption $x = \mathrm{pred}(\mathrm{succ}(y)) \vee x \neq \mathrm{pred}(\mathrm{succ}(y))$ we get $x = y \vee x \neq y$, which implies $\mathrm{succ}(x) = \mathrm{succ}(y) \vee \mathrm{succ}(x) \neq \mathrm{succ}(y)$ using a the substitution rule for equality and the injectivity of the successor function. □

In fact, if we take $\perp$ to be defined by $0 = 1$, we can base PRA on minimal logic.

Proposition 9.1.5. PRA *over minimal logic proves* $0 = 1 \to A$ *for every* A.

Proof As in the proof of Proposition 2.7.2, it suffices to do this for atomic formulas, which are of the form $u = v$. For any pair of terms u and v, define a function by

$$f(0) = u, \quad f(\mathrm{succ}(x)) = v.$$

In particular, $f(1) = v$, so if $0 = 1$, we have $u = v$. □

Implicit in the proof of Proposition 9.1.4 is the following principle of double induction, which we will put to good use in the next section.

Proposition 9.1.6. *The following is a derived rule of* PRA:

$$\frac{A(0, y) \qquad A(x, 0) \qquad \begin{matrix} \overline{A(x, y)} \\ \vdots \\ A(\mathrm{succ}(x), \mathrm{succ}(y)) \end{matrix}}{A(u, v)}$$

Proof Using Proposition 9.1.3, since we have $A(x, 0)$, it suffices to derive

$$A(x, \mathrm{pred}(y)) \to A(\mathrm{succ}(x), y).$$

We can prove this by cases on y. When y is 0, this follows from the second hypothesis. If, on the other hand, we replace y by $\mathrm{succ}(y)$, this result follows from the third hypothesis. □

Exercises

9.1.1. Show that we can replace the axiom $\mathrm{succ}(x) \neq 0$ by $1 \neq 0$. So, if we define $\bot$ to be $1 = 0$, we do not need this axiom at all. (Hint: use induction on x.)

9.1.2. Prove Proposition 9.1.3 using the alternative strategy suggested in the proof above.

9.2 Bootstrapping PRA

In Section 8.2, we saw that many familiar number-theoretic functions and relations are primitive recursive, and Section 8.3 made the case that finitary mathematical objects can be represented as numbers in such a way that common functions and relations on those are primitive recursive as well. Moreover, Section 8.4 showed that primitive recursion supports more flexible forms of recursive definition.

In this section and the next, we will argue that expected properties of the functions and relations defined in Sections 8.2 and 8.3 can be proved in PRA, and that PRA supports more flexible forms of induction. The shift of focus from definability to provability is sometimes cast as a shift of focus from *extensional* correctness of the definitions to their *intensional* correctness. Whereas in previous sections we were concerned that our formal definitions pick out the right objects in the intended semantic interpretation, here we are imposing the additional syntactic requirement that these properties are provable in a particular formal framework.

The term *bootstrapping* in the title of this section refers to the notion of pulling oneself up by one's bootstraps. In computer science, it is used to describe the way in which an operating system starts up the low-level processes that are needed to support its basic functionality. This section aims to show that, in a similar way, PRA can derive enough basic logic and arithmetic to get more substantial mathematical arguments off the ground.

To start, let us show that the natural numbers with addition and multiplication form a commutative semiring with a cancellation law for addition.

Proposition 9.2.1. *All of the following are provable in* PRA:

1. $x+0=x$
2. $(x+y)+z=x+(y+z)$
3. $x+y=y+x$
4. $x \cdot 1=x$
5. $x \cdot (y+z)=x \cdot y+x \cdot z$
6. $(x \cdot y) \cdot z=x \cdot (y \cdot z)$
7. $x \cdot y=y \cdot x$
8. $x+z=y+z \to x=y$.

Proof The first claim is a defining equation for addition. To prove claim 2, use induction on z. (A useful heuristic is that when a function is defined by recursion on a certain argument, it is generally most effective to prove statements about it using induction on the same argument.) For $z=0$, we have $(x+y)+0=x+y$ and $y+0=y$ by the defining axioms for addition. For the inductive step, suppose $(x+y)+z=x+(y+z)$. Then we have

$$\begin{aligned}(x+y)+\mathrm{succ}(z) &= \mathrm{succ}((x+y)+z)\\ &= \mathrm{succ}(x+(y+z))\\ &= x+\mathrm{succ}(y+z)\\ &= x+(y+\mathrm{succ}(z)).\end{aligned}$$

Here we have used the inductive hypothesis in the second step and the definition of addition in the others.

To prove claim 3, first we use induction on y to prove the auxiliary identities $0+y=y$ and $\mathrm{succ}(x)+y=\mathrm{succ}(x+y)$. For the first of these, the case where y is 0 follows immediately by the definition of addition, and in the inductive step, we have

$$\begin{aligned}0+\mathrm{succ}(y) &= \mathrm{succ}(0+y)\\ &= \mathrm{succ}(y),\end{aligned}$$

where the second equation holds by the inductive hypothesis. For the second identity, the case where y is 0 again follows immediately from the definition of addition, and in the inductive step we have

$$\begin{aligned}\mathrm{succ}(x)+\mathrm{succ}(y) &= \mathrm{succ}(\mathrm{succ}(x)+y)\\ &= \mathrm{succ}(\mathrm{succ}(x+y))\\ &= \mathrm{succ}(x+\mathrm{succ}(y)).\end{aligned}$$

Here again the second equation follows from the inductive hypothesis. Now we can prove claim 3 by induction on x. For the base case, we already have $0+y=y$, and $y=y+0$ is one of the defining clauses for addition. In the inductive step, we have

$$\begin{aligned} x + \mathrm{succ}(y) &= \mathrm{succ}(x+y) \\ &= \mathrm{succ}(y+x) \\ &= \mathrm{succ}(y) + x. \end{aligned}$$

Claim 4 is easily proved using the defining equation for multiplication.

We prove claim 5 by induction on z. When $z = 0$ both sides are clearly equal to $x \cdot y$. In the inductive step, we have

$$\begin{aligned} x \cdot (y + \mathrm{succ}(z)) &= x \cdot \mathrm{succ}(y+z) \\ &= x \cdot (y+z) + x \\ &= (x \cdot y + x \cdot z) + x \\ &= x \cdot y + (x \cdot z + x) \\ &= x \cdot y + x \cdot \mathrm{succ}(z). \end{aligned}$$

Claims 6 and 7 are left as exercises. To prove claim 8, the cancellation law for addition, note that the injectivity of the successor function implies $x + \mathrm{succ}(z) = y + \mathrm{succ}(z) \to x + z = y + z$. Claim 8 follows by induction on z. □

Later we will obtain a cancellation rule for multiplication by a nonzero element, but first we need to develop some properties of the $\leq$ order on the natural numbers. We define $x \leq y$ to mean $x \dot{-} y = 0$. The following characterization has a more recursive character.

Proposition 9.2.2. PRA *proves the following:*

1. $0 \leq x$
2. $x \leq 0 \leftrightarrow x = 0$
3. $\mathrm{succ}(x) \leq \mathrm{succ}(y) \leftrightarrow x \leq y$.

All these follow easily from the properties of truncated subtraction established in Proposition 9.1.2. Proposition 9.2.2 is quite useful when used in conjunction with the principle of double induction expressed by Proposition 9.1.6.

Proposition 9.2.3. PRA *proves all of the following:*

1. $x \leq x$
2. $x \leq \mathrm{succ}(y) \leftrightarrow x \leq y \vee x = \mathrm{succ}(y)$
3. $x \leq y \wedge y \leq z \to x \leq z$
4. $x \leq y \wedge y \leq x \to x = y$
5. $x \leq y \vee \mathrm{succ}(y) \leq x$
6. $x \leq y \leftrightarrow x + z \leq y + z$
7. $x \leq y \to xz \leq yz$
8. $xz \leq yz \wedge z \neq 0 \to x \leq y$.

Proof Claim 1 follows from the fact that $x \dot{-} x = 0$. Claim 2 follows by the double induction principle: it is clearly true when x or y is 0, and when we get an equivalent statement when we replace x and y by $\mathrm{succ}(x)$ and $\mathrm{succ}(y)$ respectively.

To prove claim 3, use induction on z. When $z = 0$, $y \leq 0$ implies $y = 0$ and hence $x = 0$. In the inductive step, we have $x \leq y$ and, by claim 2, either $y \leq z$ or $y = \mathrm{succ}(z)$. In the first case, we have $x \le z$ by the inductive hypothesis and $x \le \mathrm{succ}(z)$ by claim 2, and in the second case we have $x \le \mathrm{succ}(z)$ by substitution.

Claims 4, 5, and 7 are easily proved using double induction. (In the base case for $y = 0$ in claim 5, use cases on x.) For claim 6, use induction on z.

For claim 8, suppose $xz \le yz$ but not $x \le y$. We need to show $z = 0$. By claim 5, we have $\mathrm{succ}(y) \le x$, and by claim 7, we have $\mathrm{succ}(y) \cdot z = yz + z \le xz$. Together with the hypothesis $xz \le yz$ and transitivity, we have $yz + z \le yz = yz + 0$. By claim 6, we have $z \le 0$, and with claim 2 of Proposition 9.2.2, we have z = 0. □

With Proposition 9.2.3, we can now prove the cancellation law for multiplication:

Proposition 9.2.4. PRA *proves* $xz = yz \wedge z \neq 0 \to x = y$.

Proof From $xz = yz$ we have both $xz \le yz$ and $yz \le xz$. From $z \neq 0$ and claim 8 of Proposition 9.2.3, we have $x \le y$ and $y \le x$, and hence $x = y$. □

The following proposition, which shows that PRA can prove facts relating truncated subtraction and the less-than-or-equal-to relation, is left as an exercise.

Proposition 9.2.5. PRA *proves the following:*

1. $x = x + y \dot{-} y$
2. $x \le y \to y \dot{-} x + x = y$
3. $x \dot{-} y = (x + z) \dot{-} (y + z)$
4. $x(y \dot{-} z) = xy \dot{-} xz$
5. $y \le x \to x \dot{-} y + z = x + z \dot{-} y$
6. $z \le y \to x \dot{-} (y \dot{-} z) = x + z \dot{-} y$
7. $x \dot{-} (y + z) = (x \dot{-} y) \dot{-} z$.

We define $x < y$ to mean $\mathrm{succ}(x) \le y$. With Proposition 9.1.6 it is not hard to show that PRA proves that this is equivalent to $x \le y \wedge x \neq y$. The following proposition, which shows that PRA proves all the expected properties of $<$, is left as an exercise as well.

Proposition 9.2.6. PRA *proves the following:*

1. $x \not< 0$
2. $0 < \mathrm{succ}(y)$
3. $\mathrm{succ}(x) < \mathrm{succ}(y) \leftrightarrow x < y$
4. $0 < y \vee 0 = y$
5. $x < \mathrm{succ}(y) \leftrightarrow x < y \vee x = y$
6. $x < y \wedge y < z \to x < z$
7. $x + z < y + z \leftrightarrow x < y$
8. $x < y \wedge w < z \to x + w < y + z$
9. $x \not< x$
10. $\neg(x < y \wedge y < x)$
11. $x < y \to \mathrm{succ}(x) < y \vee \mathrm{succ}(x) = y$
12. $x < y \vee x = y \vee y < x$
13. $z \neq 0 \to (x < y \to xz < yz)$
14. $z \neq 0 \to (xz = yz \to x = y)$
15. $x < y \wedge w < z \to xw < yz$.

Turning from arithmetic to basic logical operations, I will leave it to you to check that we can define the conditional function $\text{cond}(x, y)$ and logical functions like $\text{and}(x, y)$ and $\text{not}(x)$ and prove properties like the following:

- $\text{and}(x, y) \neq 0 \leftrightarrow x \neq 0 \wedge y \neq 0$
- $\text{not}(x) = 0 \leftrightarrow x \neq 0$
- $x \neq 0 \rightarrow \text{cond}(x, y, z) = y$
- $x = 0 \rightarrow \text{cond}(x, y, z) = z$.

Remember that we are treating any nonzero value as *true*. As in Section 8.2, we identify relations with their characteristic functions, so that reference to a relation $R(x_0, \ldots, x_{n-1})$ in the context of PRA should be interpreted as a reference to the equation $\chi_R(x_0, \ldots, x_{n-1}) = 1$, where χ_R is a function symbol. Using the logical operations, we have that any boolean combination of primitive recursive relations is primitive recursive. We have $\chi_{\leq}(x, y) = \chi_{\text{Zero}}(x \dot{-} y)$ and $\chi_{=}(x, y) = \text{and}(\chi_{\leq}(x, y), \chi_{\leq}(y, x))$, so equality is also provably equivalent to a primitive recursive relation in PRA.

Sections 8.2–8.4 have shown that bounded quantification provides a powerful means of defining new functions and relations, so it is important to know that we can reason about bounded quantifiers effectively in PRA. In first-order logic, $\forall x < y\, A(x)$ is defined to mean $\forall x\, (x < y \rightarrow A(x))$, and $\exists x < y\, A(x)$ is defined to mean $\exists x\, (x < y \wedge A(x))$. When we extend the language of PRA to first-order logic in Section 9.4, we will see that the primitive recursive relations $\forall x < y\, R(x, \vec{z})$ and $\exists x < y\, R(x, \vec{z})$ are provably equivalent to the first-order formulations. Proposition 9.2.8 below shows that the use of these quantifiers can be simulated in PRA. To prove it, it is helpful to establish some additional properties first, including properties of bounded search.

From the recursive definitions of $\forall x < y\, R(x, \vec{z})$ and $\exists x < y\, R(x, \vec{z})$, we can prove the following in PRA:

- $\forall x < 0\, R(x, \vec{z})$
- $\forall x < y + 1\, R(x, \vec{z}) \leftrightarrow \forall x < y\, R(x, \vec{z}) \wedge R(y, \vec{z})$
- $\neg \exists x < 0\, R(x, \vec{z})$
- $\exists x < y + 1\, R(x, \vec{z}) \leftrightarrow \exists x < y\, R(x, \vec{z}) \vee R(y, \vec{z})$.

By induction on y, we can therefore prove the following:

- $\forall x < y\, R(x, \vec{z}) \wedge w < y \rightarrow R(w, \vec{z})$
- $R(w, \vec{z}) \wedge w < y \rightarrow \exists x < y\, R(x, \vec{z})$
- $\forall x < y\, R(x, \vec{z}) \leftrightarrow \neg \exists x < y\, \neg R(x, \vec{z})$.

In Section 8.2, we defined the function $\min_{x<y} R(x, \vec{z})$, which returns the smallest x less than y satisfying $R(x, \vec{z})$ if there is one and y otherwise. It satisfies these defining equations:

- $\min_{x<0} R(x, \vec{z}) = 0$
- $\min_{x<y+1} R(x, \vec{z}) = \begin{cases} \min_{x<y} R(x, \vec{z}) & \text{if } R(\min_{x<y} R(x, \vec{z}), \vec{z}) \\ y + 1 & \text{otherwise.} \end{cases}$

The next proposition provides another characterization.

Proposition 9.2.7. *The following are provable in* PRA:

1. $R(w, \vec{z}) \rightarrow \min_{x<y} R(x, \vec{z}) \leq w$

2. $R(w, \vec{z}) \land w < y \to R(\min_{x<y} R(x, \vec{z}), \vec{z})$
3. $\forall x < y \neg R(x, \vec{z}) \to \min_{x<y} R(x, \vec{z}) = y$.

These can be proved simultaneously by induction on y.

Proposition 9.2.8. *The following rules are derivable in* PRA:

$$\dfrac{\begin{array}{c}\overline{x < t}\\ \vdots \\ R(x, \vec{s})\end{array}}{\forall x < t\, R(x, \vec{s})} \qquad \dfrac{\exists x < t\, R(x, \vec{s}) \quad \begin{array}{c}\overline{x < t,\, R(x, \vec{s})}\\ \vdots \\ C\end{array}}{C}$$

In both rules, we require that x is not free in t or any hypotheses other than the ones shown, and in the second rule, we require that x is not free in C.

Proof We will justify the first rule and leave the second as an exercise. By the properties of bounded quantification enumerated above, $\forall x < t\, R(x, \vec{s})$ is equivalent to $\neg \exists x < t \neg R(x, \vec{s})$, so it suffices to derive a contradiction from $\exists x < t \neg R(x, \vec{s})$. Write u for $\min_{x<t} \neg R(x, \vec{s})$. By Proposition 9.2.7, we have $\neg R(u, \vec{s})$ and $u < t$. But the premise of the rule corresponds to a proof of $x < t \to R(x, \vec{s})$, possibly from some hypotheses that do not mention x. Substituting u for x, we obtain a contradiction. □

We will sometimes need to know that bounded quantifiers respect equivalence, in the sense that given $\forall x < y\, (R(x, \vec{z}) \leftrightarrow S(x, \vec{z}))$, we also have $\forall x < y\, R(x, \vec{z}) \leftrightarrow \forall x < y\, S(x, \vec{z})$ and $\exists x < y\, R(x, \vec{z}) \leftrightarrow \exists x < y\, S(x, \vec{z})$. This follows easily from Proposition 9.2.8.

The closure of the primitive recursive relations under logical operations shows that every quantifier-free formula in the language of primitive recursive arithmetic is equivalent to a primitive recursive relation. In first-order settings, the closure of the set of atomic formulas under boolean operations and bounded quantification is called the set of *bounded* formulas. In the first-order version of PRA, we will have that every bounded formula in the language is provably equivalent to a primitive recursive relation. In both versions, we can therefore blur the distinction between bounded formulas and relations. For example, we can use induction on bounded formulas even in the quantifier-free version of PRA.

The *least-element principle* for the natural numbers says that if any number x has a property P, there is a smallest one with that property. The following proposition makes a version of that principle available to us in PRA.

Proposition 9.2.9. *For any bounded formula* $A(x)$, PRA *proves*

$$A(x) \to \exists y \leq x\, (A(y) \land \forall z < y \neg A(z)).$$

Proof If we take y to be $\min_{w<x+1} A(w)$, Proposition 9.2.7 implies that we can prove $y \leq x$, $A(y)$, and $\forall z < y \neg A(z)$. □

In Section 8.4, we were able to upgrade the ordinary principle of recursion to a principle of course-of-values recursion, in which the value of a function at a given input is computed with respect to the values at all prior inputs. The corresponding version of induction, known as *complete induction*, allows us to use all the previous values.

Proposition 9.2.10. *The following schema is derivable in* PRA:

$$\forall y \leq x\, (\forall z < y\, A(z) \to A(y)) \to A(x).$$

Proof This is equivalent to the contrapositive of the least-element principle for $\neg A(x)$, as expressed in Proposition 9.2.9. □

Proposition 9.2.10 shows that to prove that a formula A holds of an arbitrary x, it suffices to show that for any $y \leq x$, if the formula holds for values less than y, it holds at y as well. There is no need for a separate base case: when y is 0, the hypothesis that the formula holds for values less than y is trivially true.

Exercises

9.2.1. Prove claims 6 and 7 of Proposition 9.2.1.

9.2.2. Prove claims 2 and 4–7 of Proposition 9.2.3.

9.2.3. Prove Proposition 9.2.6.

9.2.4. Prove Proposition 9.2.5.

9.2.5. Complete the proof of Proposition 9.2.8.

9.3 Finite Sets and Sequences

We saw in Section 8.3 that primitive recursive mathematics starts to look more like general finitary mathematics once we interpret finite sets and sequences in a primitive recursive way. To that end, we used products of prime numbers to represent finite sets. In the same way, the formal theory PRA will start to look more like a formal theory of finitary mathematics once we have proved some basic properties of finite sets and sequences. If we are to stick with the same representation, we need to start by proving some properties of prime numbers. You can think of this as part of the process of bootstrapping PRA, though the elementary number theory we will develop along the way is a venerable part of mathematics in its own right.

Saying that a natural number $p \geq 2$ is prime amounts to saying that it has no proper divisors. A key property is that if p is any prime number and p divides a product xy, then p divides x or p divides y. This is used to show that a positive natural number can be written as a product of prime numbers in a unique way, a fact that was implicit in our coding of finite sets. Before arguing that the key property is derivable in PRA, it will help to review an informal proof that makes use of both positive and negative integers, and then think about how it can be adapted to PRA.

The *greatest common divisor* $\gcd(x, y)$ of two nonnegative integers x and y is defined to be the largest number that divides both. There are various reasonable ways of extending the function to negative integers as well, but here we can restrict our attention to nonnegative arguments. By convention, we will take $\gcd(0, 0)$ to be equal to 0. Since most of the things we want to prove about $\gcd(x, y)$ will hold trivially when either x or y is 0, we will generally restrict our attention to the case where they are strictly positive, in which case $\gcd(x, y)$ is less than or equal to both x and y.

It is an important fact that for every x and y there are integers a and b such that $ax + by = \gcd(x, y)$. In other words, we can always express the greatest common divisor of x and y as a linear combination of x and y. This clearly holds when x or y is equal to 0. To see that it holds in the case where x and y are positive, let d be the smallest positive number that can be expressed as $d = ax + by$. Then any number that divides both x and y divides d, so, in particular, $\gcd(x, y)$ divides d. To show that $d = \gcd(x, y)$, it is therefore sufficient to show that d divides both x and y.

By symmetry, it is enough to show that d divides x. Let q be the integer quotient of x divided by d and let r be the remainder, so that we have $x = qd + r$ with $0 \leq r < d$. We want to show that r is equal to 0. But we can write

$$r = x - qd = x - q(ax + by) = (1 - qa)x - bqy,$$

which is again a linear combination of x and y. Since $0 \leq r < d$, the definition of d as the smallest such linear combination that is strictly positive implies that we must have $r = 0$, as required.

To carry out the argument in PRA, one option is to interpret the integers in PRA, using a pairing function or even a pair of variables x and x' to represent the integer $x - x'$. We would then have to define arithmetic operations on the integers, derive their properties, and show how to transfer results back to the natural numbers. With a bit of cleverness, however, we can refine the proof above so that it avoids the use of negative numbers, and therefore leave the development of the theory of the integers for a rainy day.

Let us show (still informally) that, in fact, we can write $\gcd(x, y) = ax - by$ where a and b are *nonnegative*. The fact that we have switched the plus sign to a minus sign doesn't harm the claim that $\gcd(x, y)$ can be represented that way for some integers a and b; at issue is only whether we can take them to be nonnegative. But notice that we can always add a multiple of xy to both terms without altering the difference. In other words, for any u, we have $ax - by = (a + uy)x - (b + ux)y$. Clearly, we can always choose u large enough to make both coefficients positive.

To formalize the argument, it will also be helpful to have bounds on how large a and b need to be. I claim that we can write $\gcd(x, y) = ax - by$ with $b < x$ and $a \leq y$. To see this, start with any a and b satisfying the equation. If $b \geq x$, then we have $ax = by + \gcd(x, y) > xy$, which implies $a > y$. In that case, we can subtract x from b and y from a, and the equation still holds. We can iterate this until $b < x$, at which point, we have $ax = by + \gcd(x, y) < xy + x = (y + 1)x$, which implies $a < y + 1$ and hence $a \leq y$.

Now let us turn to PRA. We define $x \mid y$ to mean that there is a $z \leq y$ such that $xz = y$. If there is any z such that $xz = y$, then there is a $z \leq y$, because if $y = 0$, then $z = 0$ suffices, and if $y \neq 0$, then $x \neq 0$, and we have $z \leq y$ by the results of the previous section. We leave it as an exercise to show that PRA shows that the divisibility relation has the expected properties.

Proposition 9.3.1. PRA *proves all of the following:*

- $x \mid 0$ *for every* x.
- $0 \mid x$ *if and only if* $x = 0$.
- *If* $x \mid y$, *then* $x \mid yz$ *for any* z.
- *If* $x \mid y$ *and* $x \mid z$ *then* $x \mid y + z$.
- *If* $x \mid y + z$ *and* $x \mid y$, *then* $x \mid z$.
- *If* $x \mid y$ *and* $x \mid z$ *then* $x \mid y \dot{-} z$.
- *If* $x \mid y$, *and* $y \neq 0$, *then* $x \leq y$.

We can then define $\gcd(x, y)$ to be y if $x = 0$, and otherwise the greatest $d \leq x$ such that d divides both x and y. From the very definition, PRA can prove that $\gcd(x, y)$ divides both x and y, and hence is less than both if they are positive.

We also made use of the quotient-remainder theorem in the informal argument, so let us record the formal properties we need.

Proposition 9.3.2. *There are functions* $\mathrm{div}(x, y)$ *and* $\mathrm{mod}(x, y)$ *such that* PRA *proves that for every x and y we have*

$$x = \mathrm{div}(x, y) \cdot y + \mathrm{mod}(x, y),$$

and if $y \neq 0$, *we also have* $\mathrm{mod}(x, y) < y$. *We can further assume* $\mathrm{div}(x, 0) = 0$, *and, in that case, for every x and y,* $\mathrm{div}(x, y)$ *and* $\mathrm{mod}(x, y)$ *are unique; in other words, if q and r satisfy the same properties, then* $q = \mathrm{div}(x, y)$ *and* $r = \mathrm{mod}(x, y)$.

Exercise 8.2.4 asked you to define $\mathrm{div}(x, y)$ and $\mathrm{mod}(x, y)$, and Exercise 9.3.4 asks you to prove (or at least sketch a proof) of Proposition 9.3.2.

Proposition 9.3.3. PRA *proves the following. Assume* $x \geq y$. *Then there are numbers* $a \leq y$ *and* $b \leq x$ *such that* $ax \dot{-} by = \gcd(x, y)$.

Proof We argue in PRA. If $y = 0$ or $x = y$ we can take $a = 1$ and $b = 0$. So, henceforth we assume $0 < y < x$.

By the least-element principle, we can let d be the smallest positive number such that for some $a \leq y$ and $b \leq x$, $ax \dot{-} by = d$. Since we can take $a = 1$ and $b = 0$, we know that $d \leq x$. We let $a \leq y$ be the least element such that $ax \dot{-} by = d$ for some $b \leq x$, and we let $b \leq x$ be the least element that works for that value of a. Since $\gcd(x, y)$ divides both x and y, it divides d, and hence $\gcd(x, y) \leq d$. It suffices to show that d divides both x and y, because then we have that d divides $\gcd(x, y)$ by the definition of gcd, which implies $d \leq \gcd(x, y)$ and hence $d = \gcd(x, y)$.

I will only show that d divides x, since the proof that d divides y is similar. We can write $x = qd + r$, where q is the quotient of x and q and where the remainder r satisfies $0 \leq r < d \leq x$. It suffices to show that r can be written in the form $a'x \dot{-} b'y$ for some $a' \leq y$ and $b' \leq x$, since the definition of d then implies that $r = 0$.

Since $x = dq + r$, we have $r = x \dot{-} qd = x \dot{-} q(ax \dot{-} by) = x \dot{-} (qax \dot{-} qby)$. We now employ the trick of adding a multiple uxy of xy to both sides. If we take u to be the maximum of qa and qb, then since x and y are greater than or equal to 1, we have $uxy \geq ux \geq qax$ and $uxy \geq uy \geq qby$. We can then use the claims of Proposition 9.1.2 to calculate as follows:

$$\begin{aligned}
x \dot{-} (qax \dot{-} qby) &= (x + uxy) \dot{-} ((qax \dot{-} qby) + uxy) \\
&= (x + uxy) \dot{-} ((qax + uxy) \dot{-} qby) \\
&= (x + uxy) \dot{-} ((uxy + qax) \dot{-} qby) \\
&= (x + uxy) \dot{-} ((uxy \dot{-} qby) + qax) \\
&= (x + uxy) \dot{-} (qax + (uxy \dot{-} qby)) \\
&= (x + uxy \dot{-} qax) \dot{-} (uxy \dot{-} qby) \\
&= (1 + uy \dot{-} qa)x \dot{-} (ux \dot{-} qb)y.
\end{aligned}$$

The net effect is that we have written $r = a''x \dot{-} b''y$ for a'' and b'' that we can describe explicitly from x, y, a, and b.

We now use the other trick described above to replace a'' and b'' by values $a' \leq y$ and $b' \leq x$ so that we have $r = a'x \dot{-} b'y$. By the choice of d, this will imply $r = 0$, as required. Let a' be the least value less than or equal to a'' such that $r = a'x \dot{-} b'y$ for some $b' \leq b''$, and let b' be the least value less than or equal to b'' that has that satisfies the equation for that choice of a'. If $b' \geq x$, then $b'y \geq xy$, and $a'x = b'y + r \geq xy$, which implies $a' \geq y$. In that case, we can subtract y from a' and x from b to get a smaller pair $\bar{a}, \bar{b}$ satisfying $r = \bar{a}x \dot{-} \bar{b}y$, contrary to the definition of a' and b'. So $b' < x$, and $a'x = b'y + r < xy + x$, since we know $r < x$. This implies $a' < y + 1$, and hence $a' \leq y$, as required. □

This gives us the key property of prime numbers.

Theorem 9.3.4. PRA *proves that if* $z \mid xy$ *and* $\gcd(z, x) = 1$ *then* $z \mid y$.

Proof Suppose $z \mid xy$ and $\gcd(z, x) = 1$. By the previous lemma, depending on whether z is greater than or less than x, there are a and b satisfying $az \dot{-} bx = 1$ or $ax \dot{-} bz = 1$. Multiplying both sides by y, we have $azy \dot{-} bxy = y$ or $axy \dot{-} bzy = y$. In each case, since $z \mid xy$, z divides the left-hand side, so z divides y. □

Corollary 9.3.5. PRA *proves that if* p *is prime and* $p \mid xy$, *then* $p \mid x$ *or* $p \mid y$.

Proof Since $\gcd(p, x) \mid p$, we have $\gcd(p, x) = p$ or $\gcd(p, x) = 1$. In the first case, we have $p \mid x$, and in the second case, by the preceding theorem, we have $p \mid y$. □

Using Corollary 9.3.5, it is straightforward to prove the unique factorization theorem. It is also straightforward to prove that every natural number greater than or equal to two has a prime factor, and that for every n, any prime factor of $n! + 1$ is a prime number greater than n. This justifies the claim that the function $x \mapsto p_x$ defined in Section 8.2 enumerates the prime numbers.

In PRA, we can now define $\{x < y \mid P(x, \vec{z})\}$ as in Section 8.3, define $x \in s$ to be $p_x \mid s$, and define $\mathrm{bound}(s) = s$.

Proposition 9.3.6. PRA *proves the following:*

- *For every* x, y, *and* $\vec{z}$, $x \in \{x < y \mid P(x, \vec{z})\}$ *if and only if* $x < y$ *and* $P(x, \vec{z})$.
- *For every* x *and* s, *if* $x \in s$, *then* $x < \mathrm{bound}(s)$.

We can go on to define the operations in Proposition 8.3.4 and show, in PRA, that they have the right properties. Thanks to Propositions 9.2.9 and 9.2.10, we can freely make use of the least-element principle and the principle of complete induction for bounded formulas. An exercise below asks you to show that PRA proves the properties of the pairing function $J(x, y)$ and the projection functions $K(z)$ and $L(z)$ given by Lemma 8.3.1 and Proposition 8.3.2. We can then go on to define finite sequences and show that they have the expected properties. With the more flexible forms of induction described in the last section, this gives rise to a robust theory of finite sets, finite sequences, and other finite objects.

We leave it as an exercise to show that PRA can support the more flexible forms of recursion described in Section 8.4. As an analog of Theorem 8.4.1, we can justify the following induction principle:

Theorem 9.3.7. *Let $m(x)$ and $k(x)$ be primitive recursive functions, and let $A(x)$ be any formula in the language of* PRA. *Then the following is a derived rule of* PRA:

$$\frac{(m(k(x)) < m(x) \to A(k(x))) \to A(x)}{A(t)}$$

If the nested implication in the hypothesis is confusing, it might be helpful to express the statement in the equivalent form $m(k(x)) \not< m(x) \vee A(k(x)) \to A(x)$. The theorem says that to show that A holds of every t, it is enough to show that, for every x, $A(x)$ follows from either $A(k(x))$ or the fact that the argument in this inductive step does not decrease.

Proof It suffices to prove $m(y) < x \to A(y)$ for arbitrary x and y, because we can then substitute t for y and $m(t) + 1$ for x. By Proposition 9.1.3, it suffices to prove $m(y) < 0 \to A(y)$ and

$$(m(k(y)) < x \to A(k(y))) \to (m(y) < \mathrm{succ}(x) \to A(y)).$$

The first is immediate, so let us sketch a natural deduction proof of the second. Assume $m(k(y)) < x \to A(k(y))$ and $m(y) < \mathrm{succ}(x)$. We want $A(y)$. Substituting y for x in the hypothesis of the rule, it suffices to show $m(k(y)) < m(y) \to A(k(y))$. So assume $m(k(y)) < m(y)$. With $m(y) < \mathrm{succ}(x)$, we have $m(k(y)) < x$, and hence $A(k(y))$, as required. □

The exercises below ask you to use this to show that in PRA we have the following analogue to the recursion principle given by Theorem 8.5.2:

Theorem 9.3.8. *Let $u(x)$ and $m(x)$ be any primitive recursive functions, and let $A(x)$ be any formula. Then the following is a derived rule of* PRA:

$$\frac{\forall x' \in u(x)\ (m(x') < m(x)) \qquad \forall x' \in u(x)\ A(x') \to A(x)}{A(t)}$$

In words, to show that $A(t)$ holds for arbitrary t, show that for every x, $A(x)$ follows from $A(x')$ for the finite set of x' in $u(x)$, all of which are smaller than x according the measure m.

Exercises

9.3.1. Show that PRA proves Lemma 8.3.1 and Proposition 8.3.2.

9.3.2. Define a representation of integers as natural numbers, show that addition, subtraction, multiplication, and the order relation are primitive recursive, and show that PRA proves that with these the integers form an ordered ring.

9.3.3. Prove Proposition 9.3.1.

9.3.4. Prove (or at least sketch a proof) of Proposition 9.3.2.

9.3.5. Identify the parts of Proposition 9.2.5 used to justify the calculation in the proof of Proposition 9.3.3.

9.3.6. Represent finite sequences as described in Section 8.3, namely, as a pair (n, α) where n is the length of the sequence and α is a finite set of ordered pairs (i, s_i) mapping each $i < n$ to the ith element of the sequence. Show that, with such a representation, it is possible to define length, elt, and append such that PRA can prove that they have the following properties, where $(x)_i$ abbreviates $\text{elt}(x, i)$:

a. $\text{length}(0) = 0$
b. $\text{length}(\text{append}(x, y)) = \text{length}(\mathrm{x}) + 1$
c. $i \geq \text{length}(\mathrm{x}) \to (x)_i = 0$
d. $i < \text{length}(\mathrm{x}) \to (\text{append}(x, y))_i = (x)_i$
e. $(\text{append}(x, y))_{\text{length}(x)} = y$.

9.3.7. With this representation, define the function $\text{concat}(s, t)$ that concatenates two sequences, and show that PRA proves that it has some of the expected properties.

9.3.8. Show that PRA can justify the more general forms of recursion described in Sections 8.4 and 8.5.

9.3.9. Prove Theorem 9.3.8.

9.4 First-Order PRA

In this section and the next, we consider two variations on primitive recursive arithmetic. On the surface, the first is more expressive and the second is less expressive than the version we have already seen. To distinguish the three versions, I will call the one presented in Section 9.1 *quantifier-free* PRA, the one presented in this section *first-order* PRA, and the one presented in the next section *equational* PRA. We will see that, in an appropriate sense, all three have the same logical strength.

First-order PRA is a theory in the language of first-order logic. In that setting, we can dispense with the substitution rule and replace the quantifier-free axioms with their universal closures. We can, moreover, express the principle of induction as follows:

$$A(0) \wedge \forall x\,(A(x) \to A(\text{succ}(x))) \to \forall x\, A(x),$$

where $A(x)$ is any quantifier-free formula. We will show that first-order PRA is conservative over quantifier-free PRA, which is to say, whenever first-order PRA proves a quantifier-free formula, quantifier-free PRA proves it as well. In fact, the following theorem says something stronger:

Theorem 9.4.1. *Suppose classical first-order* PRA *proves*

$$\forall \vec{x}\, \exists y_0, \ldots, y_{k-1}\, A(\vec{x}, y_0, \ldots, y_{k-1}),$$

where A is a bounded formula. Then there is a sequence of function symbols $f_0, \ldots, f_{k-1}$ such that quantifier-free PRA *proves*

$$A(\vec{x}, f_0(\vec{x}), \ldots, f_{k-1}(\vec{x})).$$

We will see in Chapter 10 that sentences of the kind appearing in the hypothesis are called Π_2 *sentences*. Theorem 9.4.1 says that, for theorems of that form, it is safe to blur the distinction between the first-order and the quantifier-free versions of PRA.

We will use Herbrand's theorem to prove Theorem 9.4.1. To that end, we need the following:

Lemma 9.4.2. *First-order* PRA *has a universal axiomatization.*

In other words, there is a set Γ of universal sentences whose consequences in first-order logic are exactly the theorems of first-order PRA.

Proof All the axioms of PRA are already universal, except for the induction axioms. It therefore suffices to replace these axioms with universal ones that have the same consequences in the first-order theory.

Given $A(x)$, we can define the function

$$f(x) = \min_{y<x} (A(y) \wedge \neg A(y+1)).$$

Replace the induction axiom for A with the universal closure of the following:

$$\forall x\, (A(0) \wedge (A(f(x)) \to A(f(x)+1)) \to A(x)).$$

This sentence is universal. It is easy to see that, in first-order logic, this implies the induction axiom: if we assume $A(0) \wedge \forall x\, (A(x) \to A(x+1))$, then for any x, the antecedent of the implication holds, and hence also $A(x)$. Hence it is suffices to show that the new induction axiom is, in turn, provable in the original formulation of first-order PRA.

In first-order logic, the new induction axiom is equivalent to

$$\forall x\, (A(0) \wedge \neg A(x) \to A(f(x)) \wedge \neg A(f(x)+1)).$$

Prove this by induction on x in first-order PRA. When $x = 0$, the implication is trivially true because the antecedent is false. Now suppose it is true for x. We need to show

$$A(0) \wedge \neg A(x+1) \to A(f(x+1)) \wedge \neg A(f(x+1)+1).$$

Arguing in first-order PRA, assume $A(0) \wedge \neg A(x+1)$. There are two cases, depending on whether $A(x)$ holds. If $\neg A(x)$, then we have $A(0) \wedge \neg A(x)$, and hence $A(f(x)) \wedge \neg A(f(x)+1)$ by the inductive hypothesis; then by the definition of f, we have $f(x+1) = f(x)$. Otherwise, we have $A(x) \wedge \neg A(x+1)$. If $A(f(x)) \wedge \neg A(f(x)+1)$, then again by the definition of f, $f(x+1) = f(x)$ and we are done. Otherwise, by the definition of f, $f(x+1) = x$, and again we are done. □

The argument we have just given also shows that quantifier-free PRA can prove the new induction axiom when it is expressed without the universal quantifiers in front. Conversely, it is not hard to show that from the axiom, we can derive the usual induction rule of PRA using quantifier-free logic. So this provides us with another formulation of quantifier-free PRA as well.

Proof of Theorem 9.4.1 Suppose PRA proves the given formula. By Corollary 7.1.3, there are sequences of terms $t_{i,j}$ such that

$$\bigvee_{i<n} A(\vec{x}, t_{i,0}, \ldots, t_{i,k-1})$$

has a quantifier-free proof from instances of the universal axioms of first-order PRA, and hence in quantifier-free PRA. Use definition by cases to define functions $f_0, \ldots, f_{k-1}$ by

$$f_j(\vec{x}) = \begin{cases} t_{0,j} & \text{if } A(\vec{x}, t_{0,0}, \ldots, t_{0,k-1}) \\ t_{1,j} & \text{else if } A(\vec{x}, t_{1,0}, \ldots, t_{1,k-1}) \\ \vdots & \vdots \\ t_{n-1,j} & \text{otherwise.} \end{cases}$$

In other words, the sequence of functions $f_0, \ldots, f_{k-1}$ just picks out the first sequence $t_{i,0}, \ldots, t_{i,k-1}$ that works. □

Exercise

9.4.1. Give a model-theoretic proof of the fact that PRA is universally axiomatizable. (Hint: by Theorem 5.6.7, it suffices to show that the class of models of PRA is closed under submodels.)

9.5 Equational PRA

We now move in the other direction and consider a formulation of primitive recursive arithmetic that, a priori, is less expressive than the one presented in Section 9.1. Specifically, we will consider PRA as a purely equational theory. The underlying language remains the same: we include a function symbol for each primitive recursive definition. But now we dispense with all the propositional connectives in addition to the quantifiers. The only statements we can make in the system are equations $u = v$ between terms in the language. We start with defining equations for the primitive recursive functions and we use rules to derive new equations.

More precisely, we start with all substitution instances of the defining equations for the primitive recursive functions and add the rules of equational logic, as described in Section 4.3. We replace the induction principle of quantifier-free PRA with the following, which says that if two functions satisfy the same primitive recursive definition, they are always equal:

From
- $f_0(0, \vec{z}) = f_1(0, \vec{z})$
- $f_0(\mathrm{succ}(x), \vec{z}) = h(x, f_0(x, \vec{z}), \vec{z})$
- $f_1(\mathrm{succ}(x), \vec{z}) = h(x, f_1(x, \vec{z}), \vec{z})$

conclude $f_0(t, \vec{z}) = f_1(t, \vec{z})$.

Here f_0, f_1, and h are arbitrary primitive recursive functions and t is any term. Since for any term $u(x, \vec{z})$ there is a function symbol $f(x, \vec{z})$ such that $f(x, \vec{z}) = u(x, \vec{z})$ is provable from the defining equations for the primitive recursive functions, we can use the rule above with f_0 and f_1 replaced by arbitrary terms u_0 and u_1. In particular, the rule is still valid if we replace $\vec{z}$ by arbitrary terms. Since our axioms and rules are closed under substitution, we can even dispense with the substitution rule in equational logic.

To see how the new induction principle serves as a substitute for the one in quantifier-free PRA, suppose we define addition as in Section 8.1. To prove $0 + x = x$, it suffices to show that both sides satisfy the same defining equations:

- $0 + 0 = 0$
- $0 + \mathrm{succ}(x) = \mathrm{succ}(0 + x)$
- $\mathrm{succ}(x) = \mathrm{succ}(x)$.

The first two equations hold by the defining equations for addition and the third holds by reflexivity. Similarly, to show $x + y = y + x$, it suffices to show

- $0 + x = x + 0$
- $\mathrm{succ}(x) + y = \mathrm{succ}(x + y)$
- $y + \mathrm{succ}(x) = \mathrm{succ}(y + x)$.

The first follows from the previous identity and the defining equation for addition, and the third follows from the defining equation for addition. To prove the second, using the induction principle for y, it suffices to show

- $\mathrm{succ}(x) + 0 = \mathrm{succ}(x + 0)$
- $\mathrm{succ}(x) + \mathrm{succ}(y) = \mathrm{succ}(\mathrm{succ}(x) + y)$
- $\mathrm{succ}(x + \mathrm{succ}(y)) = \mathrm{succ}(\mathrm{succ}(x + y))$.

These all follow from the defining equations for addition.

Our goal is to prove the following:

Theorem 9.5.1. *If quantifier-free* PRA *proves* $s = t$*, then equational* PRA *proves* $s = t$ *as well.*

This is the precise sense in which quantifier-free PRA and equational PRA have the same strength. We already know that every quantifier-free formula $A(\vec{x})$ has a primitive recursive characteristic function $\chi_A(\vec{x})$ such that quantifier-free PRA proves $A(\vec{x}) \leftrightarrow \chi_A(\vec{x}) = 1$. Theorem 9.5.1 implies that whenever quantifier-free PRA proves $A(\vec{x})$, equational PRA proves $\chi_A(\vec{x}) = 1$. We also know that every bounded formula in the language of first-order PRA is provably equivalent to a primitive recursive relation, so whenever first-order PRA proves a bounded formula $A(\vec{x})$, equational PRA proves $\chi_A(\vec{x}) = 1$ as well. In other words, first-order, quantifier-free, and equational PRA prove the same bounded statements modulo those identifications. In fact, they prove the same Π_2 statements in the sense of Theorem 9.4.1.

To prove Theorem 9.5.1, we will show that whenever quantifier-free PRA proves a formula $A(\vec{x})$, equational PRA proves $\chi_A(\vec{x}) = 1$. Since each equation $s = t$ is translated to the equation $\chi_=(s, t) = 1$, we also need to show that equational PRA can recover $s = t$ from the latter. In fact, we will show that equational PRA can derive each from the other, which tells us right away that it can prove the translations of all the equational axioms of quantifier-free PRA. As a result, we only have to worry about propositional logic, equality, and induction.

We will carry out the proof of Theorem 9.5.1 in three steps:

1. First, we will deal with propositional logic. We will show that equational PRA proves all instances of tautologies and that it simulates modus ponens.
2. Then we will deal with equality. We will show that equational PRA can simulate the axioms and rules for equality, and that it can prove $s = t$ from $\chi_=(s, t) = 1$ and vice versa.
3. Finally, we will show that equational PRA simulates the induction rule of quantifier-free PRA.

Along the way, we will need to know that equational PRA can prove some basic identities, including $\max(x, y) = \max(y, x)$. Surprisingly, establishing the latter will be most difficult thing we will do. We will make use of the following principle of proof by cases:

$$\frac{t_0(0) = t_1(0) \qquad t_0(\mathrm{succ}(x)) = t_1(\mathrm{succ}(x))}{t_0(u) = t_1(u)}$$

Here t_0, t_1 and u are arbitrary terms, possibly with variables other than the one shown. This becomes an instance of the first induction rule in equational PRA if we add $t_1(\mathrm{succ}(x)) = t_1(\mathrm{succ}(x))$ as a third premise.

We define $\mathrm{pred}(x)$, $x \dot{-} y$, $\max(x, y)$, and χ_{Zero} exactly as in Section 8.2. I leave it as an exercise for you to check that the identities in Proposition 9.1.2 and claims 1–7 of Proposition 9.2.1 are provable in equational PRA using the first induction rule. Remember that we define $\max(x, y)$ with the equation $\max(x, y) = x + (y \dot{-} x)$. From that definition, we have

- $\max(0, y) = y$
- $\max(x, 0) = x$
- $\max(\mathrm{succ}(x), \mathrm{succ}(y)) = \mathrm{succ}(\max(x, y))$.

We now show that the identity $\max(x, y) = \max(y, x)$ follows from this.

Lemma 9.5.2. *Equational* PRA *proves* $\max(x, y) = \max(y, x)$.

Proof First, we prove

$$\max(x, \mathrm{succ}(y)) = \mathrm{succ}(\max(\mathrm{pred}(x), y))$$

by cases on x: when x is 0, both sides are equal to $\mathrm{succ}(y)$, and if we replace x by $\mathrm{succ}(x)$, we have

$$\begin{aligned} \max(\mathrm{succ}(x), \mathrm{succ}(y)) &= \mathrm{succ}(\max(x, y)) \\ &= \mathrm{succ}(\max(\mathrm{pred}(\mathrm{succ}(x)), y)), \end{aligned}$$

as required. Define the function $\mathrm{nz}(x)$ by $\mathrm{nz}(0) = 0$, $\mathrm{nz}(x + 1) = 1$, so that $\mathrm{nz}(x)$ is the characteristic function for the property of being nonzero. Proof by cases yields the identity $x = \mathrm{pred}(x) + \mathrm{nz}(x)$. We can now prove the identity

$$\max(x, y) = \max(\mathrm{pred}(x), \mathrm{pred}(y)) + \mathrm{nz}(x + y)$$

by cases on y. When y is 0, the left-hand size is x, and the right-hand side reduces to $\mathrm{pred}(x) + \mathrm{nz}(x)$, which is equal to x as well. Replacing y with $\mathrm{succ}(y)$, the right-hand side reduces as follows:

$$\begin{aligned} \max(\mathrm{pred}(x), \mathrm{pred}(\mathrm{succ}(y))) &+ \mathrm{nz}(x + \mathrm{succ}(y)) \\ &= \max(\mathrm{pred}(x), y) + \mathrm{nz}(\mathrm{succ}(x + y)) \\ &= \max(\mathrm{pred}(x), y) + 1 \\ &= \mathrm{succ}(\max(\mathrm{pred}(x), y)), \end{aligned}$$

which is equal to $\max(x, \mathrm{succ}(y))$ by the previous identity. Now define $f(x, y, z)$ by

$$f(x, y, 0) = 0, \quad f(x, y, z + 1) = f(x, y, z) + \mathrm{nz}((x \dot{-} z) + (y \dot{-} z)).$$

Using the induction rule, we have

$$\max(x, y) = \max(x \dot{-} z, y \dot{-} z) + f(x, y, z)$$

since they are equal at 0 and both sides are unchanged when we replace z by $z + 1$. Plugging in x for z, we have

$$\max(x, y) = \max(0, y \dot{-} x) + f(x, y, x) = y \dot{-} x + f(x, y, x).$$

Switching x and y in the equation above yields

$$\max(y, x) = \max(y \dot{-} z, x \dot{-} z) + f(y, x, z)$$

and again plugging in x for z yields $\max(y, x) = y \dot{-} x + f(y, x, x)$. But since addition is commutative, we can use the induction rule on z to show $f(x, y, z) = f(y, x, z)$. Thus we have $\max(x, y) = \max(y, x)$. □

To deal with propositional logic, we define the boolean operations $\mathrm{and}(x, y)$, $\mathrm{or}(x, y)$, $\mathrm{not}(x)$, and $\mathrm{implies}(x, y)$ as in Section 8.2. We interpret 1 and 0 as true and false, respectively. Using proof by cases iteratively, we have the following:

Lemma 9.5.3. *Suppose $b_0(\vec{x})$ and $b_1(\vec{x})$ are any two terms such that equational* PRA *proves $b_0(\vec{t}) = b_1(\vec{t})$ for every tuple $\vec{t}$ of terms such that each t_i is either* 0 *or* $\mathrm{succ}(x_i)$. *Then equational* PRA *proves $b_0(\vec{x}) = b_1(\vec{x})$.*

Say that $b(\vec{x})$ is a *boolean expression* if it is built up from the variables $\vec{x}$, 0, and 1 using the boolean operations. Any boolean expression other than a variable evaluates to either 0 or 1 when each input argument is set to either 0 or an expression of the form $\mathrm{succ}(t)$. As a result, we have:

Proposition 9.5.4. *Suppose the boolean expression $b(\vec{x})$ is a tautology. Then equational* PRA *proves $b(\vec{x}) = 1$.*

Equational PRA also proves $\mathrm{implies}(1, y) = \mathrm{nz}(y)$ by cases, so the following is a derived rule:

$$\frac{s = 1 \qquad \mathrm{implies}(s, t) = 1}{\mathrm{nz}(t) = 1}$$

By cases, equational PRA proves $\mathrm{nz}(\mathrm{and}(x, y)) = \mathrm{and}(x, y)$ and similarly for the other boolean connectives. So equational PRA proves $\mathrm{nz}(t) = t$ for any boolean expression t, unless t is a variable. But formulas in the language of quantifier-free PRA are boolean combinations of equations, and we will see that when t is of the form $\chi_=(s, t)$, equational PRA also proves $\mathrm{nz}(t) = t$. As a result, the rule above is just the translation of modus ponens.

To sum up, if $A(\vec{x})$ is any instance of a tautology in the language of quantifier-free PRA, equational PRA proves $\chi_A(\vec{x}) = 1$, and if equational PRA proves $\chi_{A \to B}(\vec{x}) = 1$ and $\chi_A(\vec{x}) = 1$, it proves $\chi_B(\vec{x}) = 1$. This takes care of propositional logic.

We next turn to the second task enumerated above. To interpret equality, we define $\chi_{\leq}(x, y)$ to be $\chi_{\mathrm{Zero}}(y \dot{-} x)$ and we define $\chi_=(x, y)$ to be $\mathrm{and}(\chi_{\leq}(x, y), \chi_{\leq}(y, x))$. Since this is an instance of a boolean expression, we have $\mathrm{nz}(\chi_=(x, y)) = \chi_=(x, y)$, as desired. By cases, we also have $\mathrm{and}(\chi_{\mathrm{Zero}}(u), \chi_{\mathrm{Zero}}(v)) = \chi_{\mathrm{Zero}}(u) \cdot \chi_{\mathrm{Zero}}(v)$, and hence $\chi_=(x, y) = \chi_{\leq}(x, y) \cdot \chi_{\leq}(y, x)$. We will use this identity in the proof of Lemma 9.5.5.

From $x \dot{-} x = 0$ we have $\chi_=(x, x) = 1$, which is to say, equational PRA proves the translation of the reflexivity axiom. From the commutativity of multiplication, it also proves $\chi_=(x, y) = \chi_=(y, x)$. The following is a key lemma:

Lemma 9.5.5. *Equational* PRA *proves the identity $\chi_=(x, y) \cdot t(x) = \chi_=(x, y) \cdot t(y)$ for any term $t(x)$.*

Proof By cases on y, PRA proves $\chi_{\mathrm{Zero}}(y) \cdot t(x + y) = \chi_{\mathrm{Zero}}(y) \cdot t(x)$. Substituting $y \dot{-} x$ for y, we have

$$\chi_{\leq}(y, x) \cdot t(\max(x, y)) = \chi_{\leq}(y, x) \cdot t(x).$$

Multiplying both sides by $\chi_{\leq}(x, y)$ and using the associativity of multiplication, we have

$$\chi_{=}(x, y) \cdot t(\max(x, y)) = \chi_{=}(x, y) \cdot t(x).$$

Switching x and y we have

$$\chi_{=}(y, x) \cdot t(\max(y, x)) = \chi_{=}(y, x) \cdot t(y).$$

Since $\chi_{=}(y, x) = \chi_{=}(x, y)$ and $\max(y, x) = \max(x, y)$, from these two equations we obtain $\chi_{=}(x, y) \cdot t(x) = \chi_{=}(x, y) \cdot t(y)$, as required. □

If equational PRA proves $s = t$, then it proves $\chi_{=}(s, t) = \chi_{=}(s, s) = 1$. Conversely, taking $t(x)$ to be x in Lemma 9.5.5, we have that if equational PRA proves $\chi_{=}(s, t) = 1$, it proves $s = t$. In other words, $s = t$ and $\chi_{=}(s, t)$ are inter-derivable.

To show that the equality axioms and rules of quantifier-free PRA are preserved by the translation, it suffices to show that equational PRA proves the translation of $x = y \land C(x) \to C(y)$ for every atomic formula C. The atomic formulas of quantifier-free PRA are equations, so it suffices to show that equational PRA proves

$$\text{implies}(\text{and}(\chi_{=}(x, y), \chi_{=}(s(x), t(x))), \chi_{=}(s(y), t(y))) = 1,$$

where s and t may have free variables other than the one shown. As in the proof of Lemma 9.5.5, we can replace conjunction by multiplication, and using Lemma 9.5.5 itself we have

$$\begin{aligned}\text{and}(\chi_{=}(x, y), \chi_{=}(s(x), t(x))) &= \chi_{=}(x, y) \cdot \chi_{=}(s(x), t(x)) \\ &= \chi_{=}(x, y) \cdot \chi_{=}(s(y), t(y)) \\ &= \text{and}(\chi_{=}(x, y), \chi_{=}(s(y), t(y))).\end{aligned}$$

So it suffices to show

$$\text{implies}(\text{and}(\chi_{=}(x, y), \chi_{=}(s(y), t(y))), \chi_{=}(s(y), t(y))) = 1.$$

Since equational PRA proves any instance of $\text{implies}(\text{and}(x, y), y) = 1$, we are done.

From the foregoing, we have that for each equational axiom $s = t$ of quantifier-free PRA, equational PRA proves $\chi_{=}(s, t) = 1$. From the definition of $\chi_{=}$, it also proves $\chi_{=}(\text{succ}(x), 0) = 0$, and hence $\text{implies}(\chi_{=}(\text{succ}(x), 0), 0) = 1$. Thus, we are left only with the third task on our list, namely, showing that induction is preserved. It suffices to show that equational PRA can simulate the induction rule from quantifier-free PRA, namely, from $A(0, \vec{z})$ and $A(x, \vec{z}) \to A(\text{succ}(x), \vec{z})$ conclude $A(t, \vec{z})$, for any quantifier-free formula A. To that end, it suffices to show that in equational PRA we can derive the rule

$$\frac{p(0) = 1 \qquad \text{implies}(p(x), p(x + 1)) = 1}{p(t) = 1}$$

for any term $p(x)$, including, in particular, terms of the form $\chi_A(x, \vec{z})$.

In equational PRA, if we can prove $g(x) = g(x+1)$, we can conclude $g(t) = g(0)$. To see this, note that if we set $h(x)$ to be the constant function $h(x) = g(0)$, the hypothesis implies that we have:

- $g(0) = h(0)$
- $g(x+1) = g(x)$
- $h(x+1) = h(x)$.

We can then apply the induction rule. As a result, we have the following rule:

$$\frac{g(0) = 1 \qquad g(x) = g(x+1)}{g(t) = 1}$$

The problem is that we have $\text{implies}(p(x), p(x+1)) = 1$ instead of $p(x) = p(x+1)$. The solution is to set $g(x)$ to be the characteristic function of $\forall y < x\,(p(y) = 1)$ and use that instead. In other words, define $g(x)$ by

$$g(0) = 1, \quad g(x+1) = \text{and}(g(x), p(x)).$$

Since $p(x) = 1$ follows from $g(x+1) = 1$ using propositional logic, it suffices to show $g(x) = g(x+1)$ and then use the previous rule. We use cases on x. We have

$$g(1) = \text{and}(g(0), p(0)) = \text{and}(1, 1) = 1 = g(0),$$

and

$$\begin{aligned} g(x+2) &= \text{and}(g(x+1), p(x+1)) \\ &= \text{and}(\text{and}(g(x), p(x)), p(x+1)) \\ &= \text{and}(\text{and}(g(x), p(x)), \text{implies}(p(x), p(x+1))) \\ &= \text{and}(\text{and}(g(x), p(x)), 1) \\ &= \text{and}(g(x), p(x)) \\ &= g(x+1). \end{aligned}$$

In the last calculation, the equality between the second and third lines follows from the fact that they are instances of equivalent boolean expressions, and the next equation is a consequence of the hypothesis, $\text{implies}(p(x), p(x+1)) = 1$. This completes the proof of Theorem 9.5.1.

Bibliographical Notes

General references Goodstein (1957) is a thorough and important early reference for primitive recursive arithmetic. The reduction to the equational version of PRA in Section 9.5 draws heavily from there. See also Troelstra and van Dalen (1988).

Finitistic mathematics PRA has been proposed as an analysis of the notion of *finitistic* mathematics. See the bibliographical notes to Chapter 12.

10

First-Order Arithmetic

In the last chapter, we saw that primitive recursive arithmetic can be presented as a first-order theory. In that formulation, restricting the induction principle to quantifier-free formulas is somewhat artificial. It is therefore natural to consider *first-order arithmetic*, which is essentially the extension of PRA that has induction for all first-order formulas. *Peano arithmetic*, or PA, is the version of the theory based on classical logic, and *Heyting arithmetic*, or HA, is the version based on intuitionistic logic.

PA and HA are usually formulated in a language that is considerably smaller than that of PRA. Here we will take the language to consist of a single constant symbol, 0, a unary function symbol, succ, two binary function symbols, $+$ and $\cdot$, and a single binary relation symbol, $\leq$. With this restricted language, it is not clear that we can even describe arbitrary primitive recursive functions and relations, let alone prove things about them. We will therefore begin by exploring the extent to which the language allows us to define functions and relations in the standard model, $(\mathbb{N}, 0, \mathrm{succ}, +, \cdot, \leq)$. This gives rise to the theory of *arithmetic definability*. (In this phrase, the word "arithmetic" is read as an adjective, with the emphasis on the first and third syllables.)

Once we have a sense of what the language can do, we will return to questions of provability. We will see that primitive recursive arithmetic can be interpreted in a fragment of HA (and hence PA), even when we restrict the induction principle to an important class of formulas known as the Σ_1 formulas. This justifies thinking of PA and HA as extensions of PRA.

Our study of first-order arithmetic will allow us to explore some important themes in mathematical logic. Classically, the set of arithmetic formulas is naturally stratified into a hierarchy known as the *arithmetic hierarchy*, each level more expressive than the ones below it. Similarly, PA and HA can be stratified into hierarchies of theories, each one stronger than the ones below it. From this perspective, we will explore the extent to which the notion of *truth* for a language can be described within the language itself, and the extent to which we can reason about truth in formal axiomatic theories.

10.1 Peano Arithmetic and Heyting Arithmetic

The language of first-order arithmetic includes a constant symbol, 0, a function symbol, succ, binary function symbols $+$ and $\cdot$, and a binary relation symbol $\leq$. The basic axioms are the universal closures of the following:

1. $\mathrm{succ}(x) \neq 0$
2. $\mathrm{succ}(x) = \mathrm{succ}(y) \rightarrow x = y$
3. $x + 0 = x$

4. $x + \mathrm{succ}(y) = \mathrm{succ}(x + y)$
5. $x \cdot 0 = 0$
6. $x \cdot \mathrm{succ}(y) = x \cdot y + x$
7. $x \leq y \leftrightarrow \exists z\, (z + x = y)$.

The first two axioms tell us that zero is not the successor of any number and that the successor function is injective. The next four axioms are recursive defining equations for addition and multiplication, and the last one can be viewed as a definition of the $\leq$ relation in terms of addition. Finally, we add the schema of induction,

$$\forall \vec{z}\, (A(0) \wedge \forall x\, (A(x) \to A(\mathrm{succ}(x))) \to \forall x\, A(x)),$$

where A is any formula in the language of arithmetic and $\vec{z}$ are the free variables of A other than x. Throughout this chapter, it should be understood that the formulas occurring in schemas and rules can have free variables like these, and I will generally leave them implicit.

In natural deduction, it is natural to express induction instead as a rule:

$$\frac{\Gamma \vdash A(0) \qquad \Gamma, A(x) \vdash A(\mathrm{succ}(x))}{\Gamma \vdash A(t)}$$

Here x is assumed not to be free in any formula in Γ and t can be any term. In particular, t can be a fresh variable, y, allowing us to conclude $\forall x\, A(x)$.

The structure $(\mathbb{N}, 0, \mathrm{succ}, \cdot, +, \leq)$ can be characterized as the unique structure, up to isomorphism, satisfying the basic axioms above and the property that any subset $S \subseteq \mathbb{N}$ containing zero and closed under successor is equal to all of $\mathbb{N}$. The induction schema expresses this last principle as best as a first-order theory can: we cannot quantify over all subsets of the universe, but we can assert that the principle holds for any subset of the universe that is defined by an arithmetic formula. In Section 5.5, we saw that there are nonstandard models of arithmetic, which is to say, structures elementarily equivalent to, but not isomorphic to, the standard one. Such a structure has to satisfy the first-order induction schema, but it fails to satisfy the stronger induction principle, because the set of standard numbers contains zero and is closed under successor without exhausting the universe of the model.

As usual, we define the numerals $1 = \mathrm{succ}(0)$, $2 = \mathrm{succ}(\mathrm{succ}(0))$, and so on. Using axioms 3 and 4 we have $x + 1 = \mathrm{succ}(x)$, so we can go back and forth between the two representations freely. By induction, we can prove $\forall x\, (\mathrm{succ}(x) \neq x)$, since we have $\mathrm{succ}(0) \neq 0$ from the first axiom and $\mathrm{succ}(x) \neq x \to \mathrm{succ}(\mathrm{succ}(x)) \neq \mathrm{succ}(x)$ from the second axiom. In particular, we have $1 \neq 0$. We will use $x < y$ to denote $\mathrm{succ}(x) \leq y$.

It is straightforward to show that both HA and PA prove all the facts we have shown to be provable in PRA in Propositions 9.2.1–9.2.6. For those listed in Proposition 9.2.1, the same proofs work in HA and hence PA. For proofs involving inequalities, there is a slight mismatch in that in PRA the relation $x \leq y$ is defined via truncated subtraction. But the exercises below ask you to show that HA can carry out the principle of induction described by Proposition 9.1.6 as well as prove Proposition 9.2.2, and it is not hard to check that the other facts follow from those. In fact, in Section 10.3, we will see that the proofs go through in much weaker subsystems of HA.

With the less-than-or-equal-to relation in hand, we can show that HA, and hence PA, proves two important principles. The first, the schema of *complete induction*, also known as *strong induction* or *course of values induction*, is the first-order analogue of the version given for PRA in Proposition 9.2.10. In words, it says that to prove that a property A holds of every natural number, it suffices to show that whenever it holds up to some number x, it holds at x as well.

Proposition 10.1.1. *For every formula $A(x)$,* HA *proves*

$$\forall x\,(\forall y < x\, A(y) \to A(x)) \to \forall x\, A(x).$$

Proof Reason in HA. Assume the hypothesis, $\forall x\,(\forall y < x\, A(y) \to A(x))$. Let $B(z)$ be the formula $\forall y < z\, A(y)$, and let us use induction on z to show $\forall z\, B(z)$. Trivially, we have $B(0)$, because HA proves $y \not< 0$. In the induction step, suppose we have $\forall y < z\, A(y)$. By the main hypothesis, we have $A(z)$, and since $y < z + 1$ implies $y < z \vee y = z$, we have $\forall y < z + 1\, A(y)$, as required.

To show $\forall x\, A(x)$, let x be arbitrary. Instantiating z in $B(z)$ to $x + 1$, we have $x < x + 1 \to A(x)$, and hence $A(x)$. □

In Proposition 10.1.1, the formula $A(x)$ can have free variables other than the one shown, in which case what we have really proved is the general statement that quantifies universally over these variables. The same is true of the next principle, called the principle of *collection.* Collection says that if, for every $x < w$, there is a value y with a certain property, we can find a uniform bound on these ys. Intuitively, you can imagine picking a value y for each such x, and then taking the bound to be any number bigger than their maximum.

Theorem 10.1.2. *For every formula $A(x, y)$,* HA *proves the principle of collection:*

$$\forall w\,(\forall x < w\, \exists y\, A(x, y) \to \exists z\, \forall x < w\, \exists y < z\, A(x, y)).$$

Proof Again, reason in HA. It is possible to use induction on w, but the proof we will give is more parsimonious, in a sense that will be made precise in Theorem 10.3.7. Let w be arbitrary, assume $\forall x < w\, \exists y\, A(x, y)$, and let us show by induction on w' that the following formula holds for every w':

$$w' \leq w \to \exists z\, \forall x < w'\, \exists y < z\, A(x, y).$$

Instantiating w' by w then yields the desired conclusion.

When $w' = 0$, the conclusion holds trivially with $z = 0$. Assume the claim holds for w', and suppose $w' + 1 \leq w$. Then $w' \leq w$, so by the inductive hypothesis there is a z such that $\forall x < w'\, \exists y < z\, A(x, y)$. Also, $w' < w$, so by the main hypothesis we have a y' such that $A(w', y')$. Let z' be such that $z' \geq z$ and $z' > y'$. (For example, we can take $z' = z + y' + 1$.) It suffices to show $\forall x < w' + 1\, \exists y < z'\, A(x, y)$. So suppose $x < w' + 1$. Then $x < w'$ or $x = w'$. In the first case, we have $\exists y < z\, A(x, y)$, and in the second case, we have $A(x, y')$. In either case, we have $\exists y < z'\, A(x, y)$. □

Classically, complete induction is equivalent to the least-element principle.

Proposition 10.1.3. *For every formula $A(x)$,* PA *proves the least-element principle:*

$$\exists x\, A(x) \to \exists x\,(A(x) \wedge \forall y < x\, \neg A(y)).$$

Proof Taking the contrapositive, the least-element principle is classically equivalent to

$$\forall x\,(\neg A(x) \vee \neg \forall y < x\, \neg A(y)) \to \forall x\, \neg A(x),$$

and hence

$$\forall x\,(\forall y < x\, \neg A(y) \rightarrow \neg A(x)) \rightarrow \forall x\, \neg A(x).$$

This is just the principle of complete induction for $\neg A(x)$. □

The least-element principle provides a convenient means of introducing new functions in definitional extensions of PA: if we can prove that for every tuple $\vec{x}$ there is some y satisfying $A(\vec{x}, y)$, then we have that there is a least such y, and it is clear that this least y is unique. We can then use the means described in Section 4.7 to introduce a new function $f(\vec{x})$ denoting such an x. Using the translation described there, we can even extend the schema of induction to the new language.

In the first half of the twentieth century, it was commonly assumed that PA, based on classical logic, is stronger than HA. We will see in the chapters to come that HA can be given a computational interpretation, whereas the law of the excluded middle in PA seems to require decisions that depend on quantifiers ranging over an infinite totality, the natural numbers. The following result shows that, in a sense, this extra strength is illusory. Historically, it was the original application of the double-negation translation.

Theorem 10.1.4. *If* PA *proves a formula* A, *then* HA *proves its double-negation translation,* A^{N}. *Hence* PA *is conservative over* HA *for negative formulas.*

Proof If PA proves A, then by Theorem 4.6.3, A^{N} is provable in intuitionistic (even minimal) first-order logic from the double-negation translations axioms of PA. So, for the first claim, it suffices to show that HA proves each of these translated axioms. The translation of the induction axiom for $A(x)$ is $A^{\mathrm{N}}(0) \wedge \forall x\,(A^{\mathrm{N}}(x) \rightarrow A^{\mathrm{N}}(\mathrm{succ}(x))) \rightarrow \forall x\, A^{\mathrm{N}}(x)$, which is again an induction axiom, and hence an axiom of HA. It is easy to check that the translations of the other axioms follow from their original versions in HA.

For the second claim, if A is negative, minimal logic and hence HA proves $A^{\mathrm{N}} \leftrightarrow A$. So if PA proves A, then so does HA. □

Theorem 10.4.6 extends this conservation result to a wider class of formulas.

Exercises

10.1.1. Show that if we replace $\bot$ by $0 = 1$ in the axioms of first-order arithmetic, then we can prove that $0 = 1$ implies $\forall x\,(x = 0)$, $\forall x, y\,(x = y)$, $\forall x, y\,(x \leq y)$, and hence any formula in the language without $\bot$.

10.1.2. Show that HA justifies the principle of induction in Proposition 9.1.6, and that this suffices to prove Proposition 9.2.2.

10.1.3. Convince yourself that HA proves all the facts in Propositions 9.2.1–9.2.6.

10.2 The Arithmetic Hierarchy

For the moment, let us set aside questions of provability and consider what it is possible to express in the language of arithmetic. Remember that a k-ary relation R on the natural numbers is said to be *definable* in the standard model, $\mathfrak{N}$, if there is a formula $A(x_0, \ldots, x_{k-1})$ such that for every tuple $a_0, \ldots, a_{k-1}$ of natural numbers, $R(a_0, \ldots, a_{n-1})$ holds if and only if $\mathfrak{N} \models A(a_0, \ldots, a_{k-1})$. We will say that a k-ary function f is definable if its graph is definable, where the graph of f is, by definition, the relation $f(a_0, \ldots, a_{k-1}) = b$. The question we consider now is: what relations and functions are definable in the language of arithmetic?

We will obtain more precise results by stratifying the language of arithmetic into a hierarchy of formulas. If t is any term and A is any formula, $\forall x < t\, A$ and $\exists x < t\, A$ are taken to be abbreviations for the formulas $\forall x\, (x < t \to A)$ and $\exists x\, (x < t \wedge A)$, respectively. Assuming x does not occur in t, these quantifiers are said to be *bounded*, because rather than range over all of the natural numbers, they range only over those less than t. Since $x \leq t$ is equivalent to $x < \mathrm{succ}(t)$, $\forall x \leq t\, A$ and $\exists x \leq t\, A$ are equivalent to instances of bounded quantification as well. A formula in the language of arithmetic is said to be Δ_0 if it has no unbounded quantifiers. Alternatively, the set of Δ_0 formulas can be characterized inductively as the smallest set containing atomic formulas in the language of arithmetic and closed under boolean operations and bounded quantification. The hierarchies of Σ_n and Π_n formulas are then defined simultaneously and inductively as follows:

- $\Sigma_0 = \Pi_0 = \Delta_0$.
- If A is Σ_n, then it is also Π_{n+1}.
- If A is Π_n, then it is also Σ_{n+1}.
- If A is Σ_{n+1}, then so is $\exists x\, A$.
- If A is Π_{n+1}, then so is $\forall x\, A$.

The last four clauses can be written more succinctly, using $\vec{x}$ to denote any tuple of variables, possibly empty:

- If A is Σ_n, then $\forall \vec{x}\, A$ is Π_{n+1}.
- If A is Π_n, then $\exists \vec{x}\, A$ is Σ_{n+1}.

Roughly speaking, a Σ_n (respectively Π_n) formula consists of at most n alternating blocks of unbounded quantifiers followed by a Δ_0 formula, where the first, possibly empty, block of quantifiers is existential (respectively universal). The hierarchy is cumulative in the sense that any formula that is either Σ_n or Π_n is both Σ_{n+1} and Π_{n+1}. Not every formula falls into this hierarchy, since, in the inductive clauses, unbounded quantifiers are only added to the beginning of a formula. From the point of view of classical logic, this is not a big restriction, because quantifiers can always be brought to the front of a formula.

A relation is said to be Σ_n*-definable*, or just Σ_n, if it is defined by a Σ_n formula. Similarly, a relation is Π_n if it can be defined by a Π_n formula. A relation is Δ_n if it is both Σ_n and Π_n. These notions are extended to sets of natural numbers by interpreting such a set S as the unary predicate $x \in S$. So a set S is Σ_n if there is a Σ_n formula $A(x)$ such that for every natural number a, a is in S if and only if $\mathfrak{N} \models A(a)$.

The following provides useful closure properties of these classes of relations:

Proposition 10.2.1. *For each n, the sets of Σ_n, Π_n, and Δ_n-definable relations are each closed under conjunction, disjunction, and bounded quantification. The negation of a Σ_n relation is Π_n and vice versa, so the negation of a Δ_n relation is again Δ_n.*

Proof The theorem follows from the following claims about formulas:

- If A and B are both Σ_n, then $A \wedge B$ and $A \vee B$ are both equivalent (in the standard model) to Σ_n formulas, and similarly for Π_n.

- If A is Σ_n, then $\forall x < t\, A$ and $\exists x < t\, A$ are both equivalent to Σ_n formulas, and similarly for Π_n.
- If A is Σ_n, then $\neg A$ is equivalent to a Π_n formula, and vice versa.

We will prove only the second claim because that one requires the use of the collection principle, and the proofs of the remaining claims are otherwise similar. The proof proceeds by induction on the number of unbounded quantifiers in A. It implicitly defines a recursive transformation that yields the equivalent Σ_n or Π_n formula, with the additional property that the number of unbounded quantifiers in the transformed formula remains the same.

If there are no unbounded quantifiers in A, then A is by definition Δ_0, and clearly $\forall x < t\, A$ and $\exists x < t\, A$ are as well. We will focus on the case in which A begins with an unbounded existential quantifier, since the case in which it begins with an unbounded universal quantifier can be handled in a similar way.

If A is of the form $\exists y\, B$, then it is Σ_n for some $n > 0$, and B is either Π_{n-1} or again Σ_n. Then $\exists x < t\, A$ is equivalent to $\exists y\, \exists x < t\, B$, and we can apply the inductive hypothesis to B. More interestingly, by the collection principle, $\forall x < t\, A$ is equivalent to $\exists z\, \forall x < t\, \exists y < z\, B$, where z is a fresh variable. By the inductive hypothesis, $\exists y < z\, B$ is equivalent to a formula B' that is either Π_{n-1} or Σ_n, and applying the inductive hypothesis again, $\forall x < t\, B'$ is again equivalent to such a formula B''. Then $\forall x < t\, A$ is equivalent to $\exists z\, B''$, which is Σ_n. □

In Section 10.5, we will see that there are Σ_n relations that are not Π_n, and vice versa. A formula $\exists x_0, \ldots, x_{n-1}\, B$ is equivalent to $\exists \hat{x}\, \exists x_0 < \hat{x}, \ldots, x_{n-1} < \hat{x}\, B$, and similarly for universal quantification. Using 10.2.1 we can therefore establish by induction on n the convenient fact that every Σ_n or Π_n formula is equivalent to one in which there is at most one quantifier in each block. For example, every Π_3-definable relation $R(\vec{x})$ is defined by a formula of the form $\forall y\, \exists z\, \forall w\, A(\vec{x}, y, z, w)$, where A is Δ_0.

Proposition 10.2.2. *Let $R_f(\vec{x}, y)$ be the graph of the function $f(\vec{x})$. If R_f is Σ_n, then it is also Π_n, and hence Δ_n.*

Proof If $n = 0$, the claim is immediate. For $n \geq 1$, if R_f is Σ_n, we have

$$R_f(\vec{x}, y) \equiv \forall z\, (R_f(\vec{x}, z) \rightarrow y = z) \equiv \forall z\, (\neg R_f(\vec{x}, z) \vee y = z).$$

Applying the previous proposition to the last formula shows that $R_f(\vec{x}, y)$ is Π_n. □

When the conclusion of the Proposition 10.2.2 holds and $n \geq 1$, we say that the function f is Δ_n. In Chapter 11, when we have a definition of computability in hand, will see that the Δ_1-definable functions and relations are exactly the computable ones. This characterization is robust, in the sense that it does not change if we add arbitrary computable functions to the language of arithmetic. In contrast, the notion of Δ_0-definability is sensitive to the choice of functions in the initial language. In order to obtain nicer closure properties, we will say that a function $f(\vec{x})$ is Δ_0 if its graph is Δ_0 *and* there is a term $t(\vec{x})$ in the language of arithmetic such that for every $\vec{x}$, $f(\vec{x}) < t(\vec{x})$. This last condition is equivalent to saying that $f(\vec{x})$ is bounded by a polynomial in $\vec{x}$. This qualification makes a difference: it turns out that the exponential function $f(x) = 2^x$ has a Δ_0-definable graph even though it is not bounded by a polynomial.

Suppose we expand the language of arithmetic with a new k-ary relation symbol, $U(y_0, \ldots, y_{k-1})$, and augment the standard model, $\mathfrak{N}$, so that U is interpreted by the

k-ary relation S on $\mathbb{N}$. Then an n-ary relation $R(x_0, \ldots, x_{\ell-1})$ on $\mathbb{N}$ is said to be *definable from S*, or *definable relative to S* if it can be defined by a formula in this expanded language, interpreted in the expanded model. We then define the relativized classes $\Sigma_n(S)$, $\Pi_n(S)$, and $\Delta_n(S)$ just as we defined the unrelativized versions, except that we start with formulas that are $\Delta_0(U)$, which is to say, Δ_0 in the expanded language.

Proposition 10.2.3. *Let S be a k-ary relation and let R be an ℓ-ary relation on $\mathbb{N}$.*

1. *If R is $\Delta_0(S)$ and S is Δ_m, then R is Δ_m.*
2. *If R is $\Sigma_{n+1}(S)$ (respectively $\Pi_{n+1}(S)$, $\Delta_{n+1}(S)$) and S is Δ_{m+1}, then R is Σ_{m+n+1} (respectively Π_{m+n+1}, Δ_{m+n+1}).*

Proof The first claim is easily proved by induction on $\Delta_0(U)$ formulas using the closure properties from Proposition 10.2.1. The second claim is similarly proved by induction on n. In the base case, we have to show, for example, that anything Σ_1-definable relative to a Δ_{m+1} relation is Σ_{m+1}. This follows from the fact that any Δ_{m+1} relation is also Σ_{m+1}, and that the Σ_{m+1} relations are closed under existential quantification. □

We can also relativize a definition to a function rather than a relation, and the claims of Proposition 10.2.3 continue to hold if we replace the Δ_m relation, S, with a Δ_m function, f. The proof is more subtle because the corresponding function symbol can occur nested in terms, and, moreover, these terms can be used to bound quantifiers in a Δ_0 formula. The next proposition helps smooth over these differences.

Proposition 10.2.4. *For every $n \geq 0$, the following hold:*

- *The composition of Δ_n functions is again Δ_n: if $g(y_0, \ldots, y_{k-1})$ is Δ_n and $h_i(\vec{x})$ is Δ_n for each $i < k$, then the function f defined by $f(\vec{x}) = g(h_0(\vec{x}), \ldots, h_{k-1}(\vec{x}))$ is also Δ_n.*
- *A relation $R(\vec{x})$ is Δ_n if and only if $\chi_R(\vec{x})$ is Δ_n.*
- *If $R(x_0, \ldots, x_{k-1})$ is Δ_n and $g_i(\vec{z})$ is Δ_n for each $i < k$, then so is the relation $R(g_0(\vec{z}), \ldots, g_{k-1}(\vec{z}))$.*

Proof The reason we included a bound in the definition of a Δ_0 function was to ensure that the first claim holds with $n = 0$. Suppose the graph of $g(\vec{y})$ is defined by a Δ_0 formula $A(\vec{y}, z)$ and $g(\vec{y})$ is bounded by $t(\vec{y})$, and suppose the graph of each $g_i(\vec{x})$ is defined by a Δ_0 formula $B_i(\vec{x}, y_i)$ and each $h_i(\vec{x})$ is bounded by $s_i(\vec{x})$. Then the relation $g(h_0(\vec{x}), \ldots, h_{k-1}(\vec{x})) = z$ is defined by the Δ_0 formula

$$\exists y_0 < s_0(\vec{x}) \cdots \exists y_{k-1} < s_{k-1}(\vec{x})\ (B_0(\vec{x}, y_0) \wedge \cdots \wedge B_{k-1}(\vec{x}, y_{k-1}) \wedge A(\vec{y}, z)),$$

and by the monotonicity of addition and multiplication, $g(h_0(\vec{x}), \ldots, h_{k-1}(\vec{x}))$ is bounded by $t(s_0(\vec{x}), \ldots, s_{k-1}(\vec{x}))$. The proof for Δ_m functions when $m \geq 1$ is similar; just omit the polynomial bounds.

The second claim is left as an exercise. The third claim follows from the first two but can also be proved directly. □

Proposition 10.2.4 implies that if $s(\vec{x})$ and $t(\vec{x})$ are terms built up from Δ_m functions, then they define Δ_m functions themselves, so the relations $f(\vec{x}) = g(\vec{x})$ and $f(\vec{x}) \leq g(\vec{x})$ are also Δ_m. It also implies that if $\exists x < y\, A(x, \vec{z})$ is Δ_m in y and $\vec{z}$, then $\exists x < t(\vec{z})\, A(x, \vec{z})$ is Δ_m, and similarly for $\forall x < y\, A(x, \vec{z})$. This enables us to prove the following.

Proposition 10.2.5. *Let f be a k-ary function from $\mathbb{N}$ to $\mathbb{N}$, and let R be an ℓ-ary relation.*

1. *If R is $\Delta_0(f)$ and f is Δ_m, then R is Δ_m.*
2. *If R is $\Sigma_{n+1}(f)$ (respectively $\Pi_{n+1}(f)$, or $\Delta_{n+1}(f)$) and f is Δ_{m+1}, then R is Σ_{m+n+1} (respectively Π_{m+n+1}, or Δ_{m+n+1}).*

We can, of course, relativize a definition to a sequence of relations $\vec{S}$ and functions $\vec{f}$. It is not hard to generalize Propositions 10.2.3 and 10.2.5 so that they apply to relations that are $\Sigma_n(\vec{S}, \vec{f})$, $\Pi_n(\vec{S}, \vec{f})$, and $\Delta_n(\vec{S}, \vec{f})$, respectively.

Our goal now is to show that every primitive recursive function is Δ_1. By Proposition 10.2.4.2, this implies that every primitive recursive relation is Δ_1 as well. In the base case, zero is defined by the formula $y = 0$, the successor function is defined by the formula $y = \text{succ}(x)$, and each projection $p^n_i(x_0, \ldots, x_{n-1})$ is defined by the formula $y = x_i$. By Proposition 10.2.4.3, the Δ_1-definable functions are closed under composition, so it suffices to show the following:

Proposition 10.2.6. *For every $n \geq 1$, the Δ_n-definable functions are closed under primitive recursion.*

To prove Proposition 10.2.6, we need finite sequences.

Proposition 10.2.7. *There is a Δ_0-definable function* $\text{elt}(s, i)$ *satisfying the following: for every finite sequence $(a_0, \ldots, a_{k-1})$ of natural numbers, there is a natural number s such that for every $i < k$,* $\text{elt}(s, i) = a_i$.

Proposition 10.2.7 says that there is a way of encoding finite sequences so that the decoding function $\text{elt}(s, i)$, which I will write $(s)_i$, is Δ_0-definable. The representations we will use are not very efficient; the length of the binary representation of the number s that we will obtain is exponential in the sum of the lengths of the elements a_i. With more work, one can define representations whose length is linear in that sum. See the bibliographical notes at the end of this chapter for references.

Proof of Proposition 10.2.6 from Proposition 10.2.7 Suppose $f(x, \vec{z})$ is defined from a k-ary function $g(\vec{z})$ and a $(k+2)$-ary function $h(x, w, \vec{z})$ via the equations

$$f(0, \vec{z}) = g(\vec{z})$$
$$f(x+1, \vec{z}) = h(x, f(x, \vec{z}), \vec{z}),$$

where g and h are Δ_1. Then $f(x, \vec{z}) = y$ is equivalent to the following:

$$\exists s\, ((s)_0 = g(\vec{z}) \wedge \forall w < x\, ((s)_{w+1} = h(w, (s)_w, \vec{z})) \wedge (s)_x = y).$$

In other words, we assert that there is a sequence s that represents (at least) the values $f(0, \vec{z}), f(1, \vec{z}), \ldots, f(x, \vec{z})$, by saying that $(s)_0$ is the value of $f(0, \vec{z})$ specified by g, and for each $w < x$, $(s)_{w+1}$ is the value of $f(w+1, \vec{z})$ specified in terms of h and $(s)_w$. Write this as

$$\exists s, t\, ((s)_0 = t \wedge g(\vec{z}) = t \wedge$$
$$\forall w < x\, \exists u, v\, ((s)_w = u \wedge (s)_{w+1} = v \wedge h(w, u, \vec{z}) = v) \wedge (s)_x = y).$$

Assuming the relations $g(\vec{z}) = t$ and $h(w, y, \vec{z}) = v$ are Σ_n, we see that $f(x, \vec{z}) = y$ is Σ_n, and hence Δ_n. □

Toward proving Proposition 10.2.7, we prove the following.

Proposition 10.2.8. *The pairing function $J(x, y)$ and projection functions $K(z)$ and $L(z)$ from Section 8.3 are Δ_0.*

Proof We have:

- $J(x, y) = z \equiv \exists w \leq z\, (w + w = (x + y)(x + y + 1) \wedge z = w + x)$
- $K(z) = x \equiv \exists y \leq z\, (J(x, y) = z)$
- $L(z) = y \equiv \exists x \leq z\, (J(x, y) = z)$.

Clearly $J(x, y) \leq (x + y)(x + y + 1) + x$, $K(z) \leq z$, and $L(z) \leq z$. □

Proposition 10.2.9. *There is a Δ_0 relation* mem(i, s) *satisfying the following: for every finite set X of natural numbers, there is a number s such that* mem(i, s) *holds if and only if $i \in X$.*

Proof Just as Proposition 10.2.7 says that there is a good encoding of finite sequences, Proposition 10.2.9 says that there is a good encoding of finite sets. I will write $i \in s$ for mem(i, s). In Chapter 8, we encoded the set $\{a_0, a_1, \ldots, a_{k-1}\}$ as $p(a_0) \cdot p(a_1) \cdots p(a_{k-1})$, where $p(i)$ denotes the ith prime. The problem is that defining $p(x)$ there required primitive recursion, which we do not have yet. Instead, to represent a subset of $\{0, \ldots, n-1\}$, we will define a Δ_0 function $p(s, i)$ so that, for a suitable choice of s, if we set $p_i = p(s, i)$, then $p_0, \ldots, p_{n-1}$ are pairwise relatively prime. Assuming each a_j is less than n, we will then be able to represent the set using the product $p_{a_0} \cdot p_{a_1} \cdots p_{a_{k-1}}$.

Given n, let m be the least common multiple of $\{1, \ldots, n\}$, and for each i, take $p_i = m(i + 1) + 1$. I claim that for every $i < j < n$, p_i and p_j are relatively prime. To see this, suppose $i < j < n$, and suppose u divides both p_i and p_j. Then it divides their difference, $m(j - i)$. Hence u divides

$$(j - i)p_i - (i + 1)m(j - i) = (j - i)(m(i + 1) + 1) - (i + 1)m(j - i) = j - i.$$

Since $0 < j - i < n$, u also divides m. Hence u divides $p_i - m(i + 1) = 1$, so $u = 1$.

We now represent the set s by the pair $(m, p_{a_0} \cdot p_{a_1} \cdots p_{a_{n-1}})$, using the pairing function $J(x, y)$. We define $p(s, i) = (s)_0 \cdot (i + 1) + 1$, and we define $i \in s$ to be the relation $p(s, i) \mid (s)_1$.

To see that $i \in s$ behaves as advertised, given $\{a_0, \ldots, a_{k-1}\}$, let s be as in the last paragraph. Suppose $i = a_j$ for some j. Then $p(s, i) = p_i = p_{a_j}$, and the right-hand side divides $(s)_1$, as required. Conversely, suppose $p_i = p(s, i) \mid (s)_1$. Since the elements p_j are pairwise relatively prime and p_i divides $p_{a_0} \cdot p_{a_1} \cdots p_{a_{n-1}}$, we have that p_i divides p_{a_j} for some j, and hence $i = a_j$. □

In Section 10.3, we will show that the functions and relations described in the previous proof can be derived formally in a weak fragment of first-order arithmetic.

To prove Proposition 10.2.6, we finally define elt(s, i) to be the least $j \leq s$ such that $(i, j) \in s$. In other words, we use the set $\{(0, a_0), \ldots, (k - 1, a_{k-1})\}$ to represent the sequence $(a_0, \ldots, a_{k-1})$. Then we have elt$(s, i) = j$ if and only if

$$(i, j) \in s \wedge \forall k < j\, ((i, k) \notin s),$$

showing that elt(s, i) is Δ_0. Thus we have shown:

Theorem 10.2.10. *Every primitive recursive function is Δ_1.*

Our proof of Theorem 10.2.10 shows how to assign, to each function symbol in the language of primitive recursive arithmetic, an explicit Σ_1 formula that defines the graph of the associated primitive recursive function. We will ultimately strengthen this result in two ways. First, in Section 10.4, we will show that not only do the Σ_1 definitions meet the right specifications but, moreover, that these facts are provable in a restricted fragment of first-order arithmetic. Second, in Chapter 11, we will show that every computable function is Δ_1, not just the primitive recursive functions.

Exercises

10.2.1. If we set aside the information about the complexity of the definition, Proposition 10.2.7 says there is a representation of sequences such that a function elt(s, i) with the properties indicate there can be defined in the language of arithmetic. This claim is sometimes known as the β-function lemma, and is originally due to Gödel, who used such a representation in his original presentation of the incompleteness theorems. This exercise describes an approach that is closer to Gödel's original proof, and the next exercise describes another method of proving it. For other approaches, some of which can be carried out formally in very weak subsystems of second-order arithmetic, see the notes at the end of this chapter.

Gödel's approach makes use of the *Chinese remainder theorem*:

Theorem 10.2.11. *Suppose $x_0, \ldots, x_{n-1}$ are pairwise relatively prime. Then for any tuple $a_0, \ldots, a_{n-1}$ there is a z such that $z \equiv a_i \bmod x_i$ for each $i < n$.*

a. Prove, or look up a proof, of that theorem.
b. Given $a_0, \ldots, a_{n-1}$, let $j = \max(n, a_0, \ldots, a_{n-1}) + 1$, let m be the least common multiple of $\{1, \ldots, j\}$, and for each $i < n$, let $x_i = m(i+1) + 1$. The proof of Proposition 10.2.9 shows that $x_0, \ldots, x_{n-1}$ are pairwise relatively prime. Choose z as in the statement of the Chinese remainder theorem. Show that for every i, $a_i = \text{mod}(z, x_i)$, the remainder when dividing z by x_i.
c. Define $\beta(d, i) = \text{mod}(K(d), L(d) \cdot (i+1) + 1)$. Show that β is Δ_0-definable.
d. Represent the sequence $(a_0, \ldots, a_{n-1})$ by $d = J(z, m)$, and show that for every $i < n$, $\beta(d, i) = a_i$.

10.2.2. This provides an alternative proof of Proposition 10.2.9.
a. Show that the predicate $P(x)$ which holds if x is a power of 2 is Δ_0-definable. (Hint: this holds if x has no prime divisors other than 2.)
b. Suppose we represent a binary sequence $b_0 \ldots b_{n-1}$ by the binary number $1b_0 \ldots b_{n-1}$. If x represents any such binary sequence (i.e. x is not 0), write $\ell(x)$ for n, the length of the sequence. Show that $y = 2^{\ell(x)}$ if and only if y is a power of 2 and $y \le x < 2y$. Conclude that $y = 2^{\ell(x)}$ is Δ_0-definable.
c. If x and y represent binary sequences, notice that the concatenation of those strings, $x^\frown y$, is equal to $x \cdot 2^{\ell(y)} + y - 2^{\ell(y)}$. Use this to show that $x^\frown y = z$ is Δ_0-definable.
d. Now suppose we represent a set of numbers $\{a_0, \ldots, a_{n-1}\}$ by the sequence consisting of a one followed by a_0 zeros, another one followed by a_1 zeros, and so on, with a one at the end. Then a number m is in the set represented by u if and only if there is a substring of u of the form $10 \cdots 01$ with m zeros. Show that this statement is Δ_0-definable.

10.2.3. Show that for every $n \ge 0$ the Σ_n-definable sets satisfy the *reduction property*: if A and B are Σ_n, there are Σ_n sets $A' \subseteq A$ and $B' \subseteq B$ such that $A' \cap B' = \emptyset$ and $A' \cup B' = A \cup B$. (Hint: For $n \ge 1$, suppose $x \in A$ if and only if $\exists y\, C(x, y)$ and $x \in B$ if and only if $\exists z\, D(x, z)$, where C and D are Π_n. Put x in A' if there is a y that puts it in A and no smaller z that puts it in B.)

10.2.4. Show that for every $n \geq 0$ the Π_n-definable sets satisfy the *separation property*: if A and B are disjoint Π_n sets, there is a Δ_n set C such that $A \subseteq C$ and $B \subseteq \overline{C}$.

10.2.5. Show that for every $n \geq 0$ the Σ_n-definable relations satisfy the *uniformization property*: if $R(\vec{x}, y)$ is Σ_n, there is a partial function $f(\vec{x})$ with a Σ_n graph, such that every tuple $\vec{x}$, $f(\vec{x})$ is defined if and only if there is a y satisfying $R(\vec{x}, y)$, in which case $R(\vec{x}, f(\vec{x}))$ holds. In other words, $f(\vec{x})$ chooses a y satisfying $R(\vec{x}, y)$ whenever such a y exists, and is undefined otherwise.

10.3 Subsystems of First-Order Arithmetic

In this section, we will show that PA is strong enough to support the kinds of classical reasoning about the arithmetic hierarchy described in the last section, and in the next section we will show that primitive recursive arithmetic can be interpreted in HA. We will take some care to identify the axioms that are needed along the way, so most of the results are formulated in terms of subsystems of HA and PA. If these refinements are uninteresting to you, you can ignore them, and gloss all these results in terms of provability in HA and PA.

If Γ is any class of formulas, $(\mathrm{I}\Gamma)$ denotes the schema of induction for formulas in Γ. For example, $(\mathrm{I}\Sigma_1)$ is the principle of induction for Σ_1 formulas. We use $\mathrm{I}\Gamma$ to denote the corresponding subsystem of PA, so $\mathrm{I}\Sigma_1$ denotes the subsystem of PA in which the induction principle is restricted to formulas that are Σ_1. Because every formula is classically equivalent to one that is Σ_n for some n, the theories $\mathrm{I}\Sigma_\mathrm{n}$ stratify Peano arithmetic.

The arithmetic hierarchy is less natural in the context of HA, since, intuitionistically, formulas cannot generally be put in prenex form. But we will see that the restriction $\mathrm{I}\Sigma_1^\mathrm{i}$ of $\mathrm{I}\Sigma_1$ to intuitionistic logic is a natural theory to consider, and is strong enough to interpret the axioms of primitive recursive arithmetic. In the name of parsimony, we will show that some fundamental results can even be proved in the intuitionistic version $\mathrm{I}\Delta_0^\mathrm{i}$ of $\mathrm{I}\Delta_0$. This is the weakest subsystem of first-order arithmetic that we will consider here, but weaker ones have been studied; see the notes at the end of this chapter.

If T is any theory in the language of arithmetic, we say that a formula A is Γ *in* T when it is provably equivalent to a formula in Γ. In particular, when we say that a formula A is Δ_n in T, we mean that, on the basis of T, A is provably equivalent to a formula that is Σ_n and to another that is Π_n. Saying that a relation defined by a formula A is Σ_n (respectively Π_n) in the sense of the last section is exactly the same as saying that A is Σ_n (respectively Π_n) in *true arithmetic*, that is, the full theory of the standard model. Replacing true arithmetic with other axiomatic theories brings additional information as to the principles needed to establish the relevant equivalences and properties.

Let us start by showing that basic arithmetic can be carried out in $\mathrm{I}\Delta_0^\mathrm{i}$. We will reuse a good deal of the work that we did to bootstrap primitive recursive arithmetic in Chapter 9 by observing that many of the proofs described there can be carried out in $\mathrm{I}\Delta_0^\mathrm{i}$. Note that Δ_0 induction entails that we can use proof by cases on a Δ_0 formula, that is, we can prove $\forall x\, A(x)$ by proving $A(0)$ and $A(\mathrm{succ}(x))$.

Proposition 10.3.1. $\mathrm{I}\Delta_0^\mathrm{i}$ *proves all the facts asserted to be provable in* PRA *by Propositions 9.2.1 and 9.2.2.*

Proof All the derivations described in the proof of Proposition 9.2.1 require only quantifier-free induction, and hence they carry over. The claim $0 \leq x$ follows from $0 + x = x$. The claim

$x \leq 0 \rightarrow x = 0$ follows from $z + x = 0 \rightarrow x = 0$, which is easily proved by cases on x. Finally, $\mathrm{succ}(x) \leq \mathrm{succ}(y) \leftrightarrow x \leq y$ follows from $z + \mathrm{succ}(x) = \mathrm{succ}(y) \leftrightarrow z + x = y$, which follows from the second defining axiom for addition and the injectivity of the successor function. □

Proposition 10.3.2. $\mathrm{I}\Delta_0^{\mathrm{i}}$ *proves the principle of double induction in Proposition 9.1.6, restricted to* Δ_0 *formulas. In other words, for any* Δ_0 *formula* $A(x, y)$*, from* $\forall x\, A(x, 0)$*,* $\forall y\, A(0, y)$*, and* $\forall x, y\, (A(x, y) \rightarrow A(\mathrm{succ}(x), \mathrm{succ}(y)))$*, we conclude* $\forall x, y\, A(x, y)$.

Proof Let $B(z)$ be the Δ_0 formula $\forall u \leq z\, \forall v \leq z\, A(u, v)$. We use ordinary Δ_0 induction to prove $\forall z\, B(z)$. This suffices, because we can prove $x \leq x + y$ and $y \leq x + y$, so instantiating z to $x + y$ yields $A(x, y)$.

When $z = 0$, we have $u = v = 0$, so we have $A(u, v)$ from either of the hypotheses. For the induction step, suppose we have $B(z)$, and suppose $u \leq \mathrm{succ}(z)$ and $v \leq \mathrm{succ}(z)$. We can now use cases on u and v. If either u or v is 0, we are done. Otherwise, we have $u = \mathrm{succ}(u')$, $v = \mathrm{succ}(v')$. From $\mathrm{succ}(u') \leq \mathrm{succ}(z)$ we have $u' \leq z$ and from $\mathrm{succ}(v') \leq \mathrm{succ}(z)$ we have $v' \leq z$. From the inductive hypothesis we get $A(u', v')$, and hence $A(u, v)$, as required. □

We leave it as an exercise to prove the formulas $x = y \vee x \neq y$ and $x \leq y \vee x \not\leq y$ by double induction. By Exercise 2.7.1, this suffices to show that $\mathrm{I}\Delta_0^{\mathrm{i}}$ proves the law of the excluded middle for quantifier-free formulas. Proposition 10.3.4 shows that $\mathrm{I}\Delta_0^{\mathrm{i}}$ can in fact prove the law of the excluded middle for Δ_0 formulas, but to prove that we first need to establish some basic facts about the order relation.

Proposition 10.3.3. $\mathrm{I}\Delta_0^{\mathrm{i}}$ *proves all the facts asserted to be provable in* PRA *in Propositions 9.2.3, 9.2.4, and 9.2.6.*

Proof The proofs in Section 9.2 go through. □

Proposition 10.3.4. $\mathrm{I}\Delta_0^{\mathrm{i}}$ *proves the law of the excluded middle for* Δ_0 *formulas.*

Proof We already know that $\mathrm{I}\Delta_0^{\mathrm{i}}$ proves the law of the excluded middle for atomic formulas and that the class of formulas A for which $\mathrm{I}\Delta_0^{\mathrm{i}}$ proves $A \vee \neg A$ is closed under the boolean connectives. So it only remains to show that if $\mathrm{I}\Delta_0^{\mathrm{i}}$ proves the law of the excluded middle for a Δ_0 formula A, then it also proves it for $\exists x < y\, A(x)$ and $\forall x < y\, A(x)$.

To prove $\exists x < y\, A(x) \vee \neg\exists x < y\, A(x)$, use induction on y. The base case is easy. In the inductive step, $\exists x < y + 1\, A(x)$ is equivalent to $\exists x < y\, A(x) \vee A(y)$ by Proposition 10.3.3. We have $\exists x < y\, A(x) \vee \neg\exists x < y\, A(x)$ by the inductive hypothesis, and we are assuming that we also have $A(y) \vee \neg A(y)$. Reasoning in $\mathrm{I}\Delta_0^{\mathrm{i}}$, if either $\exists x < y\, A(x)$ or $A(y)$, then $\exists x < y + 1\, A(x)$. Otherwise, we have $\neg\exists x < y\, A(x)$ and $\neg A(y)$. These imply $\neg(\exists x < y\, A(x) \vee A(y))$ in intuitionistic logic, and hence $\neg\exists x < y + 1\, A(x, y)$.

For the universal quantifier, assuming we have the law of the excluded middle for $A(x)$ (and hence for $\neg A(x)$), the foregoing shows that $\mathrm{I}\Delta_0^{\mathrm{i}}$ proves $\exists x < y\, \neg A(x) \vee \neg\exists x < y\, \neg A(x)$. Since we have the law of the excluded middle for A, $\mathrm{I}\Delta_0^{\mathrm{i}}$ proves $\neg\neg A(x) \leftrightarrow A(x)$. Using this, the first disjunct implies $\neg\forall x < y\, A(x)$ and the second implies $\forall x < y\, A(x)$, so we have $\forall x < y\, A(x) \vee \neg\forall x < y\, A(x)$, as required. □

We can now show that PA is conservative over HA for Π_1 formulas. To see this, suppose PA proves $\forall \vec{x}\, A$, where A is Δ_0. Then by Theorem 10.1.4, HA proves $\forall \vec{x}\, A^{\mathrm{N}}$. So it suffices to show that HA proves $A^{\mathrm{N}} \leftrightarrow A$. Using induction on Δ_0 formulas and Proposition 10.3.4, we can show that this is even provable in $\mathrm{I}\Delta_0^{\mathrm{i}}$. The only interesting cases are the applications of bounded quantifiers. For any formula $B(y)$, $(\forall y < x\, B(y))^{\mathrm{N}}$ is $\forall y < x\, B^{\mathrm{N}}(y)$ and $(\exists y < x\, B)^{\mathrm{N}}$ is $\neg\forall y\, \neg(y < x \wedge B^{\mathrm{N}}(y))$. If B is Δ_0, the inductive hypothesis implies that this last formula is equivalent to $\neg(\forall y < x\, \neg B^{\mathrm{N}})$, which is in turn equivalent to $\exists y < x\, B^{\mathrm{N}}(y)$.

Theorem 10.4.6 below states a stronger result, namely, that PA is conservative over HA for Π_2 formulas, but we need to develop more machinery before we can prove it. We will make use of the following proposition in Sections 10.6 and 10.7.

Proposition 10.3.5. *$\mathrm{I}\Delta_0^{\mathrm{i}}$ and $\mathrm{I}\Delta_0$ can be axiomatized by Π_1 formulas.*

Proof We can replace the eighth axiom of Q by the universal closures of $x + z = y \to x \leq y$ and $x \leq y \to \exists z \leq x\, (x + z) = y$, since these are both provable in $\mathrm{I}\Delta_0^{\mathrm{i}}$, and they imply the eighth axiom using first-order logic. Similarly, we can express the induction axioms as

$$\forall \vec{z}\, \forall x\, (A(0, \vec{z}) \wedge \forall y < x\, (A(x, \vec{z}) \to A(x+1, \vec{z})) \to A(x, \vec{z})). \qquad \square$$

We now show that closure properties on the arithmetic hierarchy are provable in suitable fragments of arithmetic. For any class of formulas Γ, we will use $(\mathrm{B}\Gamma)$ to denote the collection schema of Theorem 10.1.2,

$$\forall w\, (\forall x < w\, \exists y\, A(x, y) \to \exists z\, \forall x < w\, \exists y < z\, A(x, y)),$$

where A is restricted to formulas in Γ. In the exercises, you are asked to show that over $\mathrm{I}\Delta_0^{\mathrm{i}}$, for every $n \geq 0$, $(\mathrm{B}\Pi_n)$ is equivalent to $(\mathrm{B}\Sigma_{n+1})$. We will therefore only consider the principles $(\mathrm{B}\Sigma_{\mathrm{n}})$ for $n \geq 1$.

Proposition 10.3.6. *For every $n \geq 1$, $\mathrm{I}\Delta_0^{\mathrm{i}} + (\mathrm{B}\Sigma_{\mathrm{n}})$ proves that the Σ_n and Π_n formulas, respectively, are closed under bounded quantification, conjunction, and disjunction.*

The argument is just a formal version of Proposition 10.2.1, showing that the Σ_n collection axioms are sufficient to show that each transformation gives rise to an equivalence. The details are left as an exercise.

Theorem 10.3.7. *For every n, $\mathrm{I}\Sigma_{\mathrm{n+1}}^{\mathrm{i}}$ proves $(\mathrm{B}\Sigma_{\mathrm{n+1}})$.*

Proof By Exercise 10.3.2, it suffices to show that $\mathrm{I}\Sigma_{\mathrm{n+1}}^{\mathrm{i}}$ proves $(\mathrm{B}\Pi_{\mathrm{n}})$. Do this by induction on n. The proof of Theorem 10.1.2 shows that in order to prove the collection axiom for a formula $A(x, y)$, it suffices to prove

$$w' \leq w \to \exists z\, \forall x < w'\, \exists y < z\, A(x, y)$$

by induction on w'. When $n = 0$, A is Δ_0, and since we have the law of the excluded middle for $w' \leq w$ in $\mathrm{I}\Sigma_1^{\mathrm{i}}$, the displayed formula is equivalent to a formula that is Σ_1. In the inductive step, A is Π_n, and by the inductive hypothesis and Proposition 10.3.6, the formula is equivalent to one that is Σ_{n+1}. $\square$

In Section 10.2, we used Proposition 10.2.1 to prove Propositions 10.2.3 and 10.2.5, which show that, for every $n \geq 1$, a Σ_n formula relativized to Δ_1 relations and functions is again Σ_n. This provides a useful way of extending a theory in the language of arithmetic with new relation and function symbols while preserving the complexity of formulas. Suppose a formula $A(\vec{x})$ is provably Δ_1 in $\mathrm{I}\Sigma_{\mathrm{n}}$, and suppose we introduce a new relation symbol $R(\vec{x})$

with the following axiom:

$$\forall \vec{x}\,(R(\vec{x}) \leftrightarrow A(\vec{x})).$$

If $\mathrm{I}\Sigma_\mathrm{n}(R)$ denotes the extension of $\mathrm{I}\Sigma_\mathrm{n}$ that adds this new axiom and extends the induction schema to formulas that are $\Sigma_n(R)$, $\mathrm{I}\Sigma_\mathrm{n}(R)$ is a conservative extension of $\mathrm{I}\Sigma_\mathrm{n}$, because all the atomic formulas $R(\vec{t})$ can be replaced by $A(\vec{t})$ without increasing the complexity of the Σ_n formulas. Similarly, suppose $A(\vec{x}, y)$ is a Σ_n formula such that $\mathrm{I}\Sigma_\mathrm{n}$ proves $\forall \vec{x}\,\exists! y\, A(\vec{x}, y)$. In this case, we say that A *defines a provably* Δ_n *function*. We can then introduce a new function symbol $f(\vec{x})$ with defining axiom

$$\forall \vec{x}, y\,(f(\vec{x}) = y \leftrightarrow A(\vec{x}, y)),$$

or, equivalently, $\forall \vec{x}\, A(\vec{x}, f(\vec{x}))$. This also results in a conservative extension, since the proof of Proposition 10.2.5 tells us how to translate away references to f without increasing the complexity of Σ_n formulas. This can be seen as a refined version of Theorem 4.7.3 that tracks the complexity of the translation and allows the new function symbol to occur in induction axioms.

Over $\mathrm{I}\Delta_0^\mathrm{i}$, the collection principles $(\mathrm{B}\Sigma_1)$ are enough to carry out the translation. The next theorem sums up the situation.

Theorem 10.3.8. *Let T be a classical or intuitionistic theory in the language of arithmetic, and let $T(\vec{R}, \vec{f})$ be an extension of T by Δ_1-definable relations and functions, as described above. Then:*

- *$T(\vec{R}, \vec{f})$ is a conservative extension of T.*
- *For every $n \geq 1$, $T(\vec{R}, \vec{f})$ together with $\mathrm{I}\Delta_0^\mathrm{i} + (\mathrm{B}\Sigma_1)$ proves that every formula that is $\Sigma_n(\vec{R}, \vec{f})$ is equivalent to one that is Σ_n.*

Proof The first claim follows from the fact that we can replace the relation symbols R with their definitions and eliminate the function symbols f using Theorem 4.7.3. The second claim follows from the fact $(\mathrm{B}\Sigma_1)$ suffices to prove that the translated formulas are equivalent to formulas that are Σ_n, formalizing the arguments in the proofs of Propositions 10.2.3 and 10.2.5. □

Corollary 10.3.9. *For $n \geq 0$, let $\mathrm{I}\Sigma_\mathrm{n}(\vec{R}, \vec{f})$ be an extension of $\mathrm{I}\Sigma_\mathrm{n}$ by Δ_1-definable relations and functions, in which the induction schema is also extended to formulas that are $\Sigma_n(\vec{R}, \vec{f})$. Then $\mathrm{I}\Sigma_\mathrm{n}(\vec{R}, \vec{f})$ is a conservative extension of $\mathrm{I}\Sigma_\mathrm{n}$. The analogous statement holds for intuitionistic theories $\mathrm{I}\Sigma_\mathrm{n}^\mathrm{i}(\vec{R}, \vec{f})$ and $\mathrm{I}\Sigma_\mathrm{n}^\mathrm{i}$.*

Proof Apply the previous theorem with T equal to $\mathrm{I}\Sigma_\mathrm{n}$. By the first claim, $T(\vec{R}, \vec{f})$ is conservative over $\mathrm{I}\Sigma_\mathrm{n}$. By the second claim, every new $\Sigma_n(\vec{R}, \vec{f})$ induction axiom is equivalent to a Σ_n induction axiom. So $T(\vec{R}, \vec{f})$ coincides with $\mathrm{I}\Sigma_\mathrm{n}(\vec{R}, \vec{f})$. The same argument works in the intuitionistic case as well. □

We also have useful versions of this theorem and corollary for Δ_0-definable functions and relations, covering the case where $n = 0$. For a Δ_0-definable function, we add the requirement that the theory T proves proves $\forall \vec{x}, y\,(A(\vec{x}, y) \to y \leq t(\vec{x}))$ for some term t in the language of arithmetic. Theorem 10.6.2 below shows that this additional requirement is superfluous when T is $\mathrm{I}\Delta_0$ or $\mathrm{I}\Delta_0^\mathrm{i}$ itself.

Theorem 10.3.10. *Let T be a classical or intuitionistic theory in the language of arithmetic, and let $T(\vec{R}, \vec{f})$ be an extension of T by Δ_0-definable relations and functions. Then $T(\vec{R}, \vec{f})$ is a conservative extension of T, and for every $n \geq 0$, $T(\vec{R}, \vec{f})$ together with $\mathrm{I}\Delta_0^{\mathrm{i}}$ proves that every formula that is $\Sigma_n(\vec{R}, \vec{f})$ is equivalent to one that is Σ_n.*

Corollary 10.3.11. *If $\mathrm{I}\Delta_0^{\mathrm{i}}(\vec{R}, \vec{f})$ is an extension of $\mathrm{I}\Delta_0^{\mathrm{i}}$ by Δ_0-definable relations and functions, then $\mathrm{I}\Delta_0^{\mathrm{i}}(\vec{R}, \vec{f})$ is a conservative extension of $\mathrm{I}\Delta_0^{\mathrm{i}}$.*

Let us now return to studying the relationship between various induction principles and the collection principles.

Theorem 10.3.12. *For every n, $\mathrm{I}\Sigma_{\mathrm{n}}^{\mathrm{i}}$ proves the principle of complete induction for Σ_n formulas.*

Proof Our proof of Proposition 10.1.1 goes through in $\mathrm{I}\Sigma_{\mathrm{n}}^{\mathrm{i}}$. □

The next two theorems rely on classical logic.

Theorem 10.3.13. *Over $\mathrm{I}\Delta_0$, the following are pairwise equivalent:*

1. *the principle of Σ_n induction*
2. *the principle of Σ_n complete induction*
3. *the principle of Π_n induction*
4. *the principle of Π_n complete induction*
5. *the least-element principle for Σ_n formulas*
6. *the least-element principle for Π_n formulas.*

Proof We have seen that Σ_n induction implies Σ_n complete induction, and it is not hard to show that the former follows from the latter, and that the corresponding claims hold with Σ_n replaced by Π_n. As in Proposition 10.1.3, the least-element principle for Π_n formulas is equivalent to the contrapositive of the induction principle for Σ_n formulas and vice versa. So it suffices to show that Σ_n and Π_n induction are equivalent.

Suppose we wish to use Σ_n induction to prove induction axiom

$$A(0) \wedge \forall x\, (A(x) \to A(x+1)) \to \forall x\, A(x)$$

for a Π_n formula A. Reason in $\mathrm{I}\Sigma_{\mathrm{n}}$. The hypothesis $\forall x\, (A(x) \to A(x+1))$ is equivalent to $\forall x\, (\neg A(x+1) \to \neg A(x))$, and it suffices to show that this, together with $\exists x\, \neg A(x)$, implies $\neg A(0)$. So fix x and assume $\neg A(x)$. Then, by induction on y, we can show $\neg A(x \dot{-} y)$. More precisely, we will be able do this once we have a way of interpreting talk of truncated subtraction in $\mathrm{I}\Sigma_{\mathrm{n}}$; in the meanwhile, we can prove the following formula, which is provably equivalent to a Σ_n formula, by induction on y:

$$\forall w \leq x\, (w + y = x \to \neg A(w)).$$

Taking $y = x$ and $w = 0$ yields the desired conclusion. When $y = 0$, the formula follows from $\neg A(x)$. Assume the claim holds for y, and assume for some $w \leq x$, $w + (y+1) = x$. Then $(w+1) + y = x$, and so $w + 1 \leq x$. By the inductive hypothesis, we have $\neg A(w+1)$, and by $\forall x\, (\neg A(x+1) \to \neg A(x))$ we have $\neg A(w)$, as required. □

In the next section, we will prove a stronger result: the class of formulas $A(x)$ for which $\mathrm{I}\Sigma_{\mathrm{n}}$ proves the corresponding induction axiom is closed under negation, boolean operations, and bounded quantification. In other words, $\mathrm{I\Sigma_n}$ proves induction for formulas that are Δ_0-definable relative to arbitrary Σ_n formulas.

Theorem 10.3.14. *Over* $\mathrm{I}\Delta_0$, $(\mathrm{B}\Sigma_{n+1})$ *implies* $(\mathrm{I\Sigma_n})$.

Proof Use induction on n. When $n = 0$, there is nothing to do. Assume the result holds for n, and let $A(x)$ be a Σ_{n+1} formula $\exists y\, B(x, y)$, where B is Π_n. By the inductive hypothesis, it suffices to show that $\mathrm{I\Sigma_n} + (\mathrm{B}\Sigma_{n+2})$ proves the induction axiom for A. Reasoning in that theory, assume $A(0)$ and $\forall x\, (A(x) \to A(x+1))$. Fix x', and let us show $A(x')$. The second hypothesis is equivalent to

$$\forall x\, \exists y\, \forall y'\, (B(x, y') \to B(x+1, y)),$$

which implies

$$\forall x \le x'\, \exists y\, \forall y'\, (B(x, y') \to B(x+1, y)).$$

The formula $\forall y'\, (B(x, y') \to B(x+1, y))$ is Π_{n+1}, so by $(\mathrm{B}\Sigma_{n+2})$, we obtain a w satisfying

$$\forall x \le x'\, \exists y < w\, \forall y'\, (B(x, y') \to B(x+1, y)).$$

We also have $\exists y\, B(0, y)$, and by replacing w if necessary, we can assume $\exists y < w\, B(0, y)$. The formula $x \le x' \to \exists y < w\, B(x, y)$ is Π_n, so now we can use Π_n induction to show that it holds for every x, and, in particular, for x'. □

Exercises

10.3.1. Use Proposition 10.3.2 to show that $x = y \lor x \ne y$ and $x \le y \lor x \not\le y$ are provable in $\mathrm{I}\Delta_0^{\mathrm{i}}$.

10.3.2. Show that over $\mathrm{I}\Delta_0^{\mathrm{i}}$, for every $n \ge 0$, $(\mathrm{B}\Pi_n)$ proves $(\mathrm{B}\Sigma_{n+1})$.

10.3.3. Prove Proposition 10.3.6.

10.3.4. Spell out the proof of Theorem 10.3.14 in greater detail.

10.4 Interpreting PRA

Having established that the language of first-order arithmetic is expressive enough to interpret the language of primitive recursive arithmetic, our goal now is to show that the axioms of PA and HA are strong enough to interpret the axioms of PRA. We will show that the intuitionistic fragment $\mathrm{I\Sigma_1^i}$ suffices for that purpose, at which point we will be justified in treating subsystems of first-order arithmetic that contain $\mathrm{I\Sigma_1^i}$ as extensions of PRA.

There are two tasks before us. First, to apply Theorem 10.3.8, we need to show that if $A_f(\vec{x}, y)$ is the Σ_1 formula we used to define the graph of $f(\vec{x})$ in Section 10.2, then $\mathrm{I\Sigma_1^i}$ proves $\forall \vec{x}\, \exists! y\, A_f(\vec{x}, y)$. Second, we need to show that once we associate function symbols to their definitions in this way, $\mathrm{I\Sigma_1^i}$ proves all the axioms of PRA. Both tasks require showing that we can formalize enough mathematics in $\mathrm{I\Sigma_1^i}$ that the informal arguments we used to establish the correctness of the definitions in Section 10.2 can be carried out formally in that theory.

For zero, the successor functions, and the projections, there is not much to do. To handle a projection p_i^n, it is trivial to prove $\forall x_0, \ldots, x_{n-1}\, \exists! y\, (y = x_i)$. Handling composition is similar; I will leave it for you to check that we obtain the totality of the definition of the composition given in Section 10.2 from the totality of the definition of each component function, and the fact that composition meets its specification follows straightforwardly from the definitions and the method of expanding definite descriptions described in Section 4.7.

The fact that $\mathrm{I}\Sigma_1^{\mathrm{i}}$ proves the interpretation of each induction axiom of PRA is also straightforward. If A is any quantifier-free formula in the language of PRA, it is, in particular, $\Delta_1(\vec{f})$, where $\vec{f}$ are the function symbols involved. Once we know that each function is Δ_1-definable in $\mathrm{I}\Sigma_1^{\mathrm{i}}$, Theorem 10.3.8 tells us that A is equivalent to a Δ_1 formula, and hence we can use Σ_1 induction in $\mathrm{I}\Sigma_1^{\mathrm{i}}$ to justify the induction axiom for A. So all that is left to do is to handle the definitions of the primitive recursive functions. This, in turn, involves showing that $\mathrm{I}\Sigma_1^{\mathrm{i}}$ can carry out the reasoning about pairs, finite sets, and finite sequences that was needed in Section 10.2.

To that end, we will use Theorem 10.3.8 to introduce auxiliary functions and relations. For our purposes, it suffices to know that they are Δ_1-definable in $\mathrm{I}\Sigma_1^{\mathrm{i}}$. For the sake of parsimony, we will even show that some of them are even Δ_0-definable in $\mathrm{I}\Delta_0^{\mathrm{i}}$. To that end, notice that if $A(\vec{x}, y)$ is any Δ_0 formula and $\mathrm{I}\Delta_0^{\mathrm{i}}$ proves $\forall \vec{x}\, \exists y < t\, A(\vec{x}, y)$, then the formula $A'(\vec{x}, y)$ given by

$$A(\vec{x}, y) \wedge \forall y' < y\, \neg A(\vec{x}, y')$$

is again a Δ_0 formula defining the least y satisfying $A(\vec{x}, y)$. Moreover, $\mathrm{I}\Delta_0^{\mathrm{i}}$ proves $\forall \vec{x}\, \exists! y\; A'(\vec{x}, y)$ as well as $\forall \vec{x}, y\, (A'(\vec{x}, y) \to y < t)$, so $A'(\vec{x}, y)$ defines a Δ_0 function.

Proposition 10.4.1. *The functions* pred(x) *and* $x \dot{-} y$ *are* Δ_0*-definable in* $\mathrm{I}\Delta_0^{\mathrm{i}}$ *in such a way that* $\mathrm{I}\Delta_0^{\mathrm{i}}$ *proves their primitive recursive defining equations,* $\mathrm{pred}(0) = 0$, $\mathrm{pred}(x+1) = x$, $x \dot{-} 0 = x$, *and* $x \dot{-} (y+1) = \mathrm{pred}(x \dot{-} y)$.

Proof By cases, $\mathrm{I}\Delta_0^{\mathrm{i}}$ proves $x = 0 \vee \exists y\, (\mathrm{succ}(y) = x)$, and in the second case, it proves that y is unique. So the formula $(x = 0 \wedge y = 0) \vee (x \neq 0 \wedge \mathrm{succ}(y) = x)$ defines the graph $\mathrm{pred}(x) = y$.

By induction on x, $\mathrm{I}\Delta_0^{\mathrm{i}}$ proves $\forall y \leq x\, \exists z \leq x\, (x = y + z)$, so we can define $x \dot{-} y$ to be the least $z \leq x$ satisfying $x + z = y$ if $y \leq x$, and 0 otherwise. I leave it as an exercise to show, in both cases, that $\mathrm{I}\Delta_0^{\mathrm{i}}$ proves that the functions so defined satisfy their defining equations. □

The properties of truncated subtraction established in Proposition 9.2.5 can be established in $\mathrm{I}\Delta_0^{\mathrm{i}}$ just as they were established in PRA. Now, for any Δ_0 formula $A(\vec{x}, y)$ and term $t(\vec{x})$, we can define in $\mathrm{I}\Delta_0^{\mathrm{i}}$ the function $\max_{y < t(\vec{x})} A(\vec{x}, y)$, which returns the greatest $y < t(\vec{x})$ such that $A(\vec{x}, y)$ holds, if there is one, and $t(\vec{x})$ otherwise. Namely, we define it to be $y \dot{-} z$, where z is the least $z' \leq t(\vec{x})$ such that $A(\vec{x}, t(\vec{x}) \dot{-} z')$ holds, if there is such a z', and 0 otherwise.

By induction, $\mathrm{I}\Delta_0^{\mathrm{i}}$ shows that for every u, there is a $v \leq u \cdot (u+1)$ such that $v + v = u \cdot (u+1)$, so $\mathrm{I}\Delta_0^{\mathrm{i}}$ can define the pairing function $J(x, y)$ as in Section 10.2. We leave it as an exercise to show that the proof of Lemma 8.3.1, which asserts that for every z there exists a unique pair x, y such that $J(x, y) = z$, can be carried out in $\mathrm{I}\Delta_0^{\mathrm{i}}$. As a result, we can define the inverse functions $K(z)$ and $L(z)$ in $\mathrm{I}\Delta_0^{\mathrm{i}}$ and show that they have the expected properties.

We next transport some elementary number theory, as developed in Section 9.3 in PRA, to $\mathrm{I}\Delta_0^{\mathrm{i}}$. If we define $x \mid y$ to be $\exists z \leq y\, (x \cdot z = y)$, then $\mathrm{I}\Delta_0^{\mathrm{i}}$ can prove the properties of divisibility enumerated in Proposition 9.3.1. Similarly, define $\mathrm{div}(x, y)$ to be the largest $z \leq x$ such that

$zy \leq x$ if $y \neq 0$, and 0 otherwise. Define $\mathrm{mod}(x, y)$ to be $x \dot{-} \mathrm{div}(x, y) \cdot y$. Then, from the definition, we have

$$x = \mathrm{div}(x, y) \cdot y + \mathrm{mod}(x, y).$$

If $y \neq 0$, we also have $\mathrm{mod}(x, y) < y$, since otherwise we would have

$$\mathrm{mod}(x, y) = x \dot{-} \mathrm{div}(x, y) \cdot y \geq y$$

and hence $\mathrm{div}(x, y) \cdot y + y = (\mathrm{div}(x, y) + 1) \cdot y \leq x$ contrary to the definition of $\mathrm{div}(x, y)$. As in Proposition 9.3.2, we can prove that $\mathrm{div}(x, y)$ and $\mathrm{mod}(x, y)$ are the unique values with these properties. We define $\gcd(x, y)$ to be the greatest common divisor of x and y, with the convention that $\gcd(0, 0) = 0$. This function is bounded by x if x is nonzero and y otherwise, and hence it is bounded by $x + y$. Following the proofs of Proposition 9.3.3 and Theorem 9.3.4, we can establish that if $\gcd(x, y) = 1$ and $x \mid yz$ then $x \mid z$. In particular, if $\gcd(x, y) = 1$ and $\gcd(x, z) = 1$, then $\gcd(x, yz) = 1$.

At this stage, we part company with Section 9.3 and follow the strategy set out in Section 10.2. We also begin to avail ourselves of the added strength of Σ_1 induction.

Lemma 10.4.2. $\mathrm{I}\Sigma_1^{\mathrm{i}}$ *proves that for every n, there exists an $m > 0$ such that, for every $i < n$, $i + 1 \mid m$.*

Proof By induction on n. When $n = 0$ or $n = 1$, take $m = 1$. Given an m that satisfies the claim for $n \geq 1$, $m \cdot (n + 1)$ satisfies the claim for $n + 1$. □

The proof of Lemma 10.4.2 implicitly describes the factorial function, but we cannot yet specify the graph of that function in the language of arithmetic. But we can, in $\mathrm{I}\Delta_0^i$, define the function $\ell(n)$ to be the least m such that $i + 1 \mid m$ for every $i < n$. The relation $\ell(n) = m$ has a Δ_0 definition, and in $\mathrm{I}\Sigma_1^{\mathrm{i}}$ we can prove that for every n, there is a unique m with that property.

Proposition 10.2.9 asserted that there is a Δ_0 membership relation $i \in s$ with the property that every finite set is represented by some number s. In subsystems of arithmetic, we do not have an independent notion of "every finite set." But for our purposes, it is enough to know that the result holds for every set $\{i < n \mid A(x)\}$ defined by a Δ_0 formula $A(x)$. So we state the next proposition in those terms.

Proposition 10.4.3. *There is a Δ_0-definable relation $i \in s$ such that for every Δ_0 formula $A(x)$, $\mathrm{I}\Sigma_1^{\mathrm{i}}$ proves the following: for every n there exists an s such for every i, $i \in s$ if and only if $i < n$ and $A(i)$. Moreover, $\mathrm{I}\Delta_0^{\mathrm{i}}$ proves that $i \in s$ implies $i \leq s$.*

Proof As in the proof of Proposition 10.2.9, we would like to define $i \in s$ to be the relation $(s)_0 \cdot (i + 1) + 1 \mid (s)_1$. But to make sure the last part of the proposition holds, we also add the conjuncts $(s)_0 > 0$ and $(s)_1 > 0$. Then whenever $i \in s$ holds, $\mathrm{I}\Delta_0^{\mathrm{i}}$ can prove $i \leq (s)_0 \cdot (i + 1) + 1 \leq (s)_1 \leq s$.

For a fixed formula $A(x)$, we need to show that $\mathrm{I}\Sigma_1^{\mathrm{i}}$ proves the existence of an s with the requisite property. Fix n, let $m = \ell(n)$, and let $p_i = m(i + 1) + 1$. By the argument in the proof of Proposition 10.2.9, $\mathrm{I}\Sigma_1^{\mathrm{i}}$ proves that for $i < j < n$, $\gcd(p_i, p_j) = 1$. In $\mathrm{I}\Sigma_1^{\mathrm{i}}$, we now show by induction on k that if $k \leq n$, there is a number $v > 0$ such that for every $i < n$,

- if $i < k$ and $A(i)$ then $p_i \mid v$, and
- otherwise $\gcd(p_i, v) = 1$.

If $k = 0$, $v = 1$ works. In the inductive step, suppose v' satisfies the claim for k, and suppose $k+1 \leq n$. If $A(k)$ does not hold, then taking $v = v'$ yields the desired conclusion. If $A(k)$ holds, take $v = v' p_k$, and let us show that this works. From the inductive hypothesis, we have that if $i < k$ and $A(i)$ then $p_i \mid v$, and clearly this holds when $i = k$. To establish the second condition, suppose $i \not< (k+1)$ or $\neg A(i)$. In either case, the inductive hypothesis tells us that $\gcd(p_i, v') = 1$. By the argument above, we also have $\gcd(p_i, p_k) = 1$, and since $v = v' p_k$, we have $\gcd(p_i, v) = 1$.

Now let $k = n$, take the corresponding v, and let $s = (m, v)$. Then we have that for $i < n$, $A(i)$ holds if and only if $p_i \mid v$, that is, if and only if $i \in s$. □

To interpret sequences, we now formally define $\mathrm{elt}(s, i)$ to be the least $j \leq s$ such that $(i, j) \in s$ if there is one, and 0 otherwise. $\mathrm{I}\Delta_0^{\mathrm{i}}$ proves that this function, which we will write $(s)_i$, is total and bounded by s. The next lemma says, roughly, that we can always append an element to the end of a sequence.

Lemma 10.4.4. $\mathrm{I}\Sigma_1^{\mathrm{i}}$ *proves that for every s, n, and a, there is an s′ such that*

- $(s')_i = (s)_i$ *for every* $i < n$, *and*
- $(s')_n = a$.

Proof Let k be the maximum of s and (n, a), and by the previous proposition, let $s' = \{x \leq k \mid (x \in s \wedge (x)_0 \neq n) \vee x = (n, a)\}$. □

We can now show that $\mathrm{I}\Sigma_1^{\mathrm{i}}$ proves that the Δ_1 functions are closed under primitive recursion. Suppose $g(\vec{z})$ is a k-ary function and $h(x, w, \vec{z})$ is a $(k+2)$-ary function with Σ_1 graphs that $\mathrm{I}\Sigma_1^{\mathrm{i}}$ proves to be total. Using Theorem 10.3.8, we can assume that symbols for g and h have been added to the language of $\mathrm{I}\Sigma_1^{\mathrm{i}}$. As in the proof of Proposition 10.2.6, let $A_f(x, \vec{z}, y)$ be the formula

$$\exists s\, ((s)_0 = g(\vec{z}) \wedge \forall w < x\, (h(w, (s)_w, \vec{z}) = (s)_{w+1}) \wedge (s)_x = y).$$

By Proposition 10.3.6 and Theorem 10.3.8, $\mathrm{I}\Sigma_1^{\mathrm{i}}$ extended with the definitions of g and h proves that this is equivalent to a Σ_1 formula. Our goal is to show that it also proves $\forall x, \vec{z}\, \exists! y\, A_f(x, \vec{z}, y)$. From this, it easily follows that if we add the defining axioms $f(x, \vec{z}) = y \leftrightarrow A_f(x, \vec{z}, y)$, the expanded theory proves

$$f(0, \vec{z}) = g(\vec{z})$$
$$f(x+1, \vec{z}) = h(x, f(x, \vec{z}), \vec{z}).$$

We can dispense with uniqueness quickly. Suppose we have $A_f(x, \vec{z}, y)$ and $A_f(x, \vec{z}, y')$. Then there are sequences s and s' satisfying the specification above for y and y', respectively. By induction we can show that for every $i \leq x$, $(s)_i = (s')_i$, and hence $y = y'$.

To show existence, it suffices to show

$$\exists s\, ((s)_0 = g(\vec{z}) \wedge \forall w < x\, (h(w, (s)_w, \vec{z}) = (s)_{w+1}))$$

because we can then take $y = (s)_x$. We can prove this using Σ_1 induction on x. When x is 0, we can apply Lemma 10.4.4 to find an s such that $(s)_0 = g(\vec{z})$. Given an s that works for x, we again apply Lemma 10.4.4 to obtain an s' that works for $x+1$ by making $(s')_i$ equal to $(s)_i$ for $i \leq x$ and making $(s')_{x+1}$ equal to $h(x, (s)_x, \vec{z})$.

We have therefore shown that we can introduce all the primitive recursive functions in a definitional extension of $\text{I}\Sigma_1^{\text{i}}$ in such a way that the axioms of PRA are provable, and in such a way that, for every $n \geq 1$, Σ_n formulas with the new function symbols are provably equivalent to Σ_n formulas in the original language. Induction and collection principles involving the new function symbols are then equivalent to induction and collection principles in the original language. Thus, whenever we deal with a subsystem of first-order arithmetic that contain $\text{I}\Sigma_1^{\text{i}}$, we can act as though it includes primitive recursive arithmetic. In other words, we have shown the following:

Theorem 10.4.5. *Let* $\text{I}\Sigma_1^{\text{i}}(\text{PRA})$ *consist of first-order primitive recursive arithmetic together with the principle of induction for* Σ_1 *formulas in the new language. Then* $\text{I}\Sigma_1^{\text{i}}(\text{PRA})$ *is a conservative extension of* $\text{I}\Sigma_1^{\text{i}}$*, and similarly for classical or intuitionistic theories extending* $\text{I}\Sigma_1^{\text{i}}(\text{PRA})$ *with additional axioms and axiom schemas that are parameterized by the complexity classes* Σ_n *and* Π_n *with* $n \geq 1$.

We now put this information to good use. The translation implicit in the proof of the next theorem is called the *Friedman–Dragalin translation.*

Theorem 10.4.6. PA *is conservative over* HA *for* Π_2 *sentences.*

Proof Suppose PA proves $\forall \vec{x}\, \exists \vec{y}\, A(\vec{x}, \vec{y})$, where A is Δ_0. We can now act as though PA and HA were defined to be first-order primitive recursive arithmetic together with the full schema of induction, since if the theorem holds for that formulation, it holds for our original formulation. We can also assume $A(\vec{x}, \vec{y})$ is an atomic formula, since otherwise we can replace it by the formula $\chi_A(\vec{x}, \vec{y}) = 1$, where $\chi_A(\vec{x}, \vec{y})$ is the characteristic function of A.

By the double-negation translation, $\forall \vec{x}\, \neg \forall \vec{y}\, \neg A(\vec{x}, \vec{y})$ is provable from the axioms of HA using minimal logic. Hence $\neg \forall \vec{y}\, \neg A(\vec{c}, \vec{y})$ is also provable, where $\vec{c}$ is a tuple of fresh constants. Since the axioms of minimal logic say nothing about $\bot$, we can replace it everywhere in the proof by the sentence $\exists \vec{z}\, A(\vec{c}, \vec{z})$. When we do this, any instance of the induction schema in HA is replaced by another instance. The only other axiom of HA that mentions $\bot$ is $\forall x\, (\text{succ}(x) \neq 0)$, which is transformed to $\forall x\, (\text{succ}(x) = 0 \to \exists \vec{z}\, A(\vec{c}, \vec{z}))$. This is also provable in HA (with the language augmented by the constants $\vec{c}$). The conclusion becomes

$$\forall \vec{y}\, (A(\vec{c}, \vec{y}) \to \exists \vec{z}\, A(\vec{c}, \vec{z})) \to \exists \vec{z}\, A(\vec{c}, \vec{z}).$$

The antecedent of this implication provable in minimal first-order logic and hence in HA. So we ultimately have that HA proves $\exists \vec{z}\, A(\vec{c}, \vec{z})$. Renaming $\vec{z}$ to $\vec{y}$, replacing $\vec{c}$ by fresh variables, and generalizing, we have that HA proves $\forall \vec{x}\, \exists \vec{y}\, A(\vec{x}, \vec{y})$. □

It is reasonable to ask whether Theorem 10.4.6 extends to subsystems of arithmetic like $\text{I}\Sigma_1$ and $\text{I}\Delta_0$. In fact, these two theories are Π_2 conservative over their intuitionistic counterparts, but the proof above does not go through, since applying the double-negation translation to a formula A and replacing $\bot$ by an existential formula does not preserve the complexity of A in the arithmetic hierarchy. Instead, these two conservation theorems follow from the results of Section 10.6.

We will henceforth blur the distinction between theories like $\mathrm{I\Sigma_n}$ and $\mathrm{I\Sigma_n(PRA)}$ and act as though PRA is part of $\mathrm{I\Sigma_n}$. Our next theorem shows that $\mathrm{I\Sigma_n}$ can derive induction principles that, at first blush, seem to go beyond Σ_n induction.

Theorem 10.4.7. *Let $\Delta_0(\Sigma_n)$ be the smallest class of formulas that contains Σ_n and is closed under conjunction, disjunction, negation, and bounded quantifiers. Then $\mathrm{I\Sigma_n}$ proves the induction axiom for any formula in $\Delta_0(\Sigma_n)$.*

Although the theorem is stated in terms of provability in the language of arithmetic, we will make use of the more expressive means afforded by PRA in order to prove it. Remember that, in PRA, we have a theory of finite sets. In PRA we can also reason about the binary representation of a natural number. In particular, we can define the function $\mathrm{bit}(x, i)$, which returns the ith bit of the number x, where we take the 0th bit to be the least significant one. The following are then provable in PRA:

- $x = \sum_{i<x} 2^i \cdot \mathrm{bit}(x, i)$
- for any fixed primitive recursive function f, the statement that if $i < k$ and $f(j) \leq 1$ for every j, then $\mathrm{bit}(\sum_{j<k} 2^j \cdot f(j), i) = f(i)$.

The following theorem says that $\mathrm{I\Sigma_n}$ proves that for any Σ_n formula $A(x)$ and any y, the finite set $\{x < y \mid A(x)\}$ exists.

Theorem 10.4.8. *For every Σ_n formula $A(x)$, $\mathrm{I\Sigma_n}$ proves the following* finite separation *principle:*

$$\forall y \, \exists s \, \forall x \, (x \in s \leftrightarrow x < y \wedge A(x)).$$

Proof Given $A(x)$, let the formula $B(y, z)$ say that for every $x < y$, if $A(x)$ holds, then $\mathrm{bit}(z, x) = 1$. In $\mathrm{I\Sigma_n}$, $B(y, z)$ is equivalent to a Π_n formula. Fix y. Clearly $B(y, z)$ holds if $z = \sum_{i<y} 2^i$, so by the least-element principle for Π_n formulas, there is a least z such that $B(y, z)$ holds. For that z, we have that for every $x < y$, if $\mathrm{bit}(z, x) = 1$, then $A(x)$ holds; otherwise, for some x, we would have $\mathrm{bit}(z, x) = 1$ and $\neg A(x)$, and $z' = \sum_{i<z \wedge i \neq x} 2^i \cdot \mathrm{bit}(z, i)$ would be less than z but satisfy $B(y, z')$. So for every $x < y$ we have $A(x)$ if and only if $\mathrm{bit}(z, x) = 1$. By Proposition 9.3.6, we can let s be $\{x < z \mid \mathrm{bit}(z, x) = 1\}$. □

One reason that Theorem 10.4.8 is useful is the following:

Proposition 10.4.9. *Suppose a theory T containing* PRA *proves the finite separation principle for a formula $A(x)$. Then it proves the principle of complete induction for $A(x)$.*

Proof Argue in T. Suppose $\forall x \, (\forall y < x \, A(y) \to A(x))$. Let x be arbitrary, and by the finite separation principle, let s be the finite set of $y \leq x$ such that $A(y)$ holds. If $\neg A(x)$, then x is not in s, and considering the least y that is not in s yields a contradiction. □

This gives us a way of leveraging the strength of induction in $\mathrm{I\Sigma_n}$. Theorem 16.4.2 in Chapter 16 provides one illustration of how it can be used. Theorem 10.4.7 provides another. The best way to prove it is to generalize the finite separation principle to n-ary relations for any n. Say that a theory T proves the finite separation principle for a formula $A(x_0, \ldots, x_{n-1})$ if it proves

$$\forall y \, \exists s \, \forall x_0, \ldots, x_{n-1} \, ((x_0, \ldots, x_{n-1}) \in s \leftrightarrow x_0 < y \wedge \cdots \wedge x_{n-1} < y \wedge A(x_0, \ldots, x_{n-1})).$$

The following is left as an exercise:

Lemma 10.4.10. *The class of formulas $A(\vec{x})$ such that $\mathrm{I\Sigma_n}$ proves the finite separation principle is closed under boolean operations and bounded quantification.*

Generalizing Theorem 10.4.8 to n-ary relations and using Lemma 10.4.10 yields the following:

Theorem 10.4.11. $\mathrm{I\Sigma_n}$ *proves the finite separation principle for $\Delta_0(\Sigma_n)$ formulas.*

Theorem 10.4.7 then follows, using Proposition 10.4.9.

We saw in the last section that, over $\mathrm{I\Delta_0}$, $(\mathrm{I\Sigma}_{n+1})$ implies $(\mathrm{B\Sigma}_{n+1})$ and that $(\mathrm{B\Sigma}_{n+1})$ implies $(\mathrm{I\Sigma_n})$. We will see in Sections 10.6 and 10.7 that there is a sense in which $(\mathrm{B\Sigma}_{n+1})$ is closer to $(\mathrm{I\Sigma_n})$ than $(\mathrm{I\Sigma}_{n+1})$. But consider the following strengthening of the collection axiom, known as *strong collection*:

$$\forall u \, \exists v \, \forall x < u \, (\exists y \, A(x, y) \rightarrow \exists y < v \, A(x, y)).$$

Rather than assume $\forall x < u \, \exists y \, A(x, y)$, we ask instead for a bound on the witnesses y for only those values of x for which such a y exists. Call this schema $(\mathrm{S}\Gamma)$ when A is restricted to formulas Γ. In Exercise 10.4.6, you are asked to show the following:

Theorem 10.4.12. *For every $n \geq 0$, $(\mathrm{S\Sigma}_{n+1})$ is equivalent to $(\mathrm{I\Sigma}_{n+1})$ over $\mathrm{I\Delta_0}$.*

Exercises

10.4.1. Show that $\mathrm{I\Delta_0^i}$ proves that the predecessor function and truncated subtraction satisfy their defining equations, given the definitions used in the proof of Proposition 10.4.1.

10.4.2. Show that the proof of Lemma 8.3.1, which asserts that for every z there exists a unique pair x, y such that $J(x, y) = z$, can be carried out in $\mathrm{I\Delta_0^i}$.

10.4.3. Convince yourself that the proofs of Proposition 9.3.3 and Theorem 9.3.4 go through in $\mathrm{I\Delta_0^i}$, and hence $\mathrm{I\Delta_0^i}$ can prove that if $\gcd(x, y) = 1$ and $\gcd(x, z) = 1$ then $\gcd(x, yz) = 1$.

10.4.4. Show that PRA proves the facts about binary numbers needed in the proof of Theorem 10.4.7.

10.4.5. Prove Lemma 10.4.10 and show that Theorem 10.4.8 generalizes to n-ary relations. (Hint: use tupling and untupling for the second part.)

10.4.6. Prove Theorem 10.4.12 as follows:
a. Show that $(\mathrm{S\Pi}_n)$ implies $(\mathrm{S\Sigma}_{n+1})$.
b. Show that $(\mathrm{I\Sigma}_{n+1})$ implies $(\mathrm{S\Pi}_n)$. (Hint: Let A be Π_n, and argue in $\mathrm{I\Sigma}_{n+1}$. Fix u, and use Theorem 10.4.11 to show that there is a number representing the set $s = \{x < u \mid \exists y \, A(x, y)\}$. Show by induction on z that we have $z \leq u \rightarrow \exists v \, \forall x < z \, (x \in s \rightarrow \exists y < v \, A(x, y))$.)
c. Show that $(\mathrm{S\Pi}_n)$ implies $(\mathrm{I\Sigma}_{n+1})$. (Hint: Use induction on n, so at stage n you can use $I\Sigma_n$. Let A be Π_n, and assume $\exists y \, A(0, y)$ and $\forall x < u \, (\exists y \, A(x, y) \rightarrow \exists y \, A(x + 1, y))$. Fix u, and apply strong collection to the second hypothesis.)

10.5 Truth and Reflection

Mathematical logic deals routinely with terms, formulas, and derivations, so formalizing logic requires having formal representations of these. We have seen that primitive recursive arithmetic provides a solid foundation for reasoning about finite objects, and we should expect it to be adequate for reasoning about syntactic notions. As we will now see, this is indeed the case. In this context, the representation of syntactic objects as natural numbers is sometimes referred to as *arithmetization of syntax*.

To simplify the presentation, we will restrict our attention to the language of arithmetic itself, though there is nothing standing in the way of developing notions of terms, formulas, and derivations for an arbitrary language. Remember that the set of terms in the language of arithmetic is defined inductively as follows:

- For each i, the variable x_i is a term.
- The constant 0 is a term.
- For every term t, $\mathrm{succ}(t)$ is a term.
- If s and t are terms, then so are $(s + t)$ and $(s \cdot t)$.

Using the encodings of tuples of numbers as numbers, we can assign to each term t a natural number, $\#t$, as follows:

$$\#x_i = (0, i), \quad \#0 = (1), \quad \#\mathrm{succ}(t) = (2, \#t), \quad \#(s + t) = (3, \#s, \#t), \quad \#(s \cdot t) = (4, \#s, \#t).$$

To reason formally in PRA, we dispense with any prior notion of term and simply define a term to be any number that arises in this way. Our representation of tuples has the convenient property that each element of the tuple is numerically smaller than the tuple itself, so the characteristic function of the predicate $\mathrm{Term}(t)$ can be defined by a course-of-values recursion, as described in Section 8.4. The predicate then satisfies the following:

$$\begin{aligned}\mathrm{Term}(t) \leftrightarrow {} & (((t)_0 = 0 \wedge \mathrm{length}(t) = 2) \vee ((t)_0 = 1 \wedge \mathrm{length}(t) = 1) \vee \\ & ((t)_0 = 2 \wedge \mathrm{length}(t) = 2 \wedge \mathrm{Term}((t)_1)) \vee \\ & (((t)_0 = 3 \vee (t)_0 = 4) \wedge \mathrm{length}(t) = 3 \wedge \mathrm{Term}((t)_1) \wedge \mathrm{Term}((t)_2)))\end{aligned}$$

We can also define a primitive recursive function $\mathrm{subst}(t, s, i)$ which returns the result of substituting s for x_i in t, and prove properties of terms and substitution in PRA. In a similar way, we can define a predicate $\mathrm{Form}(A)$ for formulas and define the corresponding notion of substitution. For simplicity, we will deal with *syntactic formulas*, as defined in Section 1.6, which is to say, we will not identify formulas up to renaming of bound variables. But we could, if we wanted to: equivalence up to renaming of bound variables is primitive recursive, and we could define $\mathrm{Term}(t)$ to pick out the numerically smallest representative of each class. The operations which construct terms and formulas from smaller ones are primitive recursive. We can define functions on terms and formulas by structural recursion, by cases, or by recursion on other complexity measures. In short, we can develop a perfectly adequate theory of syntax, like the one outlined in Chapter 1, in PRA.

What about semantics? In other words, can we say, in the language of arithmetic, what it means for a sentence in the language of arithmetic to be *true* in the standard model? Here there is cause for concern. For one thing, we argued in Section 8.6 that there is no primitive recursive evaluation function for primitive recursive functions. For another, the recursive definition of satisfaction in Section 5.1 specifies the truth value of formulas with quantifiers in terms of quantification over all elements in the model, and we would not expect to be able to do that in a primitive recursive way. In this section, we will show the following:

- The notion of truth for Δ_0 sentences is primitive recursive.
- For each n, the notion of truth for Σ_n sentences is Σ_n, and analogously for Π_n formulas.

- The adequacy of the corresponding definitions can be proved in first-order primitive recursive arithmetic.
- The notion of truth for Π_n sentences is not Σ_n and vice versa, and so the set of true sentences in the language of arithmetic is not arithmetically definable.

We will also show how to use truth predicates to prove metamathematical properties of theories of arithmetic in first-order arithmetic itself. In particular, we will show that for every n, $\mathrm{I\Sigma_{n+1}}$ proves the consistency of $\mathrm{I\Sigma_n}$. In Chapter 12, we will see that this implies that $\mathrm{I\Sigma_{n+1}}$ is strictly stronger than $\mathrm{I\Sigma_n}$.

Instead of defining predicates $\mathrm{Tr}(A)$ that say that a sentence (represented by) A is true in the standard model, we will define relations $\mathrm{Sat}(A, \sigma)$ that say that a formula A is true when each variable x_i is assigned the value $(\sigma)_i$. Here σ is a (number representing a) finite sequence, and we generally adopt the convention that $(\sigma)_i = 0$ when $i \geq \mathrm{length}(\sigma)$. The predicate $\mathrm{Tr}(A)$ is just $\mathrm{Sat}(A, ())$ with A restricted to sentences. In the context of arithmetic, satisfiability can also be defined in terms of truth, since A is satisfied by σ if and only if the result of substituting the numeral denoting $(\sigma)_i$ for each x_i in A is true. But focusing on satisfiability instead of truth avoids relying on the fact that every element in the interpretation is represented by a term in the language, and it provides for a smoother definition.

Formalization of syntax is inevitably confusing because two languages are involved: there are terms and formulas in the metatheory, which is to say, terms and formulas of the theory in which the formalization takes place, and terms and formulas of the target theory, which is to say, the formalized notions of term and formula under consideration. Here we take both to be the language of arithmetic, but when we are working in the metatheory, it will be convenient to take the language of arithmetic to include the language of PRA.

If A is a formula in the language of arithmetic, it can be represented by a natural number n, and that natural number can be represented by a term t in the language of arithmetic. We will use $\ulcorner A \urcorner$ to stand for that numeral: it is the formal representation of A in the language of arithmetic. For example, we can use the notation $\ulcorner 2+2=4 \urcorner$. This denotes a term in the language of arithmetic, specifically the term we use to represent the formula $2+2=4$. This is convenient because it allows us to use familiar notation from the metatheory to describe terms denoting the corresponding objects in the formal theory under consideration.

We will use the corner-quote notation in even more flexible ways. If we want to formalize, in the language of PRA, the statement that for every pair of formulas A and B, $A \wedge B \to A$ is an axiom of first order logic, we might describe this formula by writing

$$\forall A, B\, (\mathrm{Form}(A) \wedge \mathrm{Form}(B) \to \mathrm{Ax}(\ulcorner A \wedge B \to A \urcorner)).$$

The formula would be written, more accurately,

$$\forall A, B\, (\mathrm{Form}(A) \wedge \mathrm{Form}(B) \to \mathrm{Ax}(\mathrm{implies}(\mathrm{and}(A, B), A))),$$

where $\mathrm{implies}(x, y)$ and $\mathrm{and}(x, y)$ are primitive recursive functions that construct the formula in question. The notation $\ulcorner (A \wedge B) \to A \urcorner$ allows us to use the notation of the metatheory – the usual binary connectives – to describe the formal construction. For another example, we may wish to assert, formally, that for every natural number n, some theory T of arithmetic proves the result of substituting the numeral for n into A. We might express this formally by writing

$$\forall n \, \mathrm{Prov}_T(\ulcorner A(n) \urcorner),$$

where what we mean is really

$$\forall n \, \mathrm{Prov}_T(\mathrm{subst}(A, \mathrm{numeral}(n), x)),$$

where $\mathrm{numeral}(n)$ is the primitive recursive function which for every n returns (the number representing) the corresponding numeral.

Using corner brackets in this way is sometimes called *quotation*: $\ulcorner A \urcorner$ denotes not the formula A but a *description* of A in the theory in question. Using expressions from the metatheory within the corner brackets – for example, using A to range over an arbitrary formula or n to range over an arbitrary numeral – is known as *antiquotation*. In these examples, A is a variable in the metatheory, and n denotes the construction of the numeral, which is also defined in the metatheory. In principle, we can always dispense with corner brackets in favor of formal syntactic descriptions. But, with practice, the borders between the theory and metatheory become clear, and the expressions within the corner brackets are generally more perspicuous.

We start by defining a primitive recursive function $\mathrm{eval}(t, \sigma)$ for formulas in the language of arithmetic as follows:

$$\begin{aligned}
\mathrm{eval}(\ulcorner x_i \urcorner, \sigma) &= (\sigma)_i \\
\mathrm{eval}(\ulcorner 0 \urcorner, \sigma) &= 0 \\
\mathrm{eval}(\ulcorner \mathrm{succ}(u) \urcorner, \sigma) &= \mathrm{eval}(u, \sigma) + 1 \\
\mathrm{eval}(\ulcorner u + v \urcorner, \sigma) &= \mathrm{eval}(u, \sigma) + \mathrm{eval}(v, \sigma) \\
\mathrm{eval}(\ulcorner u \cdot v \urcorner, \sigma) &= \mathrm{eval}(u, \sigma) \cdot \mathrm{eval}(v, \sigma).
\end{aligned}$$

Because subterms are numerically smaller, this is just a course-of-values recursion, as described in Section 8.4. Primitive recursive arithmetic can prove that $\mathrm{eval}(t, \sigma)$ satisfies its defining equations. For concreteness, we can assign the value 0 when t is not a valid term.

We can also define a primitive recursive function $\mathrm{Sat}_{\Delta_0}(A, \sigma)$ for Δ_0 formulas, so that $\mathrm{Sat}_{\Delta_0}(A, \sigma)$ holds if and only if one of the following is the case:

- A is of the form $u = v$, and $\mathrm{eval}(u, \sigma) = \mathrm{eval}(v, \sigma)$.
- A is of the form $u \leq v$, and $\mathrm{eval}(u, \sigma) \leq \mathrm{eval}(v, \sigma)$.
- A is of the form $B \wedge C$ and $\mathrm{Sat}_{\Delta_0}(B, \sigma)$ and $\mathrm{Sat}_{\Delta_0}(C, \sigma)$.
- A is of the form $B \vee C$ and either $\mathrm{Sat}_{\Delta_0}(B, \sigma)$ or $\mathrm{Sat}_{\Delta_0}(C, \sigma)$.
- A is of the form $B \to C$ and if $\mathrm{Sat}_{\Delta_0}(B, \sigma)$ then $\mathrm{Sat}_{\Delta_0}(C, \sigma)$.
- A is of the form $\forall x < t \, B$, and for every n less than $\mathrm{eval}(t, \sigma)$, $\mathrm{Sat}_{\Delta_0}(B, \sigma[x \mapsto n])$.
- A is of the form $\exists x < t \, B$, and for some n less than $\mathrm{eval}(t, \sigma)$, $\mathrm{Sat}_{\Delta_0}(B, \sigma[x \mapsto n])$.

You can check that these clauses give the characteristic function of Sat_{Δ_0} the form described by Theorem 8.4.2: if we measure the size of the input (A, σ) by the complexity of A, each clause determines the value of $\mathrm{Sat}_{\Delta_0}(A, \sigma)$ from a finite set of smaller inputs.

Finally, for each n, we define formulas $\mathrm{Sat}_{\Sigma_n}(A, \sigma)$ and $\mathrm{Sat}_{\Pi_n}(A, \sigma)$ by simultaneous recursion on n. The recursion on n is in the meta-metatheory, that is, ordinary mathematics.

We start with $\mathrm{Sat}_{\Sigma_0} = \mathrm{Sat}_{\Pi_0} = \mathrm{Sat}_{\Delta_0}$, and assuming Sat_{Σ_n} and Sat_{Π_n} have been defined, we define $\mathrm{Sat}_{\Sigma_{n+1}}(A, \sigma)$ so that it expresses the following:

> A is of the form $\exists \vec{x}\, B$, where B is Π_n, and there is a σ' that differs from σ only at variables among $\vec{x}$, such that $\mathrm{Sat}_{\Pi_n}(B, \sigma')$.

To make sense of this definition, it suffices to know that the syntactic operations of extracting $\vec{x}$ and B from A are primitive recursive. The definition of $\mathrm{Sat}_{\Pi_{n+1}}$ is similar. The next theorem sums up the state of affairs.

Theorem 10.5.1. *We have the following:*

- *The evaluation function for terms in the language of arithmetic is primitive recursive.*
- *The satisfaction relation for Δ_0 formulas in the language of arithmetic is primitive recursive.*
- *For each n, the satisfaction relation for Σ_n formulas is Σ_n-definable.*
- *For each n, the satisfaction relation for Π_n formulas is Π_n-definable.*

Moreover, PRA *can prove that each of these satisfies its recursive specification.*

The last claim assumes that we are working with a first-order formulation of PRA, though it works for both intuitionistic and classical logic. It implies, in particular, that PRA proves $\mathrm{Tr}_{\Sigma_n}(\ulcorner A \urcorner) \leftrightarrow A$ for each Σ_n sentence A, and similarly for Π_n sentences. But remember that, from an intuitionistic perspective, the Σ_n and Π_n hierarchies do not exhaust the set of all formulas in the language. There are other ways of stratifying the set of formulas in the intuitionistic setting, such as simply counting the number of connectives and quantifiers. We can define partial truth predicates for those hierarchies in a similar way, and establish analogues of Theorem 10.5.1 in intuitionistic PRA. For the remainder of this section, we will focus on the arithmetic hierarchy and classical theories of arithmetic, but similar results hold for intuitionistic theories and hierarchies as well.

We now contrast the positive result of Theorem 10.5.1 with a negative one.

Theorem 10.5.2. *Let $\mathrm{Tr}_{\Sigma_n}(x)$ be the predicate that holds if and only if x represents a true Σ_n sentence. Then for every $n \geq 1$, Tr_{Σ_n} is Σ_n but not Π_n. In particular, the predicate $\mathrm{Tr}(x)$ that holds if and only if x represents a true sentence in the language of arithmetic is not arithmetically definable.*

This shows that the arithmetic hierarchy is really a hierarchy. The last sentence is known as *Tarski's theorem*, and it follows from the first sentence because if $\mathrm{Tr}(x)$ were arithmetically definable, it would be Π_n for some n (and hence for some $n \geq 1$). Restricting the predicate to Σ_n sentences would then violate the first claim.

Proof The fact that Tr_{Σ_n} is Σ_n for each $n \geq 1$ follows from Theorem 10.5.1. To see that it is not Π_n, let $n \geq 1$, and suppose, aiming for a contradiction, that $\mathrm{Tr}_{\Sigma_n}(x)$ is Π_n-definable. Let $R(x, y)$ be the relation that holds if and only if x represents a sentence $A(u)$ with only the free variable shown, and $A(\bar{y})$ is true, where $\bar{y}$ is the numeral for y. $R(x, y)$ can be defined from $\mathrm{Tr}_{\Sigma_n}(x)$ using a primitive recursive function to construct $\bar{y}$ and substitute it into the formula represented by x, and so $R(x, y)$ is Π_n as well.

Let $S(x)$ be the relation $\neg R(x, x)$. Then S is Σ_n, say, defined by Σ_n formula $B(x)$. But then we have

$$S(\ulcorner B(x) \urcorner) \equiv \neg R(\ulcorner B(x) \urcorner, \ulcorner B(x) \urcorner) \equiv \neg B(\ulcorner B(x) \urcorner) \equiv \neg S(\ulcorner B(x) \urcorner),$$

which is a contradiction. □

You can view this as a diagonalization argument. In the proof, $R(x, y)$ is a *universal* Σ_n relation, in the sense that for every Σ_n predicate $S(y)$, there is a k such that $R(k, y)$ is equivalent to $S(y)$. In other words, $R(k, y)$ runs through all the Σ_n predicates $S(y)$ as k runs through the natural numbers. But now we define $S(x) \equiv \neg R(x, x)$, ensuring that for each natural number n, $S(y)$ differs from $R(k, y)$ when y is equal to k. Assuming R is Π_n, we have that S is Σ_n, a contradiction.

Just as we can define primitive recursive predicates for the representations of terms and formulas, we can do the same for proofs. For each $n \geq 0$, let $\mathrm{Pr}_{\mathrm{I}\Sigma_\mathrm{n}}(x, y)$ say that x is a proof of y in $\mathrm{I}\Sigma_\mathrm{n}$. One way to go about this is to use the formulation of provability in terms of an axiomatic proof system: x is a proof of y if and only if x is a sequence of formulas, each of which either is an axiom of first-order logic, is an axiom of $\mathrm{I}\Sigma_\mathrm{n}$, or follows from previous formulas in the sequence by one of the inference rules of first-order logic. It is not hard to spell out the details in a way that shows that the relation $\mathrm{Pr}_{\mathrm{I}\Sigma_\mathrm{n}}(x, y)$ is primitive recursive. In fact, since we can define the axioms of each theory $\mathrm{I}\Sigma_\mathrm{n}$ uniformly with a primitive recursive relation, we can take n to be a variable. We can similarly define provability for PA or intuitionistic theories like HA.

In general, given an axiomatic theory T, we will write $\mathrm{Pr}_T(x, y)$ to denote a formal expression of the statement that x is a proof of y in T, and we will write $\mathrm{Prov}_T(y)$ for $\exists x \, \mathrm{Pr}_T(x, y)$, the statement that y is provable. Theorem 10.5.3 below refers to provability in pure first-order logic with equality, which we will write $\mathrm{Prov}(y)$.

The axioms of PA are true in the standard model, and so, by soundness, any sentence provable from the axioms of PA is true in that model. Can we hope to prove that fact in PA? Theorem 10.5.2 shows that we cannot even *state* it in the language of PA, since we cannot express the notion of truth in the standard model.

Let's try again. Using a partial truth predicate Tr_{Σ_n}, we can say that every Σ_n sentence provable from the axioms of PA is true. Such a formal statement, which uses a language L to say something about provability of formulas in a language L, is known as a *reflection principle*. Can we hope to prove it in PA?

Another problem arises: there are infinitely many induction axioms of PA, and it is not clear how PA can possibly prove, uniformly, that they are all true. In fact, Theorem 10.5.3 and the second incompleteness theorem, which we will discuss in Chapter 12, imply that it cannot.

Let's try yet again. For each particular n, PA *can* prove the truth of each axiom schema $(\mathrm{I}\Sigma_\mathrm{n})$. So maybe for fixed n and m, PA can prove that every Σ_m sentence provable in $\mathrm{I}\Sigma_\mathrm{n}$ is true. The argument would go something like this: let p be a proof of a Σ_m sentence, A, from the axioms of $\mathrm{I}\Sigma_\mathrm{n}$. By induction, every line of p is true, and hence the conclusion is true.

There is yet another problem: even though we have set bounds on the complexity of the axioms and the conclusion, the proof p can make use of formulas of arbitrary complexity along the way. A partial truth predicate (or satisfaction relation) is therefore not sufficient to state the correctness of every line in an arbitrary proof. But this time, there is a solution. If there is a proof of a Σ_m formula from some axioms of $\mathrm{I}\Sigma_\mathrm{n}$, there is a cut-free proof, and the subformula property implies that the complexity of formulas that appear in any such proof is bounded by the complexity of the axioms and the conclusion.

We will obtain the result just described as a consequence of the following.

Theorem 10.5.3. *For every $n \geq 1$, $\mathrm{I}\Sigma_n$ proves the following:*

$$\forall A\,(\mathrm{Sent}(A) \wedge A \in \Pi_{n+2} \wedge \mathrm{Prov}(A) \rightarrow \mathrm{Tr}_{\Pi_{n+2}}(A)).$$

In words, whenever a Π_{n+2} sentence A is provable in first-order logic with equality, then that sentence is true.

The reflection principle in the statement of the theorem can be upgraded to an arbitrary Π_{n+2} *formula* A, replacing the conclusion by $\forall \sigma\, \mathrm{Sat}_{\Pi_{n+2}}(A, \sigma)$. This is because whenever a formula is provable, so is the result of substituting arbitrary numerals for its free variables. In the other direction, the principle is also equivalent to the version with Π_{n+2} replaced by Σ_{n+1}, for a similar reason: if first-order logic proves the Π_{n+2} formula $\forall \vec{x}\, B$, where B is Σ_{n+1}, then it also proves every substitution instance of B, and the weaker principle then implies that B is true for every value assigned to $\vec{x}$. We will prove the principle in this latter form.

Proof We reason in $\mathrm{I}\Sigma_n$. Suppose B is the Σ_{n+1} sentence $\exists \vec{y}\, C$, where C is Π_n, and B is provable in pure first-order logic with equality. Then it has a cut-free proof, say, in the one-sided calculus for first-order logic with the equality axioms described in Section 4.3. We need to show that B is true.

Number all the sequents in the proof so that later sequents are derivable from earlier ones. The natural thing to do is to prove by induction that the sequent numbered i is true for all assignments to its free variables. Unfortunately, the complexity of this statement goes beyond Σ_n induction. By the subformula property, every formula in the proof is Σ_{n+1}, but the statement "for every assignment to the free variables of the formulas occurring in sequent Γ, some formula in Γ is true" is Π_{n+2}.

The trick is that once we start adding the outermost existential quantifiers to an instance of C, we no longer care about arbitrary assignments to the free variables, and the relevant instances of the existential quantifiers occur in the proof. So, we can argue as follows. If B is true, we are done. So it suffices to assume that B is false, and prove a contradiction from that. Hence, we can assume that for every assignment of values to the variables $\vec{y}$, C is false.

We now show by induction that for every sequent Γ occurring in the proof and every assignment to the free variables, one of the formulas in Γ of complexity at most Π_n is true. This is now a Π_n statement, Since the last sequent in the proof is $\{B\}$, and that contains no Π_n formulas, we have the contradiction we are after.

The induction is now straightforward. $\mathrm{I}\Sigma_n$ proves that every instance of an equality axiom is true, and that truth is preserved by the rules for $\wedge$, $\vee$, and $\forall$. The only step that requires some thought is the $\exists$ rule. Suppose we derive $\Gamma, \exists z\, D$ from $\Gamma, D[t/z]$. If $\exists z\, D$ is Π_n, the claim follows easily from the inductive hypothesis. If $D[t/z]$ is Σ_{n+1}, again the claim follows easily from the inductive hypothesis, since it implies that one of the Π_n formulas in Γ is true. The only subtle case is where $D[t/z]$ is exactly Π_n, so that $\exists z\, D$ is Σ_{n+1}. In that case, since $\exists z\, D$ is a generalized subformula of $\exists \vec{y}\, C$, $D[t/z]$ is of the form $C[\vec{s}/\vec{y}]$ for some tuple of terms $\vec{s}$. Our initial assumption implies that $D[t/z]$ is false, and hence the inductive hypothesis again implies that some formula in Γ is true. □

You are asked to prove the following as an exercise:

Lemma 10.5.4. *For every $n \geq 1$, $\mathrm{I}\Sigma_\mathrm{n}$ can be axiomatized by Π_{n+2} sentences.*

Corollary 10.5.5. *For every $n \geq 0$, $\mathrm{I}\Sigma_{n+1}$ proves that whenever $\mathrm{I}\Sigma_\mathrm{n}$ proves a Π_{n+3} sentence A, then A is true.*

Proof Argue in $\mathrm{I}\Sigma_{n+1}$. Suppose $\mathrm{I}\Sigma_\mathrm{n}$ proves a Π_{n+3} sentence A. Then pure logic proves $B \to A$, where B is a conjunction of Π_{n+2} axioms of $\mathrm{I}\Sigma_\mathrm{n}$. This formula is equivalent to one that is Π_{n+3}, and hence true. Since B is true, A is true as well. □

A first-order theory T is said to be *finitely axiomatizable* if there is a finite set of axioms Γ such that $T = \{A \mid \Gamma \vdash A\}$. The next theorem is another application of partial truth predicates.

Theorem 10.5.6. *For every $n \geq 1$, $\mathrm{I}\Sigma_\mathrm{n}$ is finitely axiomatizable.*

In the statement of Theorem 10.5.6, we implicitly assume that $\mathrm{I}\Sigma_\mathrm{n}$ is expressed in the language of arithmetic rather than the language of PRA, since there are infinitely many function symbols in the language of PRA and there is no way to describe them all with finitely many axioms. Nonetheless, our proof of Theorem 10.5.6 makes use of the truth predicates that we have developed in the language of PRA.

Proof In $\mathrm{I}\Sigma_1(\mathrm{PRA})$, we can use a Σ_n definition to replace the Σ_n induction schema with a single induction axiom:

$$\forall \vec{z}\,(\mathrm{Tr}_{\Sigma_n}(\ulcorner A(0,\vec{z})\urcorner) \wedge \forall x\,(\mathrm{Tr}_{\Sigma_n}(\ulcorner A(x,\vec{z})\urcorner) \to \mathrm{Tr}_{\Sigma_n}(\ulcorner A(x+1,\vec{z})\urcorner)) \to \forall x\,\mathrm{Tr}_{\Sigma_n}(\ulcorner A(x,\vec{z})\urcorner)).$$

By Theorem 10.5.1, we have that for every formula $A(x,\vec{z})$, PRA can prove $\forall x, \vec{z}\,(A(x,\vec{z}) \leftrightarrow \mathrm{Tr}_{\Sigma_n}(\ulcorner A(x,\vec{z})\urcorner))$. To obtain that result, it suffices to know that PRA can prove the equations that characterize the behavior of the evaluation function for terms and the recursive conditions specifying the satisfaction relation. So now consider the induction axiom above together with a finite fragment of PRA that suffices to prove those equations and equivalences, as well as the usual universal axioms for 0, succ, $+$, $\cdot$, and $\leq$. This provides a finite set of consequences Γ of $\mathrm{I}\Sigma_\mathrm{n}(\mathrm{PRA})$ such that Γ proves all the axioms of $\mathrm{I}\Sigma_\mathrm{n}$. The only problem is that Γ is not in the language of arithmetic; it includes symbols from the language of PRA.

But we know that $\mathrm{I}\Sigma_\mathrm{n}(\mathrm{PRA})$ is interpretable in $\mathrm{I}\Sigma_\mathrm{n}$ in such a way that the axioms of $\mathrm{I}\Sigma_\mathrm{n}(\mathrm{PRA})$, and hence all their consequences, are provable. So let Γ' be the translations of Γ to the language of arithmetic. Then every sentence in Γ' is provable in $\mathrm{I}\Sigma_\mathrm{n}$, and, by the argument above, every axiom of $\mathrm{I}\Sigma_\mathrm{n}$ is provable from Γ'. So Γ' axiomatizes $\mathrm{I}\Sigma_\mathrm{n}$. □

Exercises

10.5.1. Show that for every $n \geq 1$ there is a function $f(x)$ with a Π_{n+1}-definable graph that does not have a Σ_{n+1} definable graph. (Hint: use the results in this section to generalize Exercise 11.3.2.)

10.5.2. Show that for every $n \geq 1$, $\mathrm{I}\Sigma_{\mathrm{n}}$ proves that if $A(x)$ is an arbitrary Σ_n formula, then, for any assignment of values to the free variables of A, the induction axiom $A(0) \wedge \forall x\,(A(x) \rightarrow A(\mathrm{succ}(x))) \rightarrow \forall x\, A(x)$ is true.

10.5.3. Convince yourself that PRA proves the cut elimination theorem.

10.5.4. Prove Lemma 10.5.4. Remember that by Proposition 10.3.5, $\mathrm{I}\Delta_0$ can be axiomatized by Π_1 sentences.

a. Show that for every $n \geq 0$, over $\mathrm{I}\Sigma_{\mathrm{n}}$, any instance of $(\mathrm{B}\Sigma_{n+1})$ is equivalent to a Π_{n+3} sentence.

b. Show that for every $n \geq 0$, over $\mathrm{I}\Sigma_{\mathrm{n}} + (\mathrm{B}\Sigma_{n+1})$, any instance of $(\mathrm{I}\Sigma_{\mathrm{n+1}})$ is equivalent to a Π_{n+3} sentence. Use the fact that the induction principle for A is equivalent to $\forall \vec{z}, x\,(A(0) \wedge \forall y < x\,(A(y) \rightarrow A(y+1)) \rightarrow A(x))$.

c. Prove the desired result by induction on n. To axiomatize $\mathrm{I}\Sigma_{n+1}$, first add a Π_{n+2} axiomatization of $\mathrm{I}\Sigma_{\mathrm{n}}$, then a Π_{n+3} formulation of $(\mathrm{B}\Sigma_{n+1})$, and finally a Π_{n+3} formulation of $(\mathrm{I}\Sigma_{n+1})$.

10.5.5. Show that for each finitely axiomatized fragment T of PRA, PRA proves a reflection principle for T. In particular, PRA proves the consistency of each of its finite fragments.

10.6 Conservation Results via Cut Elimination

Let T_1 be a theory in a language L_1 and let T_2 be a theory in a bigger language L_2. According to Definition 4.7.2, T_2 is said to be a *conservative extension* of T_1 if whenever T_2 proves a formula, T_1 proves it as well. If Γ is a class of formulas in L_1, then T_2 is said to be *conservative over* T_1 *for formulas in* Γ if the preceding statement holds when restricted to formulas in Γ. For example, we know that PA is conservative over HA for negative formulas, and also for Π_2 formulas. Sometimes one extends this locution to allow for a natural identification of formulas in Γ with counterparts in the language of T_1, for example, when we identify formulas in the language of arithmetic with their natural translations to the language of PRA. We can think of such a conservation theorem as providing a *reduction* of T_2 to T_1 for the class of formulas Γ. Double-negation translations serve to reduce classical theories to constructive ones in this sense.

We now state three theorems about subsystems of first-order arithmetic, of which the first and third are conservation theorems. The first and second are analogous to Herbrand's theorem (Theorem 7.1.1) because they show that we can extract additional information from a formal derivation.

Theorem 10.6.1. $\mathrm{I}\Sigma_1$ *is conservative over first-order* PRA *for* Π_2 *formulas. In fact, if* $\mathrm{I}\Sigma_1$ *proves* $\forall \vec{x}\, \exists y\, A(x, y)$*, where A is primitive recursive, there is a function symbol f such that quantifier-free* PRA *proves* $A(\vec{x}, f(\vec{x}))$.

Theorem 10.6.2 (Parikh's theorem). *If* $\mathrm{I}\Delta_0$ *proves* $\forall \vec{x}\, \exists y\, A(\vec{x}, y)$*, where A is* Δ_0*, there is a term t such that* $\mathrm{I}\Delta_0$ *proves* $\forall \vec{x}\, \exists y < t\, A(\vec{x}, \vec{y})$.

Theorem 10.6.3. *For every $n \geq 1$, the theory* $\mathrm{B}\Sigma_{n+1}$ *is conservative over* $\mathrm{I}\Sigma_{\mathrm{n}}$ *for* Π_{n+2} *formulas.*

The statements of Theorems 10.6.1 and 10.6.2 can both be extended to the case where there is more than one existential quantifier, using the fact that the sentence $\forall \vec{x}\, \exists y_0, \ldots, y_{n-1}\, A(\vec{x}, \vec{y})$ is equivalent to $\forall \vec{x}\, \exists \hat{y}, y_0 < \hat{y}, \ldots, y_{n-1} < \hat{y}\, A(\vec{x}, \vec{y})$. Our proof of Theorem 10.6.2 will not be sensitive to the details of $I\Delta_0$, and we will see that the theorem continues to hold for extensions of $I\Delta_0$ satisfying mild requirements.

We will use the cut elimination theorem to prove Theorem 10.6.1, and we will sketch proofs of the other two theorems. In the next section, we will consider model-theoretic approaches instead. But first note that Theorems 10.6.1 and 10.6.2 bear on the relationship between classical and constructive versions of the theories involved.

Theorem 10.6.4. $I\Sigma_1$ *is conservative over* $I\Sigma_1^i$ *for* Π_2 *sentences.*

Proof If $I\Sigma_1$ proves $\forall \vec{x}\, \exists y\, A(x, y)$, where A is Δ_0, then by Theorem 10.6.1, there is a function symbol f such that PRA, and hence $I\Sigma_1^i$, proves $A(\vec{x}, f(\vec{x}))$. □

Theorem 10.6.5. $I\Delta_0$ *is conservative over* $I\Delta_0^i$ *for* Π_2 *sentences.*

Proof If $I\Delta_0$ proves $\forall \vec{x}\, \exists y\, A(\vec{x}, y)$, where A is Δ_0, then by Theorem 10.6.2 $I\Delta_0$ proves $\forall \vec{x}\, \exists y < t\, A(\vec{x}, y)$ for some term t. By the double-negation translation and Proposition 10.3.4, this is also provable in $I\Delta_0^i$. □

We now turn to the proof of Theorem 10.6.1. The second claim of that theorem follows from the first claim by Theorem 9.4.1, but we will prove the second claim directly. Our first task is to choose a convenient sequent formulation to which we can apply the cut elimination theorem. In the two-sided calculus, it is natural to express induction as a rule:

$$\frac{\Gamma \vdash \Delta, A(0) \qquad \Gamma, A(x) \vdash \Delta, A(x+1)}{\Gamma \vdash \Delta, A(t)}$$

This assumes x is not free in Γ, Δ. Alternatively, we can use the following:

$$\frac{\Gamma, A(x) \vdash \Delta, A(x+1)}{\Gamma, A(0) \vdash \Delta, A(t)}$$

But to simplify the proof we will use a one-sided calculus and assume that induction is formulated as follows:

$$\frac{\Gamma, A(0) \qquad \Gamma, \sim A(x), A(x+1)}{\Gamma, A(t)}$$

For $I\Sigma_1$, $A(x)$ is required to be a Σ_1 formula, possibly with parameters other than x. It is more convenient to work with $I\Sigma_1(\mathrm{PRA})$, which we take to be axiomatized by the universal axioms of PRA together with this last formulation of the induction rule. Moreover, using sequencing operations or a single bound to combine the existential quantifiers, we can assume that $A(x)$ is of the form $\exists y\, B(x, y)$, where B is quantifier-free.

Theorem 10.6.4 follows immediately from the next lemma, which is analogous to Herbrand's theorem. Thanks to the fact that we have definition by cases in primitive recursive arithmetic, we can get by with a single function providing a witness to each existential quantifier.

Lemma 10.6.6. *Suppose* $I\Sigma_1(\mathrm{PRA})$ *proves a sequent of the form*

$$\Gamma, \forall z\, B_0(z), \ldots, \forall z\, B_{\ell-1}(z), \exists y\, A_0(y), \ldots, \exists y\, A_{k-1}(y),$$

where all the free variables of this sequent are among $\vec{x}$, *and all the formulas in* $\Gamma \cup \{A_0, \ldots, A_{k-1}, B_0, \ldots, B_{\ell-1}\}$ *are quantifier-free. Then there are function symbols* $f_0, \ldots, f_{k-1}$ *such that* PRA *proves*

$$\Gamma, B_0(z_0), \ldots, B_{\ell-1}(z_{\ell-1}), A_0(f_0(\vec{x}, \vec{z})), \ldots, A_{k-1}(f_{k-1}(\vec{x}, \vec{z})).$$

Proof Appealing to a version of the cut elimination theorem, Theorem 6.7.2, we can assume that the only cuts in the proof in $I\Sigma_1(PRA)$ are cuts on formulas that are principal in an axiom or an instance of the induction rule. By the subformula property, every sequent in the proof is of the form described in the hypothesis of Lemma 10.6.6, so we can use induction on derivations. Notice that although the conclusion is stated in terms of the sequent calculus, the conclusion is really only about provability in PRA, and we can reason informally or in terms of any proof system that is convenient.

So let d be such a derivation of a sequent of the form described in the hypothesis of Lemma 10.6.6, such that the only cuts are of the form just mentioned. There are a number of possibilities for the last inference of d: it can be

- a quantifier-free axiom of the sequent calculus or $I\Sigma_1(PRA)$,
- a $\wedge$ or $\vee$ rule applied to quantifier-free premises,
- an $\exists$ rule, applied to a quantifier-free premise,
- a $\forall$ rule, applied to a quantifier-free premise,
- a cut where one of the cut formulas is a quantifier-free or existential formula, or
- an instance of the induction rule.

All cases but the last are fairly straightforward. Suppose, for example, the last inference is an instance of the $\wedge$ rule. Then the conclusion of this inference is of the form

$$\Gamma', \forall z\, B_0(z), \ldots, \forall z\, B_{\ell-1}(z), \exists y\, A_0(y), \ldots, \exists y\, A_{k-1}(y), C \wedge D$$

whereas the premises to this inference are

$$\Gamma', \forall z\, B_0(z), \ldots, \forall z\, B_{\ell-1}(z), \exists y\, A_0(y), \ldots, \exists y\, A_{k-1}(y), C$$

and

$$\Gamma', \forall z\, B_0(z), \ldots, \forall z\, B_{\ell-1}(z), \exists y\, A_0(y), \ldots, \exists y\, A_{k-1}(y), D.$$

For the rest of this proof it will be convenient to write $B(\vec{z})$ for $B_0(z_0) \vee \cdots \vee B_{\ell-1}(z_{\ell-1})$, where $z_0, \ldots, z_{\ell-1}$ are fresh variables. By the inductive hypothesis, there are function symbols $g_0, \ldots, g_{k-1}$ and $h_0, \ldots, h_{k-1}$ and proofs in PRA of

$$\Gamma', B(\vec{z}), A_0(g_0(\vec{x}, \vec{z})), \ldots, A_{k-1}(g_{k-1}(\vec{x}, \vec{z})), C$$

and

$$\Gamma', B(\vec{z}), A_0(h_0(\vec{x}, \vec{z})), \ldots, A_{k-1}(h_{k-1}(\vec{x}, \vec{z})), D.$$

Using weakening and the $\wedge$ rule we have a proof of

$$\Gamma', B(\vec{z}), A_0(g_0(\vec{x}, \vec{z})), A_0(h_0(\vec{x}, \vec{z})), \ldots, A_{k-1}(g_{k-1}(\vec{x}, \vec{z})), A_{k-1}(h_{k-1}(\vec{x}, \vec{z})), C \wedge D.$$

For each i, use definition by cases to define the function f_i such that

$$f_i(\vec{x}, \vec{z}) = \begin{cases} g_i(\vec{x}, \vec{z}) & \text{if } A_i(g_i(\vec{x}, \vec{z})) \\ h_i(\vec{x}, \vec{z}) & \text{otherwise.} \end{cases}$$

Since we can prove $A_i(g_i(\vec{x}, \vec{z})) \vee A_i(h_i(\vec{x}, \vec{z})) \to A_i(f_i(\vec{x}, \vec{z}))$ from the defining equations for f_i, we have a proof of

$$\Gamma', B(\vec{z}), A_0(f_0(\vec{x}, \vec{z})), \ldots, A_{k-1}(f_{k-1}(\vec{x}, \vec{z})), C \wedge D,$$

as desired.

Handling the cut rule is similar, and for the other rules, there is almost nothing to do. So the only really interesting case is where the last inference is an instance of the Σ_1 induction rule. In that case, the conclusion is of the form

$$\Gamma, \forall z\, B_0(z), \ldots, \forall z\, B_{\ell-1}(z), \exists y\, A_0(y), \ldots, \exists y\, A_{k-1}(y), \exists y\, C(y, q)$$

for some term q, and the premises are of the form

$$\Gamma, \forall z\, B_0(z), \ldots, \forall z\, B_{\ell-1}(z), \exists y\, A_0(y), \ldots, \exists y\, A_{k-1}(y), \exists y\, C(y, 0)$$

and

$$\Gamma, \forall z\, B_0(z), \ldots, \forall z\, B_{\ell-1}(z), \exists y\, A_0(y), \ldots, \exists y\, A_{k-1}(y), \forall w \sim C(w, u), \exists y\, C(y, u+1)$$

for a some u. By the inductive hypothesis, we obtain proofs of

$$\Gamma, B(\vec{z}), A_0(g_0(\vec{x}, \vec{z})), \ldots, A_{k-1}(g_{k-1}(\vec{x}, \vec{z})), C(\hat{g}(\vec{x}, \vec{z}), 0) \tag{10.1}$$

and

$$\Gamma, B(\vec{z}), A_0(h_0(\vec{x}, \vec{z}, w, u)), \ldots, A_{k-1}(h_{k-1}(\vec{x}, \vec{z}, w, u)), \sim C(w, u), C(\hat{h}(\vec{x}, \vec{z}, w, u), u+1). \tag{10.2}$$

Use primitive recursion to define

$$\hat{f}(0, \vec{x}, \vec{z}) = \hat{g}(\vec{x}, \vec{z})$$
$$\hat{f}(u+1, \vec{x}, \vec{z}) = \hat{h}(\vec{x}, \vec{z}, \hat{f}(u, \vec{x}, \vec{z}), u).$$

Then, for each $i < k$ use primitive recursion to define

$$f_i(0, \vec{x}, \vec{z}) = g_i(\vec{x}, \vec{z})$$
$$f_i(u+1, \vec{x}, \vec{z}) = \begin{cases} f_i(u, \vec{x}, \vec{z}) & \text{if } A_i(f_i(u, \vec{x}, \vec{z})) \\ h_i(\vec{x}, \vec{z}, \hat{f}(u, \vec{x}, \vec{z}), u) & \text{otherwise.} \end{cases}$$

Now let $D(u)$ be the formula

$$\bigvee \Gamma \vee B(\vec{z}) \vee A_0(f_0(u, \vec{x}, \vec{z})) \vee \cdots \vee A_{k-1}(f_{k-1}(u, \vec{x}, \vec{z})) \vee C(\hat{f}(u, \vec{x}, \vec{z}), u).$$

From (10.1) we have a proof of $D(0)$ in PRA. Moreover, by substituting $\hat{f}(u, \vec{x}, \vec{z})$ for w in (10.2) we obtain a proof of $D(u) \to D(u+1)$ in PRA. Using induction in PRA, we can therefore conclude $D(q)$. But then $f_0, \ldots, f_{k-1}, \hat{f}$ satisfy the conclusion of Lemma 10.6.6, once we replace the first argument u of each with the term q. □

We now turn to Parikh's theorem, Theorem 10.6.2. We will need the following:

Lemma 10.6.7. *Given any two terms r and s with at most the free variables shown, there is a term t such that $\mathrm{I}\Delta_0^{\mathrm{i}}$ proves $r \leq t$ and $s \leq t$. Given any two terms r and $s(y)$, where y does not occur in r, there is a term t such that $\mathrm{I}\Delta_0$ proves $\forall y < r\, (s(y) \leq t)$.*

Proof For the first claim, taking t to be $r+s$ works. Since the functions in the language of arithmetic are monotone, in the second claim we can take t to be the term $s(r)$. □

By Proposition 10.3.5, $I\Delta_0$ can be axiomatized by Π_1 formulas. Our proof uses only this fact and Lemma 10.6.7, so Theorem 10.6.2 holds more generally for any theory with these properties. In particular, we can add a symbol for any monotone function that can be axiomatized in a Π_1 way. To prove the theorem, use a one-sided sequent formulation of $I\Delta_0$, where as axioms we use all weakenings of Δ_0 instances of the Π_1 axioms.

Proof Suppose $I\Delta_0$ proves $\forall\vec{x}\,\exists y\, A(\vec{x}, y)$. Then, by cut elimination, there is a proof of $\exists y\, A(\vec{x}, y)$ in which the only cuts are on formulas that are principal in the axioms. In particular, every formula in the proof is either Δ_0 or the formula $\exists y\, A(\vec{x}, y)$ itself. So it suffices to show that whenever there is such a proof of a sequent $\Gamma, \exists y\, A(\vec{x}, y)$ in which every formula in Γ is Δ_0, then there is a term t whose free variables are among the free variables in that sequent, together with a proof of $\Gamma, \exists y < t\, A(\vec{x}, y)$. This can be proved by a straightforward induction on derivations.

First, suppose the last inference is an $\wedge$ rule:

$$\frac{\Gamma, \exists y\, A, B \qquad \Gamma, \exists y\, A, C}{\Gamma, \exists y\, A, B \wedge C}$$

From the inductive hypothesis, we have terms t_0 and t_1 and proofs of $\Gamma, \exists y < t_0\, A, B$ and $\Gamma, \exists y < t_1\, A, C$. From Lemma 10.6.7 we obtain a term $\hat{t}$ such that $I\Delta_0$ proves $t_0 \leq \hat{t}$ and $t_1 \leq \hat{t}$. Replacing t_0 and t_1 by $\hat{t}$ in the inductively obtained sequents and applying the $\wedge$ rule yields the desired conclusion.

The case where the last inference is an axiom, an $\vee$ rule, an $\exists$ rule, or a cut are all handled similarly. So suppose the last inference is a $\forall$ rule. In that case, it must introduce a bounded quantifier:

$$\frac{\Gamma, \exists y\, A, w \not< r \vee B}{\Gamma, \exists y\, A, \forall w < r\, B}$$

The eigenvariable condition tells us that w is not free in any other formula, and the definition of bounded quantification tells us that w is not free in r. From the inductive hypothesis, we get a term $t(w)$ and a proof of $\Gamma, \exists y < t(w)\, A, w \not< r \vee B$. By Lemma 10.6.7, we obtain a term $\hat{t}$ such that $w < r$ implies $t(w) \leq \hat{t}$. Then $I\Delta_0$ can prove the sequent $\Gamma, \exists y\, A, \forall w < r\, B$ as follows: if $\forall w < r\, B$ holds, we are done; otherwise, choose $w < r$ such that $\neg B$, and for that w, the inductive hypothesis tells us we have $\Gamma, \exists y < \hat{t}\, A$. □

Finally, I will a sketch a proof of Theorem 10.6.3, which says that for every n, $B\Sigma_{n+1}$ is conservative over $I\Sigma_n$. By Theorem 10.3.14, we can axiomatize $B\Sigma_{n+1}$ as $I\Delta_0$ together with the collection axioms ($B\Sigma_{n+1}$). In a sequent formulation, we express collection as the following rule:

$$\frac{\Gamma, x \not< w, \exists y < z\, A(x, y)}{\Gamma, \exists z\, \forall x < w\, \exists y < z\, A(x, y)}$$

Here we assume that x is not free in any formula in Γ and A is Π_n. The idea is that the antecedent of a collection axiom, $\forall x < w\, \exists y < z\, A(x, y)$, follows from the formula in the premise using the rules for $\vee$ and $\forall$. If we take Γ to be $\{\sim\forall x < w\, \exists y < z\, A(x, y)\}$, the premise is derivable in $I\Delta_0$, so this is equivalent to adding the collection axioms. With this formulation, we can apply an analogue of the general cut elimination theorem, Theorem 6.7.2. So if $\mathrm{B}\Sigma_{n+1}$ proves a Π_{n+2} theorem $\forall x\, \exists y\, A(x, y)$, where A is Π_n, there is a derivation of the sequent $\{\exists y\, A(x, y)\}$ in which the only cuts are on formulas that are principal in an axiom or the collection rule.

In the case where $n = 0$, the conservativity of $\mathrm{B}\Sigma_1$ over $\mathrm{I}\Delta_0$ is a consequence of the following:

Lemma 10.6.8. *If* $\mathrm{B}\Sigma_1$ *proves a sequent of the form*

$$\Gamma, \forall z\, B_0(z), \ldots, \forall z\, B_{\ell-1}(z), \exists y\, A_0(y) \vee \cdots \vee \exists y\, A_{k-1}(y),$$

where all the formulas in $\Gamma \cup \{A_0, \ldots, A_{k-1}, B_0, \ldots, B_{\ell-1}\}$ *are* Δ_0 *and the free variables are among* $\vec{x}$, *there are terms* $t_0(\vec{x}, \vec{z}), \ldots, t_{k-1}(\vec{x}, \vec{z})$ *such that* $\mathrm{I}\Delta_0$ *proves*

$$\Gamma, B_0(z_0), \ldots, B_{\ell-1}(z_{\ell-1}), \exists y < t_0(\vec{x}, \vec{z})\, A_0(y) \vee \cdots \vee \exists y < t_{k-1}(\vec{x}, \vec{z})\, A_{k-1}(y).$$

In fact, this directly gives us the result of the stronger statement obtained by applying the conclusion of Theorem 10.6.2 to Theorem 10.6.3. The proof is similar to the proof of Theorem 10.6.2, so instead we turn to the more difficult case where $n > 0$. In that case, the theorem is a consequence of the following lemma:

Lemma 10.6.9. *Suppose* $n > 0$ *and* $\mathrm{B}\Sigma_{n+1}$ *proves a sequent of the form*

$$\Gamma, \forall z\, B_0(z), \ldots, \forall z\, B_{\ell-1}(z), \exists y\, A_0(y) \vee \cdots \vee \exists y\, A_{k-1}(y),$$

where all the free variables of this sequent are among $\vec{x}$, *and all the formulas in* $\Gamma \cup \{A_0, \ldots, A_{k-1}\}$ *are* Π_n *and* $B_0, \ldots, B_{\ell-1}$ *are* Σ_n. *Then* $\mathrm{I}\Sigma_{\mathrm{n}}$ *proves*

$$\forall u\, \exists v\, \forall \vec{x} < u, \vec{z} < u\, (\bigvee \Gamma \vee B_0(z_0) \vee \cdots \vee B_{\ell-1}(z_{\ell-1}) \vee \exists y < v\, A_0(y), \ldots, \exists y < v\, A_{k-1}(y)).$$

Proof The lemma can be proved by induction on derivations of the restricted form given by the cut elimination theorem. The case where the last rule is the collection rule is almost immediate. The only case that is tricky is the $\forall$ rule:

$$\frac{\Gamma, \forall z\, B_0(z), \ldots, \forall z\, B_{\ell-1}(z), \exists y\, A_0(y), \ldots, \exists y\, A_{k-1}(y), D(w)}{\Gamma, \forall z\, B_0(z), \ldots, \forall z\, B_{\ell-1}(z), \exists y\, A_0(y), \ldots, \exists y\, A_{k-1}(y), \forall w\, D(w)}$$

If $D(w)$ is Σ_n, then it is grouped with the formulas $\forall z\, B_0(z), \ldots, \forall z\, B_{\ell-1}(z)$ and covered by the inductive hypothesis. Otherwise, it is Π_n. Let $B(\vec{x}, \vec{z})$ abbreviate $\bigvee \Gamma \vee \bigvee_i B_i(z_i)$ and let $A(\vec{y})$ abbreviate $\bigvee_j A_j(y_j)$. Then, by the inductive hypothesis, we can prove

$$\forall u\, \exists v\, \forall \vec{x} < u, \vec{z} < u, w < u\, (B(\vec{x}, \vec{z}) \vee \exists \vec{y} < v\, A(\vec{y}) \vee D(w)),$$

and hence

$$\forall u\, \exists v\, \forall \vec{x} < u, \vec{z} < u\, (B(\vec{x}, \vec{z}) \vee \exists \vec{y} < v\, A(\vec{y}) \vee \forall w < u\, D(w)).$$

We want to conclude the the formula in which $\forall w < u\, D(w)$ is replaced by $\forall w\, D(w)$. Remember that all the free variables of D are among $\vec{x}$. Given u, use the strong collection principle from Theorem 10.4.12 to obtain a u' such that we have

$$\forall \vec{x} < u\, (\exists w\, \neg D(w) \leftrightarrow \exists w < u'\, \neg D(w)),$$

and hence $\forall \vec{x} < u\, (\forall w\, D(w) \leftrightarrow \forall w < u'\, D(w))$. Without loss of generality, we can assume $u' \geq u$ by making u' larger if necessary. Apply the inductive hypothesis with u' in place of u to obtain a v satisfying

$$\forall \vec{x} < u', \vec{z} < u'\, (B(\vec{x}, \vec{z}) \vee \exists \vec{y} < v\, A(\vec{y}) \vee \forall w < u'\, D(w)).$$

We need $\forall \vec{x} < u, \vec{z} < u\, (B(\vec{x}, \vec{z}) \vee \exists \vec{y} < v\, A(\vec{y}) \vee \forall w\, D(w))$. Given $\vec{x} < u$ and $\vec{z} < u$, argue as follows: if $B(\vec{x}, \vec{z}) \vee \exists \vec{y} < v\, A(\vec{y})$, we are done. Otherwise, since $u \leq u'$, we have $\forall w < u'\, D(w)$, and hence $\forall w\, D(w)$, as required. □

Exercises

10.6.1. Show that the induction rules formulated at the beginning of this section are equivalent to the induction axioms, in that the two formulations prove the same theorems.

10.6.2. Fill in the details of the proof of Theorem 10.6.2.

10.6.3. Fill in the details of the proof of Theorem 10.6.3.

10.6.4. Show that PRA doesn't prove all the axioms of $I\Sigma_1$ outright. (Hint: Exercise 10.5.5 implies $I\Sigma_1$ proves the consistency of every finitely axiomatized fragment of PRA. Use Theorem 10.5.6 and the second incompleteness theorem, Theorem 12.4.1.)

10.7 Conservation Results via Model Theory

In this section, we will consider model-theoretic proofs of the conservation results of the last section, Theorems 10.6.1, 10.6.2, and 10.6.3. A conservation result has the general form "if T_2 proves A, where A is in Γ, then T_1 proves A as well." This is equivalent to saying that if T_1 doesn't prove A, then T_2 doesn't prove it either. Modulo the soundness and completeness of first-order logic, this is equivalent to saying that if $T_1 \cup \{\neg A\}$ has a model, then so does $T_2 \cup \{\neg A\}$. We can therefore prove such a conservation result by showing how to transform a model of T_1 to a model of T_2 in such a way that the truth of $\neg A$ is preserved. Like our model-theoretic proof of the classical cut elimination theorem in Section 6.3, such proofs can be mysterious: they imply that a proof in T_2 can be translated to a proof in T_1 without exhibiting a translation, or even providing any explicit information as to how long the proof in the weaker system may be.

I will present a proof of Theorem 10.6.1, the conservation of $I\Sigma_1$ over PRA, that is close in spirit to the syntactic proof presented in the last section but has the advantage of abstracting away the details of any specific deductive system. If $\mathfrak{M}$ is a structure for a language L, recall from the end of Section 5.6 that a *type with parameters from* $\mathfrak{M}$ is a set of formulas of $L(\mathfrak{M})$ with finitely many variables, and such a type Γ is *realized in* $\mathfrak{M}$ if there is an assignment of elements of the universe of $\mathfrak{M}$ to the variables that makes every sentence in Γ true. A type is *universal* if all the formulas are universal, and a type is *principal* if, in fact, it consists of a single formula. The *universal diagram* of $\mathfrak{M}$ is the set of universal sentences of $L(\mathfrak{M})$ that are true in $\mathfrak{M}$.

Definition 10.7.1. Let $\mathfrak{M}$ be a structure for a language L. $\mathfrak{M}$ is *Herbrand saturated* if every principal universal type consistent with the universal diagram of $\mathfrak{M}$ is realized in $\mathfrak{M}$.

A formula is said to be $\exists\forall$ if it is of the form $\exists\vec{x}\,\forall\vec{y}\,A(\vec{x},\vec{y})$, where A is quantifier-free and either of the tuples $\vec{x}$ and $\vec{y}$ can be empty. Definition 10.7.1 is equivalent to saying that any $\exists\forall$ sentence of $L(\mathfrak{M})$ that is consistent with the universal diagram of $\mathfrak{M}$ is true in $\mathfrak{M}$. Remember that, according to Definition 5.3.3, a model $\mathfrak{M}$ is said to be *saturated* if the following holds: whenever Γ is a type of $L(\mathfrak{M})$ involving a set of parameters of cardinality less than that of the universe of $\mathfrak{M}$, and Γ is consistent with the complete diagram of $\mathfrak{M}$, then Γ is realized in $\mathfrak{M}$. In contrast, Herbrand saturation only requires that principal universal types are realized; but to be realized, the type only has to be consistent with the *universal* diagram of $\mathfrak{M}$.

Theorem 10.7.2. *Every consistent universal theory has an Herbrand saturated model.*

Proof Let L be the language of T. For simplicity, I will assume that L is countable. As usual, using Zorn's lemma or a transfinite iteration yields the more general case.

Let L_ω denote a new language with an additional sequence of new constant symbols $c_0, c_1, c_2, \ldots$. Let $C_1(\vec{x}_1,\vec{y}_1)$, $C_2(\vec{x}_2,\vec{y}_2), \ldots$ enumerate the quantifier-free formulas of the new language. Recursively construct an increasing sequence of sets S_i of universal sentences, as follows. First, let S_0 be a set of universal axioms for T. At stage $i+1$, try to satisfy $\forall\vec{y}_{i+1}\,C_{i+1}(\vec{x}_{i+1},\vec{y}_{i+1})$: pick a new sequence of constants $\vec{c}$ that do not occur in S_i or C_{i+1}, and let

$$S_{i+1} = \begin{cases} S_i \cup \{\forall\vec{y}_{i+1}\,C_{i+1}(\vec{c},\vec{y}_{i+1})\} & \text{if this is consistent} \\ S_i & \text{otherwise.} \end{cases}$$

By induction, each S_i is consistent, and hence so is their union, S_ω. Let $\mathfrak{N}$ be a model of S_ω, and let $\mathfrak{M}$ be the submodel of $\mathfrak{N}$ whose universe is generated by the closed terms of L_ω; that is, $|\mathfrak{M}| = \{t^{\mathfrak{N}} \mid t \text{ a closed term of } L_\omega\}$. Since S_ω is a set of universal sentences, $\mathfrak{M}$ is also a model of S_ω, and therefore a model of T.

Note that every element of the universe of $\mathfrak{M}$ is denoted by one of the constants c_j. This is true because each element of the universe of $\mathfrak{M}$ is denoted by a term t in L_ω; if we pick i such that C_i is the formula $x = t$, then for some constant c the formula $c = t$ is in S_{i+1}.

Now it is not difficult to show that $\mathfrak{M}$ is Herbrand saturated. Suppose $\mathfrak{M} \not\models \exists\vec{x}\,\forall\vec{y}\,A(\vec{x},\vec{y},\vec{a})$, where A is quantifier-free and $\vec{a}$ is a sequence of parameters from $\mathfrak{M}$. We need to show that this sentence is inconsistent with the universal diagram of $\mathfrak{M}$. Let $\vec{d}$ be a sequence of constants in L_ω denoting the elements $\vec{a}$, choose i such that C_{i+1} is the formula $A(\vec{x},\vec{y},\vec{d})$, and let $\vec{c}$ be the constants used at stage $i+1$ in the construction. Then $\mathfrak{M} \not\models \forall\vec{y}\,A(\vec{c},\vec{y},\vec{d})$, and so, by the construction, this sentence is inconsistent with S_i. Since $\vec{c}$ does not occur in S_i, the formula $\exists\vec{x}\,\forall\vec{y}\,A(\vec{x},\vec{y},\vec{d})$ is also inconsistent with S_i. But, renaming $\vec{d}$ and the constants in S_i to the constants of $L(\mathfrak{M})$ that name the same elements, S_i is a subset of the universal diagram of $\mathfrak{M}$. □

Modifying the proof to use a transfinite iteration would enable us to realize arbitrary universal types, not just the principal ones.

If $\mathfrak{M}$ is any model and S is a finite subset of its universal diagram, then S is also satisfied by the submodel of $\mathfrak{M}$ generated by the elements mentioned in S. This can be used to show that the restriction to universal theories in Theorem 10.7.2 is necessary. For example, consider the theory T of dense linear orders, as described in Section 5.5, with the additional assumption that there exist at least two points. If $\mathfrak{M}$ is a model of T, then the $\exists\forall$ sentence asserting the existence of two points with nothing between them is consistent with the universal diagram of $\mathfrak{M}$ but is inconsistent with T, and hence false in $\mathfrak{M}$.

The following theorem describes a feature of Herbrand saturated models that makes them useful: any $\forall\exists$ sentence true in such a model is witnessed, in a strong way, by a finite set of terms with parameters.

Theorem 10.7.3. *Let $\mathfrak{M}$ be an Herbrand saturated structure for a language L. Suppose $\mathfrak{M} \models \forall\vec{x}\,\exists\vec{y}\,A(\vec{x},\vec{y},\vec{a})$, where $A(\vec{x},\vec{y},\vec{z})$ is a quantifier-free formula in L, and $\vec{a}$ is a sequence of parameters from $\mathfrak{M}$. Then there are a universal formula $B(\vec{z},\vec{w})$ with the free variables shown and sequences of terms $\vec{t}_0(\vec{z},\vec{w}),\ldots,\vec{t}_{k-1}(\vec{z},\vec{w})$, such that $\mathfrak{M} \models \exists\vec{w}\,B(\vec{a},\vec{w})$, and*

$$\models B(\vec{z},\vec{w}) \to A(\vec{x},\vec{t}_0(\vec{x},\vec{z},\vec{w}),\vec{z}) \vee \cdots \vee A(\vec{x},\vec{t}_{k-1}(\vec{x},\vec{z},\vec{w}),\vec{z}).$$

Note that the last formula is valid, and hence provable in pure first-order logic with equality. In particular, the conclusion of the theorem implies that there is a sequence of parameters $\vec{b}$ such that $\forall\vec{x}\,(A(\vec{x},\vec{t}_0(\vec{x},\vec{a},\vec{b}),\vec{a}) \vee \cdots \vee A(\vec{x},\vec{t}_{k-1}(\vec{x},\vec{a},\vec{b}),\vec{a}))$ is true in $\mathfrak{M}$. The proof of the theorem is essentially just an application of Herbrand's theorem.

Proof If $\exists\vec{x}\,\forall\vec{y}\,\neg A(\vec{x},\vec{y},\vec{a})$ is not true in $\mathfrak{M}$, then it is inconsistent with the universal diagram of $\mathfrak{M}$. Hence there is a universal formula $B(\vec{z},\vec{w})$ of L, and a sequence of parameters $\vec{b}$ from $\mathfrak{M}$, such that $\mathfrak{M} \models B(\vec{a},\vec{b})$ and $\models B(\vec{a},\vec{b}) \to \exists\vec{y}\,A(\vec{x},\vec{y},\vec{a})$. Replace the constants $\vec{a}$ and $\vec{b}$ by variables $\vec{z}$ and $\vec{w}$, note that the resulting formula is equivalent to an existential sentence, and apply Herbrand's theorem. □

Finally, the following theorem provides us with a recipe for proving conservation theorems.

Theorem 10.7.4. *Let T_2 be a universal theory and let T_1 be a theory in the language of T_2. If every Herbrand saturated model of T_2 is also a model of T_1, then every $\forall\exists$ sentence provable in T_1 is also provable in T_2.*

Proof Suppose every Herbrand saturated model of T_2 is a model of T_1. Let $A(\vec{x},\vec{y})$ be a quantifier-free formula in the language of T_2 with the free variables shown, and suppose that T_2 does not prove $\forall\vec{x}\,\exists\vec{y}\,A(\vec{x},\vec{y})$. We will show that T_1 does not prove it either.

The second assumption implies that $T_2 \cup \{\forall\vec{y}\,\neg A(\vec{d},\vec{y})\}$ is a consistent universal theory, where $\vec{d}$ is a sequence of new constants. By Proposition 10.7.2, there is an Herbrand saturated model of this theory; but then the reduct of this model to the language of T_2 is an Herbrand saturated model of T_2 satisfying $\exists\vec{x}\,\forall\vec{y}\,\neg A(\vec{x},\vec{y})$. By our hypothesis, this is also a model of T_1, in which $\forall\vec{x}\,\exists\vec{y}\,A(\vec{x},\vec{y})$ is false. □

Returning to $I\Sigma_1$, we will use this new framework to translate our syntactic proof of the fact that $I\Sigma_1$ is conservative over PRA for Π_2 sentences into a model-theoretic proof.

Model-theoretic proof of Theorem 10.6.4 Let $\mathfrak{M}$ be an Herbrand saturated model of PRA. By Proposition 10.7.4, we only need to show that $\mathfrak{M}$ satisfies the schema of Σ_1 induction. Over PRA, every Σ_1 formula is equivalent to one of the form $\exists y\,A(x,y,\vec{z})$, where A is quantifier-free, so it suffices to consider induction for formulas of that form.

To that end, suppose $\vec{a}$ is a sequence of parameters in $\mathfrak{M}$, and $\mathfrak{M}$ satisfies the induction hypotheses:

- $\exists y\, A(0, y, \vec{a})$
- $\forall x\, (\exists y\, A(x, y, \vec{a}) \rightarrow \exists y\, A(x + 1, y, \vec{a}))$.

We need to show that $\mathfrak{M}$ satisfies $\forall x\, \exists y\, A(x, y, \vec{a})$.

The second formula is equivalent to $\forall x, y\, \exists y'\, (A(x, y, \vec{a}) \rightarrow A(x + 1, y', \vec{a}))$. Using Theorem 10.7.3 and the fact that PRA supports definition by cases, we have that there are parameters $\vec{b}$ and c, and a function symbol $g(x, y, \vec{z}, \vec{w})$, such that $\mathfrak{M}$ satisfies the following:

- $M \models A(0, c, \vec{a})$.
- $\mathfrak{M} \models \forall x, y\, (A(x, y, \vec{a}) \rightarrow A(x + 1, g(x, y, \vec{a}, \vec{b}), \vec{a}))$.

Let $h(x, \vec{z}, v, \vec{w})$ be the function symbol of PRA with defining equations

$$h(0, \vec{z}, v, \vec{w}) = v$$
$$h(x + 1, \vec{z}, v, \vec{w}) = g(x, h(x, \vec{z}, v, \vec{w}), \vec{z}, \vec{w}).$$

Then $\mathfrak{M}$ satisfies

- $A(0, h(0, \vec{b}, c, \vec{a}), \vec{a})$ and
- $\forall x\, (A(x, h(x, \vec{b}, c, \vec{a}), \vec{a}) \rightarrow A(x, h(x + 1, \vec{b}, c, \vec{a}), \vec{a}))$.

Since $\mathfrak{M}$ is a model of PRA and hence satisfies quantifier-free induction, we have $\mathfrak{M} \models \forall x\, A(x, h(x, \vec{a}, c, \vec{b}), \vec{a})$, and hence $\mathfrak{M} \models \forall x\, \exists y\, A(x, y, \vec{a})$, as desired. □

Let us now turn to model-theoretic proofs of our other two conservation results. To prove Parikh's theorem, if $\mathfrak{M}$ is a model in the language of arithmetic, say $I \subseteq |\mathfrak{M}|$ is an *initial segment* of $\mathfrak{M}$ if it is closed downward, in the sense that if a is in I and $\mathfrak{M} \models b \leq a$ then b is in I as well. Say that in initial segment I is a *cut* in $\mathfrak{M}$ if I contains 0 and closed under the interpretations of succ, $+$, and $\cdot$. In that case, we can think of I as a substructure of $\mathfrak{M}$, in which the interpretations of succ, $+$, and $\cdot$ are the restrictions of those in $\mathfrak{M}$.

Clearly there are no proper cuts in the standard model of arithmetic, but the standard numbers themselves are a cut in any nonstandard model. More generally, if c is any nonstandard element in a model of $\mathrm{I}\Delta_0$, then the set $I = \{a \mid \text{for some standard } n,\ a \leq c^n\}$ is a cut, which may or may not be proper. By restricting the interpretation of the language of arithmetic from $\mathfrak{M}$ to I, we can consider I to be a model as well.

Proposition 10.7.5. *Let I be any cut in a model $\mathfrak{M}$ of* $\mathrm{I}\Delta_0$.

1. *For any Δ_0 formula $A(\vec{x})$ and any tuple $\vec{a} \in I$, $\mathfrak{M} \models A(\vec{a})$ if and only if $I \models A(\vec{a})$.*
2. *For any Π_1 formula $B(\vec{x})$ and any tuple $\vec{a} \in I$, if $\mathfrak{M} \models B(\vec{a})$ then $I \models B(\vec{a})$.*
3. *I is a model of* $\mathrm{I}\Delta_0$.
4. *If I is proper, then I is a model of* $\mathrm{B}\Sigma_1$.

Proof The first claim follows from an easy induction on formulas, and the second claim follows easily from the first. The third claim follows from Proposition 10.3.5, which says that $\mathrm{I}\Delta_0$ has a Π_1 axiomatization.

For the last claim, let suppose $b \notin I$, and suppose $I \models \forall x < a\, \exists y\, A(x, y)$, where A is a Δ_0 formula with parameters from I and a is in I. Then $\mathfrak{M} \models \forall x < a\, \exists y < b\, A(x, y)$, and since $\mathfrak{M}$ is a model of $\mathrm{I}\Delta_0$, there is a least such b. This b cannot be in $|\mathfrak{M}| \setminus I$, since if it is, then $b - 1$ is also in $|\mathfrak{M}| \setminus I$ and has the same property. So b is in I, and $I \models \forall x < a\, \exists y < b\, A(x, y)$. □

This provides us with another proof of Parikh's theorem, which says that for every Δ_0 formula $A(\vec{x}, y)$, if $\mathrm{I}\Delta_0$ proves $\forall \vec{x}\, \exists y\, A(\vec{x}, y)$, then there is a term $t(\vec{x})$ such that $\mathrm{I}\Delta_0$ proves $\forall \vec{x}\, \exists y < t(\vec{x})\, A(\vec{x}, y)$.

Model-theoretic proof of Theorem 10.6.2 Suppose that A is a Δ_0 formula and $\mathrm{I}\Delta_0$ does not prove $\forall \vec{x}\, \exists y < t\, A(\vec{x}, y)$ for any term $t(\vec{x})$. Let $\vec{c}$ be a tuple of new constants, and consider the set

$$\Gamma = \mathrm{I}\Delta_0 \cup \{\forall y < t(\vec{c})\, \neg A(\vec{c}, y) \mid t(\vec{x}) \text{ is any term}\}.$$

I claim that Γ is consistent. Otherwise, there is a finite set $\{t_0(\vec{x}), \ldots, t_{n-1}(\vec{x})\}$ such that $\mathrm{I}\Delta_0$ proves

$$\exists y < t_0(\vec{c})\, A(\vec{c}, y) \vee \cdots \vee \exists y < t_{n-1}(\vec{c})\, A(\vec{c}, y).$$

Setting $t(\vec{x}) = t_0(\vec{x}) + \ldots + t_{n-1}(\vec{x})$ we have that $\mathrm{I}\Delta_0$ proves $\exists y < t(\vec{c})\, A(\vec{c}, y)$. Since $\vec{c}$ are fresh constants, it also proves $\forall \vec{x}\, \exists y < t(\vec{x})\, A(\vec{x}, y)$, contrary to the hypothesis.

So Γ is consistent, and hence has a model, $\mathfrak{M}$. Let

$$I = \{a \in |\mathfrak{M}| \mid \mathfrak{M} \models a < t(\vec{c}) \text{ for some term } t(\vec{x})\}.$$

Then I is a cut and $I \models \forall y\, \neg A(\vec{c}, y)$, so $I \models \exists \vec{x}\, \forall y\, \neg A(\vec{x}, y)$. Hence $\mathrm{I}\Delta_0$ does not prove $\forall \vec{x}\, \exists y\, A(\vec{x}, y)$. □

As with the syntactic proofs of Theorems 10.6.2 and 10.6.3 in the last section, it is not hard to modify the model-theoretic proof of Parikh's theorem to show that $\mathrm{B}\Sigma_1$ is conservative over $\mathrm{I}\Delta_0$ for Π_2 sentences. Suppose $\mathrm{I}\Delta_0$ does not prove $\forall \vec{x}\, \exists y\, A(\vec{x}, y)$. Let Γ' be the set Γ defined above together with set $\{b > t(\vec{c}) \mid t \text{ is any term}\}$ for another fresh constant b. Then Γ' is still consistent, and if we apply the completeness theorem to Γ' instead of Γ we get a model $\mathfrak{M}$ with an element b that is not in I. By Proposition 10.7.5, I is a model of $\mathrm{B}\Sigma_1 \cup \{\neg \forall \vec{x}\, \exists y\, A(\vec{x}, y)\}$.

To extend the result to $\mathrm{B}\Sigma_{n+1}$ for $n > 0$, say that a substructure $\mathfrak{A}$ of a model $\mathfrak{B}$ of $\mathrm{I}\Delta_0$ is a *Σ_k-elementary substructure* of $\mathfrak{B}$ if for every Σ_k formula $A(\vec{x})$ and tuple $\vec{a}$ in $|\mathfrak{A}|$, if $\mathfrak{B} \models A(\vec{a})$, then $\mathfrak{A} \models A(\vec{a})$. The following is a slight variation on Proposition 5.6.1:

Proposition 10.7.6. *Suppose $\mathfrak{M}$ is a model of $\mathrm{I}\Delta_0$ and I is a cut in $\mathfrak{M}$. Then for any $n \geq 1$, I is an Σ_n-elementary substructure of $\mathfrak{M}$ if and only if the following holds: if $A(\vec{x}, y)$ is any Π_{n-1} formula and $\vec{a}$ is any tuple of elements from $|\mathfrak{A}|$, then $I \models \exists y\, A(\vec{a}, y)$ if and only if $\mathfrak{M} \models \exists y\, A(\vec{a}, y)$.*

The proof is once again a routine induction on formulas, where in the base case we use the fact that I and $\mathfrak{M}$ agree in Δ_0 formulas.

Lemma 10.7.7. *Suppose $n \geq 1$, $\mathfrak{M}$ is a model of $\mathrm{I}\Sigma_n$, and I is a proper cut in $\mathfrak{M}$. If I is a Σ_n-elementary substructure of $\mathfrak{M}$, then I is a model of $\mathrm{B}\Sigma_{n+1}$.*

Proof The proof is essentially the same as the proof of claim 4 of Proposition 10.7.5. Suppose $I \models \forall x < a\, \exists y\, A(x, y)$, where $A(x, y)$ is a Π_n formula with parameters from I and a is in I. Let b be any element of $|\mathfrak{M}| \setminus I$. Because I and $\mathfrak{M}$ agree on the truth of $A(u, v)$ when

u and v are in I, $\mathfrak{M} \models \forall x < a\, \exists y < b\, A(x, y)$. By Σ_n induction, there is a least such b, which has to be in I. □

Model-theoretic proof of Theorem 10.6.1 Suppose $\mathrm{I}\Sigma_n$ does not prove a sentence $\forall x\, \exists y\, A(x, y)$, where A is Π_n. Let b, c, $d_0, d_1, d_2, \ldots$ be new constants, and let $(B_i(x, y))_{i \in \mathbb{N}}$ be an enumeration of Π_n formulas so that every formula occurs infinitely many times on the list. Consider the set

$$\Gamma = \mathrm{I}\Delta_0 \cup \{\forall y\, \neg A(c, y)\} \cup \{b > d_i \wedge d_{i+1} > d_i^2 \wedge \forall x < d_i\, (\exists y\, B(x, y) \leftrightarrow \exists y < d_{i+1}\, B(x, y))\}_{i \in \mathbb{N}}.$$

Γ is consistent because, by Theorem 10.4.12, $\mathrm{I}\Sigma_n$ can use strong collection to prove the existence of witnesses to any finite subset of the last set in this union. Let $\mathfrak{M}$ be a model of Γ, and let $I = \{a \in |\mathfrak{M}| \mid \text{for some } i, \mathfrak{M} \models a < d_i\}$. Then it is not hard to check that I is a cut in $\mathfrak{M}$, I is a Σ_n-elementary substructure of $\mathfrak{M}$, and b is not in I. By Lemma 10.7.7, I is a model of $\mathrm{B}\Sigma_n \cup \{\exists x\, \forall y\, \neg A(x, y)\}$. □

Exercises

10.7.1. Justify the comment after Definition 10.7.1: show that a structure $\mathfrak{M}$ is Herbrand saturated if every $\exists\forall$ sentence of $L(\mathfrak{M})$ that is consistent with the universal diagram of $\mathfrak{M}$ is true in $\mathfrak{M}$.

10.7.2. Show that Theorem 10.7.2 implies (and is implied by) the statement that every model $\mathfrak{M}$ has a Σ_1-elementary extension that is Herbrand saturated. (Hint: let T be the universal diagram of $\mathfrak{M}$.)

10.7.3. Spell out the details of the model-theoretic proof of Theorem 10.6.1.

Bibliographical Notes

General references Hájek and Pudlák (1998), Kaye (1991), and Buss (1998b) are general references for subsystems of classical first-order arithmetic. Hájek and Pudlák (1998), Buss (1998a), and Krajíček (1995) also describe theories of arithmetic that are weaker than the ones discussed here.

Representing sequences There are various methods of representing sequences in the language of arithmetic, and all of the sources just listed offer ways of doing it. The method given in Exercise 10.2.2 can be found in Smullyan (1992). Hájek and Pudlák (1998) and Buss (1998a) provide representations whose correctness can be proved in weak theories of arithmetic, and they both establish the fact, mentioned in Section 10.2, that the graph of the exponential function has a Δ_0 definition.

Subsystems of first-order arithmetic Hájek and Pudlák (1998) provides more information about axiomatic principles like the ones considered in Section 10.3 and the relationships between them. Instead of saying that a theory T proves the finite separation principle for a formula $A(\vec{x})$, Hájek and Pudlák say that $A(\vec{x})$ is *piecewise coded* in T.

Intuitionistic arithmetic This chapter favors the metatheory of classical systems over intuitionistic systems, but the situation is reversed in Chapter 14, which focuses on computational interpretations. See Troelstra and van Dalen (1988) for more information about intuitionistic first-order arithmetic. Burr (2000) provides an analogue of the arithmetic hierarchy that is better suited to intuitionistic theories.

Conservation results The model-theoretic proofs of Theorems 10.6.2 and 10.6.3 in Section 10.7 are essentially the ones presented in Hájek and Pudlák (1998), which also offers a second model-theoretic proof of Theorem 10.6.3 and a model-theoretic proof of 10.6.1 that is different from the one presented here.

The proof-theoretic proofs of all three conservation theorems are essentially the ones described in Buss (1998b). An alternative approach to establishing the conservation of $I\Sigma_1$ over PRA is to use either the Dialectica interpretation or modified realizability together with an independent reduction of $I\Sigma_1$ to its intuitionistic counterpart. Both approaches require showing that the type 1 functionals of a subset of the primitive recursive functionals of finite type define primitive recursive functions. (See Chapter 14 and Avigad and Feferman (1998).)

Because the syntactic proofs of Theorems 10.6.4, 10.6.2, and 10.6.3 rely on the cut elimination theorem, they leave open the possibility that proofs can grow much longer. (See Exercise 6.4.4 and the notes at the end of Chapter 6.) In fact, for Theorems 10.6.4 and 10.6.2, this speedup is unavoidable. See Pudlák (1998) and Section 16.1 for a discussion of such phenomena. For a method of interpreting $I\Sigma_1$ in $I\Sigma_1^i$ without a substantial increase in length of proof, see Coquand and Hofmann (1999) and Avigad (2004). Exercise 10.5.5 can be used to show that first-order PRA can be translated to quantifier-free PRA without substantial increase in length: if first-order PRA proves a quantifier-free formula, quantifier-free PRA can prove that there is a cut-free derivation of every closed instance, and then, using a partial evaluation function, establish that every line in that derivation is true.

11

Computability

In Chapter 8, we defined the primitive recursive functions and showed that many common functions are among them. It is therefore reasonable to ask: do the primitive recursive functions exhaust the computable functions? Section 8.6 provided a negative answer to this question: if we systematically enumerate the unary primitive recursive functions $f_0, f_1, f_2, \ldots$ and define $g(x) = f_x(x) + 1$, we expect $g(x)$ to be computable even though it differs from each function f_x on input x.

This argument is disheartening because it points to a difficulty in obtaining *any* precise definition of the set of computable functions. After all, any such definition should tell us how we can describe a computable function, and if we are systematic enough, we should be able to enumerate all the descriptions. We can then repeat the argument above and obtain a computable function that is not on the list.

We will see that the solution is to focus on *partial* functions. We will allow for the possibility that the computations we describe may be undefined for some inputs, corresponding to the situation where a program fails to terminate. In the next section, we will define a class of partial functions, and, by fiat, declare these to be the computable ones. This class will contain all the primitive recursive functions and will be immune to the diagonalization argument above. From the definition, it should be compelling that every partial computable function conforms to an intuitive conception of what it means to be computable. It is less clear at the outset that the class of partial computable functions includes all the partial functions that we might ever consider to be computable. The claim that the formal definition accurately captures the intuitive notion is known as the *Church–Turing thesis*, and Section 11.6 discusses some of the evidence in its favor.

11.1 The Computable Functions

We will now define a collection of partial functions from the natural numbers to the natural numbers, of various arities, conventionally known as the *partial computable functions* or *partial recursive functions*. It would be more appropriate to call them the *computable partial functions*, but the former terminology has become entrenched, and we will stick with it.

The set of partial computable functions is defined, like the set of primitive recursive functions, as the closure of a set of initial function under certain operations. We will start with the same set of initial functions that we used to define the primitive recursive functions, namely, zero, successor, and the projection functions. To complete the definition, we need to do two things:

1. Modify the definition of composition and definition by primitive recursion to accommodate partial functions as well as total functions.
2. Add a new closure operation, one that can, in particular, give rise to partial functions.

The first is straightforward. If f and g are partial functions, we will write $f(x)\downarrow$ to express that f is defined at x, which is to say, that x is in the domain of f. We will write $f(x)\uparrow$ to express that f is not defined at x. We will use $f(x)\simeq g(x)$ to mean that either $f(x)$ and $g(x)$ are both undefined, or they are both defined and equal. This is often known as *Kleene equality*. If f is a partial function, we may write $f(x)\simeq y$, $f(x)\downarrow = y$, or simply $f(x)=y$ to say that f is defined at x and has the value y there.

We will adopt the convention that if g and $h_0, \ldots, h_{k-1}$ are all partial functions, then $g(h_0(\vec{x}), \ldots, h_{k-1}(\vec{x}))$ is defined if and only if each h_i is defined at $\vec{x}$, and g is defined at $h_0(\vec{x}), \ldots, h_{k-1}(\vec{x})$. We can therefore use the notation $t(\vec{x})\downarrow$, $t(\vec{x})\uparrow$, and $s(\vec{x})\simeq t(\vec{x})$ for terms as well. With this understanding, the definitions of composition and primitive recursion for partial functions are the same as the definitions of composition and primitive recursion for total functions presented in Section 8.1, except that "$=$" is replaced by "$\simeq$" to allow for the possibility that the specified value is undefined.

A priori, these definitions have nothing to do with computation. To link the formal definitions with an intuitive notion of computability, think of the values at which a computable function is undefined as being values at which the algorithm or program that computes it fails to terminate. With this intuition, the algorithm for computing a composition $g(h_0(\vec{x}), \ldots, h_{k-1}(\vec{x}))$ is to compute each $h_i(\vec{x})$ in turn and then apply g to the result. If any of the computations fail to terminate, so does the computation of the composition as a whole. A similar story applies to definition by primitive recursion: to compute $f(x, \vec{z})$, we follow the defining schemas to compute each of $f(0, \vec{z})$, $f(1, \vec{z})$, ... in turn. If any of these computations fails to terminate before we reach $f(x, \vec{z})$, the value of the latter is undefined.

What we will add to the definition of the primitive recursive functions to obtain partial functions is the *unbounded search* operator. If $f(x, \vec{z})$ is any partial function on the natural numbers, define $\mu x\, f(x, \vec{z})$ to be

> the least x such that $f(0, \vec{z}), \ldots, f(x, \vec{z})$ are all defined and $f(x, \vec{z})=0$, if such an x exists, and undefined otherwise.

This defines $\mu x\, f(x, \vec{z})$ uniquely.

The wording of the definition of $\mu x\, f(x, \vec{z})$ is delicate, but it fits the intuitive computational model. To compute $\mu x\, f(x, \vec{z})$, we compute $f(0, \vec{z})$, $f(1, \vec{z})$, and so on, until we find a value of x such that $f(x, \vec{z})$ is 0. If any of the intermediate computations fails to terminate, the value of $\mu x\, f(x, \vec{z})$ is undefined. It may turn out that all of $f(0, \vec{x})$, $f(1, \vec{x})$, ... are defined but none of them are equal to 0, in which case, $\mu x\, f(x, \vec{z})$ is again undefined.

If $R(x, \vec{z})$ is any relation, $\mu x\, R(x, \vec{z})$ is defined to be $\mu x\,(1 \dot{-} \chi_R(x, \vec{z}))$. In other words, $\mu x\, R(x, \vec{z})$ returns the least value of x such that $R(x, \vec{z})$ holds. So, if $f(x, \vec{z})$ is a total function, $\mu x\, f(x, \vec{z})$ is the same as $\mu x\,(f(x, \vec{z})=0)$. But the definition of $\mu x\, f(x, \vec{z})$ is more general since it allows for the possibility that $f(x, \vec{z})$ is not total, whereas the characteristic function of a relation is always total.

Definition 11.1.1. The set of *partial computable functions* is the smallest set of partial functions from the natural numbers to the natural numbers (of various arities) containing zero, successor, and projections, and closed under composition, primitive recursion, and unbounded search.

Of course, some of the partial computable functions happen to be total.

Definition 11.1.2. The set of *computable functions* is the set of partial computable functions that are total.

A computable function is sometimes called a *total computable function* to emphasize that it is defined everywhere, and I may adopt this terminology on occasion. But, in general, when I write "computable function" I mean "total computable function," and I will always use the word "partial" in the more general case.

As with the primitive recursive functions, we say that a set S or a relation $R(\vec{x})$ is computable if its characteristic function is computable.

Exercises

11.1.1. Let $f(x)$ be the partial function that is equal to 0 if x is prime, and is undefined otherwise. Show that f is a partial computable function.

11.1.2. Let $g(x)$ be the partial function that is equal to x if x is prime, and is undefined otherwise. Show that g is a partial computable function.

11.2 Computability and Arithmetic Definability

Suppose $f(x_0, \ldots, x_{k-1})$ is a partial function from $\mathbb{N}^k$ to $\mathbb{N}$. The *graph* of f is defined to be the relation $R(\vec{x}, y)$ that holds if and only if $f(\vec{x})$ is defined and equal to y. The following theorem represents an important correspondence between computability and arithmetic definability.

Theorem 11.2.1. *A partial function $f(\vec{x})$ is computable if and only if it has a Σ_1-definable graph.*

Proof Since the set of partial computable functions is defined to be the smallest set containing zero, successor, and projections and closed under composition, primitive recursion, and unbounded search, we can show that every partial computable function has a Σ_1 computable graph by induction on the set of partial computable functions. The graph of the constant 0 is defined by the formula $y = 0$, the graph of the successor function is defined by $y = \mathrm{succ}(x)$, and the graph of the projection function $p_i^n(\vec{x})$ is defined by the formula $y = x_i$.

Proposition 10.2.4 implies that the set of functions with Σ_1-definable graphs is closed under composition, and Proposition 10.2.6 implies that it is closed under primitive recursion. Essentially the same proofs work for partial functions as well. To show closure under composition, suppose the graph of the partial function $g(y_0, \ldots, y_{k-1})$ is defined by a Σ_1 formula $A(\vec{y}, z)$, and for every $i < k$, $B_i(\vec{x}, y_i)$ is a Σ_1 formula defining the graph of the partial function $h_i(\vec{x})$. Then the relation $g(h_0(\vec{x}), \ldots, h_{k-1}(\vec{x})) \simeq z$ is defined by the formula

$$\exists y_0, \ldots, y_{k-1} \; (B_0(\vec{x}, y_0) \wedge \cdots \wedge B_{k-1}(\vec{x}, y_{k-1}) \wedge A(\vec{y}, z)).$$

By the closure properties derived in Section 10.2, this is equivalent to a Σ_1 formula. Closure under primitive recursion is handled similarly.

Finally, to show closure under unbounded search, suppose $f(\vec{x}) \simeq \mu y \, g(y, \vec{x})$, where $g(y, \vec{x}) \simeq w$ is defined by the Σ_1 formula $A(y, \vec{x}, w)$. By the definition of unbounded search, we have that $f(\vec{x}) \simeq z$ is defined by the formula

$$\exists s\,(\forall i < z\,(A(i, \vec{x}, (s)_i) \wedge (s)_i \neq 0) \wedge A(z, \vec{x}, 0)).$$

This is equivalent to a Σ_1 formula, so we have established the forward direction of the theorem.

For the other direction, suppose the relation $f(\vec{x}) \simeq y$ is defined by the Σ_1 formula $\exists u\, A(\vec{x}, y, u)$, where A is Δ_0. Then, in particular, the relation defined by A is primitive recursive, and we can compute $f(\vec{x})$ by searching for a pair (y, u) satisfying $A(\vec{x}, y, u)$. In other words, we can write

$$f(\vec{x}) \simeq (\mu u\, A(\vec{x}, (u)_0, (u)_1))_0,$$

which shows that f is a partial computable function. □

By Proposition 10.2.2, the graph of a total function is Σ_1 if and only if it is Δ_1. Exercise 11.3.2 asks you to show that this is sharp, by exhibiting a function with a Π_1 computable graph that is not computable. Combining Proposition 10.2.2 with Theorem 11.2.1 yields the following characterization of computable sets, relations, and functions.

Corollary 11.2.2. *A set or relation is computable if and only if it is Δ_1-definable, and a function is computable if and only if it has a Δ_1-definable graph.*

We will put Theorem 11.2.1 to good use by deriving a number of important properties of the set of partial computable functions, properties that abstract away the details of any particular computational model. To start, we establish the existence of a *universal* partial computable function.

Theorem 11.2.3. *There is a partial computable function* Univ(e, x) *with the property that for every partial computable function f, there is an e such that $f(x) \simeq$* Univ(e, x) *for every x.*

Fixing such a function Univ(e, x), we will write $\varphi_e(x)$ for the function which maps x to Univ(e, x). Then $\varphi_0(x), \varphi_1(x), \varphi_2(x), \ldots$ enumerates the unary partial computable functions, and Univ(e, x) computes them all uniformly, in the sense that for every e and x it returns the value of the eth function on input x. (It really returns the value of the $(e+1)$st function on input x, but with such enumerations it is convenient to call φ_0 the zeroth function, and so on.)

Proof Let $R(e, x, y, z)$ hold if e represents a Δ_0 formula $A(u, v, w)$ with at most the three variables shown and A holds of x, y, and z. By the results of Section 10.5, $R(e, x, y, z)$ is primitive recursive.

Let $T(e, x, s)$ be the relation $R(e, x, (s)_0, (s)_1)$, let $U(s) = (s)_0$, and define Univ by

$$\text{Univ}(e, x) \simeq U(\mu s\, T(e, x, s)).$$

The relation T and the function U are both primitive recursive, so Univ(e, x) is clearly computable.

Let $f(x)$ be a partial computable function. Then f has a Σ_1 graph, given, say, by the formula $\exists z\, A(x, y, z)$, where A is Δ_0. Let e represent A. Then $f(x) \simeq y$ if and only if there is a z such that $R(e, x, y, z)$ holds, which happens if and only if there is an s such that $T(e, x, s)$ holds and $(s)_0 = y$. But this means for every x we have

$$f(x) \simeq U(\mu s\, T(e, x, s))) \simeq \text{Univ}(e, x),$$

as required. □

The proof we have just given also establishes the following important fact:

Theorem 11.2.4. *There are a primitive recursive predicate $T(e, x, s)$ and a primitive recursive function $U(s)$ such that for every e, $\varphi_e(s) \simeq U(\mu s\, T(e, x, s))$.*

Theorem 11.2.4 is known as *Kleene's normal form theorem* and the predicates T and U are often referred to as *Kleene's T and U*, after Stephen Kleene, who introduced that notation. From the point of view of computability theory, think of the primitive recursive functions and relations as being those that are very explicitly computable. Theorems 11.2.3 and 11.2.4 say that every partial computable function can be described in terms searching for an explicit piece of data and then reading off the result.

In Section 11.6, we will give an equivalent characterization of computability in terms of Turing machines. With that in mind, we can equally well think of $\varphi_e(x)$ as the function computed by Turing machine e. In those terms, $T(e, x, s)$ says that s codes a halting computation sequence of Turing machine e on input x, and $U(s)$ reads off the result. More abstractly, think of the index e in φ_e as denoting a computer program in some language that is sufficiently rich to represent all the computable functions. Then $\varphi_e(x)$ is the function computed by that program, $T(e, x, s)$ says that s is a complete record of the computation that is performed on input x, and $U(s)$ returns the result.

The fact that we can represent a finite sequence of natural numbers as a single number means that the restriction to unary functions is inessential. If we define

$$\varphi_e^n(x_0, \ldots, x_{n-1}) = \varphi_e((x_0, \ldots, x_{n-1})),$$

then the sequence $\varphi_0^n, \varphi_1^n, \varphi_2^n, \ldots$ enumerates all the n-ary functions, since an n-ary partial function $f(x_0, \ldots, x_{n-1})$ is computable if and only if the function $g(s) \simeq f((s)_0, \ldots, (s_{n-1}))$ is. Moreover, the function $\text{Univ}^n(e, x_0, \ldots, x_{n-1})$ defined by $\text{Univ}^n(e, \vec{x}) \simeq \text{Univ}(e, (\vec{x}))$ computes all the n-ary functions uniformly, so we have a universal n-ary partial computable function as well. Below, I will omit the superscript and write, for example, $\varphi_e(x, y, z)$ for the eth ternary partial computable function.

Our final abstraction says something that is finicky and intuitively clear, but once again hides the details of a specific model of computation and allows us to proceed more abstractly later on.

Theorem 11.2.5 (The s-m-n theorem). *For each pair of natural numbers n and m, there is a primitive recursive function s_n^m such that for every sequence*

$$e, a_0, \ldots, a_{m-1}, y_0, \ldots, y_{n-1},$$

we have

$$\varphi^n_{s_n^m(e, a_0, \ldots, a_{m-1})}(y_0, \ldots, y_{n-1}) \simeq \varphi_e^{m+n}(a_0, \ldots, a_{m-1}, y_0, \ldots, y_{n-1}).$$

Think of e as a program for an $(m+n)$-ary function. Then s_n^m takes such a program, e, together with fixed inputs $a_0, \ldots, a_{m-1}$, and returns a program, $s_n^m(e, a_0, \ldots, a_{m-1})$, for the n-ary function of the remaining arguments. The first program, e, expects $m + n$ inputs. The second program, $s_n^m(e, \vec{a})$, fixes the first m inputs to $a_0, \ldots, a_{m-1}$, and computes the resulting function of $y_0, \ldots, y_{n-1}$.

To prove the s-m-n theorem rigorously, we have no choice but to unpack the definition of $\text{Univ}(x, e)$. The relation $\text{Univ}(e, x) \simeq w$ is given by a Σ_1 formula. Hence, the relation

$$\text{Univ}(e, (x_0, \ldots, x_{m-1})^\frown s) \simeq w$$

is also given by a Σ_1 formula $\exists z\, A(e, x_0, \ldots, x_{m-1}, s, w, z)$, whose free variables are $e, x_0, \ldots, x_{m-1}, s$, and w. Define the function $s^m_n(e, a_0, \ldots, a_{n-1})$ so that it substitutes constants for $e, a_0, \ldots, a_{m-1}$ into A and returns a number e' representing the resulting Δ_0 formula $A'(s, w, z)$. Then we have

$$\varphi^n_{s^m_n(e,\vec{a})}(\vec{y}) \simeq \varphi_{s^m_n(e,\vec{a})}((\vec{y})) \simeq \text{Univ}(s^m_n(e, \vec{a}), (\vec{y})) \simeq \text{Univ}(e, (\vec{a})^\frown (\vec{y})) \simeq \varphi^{m+n}_e(\vec{a}, \vec{y}),$$

as required.

Exercises

11.2.1. a. Consider the function g, defined by

$$g(x) \simeq \begin{cases} f_x(x) & \text{if this is defined} \\ \text{undefined} & \text{otherwise.} \end{cases}$$

Is g a partial computable function? Is g total?

b. Consider the function h, defined by

$$h(x) \simeq \begin{cases} f_x(x) & \text{if this is defined} \\ 0 & \text{otherwise.} \end{cases}$$

Is h a partial computable function? Is h total?

11.2.2. We actually proved the following strengthening of Kleene's normal form theorem (Theorem 11.2.4): there is a primitive recursive relation $T(e, x, s)$ such that for each partial computable function $f(x)$, there is a natural number e, such that, for every x, $f(x) \simeq (\mu s\, T(e, x, s))_0$. In other words, we can take the output function $U(x)$ to simply return the first element of the sequence coded by s. Show that this strengthening can be proved directly from the statement of Theorem 11.2.4.

11.2.3. Use the s-m-n theorem to show that there is a primitive recursive function $g(n)$ such that for every n, $\varphi_{g(n)}(x)$ is the constant function, $\varphi_{g(n)}(x) = n$. Use it also to show that there is a primitive recursive function $h(n)$ such that for every n, $\varphi_{h(n)}(x) = n \cdot x$.

11.2.4. Show that an infinite set is computable if and only if it is the image of a strictly increasing computable function. In other words, given that the characteristic function $\chi_A(x)$ of an infinite set A is computable, show how to define a computable strictly increasing function f whose image is A, and, conversely, given such an f, show how to compute χ_A.

11.3 Undecidability and the Halting Problem

At this point, we have flexible means of showing that various functions and relations are computable. The fact that every primitive recursive function is computable gives us good stock of functions to start with, and the ability to define functions using recursion and unbounded search allows us to define more functions easily.

The goal of this section is to develop means of showing that certain functions and sets are *not* computable. This chapter opened with a heuristic argument that the primitive recursive functions cannot exhaust the computable functions because we can diagonalize against an enumeration of primitive recursive functions to get a computable function that is not in the enumeration. What happens when we try to apply the argument to our enumeration $\varphi_0, \varphi_1, \varphi_2, \ldots$ of partial computable functions? Define

$$f(x) \simeq \varphi_x(x) + 1 \simeq \text{Univ}(x, x) + 1.$$

Then f is a partial computable function, since it can be defined in terms of Univ, successor, and projections using composition. Hence it is equal to φ_e for some e. In particular we have

$$\varphi_e(e) \simeq f(e) \simeq \varphi_e(e) + 1.$$

This is not a contradiction. It simply means that $\varphi_e(e)$ is undefined, an interesting bit of trivia, but not much more than that.

But we can use the argument to show that there is no universal *total* computable function $\text{Univ}'(e, x)$. In other words, we can show that there no total computable function $\text{Univ}'(e, x)$ that computes all the unary total computable functions as e is fixed to different values. After all, assuming such a function exists, setting $f(x) = \text{Univ}'(x, x) + 1$ yields a contradiction.

Define the function

$$f(e, x) = \begin{cases} 1 & \text{if Univ}(e, x) \text{ is defined} \\ 0 & \text{otherwise.} \end{cases}$$

Notice that, by definition, $f(e, x)$ is a total function. The fact that there is no universal computable function shows that $f(e, x)$ cannot be computable. If it were, we could define a total computable function Univ$'$ as follows:

$$\text{Univ}'(e, x) = \begin{cases} \text{Univ}(e, x) & \text{if } f(e, x) = 1 \\ 0 & \text{otherwise.} \end{cases}$$

Intuitively, the computation of $\text{Univ}'(e, x)$ starts by asking f whether $\varphi_e(x)$ is defined. If it is, Univ$'$ calls Univ to compute it and returns the result. If it isn't, Univ$'$ safely returns 0.

We have to be careful: strictly speaking, the definition above is not enough to show that the computability of f implies the computability of Univ. We generally interpret a definition by cases in terms of the conditional function of Section 8.2, so the straightforward way of writing the right-hand side is as $\text{cond}(f(e, x), \text{Univ}(e, x), 0)$. But by the semantics of composition, this expression is undefined if any of the arguments are undefined, independent of the value of $f(e, x)$. The trick is to avoid the composition and define the conditional directly using primitive recursion,

$$g(e, x, 0) \simeq 0$$
$$g(e, x, y + 1) \simeq \text{Univ}(e, x).$$

We can then define $\text{Univ}'(e, x) = g(e, x, f(e, x))$. Now the semantics of primitive recursion works as we want: if $f(e, x) = 0$, then $g(e, x, f(e, x)) = 0$, and otherwise $g(e, x, f(e, x)) = \text{Univ}'(e, x)$. So, if $f(e, x)$ is computable, $\text{Univ}'(e, x)$ is computable as well, a contradiction.

A function that returns either 1 or 0 – that is, the characteristic function of a set or relation – is often called a *decision problem*, since we think of it as providing yes/no answers to a question. If the function is computable, one often says that the associated set or relation is *solvable* or *decidable*. If we interpret the argument e to our universal computable functions as a program or Turing machine, $f(e, x)$ essentially decides whether the program or Turing machine e terminates on input x. For that reason, f is known as the *halting problem*, and we have just shown that it is undecidable. For the record, here is a more direct proof.

Theorem 11.3.1. *The halting problem is undecidable.*

Proof Suppose $f(e, x)$ is computable. Define the function g by

$$g(x) = \begin{cases} 0 & \text{if } f(x, x) = 0 \\ \text{undefined} & \text{otherwise.} \end{cases}$$

As above, we can show that $g(x)$ is a partial computable function, and so it is equal to φ_e for some e. But then we have:

$$\begin{aligned} g(e) \downarrow &\Leftrightarrow f(e, e) = 0 \\ &\Leftrightarrow \varphi_e(e) \uparrow \\ &\Leftrightarrow g(e) \uparrow. \end{aligned}$$

The first equivalence appeals to the definition of g, and the second one appeals to the definition of f. So we have $g(e) \downarrow$ if and only if $g(e) \uparrow$, a contradiction. □

Note, incidentally, that a set A is decidable if and only if its complement, $\overline{A}$, is decidable: a decision procedure for one is a decision procedure for the other, modulo switching yes and no. As long as we are diagonalizing, we may as well establish a fact that will be useful to us in Section 12.2.

Proposition 11.3.2. *There is no universal computable relation $R(x, y)$. In other words, there is no computable relation $R(x, y)$ with the property that for every computable unary predicate $S(y)$, there is an e such that for every y, $S(y)$ holds if and only if $R(e, y)$ holds.*

Proof Given a computable relation $R(x, y)$, the predicate $S(y) \equiv \neg R(y, y)$ is also computable. Given any e, $S(e)$ holds if and only if $\neg R(e, e)$ holds, so R is not universal. □

In terms of definability, this says that there is no Δ_1 binary relation that represents all the Δ_1 unary relations, providing another perspective on the $n = 1$ case of Theorem 10.5.2.

Once we know that one set is undecidable, we can use that fact to show that other sets are undecidable as well. For example, consider the set $S = \{e \mid \varphi_e(0) \downarrow\}$, the set of indices for partial computable functions that are defined on input 0. A priori, this is an easier question to answer than determining whether a partial function halts at an arbitrary input, x, but not much. With some cleverness, we can show that this set is not computable either. To that end, it will be helpful to introduce the following notion.

Definition 11.3.3. A *many-one reduction* from one set A to a set B is a computable function f such that for every x, $x \in A$ if $f(x) \in B$. A set A is *many-one reducible* to B if there is a many-one reduction from A to B.

We will write $A \leq_m B$ to denote that A is many-one reducible to B. Saying that A is many-one reducible to B amounts to saying that there is a procedure for translating any question about membership in A to a question about membership in B in such a way that the answer to the second is the same as the answer to the first. The following proposition supports the reading of $A \leq_m B$ as saying that solving A is no harder than solving B:

Proposition 11.3.4. *For any sets A, B, and C:*

1. $A \leq_m A$.
2. *If* $A \leq_m B$ *and* $B \leq_m C$ *then* $A \leq_m C$.
3. *If B is computable and* $A \leq_m B$ *then A is computable.*
4. *If A is undecidable and* $A \leq_m B$ *then B is undecidable.*

Proof For 1, the identity function is a many-one reduction of A to A. For 2, if f is a many-one reduction of A to B and g is a many-one reduction of B to C, then $g \circ f$ is a many-one reduction of A to C, because x is in A if and only if $f(x)$ is in B, which happens if and only if $g(f(x))$ is in C.

For 3, if f is a many-one reduction of A to B, then $\chi_A(x) = \chi_B(f(x))$. Since f is computable, the left-hand side is computable if and only if the right-hand side is. Claim 4 is the contrapositive of claim 3. □

Many-one reducibility is a fairly restrictive notion of reduction: one has to translate a single membership question for A to a single membership question for B in such a way that the answers are the same. The terminology stems from the fact that we can also consider the notion of one-one reducibility, in which the reducing function is required to be injective. In Section 11.8, we will consider a more liberal notion of reducibility, *Turing reducibility*, which also enjoys the properties enumerated in Proposition 11.3.4. For the time being, many-one reducibility will suffice for our purposes.

We have already used the notion of reducibility implicitly. Our proof of Theorem 11.3.1, the undecidability of the halting problem, shows that the *diagonal set* $K = \{x \mid \varphi_x(x) \downarrow\}$ is undecidable. But that is reducible to the set $H = \{(e, x) \mid \varphi_e(x) \downarrow\}$ via the map $x \mapsto (x, x)$, so the latter is undecidable as well.

Reducing either of these to $S = \{e \mid \varphi_e(0) \downarrow\}$ is tricker. Suppose we want to reduce H to it. To appreciate the challenge, imagine you have a really smart friend that can answer any question about S, that is, tell you whether a partial computable function φ_e is defined at 0. Your friend's talents are quirky, though, and they do not extend to answering questions about any other input. You really want to know whether a particular partial computable function, φ_e, is defined at x. What can you do? You can get the answer you want by designing another partial computable function $\varphi_{e'}(y)$ that simply ignores its input and computes $\varphi_e(x)$. Now ask your friend whether $\varphi_{e'}$ is defined at 0, and you have your answer.

Proposition 11.3.5. *The set* $H = \{(e, x) \mid \varphi_e(x) \downarrow\}$ *is many-one reducible to the set* $S = \{e \mid \varphi_e(0) \downarrow\}$*, and also to the set* $K = \{x \mid \varphi_x(x) \downarrow\}$.

This implies that S is also undecidable. And since S and K are easily many-one reducible to H, by transitivity, each of S, K, and H is many-one reducible to either of the others.

Proof Let $f(e, x, y) \simeq \mathrm{Univ}(e, x) \simeq \varphi_e(x)$. Since f is a computable three-place function, it is equal to $\varphi_k(e, x, y)$ for some k. We can use the s-m-n theorem to fix the first two inputs to φ_k. Specifically, let $u(p) = s^2_1(k, (p)_0, (p)_1)$. Then we have:

$$\begin{aligned} p \in H &\Leftrightarrow \varphi_{(p)_0}((p)_1) \downarrow \\ &\Leftrightarrow f((p)_0, (p)_1, 0) \downarrow \\ &\Leftrightarrow \varphi_k((p)_0, (p)_1, 0) \downarrow \\ &\Leftrightarrow \varphi_{s^2_1(k,(p)_0,(p)_1)}(0) \downarrow \\ &\Leftrightarrow \varphi_{u(p)}(0) \downarrow \\ &\Leftrightarrow u(p) \in S. \end{aligned}$$

This shows that H is many-one reducible to S. Since f does not depend on its third argument, we can replace 0 by $u(p)$ everywhere and S by K in the last line. This shows that H is many-one reducible to K. □

The argument we have just seen shows something much more general. To reduce the halting problem to S, given a pair (e, x), we compute the index $u((e, x))$ of a partial computable function that acts like a function in S if $\varphi_e(x)$ is defined and is always undefined otherwise. As a result, asking whether p is in H is equivalent to asking whether $u(p)$ is in S. Very few specific details of S came into the argument. Abstracting them away yields a theorem known as *Rice's theorem*, which, in essence, states that *no* nontrivial property of partial computable functions is decidable. The theorem is breathtakingly general.

Theorem 11.3.6. *(Rice's theorem) Let C be any set of partial computable functions, and let $A = \{e \mid \varphi_e \in C\}$. If A is computable, then either C is $\emptyset$ or C is the set of all the partial computable functions.*

Before proving the theorem, let us consider an alternative formulation. Say a set A is an *index set* if for every e and e', if $\varphi_e = \varphi'_{e'}$, then $e \in A$ if and only if $e' \in A$. Think of A as a set of indices of partial computable functions, where membership in A depends only on the partial function that is computed by e. Given C and A as in the statement of Rice's theorem, it is clear that A is an index set. Conversely, given an index set A, if we define C to be the set of partial functions computed by indices in A, then $A = \{e \mid \varphi_e \in C\}$. In other words, there is a one-to-one correspondence between sets of partial computable functions and index sets, and Rice's theorem can be stated in this equivalent form:

Theorem 11.3.7. *No nontrivial index set is decidable.*

To appreciate the force of this theorem, remember that we can think of indices as programs in a programming language. Some questions about programs are clearly decidable: does the program have more than 1000 symbols? Does it have more than 50 lines? Does it contain a "print" statement? These are questions about *syntax*. Rice's theorem deals, in contrast, with *semantic* questions: does the program halt on empty input? Does it ever output 0? Does it ever output an odd number? The theorem says that no such question is decidable, unless the answer is always yes or always no.

Proof We prove the second formulation. Let A be an index set, and suppose neither A nor its complement is empty. We will show that either the set S of Proposition 11.3.5 or its complement is reducible to A, which implies that S is undecidable.

Let b be an index for the function that is nowhere defined. Without loss of generality, switching A with its complement if necessary, we can assume that b is in the complement of A. Let a be any element of A. Define

$$h(e, x) \simeq \begin{cases} \varphi_a(x) & \text{if } \varphi_e(0) \downarrow \\ \text{undefined} & \text{otherwise.} \end{cases}$$

For example, we can write $h(e, x) \simeq p_1^2(\mathrm{Univ}(e, 0), \varphi_a(x))$, remembering that the composition of functions is defined at an input if and only if both arguments are. By the s-m-n theorem, there is a primitive recursive function $k(e)$ such that $\varphi_{k(e)}(x) \simeq h(e, x)$ for every x. But now notice that $\varphi_{k(e)} = \varphi_a$ if $e \in S$ and $\varphi_{k(e)} = \varphi_b$ otherwise. Since A is an index set, we therefore have $e \in S$ if and only if $k(e) \in A$, which shows that k is a reduction of S to A. □

Rice's theorem is very powerful. The following illustrates some of its consequences.

Corollary 11.3.8. *The following sets are undecidable:*

- $\{e \mid$ *0 is in the image of* $\varphi_e\}$
- $\{e \mid \varphi_e$ *is total*$\}$
- $\{e \mid \varphi_e$ *is the constant* 0 *function*$\}$
- $\{e \mid$ *there is an even number in the image of* $\varphi_e\}$

Exercises

11.3.1. Let $f(x) \simeq \mathrm{Univ}(x, x) + 1$. Show that there is no way of extending f to a total computable function. In other words, show that there is no total computable function $g(x)$ such that whenever $f(x)$ is defined, $g(x) = f(x)$.

11.3.2. Show that there is a function $f(x)$ that has a Π_1-definable graph but is not computable. (Hint: let $f(x)$ return the least s such that $T(x, x, s)$ holds, if there is one, and zero otherwise.)

11.3.3. Show that there is a function f that is not primitive recursive but has a primitive recursive graph. In other words, show that there is a function f that is not primitive recursive, but such that the relation $y = f(x)$ is primitive recursive. (Hint: let $g(x)$ be any computable function that is not primitive recursive, and make use of Kleene's T.)

11.3.4. Remember our enumeration of the unary primitive recursive functions, $f_0, f_1, f_2, \ldots$, where f_i is the function computed by notation i. Notice that the function

$$g(x, s) = \begin{cases} 1 & \text{if } T(x, 0, s) \\ 0 & \text{otherwise} \end{cases}$$

is primitive recursive. For this problem, you can assume that there is a computable function k such that for every x and s,

$$f_{k(x)}(s) = g(x, s).$$

Roughly, $k(x)$ returns the notation for a function that composes g with the constant function $h(y) \equiv x$ in the first argument.

This problem shows that there is no algorithm to determine whether f_i is the constant zero function, and also no algorithm to determine whether f_i and f_j are the same function.

a. Let A be the set $\{x \mid f_x$ is the constant zero function$\}$ and let S be the set $\{e \mid \varphi_e(0) \downarrow\}$. Show that the function k defined above is a many-one reduction of S to the complement of A.
b. Conclude that A is not computable.
c. Use this to show that the set $B = \{(x, y) \mid f_x = f_y\}$ is not computable, by showing that A is reducible to B.

11.4 Computably Enumerable Sets

We have characterized the sets H, S, and K of Proposition 11.3.5 in negative terms by showing that they are undecidable. The goal of this section is to provide a positive characterization.

Definition 11.4.1. A set S of natural numbers is *computably enumerable*, or *c.e.* for short, if it is empty or the image of a computable function.

If S is the image of the function f, we can write $S = \{f(0), f(1), f(2), \ldots\}$. This provides the sense in which f enumerates the elements of S. Remember that an arbitrary set S is countable if it is empty or the image of a function $f : \mathbb{N} \to S$. (See Proposition A.3.1.) Every set of natural numbers is countable, but not not every set of natural numbers is the image of a *computable* function f. Thus computable enumerability is an effective (i.e. computable) version of countability.

Theorem 11.4.2. *Let S be a set of natural numbers. The following are equivalent.*

1. *S is empty or the image of a primitive recursive function.*
2. *S is computably enumerable.*
3. *S is the image of a partial computable function.*
4. *S is Σ_1-definable.*
5. *$S = \{x \mid \exists y\, R(x, y)\}$ for some computable relation R.*
6. *S is the domain of a partial computable function.*

Proof Clearly 1 implies 2 and 2 implies 3. To see that 3 implies 4, if S is the image of the partial computable function φ_e, then

$$S = \{y \mid \exists x\, (\varphi_e(x) \simeq y)\}.$$

By Theorem 11.2.1, the relation $\varphi(e) \simeq y$ is Σ_1, so S is Σ_1.

Clearly 4 implies 5, since any Δ_0 formula defines a computable relation. Given S and R as in 5, define the partial computable function f by

$$f(x) \simeq \mu y\, R(x, y).$$

Then the domain of f is S, so 5 implies 6.

To close the chain of equivalences, it suffices to show that 6 implies implies 1. So suppose S is the domain of φ_e. Then $S = \{x \mid \exists s\, T(e, x, s)\}$. If S is empty, then 2 holds, so we can assume that S has at least one element, a. Define

$$g(u) = \begin{cases} (u)_0 & \text{if } T(e, (u)_0, (u)_1) \\ a & \text{otherwise.} \end{cases}$$

Then g is primitive recursive and the image of g is S: if $T(e, x, s)$ holds, then $x = g((x, s))$, and, conversely, if $T(e, (u)_0, (u)_1))$ holds, then $g(u) = (u)_0$ is in S. □

In the last step of the proof, we can think of the primitive recursive function g as playing a waiting game. It searches for pairs (x, s) where s is a witness to the fact that $\varphi_e(x) \downarrow$, and whenever it finds one, it outputs x. Otherwise, it stalls by outputting a. You can think of this argument in computational terms: to enumerate the domain of φ_e, carefully simulate the computations of $\varphi_e(0), \varphi_e(1), \varphi_e(2), \ldots$ running in parallel. Put the programs in a list in such a way that each one is visited infinitely often, and at step i, carry out another step of the program at position i on the list. Whenever you see a computation halt, you can output an element in the domain of φ_e, and if you are systematic enough, every element of the domain of φ_e will eventually be discovered in this way.

Combining the characterization of c.e. sets in terms of Σ_1 definability with the results of Section 10.2, we have the following:

Corollary 11.4.3. *For any function $f(\vec{x})$, the following are equivalent:*

- *f is computable.*
- *The graph of f is computable.*
- *The graph of f is computably enumerable.*

Corollary 11.4.4. *A set S is computable if and only if S and $\overline{S}$ are computably enumerable.*

Corollary 11.4.5. *If S and T are computably enumerable, then so are $S \cap T$ and $S \cup T$.*

These three corollaries can also be proved directly, without passing through arithmetic definability. For example, to prove the harder direction of Corollary 11.4.4, suppose S and $\overline{S}$ are computably enumerable. Let S be the domain of φ_d and let $\overline{S}$ be the domain of φ_e. Define h by

$$h(x) = \mu s\,(T(e, x, s) \vee T(f, x, s)).$$

In words, on input x, h searches for either a halting computation of $\varphi_d(x)$ or a halting computation of $\varphi_e(x)$. If x is in S, it will succeed in the first case, and if x is in $\overline{S}$, it will succeed in the second case. So h is a total computable function. But now we have that for every x, $x \in S$ if and only if $T(e, x, h(x))$, i.e. if φ_e is the one that is defined. Since $T(e, x, h(x))$ is a computable relation, S is computable. We can summarize the proof as follows: to decide S, on input x search for halting computations of φ_e and φ_f. One of them is bound to halt; if it is φ_e, then x is in S, and otherwise, x is in $\overline{S}$. Exercise 11.4.1 asks you to give similarly direct proofs of Corollaries 11.4.3 and 11.4.5.

The existence of so many equivalent ways of describing the computably enumerable sets is evidence that the concept is robust. The last characterization can be viewed as saying that a set S is computably enumerable if there is a curious sort of decision procedure for S: on input x, the procedure returns a yes answer if x is in S, but never returns if the answer is no. So if the answer is no, you never find out for sure; as long as you wait, there is always the possibility that the computation has yet to halt. For that reason, computably enumerable sets are also sometimes called *semidecidable.*

Since the halting problem is Σ_1-definable, it is a computably enumerable set. We can view this fact through the lens of equivalences offered by Theorem 11.4.2. For example, there is a natural semidecision procedure for K: on input (e, x), start the computation of $\varphi_e(x)$ and wait for it to halt. Or, using the technique suggested at the end of the proof of Theorem 11.4.2, we can enumerate the set of pairs (e, x) by systematically running all the computations in parallel.

The notion of many-one reducibility provides a sense in which the halting problem has maximum difficulty for a c.e. set.

Proposition 11.4.6. *Let A and B be sets of natural numbers. If $A \leq_m B$ and B is computably enumerable, then so is A.*

Proof If f is a many-one reduction and A is the domain of the partial computable function g, then $g \circ f$ is a partial computable function, and B is the domain of $g \circ f$. □

Definition 11.4.7. A set A is a *complete computably enumerable set* (under many-one reducibility) if it is c.e. and every other c.e. set is many-one reducible to it.

Theorem 11.4.8. *The halting problem, $H = \{(e, x) \mid \varphi_e(x) \downarrow\}$, is a complete c.e. set, as are the sets $S = \{e \mid \varphi_e(0) \downarrow\}$ and $K = \{x \mid \varphi_x(x) \downarrow\}$.*

Proof We already know that H is c.e. To see that it is complete, let A be any c.e. set. Then for some e, A is the domain of φ_e, and for every x we have $x \in A$ if and only if $(e, x) \in H$. Thus the map $x \mapsto (x, e)$ is a many-one reduction.

Since $S = \{e \mid \exists s\, T(e, 0, s)\}$ and $K = \{x \mid \exists s\, T(x, x, s)\}$, they are c.e. By Proposition 11.3.5, H is reducible to either of these, so by the transitivity of many-one reducibility, S and K are complete. □

The notion of completeness generalizes.

Definition 11.4.9. Let $\mathcal{S}$ be a collection of sets of natural numbers. An set A is *hard* for $\mathcal{S}$ under many-one reducibility if for every B in $\mathcal{S}$, $A \leq_m B$. If A is also an element of $\mathcal{S}$, then A is also said to be *complete* for $\mathcal{S}$ under many-one reducibility.

The truth definitions for the arithmetic hierarchy, described in Section 10.5, provide examples of complete sets.

Proposition 11.4.10. *For every $n \geq 1$, the set of true Σ_n sentences is a complete Σ_n set under many-one reduction.*

Exercises

11.4.1. Prove Corollaries 11.4.3 and 11.4.5 directly, using any of the characterizations of c.e. sets given by Theorem 11.4.2 other than the characterization in terms of Σ_1 definability.

11.4.2. Show that the set $\{e \mid \varphi_e$ is not injective$\}$ is computably enumerable. Here, a partial function f is said to be injective if for every x and y, if $x \neq y$ and $f(x)$ and $f(y)$ are both defined, then $f(x) \neq f(y)$.

11.4.3. Prove that any infinite computably enumerable set is the image of an injective computable function.

11.4.4. Show that every infinite computably enumerable set contains an infinite computable subset.

11.4.5. Let A be the set $A = \{2n \mid n \in K\} \cup \{2n + 1 \mid n \in \overline{K}\}$ where K is the set $\{x \mid \varphi_x(x) \downarrow\}$ and $\overline{K}$ is the complement of K. Show that K is many-one reducible to both A and $\overline{A}$. Conclude that neither A nor $\overline{A}$ is computably enumerable.

11.4.6. Show that the set $A = \{e \mid \varphi_e$ is total$\}$ is not c.e. (Hint: assume that A is the image of a recursive function, and diagonalize.) In fact, show that A is a Π_2-definable set that is complete for the class of Π_2-definable sets under many-one reduction.

11.4.7. For every e, define the set W_e to be $\{x \mid \varphi_e(x) \downarrow\}$, so $W_0, W_1, W_2, \ldots$ is a list of all the computably enumerable sets. Let $A = \{e \mid W_e = \emptyset\}$.

a. Use Rice's theorem to give a quick proof that A is not computable.
b. Show that the complement of A is computably enumerable.
c. Show that the complement of A is a complete computably enumerable set.

11.4.8. Two disjoint sets A and B of numbers are *computably inseparable* if there is no computable set C such that $A \subseteq C$ and $B \subseteq \overline{C}$. (See also Exercise 10.2.4.) Here $\overline{C}$ denotes the complement of C. This condition is stronger than saying that neither A nor B is computable; the statement implies that not only are they not computable, but there is no computable set separating them.
Let $A = \{x \mid \varphi_x(x) \downarrow = 0\}$ and let $B = \{x \mid \phi_x(x) \downarrow = 1\}$.
a. Show that A and B are computably enumerable.
b. Show that A and B are computably inseparable. (Hint: suppose C is a computable separation, and let φ_e be the characteristic function of C.)

11.4.9. Prove Proposition 11.4.10.

11.5 The Recursion Theorem

Remember our proof of the undecidability of the halting problem, Theorem 11.3.1. Assuming we had a decision procedure $f(e, x)$ that could tell us whether $\varphi_e(x)$ is defined, we obtained a function $g(x)$ that is defined if and only if $\varphi_x(x)$ is undefined. Since $g(x)$ is computable, it is equal to φ_e for some e. But then $g(e) \simeq \varphi_e(e)$ is defined if and only if $\varphi_e(e)$ is undefined, a contradiction. Defining the function $g(x)$ in terms of $\varphi_x(x)$ sets up a diagonal argument, and plugging in the index e for g results in a function whose behavior is, in a sense, self-referential: the value of g at e depends on the value of $\varphi_e(e)$, which is, in turn, the value of g at e.

In this proof by contradiction, the diagonal construction self destructs, since we ultimately conclude that the posited function g does not exist. In this section, we will see that diagonalization can also be used in positive ways to obtain partial computable functions meeting self-referential specifications. We will prove a striking result known as the *recursion theorem*, or the *fixed-point theorem*, that encapsulates this idea.

The next lemma provides two equivalent ways of stating the recursion theorem. The trivial sense in which they are equivalent is that they are both true, but the point is that each one can easily be derived from the other, providing complementary perspectives on the theorem.

Lemma 11.5.1. *The following statements are equivalent:*

1. *For every partial computable function $g(x, y)$, there is an index e such that for every y,*

$$\varphi_e(y) \simeq g(e, y).$$

2. *For every computable function $f(x)$, there is an index e such that for every y,*

$$\varphi_e(y) \simeq \varphi_{f(e)}(y).$$

Proof To see that 1 implies 2, given f, define g by $g(x, y) \simeq \mathrm{Univ}(f(x), y)$. By 1, there is an index e such that for every y,

$$\begin{aligned} \varphi_e(y) &\simeq \mathrm{Univ}(f(e), y) \\ &\simeq \varphi_{f(e)}(y). \end{aligned}$$

Whereas the passage from 1 to 2 uses a universal function to move an argument into a subscript, the passage from 2 to 1 uses the s-m-n theorem to go the other way. Given g, use the s-m-n theorem to get f such that for every x and y, $\varphi_{f(x)}(y) \simeq g(x, y)$. By 2, there is an index e such that for every y,

$$\begin{aligned}\varphi_e(y) &\simeq \varphi_{f(e)}(y)\\ &\simeq g(e, y).\end{aligned}$$

□

Before proving that these statements are true, it is worthwhile to reflect on how surprising they are. Thinking of e as a computer program, the first statement says that, given any computable function $g(x, y)$, there is a program e that, for any input y, computes the value of g when it is applied to the program e itself and y. The second statement says that, given any crazy transformation of computer programs – like writing all the lines backward, or deleting every other letter – there is a program whose behavior is unchanged by the transformation.

Theorem 11.5.2. *The two statements in Lemma 11.5.1 are true.*

Proof It suffices to prove the first statement. By the s-m-n theorem, we can find a primitive recursive function diag(x) such that for every x and y,

$$\varphi_{\text{diag}(x)}(y) \simeq \text{Univ}^2(x, x, y) \simeq \varphi_x(x, y).$$

In other words, diag(x) is a program that, on input y, runs program x on inputs x and y.

Now, given a partial computable function $g(x, y)$, let k be an index for the partial computable function $g(\text{diag}(x), y)$, so that we have, for every y,

$$\varphi_k(x, y) \simeq g(\text{diag}(x), y).$$

Finally, let $e = \text{diag}(k)$. To show that e has the magical properties postulated, we need only unwrap definitions. For every y, we have

$$\varphi_e(y) \simeq \varphi_{\text{diag}(k)}(y) \simeq \varphi_k(k, y) \simeq g(\text{diag}(k), y) \simeq g(e, y),$$

as required. □

The proof is mysterious, but here is one way to think about how it works. Imagine you have been assigned the task of writing a computer program that prints itself out. Suppose, moreover, that your favorite programming language has a procedure diag(x) that takes, as input, a string of characters, and returns the result of substituting a quoted version of that string for the first occurrence of the letter X. For example, when passed the string

hello X world

the function diag(x) returns the following string:

hello "hello X world" world

Then this short program meets the specification:

print(diag("print(diag(X))"))

After all, applying the function diag(x) to the string "print(diag(X))" yields the program itself. If you are not fortunate enough to have a programming language with a built-in function like diag(x), you need to write one, add it to the program, and include it in the string that is diagonalized. The proof of Theorem 11.5.2 implements this general idea, though it also takes into account the argument y. To see how this additional parameter changes things, think about how you can write a program that, when passed a natural number y, prints itself out y times.

The recursion theorem has some whimsical applications. For example, taking $g(x, y) = x + y$ yields an index e such that for every x, $\varphi_e(x) = e + x$. It also underwrites arbitrary recursive definitions, which are commonly used in programming. For example, given the primitive recursive functions $\mathrm{mod}(x, y)$ that returns the remainder when dividing x by y (or x if y is 0), we can define the greatest common divisor function by writing

$$\mathrm{gcd}(x, y) \simeq \begin{cases} x & \text{if } y = 0 \\ \mathrm{gcd}(y, \mathrm{mod}(x, y)) & \text{otherwise.} \end{cases}$$

To see that this describes a partial computable function, define $g(e, u)$ by

$$g(e, u) \simeq \begin{cases} (u)_0 & \text{if } (u)_1 = 0 \\ \varphi_e(((u)_1, \mathrm{mod}((u)_0, (u)_1))) & \text{otherwise.} \end{cases}$$

Then $g(e, u)$ computes desired value of $\mathrm{gcd}((u)_0, (u)_1)$, assuming, optimistically, that $\varphi_e(u)$ does the same. Applying the recursion theorem to g yields a function $\varphi_e(u)$ that meets the recursive specification on the right-hand side, and setting $\mathrm{gcd}(x, y) \simeq \varphi_e((x, y))$ completes the definition.

In contrast to our primitive recursive definition of $\mathrm{gcd}(x, y)$, the appeal to the recursion theorem only guarantees the existence of a *partial* function meeting the specification we have given. But given that the second argument, y, decreases in the recursive call, we can use induction on y to prove that $\mathrm{gcd}(x, y)$ is in fact total. Thus the recursion theorem justifies arbitrary recursive definitions and separates the task of proving termination, or totality. But the theorem does even more than that, since it enables us to specify the behavior of a program, e, in terms of e itself, not just the partial function φ_e that it computes.

Exercises

11.5.1. Use the recursion theorem to show that there is an index e such that the domain of φ_e is the set $\{e\}$.

11.5.2. This exercise provides a short proof of Rice's theorem. Let A be any nontrivial index set and suppose A is computable. Let a be any element of A, and let b be any element in the complement of A. Use the recursion theorem to find an index e such that for every x,

$$\varphi_e(x) \simeq \begin{cases} \varphi_a(x) & \text{if } e \notin A \\ \varphi_b(x) & \text{if } e \in A. \end{cases}$$

Show that this yields a contradiction.

11.5.3. A set A is said to be self-dual if it is reducible to its complement, i.e. $A \leq_m \overline{A}$.

a. Use the second version of the recursion theorem to show that no index set is self-dual. (Hint: suppose f is a reduction of A to $\overline{A}$ and the fixed-point theorem to f.)

b. Use this to give a short proof of Rice's theorem. (Hint: if A is a computable set that is not empty and not equal to $\mathbb{N}$, show that A is reducible to $\overline{A}$.)

11.5.4. Let S be the set $\{e \mid \varphi_e(0) \downarrow\}$. Use the recursion theorem to show that S is not computable. (Hint: assuming it is computable, find an index e such that φ_e is the constant 0 function if e is not in S, and undefined everywhere if e is in S.)

11.6 Turing Machines

In Section 11.1, we defined the set of partial and total computable functions from the natural numbers to the natural numbers in terms of a generating schema, starting with zero, successor, and projections, and closing under composition, primitive recursion, and unbounded search. The definition was meant to capture the intuitive notion of what it means for a partial function to be algorithmically computable, that is, something that we can calculate in principle, setting aside concerns about the length of time and amount of paper we might need to carry out the computation.

It should seem plausible that any function meeting the formal definition should qualify as computable in the intuitive sense, because we can imagine calculating the values of a partial computable function by hand. *Church's thesis* is the claim that the converse also holds, in other words, that the formal definition adequately captures the intuitive notion. Turing himself considered three types of evidence that speak in favor of identifying the two. Slightly paraphrased, they are:

1. a direct appeal to intuition
2. a proof of the equivalence of two definitions, in case the new definition has a greater intuitive appeal
3. giving examples of large classes of functions that are computable according to the formal definition.

Chapter 8 and this chapter provide ample evidence of the third kind. We now turn to the first two.

Turing himself provided a model of computation that can be justified by a more direct appeal to intuition. He asked us to imagine someone sitting at a table with a sheet of paper, calculating according to an algorithmic recipe, which is a finite list of instructions. For the purposes of calculation, it does not seem harmful to require that the paper is divided into squares, and that the calculator proceeds by writing or changing the symbols in those squares, according to the recipe. All the following assumptions seem to be justified by our intuitive conception of what it means to calculate:

1. The calculator uses only finitely many symbols. (Otherwise, there will be symbols that are arbitrarily close to each other.)
2. The calculator can only remember a finite amount of information.
3. The calculator can only survey a finite portion of the paper at any given time but can shift attention to other portions of the sheet of paper, as required by the algorithm.

We also assume that the calculator has access to as much time and paper as is needed to complete the computation.

At each step in the computation, the calculator has to decide what to do based on the contents of their memory (which we will refer to as the *state* of the calculator) as well as the contents of the part of the paper that is within the calculator's scope. Based on that information, a matching instruction will tell the calculator how to execute the next step of the calculation. Executing a step may require changing some of the symbols in the squares under observation and shifting focus to a configuration nearby. (By "nearby" we assume that there is a fixed bound on how far the calculator can shift attention in a single step.)

Turing made the case that this notion of computation can be given a precise mathematical specification. He also argued that anything that can be computed according to this model can also be computed according to the similar but more restrictive model that we now call a *Turing machine*. Thus we have an argument that anything that is computable in the intuitive sense can be computed by such a machine.

The Turing machine model of computation is historically important, not only for the role it plays in providing a philosophical analysis of the notion of computation, but also for its practical implications. Our characterization of computation in terms of recursion was based on what is called a *denotational semantics*: we described the meaning of a *program* – in this case, a sequence of function definitions – in terms of the partial function it defines. The description of a Turing machine draws instead on what is called an *operational semantics*, in which we view a computation as a sequence of states, where each successive state is determined by the program. A Turing machine is an abstract representation of such a program, and the semantics specifies the partial function on the natural numbers that it computes.

The following informal description is needed to make sense of the definition. Think of a Turing machine as a machine that can be in any of a finite set of *states*, one of which is designated as the *start* state. The machine does its work on a two-way infinite *tape*, divided into squares, each containing one of a finite set of *symbols*. The machine also has a *tape head* that travels back and forth along the tape reading one square at a time. At any point in the computation, all but finitely many squares contain the *blank* symbol. The fact that the tape is infinite in both directions means that there is no bound to how far to the left or right the tape head can move as the computation progresses.

A Turing machine's program consists of a finite list of instructions, each of which specifies what to do in a particular state with a particular symbol under the tape head. An instruction can tell the machine to change the symbol under the tape head, or move left, or move right. At the same time, executing an instruction also changes the machine's state, as specified by the instruction. The computation begins with the Turing machine in the start state, with an input string on the tape and the tape head at the beginning of that string.

If there are n states, there is no harm in assuming that the states are numbered from 0 to $n-1$, and similarly for the symbols. It also does not hurt to assume that the start state is numbered 0, as is the blank symbol. Thus, abstractly, a Turing machine is specified by the number of states, the number of symbols, and the list of instructions.

Definition 11.6.1. A *Turing machine* consists of a triple (n, m, δ) where

- $n \geq 1$ is a natural number (the number of states);
- $m \geq 1$ is a natural number (the number of symbols); and
- δ is a *partial function* from the set $\{0, \ldots, n-1\} \times \{0, \ldots, m-1\}$ to the set $\{0, \ldots, m+1\} \times \{0, \ldots, n-1\}$ (the instructions).

If $\delta(i, j) = (k, \ell)$, then k represents the action that the machine should perform and ℓ is the next state. If $k < m$, then k is a symbol, and so we interpret the instruction as specifying that the machine should replace the symbol under the tape head with k. If $k = m$, we interpret the instruction as *move left*, and if $k = m + 1$, we interpret the instruction as *move right*.

We will characterize a computation as a sequence of configurations of the Turing machine, where each configuration specifies the current state, the contents of the tape, and the position of the tape head.

Definition 11.6.2. If M is a Turing machine, a *configuration* of M is a 4-tuple (i, j, r, s) satisfying the following:

- i is a state, i.e. a natural number less than the number of states of M.
- j is a symbol, i.e. a natural number less than the number of symbols of M.
- r is a finite sequence of symbols, $(r_0, \ldots, r_{k-1})$.
- s is a finite sequence of symbols, $(s_0, \ldots, s_{\ell-1})$.

The first two items, i and j, represent the current state and the symbol under the tape head, r represents the symbols to the left of the tape head (say, in reverse order), and s represent the symbols to the right of the tape head. By convention, the squares on the tape beyond those represented by j, r, and s are assumed to be blank, so we think of the result of adding zeros to the end of r or s, or deleting zeroes from the end of r or s, as describing the same machine configuration.

Suppose (i, j, r, s) is a configuration of a machine M. Call this a *halting configuration* if no instruction applies, i.e. the pair (i, j) is not in the domain of δ. Otherwise, the *configuration after c according to M* is obtained as follows:

- If $\delta(i, j) = (k, l)$, where k is a symbol, the desired configuration is (l, k, r, s).
- If $\delta(i, j) = (m, l)$, an instruction to move left, the desired configuration is (l, j', r', s'), defined as follows. If r is not empty, j' is the first symbol of r and r' is the rest of r. If r is empty, then j' is 0 and r' is empty. Either way, s' consists of j prepended to s.
- If $\delta(i, j) = (m + 1, l)$, an instruction to move right, the desired configuration is (l, j', r', s'), defined as follows. If s is not empty, then j' is the first symbol in s and s' is the rest of s. If s is empty, j' is 0 and s' is empty. Either way, r' consists of j prepended to r.

Now suppose M is a Turing machine and u is a sequence of symbols of M. The *start configuration for M with input u* is the configuration $(0, i, (), u')$, where i is the first symbol in u, u' is the rest of u, and $()$ denotes the empty sequence. (If u is empty, take i to be 0.) This corresponds to the configuration where the machine is in state 0 and u written on the input tape, with the head at the beginning of the string.

Definition 11.6.3. Let M be a Turing machine and s a sequence of symbols of M. A *partial computation sequence* for M on input u is a sequence of configurations $c_0, c_1, \ldots, c_k$ such that:

- c_0 is the start configuration for M with input u, and
- for each $i < k$, c_{i+1} is the configuration after c_i, according to M.

A *halting computation sequence* for M on input u is a partial computation sequence where the last configuration is a halting configuration. M *halts* on input u if and only if there is a halting computation of M on input u.

We are almost done. Suppose we want to compute an υ-ary partial function from $\mathbb{N}$ to $\mathbb{N}$. We need to assume that the Turing machine has at least one non-blank symbol, that is, $m \geq 2$.

We will use a string of k ones to represent the natural number k, and we will represent a v-tuple $k_0, \ldots, k_{v-1}$ by writing these strings separated by blanks. So, for example, the input 3, 5, 4 is represented by 1110111110111 and the input 3, 0, 4 is represented by 111001111.

Definition 11.6.4. Let M be a Turing machine. Then the *v-ary partial function computed by M* is defined as follows: if M halts on the input string representing $\vec{k}$, then the value of the partial function at $\vec{k}$ is the length of the longest string of 1s ending immediately to the left of the tape head in the halting configuration. If M does not halt on that input, then the partial function is undefined at $\vec{k}$.

With that, we can state the main theorem of this section:

Theorem 11.6.5. *A partial function from $\mathbb{N}$ to $\mathbb{N}$ is computable if and only if it is computed by some Turing machine.*

Although this equivalence is historically and conceptually important, nothing else we will do depends on it. There are no conceptual hurdles involved, but filling in the details carefully requires some work. I will therefore only outline the general strategy here.

To see that every Turing computable partial function is a partial computable function, first observe that it is straightforward to represent Turing machines, configurations, and computation sequences as natural numbers in such a way that all the operations and relations defined are primitive recursive. Our definitions depend only on basic operations and relations on natural numbers and finite sequences, all of which we have shown to be primitive recursive in Chapter 8. So it should seem plausible that the relation $T(e, x, s)$ that says "s is a halting computation sequence for Turing machine e on input x" is primitive recursive, as is the function $U(s)$ that returns the output of that computation. Thus every Turing computable partial function $f(x)$ can be defined by $f(x) \simeq U(\mu s\, T(e, x, s))$, where e represents the Turing machine that computes it.

In the other direction, we need to show that every partial computable function is Turing computable, which requires reasoning about Turing machines and what they can do. It suffices to show that the set of Turing computable partial functions contains zero, successor, and the projections, and is closed under composition, primitive recursion, and unbounded search. This is enough because the set of partial computable functions is the smallest such set.

To that end, it is helpful to adopt a graphical representation of Turing machines in which states are represented by nodes and an instruction $(i, j) \mapsto (k, \ell)$ is represented by an arrow labeled (j, k) between nodes i and ℓ. Remember that this instruction says that when in state i scanning symbol j, the machine should perform action k and move to state ℓ. We also adopt the convention of using arrows to represent the move-left and move-right commands. For example, Figure 11.1 describes a Turing machine with two states, two symbols, and instructions δ defined by $\delta(0, 1) = (3, 0)$, $\delta(0, 0) = (1, 1)$, and $\delta(1, 1) = (3, 1)$. Here the number 3 represents the move-right command. When started on a string of 1s, this Turing machine simply scans to the first blank space after the string, replaces it with another 1, and then moves the tape head one square to the right. In other words, it computes the successor function. The constant zero function is also easy to compute: the Turing machine simply halts right away.

For a slightly more elaborate example, the Turing machine in Figure 11.2 computes addition. When started on the leftmost symbol of a string 1^m01^n on an otherwise blank tape, it replaces this input by 1^{m+n} and halts on a blank symbol just after the last 1. Keep in mind

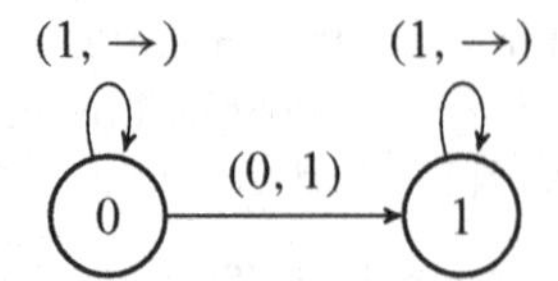

Figure 11.1 A Turing machine computing the successor function.

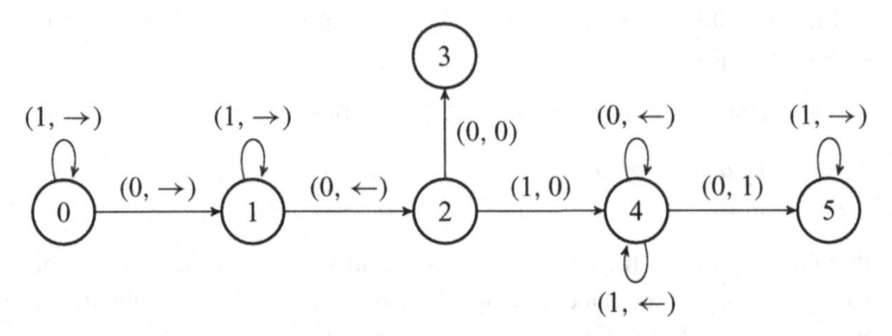

Figure 11.2 A Turing machine computing addition.

that m or n may be zero. The machine scans past the first block of 1s until it reaches a 0. It then moves to the right one square and scans past the second block of 1s until it reaches a 0, and then moves to the left. At that point, if it sees a 0, the second block is empty, and there is nothing to be done; it leaves a 0 there and halts. Otherwise, the machine need only replace the 1 by a 0, scan back to the 0 between the blocks and replace it by a 1, and then scan to the end and halt. The exercises below encourage you to try your hand at showing, similarly, that multiplication and the projection functions $p_i^n(x_0, \ldots, x_{n-1}) = x_i$ are Turing computable.

Suppose an ℓ-ary partial function $f(\vec{x})$ is defined by composing a k-ary partial function $g(y_0, \ldots, y_{k-1})$ with k-many ℓ-ary partial functions $h_0(\vec{x}), \ldots, h_{k-1}(\vec{x})$. Given Turing machines computing g and $h_0, \ldots, h_{k-1}$, there is a natural strategy for designing a Turing machine that computes f: run the Turing machines for $h_0, \ldots, h_{k-1}$ in turn, copying the inputs as needed, and then run the Turing machine for g. But this runs up against a general problem that arises when reasoning about computational processes that share a global state: with unrestricted access to the tape, any Turing machine can overwrite the results obtained by the ones before it. So to prove that the Turing computable partial functions are closed under composition, we need a stronger inductive hypothesis. Say a Turing machine computing a partial function $h(\vec{x})$ is *nice* if, when started at the beginning of an input string representing the input $\vec{x}$ and with nothing on the tape to the right of it, if $h(\vec{x})$ is defined, the Turing machine computes the value of $h(\vec{x})$ independent of what appears to the left of the input string, does not change anything to the left of the input string, and ultimately terminates with the output $h(\vec{x})$ on the tape at the same position as the input, with nothing to the right of it and the tape head one square to the right of the end of the output. One then inductively proves the claim that every partial computable function is computed by a nice Turing machine.

Using niceness, one can show closure under composition, unbounded search, and primitive recursion. Of these, handling primitive recursion is the most finicky. An alternative is to use a description of the partial computable functions that avoids it. The following characterization is quite nice:

Proposition 11.6.6. *The set of partial computable functions is the smallest set of partial functions containing zero, successor, projections, addition, multiplication, and truncated subtraction, and closed under composition and unbounded search.*

It is clear that all the functions obtained this way are computable. For the other direction, we can use the fact that for every computable function $f(\vec{x})$, there is a Δ_0 formula $A(\vec{x}, y)$ that such for every $\vec{x}$, $f(\vec{x}) \simeq L(\mu y\, A(\vec{x}, y))$, the result of searching a y satisfying $A(\vec{x}, y)$ and returning the first element. Here L is the projection function defined in Section 8.3 and shown to be Δ_0-definable by Proposition 10.2.8. It therefore suffices to show that the set of functions described by Proposition 11.6.6 contains the characteristic function of any Δ_0-definable relation. You are asked to spell out the details in Exercise 11.6.2.

Exercises

11.6.1. Show that the following functions are computable, by exhibiting Turing machines that compute them:
  a. $f(x) = 2x$
  b. the projection functions p_i^n
  c. multiplication, $g(x, y) = x \cdot y$
  d. truncated subtraction, $h(x, y) = x \dot{-} y$.

11.6.2. Prove Proposition 11.6.6 as follows.
  a. Show that for any partial computable function $f(\vec{x})$ there is a Δ_0 formula $A(\vec{x}, y)$ that such for every $\vec{x}$, $f(\vec{x}) \simeq L(\mu y\, A(\vec{x}, y))$.
  b. Let $\mathcal{S}$ be the class of functions described in Proposition 11.6.6. Show that the characteristic function $\chi_=(x, y)$ of equality and the characteristic function $\chi_<(x, y)$ of the less-than relation are in $\mathcal{S}$.
  c. Show that the usual boolean operations and(x, y) and not(x) are in $\mathcal{S}$.
  d. Show that if $f(\vec{x}, y)$ is in $\mathcal{S}$ then so is the characteristic function of the relation $R(\vec{x}, z)$ defined by $\exists y < z\, (f(\vec{x}, y) = 1)$. (Hint: show how to do a bounded search using functions in $\mathcal{S}$.)
  e. Conclude that the characteristic function of any Δ_0-definable relation is in $\mathcal{S}$, and use this to establish Proposition 11.6.6.

11.6.3. Show that if a function is computable, it can be computed by a Turing machine with at most two states.

11.6.4. Define the notion of a Turing machine with additional work tapes, and show that any such Turing machine can be simulated by a Turing machine with only one work tape, as we have defined them.

11.7 The Lambda Calculus

In this section, we will consider yet another equivalent characterization of the partial computable functions, in terms of a system of symbolic expressions known as the *lambda calculus*. The syntax of the lambda calculus is based on a notation for describing functions. Instead of saying "let f be the function defined by $f(x) = x + 5$," we can say, "let f be the function $\lambda x.\, x + 5$." In this example, $\lambda x.\, x + 5$ is just a name for the function that adds five to its argument. As with quantifiers, the variable introduced by the lambda is bound, and we identify expressions that differ only in the names of their bound variables. So we consider the expression above the same as the expression $\lambda y.\, y + 5$.

This way of defining a function from a symbolic expression is known as *lambda abstraction*. The companion to lambda abstraction is *application*. Assuming we have a function f (say, defined on the natural numbers), we can apply it to any value, like 2, to obtain what is conventionally denoted $f(2)$. In the lambda calculus, application is expressed instead by writing the function and then the argument, as in $f\ 2$, and parentheses are only needed to apply a function to a compound expression. The combination of lambda abstraction and application gives rise to a notion of *reduction*. For example, the expression $(\lambda x.\, x + 5)\ 2$ can be reduced to $2 + 5$, by substituting 2 for the bound variable x in the body of the lambda expression.

Up to this point, our use of lambda abstraction has done little more than introduce notation for conventional mathematical notions. The lambda calculus represents a more radical departure from the set-theoretic viewpoint. In that framework, everything denotes a function, and anything can be applied to anything else. For example, if t is any term in the lambda calculus, $t\,t$ is another term, informally denoting the result of applying t to itself. More precisely, the framework we will describe in this section is known as the *untyped lambda calculus*. Chapter 13 introduces a variant, the *simply typed lambda calculus*, which is, in contrast, straightforward to interpret in conventional set-theoretic terms.

In computer science, the lambda calculus has become a paradigm model of a computational system. Providing a semantics for the lambda calculus is tricky, so for the moment it is best to think of it as nothing more than as a collection of expressions together with rules for computing with them. Starting with a sequence of variables $x, y, z, \ldots$ and constant symbols $a, b, c, \ldots$, the set of terms is defined inductively, as follows:

- Each variable and constant is a term.
- If s and t are terms, so is $(s\, t)$.
- If t is a term and x is a variable, then $(\lambda x.\, t)$ is a term.

Here we will focus on the *pure* lambda calculus, in which there are no constants.

Terms that differ only up to renaming of their bound variables are said to be *α-equivalent*, and we take α-equivalent expressions to be syntactically identical. We define the set of free variables in a term, notions of substitution, and the complexity of a term as in Section 1.6. As with formulas in first-order logic, we can can carry out proofs by induction on complexity and define functions by recursion on complexity. In particular, we can define functions by recursion on the structure of a lambda term, as long as we make sure that the definition is independent of the variable x we choose to represent the body of the term $\lambda x.\, t$.

We adopt the following conventions for dropping parentheses. Applications associate to the left, so that if r, s, t, and u are terms of suitable type, then $r\,s\,t\,u$ is short for $(((r\,s)\,t)\,u)$. A lambda abstraction is assumed to have the widest scope possible, so that, for example, $\lambda x.\, r\,s\,t$ is short for $\lambda x.\, (r\,s\,t)$. For another example, $\lambda x, y.\, x\,x\,y\,x\,\lambda z.\, x\,z$ abbreviates

$$\lambda x.\, \lambda y.\, ((((x\,x)\,y)\,x)\,\lambda z.\, (x\,z)).$$

Note that this convention stands in contrast to our convention for giving quantifiers the narrowest scope possible, which is common in the traditional literature in mathematical logic.

In computer science, it is more common to give binders a wide scope and use a period to delimit the bound variables, as we have done here.

If t and s are lambda terms and x is any variable, $t[s/x]$ denotes the result of substituting s for x in t, renaming bound variables, if necessary, to avoid capture. For example,

$$(\lambda w.\, x\, x\, w)[y\, y\, z/x] = \lambda w.\, (y\, y\, z)(y\, y\, z)\, w.$$

An expression of the form $(\lambda x.\, t)\, s$ is called a *β-redex* and is said to *β-contract to* $t[s/x]$. More generally, if it is possible to reduce t_0 to t_1 by β-contracting some subterm, we say that t_0 β-reduces to t_1 in one step, and if there is a sequence of β-reductions from t_0 to t_1 (possibly none), t_0 is said to β-reduce to t_1. If no one-step reduction applies to t, t is said to be in *normal form*.

These notions will defined more precisely in Section 13.1 in the context of the simply typed lambda calculus. This informal presentation will suffice for now. Since β-reduction is the only notion of reduction we will consider in this section, I will generally leave off the β.

Let us consider some examples. We have

$$\begin{aligned}(\lambda x.\, x\, x\, y)\, \lambda z.\, z &\to_1 (\lambda z.\, z)(\lambda z.\, z) y \\ &\to_1 (\lambda z.\, z)\, y \\ &\to_1 y.\end{aligned}$$

Reducing a term can make it longer:

$$\begin{aligned}(\lambda x.\, x\, x\, y)\, (\lambda x.\, x\, x\, y) &\to_1 (\lambda x.\, x\, x\, y)\, (\lambda x.\, x\, x\, y)\, y \\ &\to_1 (\lambda x.\, x\, x\, y)\, (\lambda x.\, x\, x\, y)\, y\, y \\ &\to_1 \ldots\end{aligned}$$

It can also leave a term unchanged:

$$(\lambda x.\, x\, x)(\lambda x.\; x\, x) \to_1 (\lambda x.\, x\, x)(\lambda x.\, x\, x).$$

Some terms can be reduced in more than one way. For example,

$$(\lambda x.\, (\lambda y.\, y\, x)\, z)\, v \to_1 (\lambda y.\, y\, v)\, z$$

by contracting the outermost application, and

$$(\lambda x.\, (\lambda y.\, y\, x)\, z)\, v \to_1 (\lambda x.\, z\, x)\, v$$

by contracting the innermost one. Notice that both terms further reduce to the same term, $z\, v$. This is not a coincidence but rather an important property of the lambda calculus known as the *Church–Rosser property*, or *confluence*.

Theorem 11.7.1. *Let s, t_0, and t_1 be terms such that $s \to t_0$ and $s \to t_1$. Then there is a term u such that $t_0 \to u$ and $t_1 \to u$.*

This theorem is tricky to prove. It is not hard to show that any two one-step reductions can be reconciled, in the sense that if $s \to_1 t_0$ and $s \to_1 t_1$, there is a u such that $t_0 \to u$ and $t_1 \to u$. The problem is that the reductions to u may require multiple steps, and in Section 13.3 we will see that this property is not strong enough to imply confluence. The solution is to define a stronger notion, *parallel reduction*, that describes the possibility of doing any number of one-step reductions in parallel.

Definition 11.7.2. The *parallel reduction* relation $s \Rightarrow t$ is defined inductively as follows:

- $x \Rightarrow x$ for every variable x.
- If $s \Rightarrow t$, then $\lambda x.\, s \Rightarrow \lambda x.\, t$.
- If $s \Rightarrow t$ and $u \Rightarrow v$ then $s\,u \Rightarrow t\,v$.
- If $s \Rightarrow t$ and $u \Rightarrow v$ then $(\lambda x.\, s)\, t \Rightarrow u[v/x]$.

To each term t, we can also define the result t^* of doing *all* possible one-step reductions that can be carried out in parallel:

- $x* = x$ and $c^* = c$
- $(\lambda x.\, t)^* = \lambda x.\, t^*$
- $(ts)^* = t^*\, s^*$, if t is not a lambda abstraction
- $((\lambda x.\, t)\, s)^* = t^*[s^*/x]$.

In the exercises below, you are asked to prove the following two lemmas and show that Theorem 11.7.1 follows.

Lemma 11.7.3. *Parallel reduction satisfies the following:*

- $t \Rightarrow t$ *for every* t.
- *If* $s \rightarrow_1 t$ *then* $s \Rightarrow t$.
- *If* $s \Rightarrow t$ *then* $s \rightarrow t$.
- *If* $s \Rightarrow t$ *and* $u \Rightarrow v$, *then* $s[u/x] \Rightarrow t[v/x]$.

Lemma 11.7.4. *If* $s \Rightarrow t$ *then* $s \Rightarrow t^*$.

Theorem 11.7.1 has the following important corollary.

Corollary 11.7.5. *If t has a normal form, then this normal form is unique.*

Proof If $t \rightarrow u_0$ and $t \rightarrow u_1$, then by Theorem 11.7.1 there is a term v such that u_0 and u_1 both reduce to v. If u_0 and u_1 are both in normal form, this can only happen if u_0 and u_1 are both equal to v. □

We will say that two terms s and t are *β-equivalent*, or just *equivalent*, if there is some u such that $s \rightarrow u$ and $t \rightarrow u$. This is written $s \equiv t$. Using Theorem 11.7.1, you can check that $\equiv$ is an equivalence relation with the additional property that for every s and t, if $s \rightarrow t$ or $t \rightarrow s$ then $s \equiv t$. Exercise 13.1.4 shows that it is the smallest equivalence relation with this property.

In order to talk about computation on the natural numbers in terms of the lambda calculus, we need a way of representing the natural numbers.

Definition 11.7.6. For each natural number n, define the *numeral* $\overline{n}$ to be the lambda term $\lambda x, y.\, x\,(x\,(x\, \cdots\, (x\, y) \cdots))$, where there are n-many xs in all.

The terms $\overline{n}$ are *iterators*: for any function f, $\overline{n}\, f$ denotes the function that maps any value y to $f^n(y)$. Note that each numeral is in normal form.

Definition 11.7.7. Let $f(n_0, \ldots, n_{k-1})$ be an k-ary partial function from $\mathbb{N}$ to $\mathbb{N}$. Say a lambda term t *represents* f if for every sequence of natural numbers $n_0, \ldots, n_{k-1}$,

$$t\, \overline{n_0}\, \cdots\, \overline{n_{k-1}} \rightarrow \overline{f(n_0, \ldots, n_{k-1})}$$

if $f(n_0, \ldots, n_{k-1})$ is defined, and $t\, \overline{n_0}\, \cdots\, \overline{n_{k-1}}$ has no normal form otherwise.

As with numerals in the language of arithmetic, we will henceforth omit the bar when the context makes it clear that we are interpreting a natural number as a lambda term.

The following theorem shows that the lambda calculus provides yet another model of computation.

Theorem 11.7.8. *A function f is a partial computable function if and only if it is represented by a term in the lambda calculus.*

This theorem is not used elsewhere in this book, and, as with Turing machines, I will only sketch the equivalence. By now, the strategy for proving the "if" direction should be clear: we describe a primitive recursive relation $T(e, x, s)$ that says that s represents a reduction sequence that starts with e applied to the numeral for x and ends with a numeral, and we describe a function $U(s)$ the decodes the numeral at the end of the sequence. Then any lambda computable function can be expressed as $f(x) \simeq U(\mu s\, T(e, x, s))$ for some e.

For the other direction, we need to show how to design lambda terms to represent the partial computable functions. Let us start with the primitive recursive functions. We already have 0, and the successor function is represented by the term $S = \lambda u, x, y.\, x\,(u\,x\,y)$ because for any numeral n, $S\,n$ reduces to $\lambda x, y.\, x\,(n\,x\,y)$, which in turn reduces to $n+1$. Projections are represented by the terms P_i^n given by $\lambda \vec{x}.\, x_i$. Closure under composition is similarly straightforward. If f is defined by composition from $g, h_0, \ldots, h_{k-1}$ and these are represented by $G, H_0, \ldots, H_{k-1}$, respectively, then f is represented by the term $\lambda \vec{x}.\, G\,(H_0\,\vec{x}) \cdots (H_{k-1}\,\vec{x})$.

Primitive recursion, however, requires more work. Assuming a function f is defined from g and h by primitive recursion and that these are represented by terms G and H respectively, it suffices to find a term F such that we have the following for every natural number n:

$$F\,0\,\vec{z} \equiv G\,\vec{z}$$
$$F\,(n+1)\,\vec{z} \equiv H\,n\,(F\,n\,\vec{z})\,\vec{z}$$

Using induction and the Church–Rosser property, we can then prove that for every natural number n and tuple of numerals $\vec{m}$, $F\,n\,\vec{m}$ normalizes to the right answer. We can use lambdas to avoid worrying about the parameters $\vec{z}$: to solve the problem above, it suffices to find a term F satisfying

$$F\,0 \equiv G$$
$$F\,(n+1) \equiv H'\,n\,(F\,n)$$

for every natural number n, where H' is the term $\lambda u, v, \vec{z}.\, H\,u\,(v\,v\,\vec{z})\,\vec{z}$. So we will focus on solving the second pair of equivalences, renaming H' to H.

The intuition behind our strategy is to use the numeral n as an iterator to compute the pairs of values

$$(0, F\,0),\ (1, F\,1),\ (2, F\,2),\ \ldots,$$

where the pairs are represented by suitable lambda terms. First, define the lambda term K to be $\lambda x, y.\, x$, so for every t, $K\,t$ describes the constant function $\lambda y.\, t$. If we then define the lambda term D to be $\lambda x, y, z.\, z\,(K\,y)\,x$, we have that for every pair of terms s and t,

$$D\,s\,t\,0 \to 0\,(K\,t)\,s \to s$$
$$D\,s\,t\,1 \to 1\,(K\,t)\,s \to K\,t\,s \to t.$$

So $D\,s\,t$ represents the pair (s, t), and applying $D\,s\,t$ to 0 and 1 respectively extracts the two components. In what follows, I will suggestively write (s, t) for $D\,s\,t$ and $(t)_0$ and $(t)_1$ for $t\,0$ and $t\,1$.

Given G and H, let T be the term

$$\lambda u.\,(S\,(u)_0, H\,(u)_0\,(u)_1).$$

For any number n and term t, $T\,(n, t)$ reduces to $(n+1, H\,n\,t)$. So T is just the right thing to take an expression equivalent to $(n, F\,n)$ to another expression equivalent to $(n+1, F\,(n+1))$. Define F to the term

$$\lambda u.\,(u\,T\,(0, G))_1,$$

so that $F\,n$ applies T n-many times to the pair $(0, G)$ and then returns the second component.

To see $F\,0 \equiv G$, we simply compute:

$$F\,0 \to (0, T\,(0, G))_1 \to (0, G)_1 \to G.$$

As a result, we also have $0\,T\,(0, G) \equiv (0, F\,0)$, since the left-hand side reduces to $(0, G)$. We will show that for any n, the equivalence $n\,T\,(0, G) \equiv (0, F\,n)$ implies both of the following:

- $(n+1)\,T\,(0, G) \equiv (n+1, F\,(n+1))$
- $F\,(n+1) \equiv H\,n\,(F\,n)$.

Induction and the first implication yield $n\,T\,(0, G) \equiv (0, F\,n)$ for every n, and then the second implication gives us $F\,(n+1) \equiv H\,n\,(F\,n)$ for every n, which is what we want.

Assuming $n\,T\,(0, G) \equiv (n, F\,n)$, we have the following:

$$\begin{aligned}(n+1)\,T\,(0, G) &\equiv T\,(n\,T\,(0, G))\\ &\equiv T\,(n, F\,n)\\ &\equiv (n+1, H\,n\,(F\,n)).\end{aligned}$$

From the definition of F, we then get this:

$$\begin{aligned}F\,(n+1) &\equiv ((n+1)\,T\,(0, G))_1\\ &\equiv (n+1, H\,n\,(F\,n))_1\\ &\equiv H\,n\,(F\,n).\end{aligned}$$

Replacing $H\,n\,(F\,n)$ by $F\,(n+1)$ in the first equivalence, we get $(n+1)\,T\,(0, G) \equiv (n+1, F\,(n+1))$, as required.

We have thus shown that every primitive recursive function is represented by a lambda term. To extend the result to arbitrary partial computable functions, we need a way of doing unbounded search. The following proposition is a version of the fixed-point theorem for the lambda calculus. Compare the proof to that of Theorem 11.5.2.

Proposition 11.7.9. *There is a lambda term Y such that for any lambda term t, $Y\,t$ reduces to $t\,(Y\,t)$.*

Proof If we let Y' be the term

$$\lambda x.\,((\lambda y.\,x\,(y\,y))\,(\lambda y.\,x\,(y\,y))),$$

then both $Y' t$ and $t\,(Y' t)$ reduce to $t\,((\lambda y.\, t\,(y\,y))\,(\lambda y.\, t\,(y\,y)))$, which shows that $Y' t$ and $t\,(Y' t)$ are equivalent. The term Y' is known as *Curry's combinator*. But Proposition 11.7.9 asserts something stronger, namely, that Yt reduces to $t\,(Y\,t)$. Taking Y to be *Turing's combinator*,

$$(\lambda y, x.\, x\,(y\,y\,x))\,(\lambda y, x.\, x\,(y\,y\,x)),$$

yields that stronger claim. □

We can use Y to carry out unbounded search. Let $h(x)$ be the characteristic function of some predicate, meaning that it always returns 0 or 1. Suppose $h(x)$ is represented by the lambda term H. Suppose, moreover, we know that we know that there is at least one value of x such that $h(x) = 1$, and we want to find the smallest one. Using the Y combinator, we can find a term V such that for every term u,

$$V \equiv \lambda u.\, D\,(V\,(S\,u))\,u\,(H\,u).$$

Specifically, we take V to be $Y\,(\lambda x, u.\, D\,(x\,(S\,u))\,u\,(H\,u))$, which, by Proposition 11.7.9, reduces to $\lambda u.\, D\,(V\,(S\,u))\,u\,(H\,u)$. Then we have that for any numeral n, $V\,n$ reduces to $V\,(n + 1)$ if $H\,n$ reduces to 0, and n if it reduces to 1. So $V\,0$ carries out the search we want. If we replace H by a variable h in the definition of V, then $\lambda h.\, V\,0$ works generically when applied to any H.

Now, suppose $f(x)$ is a partial computable function. Then for some e, we have that for every x, $f(x) \simeq U(\mu s\, T(e, x, s))$, making use of Kleene's T and U. Let U and T also denote the lambda terms that represent U and the characteristic function of T, let H be the term $T\,e\,x$, and let F be the term $\lambda x.\, U\,(V\,0)$. Then we have shown that if $f(x)$ is total, then for every n, $F\,n$ reduces to $f(n)$. In other words, we have shown that every total computable function is representable by a lambda term.

What happens if f is partial? In that case, we need to know that for any value of n, if there is no value of s such that $T(e, n, s) = 1$, then the term $F\,n$ has no normal form. In fact, this is true, but proving it requires more work. It is clear that the natural way of trying to reduce $F\,n$, by iteratively computing T and then moving on to the next element in the search, never terminates. But that leaves the possibility that there might be some other bizarre way to reduce $F\,n$ to a normal form. Ruling out that possibility requires a careful analysis of the structure of lambda terms and reductions, which would take us too far afield.

In short, we have sketched a proof that every total computable function is representable in the lambda calculus, and, modulo the nagging concern just described, a proof that every partial computable function is representable there as well. It is surprising that the general notion of computation can be understood in terms of such a simple notion of reduction. In Chapter 13, we will consider an even simpler calculus, combinatory logic, with two constants K and S, and two reduction rules:

- $K\,s\,t \to s$
- $S\,r\,s\,t \to r\,t\,(s\,t)$.

In Chapter 13, in the setting of the simply typed lambda calculus, we will see that lambda terms can be translated to terms in combinatory logic. These considerations carry over to the untyped setting as well, and arguments similar to the ones presented here show that every partial computable function can be represented in combinatory logic as well. This means that computation can be understood in terms of nothing more than reducing terms involving K and S, a truly remarkable fact.

Exercises

11.7.1. Reduce the following terms to normal form. It might help to rename bound variables, to avoid confusion.
a. $(\lambda x, y, z. x y (y z)) (\lambda y. y y) (\lambda x. w) (\lambda x. x)$
b. $(\lambda z. (\lambda x. x x z) (\lambda x. y)) w w$

11.7.2. Find a lambda term t such that t has a normal form but there is also an infinite sequence of one-step reductions starting from t.

11.7.3. Prove Lemmas 11.7.3 and 11.7.4, and show that Theorem 11.7.1 follows.

11.7.4. Present explicit lambda terms that represent addition and multiplication.

11.7.5. Prove that every computable function is representable in the lambda calculus, using the characterization of computable functions given by Proposition 11.6.6. (This is not a big win. Representing truncated subtraction requires representing the predecessor function, which, in turn, requires something similar to the method we used to handle primitive recursion.)

11.8 Relativized Computation

Given that there are sets and functions that are not computable, it is often interesting to consider what can be computed *relative to* such data. Since we can identify a set with its characteristic function and, conversely, a function with its graph, in the name of parsimony we could restrict attention to one or the other. But sometimes one formulation is more natural and sometimes the other, so we will pass freely between the two.

Let $f_0, \ldots, f_{n-1}$ be a tuple of functions from $\mathbb{N}$ to $\mathbb{N}$ of various arities. The set of *partial computable functions relative to* $\vec{f}$ is the smallest set of partial functions containing zero, successor, projections, and each f_i, and closed under composition, primitive recursion and unbounded search. The intuition is that a partial function is computable relative to $\vec{f}$ if there is a recipe for computing it given the ability to make arbitrary calls to the functions in $\vec{f}$. By definition, a partial function is computable relative to a tuple of functions $\vec{f}$ and a tuple of sets $\vec{X}$ if it is computable relative to $\vec{f}$ and the characteristic functions of $\vec{X}$. Relativized computation is sometimes understood as computability relative to an *oracle*: although $\vec{f}$ and $\vec{X}$ may not be computable, we imagine having an external source that somehow answers our queries about membership in a set X_i or the value of a function f_j on some input.

We can similarly talk about sets that are computable or computably enumerable relative to $\vec{f}$ and $\vec{X}$. Section 10.2 introduced the notion of arithmetic definability relative to a tuple of functions and sets. The common use of the term "relative" is not a coincidence. The following is a straightforward generalization of Theorem 11.2.1, Corollary 11.2.2, and the equivalence between clauses 2 and 4 of Theorem 11.4.2:

Proposition 11.8.1. *Let $\vec{f}$ be any tuple of functions and $\vec{X}$ be any tuple of sets.*

1. *A partial function g is computable relative to $\vec{f}$ and $\vec{X}$ if and only if the graph of g is Σ_1-definable relative to $\vec{f}$ and $\vec{X}$.*

2. *A total function g is computable relative to $\vec{f}$ and $\vec{X}$ if and only if the graph of g is Δ_1-definable relative to $\vec{f}$ and $\vec{X}$.*
3. *A relation R is computable relative to $\vec{f}$ and $\vec{X}$ if and only if R is Δ_1-definable relative to $\vec{f}$ and $\vec{X}$.*
4. *A relation R is computably enumerable relative to $\vec{f}$ and $\vec{X}$ if and only if Y is Σ_1-definable relative to $\vec{f}$ and $\vec{X}$.*

It is common to say that a set A is computable *in* X when A is computable relative to X, and similarly for other variations on relative definability.

We can also characterize computability relative to a function $f(\vec{x})$ in the Turing machine model by adding a new *oracle tape* and an *oracle tape head.* If the program places input $\vec{x}$ on the oracle tape and enters a special *query state*, then at the next step, the machine enters a special *response state*, with the value of $f(\vec{x})$ on the oracle tape and the oracle tape head reset to the beginning of that input. Computability relative to multiple sets and functions can be defined by adding more oracle tapes, or representing such tuples by a single function.

Relative computability gives rise to another natural notion of reduction:

Definition 11.8.2. A set X is *Turing reducible to* Y, write $X \leq_T Y$, if χ_X is computable relative to Y.

In other words, X is Turing reducible to Y if there is a Turing machine that, on input z, determining whether z is a member of X after making arbitrarily many queries to an oracle for Y. The following are straightforward.

Proposition 11.8.3. *If $X \leq_T Y$ and Y is computable, then X is computable.*

Proposition 11.8.4. *The relation $\leq_T$ is reflexive and transitive.*

Proposition 11.8.3 follows from Proposition 10.2.3, given the relationship between computability and Δ_1-definability. The transitivity claim in Proposition 11.8.4 follows from a relativized version thereof. But Propositions 11.8.3 and 11.8.4 can also be proved directly in terms of the other descriptions of relativized computation. For example, it is not hard to show from the definition of relativized computability that if a function f is computable relative to g, then so is any function computable relative to f. In terms of Turing machine, you should imagine replacing queries about f in an oracle computation by calls to a subroutine that computes f from an oracle for g.

Many-one reducibility is stricter than Turing reducibility in the sense that $A \leq_m B$ implies $A \leq_T B$ but not conversely. The difference is that in deciding membership in A, a many-one reduction from B is only allowed to query B once, and it has to report the same answer. So, for example, whereas the halting problem is Turing reducible to its complement, it is not many-one reducible to it. As preorders, $\leq_m$ and $\leq_T$ induce equivalence relations: A is many-one equivalent to B, written $A \equiv_m B$, if $A \leq_m B$ and $B \leq_m A$. Turing equivalence, $A \equiv_T B$, is defined analogously. The preorders induce a partial order on the equivalence classes, known as the *many-one degrees* and *Turing degrees*, respectively, and these are structures of interest in the theory of computability.

Given two sets A and B, define $A \oplus B$ to be the set $\{2n \mid n \in A\} \cup \{2n+1 \mid n \in B\}$. This operation descends to both the many-one and Turing degrees, in the sense that if A_1 is equivalent to A_2 and B_1 is equivalent to B_2, then $A_1 \oplus B_1$ is equivalent to $A_2 \oplus B_2$. The exercises below ask you to show that the equivalence class of $A \oplus B$ is the least upper bound of A and B in both the lattice of many-one degrees and the lattice of Turing degrees. Similarly, if A is any set, A' denotes the halting problem relative to A, and operation that carries over to degrees. If $\mathbf{a}$ is any Turing degree, then $\mathbf{a} <_T \mathbf{a}'$, and similarly for many-one degrees.

Up to this point, we have considered computability relative to fixed data, $\vec{f}$ and $\vec{X}$. But our descriptions are uniform enough that we can also view $\vec{f}$ and $\vec{X}$ are *arguments* to the functions we describe. We can therefore say what it means for a function $F(\vec{X}, \vec{f}, \vec{y})$ to be computable, where F takes arbitrary sets $\vec{X}$, functions $\vec{f}$, and natural numbers $\vec{y}$ as arguments and returns a natural number. According to terminology that is explained in Chapter 13, these are called *type 2 functionals*.

For example, restricting attention to unary functions for notational simplicity, to define the partial computable type 2 functionals we start with zero, successor, projections, and the evaluation functions

$$F(f_0, \ldots, f_{n-1}, y_0, \ldots, y_{m-1}) = f_i(y_j).$$

We then define the set of partial computable type 2 functionals to be the smallest set of functionals containing these and closed under composition, primitive recursion, and unbounded search. The computable type 2 functionals, as usual, are the ones that happen to be total. In terms of definability, a partial computable type 2 functional is one whose graph $F(\vec{X}, \vec{f}, \vec{y}) \simeq z$ is definable by a Σ_1 formula with additional predicate symbols $\vec{X}$ and function symbols $\vec{f}$, as well as the variables $\vec{y}$ and z. Any total computable type 2 functional is Δ_1-definable in this way. In terms of Turing machines, the partial computable type 2 functionals are the ones that are computable Turing machines with an oracles for their set and function arguments.

In Chapter 15, we will consider the language of *second-order arithmetic*, in which formulas can have second-order variables and quantifiers that range over functions and relations on the natural numbers. This opens the door to the study of definability of not only sets, relations, and functions on the natural numbers, but also sets of sets, relations between sets, sets of functions, and so on. For example, a collection $\mathcal{X}$ of sets of natural numbers is definable by a formula $A(X)$ in the language of second-order arithmetic if, for every X, $X \in \mathcal{X}$ if and only if $\mathfrak{N} \models A(X)$. Here $\mathfrak{N}$ denotes the standard model of the natural numbers and the symbol $\models$ denotes the standard semantics for second-order logic. For formulas with only first-order quantifiers, we can alternatively formulate definability in terms of the language of first-order arithmetic, provided we extend it to include additional predicate symbols $\vec{X}$ and function symbols $\vec{f}$.

The usual enumeration $\varphi_0, \varphi_1, \varphi_2, \ldots$ of partial computable functions can be extended to an enumeration $\varphi_0^X, \varphi_1^X, \varphi_2^X, \ldots$ of partial computable computable functions relative to X. We can take φ_e^X to be the partial function defined by the Σ_1 formula with parameter X represented by e, or, alternatively, the partial function computed by Turing machine e with an oracle for X. We can once again treat X as a variable, so that $F(X, y) \simeq \varphi_e^X(y)$ is the eth type 2 partial computable functional with arguments X and y. Given the correspondences between sets and functions and the pairing operation $\oplus$, we can generally pass between computability relative to a set and computability relative to arbitrary tuples of sets and functions freely. As a result, we can just as well enumerate partial functions computable relative to tuples $\vec{X}$ and $\vec{f}$.

We should expect that if the value of a partial computable function $\varphi_e^X(y)$ is defined at y, the computation can only query finitely many values of X. Similarly, and more or less equivalently, if a Σ_1 formula $\exists y\, C(y, X)$ is true, a finite initial segment of X is enough to make it true, in the sense that $\exists y\, C(y, Z)$ is also true for any Z that agrees with X on sufficiently many values. Theorems 11.8.5 and 11.8.6 make these intuitions precise. If X is a set and σ is a finite binary sequence, say that *σ is an initial segment of X*, written $\sigma \subset X$, if the following holds: for every i less than $\mathrm{length}(\sigma)$, if i is in X, then $(\sigma)_i = 1$, and if i is not in X, then $(\sigma)_i = 0$.

Theorem 11.8.5. *Let $A(X)$ be any Σ_1 formula relative to a predicate variable X. Then there is a Δ_0 formula $B(\sigma)$ such that for every set X, $A(X)$ is true in the standard interpretation if and only if for some $\sigma \subset X$, $B(\sigma)$ is true. We can moreover assume that $B(\sigma)$ has the additional property that $B(\sigma)$ and $\tau \supseteq \sigma$ imply $B(\tau)$.*

The last condition is equivalent to saying that the set $\{\sigma \in \{0, 1\}^{\mathbb{N}} \mid \neg B(\sigma)\}$ is a tree on $\{0, 1\}$.

Proof Intuitively, $B(\sigma)$ says that σ contains enough information about X to verify that $A(X)$ is true. Note that $A(X)$ may have free variables $\vec{z}$ distinct from σ, in which case $B(\sigma)$ can have the same free variables and the equivalence stated in the theorem is asserted to hold for all assignments to these variables.

We first prove the claim for Δ_0 formulas $A(X)$ in negation normal form, using induction on the set of such formulas. If $A(X)$ is an atomic formula or negated atomic formula that doesn't mention X, we take $B(\sigma)$ to be the same formula. If $A(X)$ is the formula $X(t)$, we take $B(\sigma)$ to be the formula $t < \mathrm{length}(\sigma) \wedge (\sigma)_i = 1$, and if $A(X)$ is $\neg X(t)$, we take $B(\sigma)$ to be $t < \mathrm{length}(\sigma) \wedge (\sigma)_i = 0$. If $A(X)$ is $A_0(X) \wedge A_1(X)$, we take $B(\sigma)$ to $B_0(\sigma) \wedge B_1(\sigma)$, for B_0 and B_1 obtained from the inductive hypothesis. Disjunction is handled similarly.

Suppose $A(X)$ is $\forall y < t\, A'(y, X)$. By the inductive hypothesis, we obtain a $B'(y, \sigma)$ corresponding to $A'(y, X)$ with the stated properties. Let $B(\sigma)$ be the formula $\forall y < t\, B'(y, \sigma)$. To see that this works, suppose $A(X)$ holds. Then for every $y < t$, $A'(y, X)$ holds, and so for each $y < t$, there is a $\sigma_y \subset X$ such that $B'(y, \sigma_y)$ holds. Let σ be the longest of these initial segments; then $B'(y, \sigma)$ holds for each $y < t$ as required. The converse direction is straightforward, and it is immediate that if $B'(y, \sigma)$ is preserved under extensions of σ for every y, then $B(\sigma)$ is preserved under extensions of σ. The case where $A(X)$ begins with a bounded existential quantifier is similar.

Finally, suppose $A(X)$ is $\exists y\, A'(y, X)$, where A' is Δ_0. Let $B'(y, \sigma)$ satisfy the statement of the theorem for A'. We can absorb the existential quantifier in A into σ by taking $B(\sigma)$ to be $\exists y < \mathrm{length}(\sigma)\, B'(y, \sigma)$. Clearly $B(\sigma)$ implies $\exists y\, A'(y, X)$. On the other hand, if $\exists y\, A'(y, X)$ holds, taking a $\sigma \subset X$ such that $B'(y, \sigma)$ and $\mathrm{length}(\sigma) > y$ shows that the converse also holds. □

Theorem 11.8.6. *For every partial computable function $\varphi_e^X(y)$ relative to X, there is a computable relation $R(\sigma, y, z)$ such that $\varphi_e^X(y) \downarrow = z$ if and only if for some $\sigma \subset X$, $R(\sigma, y, z)$ holds. We can moreover assume that $R(\sigma, y, z)$ implies $z < \mathrm{length}(\sigma)$, and if $R(\sigma, y, z)$ and $\tau \supseteq \sigma$ then $R(\tau, y, z)$.*

Proof Apply Theorem 11.8.5 to the Σ_1 statement $\varphi_e^X(y) \downarrow = z$, let $B(\sigma, y, z)$ be the corresponding Δ_0 formula, and let $R(\sigma, y, z)$ say that $B(\sigma, y, z)$ holds and $z < \mathrm{length}(\sigma)$. □

In the next section, we will write $\varphi_e^\sigma(y) \downarrow = z$ to indicate that $R(\sigma, y, z)$ holds, which we can interpret as saying that σ provides enough information to see that $\varphi_e^X(y)$ is defined and equal to z. If σ is a finite sequence of natural numbers and f is a function from natural numbers to natural numbers, we say that σ is an *initial segment of* f, written $\sigma \subset f$, if $(\sigma)_i = f(i)$ for every $i < \text{length}(\sigma)$. In Chapters 15 and 16, we will need the following variant of Theorem 11.8.5.

Theorem 11.8.7. *Let $A(f)$ be any Σ_1 formula relative to a variable f ranging over unary functions from $\mathbb{N}$ to $\mathbb{N}$. Then there is a Δ_0 formula $B(\sigma)$ such that for every f, $A(f)$ is true in the standard interpretation if and only if for some $\sigma \subset f$, $B(\sigma)$ is true. We can moreover assume that $B(\sigma)$ has the additional property that $B(\sigma)$ and $\tau \supseteq \sigma$ imply $B(\tau)$.*

This time, the last condition says that the set $\{\sigma \in \{0, 1\}^{\mathbb{N}} \mid \neg B(\sigma)\}$ is a tree on $\mathbb{N}$ instead of a tree on $\{0, 1\}$. The proof of Theorem 11.8.7 is similar to that of Theorem 11.8.5, though it is slightly complicated by the fact that terms t appearing in A may involve f. To handle this, show by induction on terms that for any such t, there is an initial segment $\sigma \subset f$ such that the values of t are determined by σ. The details are left to you as an exercise.

The set $\{0, 1\}^{\mathbb{N}}$ equipped with the product topology is known as *Cantor space* and the set $\mathbb{N}^{\mathbb{N}}$ equipped with the product topology is known as *Baire space*. (See Section A.4.) We can also view Cantor space as the set of subsets of $\mathbb{N}$ if we identify such subsets with their characteristic functions. Despite the similarity between Theorems 11.8.5 and 11.8.7, there are important differences between set parameters and function parameters in a Σ_1 formula, stemming from the fact that Cantor space is compact but Baire space is not. The exercises below ask you to show if $B(X)$ is a Σ_1 formula, possibly with free variables other than X, then $\forall X\, B(X)$ is equivalent in the standard model to a Σ_1 formula. In contrast, any formula of the form $\forall X\, A(X)$, where A is arithmetic, is equivalent to one of the form $\forall f\, B(f)$, for some Σ_1 formula B. In other words, a universal quantifier over function variables followed by a Σ_1 formula can be used to express any arithmetically definable relation, whereas allowing universal quantifiers over set variables does not extend Σ_1 definability.

Once again, Theorems 11.8.5 and 11.8.6 can be extended to tuples of sets $\vec{X}$ and functions $\vec{f}$. Consider the product topologies on $\{0, 1\}^{\mathbb{N}}$ and $\mathbb{N}^{\mathbb{N}}$, where we start with the discrete topology on $\{0, 1\}$ and $\mathbb{N}$, respectively. The fact that a computable functional $F(X, f, y)$ of a set X and a unary function f depends on only finitely many values of X and f implies that it is continuous when we view X and f as ranging over these spaces. Similarly, Theorem 11.8.5 says that Σ_1-definable sets of sets and functions are open in the respective topologies. For that reason, type 2 computable functions are sometimes said to be *effectively continuous* functions, and Σ_1-definable collections of sets and functions are sometimes said to be *effectively open*. We will return to the topic of the definability of collections of sets and functions in Section 15.5.

The Turing jump can be iterated, and it is common to write $A^{(n)}$ for the nth jump of A. Theorems 11.8.5 and 11.8.6 are useful in proving *Post's theorem*, which establishes a fundamental correspondence between the arithmetic hierarchy and the iterated jump operator.

Theorem 11.8.8 (Post's theorem). *The following hold for every n:*

1. *A set A is Σ_{n+1}-definable if and only if it is computably enumerable relative to some Π_n set.*
2. *The set $\emptyset^{(n)}$ is a complete Σ_n-definable set under many-one reducibility.*

In fact, the following relativized versions hold for any set of natural numbers B and every n:

1. *A set A is Σ_{n+1}-definable relative to B if and only if it is computably enumerable relative to some set that is Π_n-definable relative to B.*
2. *The set $B^{(n)}$ is complete for the class of sets that are Σ_n-definable relative to B under many-one reducibility.*

Keep in mind that being computably enumerable relative to a Π_n-definable set is the same as being computably enumerable relative to a Σ_n-definable set, since every set is Turing reducible to its complement. As a corollary of Post's theorem, we have that a set A is Σ_{n+1}-definable relative to B if and only if it is computably enumerable in $B^{(n)}$, and a set A is Δ_{n+1}-definable relative to B if and only if it is Turing reducible to $B^{(n)}$. Post's theorem is proved in most introductions to computability theory, but the exercises below encourage you to try it yourself.

Exercises

11.8.1. Outline a proof of Proposition 11.8.1.

11.8.2. Prove Propositions 11.8.3 and 11.8.4 using the description of relative computability in terms of the relativized partial computable functions.

11.8.3. Prove that $A \oplus B$, defined above, descends to a function on both many-one degrees and the Turing degrees, and represents the least upper bound of the degrees of A and B in each lattice.

11.8.4. Prove Theorem 11.8.7.

11.8.5. Show that if $B(X)$ is a Σ_1 formula, possibly with free variables other than X, then $\forall X\, B(X)$ is equivalent to a Σ_1 formula. (Hint: use Kőnig's lemma.)

11.8.6. Show that any formula of the form $\forall X\, A(X)$, where A is arithmetic, is equivalent to one of the form $\forall f\, B(f)$ for some Σ_1 formula B. (Hint: Skolemize and use a pairing operation.)

11.8.7. Let $F(X, y)$ be a computable functional. Show that the function $g(y) = \max_{X \subseteq \mathbb{N}} F(X, y)$ is computable.

11.8.8. Show that a set of natural numbers is Δ_2-definable if and only if it is computable from the halting problem, $\emptyset'$.

11.8.9. The *Shoenfield limit lemma* says that a set A is computable from $\emptyset'$ if and only if there is a computable function $g(n, s)$ taking values in $\{0, 1\}$ such that:

a. for every n, $\lim_{s\to\infty} g(n, s)$ exists, and
b. for every n, $n \in A$ iff $\lim_{s\to\infty} g(n, s) = 1$.

Prove this. (Hint: The statement "$\lim_{s\to\infty} g(n, s)$ exists" means that there is some s_0 such that either $g(n, s) = 1$ for every $s \geq s_0$ or $g(n, s) = 0$ for every $s \geq s_0$. For the "if" direction, describe an algorithm for computing A from g, using $\emptyset'$ to answer Σ_1 questions. For the "only if" direction, use Theorem 11.8.5 to obtain a computable relation R such that, for every n, $n \in A$ if and only if there is some $\sigma \subset \emptyset'$ such that $R(n, \sigma)$, and R moreover satisfies the relevant monotonicity property. For each s, define σ_s to be a finite binary sequence of length s which approximates $\emptyset'$, so that $(\sigma_s)_i = 1$ if and only if Turing machine i halts on input 0 in at most s steps. Define $g(n, s) = 1$ if $R(n, \sigma_s)$, and $g(n, s) = 0$ otherwise. Show that this g works.)

11.8.10. Prove Theorem 11.8.8 by induction on n, using Theorems 11.8.5 and 11.8.6.

11.9 Computability and Infinite Binary Trees

Let T be a binary tree, that is, a set of finite sequences of elements of $\{0, 1\}$ that is closed under initial segments. Kőnig's lemma says that if T is infinite, it has an infinite path, a fact which expresses the compactness of Cantor space. (See Exercise 1.4.2.) A number of important mathematical constructions can be viewed as the result of finding an infinite branch through such a tree. In Section 6.3, we used a path through a tree to construct a model of a consistent set of sentences in our proof of the completeness theorem for the classical cut-free sequent calculus. Other constructions in Chapter 5 can be expressed in these terms.

Since any finite binary sequence can be represented by a natural number, a binary tree T can be represented by a set of numbers. We can conveniently identify any path through T, viewed as a binary sequence indexed by $\mathbb{N}$, with the characteristic function of a set P. P represents a path through T in this sense if and only for every $\sigma \subset P$, σ is in T. We can therefore ask: if T is a computable tree, does it necessarily have a computable path? The next theorem shows that the answer is no.

Theorem 11.9.1. *There is a computable infinite binary tree with no computable path.*

To prove this, we will make use of the next definition and proposition.

Definition 11.9.2. If A and B are disjoint sets, C is a *separation* of A from B if $A \subseteq C$ and $B \subseteq \overline{C}$. Two disjoint sets A and B are *computably inseparable* if there is no computable set separating them.

If A and B are disjoint sets and A is computable, then A itself is a computable separation of A from B. Similarly, if B is computable, then $\overline{B}$ is a computable separation of A from B. Saying that A and B are computably inseparable says that not only is neither of them is computable but there is, moreover, no way of squeezing a computable set between them.

Proposition 11.9.3. *There exist disjoint computably enumerable sets A and B such that A and B are computably inseparable.*

Proof Let $A = \{x \mid \varphi_x(x) \downarrow = 0\}$ and let $B = \{x \mid \varphi_x(x) \downarrow = 1\}$. If C is a separation of A from B with characteristic function φ_e, then $e \in C$ implies $\varphi_e(e) = 1$, which implies $e \in B$ and hence $e \notin C$, a contradiction. Similarly, $e \notin C$ implies $\varphi_e(e) = 0$, which implies $e \in A$ and hence $e \in C$, again a contradiction. □

Proposition 11.9.3 was foreshadowed by Exercise 11.4.8, and Lemma 12.2.8 in the next chapter provides another important example of a pair of computably enumerable sets that are computably inseparable.

Proof of Theorem 11.9.1 Let A and B be disjoint computably enumerable sets that are computably inseparable. Let $R(x, y)$ and $S(x, y)$ be computable relations such that $A = \{x \mid \exists y\, R(x, y)\}$ and $B = \{x \mid \exists y\, S(x, y)\}$. Construct a tree T as follows: put a finite sequence σ in T if and only if

$$\forall x < \text{length}(\sigma)\ ((\exists y < \text{length}(\sigma)\, R(x, y) \to (\sigma)_x = 1) \land$$
$$(\exists y < \text{length}(\sigma)\, S(x, y) \to (\sigma)_x = 0)).$$

In words, σ is in T if and only if looks like an initial segment of a separation of A from B as far as witnesses up to length(σ) are concerned. T is computable, and the assumption that

A and B are disjoint imply that for every y, there exists an element σ of T of length y. So T is infinite. To see that T has no computable path, let P be any path through T. If a natural number x is in A, then there is some y such that $R(x, y)$ holds; if σ is the initial segment of the characteristic function of P of length $y+1$, the fact that σ is in T implies $(\sigma)_x = 1$, so x is in P. Similarly, if x is in B for such a σ, we have $(\sigma)_x = 0$, so x is not in P. So P is a separation of A from B, and hence it is not computable. □

On the positive side, the proof of Kőnig's lemma in Section 1.4 shows that if T is an infinite binary tree, there is a path P that can be computed from T'. Saying that there are only finitely many nodes τ extending σ in T is equivalent to saying that for some n, there are no binary sequences in T of length n that extend σ:

$$\exists n\,(\forall \tau \in \{0, 1\}^n\,(\tau \supseteq \sigma \to \tau \notin T)).$$

The expression in parentheses is computable relative to T, since it only requires checking the 2^n many binary sequences of length T. The expression above is therefore Σ_1 in T, and so the set

$$S = \{\sigma \mid \text{there are only finitely many } \tau \text{ extending } \sigma \text{ in } T\}$$

is computable from T'. Now define the sequence $(\sigma_i)_{i\in\mathbb{N}}$ recursively by $\sigma_0 = ()$ and

$$\sigma_{i+1} = \begin{cases} \sigma_i{}^\frown 0 & \text{if } \sigma_i{}^\frown 0 \notin S \\ \sigma_i{}^\frown 1 & \text{otherwise.} \end{cases}$$

Since T is infinite, this defines a path through T, and the sequence $(\sigma_i)_{i\in\mathbb{N}}$ is computable from S and hence from T'.

So, up to Turing reducibility, finding a path through an infinite binary tree T is no harder than computing T'. What about the converse? Given the ability to query a set X, is it possible to compute a tree T such that X' is computable from any path through T? A remarkable theorem known as the *low basis theorem* implies that the answer is no. First, we need the following.

Definition 11.9.4. A set A of natural numbers is *low* if $A' \leq_T \emptyset'$.

A priori, if a set A is computable from $\emptyset'$, then the halting problem A' relative to A can be Turing equivalent to $\emptyset''$. A set A is low if not only is A computable from $\emptyset'$ but, moreover, A' is also computable from $\emptyset'$. Since for any A we have $\emptyset' \leq_T A'$, a set A is low if A' is as low in the degree hierarchy as it can possibly be. Clearly every computable set is low. But Theorem 11.9.5 implies that every computable infinite binary tree has a low path, which in turn implies that there are non-computable sets that are low as well.

More generally, a set A of natural numbers is *low in B* if $A' \leq_T B'$, and the following generalization to arbitrary trees T is no harder to prove.

Theorem 11.9.5. *Let T be any infinite binary tree. Then there is a path P that is low in T.*

Proof Using the notation introduced at the end of the previous section, it will be convenient to take P' to be given by $\{e \mid \exists \sigma \subset P\,(\varphi_e^\sigma(0)\downarrow)\}$, the set of indices of partial computable functions relative to P that are defined on input 0. Remember that if $\varphi_e^\sigma(0)\downarrow$ and $\tau \supseteq \sigma$, then

$\varphi_e^\tau(0)\downarrow$; in words, if σ provides enough information about P to guarantee that φ_e^P is defined, then so does any τ that extends σ.

Given an infinite binary T, our goal is to build a path P through T carefully so that P' is Turing reducible to T'. As before, we will build P as a limit of finite sequences $\sigma_0, \sigma_1, \sigma_2, \ldots$, where $\sigma_0 = ()$ and each σ_{i+1} extends σ_i by 0 or 1, in such a way that at each step we ensure that there are infinitely many nodes of T extending σ_{i+1}. We have seen that we can do this computably using T' as an oracle. But, as we go, we will also construct a sequence of infinite binary trees

$$T = T_0 \supseteq T_1 \supseteq T_2 \supseteq \cdots$$

and constrain each choice of σ_i to be in T_i, in a way that guarantees that P' will also be computable from T'.

Start with $T_0 = T$ and $\sigma_0 = ()$. To define T_{e+1} and σ_{e+1} do the following:

1. Ask whether there are infinitely many ρ in T_e such that $\varphi_e^\rho(0)\downarrow$ does not hold, that is, infinitely many ρ that do not witness the fact that $\varphi_e^P(0)$ is defined for every $P \supset \rho$. If there are, temporarily let $\hat{T} = \{\rho \in T_e \mid \varphi_e^\rho(0) \not\downarrow\}$, so that $\hat{T}$ is an infinite subtree of T_e and $\varphi_e^P(0)$ is guaranteed to be undefined for any path P through $\hat{T}$. Otherwise, at this stage we know that $\varphi_e^P(0)\downarrow$ holds for any infinite path P through T_e, so we set $\hat{T} = T_e$.
2. If $\sigma_e \frown 0$ has infinitely many descendants in $\hat{T}$, set $\sigma_{e+1} = \sigma_e \frown 0$, and otherwise set $\sigma_{e+1} = \sigma_e \frown 1$. Define T_{e+1} to be the set of ρ in $\hat{T}$ that are compatible with σ_{e+1}, that is, satisfy $\rho \subseteq \sigma_{e+1}$ or $\rho \supseteq \sigma_{e+1}$.

We let P be the limit of the σ_is, that is, the unique set such that for each i, $\sigma_i \subset P$.

We need to show that P' is computable from T'. To do that, we represent the construction above as a sequence of pairs $(\sigma_0, \alpha_0), (\sigma_1, \alpha_1), (\sigma_2, \alpha_2), \ldots$, where each α_e records entire sequence of yes/no answers to the first query before stage e. Then for every e, we have

$$T_e = \{\rho \in T \mid \rho \text{ is compatible with } \sigma_e \text{ and for every } i < e, \text{ if } (\alpha_e)_i = 1, \text{ then } \varphi_e^\rho(0) \not\downarrow\}.$$

The sequence $(\sigma_0, \alpha_0), (\sigma_1, \alpha_1), (\sigma_2, \alpha_2), \ldots$ is computable from T', since each of 1 and 2 can be phrased as the negation of an existential query relative to T. Moreover, for every e, $e \in P'$ if and only if $(\alpha_{e+1})_e = 0$, so P' is computable from T'. $\square$

We can extract more information from the construction. Each (σ_e, α_e) is a pair of approximations to (P, P') in the sense that $\sigma_e \subset P$ and $\alpha_e \subset P'$. At stage e, the tree T_e is computed from T using σ_e and α_e, so we can write it as $T(e, \sigma_e, \alpha_e)$ to indicate that dependence. Define I_1 to be the set of triples (e, σ, α) such that the answer to the first query is no:

$$I_1 = \{(e, \sigma, \alpha) \mid \sigma \in \{0, 1\}^e \text{ and } \alpha \in \{0, 1\}^e \text{ and } T(e, \sigma, \alpha) \cap \{\rho \mid \varphi_e^\rho(0) \not\downarrow\} \text{ is finite}\}.$$

Then I_1 is $\Sigma_1(T)$ and hence many-one reducible to T'. Similarly, the tree $\hat{T}$ depends on $T(e, \sigma_e, \alpha_e)$ and the answer to query 1, which is the last bit of α_{e+1}. As a result, we can write it as $\hat{T}(e, \sigma_e, \alpha_{e+1})$. Let I_2 be the set of triples (e, σ, α) such that the answer to the second query is no:

$$I_2 = \{(e, \sigma, \alpha) \mid \sigma \in \{0, 1\}^e \text{ and } \alpha \in \{0, 1\}^{e+1} \text{ and } \hat{T}(e, \sigma, \alpha) \cap \{\rho \mid \rho \supseteq \sigma \frown 0\} \text{ is finite}\}.$$

Then I_2 is also $\Sigma_1(T)$ and hence many-one reducible to T'. For every x, let $I_{1,x}$ be the restriction of I_1 to the finite set of triples where $e \leq x$ and similarly for $I_{2,x}$. Then the construction shows that for every e, σ_{e+1} and α_{e+1} can be computed in a primitive recursive way from finite representations of $I_{1,e}$ and $I_{2,e}$. This is stronger than saying that P' is computable from T'; it says that to determine whether or not an element x is in P', it suffices to know a finite initial segment of T' that can be determined in advance, since it depends only on x. In computability theory, this is known as a *truth-table reduction*, and the strengthening of Theorem 11.9.5 that makes this stronger claim about P' is known as the *superlow basis theorem*. We will make use of this extra information in Section 16.4.

The proof of Theorem 11.9.1 shows that if A and B are any two disjoint sets computably enumerable in X, there is an infinite binary tree T computable in X such that from any path P through T we can compute a separation of A from B. The following lemma provides a reduction in the other direction.

Lemma 11.9.6. *Let T be any infinite binary tree. Then there are a pair of disjoint sets A and B computably enumerable in T such that a path through T is uniformly computable from any separation of A from B.*

Proof Let A be the set

$$\{\sigma \mid \exists \tau \supseteq \sigma^\frown 0\,(\tau \in T \wedge \forall \rho\,(\mathrm{length}(\rho) = \mathrm{length}(\tau) \wedge \rho \supseteq \sigma^\frown 1 \to \rho \notin T))\}$$

and let B be the set

$$\{\sigma \mid \exists \tau \supseteq \sigma^\frown 1\,(\tau \in T \wedge \forall \rho\,(\mathrm{length}(\rho) = \mathrm{length}(\tau) \wedge \rho \supseteq \sigma^\frown 0 \to \rho \notin T))\}.$$

Roughly, $\sigma \in A$ means that the left descendants of σ in T survive longer than the right descendants, and $\sigma \in B$ means that the right descendants of σ survive longer than the left descendants. (If σ is in neither, then either both sets of descendants disappear at the same time, or both go on forever.) If C is a separation of A from B, use C to compute a path through T as follows: let $\sigma_0 = ()$ and at stage i set σ_{i+1} equal to $\sigma_i^\frown 0$ if $\sigma_i \in C$ and $\sigma_i^\frown 1$ if $\sigma_i \notin C$. □

Having shown a correspondence between separations of disjoint computably enumerable sets on the one hand and paths through computable infinite binary trees on the other, we now construct universal objects of both kinds. A relativization of Proposition 11.9.3 shows that for every X, the sets $\{(e,x) \mid \varphi_e^X(x)\downarrow = 0\}$ and $\{(e,x) \mid \varphi_e^X(x)\downarrow = 1\}$ cannot be separated by a set computable in X. The next theorem shows that, in a sense, they are the most extreme example of this phenomenon.

Theorem 11.9.7. *Let $A = \{(e,x) \mid \varphi_e^X(x)\downarrow = 0\}$ and $B = \{(e,x) \mid \varphi_e^X(x)\downarrow = 1\}$. Then A and B are universal in the following sense: if A_0 and B_0 are any two sets that are disjoint and computably enumerable in X, then for any separation C of A from B there is a separation C_0 of A_0 from B_0 such that $C_0 \leq_m C$.*

Proof Suppose $A_0 = \{x \mid \exists y\, R(x,y,X)\}$ and $B_0 = \{x \mid \exists y\, S(x,y,X)\}$. On input x, let $\varphi_e^X(x)$ search for the least y such that $R(x,y,X)$ or $S(x,y,X)$ holds, and, assuming it finds one, let it return 0 if $R(x,y,X)$ holds and 1 otherwise. If x is in A_0 then (e,x) is in A and if x is in B_0 then (e,x) is in B. So if C is separation of A from B, then $C_0 = \{x \mid (e,x) \in C\}$ is a separation of A_0 from B_0, and the map $x \mapsto (e,x)$ is a many-one reduction of C_0 to C. □

The proof shows that, moreover, C_0 can be computed uniformly in C and the descriptions of A_0 and B_0. In other words, there is a primitive recursive function s such that given, say, indices k and ℓ for the relations R and S, $\varphi^C_{s(k,\ell)}$ is a separation of A_0 from B_0, for any separation C of A from B. Combining Theorem 11.9.7 with the correspondence we have established between paths and separations, we obtain a universal computable infinite binary tree.

Theorem 11.9.8. *For any set X, there is an infinite binary tree T computable in X that is universal in the following sense: if T_0 is any infinite binary tree computable in X, a path P_0 through T_0 can be computed uniformly from an index for the characteristic function of T_0 and any path P through T.*

Proof By Theorem 11.9.7 and Lemma 11.9.6, there are disjoint sets A and B computably enumerable in X such that the path through any infinite binary tree T_0 computable in X can be computed uniformly from an index for the characteristic function of T_0 and a separation of A from B. The proof of Theorem 11.9.1 yields a tree T computable in X such that a separation of A from B can be computed from any path P through T. □

We can do even better: we can use a single infinite binary tree to obtain paths through all the computable infinite binary trees, paths through all the infinite binary trees that are computable relative to those paths, paths through all the infinite binary trees that are computable relative to *those* paths, and so on. Chapter 16 will introduce the notion of an *ω-model of* WKL_0, which is a collection $\mathcal{S}$ of subsets of $\mathbb{N}$ with the following properties:

- Whenever X and Y are in $\mathcal{S}$, so is $X \oplus Y$.
- Whenever X is in $\mathcal{S}$ and $Y \leq_T X$, then Y is in $\mathcal{S}$.
- Whenever T is an infinite binary tree in $\mathcal{S}$, there is a P in $\mathcal{S}$ such that P is a path through T.

Equivalently, $\mathcal{S}$ is an ω-model of WKL_0 if it satisfies the following:

- Whenever X and Y are in $\mathcal{S}$, so is $X \oplus Y$.
- Whenever X is in $\mathcal{S}$ and A and B are disjoint sets computably enumerable in X, there is a C in $\mathcal{S}$ such that C is a separation of A from B.

We can take any set U to represent an infinite sequence of sets by defining $x \in (U)_i$ to mean $(i, x) \in U$. The following theorem shows that for any set X, we can compute an ω-model of WKL_0 containing X from any path through a certain tree that is computable in X.

Theorem 11.9.9. *For any set X, there is a tree T computable in X such that if P is any path through T, then the set $\mathcal{S} = \{(P)_i \mid i \in \mathbb{N}\}$ is an ω-model of* WKL_0 *containing X.*

Proof By Theorem 11.8.6 there is a primitive recursive relation $R(x, y, \sigma)$ such that if we let $U(x, y, Z)$ be $\exists \sigma \subset Z\, R(x, y, \sigma)$, then $U(x, y)$ is a universal Σ_1 relation relative to Z. If for every x we define W^Z_x to be the set $\{y \mid U(x, y, Z)\}$, then $W^Z_0, W^Z_1, \ldots$ enumerates the computably enumerable sets relative to Z.

We will define an infinite binary tree T, computable in X, such that if P is any path through T, the following hold:

- $(P)_0 = X$.
- For every e, if $e = (i, j)$, then $(P)_{2e+1} = (P)_i \oplus (P)_j$.
- For every e, if $e = (u, v, k)$ and $W^{(P)_k}_u$ and $W^{(P)_k}_v$ are disjoint, then $(P)_{2e+2}$ is a separation of $W^{(P)_k}_u$ from $W^{(P)_k}_v$.

Our strategy will be to define T to be a set of sequences σ such that, considering only data represented by numbers less than length(σ), it is possible that σ is the initial segment of a set P satisfying those requirements. Then if every initial segment of P is in T, P will satisfy the requirements outright.

To establish the last requirement, as in the proof of Theorem 11.9.7, it suffices to know that if ρ is any initial segment of $(P)_k$, then for every z, if $R(u, z, \rho)$ holds but not $R(v, z, \rho)$ then z is in $(P)_k$, and if $R(v, z, \rho)$ holds but not $R(u, z, \rho)$ then z is not in $(P)_k$. Expanding the definition of $\oplus$ and the encoding of sequences of sets, to establish the three conditions above, it is therefore sufficient that P satisfies the following:

- For every y, $(0, y) \in P$ if and only if $y \in X$.
- For every e, i, j such that $e = (i, j)$ and for every y, $(2e + 1, 2y) \in P$ if and only if $(i, y) \in P$, and $(2e + 1, 2y + 1) \in P$ if and only if $(j, y) \in P$.
- For every e, u, v, k such that $e = (u, v, k)$, for every binary sequence $\rho \subset (P)_k$, and for every z, if $R(u, z, \rho)$ holds but not $R(v, z, \rho)$ then z is in $(P)_{2e+1}$, and if $R(v, z, \rho)$ holds but not $R(u, z, \rho)$ then z is not in $(P)_{2e+1}$.

Let T be the set of binary sequences σ that satisfy the following three conditions:

- For every $y < \text{length}(\sigma)$, if $(0, y) < \text{length}(\sigma)$, $\sigma_{(0,y)} = 1$ if and only if $y \in X$.
- For every e, i, j, and y less than $\text{length}(\sigma)$, if $e = (i, j)$ and if $(2e + 1, 2y)$, (i, y), and (j, y) are all less than $\text{length}(\sigma)$, then $\sigma_{(2e+1,2y)} = 1$ if and only if $\sigma_{(i,y)} = 1$ and $\sigma_{(2e+1,2y+1)} = 1$ if and only if $\sigma_{(j,y)} = 1$.
- For any e, u, v, and k less than $\text{length}(\sigma)$, suppose $e = (u, v, k)$. For any binary sequence ρ less than $\text{length}(\sigma)$, suppose moreover that for every $\ell < \text{length}(\rho)$, we have $(k, \ell) < \text{length}(\sigma)$, and $(\rho)_l = 1$ if and only if $\sigma_{(k,\ell)} = 1$. Then for any $z < \text{length}(\sigma)$ such that $(z, 2e + 1) < \text{length}(\sigma)$, if $R(u, z, \rho)$ holds but not $R(v, z, \rho)$ then $(\sigma)_{(2e+1,z)} = 1$, and if $R(v, z, \rho)$ holds but not $R(u, z, \rho)$ then $(\sigma)_{(2e+1,z)} = 0$.

Since all the quantifiers are bounded, T is computable relative to X. It takes a bit of concentration to show that T is a tree; in the third clause, making σ longer only increases the constraints on the values $(\sigma)_i$ by adding more tuples $e = (u, v, k)$, more binary sequences $\rho \subseteq \sigma$, and more values z that need to be considered. T is infinite, because for every n, the constraints on a sequence σ of length n are consistent with one another, so there is always at least one sequence σ that meets those constraints. Finally, it is not hard to see that any path P through T satisfies the requirements we have set: to check each condition, we only need to take an initial segment σ of P long enough and apply the corresponding condition in the definition of T. □

Exercises

11.9.1. Show that we can simplify the construction in the proof of Theorem 11.9.5 by omitting the second query, building a descending sequence of trees $T_0 \supseteq T_1 \supseteq \cdots$ and letting P be any path through the intersection of the T_is. By compactness, such a path exists, and the fact that P' is computable from T' independent of the choice of P means that P itself is computable from T' and is determined uniquely by the construction.

11.9.2. Let T be a consistent theory. Use the low basis theorem to show that there is a complete, consistent theory $\hat{T} \supseteq T$ such that $\hat{T}$ is low in T.

11.9.3. Let T be a consistent theory, and take a set of numbers M to represent a model of T if it encodes both a structure satisfying T as well as a satisfaction function that evaluates the truth of formulas in the language under a given assignment to their free variables. Emulate the proof of the completeness theorem for first-order logic, Theorem 5.1.4, to show that T has such a model M that is low in T.

11.9.4. Fill in the details of the proof of Theorem 11.9.9. Specifically, show that the set T defined there is a tree and that any path through T has the properties claimed.

Bibliographical Notes

Computability There are a number of good introductions to the theory of computability, including Cooper (2004), Cutland (1980), Davis (1958), Soare (2016), and Weber (2012). Soare (1987) focuses on the theory of computably enumerable sets and reducibility. The models of computation discussed here – Turing machines, the recursive functions, and the lambda calculus – were all introduced in the 1930s, and many of the original papers are collected in Davis (2004).

Theorem 11.5.2, the recursion theorem, is often called, more precisely, *Kleene's second recursion theorem*. Moschovakis (2010) provides an overview of the theorem and its applications.

Undecidability Since the discovery of the undecidability of the halting problem, a number of problems have been shown to be undecidable, including problems in combinatorics, number theory, computer science, and linguistics. See Poonen (2014) for a thorough survey, and see Matiyasevich (1993) for the undecidability of problems in number theory. Chapter 12 deals with decidability and undecidability phenomena related to logic.

Turing machines Davis (1958) provides a thorough mathematical treatment of Turing machines. Turing's original paper can be found in Davis (2004).

The lambda calculus and combinators References for the lambda calculus include Barendregt (1981), Hindley and Seldin (2008), and Sørensen and Urzyczyn (2006). All of these cover the confluence of the lambda calculus and combinatory logic, the representability of partial computable functions in these systems, and the semantics of these systems. Wolfram (2021) provides an exploration of combinators and their history, as well as a detailed bibliography. See also Chapter 13.

Computable functionals The theory of computable functionals whose arguments are numbers, sets of numbers, and functions from numbers to numbers is sometimes called the theory of *type 2* computability. This is the main focus of Weihrauch (2000), but insofar as it forms the basis for computable treatments of infinitary mathematical objects, the references under *computability in mathematics* below are relevant.

One can study computable functionals $F(f)$ of *computable* arguments f in at least two ways. The first is to view them as the restriction of computable type 2 functionals to the domain of computable functions. The second is to consider partial computable functions $g(e)$ on indices with the properties that $g(e)$ is defined whenever e is the index of a total computable function and that $g(e) = g(e')$ whenever e and e' are indices for the same total computable function. It is straightforward to show that if a functional $F(f)$ is computable in the first sense, it is computable in the second; simply define $g(e) = F(\varphi_e)$. A theorem by Kreisel, Shoenfield, and Lacomb asserts that the converse holds. See Moschovakis (2010) for details.

The low basis theorem For a survey of the low basis theorem and related results, see Diamondstone et al. (2010). This also includes a discussion of superlow basis theorem; see also Hájek and Pudlák (1998). For more on truth-table reducibility, see Soare (1987, 2016), or Downey and Hirschfeldt (2010).

Computability in mathematics Given that the modern notions of set and function include far more than the computable ones, we can inquire as to the extent to which the sets and functions that arise in everyday mathematics are computable, and the extent to which we can develop mathematics in computable terms. *Computable analysis* treats various branches of analysis in this way, including real and complex analysis, functional analysis, and measure theory, which generally relies on limiting processes and other infinitary constructions. See Downey and Hirschfeldt (2010), Nies (2009), Pour-El and Richards (1989), and Weihrauch (2000), as well as Avigad and Brattka (2014) for a survey. See also Section 16.3. Hirschfeldt (2015) provides a good overview of work on the analysis of infinitary combinatorics in computable terms.

12

Undecidability and Incompleteness

We have seen that PA proves lots of things about the natural numbers. Can it prove all the true statements that can be expressed in its language? A theory T is said to be *complete* if, for every sentence A, either $T \vdash A$ or $T \vdash \neg A$. In other words, T is complete if it either proves or refutes each sentence in its language. Since the axioms of PA are true of the natural numbers, it proves only true sentences. Our question is therefore tantamount to asking whether PA is a complete theory.

This notion of completeness is purely syntactic, expressed in terms of provability alone, without reference to semantic notions. It should therefore be distinguished from the notion of a proof system being complete with respect to a semantics, which was the center of attention in Chapters 3 and 5. But the two notions are related: if T is a theory that is true of a model, $\mathfrak{M}$, then T is a complete theory if and only if provability in T is complete for the semantics of being true in $\mathfrak{M}$. In particular, asking whether PA is complete is equivalent to asking whether it is the complete theory of the structure $(\mathbb{N}, 0, \mathrm{succ}, +, \cdot, \leq)$.

I will sketch three arguments that show that the answer is no. The first argument diagonalizes over the set of computable functions whose totality is provable in PA. We know that every computable function has a Σ_1 graph $\exists z\, A(x, y, z)$, where A is Δ_0. Define a computable function $f(u)$ as follows: if u represents a proof of a sentence of the form $\forall x\, \exists y, z\, A(x, y, z)$ in PA, let $f(u)$ be $(\mu s\, A(u, (s)_0, (s)_1))_0 + 1$, and let $f(u)$ be 0 otherwise. In other words, given a guarantee that it will succeed, $f(u)$ finds the least pair (y, z) satisfying $A(u, y, z)$ and returns $y + 1$.

By definition, $f(u)$ is a total computable function, so its graph is defined by a Σ_1 formula $\exists z\, A(x, y, z)$. Since f is total, the sentence $\forall x\, \exists y, z\, A(x, y, z)$ is true. I claim that it is not provable in PA. To see this, suppose u represents such a proof. Since $\exists z\, A(x, y, z)$ defines the graph of f, $f(u)$ is the unique value of y for which $\exists z\, A(u, y, z)$ holds. By definition, however, $f(u)$ returns $y + 1$, a contradiction. We have therefore shown that $\forall x\, \exists y, z\, A(x, y, z)$ is a true sentence that cannot be proved in PA, and since it is true, PA cannot refute it either. So PA is incomplete.

The second argument hinges on the undecidability of the halting problem and is even shorter. We know that Kleene's T predicate can be defined in PA, and for every e and x, the eth computable function φ_e is defined at x if and only if $\exists s\, T(e, x, s)$ is true. If PA were complete, for every e and x it would either prove or refute $\exists s\, T(e, x, s)$, and we could decide which is the case by systematically searching through proofs in PA until we found one or the other. Since the halting problem is undecidable, PA is necessarily incomplete.

The third argument unwraps our proof of the undecidability of the halting problem to extract more information. Let $\varphi_e(x)$ be the partial computable function which, on input x, searches for a proof in PA that $\varphi_x(x)$ is undefined. If it finds such a proof, let $\varphi_e(x)$ return 0.

If it never finds such a proof, $\varphi_e(x)$ is undefined. Does PA prove that $\varphi_e(e)$ is undefined? If it does, then, by definition, $\varphi_e(e)$ is defined, a contradiction. So PA doesn't prove that $\varphi_e(e)$ is undefined. But then, by definition, $\varphi_e(e)$ *is* undefined, and once again we have found a true sentence that is not provable.

These arguments rely on the fact that we can make statements about the behavior of computable functions in the language of PA, and that PA can only prove true statements. In this chapter, we will refine these arguments so that they apply more generally to any theory T that satisfies some meager hypotheses, namely, that it is consistent, it has a computable set of axioms, and it can prove some very elementary facts of arithmetic. The result is a strong formulation of Gödel's *first incompleteness theorem*.

We will consider two proofs of that theorem. The first strengthens the second argument we have just seen and highlights interesting correspondences between provability and computability. Our second proof is structurally similar to the third argument, but it is phrased in terms of provability rather than computability and it is closer to Gödel's original proof. Whereas the third argument above involves a statement, "$\varphi_e(e)$ is undefined," whose provability is equivalent to statement that $\varphi_e(e)$ is defined, Gödel's argument involves a statement, "A is not provable," whose provability implies its own falsity.

We will also discuss Gödel's *second incompleteness theorem*. The first incompleteness theorem is often interpreted, informally, as the statement that any reasonable theory of mathematics is incomplete. In a similar way, the second incompleteness theorem is often interpreted as saying that no reasonable theory of mathematics can prove its own consistency.

Some historical context will help explain the significance of this result. The nineteenth century brought fundamental changes to mathematics, including the development of liberal ways of reasoning about sets, functions, and the infinite. At the turn of the twentieth century, it became clear that overly naive ways of using these notions lead to contradictions. In the early 1920s, David Hilbert proposed a program of *Beweistheorie*, or *Proof Theory*, that was designed to respond to the set-theoretic paradoxes and defend the use of the new methods. The program aimed to represent these methods with formal axiomatic systems, and then prove that those systems are consistent – free from contradiction – using only explicit, *finitistic* reasoning. By establishing that no reasonable theory of mathematics can establish its own consistency, the second incompleteness theorem showed, a fortiori, that the consistency of such a theory cannot be established using only a small finitistic part. Put in other terms, the second incompleteness theorem implies that a body of finitistic methods cannot even establish its own consistency, let alone that of any theory that properly contains it. As a result, the second incompleteness theorem dealt a strong blow to Hilbert's program. We will provide a precise statement of the theorem in Section 12.4.

12.1 Computability and Representability

We have already developed a number of connections between logic and computability. We have seen that computable functions can be described in formal axiomatic theories like primitive recursive arithmetic and first order arithmetic, enabling us to use such systems to reason about computable sets, functions, and structures. In the other direction, Section 10.5 showed us how to represent the objects of logic – terms, formulas, and proofs – in such a way that we can compute with them.

This section establishes another connection between logic and computation. We will describe a minimal notion of what it means to *represent* a relation or a function on the natural numbers in a formal axiomatic theory, and then we will show that a weak theory of arithmetic known as *Robinson's Q* suffices to represent all the computable functions and relations. The two most important features of Q are, first, that it has a finite set of axioms, and, second, that these axioms express only a few meager properties of arithmetic. We will later show that Gödel's first incompleteness theorem applies to any theory that includes or interprets Q. The fact that Q is so weak means that the incompleteness theorem applies to any theory that can express and prove just a bit of arithmetic, making the statement of the incompleteness theorem quite strong.

Q is the theory in the language of first-order arithmetic based on classical first-order logic and the universal closures of the following axioms:

1. $\mathrm{succ}(x) = \mathrm{succ}(y) \to x = y$
2. $0 \neq \mathrm{succ}(x)$
3. $x \neq 0 \to \exists y\, (x = \mathrm{succ}(y))$
4. $x + 0 = x$
5. $x + \mathrm{succ}(y) = \mathrm{succ}(x + y)$
6. $x \cdot 0 = 0$
7. $x \cdot \mathrm{succ}(y) = x \cdot y + x$
8. $x \leq y \leftrightarrow \exists z\, (z + x = y)$.

Most of these are familiar from Section 10.1 as axioms of Peano Arithmetic and Heyting Arithmetic. The only one that is new is axiom 3, which says that every number other than zero has a predecessor. Since this is easily provable in PA and HA, we can think of first-order arithmetic as Q together with the full schema of induction. Conversely, we can think of Q as first-order arithmetic without the induction schema.

Giving up induction means giving up a lot. Exercise 12.1.1 asks you to show that Q can't even establish trivial statements like $\forall x\, (0 + x = x)$ and $\forall x\, (\mathrm{succ}(x) \neq x)$, since there is a model of Q in which these statements are false. The characterization of $\leq$ given by axiom 8 has been chosen carefully, given the fact that Q also doesn't prove $x + z = z + x$.

So what is Q good for? Well, it is good at proving concrete statements like $2 + 2 = 4$. In fact, the next few lemmas will enable us to show that it proves any true Σ_1 sentence. Remember that if n is a natural number, $\overline{n}$ denotes the formal numeral denoting n in the standard model. Although we generally leave the bar over n implicit, we need to make it explicit in the formulation of the next lemma.

Lemma 12.1.1. *Let m and n be natural numbers.*

1. $Q \vdash \mathrm{succ}(x) = \mathrm{succ}(y) \leftrightarrow x = y$.
2. *If* $n \neq m$, $Q \vdash \overline{n} \neq \overline{m}$.
3. $Q \vdash \overline{m} + \overline{n} = \overline{m + n}$.
4. $Q \vdash \overline{m} \cdot \overline{n} = \overline{m \cdot n}$.

Proof In claim 1, the forward direction is axiom 1 of Q and the reverse direction is an instance of substitution for equality.

For claim 2, assume without loss of generality that $n < m$, and use induction on n. If n is 0, $m = k + 1$ for some k, and $\overline{m}$ is $\mathrm{succ}(\overline{k})$. Then Q proves $\overline{n} \neq \overline{m}$ by axiom 2. Otherwise,

$n = \text{succ}(k)$ and $m = \text{succ}(\ell)$ for some k and ℓ. By the inductive hypothesis, Q proves $\overline{k} \neq \overline{\ell}$, and hence it proves $\overline{n} \neq \overline{m}$ by axiom 1.

For claim 3, use induction on n. When $n = 0$, we use axiom 4. In the inductive step, Q proves $\overline{m} + \overline{n+1} = \text{succ}(\overline{m} + \overline{n}) = \text{succ}(\overline{m+n})$ using axiom 5, the inductive hypothesis, and the substitution rule for equality. The right-hand side is by definition $\overline{m + (n+1)}$.

The proof of claim 4 is similar to the proof of claim 2. □

This shows that Q is strong enough to evaluate closed terms.

Corollary 12.1.2. *Let s and t be closed terms in the language of arithmetic, and let $\mathfrak{N}$ denote the standard interpretation* $(\mathbb{N}, 0, \text{succ}, +, \cdot, \leq)$. *If $s^{\mathfrak{N}} = t^{\mathfrak{N}}$, then $Q \vdash s = t$, and if $s^{\mathfrak{N}} \neq t^{\mathfrak{N}}$, then $Q \vdash s \neq t$.*

Proof By induction on closed terms and Lemma 12.1.1, we have that Q proves $t = \overline{t^{\mathfrak{N}}}$ for every closed term t. If $s^{\mathfrak{N}} = t^{\mathfrak{N}}$, then Q proves $\overline{s^{\mathfrak{N}}} = \overline{t^{\mathfrak{N}}}$ by reflexivity, and if $s^{\mathfrak{N}} \neq t^{\mathfrak{N}}$, then Q proves $\overline{s^{\mathfrak{N}}} \neq \overline{t^{\mathfrak{N}}}$ by claim 1 of Lemma 12.1.1. □

We now return to writing n instead of $\overline{n}$ when it is clear that n should be interpreted as a term. The next lemma allows us to extend Corollary 12.1.2 to the less-than-or-equal-to relation. Remember that saying that Q proves a formula with free variables is equivalent to saying that it proves its universal closure.

Lemma 12.1.3. *1.* $Q \vdash \text{succ}(x) \leq \text{succ}(y) \leftrightarrow x \leq y$.
2. $Q \vdash 0 \leq x$.
3. $Q \vdash \text{succ}(x) \not\leq 0$.
4. For every m and n, if $m \leq n$, then $Q \vdash m \leq n$.
5. For every m and n, if $m \not\leq n$, then $Q \vdash m \not\leq n$.

Proof For claim 1, using axioms 8 and 5, Q proves

$$\text{succ}(x) \leq \text{succ}(y) \leftrightarrow \exists z\, (z + \text{succ}(x) = \text{succ}(y)) \leftrightarrow \exists z\, (z + x = y) \leftrightarrow x \leq y.$$

This is why $x \leq y$ is defined by $\exists z\, (z + x = y)$ rather than $\exists z\, (x + z = y)$ in axiom 8. Claim 2 is proved similarly using axioms 8 and 4, and claim 3 is proved using axioms 8, 5, and 2.

For claim 4, suppose $n = k + m$. By claim 1 and induction on m in the metatheory, Q proves $m \leq n \leftrightarrow 0 \leq k$, and Q proves $0 \leq k$ by claim 2. For claim 5, if $m \not\leq n$, then $m = n + k + 1$ for some k, and Q proves $m \leq n \leftrightarrow \text{succ}(k) \leq 0$. The conclusion then follows from claim 3. □

Corollary 12.1.4. *For any two closed terms s and t, if $s^{\mathfrak{N}} \leq t^{\mathfrak{N}}$, then $Q \vdash s \leq t$, and if $s^{\mathfrak{N}} \not\leq t^{\mathfrak{N}}$, then $Q \vdash s \not\leq t$.*

The proof of Corollary 12.1.4 is similar to that of Corollary 12.1.2.

Remember from Section 10.2 that the set of Δ_0 formulas is the smallest set of formulas in the language of arithmetic containing all the atomic formulas and closed under boolean connectives and bounded quantification, and that any Σ_1 formula is obtained by prefixing existential quantifiers to a Δ_0 formula. Toward showing that every true Σ_1 sentence is provable in Q, we need to show that every true Δ_0 sentence is provable in Q. Corollaries 12.1.2 and 12.1.4 provide the base cases for a proof by induction, but we will have to deal with the fact that Δ_0 formulas are closed under bounded quantification. The next lemma is designed with that in mind.

Axiom 3 of Q implies that every natural number is either 0 or a successor. So, even though we do not have proof by induction, we can carry out proof by cases on these two possibilities.

Lemma 12.1.5. *For any natural number n, Q proves the following:*

1. $x \leq n \leftrightarrow x = 0 \vee \cdots \vee x = n$.
2. $x \leq n \vee n + 1 \leq x$.

Proof Claim 1 can be proved by induction on n. In the base case, Q proves $x \leq 0 \leftrightarrow x = 0$ by cases on x, using claim 3 of Lemma 12.1.3 and axiom 2 of Q. In the inductive step, use cases on x. When $x = 0$, the claim holds using claim 2 of Lemma 12.1.3. When $x = \mathrm{succ}(y)$, the claim reduces to the inductive hypothesis using claim 1 of Lemma 12.1.3.

Claim 2 can also be proved by induction on n. When n is 0, use cases on x and Lemma 12.1.3. In the induction step, again use cases on x; when we replace x by $\mathrm{succ}(x)$, the provability of $\mathrm{succ}(x) \leq n + 1 \vee n + 2 \leq \mathrm{suc}(x)$ follows from the inductive hypothesis using Lemma 12.1.3. □

Below it will be convenient to know that Q proves a weaker version of claim 2, namely, $x \leq n \vee n \leq x$. This follows from claim 2 and claim 1 together, though as an exercise you can also prove it directly by induction on n.

Lemma 12.1.6. *Q proves every true Δ_0 sentence.*

Proof We prove by induction that for every Δ_0 sentence A, if A is true, then Q proves A, and if A is false, then Q proves $\neg A$. More precisely, we will use induction on the number of (bounded) quantifiers occurring in A and a secondary induction on the depth of A. Corollaries 12.1.2 and 12.1.4 handle the base case where A is an atomic formula.

Closure under the propositional connectives is straightforward. For example, suppose A is the sentence $B \wedge C$. If A is true, then B and C are both true, and by the inductive hypothesis Q proves both B and C, and hence $B \wedge C$. If $B \wedge C$ is false, then either B or C is false, in which case, Q proves either $\neg B$ or $\neg C$. Each of these implies $\neg(B \wedge C)$ in first-order logic.

Suppose A is the sentence $\forall x \leq t\, B(x)$. Then t is a closed term, and by Corollary 12.1.2 and 12.1.5, Q proves $A \leftrightarrow (B(0) \wedge \cdots \wedge B(t^{\mathfrak{N}}))$. By the inductive hypothesis, Q proves the right-hand side if it is true and Q proves its negation it is false, and hence the same holds for A. The case where A is the sentence $\exists x \leq t\, B(x)$ is similar, with $\wedge$ replaced by $\vee$. □

Theorem 12.1.7. *Q proves every true Σ_1 sentence.*

Proof Let A be the sentence $\exists x_0, \ldots, x_{k-1}\, B(x_0, \ldots, x_{k-1})$, where B is Δ_0. If A is true, then for some $n_0, \ldots, n_{k-1}$, the sentence $B(n_0, \ldots, n_{k-1})$ is also true. By Lemma 12.1.6, Q proves $B(n_0, \ldots, n_{k-1})$, and hence A. □

Let T be any theory in a language that includes 0 and the successor function, $\mathrm{succ}(x)$.

Definition 12.1.8. An k-ary relation R on the natural numbers is *representable in T* if there is a formula $A(x_0, \ldots, x_{k-1})$ such that for every tuple $n_0, \ldots, n_{k-1}$,

- if $R(n_0, \ldots, n_{k-1})$ holds, then $T \vdash A(n_0, \ldots, n_{k-1})$, and
- if $R(n_0, \ldots, n_{k-1})$ doesn't hold, then T proves $\neg A(n_0, \ldots, n_{k-1})$.

Saying that a relation R is representable in T does not imply that T can prove any interesting general statements about R. It only says that T can settle every specific instance. Corollaries 12.1.2 and 12.1.4 say that equality and the less-than-or-equal-to relation on $\mathbb{N}$ are representable in Q, and Lemma 12.1.6 says, more generally, that every Δ_0 relation is representable in Q. We now prove something stronger.

Theorem 12.1.9. *Every computable relation is representable in Q.*

Proof If $R(x_0, \ldots, x_{k-1})$ is computable, it is Δ_1-definable, so there is a Δ_0 formula $B(x_0, \ldots, x_{k-1}, y)$ such that $R(n_0, \ldots, n_{k-1})$ holds of a tuple $\vec{n}$ of natural numbers if and only if $\exists y\, B(n_0, \ldots, n_{k-1}, y)$ is true, and there is a Δ_0 formula $C(x_0, \ldots, x_{k-1}, z)$ such that $\neg R(n_0, \ldots, n_{k-1})$ holds if and only if $\exists z\, C(n_0, \ldots, n_{k-1}, z)$ is true. By Theorem 12.1.7, if $R(\vec{n})$ holds, Q proves the first sentence, and if $\neg R(\vec{n})$ holds, Q proves the second. The challenge is to write down a single formula $A(x_0, \ldots, x_{k-1})$ which makes use of both Σ_1 formulas to satisfy Definition 12.1.8.

The trick is to take $A(\vec{x})$ to be the formula $\exists y\, (B(\vec{x}, y) \wedge \neg \exists z \leq y\, C(\vec{x}, z))$. Roughly, $A(\vec{x})$ says that there is a witness to the truth of $R(\vec{x})$ and no smaller witness to its falsehood. Let us show that A has the right properties.

For any tuple $n_0, \ldots, n_{k-1}$, if $R(n_0, \ldots, n_{k-1})$ holds, then $A(n_0, \ldots, n_{k-1})$ is a true Σ_1 sentence, and hence by Theorem 12.1.7 it is provable in Q. So suppose $R(n_0, \ldots, n_{k-1})$ doesn't hold. Then for every i, Q proves $\neg B(n_0, \ldots, n_{k-1}, i)$, and there is an m such that Q proves $C(n_0, \ldots, n_{k-1}, m)$. The sentence $\neg A(n_0, \ldots, n_{k-1})$ is equivalent to

$$\forall y\, (B(n_0, \ldots, n_{k-1}, y) \to \exists z \leq y\, C(n_0, \ldots, n_{k-1}, z)),$$

so it suffices to show that Q proves that.

By the comment after Lemma 12.1.5, Q proves $y \leq k \vee k \leq y$. Splitting on cases, in the first case we have $y = 0 \vee \cdots \vee y = k$. In each of these cases, Q proves $\neg B(n_0, \ldots, n_{k-1}, y)$. In the second case, Q proves $k \leq y \wedge C(n_0, \ldots, n_{k-1}, k)$. Hence in both cases Q proves

$$B(n_0, \ldots, n_{k-1}, y) \to \exists z \leq y\, C(n_0, \ldots, n_{k-1}, z),$$

which is what we need. □

Definition 12.1.10. A k-ary function f on the natural numbers is *representable* in a theory T if there is a formula $A(x_0, \ldots, x_{k-1}, y)$ such that for every tuple $n_0, \ldots, n_{k-1}$, Q proves

$$A(n_0, \ldots, n_{k-1}, y) \leftrightarrow y = f(n_0, \ldots, n_{k-1}).$$

The exercises ask you to check that the last requirement is equivalent to saying that Q proves $A(\vec{n}, f(\vec{n}))$ and $\forall y\, (A(\vec{n}, y) \to y = f(\vec{n}))$. This formulation is sometimes more convenient.

Theorem 12.1.11. *Every computable function is representable in Q.*

Proof If $f(\vec{x})$ is computable, its graph $f(\vec{x}) = y$ is computable, and hence representable in Q by a formula $B(\vec{x}, y)$. Then for every $\vec{n}$, Q proves $B(\vec{n}, f(\vec{n}))$, and $f(\vec{n})$ is the only m such that Q proves $B(\vec{n}, m)$. This doesn't necessarily mean that Q can prove the characterization above for B, however, so again the challenge is to come up with a description $A(\vec{x}, y)$ that has the right property.

Take $A(\vec{x}, y)$ to be the formula $B(\vec{x}, y) \wedge \forall z \leq y\, (B(\vec{x}, y) \to y = z)$, which says that y is the smallest value satisfying $B(\vec{x}, y)$. Then for every tuple $\vec{n}$, Q proves $A(\vec{n}, f(\vec{n}))$ since, by

Lemma 12.1.5, the second part of the formula is equivalent to a conjunction over the cases $z=0, \ldots, z=f(\vec{n})$. So Q proves $y=f(\vec{n}) \rightarrow A(\vec{n}, y)$.

For the other direction, reason in Q. Suppose we have $A(\vec{n}, y)$, which is to say, we have $B(\vec{n}, y)$ and $\forall z \leq y\,(B(\vec{n}, z) \rightarrow z=y)$. We need to show $y=f(\vec{n})$. By Lemma 12.1.5, Q proves $y \leq f(\vec{n}) \vee f(\vec{n}) \leq y$. In the first case, we have $y=0 \vee \cdots \vee y=f(\vec{n})$. In all these subcases but the last, Q proves $\neg B(\vec{n}, y)$, resulting in a contradiction. So, in the first case, we have $y=f(\vec{n})$. In the second case, $f(\vec{n}) \leq y$, and substituting $f(\vec{n})$ for z in the second hypothesis yields $B(\vec{n}, f(\vec{n})) \rightarrow f(\vec{n})=y$. Q proves $B(\vec{n}, f(\vec{n}))$, so, once again, we have $y=f(\vec{n})$. □

Definitions 12.1.8 and 12.1.10 make only positive assertions about provability, so it is clear that if a relation or function is representable in Q, then it is representable in any theory T that contains Q. Conversely, suppose T is any consistent theory with a computable set of axioms. Then the relation $\Pr_T(x, y)$ which holds if and only if x is a proof of y from the axioms of T in first-order logic is computable. If a relation $R(\vec{x})$ is representable in T, then it is clearly computable: assuming R is represented by $A(\vec{x})$, on input $\vec{n}$, search simultaneously for a proof of $A(\vec{n})$ and a proof of $\neg A(\vec{n})$, and the one you find determines whether or not $R(\vec{n})$ holds. A similar argument shows that any function representable in T is also computable. Thus we have yet another characterization of what it means for a function or relation to be computable.

Theorem 12.1.12. *A function or relation on the natural numbers is computable if and only if it is representable in a consistent, computably axiomatized theory extending Q.*

I will close this section with an observation that will be helpful to us in Section 12.4: everything we have done in this section deals with explicit properties of syntax (terms, formulas, and proofs), and so all of the results here can be derived straightforwardly in first-order PRA. We will need this one in particular:

Proposition 12.1.13. PRA *proves that every true Σ_1 sentence is derivable in Q.*

Exercises

12.1.1. Consider the following structure for the language of arithmetic. The universe consists of the usual natural numbers, $\mathbb{N}$, plus two additional objects, α and β. Interpret successor, addition, and multiplication with the following tables:

x	$\mathrm{succ}(x)$
m	$m+1$
α	α
β	β

$+$	n	α	β
m	$m+n$	β	α
α	α	β	α
β	β	β	α

$\cdot$	0	$n \neq 0$	α	β
m	0	mn	α	β
α	0	β	β	β
β	0	α	α	α

Here m and n denote elements of $\mathbb{N}$, and in the column labeled $n \neq 0$ in the multiplication table, n is any nonzero element of $\mathbb{N}$.

a. Show that all the axioms of Q are true in this model, assuming $\leq$ is defined so as to satisfy the last axiom.

b. Show that none of these sentences are true in this model:

- $\forall x\,(x \neq \mathrm{succ}(x))$
- $\forall x\,(0+x=x)$

- $\forall x\,(x \leq x)$
- $\forall x, y\,(x + y = y + x)$
- $\forall x, y, z\,(x + (y + z) = (x + y) + z)$
- $\forall x, y, z\,(x \cdot (y \cdot z) = (x \cdot y) \cdot z)$
- $\forall x, y, z\,(x \cdot (y + z) = x \cdot y + x \cdot z)$.

12.1.2. Fill in details of the proofs of Lemmas 12.1.1 and Lemma 12.1.3.

12.1.3. Claim 2 of Lemma 12.1.5 implies that Q proves $x \leq n + 1 \leftrightarrow x \leq n \vee x = n + 1$. Prove this instead directly, using induction on n.

12.1.4. Verify the claim after Definition 12.1.10.

12.1.5. Verify that the proof of Theorem 12.1.11 shows more generally that if the graph of a function f is representable in a theory T containing Q, then f is representable in T. Show also that if the characteristic function of a relation R is representable in a theory T containing Q, then so is R.

12.1.6. Our informal proof of Theorem 12.1.12 described algorithms for computing functions and relations given the fact that they are representable in a computably axiomatized theory. Describe these computations more formally, using the notation established in Chapter 11.

12.2 Incompleteness via Undecidability

Section 12.1 established an important correspondence between computability and provability. We will now exploit this correspondence to derive a number of important metamathematical results having to do with undecidability and incompleteness of formal axiomatic theories.

When I refer to Q as a theory, I mean the set of sentences provable from the eight axioms presented in the last section. So saying that Q proves A and saying $A \in Q$ mean the same thing. Whenever we talk about the decidability or undecidability of an axiomatic theory, we are implicitly assuming that we have fixed representations of formulas as natural numbers such that the usual syntactic operations and primitive recursive, and such that their basic properties are provable in primitive recursive arithmetic.

Proposition 12.2.1. *Q is a complete computably enumerable set under many-one reduction.*

Proof Since $Q = \{y \mid \exists x\, \mathrm{Pr}_Q(x, y)\}$ and $\mathrm{Pr}_Q(x, y)$ is computable, it is clear that Q is computably enumerable.

Let $S = \{e \mid \varphi_e(0)\downarrow\}$. To show that Q is complete, by Proposition 11.3.5 it suffices to show that S is many-one reducible to it. Since Kleene's T is primitive recursive, it is representable in Q, say by the formula $A_T(e, x, s)$. I claim that for every e, $e \in S$ if and only if $Q \vdash \exists s\, A_T(e, 0, s)$. This implies that the map $e \mapsto \exists s\, A_T(e, 0, s)$ is the desired reduction.

To prove the claim, suppose $e \in S$. Then $\varphi_e(0)$ is defined, and so for some $k \in \mathbb{N}$, $T(e, 0, k)$ holds. By representability, Q proves $A_T(e, 0, k)$, and hence Q proves $\exists s\, A_T(e, 0, s)$. Conversely, suppose Q proves $\exists s\, A_T(e, 0, s)$. Since Q is a true theory of arithmetic, the sentence is true in the standard interpretation. So, for some s, $T(e, 0, s)$ holds, and hence $\varphi_e(0)$ is defined. $\square$

The previous proof relies on the fact that Q is a true theory of arithmetic. We can weaken this requirement.

Definition 12.2.2. A theory T in a language with 0 and $\mathrm{succ}(x)$ is said to be *ω-consistent* if for every formula $A(x)$ with the free variable shown, it does not simultaneously prove $\exists x\, A(x)$ and $\neg A(n)$ for every n.

Here we think of T as being a theory of the natural numbers. It is *ω-inconsistent* if it proves the existence of a number with some property but refutes each particular instance. This is not quite as bad as being inconsistent, and we will see that there are consistent theories that are ω-inconsistent. But clearly it is a bad property for a theory of the natural numbers to have.

Proposition 12.2.3. *Let T be any computably axiomatized, ω-consistent theory extending Q. Then T is a complete computably enumerable set.*

Proof Since we can write $T = \{y \mid \exists x\, \mathrm{Pr}_T(x, y)\}$, it is clear that T is computably enumerable. As in the proof of Proposition 12.2.1, it suffices to show that for every e, T proves $\exists s\, A_T(e, 0, s)$ if and only of $\varphi_e(0)$ is defined.

One direction is unchanged: it $\varphi_e(0)$ is defined, then there is an s such that $T(e, 0, s)$ holds, and so Q, and hence T, proves $\exists s\, A_T(e, 0, s)$. For the other direction, suppose T proves $\exists s\, A_T(\bar{e}, 0, s)$. If $\varphi_e(0)$ is undefined, then for every k, $T(e, 0, k)$ does not hold. Since A_T represents T, this implies that Q, and hence T, proves $\neg A_T(e, 0, k)$ for every k. This contradicts the assumption that T is ω-consistent. So $\varphi_e(0)$ is defined. □

Provability in any theory satisfying the hypotheses of Proposition 12.2.3 is therefore a lot like the halting problem. In particular, it is undecidable. In the previous proof, we only needed to assume that T is computably axiomatized to show that it is computably enumerable; the proof shows that the halting problem is reducible to any ω-consistent theory extending Q. And, as far as undecidability is concerned, we can weaken the assumption that T is ω-consistent.

Lemma 12.2.4. *Let T be any consistent theory extending Q. Then T is undecidable.*

Proof Proposition 11.3.2 established that there is no universal computable binary relation $R(x, y)$, that is, no computable relation with the property that as we set the first input to $0, 1, 2, \ldots$ we obtain all the computable unary relations. It is therefore sufficient to show that the existence of a consistent, decidable theory extending Q implies the existence of such a relation.

So, suppose T is consistent, decidable, and extends Q. Let $R(x, y)$ be defined to hold if and only if

x codes a formula $A(u)$ and T proves $C(\bar{y})$.

Since T is decidable, R is computable. Let us show that R is universal. If $S(y)$ is any computable relation, then it is representable in Q, and hence T, by a formula $A(u)$. Let k be the number coding $A(u)$. Then for every n, we have

$$\begin{aligned} S(n) &\to T \vdash A(n) \\ &\to R(k, n) \end{aligned}$$

and

$$\begin{aligned}\neg S(n) &\rightarrow T \vdash \neg A(n)\\ &\rightarrow T \nvdash A(n)\\ &\rightarrow \neg R(k, n).\end{aligned}$$

That is, for every y, $S(y)$ is true if and only if $R(k, y)$ is. So R is universal, and we have the contradiction we were looking for. □

We are now just a step away from the first incompleteness theorem.

Proposition 12.2.5. *Any complete, computably axiomatized theory is decidable.*

Proof If T is inconsistent, then T proves everything, and there is an easy decision procedure for T: just say yes to every input.

Otherwise, we can decide T as follows. Given a sentence A as input, search simultaneously for a proof of A and a proof of $\neg A$. Since T is complete, the search will find a proof of one or the other. If it finds a proof of A, then T proves A, and if it find a proof of $\neg A$, then, by consistency, T does not prove A. □

Putting Lemma 12.2.4 together with Proposition 12.2.5 together immediately yields the first incompleteness theorem.

Theorem 12.2.6. *There is no complete, consistent, computably axiomatized theory that contains Q.*

The theorem says that a theory extending Q cannot simultaneously be complete, consistent, and computably axiomatized. Exercise 12.2.4 asks you to show that there are theories extending Q satisfying any two of these three properties.

Theorem 12.2.6 tells us that no reasonable theory of mathematics is complete. The requirement that a "reasonable theory" has a computable set of axioms should seem unobjectionable: it is hard to see how one could claim to have a mathematical theory without being able to tell whether or not something is an axiom. The restriction to theories in the language of arithmetic is not essential. We will see below that the result carries over straightforwardly to theories in other languages, as long as they are strong enough to interpret the axioms of Q.

We will now set aside the question of completeness, for the rest of this section, and turn to the question of decidability. Lemma 12.2.4 says that any consistent theory that extends Q is undecidable. We can do better: the next theorem asserts that any consistent theory that is *compatible* with Q is undecidable, whether it extends it or not.

Theorem 12.2.7. *Let T be any theory in the language of arithmetic such that $T \cup Q$ is consistent. Then T is undecidable.*

To prove Theorem 12.2.7, remember that, according to Definition 11.9.2, two disjoint sets A and B are *computably inseparable* if there is no set C such that $A \subseteq C$ and $B \subseteq \overline{C}$.

Lemma 12.2.8. *Let $Q' = \{A \mid Q \vdash \neg A\}$ be the set of sentences whose* negations *are provable in Q. Then Q and Q' are computably inseparable.*

Proof Suppose C is a computable set such that $Q \subseteq C$ and $Q' \subseteq \overline{C}$. Let $R(x, y)$ be the relation

x codes a formula $A(u)$ and $A(\overline{y})$ is in C.

I will show that $R(x, y)$ is a universal computable relation, yielding a contradiction.

Suppose $S(y)$ is computable. Then it is represented in Q by a formula $A(u)$ in Q. Let k be a natural number representing $A(u)$. Then

$$\begin{aligned} S(n) &\to Q \vdash A(n) \\ &\to A(n) \in C \\ &\to R(k, n), \end{aligned}$$

and

$$\begin{aligned} \neg S(n) &\to Q \vdash \neg A(n) \\ &\to A(n) \in Q' \\ &\to A(n) \notin C \\ &\to \neg R(k, n). \end{aligned}$$

So for every n, $S(n)$ is equivalent to $R(k, n)$. Since $S(y)$ is an arbitrary computable predicate, $R(x, y)$ is a universal computable relation. □

Proof of Theorem 12.2.7 Remember that Q has a finite set of axioms. Taking their conjunction, we can replace these by a single axiom, E.

Suppose T is a decidable theory consistent with Q. Let $C = \{A \mid T \vdash E \to A\}$. I claim that C is a computable separation of Q from Q', contradicting the previous lemma.

If A is in Q, then A is provable from the axioms of Q; by the deduction theorem, there is a proof of $E \to A$ in first-order logic. So A is in C. On the other hand, if A is in Q', then there is a proof of $E \to \neg A$ in first-order logic. If T also proves $E \to A$, then T proves $\neg E$, in which case $T \cup Q$ is inconsistent. But we are assuming $T \cup Q$ is consistent, so T does not prove $E \to A$, and so A is not in C. Hence C is a computable separation of Q from Q'. □

Corollary 12.2.9. *First-order logic for the language of arithmetic (that is, the set of valid first-order formulas) is undecidable.*

Proof First-order logic is the set of consequences of the empty set, which is consistent with Q. □

We can remove the dependence on the language of arithmetic. Let L_1 and L_2 be two languages. An *interpretation* of L_1 in L_2 consists of

- a formula $U(x)$ of L_2 (intuitively, defining the intended universe of T_1),
- a formula $A_R(\vec{x})$ of L_2 for each relation symbol of L_1, and
- a formula $A_f(\vec{x}, y)$ of L_2 for each function symbol $f(\vec{x})$ of L_1.

This gives rise to a translation of first-order formulas in the language of L_1 to first-order formulas in the language of L_2 in which quantifiers in L_1 are interpreted as quantifiers ranging over $U(x)$ in L_2, and the formulas A_R and A_f are used to interpret the relation and function symbols. If Γ is a set of axioms in the language of L_1 and T_2 is a theory in the language of L_2, an interpretation of L_1 in L_2 is said to be an interpretation of Γ in T_2 if the following hold:

- For each k-ary function symbol f, T_2 proves

$$\forall x_0, \ldots, x_{k-1} \, (U(x_0) \wedge \cdots \wedge U(x_{k-1}) \rightarrow \exists! y \, (U(y) \wedge A_f(\vec{x}, y))).$$

- T_2 proves the translation of each element of Γ.

In that case we have:

Theorem 12.2.10. *If T is any theory that is consistent with a theory that interprets the axioms of Q, then T is undecidable.*

The proof of Theorem 12.2.7 carries over straightforwardly, since a counterexample would give rise to a universal computable relation. Notice that if T satisfies the antecedent of the theorem, then so does any theory $T' \subseteq T$.

Let $\mathfrak{M}$ be a model in the language of L_2 and let Γ be a set of sentences in L_1. An interpretation of L_1 in L_2 is said to interpret Γ in $\mathfrak{M}$ if, in $\mathfrak{M}$, each formula A_f defines the graph of a function and the sentences in Γ are true of the model defined by the formulas in the interpretation. By soundness and completeness, the antecedent to Theorem 12.2.10 is equivalent to saying that T has a model which interprets a model of Q.

Theorem 12.2.11. *Suppose T has a model which interprets a model of Q. Then T is undecidable.*

A theory T with the property that every theory $T' \subseteq T$ is undecidable is said to be *hereditarily undecidable*. Theorems 12.2.10 and 12.2.11 imply that any theory that is consistent with an interpretation of Q, or, equivalently, has a model that interprets Q, has this property. A theory T that is undecidable and cannot be extended to a decidable theory is said to be *essentially undecidable*. Theorems 12.2.10 and 12.2.11 imply that any consistent theory that contains or interprets Q is both essentially and hereditarily undecidable.

These results are very powerful. They imply the following.

Corollary 12.2.12. *There is no decidable theory consistent with axiomatic set theory.*

Corollary 12.2.13. *First-order logic in a language with one binary relation is undecidable.*

The second corollary follows from the first because axiomatic set theory can be presented in a language with a single binary relation, $\in$. Even the equality symbol can be defined in terms of that, so the corollary still holds if we interpret "first-order logic" as first-order logic without equality.

The methods also yield interesting undecidability results for algebraic structures. A landmark result by Julia Robinson shows that there are first-order formulas defining the integers, as well as the order on the integers, in the structure $(\mathbb{Q}, 0, 1, +, \cdot)$ of the rational numbers as a field. By Theorem 12.2.11, this immediately implies that the theory of this structure is hereditarily undecidable. As a corollary, the theory of fields is undecidable. But in Section 12.5, we will see that it is not essentially undecidable: there are decidable extensions of the theory of fields.

Exercises

12.2.1. Show that Q together with the axiom $\exists x \, (\mathrm{succ}(x) = x)$ is consistent but not ω-consistent. (Hint: use Exercise 12.1.1.)

12.2.2. A sentence A in the language arithmetic is said to be *consistent with Q* if the set of axioms $Q \cup \{A\}$ is consistent. This is equivalent to saying Q does not prove $\neg A$. Show that the set of sentences that are consistent with Q is not computably enumerable.

12.2.3. Suppose T has a computably enumerable set of axioms Γ; in other words, there is a computably enumerable set Γ such that $T = \{A \mid \Gamma \vdash A\}$. Show that T has a computable set of axioms. This result is known as *Craig's theorem*.

(Hint: Use Exercise 11.2.4, which shows that an infinite set is computable if it is enumerated by an increasing computable function. Given a function enumerating Γ, convert it to a function enumerating a set Γ' that proves the same theorems.)

12.2.4. The incompleteness theorem says that no theory extending Q is complete, consistent, and computably axiomatized. Show that there are theories extending Q satisfying any two of these three properties.

12.3 Incompleteness via Self-Reference

The theory of computability is needed to state the incompleteness theorem in a strong way, since we cannot say what it means to be computably axiomatized without being able to say what it means to be computable. Moreover, in Section 12.2, we cast the diagonalization argument in terms of the nonexistence of a universal computable relation. Gödel's original proof carried out the diagonalization in terms of logic rather than computability. He used a self-referential sentence, now known as the Gödel sentence, that says "I am not provable." The argument turns on a proof-theoretic rather than a computability-theoretic version of the fixed-point theorem, which is interesting in its own right.

Theorem 12.3.1 (Gödel's fixed-point lemma). *Let $B(x)$ be any formula in the language of arithmetic with at most the free variable shown. Then there is a sentence A such that Q proves $A \leftrightarrow B(\ulcorner A \urcorner)$.*

Recall from Section 10.5 that $\ulcorner A \urcorner$ denotes the numeral corresponding to the natural number representing A, assuming a reasonable encoding of syntax. So the sentence A in Lemma 12.3.1 says "this sentence satisfies B," much the way that the Epimenides sentence says "this sentence is false." Unlike natural language, the language of arithmetic does not provide an easy way for a sentence to refer to itself, though our proof of the computability-theoretic fixed-point theorem in Section 11.5 accomplished a similar task. To understand how the idea plays out in the current setting, consider the following version of the Epimenides paradox, due to W. V. O. Quine:

> "Yields falsehood when preceded by its quotation" yields falsehood when preceded by its quotation.

This sentence is not directly self-referential; it simply makes a statement about a syntactic object, namely, the phrase between the quotation marks. But when we put a quoted version of the phrase in front of itself, we get the original sentence. So, in this cleverly indirect way, the sentence asserts that it is false.

The idea behind the proof of the lemma is to construct a sentence that says

> "X, when substituted for the variable in itself, satisfies B," when substituted for the variable in itself, satisfies B.

To that end, let diag(x) be the primitive recursive function that, given a formula $C(x)$ with one free variable, returns the sentence $C(\ulcorner C(x) \urcorner)$. If diag($x$) were represented by a symbol in the language of arithmetic, we could take A to be $B(\text{diag}(\ulcorner B(\text{diag}(x)) \urcorner))$, because the diagonalization of the formula in corner brackets would then be A itself. Instead, using the fact that every computable function is representable in Q, let $C_{\text{diag}}(x, y)$ represent diag(x). Let $D(x)$ be the formula defined by

$$D(x) \equiv \exists y\, (C_{\text{diag}}(x, y) \wedge B(y))$$

and let A be the sentence $D(\ulcorner D(x) \urcorner)$. Then, by representability, Q proves

$$\forall y\, (C_{\text{diag}}(\ulcorner D(x) \urcorner, y) \leftrightarrow y = \ulcorner D(\ulcorner D(x) \urcorner) \urcorner).$$

The sentence $D(\ulcorner D(x) \urcorner)$ is, by definition,

$$\exists y\, (C_{\text{diag}}(\ulcorner D(x) \urcorner, y) \wedge B(y)).$$

Hence Q proves $D(\ulcorner D(x) \urcorner) \leftrightarrow B(\ulcorner D(\ulcorner D(x) \urcorner) \urcorner)$, which is, by definition, the sentence $A \leftrightarrow B(\ulcorner A \urcorner)$. This concludes the proof of Lemma 12.3.1.

We can now describe Gödel's proof of the first incompleteness theorem. Let T be a computably axiomatized theory extending Q, let $\text{Pr}_T(x, y)$ represent the relation "x is a proof of y in T, " and let $\text{Prov}_T(y)$ be the formula $\exists x\, \text{Pr}_T(x, y)$. The fixed-point theorem yields a sentence A such that Q, and hence T, proves

$$A \leftrightarrow \neg \text{Prov}_T(\ulcorner A \urcorner).$$

In other words, A says "I am not provable." This is known as the *Gödel sentence* for T.

Proposition 12.3.2. *If T is consistent, it does not prove A. If T is ω-consistent, it does not prove $\neg A$.*

Proof First we show that if T proves A, it is inconsistent. Suppose T proves A. Then there is a proof of A, represented by a natural number, d. By representability, T proves $\text{Pr}_T(d, \ulcorner A \urcorner)$, and hence $\text{Prov}_T(\ulcorner A \urcorner)$. But since A is equivalent to $\neg\text{Prov}_T(\ulcorner A \urcorner)$, T proves that as well, and hence it is inconsistent.

Next, we show that if T proves $\neg A$, it is ω-inconsistent. Suppose T proves $\neg A$. Then T proves $\exists x\, \text{Pr}_T(x, y)$. If T is inconsistent, then it is ω-inconsistent, and we are done. Otherwise, T does not prove A. By representability, then, T proves $\neg\text{Pr}_T(d, \ulcorner A \urcorner)$ for every d, and so T is ω-inconsistent. □

We have shown that if T is any computably axiomatized, ω-consistent theory extending Q, then there is a sentence A such that T proves neither A nor $\neg A$ – in other words, T is incomplete. This is only slightly weaker than our formulation of the first incompleteness theorem, Theorem 12.2.6, in that it assumes that T is ω-consistent rather than just consistent.

A slight modification of Gödel's argument yields Theorem 12.2.6 on the nose. It uses a method similar to one used in the proof of Theorem 12.1.9. The idea is to modify the provability predicate slightly. For convenience, let us act as though the language of arithmetic has a function symbol neg(x) that returns the negation of a formula; the use of that symbol can be eliminated using representability, just as we avoided the use of a function diag(x) in the proof of the fixed-point lemma above. Define an alternative version of the provability predicate:

$$\text{Prov}'(y) \equiv \exists x\, (\text{Pr}_T(x, y) \wedge \forall z \leq x\, \neg \text{Pr}_T(x, \text{neg}(y))).$$

Let us refer to the property expressed by Prov′ as being *shmovable* instead of provable. In words, a sentence C is shmovable if there is a proof of C and no smaller proof of $\neg C$. Of course, if T is consistent, provability and shmovability coincide. But, as we will see in the next section, T does not know that it is consistent, and so it cannot prove the two formulas equivalent. We can use the fixed-point lemma to find a sentence A such that Q proves $A \leftrightarrow \text{Prov}'(\ulcorner A \urcorner)$. Such a sentence is known as the *Rosser sentence* for T. Exercise 12.3.3 asks you to show that if T is consistent, then it proves neither A nor $\neg A$.

The fixed-point lemma also yields a short proof of Tarski's theorem, Theorem 10.5.2, which says that the set of true sentences in the language of arithmetic is not arithmetically definable. For any formula $B(x)$, the fixed-point lemma yields a sentence A such that Q proves $A \leftrightarrow \neg B(\ulcorner A \urcorner)$. This means that A holds in the standard model if and only if $B(\ulcorner A \urcorner)$ does not, so $B(x)$ cannot define the set of true sentences in the language of arithmetic.

Exercises

12.3.1. Given two formulas $B_1(y)$ and $B_2(y)$, show that there are sentences A_1 and A_2 such that Q proves $A_1 \leftrightarrow B_1(\ulcorner A_2 \urcorner)$ as well as $A_2 \leftrightarrow B_2(\ulcorner A_1 \urcorner)$. (Hint: in fact, you can let A_2 *be* the formula $B_2(\ulcorner A_1 \urcorner)$ for a suitable A_1.)

12.3.2. Let T be a consistent, computably axiomatized extension of Q, and let A be the Gödel sentence for T.
a. Show that $T \cup \{\neg A\}$ is consistent.
b. Show that $T \cup \{\neg A\}$ is not ω-consistent.

12.3.3. Let T be a consistent, computably axiomatized theory extending Q. Let A be the Rosser sentence for T. Show that T doesn't prove A and that it also doesn't prove $\neg A$.

12.3.4. Let T be a consistent, computably axiomatized theory extending Q, and let the formula $\text{Pr}_T(x, y)$ represent the relation "x is a proof of y in T." Assume T has a function symbol representing exponentiation with base 2, so that T can prove $2^{\overline{n}} = \overline{2^n}$ for every n. Let t be the following term:

$$\overline{2}\hat{}(\overline{2}\hat{}(\overline{2}\hat{}(\overline{2}\hat{}(\overline{2}\hat{}(\overline{2}\hat{}(\overline{2}\hat{}(\overline{2}\hat{}(\overline{2}\hat{}(\overline{2}\hat{}(\overline{2}\hat{}0)))))))))).$$

So t denotes a very big number! Call this number n. Let $\text{SProv}_T(y)$ be the formula $\exists x\, (x < t \wedge \text{Pr}_T(x, y))$, so $\text{SProv}_T(y)$ says "there is a short proof of y," that is, a proof of y that is represented by a number less than a n. By the fixed-point lemma, there is a sentence A such that T proves $A \leftrightarrow \neg \text{SProv}_T(\ulcorner A \urcorner)$, so A says "there is no short proof of me." Show that T proves A but there is no short proof of A in T, that is, no proof of A that is represented by a number less than n.

12.4 The Second Incompleteness Theorem

Roughly speaking, the second incompleteness theorem says that no reasonable theory of mathematics can prove its own consistency. This claim differs from the first incompleteness theorem in that, in order to state it precisely, we have to say what it means for a theory to prove its own consistency. In other words, for the theory T in question, we have to express "T is consistent" as a formal sentence in the language of T, and then whether the result is provable in T will depend on the expression we choose.

For simplicity, let us assume that T extends Q, although everything we will say can be generalized to theories that interpret Q instead. We will express Con_T as the sentence $\neg\mathrm{Prov}_T(\ulcorner \bot \urcorner)$, which says that T does not prove $\bot$, assuming $\mathrm{Prov}_T(y)$ expresses the fact that y is provable in T. This only pushes the problem back to determining when $\mathrm{Prov}_T(y)$ is an adequate expression of the statement "y is provable in T." In order to prove the second incompleteness theorem as generally as possible, we will identify minimal conditions on $\mathrm{Prov}_T(y)$ that suffice to establish the conclusion.

A short sketch of the proof will help us isolate the necessary conditions. Suppose $\mathrm{Prov}_T(y)$ expresses provability in T, and let A be the Gödel sentence with respect to $\mathrm{Prov}_T(y)$. The first incompleteness theorem says that if T is consistent, then T does not prove A. Gödel's proof, though clever, is not very difficult, and with mild assumptions on T and the provability predicate Prov_T, we should be able to formalize the argument in T. In other words, T should be able to prove

$$\mathrm{Con}_T \to \neg\mathrm{Prov}_T(\ulcorner A \urcorner).$$

But since $\neg\mathrm{Prov}_T(\ulcorner A \urcorner)$ is equivalent to A, T can then prove $\mathrm{Con}_T \to A$. By the first incompleteness theorem, T cannot prove A, so T cannot prove Con_T.

To prove the first incompleteness theorem, it was enough to assume that $\mathrm{Prov}_T(y)$ is defined as $\exists x\, \mathrm{Pr}_T(x, y)$ for some formula $\mathrm{Pr}_T(x, y)$ that represents provability in T. This is not enough to prove the second incompleteness theorem. We will assume instead that the following hold for all sentences A and B:

1. If $T \vdash A$, then $T \vdash \mathrm{Prov}_T(\ulcorner A \urcorner)$.
2. $T \vdash \mathrm{Prov}_T(\ulcorner A \to B \urcorner) \to (\mathrm{Prov}_T(\ulcorner A \urcorner) \to \mathrm{Prov}_T(\ulcorner B \urcorner))$.
3. $T \vdash \mathrm{Prov}_T(\ulcorner A \urcorner) \to \mathrm{Prov}_T(\ulcorner \mathrm{Prov}_T(\ulcorner A \urcorner) \urcorner)$.

These are sometimes called the Hilbert–Bernays–Löb provability conditions.

I will now try to convince you that these are reasonable requirements for a formal notion of provability. I will also try to convince you that these properties are easily established in PRA for a straightforward formalization of the notion of provability in first-order logic from a computable set of axioms. Finally, I will state and prove the second incompleteness theorem in terms of any provability predicate satisfying these conditions.

Let us consider each of the conditions in turn. The first claim will hold if provability is defined in terms of the existence of a proof, assuming T can recognize a proof when it sees it. In other words, it will hold whenever $\mathrm{Prov}_T(y)$ is defined as $\exists x\, \mathrm{Pr}_T(x, y)$, if $\mathrm{Pr}_T(x, y)$ represents provability in T in the sense of Definition 12.1.8. This is a minimal assumption: if T cannot recognize explicit proofs as such, it is hard to make the case that $\mathrm{Prov}_T(y)$ is an adequate definition of provability within the theory T.

Condition 2 says that if T proves sentences $A \to B$ and T proves A, then it proves B. In other words, it says that T knows that the notion of provability is closed under modus ponens. Once again, this is generally easy to establish and it describes a fundamental property of provability. If we take a proof to be a sequence of sentences in a Hilbert-style axiomatization of first-order logic, we only need to know that T can prove that concatenating proofs of $A \to B$ and A and then applying modus ponens yields a proof of B. Similarly, if we define provability in terms of natural deduction, we only need that T can show that we can form a new proof using the implication elimination rule.

The third condition is the most subtle. You can think of it as an internalized version of the first claim, asserting that the first claim is provable in T. In other words, it says that T proves "if I can prove a sentence A, then I can prove that I can prove it." This requires that, to some extent, T can reflect on its own abilities. The formal argument in T might go something like this: "Suppose I can prove A. Then there is a proof d of A, and just by running through all the steps, I can prove that d is a proof of A. Hence I can prove that I can prove A." If $\mathrm{Prov}_T(y)$ is a Σ_1 formula, Proposition 12.1.13, which says that PRA proves that every true Σ_1 sentence is provable in Q, is relevant. If T extends or interprets PRA and we use any Σ_1 formula of provability, then that formulation automatically satisfies the third condition. Proposition 12.1.13 is even more than we need, however: it makes a statement about arbitrary Σ_1 sentences, whereas condition 3 refers only to instances of a *single* Σ_1 formula. Instead of using induction and recursion in T, T can painstakingly describe the relevant proofs for each subformula of $\mathrm{Prov}_T(y)$ in turn. So again all that is needed is that T can manage a modicum of syntactic reasoning.

The following is our formulation of the second incompleteness theorem:

Theorem 12.4.1. *Let T be any computably axiomatized theory that interprets Q, let* Prov_T *satisfy conditions 1–3, and let* Con_T *be the sentence* $\neg\mathrm{Prov}_T(\ulcorner \bot \urcorner)$. *If T is consistent, then T doesn't prove* Con_T.

Proof Let A be the Gödel sentence for Prov_T, so T proves $A \leftrightarrow \neg\mathrm{Prov}_T(\ulcorner A \urcorner)$. Condition 1 implies that T does not prove A: otherwise, it proves both $\mathrm{Prov}_T(\ulcorner A \urcorner)$ and $\neg\mathrm{Prov}_T(\ulcorner A \urcorner)$, and hence is inconsistent. It therefore suffices to show that T proves $\mathrm{Con}_T \to A$.

We need only fill out the argument sketched at the beginning of this section. We will repeatedly use the fact that if T proves a sentence of the form $B \to C$, then, by conditions 1 and 2, it also proves $\mathrm{Prov}_T(\ulcorner B \urcorner) \to \mathrm{Prov}_T(\ulcorner C \urcorner)$.

Since T proves $A \to \neg\mathrm{Prov}_T(\ulcorner A \urcorner)$, it also proves

$$\mathrm{Prov}_T(\ulcorner A \urcorner) \to \mathrm{Prov}_T(\ulcorner \neg\mathrm{Prov}_T(\ulcorner A \urcorner) \urcorner).$$

Since Q, and hence T, proves $\neg\mathrm{Prov}_T(\ulcorner A \urcorner) \to (\mathrm{Prov}_T(\ulcorner A \urcorner) \to \bot)$, T also proves

$$\mathrm{Prov}_T(\ulcorner \neg\mathrm{Prov}_T(\ulcorner A \urcorner) \urcorner) \to \mathrm{Prov}_T(\ulcorner \mathrm{Prov}_T(\ulcorner A \urcorner) \to \bot \urcorner).$$

(In fact, we have defined $\neg B$ to be $B \to \bot$, so the antecedent and conclusion of the implication are syntactically the same. But spelling out the step shows that our argument does not depend on the precise syntactic formulation of negation.) From 2, we also have

$$\mathrm{Prov}_T(\ulcorner \mathrm{Prov}_T(\ulcorner A \urcorner) \to \bot \urcorner) \to (\mathrm{Prov}_T(\ulcorner \mathrm{Prov}_T(\ulcorner A \urcorner) \urcorner) \to \mathrm{Prov}_T(\ulcorner \bot \urcorner)).$$

Chaining these three implications yields

$$\mathrm{Prov}_T(\ulcorner A \urcorner) \to (\mathrm{Prov}_T(\ulcorner \mathrm{Prov}_T(\ulcorner A \urcorner) \urcorner) \to \mathrm{Prov}_T(\ulcorner \bot \urcorner)).$$

By condition 3, we have that T also proves

$$\mathrm{Prov}_T(\ulcorner A \urcorner) \to \mathrm{Prov}_T(\ulcorner \mathrm{Prov}_T(\ulcorner A \urcorner) \urcorner).$$

Combining these, we have $\text{Prov}_T(\ulcorner A \urcorner) \to \text{Prov}_T(\ulcorner \bot \urcorner)$, which implies the contrapositive, $\neg\text{Prov}_T(\ulcorner \bot \urcorner) \to \neg\text{Prov}_T(\ulcorner A \urcorner)$. By the definition of the Con_T and the fact that A is the Gödel sentence, this is equivalent to $\text{Con}_T \to A$. □

Formal provability and consistency have surprising properties. For example, in the exercises below you are asked to show that there are consistent theories that can prove that they are inconsistent. Such theories are false in the standard interpretation, but that does not make them inconsistent. For another example, we might wonder about the conditions under which a theory T can prove $\text{Prov}_T(\ulcorner A \urcorner) \to A$, that is, the conditions under which T can establish that the provability of a sentence implies its truth. The following theorem, known as Löb's theorem, shows that in general a theory *cannot* prove this, unless it already proves A itself.

Theorem 12.4.2. *Suppose T is a theory extending (or interpreting) Q, and* $\text{Prov}_T(y)$ *is a provability predicate satisfying conditions 1–3. If T proves* $\text{Prov}_T(\ulcorner A \urcorner) \to A$*, then T proves A.*

This is known as Löb's theorem. To motivate the argument, consider the following proof that Santa Claus exists:

1. Let B be the sentence, "If B is true, then Santa Claus exists."
2. Suppose B is true.
3. Then what it says is true; i.e. if B is true, then Santa Claus exists.
4. Since we are assuming B is true, we can conclude that Santa Claus exists.
5. So, we have shown: "If B is true, then Santa Claus exists."
6. But this is just the statement B. So we have shown that B is true.
7. But then, by the argument above, Santa Claus exists.

We obtain a more sober proof of Löb's theorem by replacing "is true" by "is provable," replacing "Santa Claus exists" by any formula A such that T proves $\text{Prov}_T(\ulcorner A \urcorner) \to A$, and appealing to the fixed-point lemma to obtain a sentence B such that T proves

$$B \leftrightarrow (\text{Prov}_T(\ulcorner B \urcorner) \to A).$$

Then each of the following is provable in T:

1. $B \to (\text{Prov}_T(\ulcorner B \urcorner) \to A)$
2. $\text{Prov}_T(\ulcorner B \urcorner) \to \text{Prov}_T(\ulcorner \text{Prov}_T(\ulcorner B \urcorner) \to A \urcorner)$
3. $\text{Prov}_T(\ulcorner B \urcorner) \to (\text{Prov}_T(\ulcorner \text{Prov}_T(\ulcorner B \urcorner) \urcorner) \to \text{Prov}_T(\ulcorner A \urcorner))$
4. $\text{Prov}_T(\ulcorner B \urcorner) \to \text{Prov}_T(\ulcorner \text{Prov}_T(\ulcorner B \urcorner) \urcorner)$
5. $\text{Prov}_T(\ulcorner B \urcorner) \to \text{Prov}_T(\ulcorner A \urcorner)$
6. $\text{Prov}_T(\ulcorner B \urcorner) \to A$
7. B
8. $\text{Prov}_T(\ulcorner B \urcorner)$
9. A.

This completes the proof. The fact that the argument can be used to establish anything at all is sometimes known as *Curry's paradox*.

Löb's theorem provides a short proof of the first incompleteness theorem for any theory T having a provability predicate satisfying conditions 1–3: if T proves Con_T, then it proves $\text{Prov}_T(\ulcorner \bot \urcorner) \to \bot$, and hence, by Theorem 12.4.2, it also proves $\bot$.

Exercises

In the exercises that follow, let T be a consistent, computably axiomatized theory in a language extending (or interpreting) Q, and let $\mathrm{Prov}_T(y)$ be a formula satisfying the Hilbert–Bernays–Löb provability conditions.

12.4.1. Let T' be the theory T together with the additional axiom $\neg\mathrm{Con}_T$. Suppose $\mathrm{Prov}_{T'}(y)$ is a formula such that T' can prove $\forall y\,(\mathrm{Prov}_T(y) \to \mathrm{Prov}_{T'}(y))$, and suppose $\mathrm{Con}_{T'}$ is defined, as usual, as $\neg\mathrm{Prov}_{T'}(\ulcorner\bot\urcorner)$.

a. Show that T' is consistent.
b. Show that T' proves $\neg\mathrm{Con}_{T'}$.

In other words, T' is a consistent theory that proves that it is inconsistent.

12.4.2. In the previous problem, show that if T' proves that $\mathrm{Prov}_{T'}(y)$ is equivalent to a formula of the form $\exists x\,\mathrm{Pr}_{T'}(x, y)$, where $\mathrm{Pr}_{T'}(x, y)$ represents the relation "x is a proof of y in T'," then T' is a consistent theory that is not ω-consistent.

12.4.3. Show that if A is any sentence at all, then T does not prove $\neg\mathrm{Prov}(\ulcorner A\urcorner)$.

12.4.4. Let $\mathrm{Prov}'_T(y)$ be the formula $\mathrm{Prov}_T(y) \wedge y \neq \ulcorner\bot\urcorner$, and let Con'_T be the sentence $\neg\mathrm{Prov}'_T(\ulcorner\bot\urcorner)$.

a. Show that T proves Con'_T.
b. Show that the first and third of the Hilbert–Bernays–Löb provability conditions hold for T.
c. Find specific sentences A and B for which the second condition fails.

12.4.5. Justify each step in the proof of Löb's theorem, Theorem 12.4.2.

12.4.6. Let A and B be sentences T proves $\mathrm{Prov}_T(\ulcorner A\urcorner) \to B$ and $\mathrm{Prov}_T(\ulcorner B\urcorner) \to A$. Show that T proves both A and B.

12.4.7. Suppose $\mathrm{Tr}(y)$ is any formula in the language of arithmetic, and let $\hat{T}$ be Q together with an axiom $\mathrm{Tr}(\ulcorner A\urcorner) \leftrightarrow A$ for every sentence A.

a. Show $\mathrm{Tr}(y)$ satisfies the Hilbert–Bernays–Löb provability conditions with respect to $\hat{T}$.
b. Use Löb's theorem to conclude that $\hat{T}$ is inconsistent.
c. Show that this yields another proof of Tarski's theorem (Theorem 10.5.2).

12.4.8. Consider the following four statements:

a. If T proves A, then T proves $\mathrm{Prov}_T(\ulcorner A\urcorner)$.
b. T proves $A \to \mathrm{Prov}_T(\ulcorner A\urcorner)$.
c. If T proves $\mathrm{Prov}_T(\ulcorner A\urcorner)$, then T proves A.
d. T proves $\mathrm{Prov}_T(\ulcorner A\urcorner) \to A$

Under what conditions are each of these statements true?

12.4.9. For another proof of Löb's theorem, suppose T proves $\mathrm{Prov}_T(\ulcorner A\urcorner) \to A$.

a. Let $P(y)$ be the formula $\mathrm{Prov}_T(\ulcorner \neg A \to y\urcorner)$. (More precisely, I mean the formula $\exists z\,(B(y, z) \wedge \mathrm{Prov}_T(z))$, where B represents the function that maps each y coding a formula C to the Gödel number $\neg A \to C$.) Show that $P(y)$ is a provability predicate for $T \cup \{\neg A\}$.
b. Show that $T \cup \{\neg A\}$ proves $\neg P(\ulcorner\bot\urcorner)$.

The second incompleteness theorem implies that $T \cup \{\neg A\}$ is inconsistent, so T proves A.

12.5 Some Decidable Theories

Theorem 12.2.10 shows that any theory that is even vaguely compatible with a theory of arithmetic is undecidable, including theories that, a priori, look like they have nothing to do with arithmetic at all. For example, the first-order theory of groups and the first-order theory of fields are both undecidable.

But that does not mean that there aren't any interesting decidable theories. Quite the contrary: some very important algebraic structures and classes of structures have theories that are decidable. In this section, we will consider a few important examples, restricting attention to classical logic. We will make use of a powerful method, *quantifier elimination*, for establishing decidability.

Definition 12.5.1. A theory T *admits quantifier elimination* if for every formula $A(\vec{x})$ in the language of T there is a quantifier-free formula $A'(\vec{x})$ such that T proves $\forall \vec{x}\, (A(\vec{x}) \leftrightarrow A'(\vec{x}))$.

Below, we will discuss explicit procedures that, given a formula $A(\vec{x})$, produce a quantifier-free equivalent $A'(\vec{x})$ and a proof of the equivalence. But it is worth noting that if T is a computably axiomatized theory that admits quantifier elimination, then, in principle, this data can always be obtained by a brute-force search. Moreover, suppose T has the additional property that for every quantifier-free *sentence* A, either T proves A or T proves $\neg A$. Then we can decide whether or not T proves a sentence A by finding a quantifier-free equivalent A' and determining whether T proves A'.

Proposition 12.5.2. *Suppose T is a computably axiomatized theory that admits quantifier elimination and that for every quantifier-free sentence A, either $T \vdash A$ or $T \vdash \neg A$. Then T is complete and decidable.*

Remember also that if T is a complete theory and is satisfied by a model $\mathfrak{M}$, then T necessarily coincides with the theory of $\mathfrak{M}$. Thus a decision procedure for T is at the same time a decision procedure for the truth or falsity of sentences in $\mathfrak{M}$.

The decision procedures that are implicit in the presentations that follow are generally inefficient, and a good deal of research in logic and computer science has been devoted to developing algorithms that work better in practice. But before looking for good algorithms, the first step is to determine that a theory is decidable in principle, and that is the only concern that we will address here.

A number of considerations simplify the effort involved in showing that a theory admits quantifier elimination. First, it is sufficient to show that every formula of the form $\exists x\, B(x, \vec{y})$, with $B(x, \vec{y})$ quantifier-free, is equivalent to a quantifier-free formula. This follows from the fact that if we can eliminate a single existential quantifier then we can iterate the process, and the fact that $\forall x\, C(x)$ is equivalent to $\neg \exists x\, \neg C(x)$.

Second, without loss of generality, we can assume that $B(x, \vec{y})$ is a conjunction of literals. After all, any quantifier-free formula $B(x)$ can be put in disjunctive normal form, and using the identity $\exists x\, (C(x) \vee D(x)) \leftrightarrow \exists x\, C(x) \vee \exists x\, D(x)$ we can move the existential quantifier through the disjunction. If each of $\exists x\, C(x)$ and $\exists x\, D(x)$ is equivalent to a quantifier-free formula, then so is $\exists x\, (C(x) \vee D(x))$.

We can also assume, without loss of generality, that x is mentioned in every conjunct, because $\exists x\, (C \wedge D(x))$ is equivalent to $C \wedge \exists x\, D(x)$ when x is not free in C. If one of the literals present is equivalent to $x = t$, we are done, because $\exists x\, (x = t \wedge C(x))$ is equivalent to $C(t)$.

Another useful trick is that we can also expand a conjunction of literals when it is helpful to have more information about the quantified variable. For any formula $C(x)$, the formula $\exists x\, B(x)$ is equivalent to the disjunction of $\exists x\,(B(x) \wedge C(x))$ and $\exists x\,(B(x) \wedge \neg C(x))$, so it suffices to know that we can eliminate the existential quantifier in each case.

A final trick we will use below is that we can sometimes show that a theory T', expressed in a larger language, is conservative over T, and that T' admits elimination of quantifiers. If quantifier-free formulas in the language T' can be translated back to quantifier-free formulas in the language of T, this shows that T admits elimination of quantifiers of well. Even in cases where quantifier-free formulas in the language of T' cannot be translated back to quantifier-free formulas in the language of T, the completeness and decidability of T' implies the completeness and decidability of T, which is often what we are after.

As a first, simple example, consider the theory of *dense linear orders*, which is axiomatized by the (universal closures of the) following:

- $x < y \wedge y < z \to x < z$
- $\neg(x < y \wedge y < x)$
- $x < y \vee x = y \vee y < x$
- $x < y \to \exists z\,(x < z \wedge z < y)$.

The first three axioms say that $<$ is a strict linear order, and the last one says that it is dense. The theory could equally well be formulated in terms of $\leq$ or in terms of both $<$ and $\leq$ without altering the results below. We will see that the choice of $<$ is convenient.

If we add an axiom $\forall x\, \exists y\,(y < x)$ asserting that there is no smallest element and another axiom $\forall x\, \exists y\,(x < y)$ asserting that there is no largest element, we obtain the theory *dense linear orders without endpoints*.

Theorem 12.5.3. *The theory of dense linear order without endpoints admits quantifier elimination. Hence it is complete and decidable, and every dense linear order without endpoints, such as* $(\mathbb{Q}, \leq)$ *and* $(\mathbb{R}, \leq)$, *has the same first-order theory.*

Proof Since there are no constant symbols in the language, the only quantifier-free formulas are boolean combinations of the propositional constant $\bot$, which are easily proved or refuted using propositional logic. So quantifier elimination implies decidability.

Let B be any quantifier-free formula. We need to show $\exists x\, B$ is equivalent to a formula that is again quantifier-free. If we put B in negation-normal form, we can replace every negated atomic formula by a quantifier-free equivalent without negations: $u \not< v$ is equivalent to $u = v \vee v < u$ and $u \neq v$ is equivalent to $u < v \vee v < u$. Putting the resulting formula in disjunctive normal form and factoring the existential quantifier through the disjunctions, we can assume that B is a conjunction of formulas of the form $x = y_i$ or $x < y_i$ or $y_i < x$, where y_i ranges over the variables appearing in B.

If any of these atomic formulas is $x < x$, $\exists x\, B(x)$ is equivalent to $\bot$. If an equation $x = y_i$ occurs in the conjunction, $\exists x\, B(x)$ is equivalent to $B(y_i)$. Otherwise, $B(x)$ can be written as a conjunction

$$\bigwedge_{i \in A} y_i < x \wedge \bigwedge_{j \in B} x > y_j.$$

If A is empty, the axiom that says that there is no greatest element shows that this formula is equivalent to $\top$. The case where B is empty is similar. Otherwise, the density axiom implies that $\exists x\, B(x)$ is equivalent to

$$\bigwedge_{i \in A, j \in B} y_i < y_j. \qquad \square$$

It turns out that the structures $(\mathbb{Q}, 0, 1, +, <)$ and $(\mathbb{R}, 0, 1, +, <)$ also have the same first-order theory in the language with 0, 1, +, and <, and it is not hard to characterize this theory axiomatically. First, we add a unary negation operation $-$ to the language. (Later we will write $x - y$ for $x + (-y)$.) The structures $(\mathbb{Q}, 0, +, -)$ and $(\mathbb{R}, 0, +, -)$ are examples of *abelian groups*, which is to say, they satisfy the following axioms:

- $\forall x, y, z\, ((x + y) + z = x + (y + z))$
- $\forall x\, (x + 0 = x)$
- $\forall x\, (x + (-x) = 0)$
- $\forall x, y\, (x + y = y + x)$.

Sometimes one adds the word "additive" to indicate that the group operation is written as addition, but of course that is only a notational distinction. To these, we add axioms asserting that $<$ is a linear order and that addition respects the order:

- $\forall x, y, z\, (x < y \to x + z < y + z)$.

These are the axioms for an *ordered abelian group*. We now add a schema with the axiom

- $\forall x\, \exists y\, (n \cdot y = x)$

for each natural number $n > 0$, where $n \cdot y$ abbreviates a sum $y + y + \ldots + y$ of n-many ys. This yields the theory of *ordered divisible abelian groups*. Finally, we add the axiom $0 < 1$ to ensure that the group consists of more than just the 0 element.

Theorem 12.5.4. *The theory of nontrivial ordered divisible abelian groups admits quantifier elimination. Hence is it complete and decidable, and every nontrivial ordered divisible abelian group has the same first-order theory.*

Proof If n is a natural number, for any term t we can interpret nt in terms of iterated addition and $-nt$ as its negation. Similarly, we can interpret n by itself as the term $n \cdot 1$. So we can express any term t involving the variables $x_0, \ldots, x_{n-1}$ in the form $a_0 x_0 + \ldots + a_{n-1} x_{n-1} + b$ where $a_0, \ldots, a_{n-1} + b$ are integers. Using the axioms of the theory, any atomic formula can be proved equivalent to one of the form $t > 0$ or $t = 0$. By the usual reductions it suffices to show that it is possible to eliminate a single existential quantifier in front of a conjunction of formulas of this form.

It will also be useful to work with expressions $a_0 x_0 + \ldots + a_{n-1} x_{n-1} + b$ in which $a_0, \ldots, a_{n-1}$ and b are rational numbers. If u and v are rational linear combinations of variables $x_0, \ldots, x_{n-1}$, let m be the least common multiple of all the denominators and interpret $u = v$ and $u < v$ as the expressions obtained by using multiplication by m to clear the denominators. The fact that the axiomatic theory proves $nx = ny \leftrightarrow x = y$ and $nx < ny \leftrightarrow x < y$ for any natural number n ensures that we can always scale equations and inequalities to simulate comparisons between terms with rational coefficients.

So consider a formula $\exists x\, B(x)$ where B is any conjunction of comparisons of the form $t = 0$ or $t > 0$, where each t contains x. Each of these can be reexpressed as $x = t'$ or $x > t'$ or $x < t'$ where t' is a rational linear combination of the other variables. If any of the conjuncts is an equality $x = t'$, the formula $\exists x\, B(x)$ is equivalent to $B(t')$, and we are done. Otherwise, we can write $B(x)$ as

$$\bigwedge_{i \in A} t'_i < x \wedge \bigwedge_{j \in B} x > t'_j.$$

where each t'_i or t'_j is a rational linear combination of the other variables. As in the proof of Theorem 12.5.3, $\exists x\, B(x)$ is equivalent to

$$\bigwedge_{i \in A, j \in B} t'_i < t'_j,$$

or simply $\top$ if either A or B is empty. □

This proof shows that we can view the theory of nontrivial ordered divisible abelian groups as a theory of equalities and inequalities between linear expressions $a_0 x_0 + \ldots + a_{n-1} x_{n-1} + b$. As a result, the theory is often known as *linear arithmetic*.

The fact that the rationals and reals are densely ordered played a key role in the proofs above. At the other extreme, the structure $(\mathbb{Z}, 0, 1, +, <)$ is a *discrete* ordered abelian group: for any integer x, there is nothing between x and $x + 1$. We will show that, nonetheless, the theory of this structure is also decidable and admits quantifier elimination in a slightly expanded language. For each integer $m > 0$, define $D_m(x)$ to be the formula $\exists y\, (m \cdot y = x)$, where once again $m \cdot y$ denotes a sum m instances of y. The predicate $D_m(x)$ says that x is divisible by m. The theory known as *Presburger arithmetic*, or *integer linear arithmetic*, has the following axioms:

- the axioms for an ordered abelian group with $0 < 1$
- the axiom $\forall x\, \neg(0 < x \wedge x < 1)$
- for each $m > 0$, an axiom

$$\forall x\, (D_m(x) \vee D_m(x + 1) \vee D_m(x + 2) \vee \cdots \vee D_m(x + (m - 1))).$$

We leave it as an exercise to show that Presburger arithmetic proves all of the following:

1. $mx = 0 \leftrightarrow x = 0$
2. $D_m(x) \wedge D_m(y) \rightarrow D_m(x + y)$
3. $D_m(-x) \leftrightarrow D_m(x)$
4. $D_m(x) \rightarrow D_n(x)$ whenever $n \mid m$
5. $D_m(x) \leftrightarrow D_{km}(kx)$ for any $k > 0$
6. $\neg(D_m(x + k) \wedge D_m(x + \ell))$ for every k and ℓ such that $0 \leq k < l < m$.

In the proof that follows, we will make use of formulas $t \equiv_m k$, defined as $D_m(t - k)$, where $0 \leq k < m$ are natural numbers. The formula $t \equiv_m k$ says that t is congruent to k modulo m. Presburger arithmetic proves all of these:

7. $x \equiv_m 0 \vee x \equiv_m 1 \vee \cdots \vee x \equiv_m m - 1$
8. $\neg(x \equiv_m k \wedge x \equiv_m \ell)$ for any $0 < k < \ell < m$

9. $x \equiv_m k \leftrightarrow (x \equiv_{m\ell} k \vee x \equiv_{m\ell} k + m \vee \cdots x \equiv_{m\ell} k + m(\ell - 1))$
10. $D_m(x+y) \leftrightarrow (x \equiv_m 0 \wedge y \equiv_m 0) \vee (x \equiv 1 \wedge y \equiv_m m-1) \vee \cdots \vee (x \equiv_m m-1 \wedge y \equiv_m 1)$.

Theorem 12.5.5. *Presburger arithmetic admits elimination of quantifiers up to the formulas $D_m(x)$. In other words, every formula in the language of Presburger arithmetic is provably equivalent to a boolean combination of atomic formulas and formulas of the form $D_m(t)$. As a result, Presburger arithmetic is decidable and complete, and hence the theory of $(\mathbb{Z}, 0, 1, +, -, <)$.*

Proof Decidability follows from the fact the Presburger arithmetic decides the truth of every sentence that is a boolean combination of atomic formulas and formulas of the form $D_m(t)$, which you are asked to prove as an exercise.

Suppose $B(x)$ is a conjunction of formulas of the following form:

- $t = 0$
- $t > 0$
- $D_m(t)$
- $\neg D_m(t)$.

By the usual reductions, it suffices to show that for any such $B(x)$, the formula $\exists x\, B(x)$ is equivalent to a boolean combination of formulas of that form. We can also assume that every term t is of the form $ax + t'$ for some integer a and some term t' that doesn't contain x.

We now show that, without loss of generality, we can assume that in each case $a = 1$, at the expense of allowing formulas $t < 0$ and well as $t > 0$. In other words, we can assume that every term that appears is of the form $x + t'$ where x is not free in t'. First, notice that we can transform $B(x)$ so that the coefficient of x is the same in every formula, using the following identities, all provable in Presburger arithmetic:

- $t = 0 \leftrightarrow kt = 0$, for any integer $k \neq 0$
- $t > 0 \leftrightarrow kt > 0$, for any $k > 0$
- $t > 0 \leftrightarrow kt < 0$, for any $k < 0$
- $D_m(t) \leftrightarrow D_{km}(kt)$.

As a result, we can express $B(x)$ as $B'(kx)$, where $B'(y)$ is a conjunction of atomic formulas $t = 0$, $t > 0$, $t < 0$, $D_m(t)$, and $\neg D_m(t)$ such that y occurs with coefficient 1 in every t. But then $\exists x\, B(x)$ is equivalent to $\exists x\, B'(kx)$, which is equivalent to $\exists y\, (D_k(y) \wedge B'(y))$. Renaming y to x, we have justified the claim.

If there is any equation in the conjunction, we can put it in the form $x = t'$, in which case $\exists x\, B(x)$ is equivalent to $B(t')$. So we can assume there are no equalities. Any formula of the form $\neg D_m(t)$ can be replaced by the equivalent formula

$$D_m(t+1) \vee D_m(t+2) \vee \cdots \vee D_m(t + (m-1)).$$

Distributing the conjunction and factoring the existential quantifier over these, we obtain conjunctions in which the formulas of the form $D_m(t)$ occur only positively.

By clause 10 above, we can replace every formula of the form $D_m(x + t')$ by a disjunction of formulas $x \equiv_m k \wedge t' \equiv_m m - k$. Distributing the rest of the conjunction over these disjuncts, pushing the existential quantifier inwards, and removing the conjuncts that don't depend on x, we are left with conjuncts of the form $x \equiv_m k$.

To sum up, we have reduced our task to eliminating the existential quantifier in a formula $\exists x\, B(x)$ where $B(x)$ is a conjunction of formulas of the form $s_i < x, x < t_j$, and $x \equiv_{m_\ell} k_\ell$, where each k_ℓ is a natural number less than m_ℓ.

We can simplify the task even further. For any two conjuncts $s < x$ and $s' < x$, we can split on the provable disjunction $s \leq s' \vee s' \leq s$. In the first case, the truth of the formula is unchanged if we drop $s < x$, and in the second case we may drop $s' < x$. Since the formulas $s \leq s'$ and $s' \leq s$ don't mention x, we can take them outside the existential quantifier and forget about them. So we can assume there is at most one conjunct of the form $s < x$, and similarly there is at most one conjunct of the form $x < t$.

Finally, using item 9 above we can replace each conjunct $x \equiv_m k$ by a disjunction involving equivalence modulo the least common multiple of all the ms that occur. Splitting across all the disjunctions, we can therefore assume that all the conjuncts of this form involve the same m, and then, by item 8 we can assume that there is at most one of them. So we have reduced our task to proving $\exists x\, B(x)$ is equivalent to a quantifier-free formula when $B(x)$ is a conjunction of at most one formula of each of the following three forms: $s < x, x < t$, and $x \equiv_m k$. If $B(x)$ consists of only one conjunct, it is easy to see that Presburger arithmetic proves $\exists x\, B(x)$. Since it proves $s + 1 \equiv_m \vee s + 2 \equiv_m k \vee \cdots \vee s + m \equiv_m k$, it proves $\exists x\, (s < x \wedge x \equiv_m k)$, and, similarly, $\exists x\, (x < t \wedge x \equiv_m k)$. The formula $\exists x\, (s < x \wedge x < t)$ is provably equivalent to $s + 1 < t$.

So all that remains is to deal with the formula $\exists x\, (s < x \wedge x < t \wedge x \equiv_m k)$. Do another case split on the formula $s + (m+1) < t \vee t \leq s + (m+1)$. In the first case, Presburger arithmetic proves $\exists x\, (s < x \wedge x < s + (m+1) \wedge x \equiv_m k)$, and hence the original formula. In the second case, it proves $x = s + 1 \vee s + x + 1 \vee \cdots \vee x = s + m + 1$, and splitting across that disjunction, in each case the equation can be used to eliminate the existential quantifier. □

Presburger arithmetic is often taken to refer to the theory of the natural numbers as an ordered additive semigroup rather than the integers. Since the set of natural numbers is definable in the integers, decidability of the former follows from decidability of the latter. The distance between the theories is not large, however. The previous proof can be rewritten so that it describes a quantifier-elimination procedure for $(\mathbb{N}, 0, 1, +, <)$, interpreting an atomic formula like $x - y = z$ as shorthand for $x = y + z$. Exercise 12.5.5 below asks you to determine how the axioms above should be modified to characterize the natural numbers.

We have seen that the theories of the structures $(\mathbb{N}, 0, 1, +, <)$, $(\mathbb{Z}, 0, 1, +, <)$, $(\mathbb{Q}, 0, 1, +, <)$, and $(\mathbb{R}, 0, 1, +, <)$ are all decidable. From Section 12.2, we know that the theories of the first three structures with multiplication, $(\mathbb{N}, 0, 1, +, \cdot, <)$, $(\mathbb{Z}, 0, 1, +, \cdot, <)$, and $(\mathbb{Q}, 0, 1, +, \cdot, <)$, are undecidable. What about the real numbers with multiplication, $(\mathbb{R}, 0, 1, +, \times, <)$, and the complex numbers with multiplication, $(\mathbb{C}, 0, 1, +, \times)$? Remarkably, these have decidable theories. I will present axiomatic characterizations of both and briefly sketch a quantifier-elimination procedure for the second one. For more information, see the bibliographical notes at the end of this chapter.

A *ring* is a structure $(R, 0, 1, +, \times)$ in which $(R, 0, +)$ is an abelian group, multiplication is associative and has an identity element 1, and multiplication distributes over addition. It is convenient to expand the language of rings to include a negation operation $-y$ to denote the additive inverse of y. A ring is *commutative* if the multiplication is commutative, and a *field* if, moreover, every nonzero element x has a multiplicative inverse x^{-1} satisfying

$x \cdot x^{-1} = x^{-1} \cdot x = 1$. In a ring or field, it is possible that for some positive natural number n, the sum $n \cdot 1$ of n copies of 1 is equal to 0. The smallest such n, if it exists, is called the *characteristic* of the ring, and if there is no such n, the ring is said to have *characteristic zero*. (The field axioms imply that if a field has positive characteristic, then that characteristic must be a prime number.)

The complex numbers are an instance of a field of characteristic zero, but what makes the structure special is that it is *algebraically closed*: every nonzero polynomial of positive degree has a root. In other words, the complex numbers are a model of the following theory, known as the *theory of algebraically closed fields of characteristic zero*:

- the axioms for a field
- for every $n > 0$, an axiom $n \cdot 1 \neq 0$
- for every $n > 0$, the following axiom:

$$\forall a_0, \ldots, a_n \, (a_n \neq 0 \rightarrow \exists x \, (a_n x^n + \cdots + a_1 x + a_0 = 0)).$$

Theorem 12.5.6. *The theory of algebraically closed fields of characteristic zero admits quantifier elimination. Hence it is complete and decidable, and equal to the theory of* $(\mathbb{C}, 0, 1, +, \times)$.

Any equation involving x can be expressed as $a_n x^n + a_{n-1} x_{n-1} + \cdots a_1 x + a_0 = 0$, where the coefficients a_i are terms involving the constant 1 and the other variables. By the usual reductions, then, we only have to worry about eliminating a single existential quantifier in front of a conjunction of equalities and disequalities

$$p_0(x) = 0 \wedge \cdots \wedge p_{k-1}(x) = 0 \wedge q_0(x) \neq 0 \wedge \cdots \wedge q_{\ell-1}(x) \neq 0,$$

where each $p_i(x)$ and $q_j(x)$ is a polynomial in x. Since the field axioms imply that a product uv is equal to zero if and only if $u = 0$ or $v = 0$, we can replace the q_js by their product, and so assume that there is at most one q_j. Also by splitting across disjunctions, we can assume that the leading coefficient of each polynomial is nonzero.

We can also assume that there is at most one p_i. This is because the theory of fields can prove a version of the quotient-remainder theorem for polynomials: given any two polynomials $f(x)$ and $g(x)$, if the degree of $g(x)$ is greater than 0, there are polynomials $q(x)$ and $r(x)$ and a constant c such that $cf(x) = g(x) \cdot q(x) + r(x)$ and the degree of $r(x)$ is less than the degree of $g(x)$. (The term c is a power of the leading coefficient of g, and is needed to compensate for the fact that we do not have a division symbol in the language. The process of obtaining c, q, and r is sometimes known as *pseudodivision*, and $q(x)$ and $r(x)$ are sometimes called a *pseudoquotient* and *pseudoremainder*, respectively.) Moreover, x is a zero of both $f(x)$ and $g(x)$ if and only if it is zero of both $g(x)$ and $r(x)$, so any pair $p_i(x)$ can $p_j(x)$ and be replaced by a pair in which the total degree is reduced. When a polynomial $p(x)$ has degree 0, then $p(x) = 0$ is of the form $a_0 = 0$ and can be moved outside the existential quantifier. We are essentially using the Euclidean algorithm to replace the polynomials $p_0(x), p_1(x), \ldots, p_{k-1}(x)$ by their greatest common divisor.

So we are reduced to showing that $\exists x \, B(x)$ is equivalent to a quantifier-free formula when it is a conjunction of at most one equation $p(x) = 0$ and one disequation $q(x) \neq 0$. If $p(x)$ is a nonzero constant, the equation is false, and if it is zero, the equation is true, and dually for $q(x)$. So, splitting on cases, we can assume that $p(x)$ and $q(x)$ are polynomials of positive degree. If $\exists x \, B(x)$ is of the form $\exists x \, (p(x) = 0)$, the theory of algebraically closed fields proves that it equivalent to $\top$. Similarly, the theory of algebraically closed fields proves $\exists x\,(q(x) \neq 0)$ is equivalent to $\top$, since it can prove that $q(x)$ has at most n roots, and, for example, the elements $0, 1, 2, 3, \ldots, n$ are all distinct.

So we only have to deal with formulas of the form $\exists x\,(p(x) = 0 \wedge q(x) \neq 0)$, which say that some root of $p(x)$ is not a root of $q(x)$. For each fixed n, the theory of algebraically closed fields can prove that any polynomial $p(x)$ of degree n can be written in the form $a(x - r_0) \cdots (x - r_{n-1})$, where $r_0, \ldots, r_{n-1}$ are its roots. So every root of $p(x)$ is a root of $q(x)$ if and only if $p(x)$ divides $q(x)^n$, where n is the degree of p. So saying that some root of $p(x)$ is not a root of $q(x)$ is equivalent to saying that dividing $q(x)^n$ by $p(x)$ leaves a nonzero remainder. This can be expressed in a quantifier-free way, using the methods described two paragraphs earlier.

The theory of the reals is more subtle. Whereas there is no useful notion of an order on the complex numbers, the usual order on the real numbers plays an essential role in reasoning about them. A polynomial $p(x)$ of odd degree always has a root, since such a polynomial changes sign at least once as x goes from negative infinity to infinity. Every positive real number has a square root. Remarkably, these two facts are enough to characterize the first-order theory of the real numbers in a language with 0, 1, $+$, $\cdot$, and $<$. The axioms for a *real closed field* are as follows:

- the field axioms
- axioms that make the structure an *ordered field*: if $x < y$ then $x + z < y + z$, and if $0 < x$ and $0 < y$ then $0 < x \cdot y$
- the axiom that every positive number has a square root: $0 < x \to \exists y\,(y \cdot y = x)$
- for every odd natural number n, an axiom asserting that every polynomial of degree n has a root, as in the theory of algebraically closed fields.

Theorem 12.5.7. *The theory of real closed fields admits quantifier elimination. Hence it is complete, decidable, and equal to the theory of* $(\mathbb{R}, 0, 1, +, \cdot, <)$.

This important result is due to Alfred Tarski. By reductions similar to the ones for algebraically closed fields, the proof boils down to showing how to eliminate an existential quantifier over a conjunction of the form

$$p_0(x) = 0 \wedge \cdots \wedge p_{k-1}(x) = 0 \wedge q_0(x) > 0 \wedge \cdots \wedge q_{\ell-1}(x) > 0.$$

But the reasoning here is more involved. Consider, for example, the statement that a quadratic polynomial has a root, $\exists x\,(a_2x^2 + a_1x + a_0 = 0)$. If a_2 is equal to 0, the expression in question is equal to $a_1x + a_0$, and has a root unless a_1 is zero and a_0 is nonzero. If a_2 is nonzero, by the quadratic formula, the polynomial has a real root if and only if it has a nonnegative discriminant, which is to say, $a_1^2 - 4a_2a_0 \geq 0$. Thus the formula above is equivalent to

$$(a_2 = 0 \wedge (a_0 = 0 \vee a_1 \neq 0)) \vee (a_2 \neq 0 \wedge a_1^2 \geq 4a_2a_0).$$

Handling arbitrary conjunctions of equations and inequalities requires reasoning formally about polynomials and their derivatives.

It may seem surprising at first that the theory of fields is undecidable, but the theory of algebraic closed fields and the theory of real closed fields, each of which has additional axioms, are decidable. But Theorem 12.2.10 provides some helpful intuitions. Pure first-order logic in a language with a single binary relation symbol is undecidable; in general, adding axioms can contribute to decidability by providing additional information about the class of structures that satisfy them. In particular, it can rule out structures in which it is possible to interpret the integers. For another example of this phenomenon, even though the theory of densely ordered abelian groups is decidable, the theory of groups is not. In fact, as noted in Section 4.4, the word problem for groups is undecidable.

There is more that can be said when it comes to pure first-order logic. Like first-order logic with a binary relation symbol, first-order logic with equality in a language with two unary functions is also undecidable. But there are also positive results: first-order logic without equality with any number of unary predicates and unary functions is decidable, as is first-order logic with equality, any number of unary predicates, and a single unary function. We can also ask about decision procedures for *fragments* of first-order logic. For example, Theorem 4.4.4 states that the universal fragment of first-order logic with equality, in any language, is decidable. With restrictions on the language, we can extend such decidability results to wider classes of formulas. For example, in a language with function symbols, the validity of $\exists\forall$ formulas, that is, formulas with a block of existential quantifiers followed by a block of universal quantifiers, is decidable. This is known as the Bernays–Schönfinkel–Ramsey class. Questions as to the decidability of fragments of classical first-order logic are known as instances of the *classical decision problem*.

Exercises

12.5.1. Show that there are exactly four complete theories that extend the theory of dense linear orders.

12.5.2. Show, by describing explicit proofs from the axioms, that the theory of ordered groups proves the following:

a. $0 < x \leftrightarrow -x < 0$

b. $nx < ny \leftrightarrow x < y$

c. $nx = ny \leftrightarrow x = y$.

12.5.3. Justify all the claims about provability in Presburger arithmetic before the statement of Theorem 12.5.5, again by describing explicit proofs.

12.5.4. Show that for every quantifier-free sentence A, Presburger arithmetic proves either A or $\neg A$.

12.5.5. Give a complete axiomatization of the structure $(\mathbb{N}, 0, 1, +, <)$, in other words, present a version of Presburger arithmetic for the natural numbers.

12.5.6. Show that if $p(x)$ is a polynomial in the language of fields, the theory of algebraically closed fields proves that $p(x)$ can be written in the form $a(x - r_0) \cdots (x - r_{n-1})$, where $r_0, \ldots, r_{n-1}$ are its roots.

12.5.7. Show that we can replace the last axiom and axiom schema in the axiomatization of real closed fields given above by an axiom schema that says that every definable subset of the reals that is nonempty and bounded has a least upper bound. In other words, for every formula $A(x)$, possibly with variables other than x, we add an axiom that says that the set defined by $A(x)$ has this property.

12.5.8. Consider first-order logic with equality without any function and relation symbols at all, that is, the pure theory of equality. For each n, let $P_n(x)$ say that there are at least n elements in the universe. Show that the pure theory of equality admits quantifier elimination up to boolean combinations of formulas of that form.

12.5.9. Provide a similar analysis of the *monadic* theory of first-order logic, that is, first-order logic with equality, no function symbols, and arbitrarily many unary relation symbols.

12.5.10. Show that first-order logic with equality and two unary function symbols is undecidable.

Bibliographical Notes

The incompleteness theorems Reading Gödel's original paper on the incompleteness theorems is enjoyable and rewarding. It can be found in his collected works, Gödel (1986–2003), in Davis (2004), and in van Heijenoort (1967). For historical background to Hilbert's program, see Feferman et al. (2010), Sieg (2013), and Zach (2003).

The incompleteness theorems and their epistemological consequences have been the subject of extensive philosophical debate. See Franzén (2005) and Shapiro (1998) for good overviews. The self-referential sentence due to Quine that was used in Section 12.3 is taken from his essay, *Ways of Paradox*, which can be found in Quine (1997). An expository gem, "Gödel's second incompleteness theorem explained in words of one syllable," can be found in Boolos (1998).

For more of the mathematics behind the incompleteness theorems, see Smith (2013), Smorynski (1977), or Smullyan (1992). For versions of the second incompleteness theorem for weak theories of arithmetic, see Hájek and Pudlák (1998).

Decidable and undecidable theories Classical sources on decidable and undecidable theories are Tarski (1968) and Robinson (1949). The fact that the theory of integers is interpretable in the theory of the rational numbers as a field is due to Robinson (1949), but there is a more recent proof in Poonen (2009). These show that the theory of fields is undecidable. For a proof that the theory of groups is undecidable, see Rotman (1995).

Kreisel and Krivine (1967) and Harrison (2009) describe decision procedures for real closed fields that are not efficient but easy to understand. van den Dries (1988) provides additional information and background on Tarski's result, which is fundamental to a field known as *real algebraic geometry*; see, for example, Bochnak et al. (1998). Marker (2002) is a good reference for model-theoretic proofs of quantifier elimination.

The decision procedures described in Section 12.5 are computationally inefficient, but they serve as a starting point for implementing more practical versions. For studies of decision procedures like linear arithmetic and integer linear arithmetic with an eye toward applications like hardware and software verification, see Kroening and Strichman (2016) or Bradley and Manna (2007). For computational aspects of the theory of the real numbers with an eye toward applications like optimization, see Basu et al. (2006) and Blekherman et al. (2013).

For the results about the classical decision problem, that is, questions as to decidability of fragments of first-order logic, see Börger et al. (2001).

13

Finite Types

In first-order logic, quantifiers range over a homogeneous universe of objects. Section 4.8 introduced many-sorted logic, in which quantifiers range over objects of different sorts. But we often want to quantify over functions and relations between such domains as well. This is, in a sense, tantamount to viewing functions and relations as objects in their own right. Once we do that, we can consider functions and relations defined on domains of other functions and relations, and then consider functions and relations on *those*, and so on.

Such domains are often called *higher types*, and our goal in this chapter is to develop ways of reasoning about them. We will focus on functions rather than predicates and relations, bearing in mind that predicates and relations can be viewed as functions that return a truth value. We will focus specifically on the collection of *finite types*, also known as the *simple types*, which arise from finitely many applications of the function-space construction. In Chapter 17, we will see that these offer a basis for formalizing a good deal of mathematics.

We will consider two closely related formalisms, the simply typed lambda calculus and simply typed combinatory logic. We will also consider some of the relationships between them. These have a lot in common with the untyped versions described in Section 11.7, but there are important differences. Most notably, we will see that terms of the simply typed calculi are *strongly normalizing*, which means that every term can be reduced to one in normal form. In order to keep this presentation self-contained, some of the definitions and conventions from Section 11.7 will be repeated here.

13.1 The Simply Typed Lambda Calculus

Every term in the simply typed lambda calculus has an associated *type*, a syntactic entity that indicates what kind of object the term denotes. The syntax of the system is therefore specified in two stages: first, we define the set of types, and then we specify the sets of terms of each type.

We start with a collection of symbols, $\mathcal{B}$, which are called the *basic types*. The set of types relative to $\mathcal{B}$ is then generated inductively by the following clauses:

- If $B \in \mathcal{B}$, then B is a type.
- If α and β are types, then so is $(\alpha \to \beta)$.
- If α and β are types, then so is $(\alpha \times \beta)$.

In a standard set-theoretic interpretation, each basic type B denotes a set of objects, $(\alpha \to \beta)$ denotes the set of functions from α to β, and $(\alpha \times \beta)$ denotes the cartesian product of α and β. For example, we might start with a single basic type, NAT, to denote the natural numbers. The type $(((\text{NAT} \to \text{NAT}) \times \text{NAT}) \to \text{NAT})$ then denotes the set of functions F that map each pair (f, n), consisting of a function from natural numbers to natural numbers and a natural number, to a natural number.

As usual, we adopt conventions for dropping parentheses. We give products higher precedence than the arrow, so that $\alpha \to \beta \times \gamma$ denotes $(\alpha \to (\beta \times \gamma))$. Products associate to the right, so that $\alpha \times \beta \times \gamma$ is interpreted as $(\alpha \times (\beta \times \gamma))$. Iterated arrows also associate to the right, so that $\alpha \to \beta \to \gamma$ denotes $(\alpha \to (\beta \to \gamma))$. We will see below why this makes sense.

We assign to each type α a natural number, its *type level*:

1. Each basic type has level 0.
2. The level of $\alpha \to \beta$ is equal to the maximum of the level of β and one plus the level of α.
3. The level of $\alpha \times \beta$ is the maximum of the levels of α and β.

So, for example, $(\mathrm{NAT} \to \mathrm{NAT}) \to \mathrm{NAT}$ has level 2, while $\mathrm{NAT} \to (\mathrm{NAT} \to \mathrm{NAT})$ has level 1. Again, we will see below see why this makes sense. An object of a type β with level n is sometimes said to be of *type n*. So an object of type $(\mathrm{NAT} \to \mathrm{NAT}) \to \mathrm{NAT}$ is of type 2 while an object of type $\mathrm{NAT} \to (\mathrm{NAT} \to \mathrm{NAT})$ is of type 1.

The set of types just defined is known as the set of *finite types* over $\mathcal{B}$. Sometimes, cartesian products are left out of the definition; we will see that this is not a substantial omission.

Before defining the set of terms, we also need to assume that we are given a set of constant symbols $\mathcal{C}$, each with an associated type. This set of constant symbols may be empty, in which case the definition below specifies the set of *pure* lambda terms. Relative to the set of constants, the set of terms is generated inductively, as follows:

- For each type α, there are infinitely many variables $x, y, z, \ldots$ of type α.
- Each constant $c \in \mathcal{C}$ is a term of the associated type.
- If t is a term of type β and x is a variable of type α, then $(\lambda x.\, t)$ is a term of type $(\alpha \to \beta)$.
- If t is a term of type $(\alpha \to \beta)$ and s is a term of type α, then $(t\, s)$ is a term of type β.
- If s and t are terms of type α and β respectively, then (s, t) is a term of type $\alpha \times \beta$.
- If t is a term of type $\alpha \times \beta$, then $(t)_0$ is a term of type α and $(t)_1$ is a term of type β.

Although I will use $x, y, z \ldots$ to denote variables of any type, I will assume that distinct variables have been chosen for each type. I will write $t \colon \alpha$ or t^α to indicate that t has type α. In the set-theoretic interpretation, if t is of type β and x is of type α, $(\lambda x.\, t)$ is interpreted as the function which maps an object a of type α to the value of t with x replaced by a. If t has type $\alpha \to \beta$ and s has type α, $(t\, s)$ denotes the result of applying t to s. Similarly, (s, t) denotes an ordered pair, and $(t)_0$ and $(t)_1$ denote the two components of a pair t. This semantics will be made precise in Section 13.6.

In the expression $\lambda x.\, t$, the lambda denotes a binding operation, like quantification in first-order logic. As a result, we will identify terms that differ only up to renaming of their bound variables, just as we did with first-order formulas in Section 1.6. In the literature on the lambda calculus, terms that differ only up to renaming of their bound variables are said to be *α-equivalent*, and we are taking α-equivalent expressions to be syntactically identical. As in Section 1.6, we define the set of free variables in a term, notions of substitution, and the complexity of a term. As with formulas in first-order logic, we can do proofs by induction on complexity and define functions by recursion on complexity. In particular, we can define functions by recursion on the structure of a term, as long as we make sure that the definition is independent of the variable x we choose to represent the body of the term $\lambda x.\, t$.

We adopt the following conventions for dropping parentheses. Applications associate to the left, so that if r, s, t, and u are terms of suitable type, then $r\,s\,t\,u$ is short for $(((r\,s)\,t)\,u)$. A lambda abstraction is assumed to have the widest scope possible, so that, for example, $\lambda x.\, r\,s\,t$ is short for $(\lambda x.\,(r\,s\,t))$. Note that this convention is opposite to the convention we have adopted for first-order quantifiers. Note also that we continue to omit the type of x when it is arbitrary or when it can be inferred from context. It is also convenient sometimes to write $\lambda x, y, z.\, t$ instead of $\lambda x.\, \lambda y.\, \lambda z.\, t$.

In ordinary mathematical notation, $t\,s$ is conventionally written $t(s)$, but the more streamlined notation is convenient. Notice that there are two distinct ways of representing functions with multiple arguments. We can represent a function f taking arguments from α and β to γ as having type $\alpha \to \beta \to \gamma$, in which case, the value at arguments a and b is $f\,a\,b$. Or we could represent it as a function f' having type $\alpha \times \beta \to \gamma$, in which case, its value at a and b is $f'\,(a, b)$. Exercise 13.1.2 shows that the two representations are easily intertranslatable. The process of turning a function of type $\alpha \times \beta \to \gamma$ into a function of type $\alpha \to \beta \to \gamma$ is known as *currying*, after Haskell Curry, and the reverse process is known as *uncurrying*. The ability to curry functions explains why product types are not essential to the typed lambda calculus. It also helps explain the assignment of type levels: since a function of type $\alpha \to \beta \to \gamma \to \delta$ can be viewed as a function of type $\alpha \times \beta \times \gamma \to \delta$ in disguise, the former should be viewed as a type 1 function.

For every type α, the term $\lambda x^{\alpha}.\, x$ denotes the identity function, which we write id_{α}. Given types α, β, and γ we can also define the term $\mathrm{comp}_{\alpha,\beta,\gamma}$ to be $\lambda x^{\beta\to\gamma}.\, \lambda y^{\alpha\to\beta}.\, \lambda z^{\alpha}.\, x\,(y\,z)$. Given $s^{\alpha\to\beta}$ and $t^{\beta\to\gamma}$, the term $\mathrm{comp}\, t\, s$ represents the composition of t and s. For yet another example, suppose we choose a single base type NAT to denote the natural numbers, a constant 0^{NAT}, and a constant $\mathrm{succ}^{\mathrm{NAT}\to\mathrm{NAT}}$ to denote the successor function. Then the term $\lambda x.\, \mathrm{succ}\,(\mathrm{succ}\, x)$ represents the function that adds 2 to its argument.

If x and s are of type α and t is of type β, then $(\lambda x.\, t)\, s$ is supposed to denote the same thing as $t[s/x]$. This allows us to simplify terms, in a sense, though you should bear in mind that such simplifications can result in terms that are larger. For example, $(\lambda x, y.\, x\,(x\,y))\, s\, t$ simplifies, in this sense, to $s\,(s\,t)$. Formally, a term of the form $(\lambda x.\, t)\, s$ is said to be a *β-redex*. The term *β-contracts* to $t[s/x]$, which we express by writing

$$(\lambda x.\, t)\, s \rhd_{\beta} t[s/x].$$

Similarly, terms of the form $(s, t)_0$ and $(s, t)_1$ are also β-redexes, and β-contract to s and t, respectively:

$$(s, t)_0 \rhd_{\beta} s, \quad (s, t)_1 \rhd_{\beta} t.$$

If s and t are terms, *s reduces to t in one step*, written $s \to_{\beta,1} t$, if t can be obtained by replacing a subterm u of s by a v such that $u \rhd_{\beta} v$. More precisely, the relation $\to_{\beta,1}$ can be defined inductively as the smallest relation satisfying the following:

- If $s \rhd_{\beta} t$ then $s \to_{\beta,1} t$.
- If $s \to_{\beta,1} t$ then $u\,s \to_{\beta,1} u\,t$.
- If $s \to_{\beta,1} t$ then $s\,v \to_{\beta,1} t\,v$
- If $s \to_{\beta,1} t$ then $\lambda x.\, s \to_{\beta,1} \lambda x.\, t$.

In each case, I am assuming that all the terms are of an appropriate type.

Define the relation of *β-reduction*, denoted $\rightarrow_\beta$, to be the reflexive-transitive closure of the one-step reducibility relation, i.e. the smallest reflexive and transitive relation including $\rightarrow_{\beta,1}$. More explicitly, the relation $\rightarrow_\beta$ is generated by the following clauses:

- $s \rightarrow_\beta s$.
- If $s \rightarrow_{\beta,1} t$ then $s \rightarrow_\beta t$.
- If $s \rightarrow_\beta t$ and $t \rightarrow_\beta u$ then $s \rightarrow_\beta u$.

Similarly, define *β-equivalence*, denoted $\equiv_\beta$, to be the smallest equivalence relation including $\rightarrow_\beta$ (or, equivalently, $\rightarrow_{\beta,1}$). The relation $\equiv_\beta$ is generated by the following clauses:

- If $s \rightarrow_\beta t$ then $s \equiv_\beta t$.
- If $s \equiv_\beta t$ then $t \equiv_\beta s$.
- If $s \equiv_\beta t$ and $t \equiv_\beta u$ then $s \equiv_\beta u$.

Other characterizations of these relations will be given in the exercises. If there is a term t' such that $t \rightarrow_{\beta,1} t'$ then t is said to be *β-reducible*, and otherwise it is said to be *β-irreducible*, *β-normal*, or *in β-normal form*.

The following is straightforward. The first claim is sometimes called *subject reduction* or *type preservation*.

Proposition 13.1.1. *Suppose $s \rightarrow_\beta t$. Then:*

1. *s and t have the same type.*
2. *Every free variable of t is a free variable of s.*
3. *$r[s/x] \rightarrow_\beta r[t/x]$.*
4. *$s[q/y] \rightarrow_\beta t[q/y]$.*

The analogous lemma holds for β-equivalence.

Proposition 13.1.2. *Suppose $s \equiv_\beta t$. Then:*

1. *s and t have the same type.*
2. *$r[s/x] \equiv_\beta r[t/x]$.*
3. *$s[q/y] \equiv_\beta t[q/y]$.*

There is another important reduction that is often considered in connection with the lambda calculus. Suppose t is a term of type $\alpha \rightarrow \beta$ and x is a variable of type α that is not free in t. Then $\lambda x.\, t\, x$ describes the function which, given a value s, applies t to s. This is, in essence, just the function denoted by t. Any term of the form $\lambda x.\, t\, x$ is called an *η-redex*, and is said to *η-contract* to t. (The Greek letter here is *eta*.) The reduction is written

$$\lambda x.\, t\, x \rhd_\eta t.$$

Analogously, for pairs, we have

$$((t)_0, (t)_1) \rhd_\eta t.$$

η-reduction is rarely considered on its own but, rather, in conjunction with β-reduction. When we combine the two, the terminology introduced above carries over as expected, yielding notions of $\beta\eta$-redex, $\beta\eta$-reduction, $\beta\eta$-equivalence, $\beta\eta$-normal form, and so on. When the appropriate notion of reduction is clear from context, it is convenient to use simply the terms *redex*, *reduces to*, *equivalent*, and so on.

The phrase *simply typed lambda calculus* is often used to describe the system without cartesian products, and terms like β- and $\beta\eta$-reduction are more commonly used to describe the systems of reductions involving lambda abstraction and application alone, rather than the extension to the systems with pairing and projection. Cartesian products are generally nice to have around, however, and we will put them to use in Chapter 14. Other extensions of the simply typed lambda calculus are discussed in Section 13.8.

Exercises

13.1.1. Consider the term $(\lambda v.\, v\, x)\, (\lambda z.\, u\, z)\, x$.

a. Show all the parentheses that are implicit, according to our conventions.

b. Determine the free variables.

c. Determine what types the variables can have, given that x has type NAT and the entire term has type NAT.

d. Reduce this term to normal form.

13.1.2. Let f be a variable of type $\alpha \to \beta \to \gamma$ and let f' be of type $\alpha \times \beta \to \gamma$. Let $F = \lambda f', a^{\alpha}, b^{\beta}.\, f', (a, b)$ and let $G = \lambda f, p^{\alpha\times\beta}.\, f\, (p)_0\, (p)_1$. Show the following:

a. $F\,(G\, f) \to_{\beta} f$.

b. $G\,(F\, f') \to_{\beta\eta} f'$.

c. $\mathrm{comp}\, F\, G \equiv_{\beta\eta} \mathrm{id}_{\alpha\to\beta\to\gamma}$.

d. $\mathrm{comp}\, G\, F \equiv_{\beta\eta} \mathrm{id}_{\alpha\times\beta\to\gamma}$.

13.1.3. Let $\to_1$ be any binary relation on a set A. Let $\to$ be the reflexive, transitive closure of $\to_1$, i.e. the intersection of all reflexive, transitive relations that include $\to_1$. Show that for any two elements a and b of A, $a \to b$ if and only if there is a sequence of elements $a = a_0, \ldots, a_k = b$ such that for each $i < k$, $a_i \to_1 a_{i+1}$.

13.1.4. Let $\to_1$ and $\to$ be as in the previous problem, and let $\equiv$ be the smallest equivalence relation including $\to_1$, i.e. the intersection of all such equivalence relations. Show that for any two elements a and b of A, $a \equiv b$ if and only if there is a sequence of elements $a = a_0, \ldots, a_k = b$ such that for each $i < k$, either $a_i \to a_{i+1}$ or $a_{i+1} \to a_i$.

13.1.5. Consider the simply typed lambda calculus with a single basic type, NAT, a constant, 0^{NAT}, another constant $\mathrm{succ}^{\mathrm{NAT}\to\mathrm{NAT}}$, and no other constants. A *numeral* in this system is a term of the form succ (succ (succ ... (succ 0))).

Show by induction on terms that if t is any closed and normal then it is either a numeral, the constant succ, or of the form $\lambda x.\, s$ or (r, s). In particular, if t is of type NAT, then it must be a numeral.

13.1.6. Define a sequence of types (i) for $i \in \mathbb{N}$ as follows:

$$(0) = \mathrm{NAT}$$
$$(i+1) = (i) \to (i).$$

For each i, define the expression $B^{(i+3)}$ of type $(i+3)$ as follows:

$$B^{(i+3)} = \lambda H, f, x.\, H\, f\, (H\, f\, x).$$

For every i and j, define terms $C_j^{(i+2)}$ by

$$C_0^{(i+2)} = \lambda f, x.\, f\, x$$
$$C_{j+1}^{(i+2)} = C_j^{(i+4)}\, B^{(i+3)}\, C_0^{(i+2)}$$

Think about why the second line makes sense: a functional of type $(i+4)$ takes an argument of type $(i+3)$ and returns an object of type $(i+3)$. That, in turn, takes an object of type $(i+2)$ and returns on object of type $(i+2)$. Similarly, in the first line, the variable f has type $(i+1)$, and x has type (i).

Define the ordinary stack-of-twos function 2^m_n by $2^m_0 = m$, $2^m_{n+1} = 2^{(2^m_n)}$. Write $f^n x$ for the nth iterate of f on x, that is, $f\ (f\ (\cdots (f\ x) \cdots)$, where f is applied n times in all.

a. Show that in the ordinary set-theoretic interpretation, if $A^{(i+2)}$ is the function satisfying $A\ f\ x = f^n x$, then $B^{(i+3)}\ A$ satisfies $B\ A\ f\ x = f^{2n}\ x$.
b. Show that for $B = B^{(i+3)}$ and $C = C_0^{(i+2)}$, the nth iterate of B on C, $B^n C$, satisfies $B^n\ C\ f\ x = f^{2^n}\ x$. (Remember that $B^n C = B(B(B \ldots (BC)))$.)
c. Show by induction on j that for every i, $C_j^{(i+3)}\ f\ x = f^{2^1_j}\ x$.

By the previous problem, assuming $C_j^{(2)}$ succ 0 has a normal form, it is the numeral corresponding to 2^1_j.

13.1.7. Prove Propositions 13.1.1 and 13.1.2.

13.2 Strong Normalization

In this section, we will see that every term in the simply typed lambda calculus has a normal form, and that there are no infinite sequences of one-step reductions, so that any reduction procedure eventually terminates. The results of the next section will show that this normal form is unique, which is to say, different sequences of one-step reductions cannot result in different normal forms.

The following definition makes sense more generally for any binary relation $\rightarrow_1$ on a set A. In that setting, I will write $\rightarrow$ for the reflexive-transitive closure of $\rightarrow_1$. When a binary relation $\rightarrow_1$ is viewed as a system of rules for rewriting expressions, it is often called a *rewrite system* or *rewriting system*. I will call elements of the underlying set *terms* and refer to the relation $\rightarrow$ as *reduces to*, since these are the cases that are of interest to us in this chapter.

Definition 13.2.1. 1. A term t is *reducible* if there is some t' such that $t \rightarrow_1 t'$, and *irreducible* otherwise. A term that is irreducible is also said to be *normal*, or *in normal form*.
2. A term t is *normalizing* or *normalizable* if there is some t' such that $t \rightarrow t'$ and t' is in normal form. In this case, we say that t' is a *normal form* of t, and that t *has a normal form*.
3. The relation $\rightarrow_1$ is *(weakly) normalizing* if every term is normalizing.
4. A term t is *strongly normalizing* or *strongly normalizable* if there is no infinite sequence of one-step reductions beginning with t, that is, every sequence of one-step reductions starting from t is finite.
5. The relation $\rightarrow_1$ is *strongly normalizing*, or *terminating*, if every term is strongly normalizing.
6. The relation $\rightarrow_1$ is *confluent*, or has the *Church–Rosser property*, if whenever $s \rightarrow u$ and $s \rightarrow v$ then there is a term t such that $u \rightarrow t$ and $v \rightarrow t$.
7. The relation $\rightarrow_1$ is *convergent* if and only if it is strongly normalizing and confluent.

It is not hard to check that if t is irreducible, then whenever $t \rightarrow t'$, we have $t = t'$. In the exercises below, however, you are asked to show that the converse does not necessarily hold.

Notice that confluence is a statement about the reflexive-transitive closure $\twoheadrightarrow$ of $\rightarrow_1$. We can also define the equivalence relation $\equiv$ to be the smallest equivalence relation containing $\rightarrow_1$, as in Section 13.1.

Proposition 13.2.2. *Suppose that the relation $\rightarrow_1$ is confluent. Then:*

1. *If a term t has a normal form, it is unique.*
2. *If s is equivalent to t, then there is some u such that $s \twoheadrightarrow u$ and $t \twoheadrightarrow u$.*
3. *If s and t are normalizing, then they have the same normal form if and only if they are equivalent.*

Proof Suppose $t \twoheadrightarrow t_0$ and $t \twoheadrightarrow t_1$, with t_0 and t_1 in normal form. By confluence, there is a term $\hat{t}$ such that $t_0 \twoheadrightarrow \hat{t}$ and $t_1 \twoheadrightarrow \hat{t}$. But since t_0 and t_1 are in normal form, we have $t_0 = \hat{t} = t_1$.

For the second claim, it suffices to show that the relation $R(s, t)$ which holds if there is a u such that $s \twoheadrightarrow u$ and $t \twoheadrightarrow u$ is an equivalence relation containing $\rightarrow_1$. It is clearly reflexive, and symmetric, and it contains $\rightarrow_1$. Confluence implies that it is transitive as well.

(Alternatively, if s is equivalent to t, then by Exercise 13.1.4, there is a sequence $s = s_0, \ldots, s_n = t$ such that for each i, either $s_i \twoheadrightarrow s_{i+1}$ or $s_{i+1} \twoheadrightarrow s_i$. Use induction on i and confluence to show that for each i, there is a term u_i such that $s_0 \twoheadrightarrow u_i$ and $s_i \twoheadrightarrow u_i$. In particular $s \twoheadrightarrow u_n$ and $t \twoheadrightarrow u_n$.)

Finally, suppose s and t are equivalent, s has normal form s', and t has normal form t'. Then s' and t' are equivalent and so, by 2, reduce to a common term u. But since s' and t' are irreducible, we have $s' = u = t'$. Conversely, if s and t reduce to a common term u, then $s \equiv u$ and $t \equiv u$, and so $s \equiv t$. □

If we interpret reduction as a form of computation, convergent reducibility relations are very well behaved. Taken together, strong normalization and confluence imply that the system of computation is independent of its implementation: strong normalization implies that every reduction sequence terminates regardless of the order in which the reductions are performed, and confluence implies that each such computation yields the same answer. In this section, we will show that the simply typed lambda calculus is strongly normalizing, and in the next section we will show that it is confluent.

Theorem 13.2.3. *β-reduction on the set of simply typed lambda terms is strongly normalizing. The same holds of $\beta\eta$-reduction.*

The theorem can be stated more precisely by saying that the relations $\rightarrow_{\beta,1}$ and $\rightarrow_{\beta\eta,1}$ are strongly normalizing. As will be clear from the proof, the theorem does not depend on the choice of basic types. For simplicity, we will consider the pure simply typed lambda calculus, without any constants. But the theorem extends immediately to the case where there are constants, since we can simply replace these constants by variables. Most of the argument is the same whether we consider β- or $\beta\eta$-reduction; the additional cases needed to handle the latter are noted explicitly.

It is not hard to show, by induction on terms, that for every term t there are only finitely many terms t' such that $t \rightarrow_{\beta,1} t'$, and similarly for $\beta\eta$-reduction. The following lemma holds for any system of one-step reductions with this property.

Lemma 13.2.4. *A term t is strongly normalizing if and only if for some n every sequence of one-step reductions starting from t has length at most n.*

Proof The backward direction is immediate. To prove the forward direction, suppose t is strongly normalizing. Build a tree with t at the root; as children of node labeled s, add a node for each term s' such that $s \to_1 s'$. Clearly this tree is finitely branching. Strong normalization implies that there is no infinite branch through the tree. By Proposition 1.4.4 (Kőnig's lemma), the tree is finite. □

In the proof below, whenever t is strongly normalizing, let $h(t)$ denote the least n such that every sequence of one-step reductions from t has length at most n. Note that if t is strongly normalizing and $t \to_1 s$, then s is strongly normalizing as well and $h(s) < h(t)$.

Before we begin the proof, I will try to convey some intuition as to how it works. Say that an object of a basic type is *computable* if it can be reduced to normal form; an object of type $\alpha \to \beta$ is computable if, when it is applied to a computable object of type α, it yields a computable object of type β; and an object t of type $\alpha \times \beta$ is computable if both $(t)_0$ and $(t)_1$ are computable. Then, by induction on terms, we can show that every term in the simply typed lambda calculus is computable, under the substitution of arbitrary computable terms for its free variables. Spelling out the details would yield a proof that every term of basic type is normalizing.

Theorem 13.2.3 says something stronger: it applies to terms of every type, and it says that terms are strongly normalizing, not just normalizing. But the proof will still follow the same general outline. We will define a stronger notion, *strong computability*, using strong normalization rather than normalization. The proof then proceeds in two steps: we will show by induction on types α that every strongly computable term of type α is strongly normalizing, and then we will show by induction on terms that every term is strongly computable.

Definition 13.2.5. For each type α, we define the set of *strongly computable* terms of type α:

- If t is a term of a basic type, then t is strongly computable if and only if it is strongly normalizing.
- If t is a term of type $\alpha \to \beta$, then t is strongly computable if and only if for every strongly computable term s of type α, $t\,s$ is strongly computable.
- If t is a term of type $\alpha \times \beta$, then t is strongly computable if and only if both $(t)_0$ and $(t)_1$ are strongly computable.

Writing the second clause symbolically, we have that $t^{\alpha \to \beta}$ is strongly computable if and only if

$$\forall s^{\alpha}\ (s \text{ strongly computable} \to t\,s \text{ strongly computable}).$$

As a result, the depth of quantifiers in the statement "t^{β} is strongly computable" grows with the complexity of β.

Before starting the proof, it is useful to have the following:

Definition 13.2.6. Let t be a term in the simply typed lambda calculus. Say that t is *neutral* if it is not of the form $\lambda x.\, s$ or (s, t).

In the literature, sometimes the terms *non-introduced* or *simple* are used instead. The notion is a technical one, designed to make the necessary inductions work. If t is a neutral term of type $\alpha \to \beta$ and s is any term of type α, then any one-step reduction in $t\,s$ has to occur in t or in s. In other words, nothing can happen at the top level. Similarly, if t is a neutral term of type $\alpha \times \beta$, then any one-step reduction in $(t)_0$ or $(t)_1$ has to occur in t. These cases are handled in one way in the proof, whereas terms of the form $(\lambda x.\, t)\, s$, $(s, t)_0$, and $(s, t)_1$ are handled in another.

We are now ready to carry out the first stage outlined above. The following lemma introduces a stronger inductive hypothesis that is needed to make the proof go through. The proof is the same for both β- and $\beta\eta$-reduction.

Lemma 13.2.7. *Let α be any type, and let t be a term of type α.*

1. *If t is strongly computable, then t is strongly normalizing.*
2. *If t is strongly computable and $t \to t'$, then t' is strongly computable.*
3. *If t is neutral and t' is strongly computable for every t' such that $t \to_1 t'$, then t is strongly computable.*

Proof Note that a special case of 3 is that if t is neutral and normal, then t is strongly computable. In particular, if 3 holds for α, then each variable of type α is strongly computable. Claims 1–3 are proved simultaneously by induction on α.

In the base case, α is a basic type. Claims 1 and 2 are immediate. For 3, if t' is strongly normalizing for each t' such that $t \to_1 t'$, then t is strongly normalizing as well, since any reduction sequence from t has to pass through some such t'.

In the induction step, α is of the form $\beta \to \gamma$ or $\beta \times \gamma$. Let us consider each claim in turn.

For 1, suppose $t^{\beta \to \gamma}$ is strongly computable. By inductive hypothesis 3 and the remark above, the variable x^β is strongly computable. By the definition of strong computability, $t\,x$ is strongly computable. By inductive hypothesis 1, $t\,x$ is strongly normalizing. But this means that t is strongly normalizing as well: if $t_0, t_1, t_2, \ldots$ were an infinite sequence of one-step reductions from t, then $t_0\,x, t_1\,x, t_2\,x, \ldots$ would be an infinite sequence of one-step reductions from $t\,x$.

Similarly, suppose $t^{\beta \times \gamma}$ is strongly computable. By the inductive hypothesis, every term of type β is strongly normalizing. If $t_0, t_1, t_2, \ldots$ were an infinite sequence of one-step reductions from t, then $(t_0)_0, (t_1)_0, (t_2)_0, \ldots$ would be an infinite sequence of one-step reductions from $(t)_0$. So t is strongly normalizing.

For 2, suppose $t \to t'$ and t is strongly computable. If t has type $\beta \to \gamma$, we need to show that for any strongly computable s of type β, $t'\,s$ is strongly computable. By the inductive hypothesis, we have that $t\,s$ is strongly computable. Since $t\,s \to t'\,s$, inductive hypothesis 2 implies $t'\,s$ is strongly computable, as required.

Similarly, if t has type $\beta \times \gamma$, we need to show that $(t')_0$ and $(t')_1$ are strongly computable. But since $(t)_0 \to (t')_0$ and $(t)_1 \to (t')_1$, the result follows from the inductive hypothesis.

For 3, suppose t is neutral and whenever $t \to_1 t'$ then t' is strongly computable. If t has type $\beta \to \gamma$, let s be strongly computable of type β; we need to show that $t\,s$ is strongly computable. By the inductive hypothesis, s is strongly normalizing, and we can use a secondary induction on $h(s)$. Since t is neutral, in one step, $t\,s$ can only reduce to terms of the following form:

- $t'\,s$, where $t \to_1 t'$
- $t\,s'$, where $s \to_1 s'$.

In the first case, our hypothesis tells us that t' is strongly computable, and hence, by the definition of strong computability, $t'\,s$ is as well. In the second case, the side inductive hypothesis on $h(s)$ tells us that $t\,s'$ is strongly computable. But now inductive hypothesis 3 for $t\,s$ tells us that $t\,s$ is strongly computable, as required.

In the case where t has type $\beta \times \gamma$, we need to show that each $(t)_i$ is strongly computable, where i is 0 or 1. Since t is neutral, in one step $(t)_i$ can only reduce to a term $(t')_i$ where $t \to_1 t'$. Our hypothesis tells us that for any such t', $(t')_i$ is strongly computable, and so by the inductive hypothesis, $(t)_i$ is strongly computable as well. □

The next three lemmas constitute the second stage of the proof of strong normalization. Lemma 13.2.10 provides a stronger inductive hypothesis needed to make the proof go through. But first, we need two auxiliary facts.

Lemma 13.2.8. *Suppose that whenever s is strongly computable, $t[s/x]$ is strongly computable. Then $\lambda x.\, t$ is strongly computable.*

Proof Notice that whenever the hypothesis of the lemma holds for t, then, in particular, taking s to be x, t is strongly computable.

Suppose the hypothesis holds for t. We need to show that for every s, if s is strongly computable, then so is $(\lambda x.\, t)\, s$. So suppose s is strongly computable. Since t and s are both strongly computable, claim 1 of Lemma 13.2.7 implies that they are strongly normalizing, so we can prove that $(\lambda x.\, t)\, s$ is strongly computable by induction on $h(s) + h(t)$.

By claim 3 of Lemma 13.2.7, it suffices to show that every one-step reduction of $(\lambda x.\, t)\, s$ is strongly computable. Under β-reduction, in one step $(\lambda x.\, t)s$ reduces to one of the following:

- $(\lambda x.\, t)\, s'$, where $s \to_1 s'$
- $(\lambda x.\, t')\, s$, where $t \to_1 t'$
- $t[s/x]$.

In the first case, the inductive hypothesis implies that $(\lambda x.\, t)\, s'$ is strongly computable, since $h(s') < h(s)$. In the second case, it is not hard to verify, using claim 2 of Lemma 13.2.7, that t' also satisfies the hypothesis of Lemma 13.2.8. So, again, the inductive hypothesis implies that $(\lambda x.\, t')\, s$ is strongly computable, since $h(t') < h(t)$. In the third case, $t[s/x]$ is strongly computable, by hypothesis.

To handle $\beta\eta$-reduction, there is one additional case to consider, when t is of the form $t'\, x$ where x is not free in t'. In that case, $(\lambda x.\, t)\, s$ also reduces in one step to $t'\, s$. But the hypothesis implies that $(t'\, x)[s/x]$ is strongly computable, and this is equal to $t'\, s$. □

Lemma 13.2.9. *Suppose that s and t are strongly computable. Then (s, t) is strongly computable.*

Proof As above, we can use induction on $h(s) + h(t)$. We need to show that $(s, t)_0$ and $(s, t)_1$ are both strongly computable, and it suffices to show that every one-step reduction of each of these is strongly computable. Under β-reduction, in one step $(s, t)_0$ reduces to either s, $(s', t)_0$ for some s' such that $s \to_1 s'$, or $(s, t')_0$ for some t' such that $t \to_1 t'$. By hypothesis, s and t are strongly computable, and by claim 2 of Lemma 13.2.8, in the respective cases s' and t' are strongly computable with $h(s') < s$ and $h(t') < t$, so the inductive hypothesis applies.

For $\beta\eta$-reduction, we also have to consider the case where $s = (p)_0$ and $t = (p)_1$. In that case, $(s, t)_i$ also reduces in one step to $(p)_i$ for $i = 0, 1$. But this is equal to either s or t, and by hypothesis these are both strongly computable. □

Lemma 13.2.10. *Suppose t is a term and $x_0, \ldots, x_{k-1}$ is a sequence of variables. Then for every sequence $s_0, \ldots, s_{k-1}$ of strongly computable terms of the same type as $x_0, \ldots, x_{k-1}$, respectively, $t[s_0/x_0, \ldots, s_{k-1}/x_{k-1}]$ is strongly computable.*

Proof By induction on terms. If t is a variable, the result is immediate.

Suppose t is of the form $\lambda x_k.\, t'$, and suppose $s_0, \ldots, s_{k-1}$ are strongly computable. Then t' has at most the free variables $x_0, \ldots, x_k$, and we need to show that the expression $\lambda x_k.\, t'[s_0/x_0, \ldots, s_{k-1}/x_{k-1}]$ is strongly computable. By Lemma 13.2.8, it suffices to show that $t'[s_0/x_0, \ldots, s_{k-1}/x_{k-1}])$ is strongly computable for each strongly computable term s_k. But this follows from the inductive hypothesis for t'.

Suppose t is of the form $q\, r$ and $s_0, \ldots, s_{k-1}$ are strongly computable. By the inductive hypothesis, $q[s_0/x_0, \ldots, s_{k-1}/x_{k-1}]$ and $r[s_0/x_0, \ldots, s_{k-1}/x_{k-1}]$ are strongly computable. Then $t[s_0/x_0, \ldots, s_{k-1}/x_{k-1}]$ is strongly computable by the definition of strong computability and substitution.

Suppose t is of the form (q, r) and $s_0, \ldots, s_{k-1}$ are strongly computable. By the inductive hypothesis applied to q and r, we have that $q[s_0/x_0, \ldots, s_k/x_{k-1}]$ and $r[s_0/x_0, \ldots, s_k/x_{k-1}]$ are strongly computable. By Lemma 13.2.9, this implies that (q, r) is strongly computable.

Finally, if t is a projection $(t')_i$ and $s_0, \ldots, s_{k-1}$ are strongly computable, the inductive hypothesis implies that $t'[s_0/x_0, \ldots, s_k/x_{k-1}]$ is strongly computable, and hence also $(t'[s_0/x_0, \ldots, s_k/x_{k-1}])_i$. □

In our presentation of the simply typed lambda calculus, we added constructions (s, t), $(t)_0$, $(t)_1$ to the inductive definition of terms. Since we have lambda abstraction and application, an alternative is to add terms pair: $\alpha \to \beta \to \alpha \times \beta$ and projections fst: $\alpha \times \beta \to \alpha$ and snd: $\alpha \times \beta \to \beta$, with β-reductions fst (pair $s\, t$) $\rhd s$ and snd (pair $s\, t$) $\rhd t$, and similarly for the η-reductions. The difference between the two presentations is not great, since with the first presentation we can define pair to be $\lambda x, y.\, (x, y)$, fst to be $\lambda p.\, (p)_0$, and snd to be $\lambda p.\, (p)_1$, and, in the other direction, interpret the terms (s, t), $(t)_0$, and $(t)_1$ as pair $s\, t$, fst t, and snd t, respectively. But this does change the notion of reduction, since, for example, pair x is irreducible in the second presentation but not the first.

As an exercise, you can try modifying the proof of strong normalization to apply to this alternative presentation. When we discuss combinatory logic in Section 13.4, we will dispense with the lambda, in which case we are forced to adopt this alternative style. In Section 13.8, we will see that the two styles of extending the calculus – with new term constructions, or new constants – parallel the two styles of extending a deductive system, namely, with rules or with new axioms.

Exercises

13.2.1. Show that for a general binary relation $\to_1$ on a set A, if a term t is irreducible, then whenever $t \to t'$, $t' = t$. Show that the converse holds if t is strongly normalizing but not in general.

13.2.2. This problem and the next provide another proof of strong normalization for the simply typed lambda calculus, due to Felix Joachimski and Ralph Matthes. For simplicity, we will stick to the $\to$ fragment of the simply typed lambda calculus, leaving out cartesian products.

Note that the set of terms can be defined inductively as follows:

- If $r_0, \ldots, r_{k-1}$ is a sequence of terms of types $\alpha_0, \ldots, \alpha_{k-1}$, respectively, and x is a variable of type $\alpha_0 \to \cdots \to \alpha_{k-1} \to \beta$, then $x\vec{r}$ is a term of type β.

- If r is a term of type β and x is a variable of type α, then $\lambda x.\, r$ is of type $\alpha \to \beta$.
- If $k \geq 1$, r is a term of type $\alpha_0 \to \cdots \to \alpha_{k-1} \to \beta$, x is a variable of type α, s is a term of type α, and $s_0, \ldots, s_{k-1}$ are terms of type $\alpha_0, \ldots, \alpha_{k-1}$, respectively, then $(\lambda x.\, r)\, s\, s_0 \ldots s_{k-1}$ is a term of type β.

The resulting set is freely generated by these claims. This way of breaking up the definition has the advantage that if we drop the third claim, we get exactly the set of terms in normal form, denoted NF.

Now define the set of terms SN inductively as follows:

- If each element of $\vec{r}$ is in SN, then $x\,\vec{r}$ is in SN.
- If r is in SN, then $\lambda x.\, r$ is in SN.
- If $r[s/x]\,\vec{s}$ is in SN, and $s \in \text{SN}$, then $(\lambda x.\, r)\, s\, \vec{s}$ is in SN.

Note that $\text{NF} \subseteq \text{SN}$, since the first two claims generate NF.

Recall that a term t is said to be *strongly normalizing* if every reduction sequence terminates, in which case $h(t)$ is defined to be the length of the longest such sequence. The proof proceeds in two steps: this problem has you show that every term in SN is strongly normalizing, and the next problem has you show that every term is in SN.

a. Show by induction on $n + m$ that if s is strongly normalizing with $h(s) \leq n$, and $r[s/x]\vec{s}$ is strongly normalizing with $h(r[s/x]\vec{s}) \leq m$, then $(\lambda x.\, r)\, s\, \vec{s}$ is strongly normalizing. (Hint: consider all the results of one-step reductions from the last term, and use the IH to show that they are all strongly normalizing.)

b. Show by induction on SN that every term in SN is strongly normalizing. In other words, show that the set of strongly normalizing terms is closed under the rules defining SN.

13.2.3. a. Show that for every α, s of type α in SN, and term r in SN, the following two claims hold:

1 If r has type $\alpha \to \beta$ for some β, then $r\,s$ is in SN.
2 If r has any type and x is a variable of type β, then $r[s/x]$ is in SN.

Do this with a primary induction on α and a secondary induction on r in SN. What this means is that you can fix α and s, consider each of the three clauses that put a term in SN, and show that the two claims hold in each case. In doing so, you can assume that the two claims hold for

- s and any term r' that is assumed to be in SN in the antecedent of the clause, and
- any pair of terms $\hat{s}$ and $\hat{r}$ where the type of $\hat{s}$ is smaller than α.

For the last clause, remember that in general $r[t/y][s/x]$ is equal to $r[s/x][t[s/x]/y]$.

b. Show that all terms are in SN, by induction on terms and part a. Use the usual inductive definition of the set of terms: every term is either a variable, a lambda abstraction, or an application.

c. Conclude, using the previous problem, that all terms are strongly normalizing.

13.3 Confluence

In Section 11.7, we sketched a proof of confluence for the untyped lambda calculus. The proof carries over straightforwardly to simply typed lambda calculus, but the fact that the latter has the strong normalization property makes it possible to provide a more direct proof. The use of the following lemma, known as *Newman's lemma*, is the key.

Lemma 13.3.1. *Let $\to_1$ be a binary relation on a set and let $\to$ be its reflexive transitive closure. Suppose $\to_1$ is strongly normalizing, and for any element t, if $t \to_1 t_0$ and $t \to_1 t_1$ then there is a $\hat{t}$ such that $t_0 \to \hat{t}$ and $t_1 \to \hat{t}$. Then $\to$ has the Church–Rosser property.*

In other words, if any two one-step reductions starting from t can be reconciled in some number of steps, and the system is strongly normalizing, then it has the Church–Rosser property. A system satisfying the hypothesis is said to be *locally confluent*. Exercise 13.3.1 asks you to show that without the extra hypothesis that the system is strongly normalizing, the conclusion need not hold.

Proof As in the last section, define $h(t)$ to be the length of the longest reduction sequence starting from t, and use induction on $h(t)$. Suppose $t \to t_0$ and $t \to t_1$. If either t_0 or t_1 is equal to t the conclusion is immediate. Otherwise, let t'_0 be the first element of a reduction sequence of t to t_0 and let t'_1 be the first element of a reduction sequence of t to t_1. By hypothesis, there is a t' such that $t'_0 \to t'$ and $t'_1 \to t'$. Since $h(t'_0) < h(t)$, the inductive hypothesis implies that there is a t''_0 such that $t_0 \to t''_0$ and $t' \to t''_0$. Similarly, since $h(t_1) < h(t)$, there is a t''_1 such that $t_1 \to t''_1$ and $t' \to t''_1$. Since $h(t') < h(t)$, there is a $\hat{t}$ such that $t''_0 \to \hat{t}$ and $t''_1 \to \hat{t}$. But then $t_0 \to \hat{t}$ and $t_1 \to \hat{t}$, as required. □

Theorem 13.3.2. *The simply typed lambda calculus, with either β- or $\beta\eta$-reduction, is confluent.*

Proof By Lemma 13.3.1, it suffices to show that if $t \to_1 t_0$ and $t \to_1 t_1$ there is a $\hat{t}$ such that $t_0 \to \hat{t}$ and $t_1 \to \hat{t}$. We will consider β-reduction first and use induction on terms t. If t is a constant or variable, there is nothing to do.

If t is of the form $\lambda x.\, s$, then we must have $t_0 = \lambda x.\, s_0$ and $t_1 = \lambda x.\, s_1$, where $s \to_1 s_0$ and $s \to_1 s_1$. By the inductive hypothesis, there is an $\hat{s}$ such that s_0 and s_1 both reduce to $\hat{s}$. But then t_0 and t_1 both reduces to $\lambda x.\, \hat{s}$.

If t is of the form $r\,s$, there are a number of ways t can reduce in one step:

1. If $t_0 = r_0\,s$ and $t_1 = r_1 s$ where $r \to_1 r_0$ and $r \to r_1$, apply the inductive hypothesis to obtain $\hat{r}$ such that r_0 and r_1 both reduce to $\hat{r}$, and hence t_0 and t_1 both reduces to $\hat{r}\,s$.
2. If $t_0 = r\,s_0$ and $t_1 = r\,s_1$ where $s \to_1 s_0$ and $s \to_1 s_1$, apply the inductive hypothesis to obtain $\hat{s}$ such that t_0 and t_1 both reduce to $r\,\hat{s}$.
3. If $t_0 = r_0\,s$ and $t_1 = r\,s_1$ where $r \to_1 r_0$ and $s \to_1 s_1$, then t_0 and t_1 both reduce to $r_0\,s_1$. The symmetric case with t_0 and t_1 switched is handled similarly.
4. If r is of the form $\lambda x.\, u$, $t_0 = (\lambda x.\, u_0)\, s$ where $u \to_1 u_0$, and $t_1 = u[s/x]$, then by Proposition 13.1.1 both t_0 and t_1 reduce to $u_0[s/x]$. The symmetric case with t_0 and t_1 switched is handled similarly.
5. If r is of the form $\lambda x.\, u$, $t_0 = (\lambda x.\, u)\, s_0$ where $s \to_1 s_0$, and $t_1 = u[s/x]$, then by Proposition 13.1.1 both t_0 and t_1 reduce to $u[s_0/x]$. Again, the symmetric case with t_0 and t_1 switched is handled similarly.

If t is of the form (r, s), then any one-step reduction has to happen in either r or s. If both happen in r or both happen in s, apply the inductive hypothesis. If one happens in r and the other in s, we simply perform the other reduction in each.

If t is of the form $(s)_0$ and both reduction happen in s, apply the inductive hypothesis. Another possibility is that s is of the form (r, u), one of t_0 or t_1 is r, and the other is $(r_0, s)_0$ or $(r, s_0)_0$, where $r \to_1 r_0$ or $s \to_1 s_0$, respectively. In that case, both terms reduces to r_0 or r, respectively. The case where t is of the form $(s)_1$ is handled similarly.

When we are considering $\beta\eta$-reduction, when t is of the form $\lambda x.\, s$, we also have to consider the case where s is of the form $s'\,x$ for some x not free in s', and one of t_0 or t_1 is s'. In one step, the other term can reduce to $\lambda x.\, s'_0 x$, where $s' \to_1 s'_0$, in which case both reduce to s'_0. Another possibility is s' is itself of the form $\lambda y. s''$, and the other term reduces to $\lambda x. s''[x/y]$, which is the same as s', since we have identified terms up to renaming of their bound variables.

Finally, when we are considering $\beta\eta$-reduction, when t is of the form (r, s), we have to consider the possibility that r and s are of the form $(p)_0$ and $(p)_1$, respectively, and one of t_0 or t_1 is p. Considering the possibilities for the other term is left as an exercise. □

Exercises

13.3.1. Show that the conclusion of Newman's lemma need not hold if the rewrite system is not strongly normalizing. (Hint: You can get by with a relation $\rightarrow_1$ on four elements. Draw a directed graph with arrows representing $\rightarrow_1$, with a cycle to make the system fail to be strongly normalizing.)

13.3.2. To establish that a system is confluent, show that it is sufficient to show the following: whenever $t \rightarrow_1 t_0$ and $t \rightarrow t_1$, then there is a $\hat{t}$ such that $t_0 \rightarrow \hat{t}$ and $t_1 \rightarrow \hat{t}$.

13.3.3. Complete the last case in the proof of Theorem 13.3.2.

13.4 Combinatory Logic

Terms in the simply typed lambda calculus differ from terms in first-order logic in the sense that lambda abstraction binds variables, whereas in first-order logic there is no binding at the level of terms. In this section, we will consider a related calculus, simply typed combinatory logic. We can view combinatory logic as a poor man's lambda calculus since it provides many of the features of the lambda calculus in a first-order framework. To simplify the presentation, we will henceforth ignore product types, though adding them is straightforward.

A combinator is, informally, an operation on functions. In the simply typed lambda calculus, given types α, β, and γ, we can define terms

$$K_{\alpha,\beta} = \lambda x^{\alpha}, y^{\beta}. x$$

and

$$S_{\alpha,\beta,\gamma} = \lambda x^{\alpha\rightarrow\beta\rightarrow\gamma}, y^{\alpha\rightarrow\beta}, z^{\alpha}. x z (y z).$$

Then K has type $\alpha \rightarrow \beta \rightarrow \alpha$ and satisfies

$$K s t = s,$$

and S has type $(\alpha \rightarrow \beta \rightarrow \gamma) \rightarrow (\alpha \rightarrow \beta) \rightarrow (\alpha \rightarrow \gamma)$ and satisfies

$$S r s t = r t (s t),$$

where r, s, and t are assumed to have the appropriate type. In combinatory logic, we dispense with lambda abstraction and instead take the constants K and S to be basic. Just as we can define K and S using lambda abstraction, we will see that there is a sense in which we can simulate lambda abstraction using K and S.

Suppose we are given a set of basic types $\mathcal{B}$, and a set of constants $\mathcal{C}$, each with an associated type over $\mathcal{B}$. The set of types in simply typed combinatory logic is the same as in the simply typed lambda calculus. The set of terms is defined inductively as follows:

- There are infinitely many variables $x, y, z, \ldots$ of each type β.
- Each constant is a term of the appropriate type.
- For each pair of types α, β, $K_{\alpha,\beta}$ is a constant of type $\alpha \to \beta \to \alpha$.
- For each triple of types α, β, γ, $S_{\alpha,\beta,\gamma}$ is a constant of type $(\alpha \to \beta \to \gamma) \to (\alpha \to \beta) \to \alpha \to \gamma$.
- If t a term is of type $\alpha \to \beta$ and s is a term of type α, then $(t\, s)$ is a term of type β.

In contrast to the simply typed lambda calculus, here there is no notion of variable binding or equivalence up to renaming of bound variables. We can therefore view terms in simply typed combinatory logic as terms in a many-sorted first-order logic with binary operations $\mathrm{app}_{\alpha,\beta}$ that map a pair of objects with types $\alpha \to \beta$ and α, respectively, to an object of type β. In other words, if t and s are terms of type $\alpha \to \beta$ and α, respectively, we can view $(t\, s)$ as an abbreviation for $\mathrm{app}_{\alpha,\beta}(t, s)$. The notion of substitution of terms for variables then carries over from first-order logic. To improve readability, I will generally omit type subscripts and superscripts on the constants and variables.

Terms of the form $K\, s\, t$ and $S\, r\, s\, t$ are called *weak redexes*, and are said to *contract* to s and $(r\, t)\, (s\, t)$, respectively. This is written as follows:

$$K\, s\, t \rhd_{\mathrm{w}} s$$
$$S\, r\, s\, t \rhd_{\mathrm{w}} r\, t\, (s\, t).$$

The relations *weakly reduces to in one step* and *weakly reduces to* are defined analogously to β-reduction.

We can define other useful combinators in terms of S and K. For example, for each type α, let I_α be the *identity combinator*, defined by

$$I_\alpha = S_{\alpha,\alpha\to\alpha,\alpha}\, K_{\alpha,\alpha\to\alpha}\, K_{\alpha,\alpha}.$$

This combinator gets its name from the fact that for each term s of type α, we have

$$I\, s = S\, K\, K\, s \to_{\mathrm{w}} K\, s\, (K\, s) \to_{\mathrm{w}} s.$$

To each term t of simply typed combinatory logic, we assign a term $\lambda x.\, t$ by recursion as follows:

- $\lambda x^\alpha.\, x = I_\alpha$
- $\lambda x^\alpha.\, t^\beta = K_{\beta,\alpha}\, t$, if x is not a free variable of t
- $\lambda x^\alpha.\, t^{\alpha\to\gamma}\, x^\alpha = t$, if x is not a free variable of t
- $\lambda x^\alpha.\, t^{\beta\to\gamma}\, s^\beta = S_{\alpha,\beta,\gamma}\, (\lambda x^\alpha.\, t)\, (\lambda x^\alpha.\, s)$, if the previous clauses do not apply.

The sense in which this definition simulates lambda abstraction is given by the following:

Proposition 13.4.1. *For every term t and variable x, we have*

$$(\lambda x.\, t)\, s \to_{\mathrm{w}} t[s/x].$$

Proof Use induction on terms. If t is the variable x, we have

$$(\lambda x.\, t)\, s = I\, s \to_{\mathrm{w}} s = t[s/x].$$

If t is any other variable or a constant, we have

$$(\lambda x.\, t)\, s = K\, t\, s \to_{\mathrm{w}} t = t[s/x].$$

This also holds if t is of the form $q\, r$ and x is not free in t. Finally, if t is of the form $q\, r$ and x is free in t, we have

$$(\lambda x.\, t)\, s = S\, (\lambda x.\, q)\, (\lambda x.\, r)\, s \to_{\mathrm{w}} (\lambda x.\, q)\, s\, ((\lambda x.\, r)\, s) \to_{\mathrm{w}} (q[s/x])\, (r[s/x]) = (q\, r)[s/x].$$

□

In Section 2.3, we saw that the deduction theorem holds for any system of propositional axioms that includes modus ponens as a rule and $\alpha \to \beta \to \alpha$ and $(\alpha \to \beta \to \gamma) \to (\alpha \to \beta) \to \alpha \to \gamma$ as axiom schemas. The analogy to the previous result is striking. Here application plays the role of modus ponens, lambda abstraction plays the role of moving a hypothesis to the antecedent of an implication, and the combinators K and S take the place of the two axiom schemas. Our construction of the identity combinator I mirrors the proof of $P \to P$ in Section 2.2. We will expand on this correspondence in Section 13.8.

Proposition 13.4.2. *The following hold:*

1. *If* $\mathrm{fv}(t)$ *denotes the set of free variables of a term* t*, then* $\mathrm{fv}(\lambda x.\, t) = \mathrm{fv}(t) \setminus \{x\}$.
2. *If* y *is not free in* t*,* $\lambda y.\, (t[y/x]) = \lambda x.\, t$.
3. *If* x *and* y *are different variables and* x *is not a free variable of* s*, then* $(\lambda x.\, t)[s/y] = \lambda x.\, (t[s/y])$.

The second claim says that α-equivalence of lambda terms translates to syntactic identity for combinators. But Exercise 13.4.2 shows that it is not, in general, the case that if $s \equiv_{\mathrm{w}} t$ then $\lambda x.\, s \equiv_{\mathrm{w}} \lambda x.\, t$. As we will see, this is a key difference between the lambda calculus and combinatory logic, and this bears on the axiomatic theories that formalize these systems. In the definition of lambda abstraction for combinatory logic, we could have left out the third clause and restricted the second clause to atomic formulas without damaging Proposition 13.4.1. But our use of the second clause makes the translation more efficient, and the third clause is needed in the proof of the third claim in Proposition 13.4.2.

The proofs of strong normalization and confluence for the simply typed lambda calculus carry over to simply typed combinatory logic. To prove strong normalization, now say that a term is *neutral* if it is not of the form $K\, s$ or $S\, r\, s$. Then we still have the property that if t is neutral of type $\alpha \to \beta$ and s is any term of type α, then any one-step reduction in $t\, s$ has to occur in either t or s. The proof of Lemma 13.2.7 goes through essentially unchanged and yields the corresponding statement for simply typed combinatory logic:

Lemma 13.4.3. *Let* α *be any type, and let* t *be a term of type* α.

1. *If* t *is strongly computable, then* t *is strongly normalizing.*
2. *If* t *is strongly computable and* $t \to t'$*, then* t' *is strongly computable.*
3. *If* t *is neutral, and whenever* $t \to_1 t'$*,* t' *is strongly computable, then* t *is strongly computable.*

We need only replace Lemma 13.2.8 with the following:

Lemma 13.4.4. *For every* α *and* β*,* $K_{\alpha,\beta}$ *is strongly computable.*

Proof If suffices to show that for every s and t of appropriate type, if s and t are strongly computable, then so is $K\,s\,t$. Since $K\,s\,t$ is neutral, by (the analogue of) Lemma 13.2.7, it suffices to show that every one-step reduction of $K\,s\,t$ is strongly computable. Since s and t are strongly computable, they are strongly normalizing, so we can use induction on $h(s)+h(t)$. In one step, $K\,s\,t$ reduces to either

- $K\,s'\,t$, where $s \rightarrow_1 s'$,
- $K\,s\,t'$, where $t \rightarrow_1 t'$, or
- s.

In the first case, $h(s') < h(s)$, and in the second case, $h(t') < h(t)$, so we can apply the inductive hypothesis. In the last case, s is strongly computable by hypothesis. □

Lemma 13.4.5. *For every α, β, and γ, $S_{\alpha,\beta,\gamma}$ is strongly computable.*

The proof is similar.

Theorem 13.4.6. *Every term of combinatorial logic is strongly computable.*

Proof A straightforward induction on terms, using Lemmas 13.4.3,13.4.4, and 13.4.5. □

The task of proving confluence is left as an exercise, as well as the task of extending these proofs to simply typed combinatory logic with products.

Exercises

13.4.1. Prove Proposition 13.4.2.

13.4.2. Let $s = S\,x\,y\,z$ and $t = x\,z\,(y\,z)$. Show that with the definition of lambda abstraction above, $s \equiv_{\mathrm{w}} t$ but $\lambda x.\, s \not\equiv_{\mathrm{w}} \lambda x.\, t$.

13.4.3. Show that in the definition of lambda abstraction, we can restrict the second clause to atomic terms (that is, variables and constants), at the expense of giving up the third clause of Proposition 13.4.2.

In greater detail, suppose, in combinatory logic, we define lambda abstraction as follows:

- $\lambda x.\, x = I$
- $\lambda x.\, t = K\,t$, if t is a constant or a variable other than y
- $\lambda x.\, (s\,t) = S(\lambda x.\, s)(\lambda x.\, t)$

assuming the types all make sense. Show that it is possible to assign types so that $(\lambda x.\, y)[K\,K/y]$ makes sense. Then compute $(\lambda x.\, y)[K\,K/y]$ and $\lambda x.\, (K\,K)$, and show that they are not weakly equivalent.

13.4.4. Show that we can replace the third clause of the definition of lambda abstraction with the following two, while preserving Propositions 13.4.1 and 13.4.2:

- $\lambda x.\, s\,x = s$ if x is not free in s
- $\lambda x.\, s\,t = S\,(\lambda x.\, s)\,(\lambda x.\, t)$, if the previous clauses do not apply.

Explain how this can shorten the lengths of terms obtained in the translation.

13.4.5. Prove Lemma 13.4.5.

13.4.6. Prove that simply typed combinatory logic is confluent.

13.4.7. Extend simply typed combinatory logic with types $\alpha \times \beta$ and constants pair: $\alpha \to \beta \to \alpha \times \beta$, fst: $\alpha \times \beta \to \alpha$, and snd: $\alpha \times \beta \to \beta$. Add the reductions fst (pair s t) $\rhd s$ and snd (pair s t) $\rhd t$. Show that the resulting system is strongly normalizing, as follows.

Extend the definition of strong computability to product types by saying that if t has type $\alpha \times \beta$, then t is strongly computable if both fst t and snd t are. Now say that a term is *neutral* if it is not of the form $K\, s$, $S\, r\, s$, fst, or snd.

a. Prove Lemma 13.4.3 for the extended system.
b. Show that fst and snd are strongly computable.
c. Show that pair is strongly computable.

A straightforward induction then shows that every term of simply typed combinatory logic is strongly computable, and hence strongly normalizing.

13.5 Equational Theories

In this section, we will discuss formal axiomatic theories for reasoning about equations between terms in the simply typed lambda calculus or simply typed combinatory logic. For both calculi, we will consider systems where, for each type α and pair of terms s, t of type α, the assertion $s =_\alpha t$ expresses that s and t are equal. We will often omit the subscripted α. Much of what I say here holds equally well for the untyped versions of these systems, modulo the fact that in the untyped systems, terms do not always have normal forms.

In both the lambda calculus and combinatory logic, we include rules expressing the reflexivity, symmetry, and transitivity of equality:

$$\frac{}{t = t} \qquad \frac{s = t}{t = s} \qquad \frac{r = s \qquad s = t}{r = t}$$

We also include rules that say that application respects equality:

$$\frac{t = t'}{ts = t's} \qquad \frac{s = s'}{ts = ts'}$$

The rules we have given so far can be viewed as equational logic with a sort for each type and a binary operation for application. But the fact that lambda abstraction binds variables means that we cannot view terms of the simply typed lambda calculus as first-order terms. For that calculus, we also add the following rule, which asserts that lambda abstraction respects equality as well:

$$\frac{s = t}{\lambda x.\, s = \lambda x.\, t}$$

This is sometimes known as the rule ξ (which is the Greek letter *xi*). Finally, we add the following axioms, which correspond to β-reduction:

$$\frac{}{(\lambda x.\, t)\, s = t[s/x]} \qquad \frac{}{(s, t)_0 = s} \qquad \frac{}{(s, t)_1 = t}$$

It is not hard to show that substitution is an admissible rule. This equational theory is known as $\lambda\beta$ because it captures the theory of β equivalence, in the following sense:

Theorem 13.5.1. *For every pair of terms s and t, $\lambda\beta$ proves $s = t$ if and only if $s \equiv_\beta t$.*

Remember that by Proposition 13.2.2, $s \equiv_\beta t$ is equivalent to saying that s and t have the same β-normal form. So Theorem 13.5.1 says, equivalently, that $\lambda\beta$ proves $s = t$ if and only if $s = t$ have the same β-normal form.

Proof By induction on derivations, it is easy to show that provably equal terms are β-equivalent. For the converse, it suffices to show that the relation $R(s, t)$ that holds if and only if $\lambda\beta$ proves $s = t$ is an equivalence relation containing $\rightarrow_{\beta,1}$, because β-equivalence is defined to be the smallest such relation. Clearly it is an equivalence relation, and rules of the calculus were chosen exactly to yield closure under the clauses defining $\rightarrow_{\beta,1}$. □

We will denote the system with additional axioms $\lambda x.\, (t\, x) = t$ and $((t)_0, (t)_1) = t$ as $\lambda\beta\eta$. Essentially the same proof shows the following:

Theorem 13.5.2. *For every pair of terms s and t, $\lambda\beta\eta$ proves $s = t$ if and only if $s \equiv_{\beta\eta} t$.*

Equivalently, $\lambda\beta\eta$ proves $s = t$ if and only if s and t have the same $\beta\eta$-normal form.

It is the rule "from $s = t$ conclude $\lambda x.\, s = \lambda x.\, t$" that precludes us from viewing either $\lambda\beta$ or $\lambda\beta\eta$ as a first-order system. Combinatory logic fares better in this regard, since it does not rely on any variable binding operations. Remember that we can view an expression $t\, s$ as syntactic sugar for an expression $\mathrm{app}_{\alpha,\beta}(s, t)$ in a many-sorted first-order logic with a sort for each type. In that case, the fourth and fifth rules above are equivalent to the usual equality rules for $\mathrm{app}(s, t)$. We obtain a system CL for combinatorial logic by adding axioms $K\, s\, t = s$ and $S\, s\, t\, u = (s\, u)\, (t\, u)$ at all appropriate types. In the version of CL with products, of course, we also add the axioms $\mathrm{fst}\ (\mathrm{pair}\, s\, t) = s$ and $\mathrm{snd}\ (\mathrm{pair}\, s\, t) = t$.

Theorem 13.5.3. *For every pair of terms s and t, CL proves $s = t$ if and only if $s \equiv_{\mathrm{w}} t$.*

Again, by Proposition 13.2.2, the conclusion of Theorem 13.5.3 is equivalent to saying that s and t have the same weak normal form. We therefore have equational theories tailor-made to reason about β-equivalence and $\beta\eta$-equivalence in the simply typed lambda calculus and weak equivalence in combinatory logic. To help clarify some of the relationships between them, we turn to a discussion of the notion of *extensionality*.

Given an irreducible term t of type $\alpha \rightarrow \beta$ and a variable x of type α, the expressions t and $\lambda x.\, (t\, x)$ are not provably equal in $\lambda\beta$, because they do not have the same β-normal form. Intuitively, however, they denote the same function, since when applied to any s of type α they both reduce to ts. Speaking informally, two functions f and g with common domain and codomain are said to be *extensionally equal* if they have equal values for every input. On the set-theoretic understanding of a function, any two functions that are extensionally equal are in fact equal. In a language with quantifiers and variables ranging over functions, this can be expressed by writing $\forall f, g\, (\forall x\, (f(x) = g(x)) \rightarrow f = g)$. This is known as the axiom of *function extensionality*.

The example in the previous paragraph shows that there are terms u_0 and u_1 such that it is possible to prove $u_0\, s = u_1\, s$ in $\lambda\beta$ for every term s (and, in particular, $u_0\, x = u_1\, x$ for a fresh variable x) but impossible to prove $u_0 = u_1$. A system with this property is said to be *intensional*. Rudolf Carnap distinguished between an expression's *intension* and its *extension* in much the same way that Frege distinguished between its *sense* and its *reference*. Two function expressions are extensionally equivalent if they take the same values for the same arguments. In that case, we say that the two expressions have the same extension, i.e. the same course of values. On the other hand, a function expression's intension, if not exactly its syntactic presentation, would be something closer to it. The distinction makes sense if we think of terms in the systems here as computer programs. We can then think of the extension of a term as its input/output behavior and its intension as the algorithm it represents.

We cannot express the axiom of function extensionality in an equational system, but we can approximate it with the following rule:

$$\frac{s\,x = t\,x}{s = t}\ \zeta$$

In this rule, which is named by the Greek letter *zeta*, we assume that the variable x is not a free variable of s or t. The rule says that we can show that s and t are equal by showing that they return equal values when applied to an arbitrary value x. Perhaps surprisingly, in the context of the simply typed lambda calculus, this rule is equivalent to η:

Proposition 13.5.4. *Over $\lambda\beta$, the rule ζ is equivalent to the axiom schema η for function types, $\lambda x.\,(t\,x) = t$.*

Proof Using the axiom for β equality we have $(\lambda x.\,(t\,x))\,y = t\,y$, from which $\lambda x.\,(t\,x) = t$ follows by ζ. Conversely, from $s\,x = t\,x$ we obtain $\lambda x.\,(s\,x) = \lambda x.\,(t\,x)$ using ξ, and hence $s = t$ using η. □

In other words, $\lambda\beta\eta$ is extensional. In contrast, adding η to CL does not allow us to derive the rule ζ. In fact, according to the definition of lambda abstraction for CL given in Section 13.4, $\lambda x.\,(t\,x)$ is syntactically identical to t when x is not free in t. As a result, adding η as an axiom does absolutely nothing. Deriving ζ from η in Proposition 13.5.4 used the ξ rule in an essential way. We could also add this rule to CL, despite the fact that λ is not a built-in construct. The fact that we have η in CL for free means that, over CL, ζ and ξ are equivalent. It is more natural to take ζ as the more basic rule, and we denote the resulting system CLζ.

Turning to products, in the simply typed lambda calculus, it is not hard to show that the η axiom for products is equivalent to the following ζ-like formulation:

$$\frac{(s)_0 = (t)_0 \qquad (s)_1 = (t)_1}{s = t}$$

In the context of CLζ, we can use the combinatory versions of either the axioms or the rules. To keep the discussion below focused, I will restrict attention to the systems with arrow types only, but it is straightforward to extend these considerations to the systems with products.

The extensional theories $\lambda\beta\eta$ and CLζ are intertranslatable. There is a natural translation mapping terms t in combinatory logic to terms t_λ in the simply typed lambda calculus, mapping K to $\lambda x, y.\,x$, mapping S to $\lambda x, y, z.\,x\,z\,(y\,z)$, mapping variables to variables, and mapping $t\,s$ to $t_\lambda\,s_\lambda$. Conversely, the simulation of the lambda operation described in the previous section maps terms t in the simply typed lambda calculus to terms t_{CL} in combinatory logic. (Since we identify terms up to renaming of bound variables in the lambda calculus, we need to check that the translation respects α-equivalence, but this follows from claim 2 of Proposition 13.4.2.) We leave the following as an exercise:

Theorem 13.5.5. *The translations $t \mapsto t_\lambda$ and $t \mapsto t_{\mathrm{CL}}$ have the following properties:*

1. *For every term t of combinatory logic, $(t_\lambda)_{\mathrm{CL}}$ is syntactically identical to t.*
2. *For every term t of the simply typed lambda calculus, $\lambda\beta\eta$ proves $(t_{\mathrm{CL}})_\lambda = t$.*
3. *If CLζ proves $s = t$, $\lambda\beta\eta$ proves $s_\lambda = t_\lambda$.*
4. *If $\lambda\beta\eta$ proves $s = t$, CLζ proves $s_{\mathrm{CL}} = t_{\mathrm{CL}}$.*

To prove the last claim, it helps to prove first that the translation $t \mapsto t_{\mathrm{CL}}$ preserves substitution, that is, $(t[s/x])_{\mathrm{CL}}$ is syntactically identical to $t_{\mathrm{CL}}[s_{\mathrm{CL}}/x]$. Note that the theorem implies that the converses of 3 and 4 also hold.

The relationship between the intensional theories $\lambda\beta$ and CL is more subtle. The problem is that the lack of extensionality makes these systems sensitive to the way a function is described. To appreciate the difference, let $f\colon \mathbb{N} \to \mathbb{N}$ be the constant 0 function, and define $g\colon \mathbb{N} \to \mathbb{N}$ to be the function such that for every x, $g(x)$ is equal to 1 if there are a, b, c, and n, all less than x, such that a, b, and c are nonzero, $n > 2$, and $a^n + b^n = c^n$, and $g(x)$ is equal to 0 otherwise. By Fermat's last theorem, g is also the constant 0 function, which is to say, f and g are extensionally equal. But the descriptions of f and g are so dramatically different, we may hesitate to view them as the same function. However, if we reject the extensional point of view, we have to think long and hard about what it means to say that two functions *are* the same. For example, if f is defined by $f(x) = x + 1$ and g is defined by $g(x) = 1 + x$, are they the same? The theories $\lambda\beta$ and CL each determine a notion of sameness based on a specific set of reductions. It is not hard to show that if CL proves $s = t$, then $\lambda\beta$ proves $s_\lambda = t_\lambda$. But it is not generally the case that if $\lambda\beta$ proves $s = t$ then CL proves $s_{\mathrm{CL}} = t_{\mathrm{CL}}$. With some effort, we can define a modified translation $t \mapsto t_{\mathrm{CL}'}$ and a strengthening CL′ of CL such that $\lambda\beta$ proves $s = t$ if and only if CL′ proves $s_{\mathrm{CL}'} = t_{\mathrm{CL}'}$; see the bibliographical notes at the end of this chapter.

Two further considerations highlight the difference between using an equational theory to reason about an expression's computational behavior and using an equational theory to reason about the functions they denote. First, it is not an easy task to come up with a notion of reduction for combinatory logic that is strongly normalizing, confluent, and also captures provable equivalence in CLζ the way that β-reduction, $\beta\eta$-reduction, and weak reduction capture provable equivalence in $\lambda\beta$, $\lambda\beta\eta$, and CL, respectively. Second, we will see in the next section that it is harder to reason about models of the simply typed lambda calculus without extensionality, given that such a model has to satisfy the ξ rule but not necessarily η. In contrast, using Theorem 13.5.5, the theory $\lambda\beta\eta$ can be understood in terms of first-order logic, which has a more straightforward model theory. So CLζ captures an extensional worldview at the expense of diverging from a straightforward computational interpretation, whereas $\lambda\beta$ is faithful to the natural computational interpretation but harder to think about in model-theoretic terms. For more information, see the bibliographical notes.

Exercises

13.5.1. Show that we can replace the combination of η and ξ with ζ in $\lambda\beta\eta$.

13.5.2. Show that the mapping $t \to t_{\mathrm{CL}}$ from the simply typed lambda calculus to combinatory logic respects α-equivalence. To that end, you will have to rely on the definition of substitution defined in Section 1.6.

13.5.3. Show that the mapping $t \to t_{\mathrm{CL}}$ also respects substitution, i.e. that $(t[s/x])_{\mathrm{CL}}$ is syntactically identical to $t_{\mathrm{CL}}[s_{\mathrm{CL}}/x]$.

13.5.4. Prove Theorem 13.5.5.

13.6 First-Order Theories and Models

We now turn to the semantics of the simply typed lambda calculus and simply typed combinatory logic. The good news is that we already know what a model of CL is, given that we can view CL as an equational theory in a many-sorted first-order logic. A first-order model of such a theory consists of a set D_α for each type α, interpretations $k_{\alpha,\beta}$ and $s_{\alpha,\beta,\gamma}$ of the constants $K_{\alpha,\beta}$ and $S_{\alpha,\beta,\gamma}$, and interpretations of the functions $\mathrm{app}_{\alpha,\beta}$ such that the axioms of combinatory logic are satisfied. In other words, using $\cdot$ to denote the interpretations of app, we require

- $k \cdot u \cdot v = u$ for every u in $D_{\alpha\to\beta}$ and v in D_α, and
- $s \cdot u \cdot v \cdot w = u \cdot w \cdot (v \cdot w)$ for every u in $D_{\alpha\to\beta\to\gamma}$, v in $D_{\alpha\to\beta}$, and w in D_α.

We will continue to restrict attention to the theories with function types only, but extending the semantics to theories with products is straightforward. A structure with sets D_α and functions interpreting $\mathrm{app}_{\alpha,\beta}$ is called an *applicative structure*, and an applicative structure with interpretations of each K and S with the properties above is called a *combinatory algebra*. (These notions are also defined for the untyped lambda calculus and untyped combinatory logic, and when there is danger of confusion, it would be more precise to call these *typed applicative structures* and *typed combinatory algebras*. Much of what I say here carries over to the untyped setting, but since we are focusing on the simply typed versions, I will drop the word "typed.") Remember that if we want these to be models of the corresponding first-order theories, we need to require that each set D_α is nonempty.

Suppose we start with basic types $A, B, C, \ldots$, and assign to each a set of elements $D_A, D_B, D_C, \ldots$. The *full set theoretic model* is the combinatory algebra that recursively interprets each type $\alpha \to \beta$ as set of all functions from the interpretation of α to the interpretation of β.

In Section 5.2 we considered the *term model* for any first-order language, which, in the many-sorted setting, assigns to each basic sort the set of all equivalence classes of closed terms of that sort, where two terms are considered equivalent if they are provably equal in the equational theory in question. Such models can always be expanded by adding additional constants. In the context of combinatory logic, the term model is usually taken to consist of all *open terms* modulo provable equivalence, essentially treating the variables as constants. In that case, we still have $[\![t]\!]_\sigma = t$ for the assignment σ that interprets each variable as itself, and hence CL proves an equation $s = t$ if and only if $s = t$ holds in the model, under this truth assignment. In other words, we have:

Theorem 13.6.1. *Let s and t be terms of combinatory logic. The following are equivalent:*

1. CL *proves* $s = t$.
2. $s = t$ *is true in every combinatory algebra, under any assignment of values to the free variables.*
3. $[\![s]\!]_\sigma = [\![t]\!]_\sigma$ *in the term model for* CL, *where σ interprets each variable as itself.*

Here and below, when we speak of an assignment of values to the free variables of combinatory logic, we implicitly mean an assignment of a value in D_α to each variable of type α. The implication from 1 to 2 expresses the soundness of CL, which follows from the fact that a combinatory algebra is exactly a first-order model of the equational theory CL. The implication from 2 to 3 is immediate, and the implication from 3 to 1 follows from the definition of the term model. Thus we have soundness and completeness of CL with respect to the semantics.

Suppose we are given a combinatory algebra $((D_\alpha), (\mathrm{app}_{\alpha\to\beta}), (k_{\alpha,\beta}), (s_{\alpha,\beta,\gamma}))$, where α, β, and γ range over types. Let u be an element of $D_{\alpha\to\beta}$ for some α and β. Then u gives rise to the function from D_α to D_β which maps any v in D_α to $u \cdot v$. Such a function is said to be *representable* in the model. There is nothing in the definition of a combinatory algebra that rules out the possibility that two distinct elements u and u' of $D_{\alpha\to\beta}$ give rise to the same function. A combinatory algebra is said to be *extensional* if that does not happen; in other words, if, for every pair of types α and β and every pair of elements u and u' in $D_{\alpha\to\beta}$, whenever $u \cdot v = u' \cdot v$ for every v in D_α, $u = u'$.

Theorem 13.6.2. *Let s and t be terms of combinatory logic. The following are equivalent:*

1. *CLζ proves $s = t$.*
2. *$s = t$ is true in every extensional combinatory algebra, under any assignment of values to the free variables.*
3. *$[\![s]\!]_\sigma = [\![t]\!]_\sigma$ in the term model for CLζ, where σ interprets each variable as itself.*

The implication from 1 to 2 expresses the soundness of CLζ with respect to extensional models. To see that any such model validates the ζ rule, suppose $[\![s\,x]\!]_\sigma = [\![t\,x]\!]_\sigma$ for any assignment σ to the free variables, where s and t have type $\alpha \to \beta$ and x is not free in s or t. We need to show that $[\![s]\!]_\sigma = [\![t]\!]_\sigma$ for any σ. By extensionality, it suffices to show $[\![s]\!]_\sigma \cdot v = [\![t]\!]_\sigma \cdot v$ for any v in D_α. But if we let σ' be the modification of σ that maps x to v, we have

$$[\![s]\!]_\sigma \cdot v = [\![s\,x]\!]_{\sigma'} = [\![t\,x]\!]_{\sigma'} = [\![t]\!]_\sigma \cdot v,$$

as required. The implication from 2 to 3 is again immediate, and the implication from 3 to 1 follows from the definition of the term model.

Recursively identifying each element $u \in D_{\alpha\to\beta}$ with the function from D_α to D_β that it represents, we have that every extensional combinatory algebra is isomorphic to a submodel of the full set-theoretic combinatory algebra over the interpretations of the basic types. This submodel contains all the interpretations of the constants k and s and is closed under function application, and, conversely, any such submodel of the full set-theoretic model is clearly an extensional combinatory algebra. Moreover, via the equivalence of CLζ and $\lambda\beta\eta$ described in the last section, every term of the simply typed lambda calculus can be interpreted in any extensional combinatory algebra, and Theorem 13.6.2 carries over:

Theorem 13.6.3. *Let s and t be terms of the simply typed lambda calculus. The following are equivalent:*

1. *$\lambda\beta\eta$ proves $s = t$.*
2. *$s = t$ is true in every extensional combinatory algebra, under any assignment of values of the right type to the free variables.*
3. *$[\![s]\!]_\sigma = [\![t]\!]_\sigma$ in the term model for $\lambda\beta\eta$, where σ interprets each variable as itself.*

What about the intensional theory $\lambda\beta$? Because λ is a binding operation, we can no longer use first-order model theory. The challenge is then to come up with a notion of model that validates the rule ξ but does not go so far as to validate η. This can be done in such a way that we get a sound and complete semantics for $\lambda\beta$, but the details are subtle, and pursuing this will take us too far afield. We will therefore stick to combinatory logic when we want to consider an intensional theory of functions in semantic terms.

Using combinatory logic also allows us to add quantifiers and stay within the realm of first-order logic. Specifically, we define *first-order* CL to be the many-sorted theory in the language of combinatory logic with the axioms $\forall x, y\, (K\, x\, y = x)$ and $\forall x, y, z\, (S\, x\, y\, z = x\, z\, (y\, z))$. In this setting, it is natural to express the axiom schema (ext) of extensionality as follows:

$$\forall x, x' : \alpha \to \beta\, (\forall y : \alpha\, (x\, y = x'\, y) \to x = x').$$

Any model of CL with at least one element in the interpretation of each type is a model of first-order CL as well, since it satisfies the axioms for K and S. Similarly, any extensional model of CL is a model of first-order CL with (ext). This yields the following:

Theorem 13.6.4. *First-order* CL *is a conservative extension of* CL, *and first-order* CL + (ext) *is a conservative extension of* CLζ.

There are easy model-theoretic proofs: if something is not provable in CL or CLζ, respectively, it is false in the term model, and hence false in a model of the corresponding first-order theory.

Notice that (ext) implies that for any two functions f and g of type $\alpha \to \beta$, $f =_{\alpha \to \beta} g$ if and only if $\forall x\, (f\, x = g\, x)$. The forward direction is just the substitution property for equality, and the converse direction is (ext). Similarly, the corresponding extensionality rule for products says that for any two objects p and q of type $\alpha \times \beta$, $p = q$ if and only if $(p)_0 = (q)_0$ and $(p)_1 = (q)_1$. As a result, in the extensional versions of the theories, we can take higher-type equality to be defined, ultimately, in terms of equality at basic types. We can even restrict the language to equality at basic types, in which case, the equality axioms consist of the usual equality axioms at the basic types, together with the substitution axioms for higher-type equality, expressed in terms of equality at basic types.

There are interesting models between term models and full set-theoretic models. For example, consider simple types over a single basic type NAT, intended to denote the natural numbers. Rather than use all the functions from $\mathbb{N} \to \mathbb{N}$, we can interpret NAT $\to$ NAT by indices for (total) computable functions, i.e. indices e such that φ_e is total. We can then interpret (NAT $\to$ NAT) $\to$ NAT to be indices e for partial computable functions φ_e that take an index, f, for any total computable function from $\mathbb{N}$ to $\mathbb{N}$ and return a natural number. Continuing in that way gives a model known as HRO, the *hereditarily recursive operations*. More precisely, define $\mathrm{HRO}_{\mathrm{NAT}}$ to be $\mathbb{N}$, and recursively define $\mathrm{HRO}_{\alpha \to \beta}$ to be the set of indices e for computable functions φ_e such that, for each element f of HRO_α, $\varphi_e(f)$ is defined and is an element of HRO_β. Application $e \cdot f$ is defined to be $\varphi_e(f)$. Choosing suitable indices to represent K and S at the relevant types, we obtain a model of CL. This model is intensional because different indices can represent the same computable function. In other words, it is not a model of (ext).

There is a general method for cutting an intensional model down to an extensional one by picking out the objects that, hereditarily, respect extensional equivalence, and then quotienting by that equivalence relation. In fact, we can define both the *hereditarily extensional* objects of each type and the necessary equivalence relations using first-order formulas $\mathrm{HE}_\alpha(x)$ and $x \equiv_\alpha y$, where the variables x and y have type α. For each type α, define formulas $\mathrm{HE}_\alpha(x^\alpha)$ and $x^\alpha \equiv y^\alpha$ in the language of CL recursively, as follows:

- If α is a basic type and x and y have type α, then $\mathrm{HE}_\alpha(x)$ is $\top$ and $x \equiv_\alpha y$ is the formula $x = y$.

- Given x and y of type α, $\mathrm{HE}_{\alpha\to\beta}(x)$ is the formula

$$\forall z, z' \, (z \equiv_\alpha z' \to x\, z \equiv_\beta x\, z')$$

and $x \equiv_{\alpha\to\beta} y$ is the formula

$$\mathrm{HE}_{\alpha\to\beta}(x) \wedge \mathrm{HE}_{\alpha\to\beta}(y) \wedge \forall z \, (\mathrm{HE}_\alpha(z) \to x\, z \equiv_\beta y\, z).$$

In any model $\mathfrak{M}$ of CL, elements satisfying the formula $\mathrm{HE}_{\alpha\to\beta}$ represent functions that respect the equivalence relations at α and β, and two such elements are themselves equivalent if they return equivalent values on all inputs. Elements satisfying $\mathrm{HE}_\alpha(x)$ are said to be *hereditarily extensional of type* α. We obtain a model $\mathfrak{M}'$ of CL + (ext) by first restricting to the hereditarily extensional elements and then quotienting by the relation defined by $\equiv_\alpha$. The fact that we can define the relations in first-order logic, however, means we get something stronger than a model-theoretic construction, namely, a syntactic interpretation of CL + (ext) in CL.

Proposition 13.6.5. *We have the following:*

- *For each* α, CL *proves that* $\equiv_\alpha$ *is an equivalence relation on the set of elements satisfying* HE_α, *and that this equivalence relation respects application:*

$$\forall x, x' : \alpha \to \beta \, \forall y, y' : \alpha \, (x \equiv x' \wedge y \equiv y' \to x\, y \equiv x'\, y').$$

- CL *proves that the combinators K and S satisfy* HE.
- *Let* α *be a type of level 1, i.e. a type of the form* $\beta_0 \to \cdots \beta_{k-1} \to \gamma$ *where* $\beta_0, \ldots, \beta_{k-1}$ *and* γ *are basic types. Then* CL *proves that every element* x *of type* α *satisfies* HE_α.
- *For every* α, CL + (ext) *proves* $\forall x^\alpha \, \mathrm{HE}(x^\alpha)$, *and* $\forall x^\alpha, y^\alpha \, (x \equiv y \leftrightarrow x = y)$.

We can thus define a translation A to A^* from the language of CL to itself, replacing quantified formulas $\forall x^\alpha \, A$ and $\exists x^\alpha \, A$ by $\forall x^\alpha \, (\mathrm{HE}(x) \to A^*)$ and $\exists x^\alpha \, (\mathrm{HE}(x) \wedge A^*)$, respectively. Proposition 13.6.5 implies the following:

Corollary 13.6.6. CL + (ext) *proves a first-order formula* A *with free variables* $x_0, \ldots, x_{n-1}$ *if and only if* CL *proves*

$$\mathrm{HE}(x_0) \wedge \cdots \wedge \mathrm{HE}(x_{n-1}) \to A^*.$$

Moreover, if A *is any formula with only equality at basic types and quantifiers and variables at level at most* 1, *then* CL *proves* $A \leftrightarrow A^*$.

In model-theoretic terms, we have the following.

Corollary 13.6.7. *Let* $\mathfrak{M}$ *be any combinatory algebra. Let* $\mathfrak{M}'$ *be the result of restricting* $\mathfrak{M}$ *to the hereditarily extensional objects and then quotienting by extensional equivalence. Then* $\mathfrak{M}'$ *is an extensional combinatory algebra, and if* A *is a sentence with only equality at basic types and quantifiers and variables at level at most* 1, *then* $\mathfrak{M}'$ *satisfies* A *if and only if* $\mathfrak{M}$ *does.*

Moreover, since $\mathfrak{M}'$ is extensional, it is isomorphic to a submodel of the full set-theoretic model over the interpretation of the basic types. If $\mathfrak{M}$ is the combinatory algebra HRO described above, the corresponding submodel of the set-theoretic model is known as the *hereditarily effective operations*, denoted HEO.

Exercise

13.6.1. Prove Proposition 13.6.5.

13.7 Primitive Recursive Functionals

Suppose we include among the basic types of the simply typed lambda calculus a type NAT intended to denote the natural numbers, with constants $0\colon \mathrm{NAT}$ and $\mathrm{succ}\colon \mathrm{NAT} \to \mathrm{NAT}$. We can add a principle of definition by primitive recursion by adding a new term construction:

- For every type α and terms $f\colon \alpha$, $g\colon \mathrm{NAT} \to \alpha \to \alpha$, and $t\colon \mathrm{NAT}$, $(R\ f\ g\ t)$ is a term of type α.

The meaning of the new term is given by its reduction rules:

$$(R\ f\ g\ 0) \rhd f$$
$$(R\ f\ g\ (\mathrm{succ}\ t)) \rhd g\ t\ (R\ f\ g\ t).$$

In the literature, this is often called ι-reduction, where the Greek letter is *iota*. In the context of combinatory logic, it is more natural to take R to be a constant of type $\alpha \to (\mathrm{NAT} \to \alpha \to \alpha) \to \mathrm{NAT} \to \alpha$. In that case, $R\ f\ g$ denotes the function h defined by

$$h\,0 = f, \quad h\,(\mathrm{succ}\,x) = g\,x\,(h\,x),$$

and the constant R is called a *recursor*. Because the return type, α, is an arbitrary type, there is no need to include side parameters in the defining schema. To define a function h satisfying

$$h\,0\,\vec{z} = u\,\vec{z}, \quad h\,(\mathrm{succ}\,x)\,\vec{z} = v\,x\,(h\,x\,\vec{z})\,\vec{z},$$

take $f = u$ and $g = (\lambda x, w, \vec{z}.\ v\,x\,(w\,\vec{z})\,\vec{z})$ in the schema above, so that $h\,0 = u$ and

$$h\,(\mathrm{succ}\,x) = (\lambda x, w, \vec{z}.\ v\,x\,(w\,\vec{z})\,\vec{z})\,x\,(h\,x) = \lambda\vec{z}.\ v\,x\,(h\,x\,\vec{z})\,\vec{z}.$$

Then $h\,0\,\vec{z} = u\,\vec{z}$ and $h\,(\mathrm{succ}\,x)\,\vec{z} = v\,x\,(h\,x\,\vec{z})\,\vec{z}$, as required.

Our proofs of strong normalization and confluence extend to both formulations. The following lemma handles strong normalization for the simply typed lambda calculus:

Lemma 13.7.1. *We have the following:*

1. *The constants* 0 *and* succ *are strongly computable.*
2. *If f, g, and t are strongly computable, then so is $(R\ f\ g\ t)$.*

Proof The constant 0 is irreducible, and hence strongly normalizing. Since NAT is a basic type, this means that 0 is strongly computable.

To handle succ, suppose t is strongly computable of type NAT, which is to say, it is strongly normalizing. We need to show that $\mathrm{succ}\,t$ is also strongly normalizing. But this follows from the fact that any infinite sequence of reductions starting from $\mathrm{succ}\,t$ would have to be of the form

$$\mathrm{succ}\,t \to_1 \mathrm{succ}\,t_0 \to_1 \mathrm{succ}\,t_1 \to_1 \mathrm{succ}\,t_2 \to_1 \cdots$$

corresponding to an infinite sequence $t, t_0, t_1, t_2, \ldots$ of one-step reductions from t.

For the second claim, suppose f, g, and t are strongly computable. Since t has type NAT, it is strongly normalizing. Since $(R\ f\ g\ t)$ is neutral, it suffices to show that every one-step reduction is strongly computable. We show this by induction on $h(f)+h(g)+h(g)+h'(t)$, where $h'(t)$ is the length of the normal form of t. These are all the possible one-step reductions of $R\ f\ g\ t$:

- $(R\ f'\ g\ t)$, where $f \rightarrow_1 f'$
- $(R\ f\ g'\ t)$, where $g \rightarrow_1 g'$
- $(R\ f\ g\ t')$, where $t \rightarrow_1 t'$
- f, if t is 0
- $g\ t\ (R\ f\ g\ t')$, if t is $\mathrm{succ}\ t'$.

In the first three cases, we have $h(f') < h(f)$, $h(g') < h(g)$, and $h(t') < h(t)$, respectively. In the fourth case, f is strongly computable by hypothesis. In the last case, notice that if t' has normal form $\hat{t}$, then t has normal form $\mathrm{succ}\ \hat{t}$, and so $h'(t') < h'(t)$. So, by the inductive hypothesis, $(R\ f\ g\ t')$ is strongly computable. Since g and t are also strongly computable by hypothesis, $g\ t\ (R\ f\ g\ t')$ is strongly computable, as required. □

We leave the task of proving confluence and extending strong normalization and confluence to combinatory logic as exercises. The full set-theoretic model in which NAT interpreted by $\mathbb{N}$ is an extensional model of the calculus, where each R is interpreted as the corresponding set-theoretic function. The interpretations of the closed terms in this model are called the *primitive recursive functionals of finite type*.

We also have that HRO is a model of the intensional theory based on combinatory logic, where each recursor R is interpreted by the function that maps recursive functions f and g to the recursive function r defined by primitive recursion in the obvious way, so that $r(0)=f$ and $r(n+1)=g \cdot n \cdot r(n)$. As a result, HEO is a model of the extensional theory.

Identifying functions from the natural numbers to the natural numbers of arbitrary arity with functionals of type level 1 (that is, of type $\mathrm{NAT} \rightarrow \cdots \rightarrow \mathrm{NAT} \rightarrow \mathrm{NAT}$), it should be clear that every primitive recursive function, in the usual sense, is a primitive recursive functional. After all, the type 1 functionals contain 0, the successor function, and projections, and are closed under composition and primitive recursion. You might guess that the converse is true, i.e. that every primitive recursive functional of type level 1 is a primitive recursive function. But this is not the case: the ability to use higher-type recursion means that we can define more functions. To see this, first notice that for each α we can use primitive recursion to define an iteration functional for functions of type $\alpha \rightarrow \alpha$:

$$\mathrm{iter}_\alpha\ f\ 0 = \mathrm{id}_\alpha$$
$$\mathrm{iter}_\alpha\ f\ (n+1) = \lambda x.\ f\ (\mathrm{iter}_\alpha\ f\ n\ x),$$

where $\mathrm{id}_\alpha = \lambda x.\, x$ is the identity function. We can then define the fast-growing hierarchy defined in Section 8.6:

$$f\ 0 = \mathrm{succ}$$
$$f\ (n+1) = \lambda x.\ \mathrm{iter}_{\mathrm{NAT} \rightarrow \mathrm{NAT}}\ (f\ n)\ (x+2)\ (x+1).$$

Then for each natural number n, $f\ n\colon \mathrm{NAT} \rightarrow \mathrm{NAT}$ is the function f_n defined in Section 8.6, and the function $f_\omega = \lambda x.\ f\ x\ x$ is a type 1 primitive recursive functional that is not primitive recursive. In fact, both instances of primitive recursion needed here are mild, returning objects of type $\mathrm{NAT} \rightarrow \mathrm{NAT}$. In a similar manner, we can show that the evaluation function $\mathrm{eval}(x, s)$ for primitive recursive functions that was described in Section 8.6 is definable in the typed calculi with primitive recursion. In Chapter 14, we will have other ways of understanding the additional strength given by higher-type primitive recursion.

We can obtain equational theories and first-order theories with recursors as in the last section. In Section 14.4, we will consider a quantifier-free theory known as *Gödel's* T, which axiomatizes the primitive recursive functionals much the way that the theory PRA axiomatizes the primitive recursive functions. We will also consider theories HA^{ω} and PA^{ω}, which extend the first-order version of CL over a single basic type NAT with constants 0, succ, and recursors R_{α}. We add the following axioms:

- $0 \neq 1$
- $R\, f\, g\, 0 = f$
- $R\, f\, g\, (x+1) = g\, x\, (R\, f\, g\, x)$
- the schema of induction:

$$A(0) \wedge \forall x\, (A(x) \to A(x+1)) \to \forall x\, A(x),$$

  where A is any formula in the language, possibly with parameters other than the ones shown.

The intuitionistic version of the resulting theory is called HA^{ω}, and the classical version is called PA^{ω}. The versions with the additional principle (ext) are denoted E-HA^{ω} and E-PA^{ω}. These clearly extend HA and PA, assuming we identify the basic symbols of those theories with suitable primitive recursive functionals. But we can say more: even with the addition of the principle of extensionality, the finite-type theories prove the same first-order theorems as their first-order counterparts.

Proposition 13.7.2. *The model* HEO *can be defined in* HA *using the predicates* HE *described in the last section.* HA *proves each axiom of* HA^{ω} *when it is interpreted as a statement about* HEO *and equality is interpreted by the relation* $\equiv$.

I will leave it to you to work out the details. The key point is that induction in HA is used to prove the totality of the recursors.

Theorem 13.7.3. PA^{ω} + (ext) *is conservative over* PA, *and* HA^{ω} + (ext) *is conservative over* HA.

Proof In each case, Proposition 13.7.2 provides an interpretation of the higher-type theory in the first-order theory that preserves first-order statements. □

So PA^{ω} and HA^{ω} are not more powerful than their first-order counterparts, but they are natural theories for reasoning about higher-type functions and functionals. We will have much more to say about these theories in Chapter 14.

Exercises

13.7.1. Prove confluence of the simply typed lambda calculus with primitive recursors.

13.7.2. Prove strong normalization of simply typed combinatory logic with primitive recursors.

13.7.3. Suppose we decide to augment the primitive recursive functionals with another type, BOOL, with constants $\top, \bot$: BOOL. Write down the type of an appropriate recursor, and write down the appropriate reduction rules.

13.7.4. Use higher types to define functions that grow even faster than Ackermann's function. (See Exercise 13.1.6.)

13.7.5. Use primitive recursive functionals to define an evaluation function for primitive recursive functions. (See Section 10.5.)

13.7.6. Spell out some of the details in the proof of Proposition 13.7.2.

13.8 Propositions as Types

We have already alluded to similarities between the simply typed lambda calculus and natural deduction for propositional logic. Recall the propositional rules for implication and products:

$$\dfrac{\begin{array}{c}\overline{A}\\ \vdots \\ B\end{array}}{A \to B} \qquad \dfrac{A \to B \qquad A}{B} \qquad \dfrac{A \qquad B}{A \wedge B} \qquad \dfrac{A \wedge B}{A} \qquad \dfrac{A \wedge B}{B}$$

These bear a distinct resemblance to the construction rules for the terms in the lambda calculus:

$$\dfrac{\begin{array}{c}\overline{x\colon \alpha}\\ \vdots \\ t\colon \beta\end{array}}{\lambda x.\, t\colon \alpha \to \beta} \qquad \dfrac{t\colon \alpha \to \beta \qquad s\colon \alpha}{t\, s\colon \beta} \qquad \dfrac{s\colon \alpha \qquad t\colon \beta}{(s, t)\colon \alpha \times \beta} \qquad \dfrac{t\colon \alpha \times \beta}{(t)_0\colon \alpha} \qquad \dfrac{t\colon \alpha \times \beta}{(t)_1\colon \beta}$$

We can modify our formulation of the simply typed calculus to make the correspondence more exact. Instead of taking each variable to come equipped with a type, define a *context* Γ to be a finite set of assignments $x\colon \alpha$ of variables to types, and write $\Gamma \vdash t\colon \alpha$ to indicate that t has type α in the context Γ. In that formulation, the rules are as follows:

$$\dfrac{\Gamma, x\colon \alpha \vdash t\colon \beta}{\Gamma \vdash \lambda x.\, t\colon \alpha \to \beta} \qquad \dfrac{\Gamma \vdash t\colon \alpha \to \beta \qquad \Gamma \vdash s\colon \alpha}{\Gamma \vdash t\, s\colon \beta}$$

$$\dfrac{\Gamma \vdash s\colon \alpha \qquad \Gamma \vdash t\colon \beta}{\Gamma \vdash (s, t)\colon \alpha \times \beta} \qquad \dfrac{\Gamma \vdash t\colon \alpha \times \beta}{\Gamma \vdash (t)_0\colon \alpha} \qquad \dfrac{\Gamma \vdash t\colon \alpha \times \beta}{\Gamma \vdash (t)_1\colon \beta}$$

If we replace $\times$ and $\to$ by the corresponding propositional connectives, the types α become propositional formulas, and the expression $t\colon \alpha$ becomes notation for a proof of α. The variable $x\colon \alpha$, representing the hypothesis that α holds, is analogous to the labels that we used in Chapter 2 to cancel hypotheses. In the sequent formulation of natural deduction in Chapter 2, we did not label hypotheses or allow multiple instances of the same formula, but we could have, and with those minor changes, the presentations agree.

The correspondence I have just described is known as the *Curry–Howard correspondence*. It identifies propositions with types and natural deduction proofs with terms in the typed lambda calculus, and is hence also known as the *propositions as types* interpretation. The correspondence is useful for transporting ideas between calculi for terms and calculi for data. For example, notions of reduction in the simply typed lambda calculus can be interpreted as reductions for natural deduction derivations. The rules for β-reduction correspond to elimination of detours:

$$\dfrac{\dfrac{\begin{array}{c}\overline{A}\\ \vdots\\ B\end{array}}{A \to B} \qquad \begin{array}{c}\vdots\\ A\end{array}}{B} \qquad \text{reduces to} \qquad \begin{array}{c}\vdots\\ A\\ \vdots\\ B\end{array}$$

and:

$$\dfrac{\dfrac{\begin{array}{c}\vdots\\ A\end{array} \qquad \begin{array}{c}\vdots\\ B\end{array}}{A \wedge B}}{A} \qquad \text{reduces to} \qquad \begin{array}{c}\vdots\\ A\end{array}$$

You should consider the corresponding η-reductions.

Proofs in normal form have properties similar to cut-free proofs. There are various relationships between cut elimination and normalization, and the two can be used in similar ways. Normalization works best with intuitionistic theories, but there are extensions to classical proof systems as well. Here we will not pursue the theory of normalization in connection with deductive systems, but the bibliographical notes include some pointers to the literature.

In the other direction, we can carry insights from proof calculi back to term calculi. The connective for disjunction $A \vee B$ in propositional logic corresponds to a *sum type* construction $\alpha + \beta$, whose elements consist of either an element of α or an element of β, tagged to indicate which is the case. In correspondence with the propositional induction and elimination rules, we have the following term constructions:

$$\dfrac{t\colon \alpha}{\mathrm{inl}\, t\colon \alpha + \beta} \qquad \dfrac{t\colon \beta}{\mathrm{inr}\, t\colon \alpha + \beta} \qquad \dfrac{r\colon \alpha \vee \beta \qquad \begin{array}{c}\overline{x\colon \alpha}\\ \vdots\\ s\colon \gamma\end{array} \qquad \begin{array}{c}\overline{y\colon \beta}\\ \vdots\\ t\colon \gamma\end{array}}{\mathrm{cases}\, xy.\, r\, s\, t\colon \gamma}$$

The constructors inl and inr, short for *introduction left* and *introduction right*, insert elements of α and β, respectively, into $\alpha + \beta$. The cases construction takes an element r of the sum type, and splits on cases: if r is a tagged element x of α, the expression returns s, and if it is a tagged element y of β, the expression returns t. The variables x and y are both bound by the cases constructor. The reduction rules are as follows:

$$\mathrm{cases}\, xy.\, (\mathrm{inl}\, u)\, s\, t \rhd s[u/x], \qquad \mathrm{cases}\, xy.\, (\mathrm{inr}\, v)\, s\, t \rhd t[v/x].$$

In Section 17.6, we will see that the sum type can be viewed as an instance of an *inductively defined type*, and so these reductions are often considered to be ι-reductions, in the same family as the reductions we associated with the recursors on the natural numbers. They can also be viewed as analogous to β-reduction. I leave it to you to work out the corresponding η rule, which is not as natural.

Normalization is subtle in the presence of disjunctions or sums. The following snippet of a proof seems to have an unnecessary detour:

$$\dfrac{\dfrac{\begin{matrix}\vdots\\ A\vee B\end{matrix}\qquad \dfrac{\dfrac{\begin{matrix}\vdots\\ A\to C\end{matrix}\quad \overline{A}}{C}\qquad \begin{matrix}\vdots\\ D\end{matrix}}{C\wedge D}\qquad \dfrac{\dfrac{\begin{matrix}\vdots\\ B\to C\end{matrix}\quad \overline{B}}{C}\qquad \begin{matrix}\vdots\\ D\end{matrix}}{C\wedge D}}{C\wedge D}}{C}$$

Specifically, the $\wedge$-introduction and $\wedge$-elimination seem gratuitous. The problem is that they are separated by the disjunction elimination. But suppose we move the elimination upward:

$$\dfrac{\begin{matrix}\vdots\\ A\vee B\end{matrix}\qquad \dfrac{\dfrac{\dfrac{\begin{matrix}\vdots\\ A\to C\end{matrix}\quad \overline{A}}{C}\qquad \begin{matrix}\vdots\\ D\end{matrix}}{C\wedge D}}{C}\qquad \dfrac{\dfrac{\dfrac{\begin{matrix}\vdots\\ B\to C\end{matrix}\quad \overline{B}}{C}\qquad \begin{matrix}\vdots\\ D\end{matrix}}{C\wedge D}}{C}}{C}$$

We can then perform a contraction in each branch:

$$\dfrac{\begin{matrix}\vdots\\ A\vee B\end{matrix}\qquad \dfrac{\begin{matrix}\vdots\\ A\to C\end{matrix}\quad \overline{A}}{C}\qquad \dfrac{\begin{matrix}\vdots\\ B\to C\end{matrix}\quad \overline{B}}{C}}{C}$$

For that reason, normalization procedures for proofs usually include such *permutative conversions*.

In the correspondence with propositional logic, the propositional constant $\bot$ corresponds to a type EMPTY, which has no elements. The construction corresponding to *ex falso* produces an element of an arbitrary type α from an element of EMPTY. (There shouldn't be any, just as there should be no proof of falsity, but just as we can obtain proofs of false from inconsistent hypotheses, we can build terms of type EMPTY within a lambda expression.) Corresponding to $\top$, one often introduces a type UNIT with a single constructor, $*:$ UNIT.

What about first-order logic? First-order formulas depend on variables, and the universal and existential quantifiers bind them. If we want to extend the Curry–Howard correspondence, then, we have to allow types to depend on variables as well. This brings us to the realm of *dependent type theory*. Given a type β that depends on the variable x of type α, the analogue of a universally quantified formula is a *Pi type*, $\Pi_{x:\,\alpha}\,\beta$. Given an element t of type $\Pi_{x:\,\alpha}\,\beta$, for any $s:\alpha$, the term $t\,s$ has type $\beta[s/x]$. In other words, t is a function whose return type depends on its argument. Similarly, the analogue of an existentially quantified formula is a *Sigma type*, $\Sigma_{x:\,\alpha}\,\beta$, whose elements consist of pairs (s, t) where s has type α and t has type $\beta[s/x]$. So elements of this type are pairs with the property that the type of the second element depends on the first.

In dependent type theory, we can introduce the rules for primitive recursion on the natural numbers as follows:

$$\frac{\Gamma \vdash f : \alpha[0/x] \qquad \Gamma, n : \text{NAT}, y : \alpha[n/x] \vdash g : \alpha[\text{succ}\, n/x]}{\Gamma \vdash \text{rec}\, ny.\, fg : \Pi_{n:\,\text{NAT}}\, \alpha[n/x]}$$

This corresponds to a natural-deduction formulation of the induction rule. We will discuss dependent type theory in greater detail in Section 17.5.

Bibliographical Notes

General references Good references for the simply typed lambda calculus and simply typed combinatory logic include Girard et al. (1989), Hindley and Seldin (2008), Sørensen and Urzyczyn (2006), and Barendregt et al. (2013). For the untyped versions, see Hindley and Seldin (2008) or Barendregt (1981). The presentation of the simply typed lambda calculus here adopts what is known as Church-style typing, which means that, by definition, every term has a unique type assigned to it. In contrast, Curry-style typing can be viewed as a means of assigning types to a suitable subset of the untyped lambda terms. See Sørensen and Urzyczyn (2006) for a comparison.

Normalization and confluence For an introduction to term rewriting, see Baader and Nipkow (1998). The presentation here draws heavily on Girard et al. (1989) and Hindley and Seldin (2008), especially for the proofs of strong normalization. The proof in Exercises 13.2.2 and 13.2.3 is taken from Joachimski and Matthes (2003).

The confluence of the simply typed lambda calculus does not follow directly from the confluence of the untyped lambda calculus. See the solution to Exercise 3.19 of Sørensen and Urzyczyn (2006), which can be found in Appendix B there, for an explanation.

Intesionality, extensionality, and semantics Hindley and Seldin (2008) is especially good about sorting out the issues raised at the end of Section 13.5. It discusses the semantics of all the theories discussed there, and Chapter 9 provides more information on the correspondences between the lambda calculus and combinatory logic.

Primitive recursive functionals of finite type It is possible to restrict the finite type recursors to obtain a subset of the primitive recursive functionals of finite type with the property that the type 1 functionals are exactly the primitive recursive functions. See Section 5 of Avigad and Feferman (1998) and an analogous construction in Section 6 of Cook and Urquhart (1993).

Propositions as types For more on the Curry–Howard correspondence, see, for example, Sørensen and Urzyczyn (2006), Girard et al. (1989), Troelstra and van Dalen (1988), or Beeson (1985). Girard et al. (1989), Troelstra and van Dalen (1988), and Troelstra and Schwichtenberg (2000) describe normalization procedures for systems with proof systems $\vee$, or, equivalently, sum types. For more on dependent type theory, see Chapter 17 and the bibliographical notes there.

14

Arithmetic and Computation

Chapter 9 introduced a formal axiomatic theory for reasoning about primitive recursive functions and relations, and Chapters 10–12 showed that the language of first-order arithmetic can be used to reason about arbitrary computable functions and relations as well. We have also seen that common operations on terms, formulas, and proofs are primitive recursive. In short, we can reason formally about computation and we can compute with the objects of formal logic.

The goal of this chapter is to develop the relationship between logic and computation in yet another way, namely, by considering the extent to which we can interpret formal derivations in computable terms. From a proof of $\forall x\, \exists y\, A(x, y)$ in a sufficiently constructive theory of the natural numbers, we would expect to be able to extract a computable function f such that $A(n, f(n))$ holds for every n. This chapters explores some of the ways that this expectation is borne out.

14.1 Realizability

Kleene's notion of *realizability* provides one way of associating computational information with formulas in the language of arithmetic. We start by defining a relation *e realizes A* between natural numbers e and sentences A in the language of arithmetic. You can think of the statement "e realizes A" as saying that e represents the computational content of A, encoding information that is implicit in a computational reading of the quantifiers and connectives. The relation is defined inductively as follows:

- If A is atomic, then e realizes A if and only if A holds.
- e realizes $A \wedge B$ if and only if e represents a pair (a, b) such that a realizes A and b realizes B.
- e realizes $A \vee B$ if and only if e represents a pair (i, c) where either $i = 0$ and c realizes A or $i \neq 0$ and c realizes B.
- e realizes $A \to B$ if and only if, given any a realizing A, $\varphi_e(a)$ is defined and realizes B.
- e realizes $\forall x\, A(x)$ if and only if for every n, $\varphi_e(n)$ is defined and realizes $A(n)$.
- e realizes $\exists x\, A(x)$ if and only if e represents a pair (n, a) such that a realizes $A(n)$.

In this definition $A(n)$ would more properly be written $A(\overline{n})$ where $\overline{n}$ is the numeral denoting n, but we will continue our practice of leaving the bar implicit. Remember that φ_e denotes the eth partial computable function, as defined in Section 11.2. Throughout this chapter, it will be convenient to take the theory HA to include primitive recursive arithmetic, so that the atomic formulas assert primitive recursive relations between primitive recursive expressions.

You can think of the statement that A is realizable as saying that A is true with respect to a computational reading of the quantifiers and logical connectives. It follows from the definition that a universal sentence is realizable if and only if it is true, in which case any natural number that is the index of a total computable function realizes it. Also, a formula $\neg A$ is realizable if and only if A is not realizable, and in that case any natural number that is an index of a total computable function realizes $\neg A$.

With respect to the classical, model-theoretic notion of truth, there are sentences that are true but not realizable. For example, using Kleene's primitive recursive T predicate, consider the sentence

$$\forall e\, \exists s\, \forall s'\, (T(e, 0, s') \rightarrow T(e, 0, s)).$$

Because it is a prenexation of the statement $\forall e\, (\exists s'\, T(e, 0, s') \rightarrow \exists s\, T(e, 0, s))$, it is straightforwardly provable in classical first-order logic. But it is not realizable: any realizer would give rise to a total computable function f with the property that $f(e)$ returns a halting computation sequence for Turing machine e on input 0 whenever such a sequence exists, thereby solving the halting problem. As a consequence, the negation of that sentence *is* realizable, and, in fact, realized by every natural number. So there are also sentences that are realizable but classically false.

It is debatable whether we should think of the statement that A is realizable as saying that A is "constructively true" in some sense. At the very least, the claim that there are statements about the natural numbers that are constructively true but classically false might give us pause. We should expect, however, that if A can be proved in a suitably constructive theory, it is realizable. For example, we have the following:

Theorem 14.1.1. *If* HA *proves a sentence A, then A is realizable.*

We will actually prove something stronger than that by formalizing the notion of realizability in the language of arithmetic itself. We can take the clauses above to assign to each formula A in the language of arithmetic another formula $e\,\mathbf{r}\,A$ whose free variables are among those of A and a new variable e, formally expressing the fact that e realizes A. The map which takes A to the formula $e\,\mathbf{r}\,A$ is defined inductively as follows:

- If A is atomic, $e\,\mathbf{r}\,A \equiv A$.
- $e\,\mathbf{r}\,(A \wedge B) \equiv (e)_0\,\mathbf{r}\,A \wedge (e)_1\,\mathbf{r}\,B$.
- $e\,\mathbf{r}\,(A \vee B) \equiv ((e)_0 = 0 \rightarrow (e)_1\,\mathbf{r}\,A) \wedge ((e)_0 \neq 0 \rightarrow (e)_1\,\mathbf{r}\,B)$.
- $e\,\mathbf{r}\,(A \rightarrow B) \equiv \forall u\,(u\,\mathbf{r}\,A \rightarrow \varphi_e(u)\downarrow \wedge\, \varphi_t(u)\,\mathbf{r}\,B)$.
- $e\,\mathbf{r}\,\forall x\, A \equiv \forall x\,(\varphi_e(x)\downarrow \wedge\, \varphi_e(x)\,\mathbf{r}\,A)$.
- $e\,\mathbf{r}\,\exists x\, A \equiv (e)_1\,\mathbf{r}\,A[(e)_0/x]$.

We represent $\varphi_e(x)\downarrow$ by the formula $\exists s\, T(e, x, s)$, and when that holds, we can interpret $A(\varphi_e(x))$ as either $\exists s\,(T(e, x, s) \wedge A(U(s)))$ or $\forall s\,(T(e, x, s) \rightarrow A(U(s)))$. You can check that the realizability interpretation commutes with substitution in the sense that $e\,\mathbf{r}\,(A[t/x])$ is the same formula as $(e\,\mathbf{r}\,A)[t/x]$ assuming that e is a fresh variable, that is, e is different from x and does not occur in A or t. As a result, there is no ambiguity in the expression $e\,\mathbf{r}\,A[t/x]$.

Now that we have formalized the realizability relation, the definition covers formulas as well as sentences. But if A has free variables $\vec{x}$ and u is a fresh variable, then $u\,\mathbf{r}\,A$ holds in the standard model with e assigned to u and $\vec{n}$ assigned to $\vec{x}$ if and only if e realizes $A[\vec{n}/\vec{x}]$. So Theorem 14.1.1 is a consequence of the following:

Theorem 14.1.2. *Suppose* HA *proves* $A(\vec{x})$. *Then there is a closed term* t *such that* HA *proves* $\forall \vec{x}\,(\varphi_t(\vec{x}) \downarrow \wedge (\varphi_t(\vec{x})\ \mathbf{r}\ A(\vec{x})))$.

In the statement of the theorem, if $\vec{x}$ is a k-tuple, $\varphi_t(\vec{x})$ denotes the evaluation of the k-ary computable function with index t. Assuming HA includes symbols for the primitive recursive functions, we will see that the term t is reasonably short and can be read off straightforwardly from a proof of $A(\vec{x})$ in HA. In particular, if A is a sentence, φ_t is a 0-ary computable description of a natural number n. Since HA can prove $\varphi_t = \overline{n}$, it also proves $\overline{n}\ \mathbf{r}\ A$. The difference is that the numeral $\overline{n}$ may be astronomically large. HA can prove the existence of numbers too big to be represented by any small term t in the language of primitive recursive arithmetic, and any number realizing such an existence theorem has to be at least as large. So it helps to keep in mind that Theorem 14.1.2 is stated in terms of computable *descriptions* of realizers rather than their representations as numerals, and it is those descriptions that we reason about in HA.

To prove Theorem 14.1.2, we will reason about the construction of the partial computable function φ_t informally. We will generally leave it implicit that the construction is represented by a term t in the language of primitive recursive arithmetic in such a way that HA can prove that φ_t has the requisite properties. We will make use of the fact that first order primitive recursive arithmetic can prove the following:

1. Each primitive recursive function is computable. In other words, for each primitive recursive function $f(\vec{x})$, there is a closed term t such that PRA proves that for every $\vec{x}$, $\varphi_t(\vec{x}) \downarrow = f(\vec{x})$.
2. The partial computable functions are uniformly closed under composition, primitive recursion, and μ. Here the word *uniformly* means that there are corresponding primitive recursive functions that carry out the operations on indices. For example, there is a primitive recursive function $u(e, f_0, \ldots, f_{k-1})$ such that PRA proves that whenever e is an index for a k-ary partial computable function and $f_0, \ldots, f_{k-1}$ are indices for ℓ-ary partial computable functions, then
$$\varphi_{u(e, f_0, \ldots, f_{k-1})}(\vec{x}) \simeq \varphi_e(\varphi_{f_0}(\vec{x}), \ldots, \varphi_{f_{k-1}}(\vec{x})).$$
3. There is a partial computable evaluation function: there is a closed term t such that PRA proves that for every e and x, $\varphi_t(e, x) \simeq \varphi_e(x)$, and similarly for evaluation of functions of higher arity.
4. The s-m-n theorem: PRA proves $\varphi_{s^m_n(e,\vec{x})}(\vec{y}) \simeq \varphi_e(\vec{x}, \vec{y})$.

Taking $n = 1$ in the last claim and taking t to be an index for the function $\vec{x} \mapsto s^m_n(e, \vec{x})$, we have that for every term e, there is term t such that for every $\vec{x}$,

$$\varphi_t(\vec{x}) \downarrow \wedge\, \varphi_{\varphi_t(\vec{x})}(y) \simeq \varphi_e(\vec{x}, y).$$

So, given a description e of a $(k+1)$-ary partial function $\varphi_e(\vec{x}, y)$, we have a description t of a k-ary function $\varphi_t(\vec{x})$ which, for every $\vec{x}$, returns an index for the unary function $\lambda y.\ \varphi_e(\vec{x}, y)$. We will make use of this below. These facts, incidentally, can be used to formalize the theory of computability introduced in Chapter 11. For example, the recursion theorem, Theorem 11.5.2, can be proved in primitive recursive arithmetic.

Before proving Theorem 14.1.2, we need some definitions. A formula A is said to be *∃-free* if it is in the smallest class of formulas containing the atomic formulas and closed under $\forall$, $\wedge$, and $\to$. (In other words, it is really $\exists$- and $\vee$-free.) Recall also that a formula is *negative* if it is in the smallest class of formulas containing negated atomic formulas and closed under $\forall$, $\wedge$, and $\to$. Since HA proves the law of the excluded middle for atomic formulas, every atomic formula A is equivalent to $\neg\neg A$. So, in HA, every $\exists$-free formula is provably equivalent to one that is negative.

A formula A in the language of arithmetic is said to be *almost negative* or *essentially ∃-free* if it is in the smallest class of formulas containing atomic formulas and formulas of the form $\exists x\, B$, where B is atomic, and closed under $\forall$, $\wedge$, and $\to$. In other words, a formula is almost negative if it does not contain $\exists$ or $\vee$, except for possibly $\exists$ before atomic formulas. Since we are taking the language of HA to include the primitive recursive functions, every Σ_1 formula $A(x)$ is equivalent to one of the form $\exists y\, (f(x,y) = 0)$, so, up to equivalence, the almost negative formulas include all the Π_2 formulas. In the proof of the following proposition, however, it is convenient to restrict to the narrow syntactic definition.

Proposition 14.1.3. *If $A(\vec{x})$ is almost negative with the free variables shown, there is a closed term t such that* HA *proves that the following are pairwise equivalent:*

1. A
2. $\varphi_t(\vec{x}) \downarrow \wedge (\varphi_t(\vec{x})\, \mathbf{r}\, A)$
3. $\exists e\, (e\, \mathbf{r}\, A)$.

In words, if A is almost negative, it is realizable if and only if it is true, in which case it is realized by a canonical realizer $\varphi_t(\vec{x})$.

Proof We define t by recursion on almost negative formulas. If A is atomic, choose t such that $\varphi_t(\vec{x})$ is identically zero. If A is of the form $\exists y\, B(\vec{x}, y)$ with B atomic, choose t such that $\varphi_t(\vec{x}) \simeq \mu y\, B(\vec{x}, y)$. If A is of the form $B \wedge C$, let t_0 and t_1 be the terms recursively assigned to B and C, and choose t such that $\varphi_t(\vec{x}) = (\varphi_{t_0}(\vec{x}), \varphi_{t_1}(\vec{x}))$. If A is of the form $B \to C$, let t' be the term assigned recursively to C, and choose t such that $\varphi_t(\vec{x})$ is an index for the constant function $\lambda u.\, \varphi_{t'}(\vec{x})$. If A is of the form $\forall y\, B(\vec{x}, y)$, let t' be the value assigned recursively to B, and choose t so that $\varphi_t(\vec{x})$ is an index for the function $\lambda y.\, \varphi_{e'}(\vec{x}, y)$.

To show that t behaves as advertised, use induction on A. Since 2 clearly implies 3, we only need to show that 1 implies 2 and that 3 implies 1. The base cases are immediate given the definition of the μ operator. I will carry out the inductive step where A is of the form $B \to C$. The others are similarly straightforward.

To prove 1 implies 2, suppose $B \to C$. Since $\varphi_t(\vec{x})$ is a syntactic operation on indices (formally, defined with an s-m-n function), it is a total function. To show that it realizes $B \to C$, suppose $u\, \mathbf{r}\, B(\vec{x})$. Then, by the inductive hypothesis, $B(\vec{x})$ is true, and hence $C(\vec{x})$. By definition, $\varphi_{\varphi_t(\vec{x})}(u) = \varphi_{t'}(\vec{x})$, where $\varphi_{t'}(\vec{x})$ is the canonical realizer of $C(\vec{x})$. By the inductive hypothesis, $\varphi_{t'}(\vec{x})$ realizes $C(\vec{x})$. So $\varphi_t(\vec{x})$ realizes $B \to C$.

To prove 3 implies 1, suppose $e\, \mathbf{r}\, B \to C$, and suppose $B(\vec{x})$. By the inductive hypothesis, there is a u such that $u\, \mathbf{r}\, B(\vec{x})$, and hence $\varphi_e(u) \downarrow$ and $\varphi_e(u)\, \mathbf{r}\, C(\vec{x})$. Again by the inductive hypothesis, this implies $C(\vec{x})$. So we have $B(\vec{x}) \to C(\vec{x})$, as required. □

A straightforward induction on formulas also shows that for every formula A, the formula $e \mathbf{r} A$ is almost negative. With Proposition 14.1.3, this implies that HA proves that a formula is realizable if and only it is realizably realizable:

$$\exists e\,(e\,\mathbf{r}\,A) \leftrightarrow \exists e'\,(e'\,\mathbf{r}\,\exists e\,(e\,\mathbf{r}\,A)).$$

Proposition 14.1.3 holds, a fortiori, for negative formulas. The preceding proof shows that the only real content to a realizer for an almost negative formula comes from an existential quantifier in the base case, so, for a negative formula, the realizer has no content at all. It would be nice if we could say that if a negative formula is realized then any number realizes it, but that is not quite true. For example, for an implication, the realizer needs to be an index for a total computable function. In Section 14.3, we will be able to make precise the intuition that realizers for negative formulas have no content.

We can now prove Theorem 14.1.2. We will show by induction on derivations that whenever a sequent $A_0, \ldots, A_{k-1} \vdash B$ is derivable in natural deduction, assuming all the free variables are among $\vec{x}$, there is an index t such that HA proves that for every $\vec{x}$ and $a_0, \ldots, a_{k-1}$, if each a_i realizes A_i, then $\varphi_t(\vec{x}, \vec{a})$ is defined and realizes B.

Let us start with the rules for the propositional connectives. Suppose the last inference is implication introduction, yielding a proof of $\Gamma \vdash A \to B$ from $\Gamma, A \vdash B$. By the inductive hypothesis, we have an index t' such that for every $\vec{x}$, $\varphi_{t'}(\vec{x}, \vec{c}, a)$ realizes B whenever a realizes A and $\vec{c}$ realizes the other hypotheses in Γ. Let $\varphi_t(\vec{x}, \vec{c})$ return an index for $\lambda a.\ \varphi_{t'}(\vec{x}, \vec{c}, a)$. Then for every $\vec{x}$, if $\vec{c}$ realizes the hypotheses in Γ, $\varphi_t(\vec{x}, \vec{c})$ realizes $A \to B$, as required. Implication elimination goes the other way: if $\varphi_{t'}(\vec{x}, \vec{c})$ is defined and realizes $A \to B$ and $\varphi_{t''}(\vec{x}, \vec{c})$ is defined and realizes A, then $\varphi_{\varphi_{t'}(\vec{x}, \vec{c})}(\varphi_{t''}(\vec{x}, \vec{c}))$ realizes B. Using a computable evaluation function, we can find a term t such that

$$\varphi_t(\vec{x}, \vec{c}) \simeq \varphi_{\varphi_{t'}(\vec{x}, \vec{c})}(\varphi_{t''}(\vec{x}, \vec{c})).$$

If $\vec{c}$ realizes the hypotheses in Γ, the right-hand side is defined and realizes B.

The other propositional connectives are handled in a similar manner. Conjunction introduction and elimination correspond to pairing and projections. The left and right introduction rules for disjunction correspond to the functions $\varphi_t(\vec{x}, \vec{c}) \simeq (0, \varphi_{t'}(\vec{x}, \vec{c}))$ and $\varphi_t(\vec{x}, \vec{c}) \simeq (1, \varphi_{t'}(\vec{x}, \vec{c}))$, respectively, where t' is the term obtained from the inductive hypothesis. For disjunction elimination, suppose t_0, t_1, and t_2 are the terms corresponding to the hypotheses of the elimination rule:

$$\frac{\Gamma \vdash A \vee B \qquad \Gamma, A \vdash C \qquad \Gamma, B \vdash C}{\Gamma \vdash C}$$

Then we use the function

$$\varphi_t(\vec{x}, \vec{c}) \simeq \begin{cases} \varphi_{t_1}(\vec{x}, \vec{c}, (\varphi_{t_0}(\vec{x}, \vec{c}))_1) & \text{if } (\varphi_{t_0}(\vec{x}, \vec{c}))_0 = 0 \\ \varphi_{t_2}(\vec{x}, \vec{c}, (\varphi_{t_0}(\vec{x}, \vec{c}))_1) & \text{otherwise.} \end{cases}$$

The *ex falso* rule is covered by the fact that there are no realizers of $\bot$.

The handling of the quantifier rules is similar. We will treat exists introduction as an example. Suppose we derive $\Gamma \vdash \exists y\, A$ from $\Gamma \vdash A[s/y]$. By the inductive hypothesis, we have a term t' such that whenever $\vec{c}$ realizes the hypotheses in Γ, $\varphi_{t'}(\vec{x}, \vec{c})$ is defined and realizes $A[s/y]$. If we choose t such that $\varphi_t(\vec{x}, \vec{c}) = (s, \varphi_{t'}(\vec{x}, \vec{c}))$, then $\varphi_t(\vec{x}, \vec{c})$ realizes the conclusion.

If we use axioms for equality and restrict the substitution axiom to atomic formulas, then all the equality axioms are negative, and so Proposition 14.1.3 implies that they have canonical realizers. The same holds of the axioms of HA other than induction. It is convenient to formulate induction as a rule:

$$\frac{\Gamma \vdash A(0) \qquad \Gamma, A(y) \vdash A(y+1)}{\Gamma \vdash A}$$

This rule has the constraint that y is not free in any formula in Γ. Suppose, inductively, that $\varphi_{t_0}(\vec{x}, \vec{c})$ is defined and realizes $A(0)$ whenever $\vec{c}$ realizes the hypotheses in Γ, and $\varphi_{t_1}(\vec{x}, y, \vec{c}, a)$ is defined and realizes $A(y+1)$ whenever $\vec{c}$ realizes the hypotheses in Γ and a realizes $A(y)$. Choose t such that we have

$$\varphi_t(\vec{x}, \vec{c}, 0) \simeq \varphi_{t_0}(\vec{x}, \vec{c})$$
$$\varphi_t(\vec{x}, \vec{c}, y+1) \simeq \varphi_{t_1}(\vec{x}, y, \vec{c}, \varphi_t(\vec{x}, \vec{c}, y)).$$

Then by induction we have that for every y, assuming $\vec{c}$ realizes the hypotheses in Γ, $\varphi_t(\vec{x}, \vec{c}, y)$ is defined and realizes $A(y)$, as required.

This completes the proof of Theorem 14.1.2. We have shown, moreover, that if HA proves a theorem A from additional hypotheses that are realized, then HA proves that A is realized as well. Using the deduction theorem, this fact can also be seen as a corollary to Theorem 14.1.2.

Theorem 14.1.4. *Suppose* $\mathrm{HA} + \Gamma + \Delta$ *proves* A*, where the free variables of* A *are among* $\vec{x}$*, there is a closed term* t_B *for each sentence* B *in* Γ *such that* HA *proves* $\varphi_{t_B}()\ \mathbf{r}\ B$*, and there is a closed term* t_C *for every sentence* C *in* Δ *such that* HA *proves* $C \to \varphi_{t_C}()\ \mathbf{r}\ C$*. Then there is a closed term* t *such that* $\mathrm{HA} + \Delta$ *proves* $\varphi_t(\vec{x}) \downarrow$ *and* $\varphi_t(\vec{x})\ \mathbf{r}\ A$*.*

The sentences in Δ have canonical realizers, and the sentences in Γ are said to be *self-realizing*. For instance, if $\mathrm{HA} \cup \{C\}$ proves A, then HA proves that any realizer for C gives rise to a realizer for A. If C is self-realizing, then $\mathrm{HA} \cup \{C\}$ proves that there is a realizer for C, and hence a realizer for A. By Proposition 14.1.3, we can add any almost negative sentence to Δ. In the next section, we will consider some additional axioms that can be added to Γ, including axioms that are strong enough to show that $\exists e\ (e\ \mathbf{r}\ A)$ and A coincide.

When A is almost negative, the conclusion to Theorem 14.1.4 implies that HA proves A as well. But we can identify a larger class of formulas A with the property that HA proves $\exists e\ (e\ \mathbf{r}\ A) \to A$. Let Π be the smallest set containing the almost negative formulas and closed under $\wedge, \vee, \forall, \exists$, and the following rule: if A is almost negative and B is in Π, then $A \to B$ is in Π. An easy induction on formulas shows that for every A in Π, HA proves $\forall e\ (e\ \mathbf{r}\ A \to A)$. As a consequence, we also have that for every A in Π, HA proves $\forall e\ (\neg A \to e\ \mathbf{r}\ \neg A)$. So we can strengthen Theorem 14.1.4 by adding negations of sentences in Π to Δ as well. In addition, we have the following:

Corollary 14.1.5. *Let* Γ *and* Δ *be as in Theorem 14.1.4. Then* $\mathrm{HA} + \Gamma + \Delta$ *is conservative over* $\mathrm{HA} + \Delta$ *for formulas in* Π*.*

Exercises

14.1.1. Show that every number realizes $\neg \forall e\ \exists s\ \forall s'\ (T(e, 0, s') \to T(e, 0, s))$.

14.1.2. Show that, with the assumption $\forall x\,(x=0 \vee x \neq 0)$, the definition of $e\,\mathbf{r}\,A \vee B$ given above is equivalent to $((e)_0 = 0 \wedge (e)_1\,\mathbf{r}\,A) \vee ((e)_1 \neq 0 \wedge (e)_0\,\mathbf{r}\,B)$. (The version used in this section stems from the fact that, with the assumption $\forall x\,(x=0 \vee x \neq 0)$, we can define disjunction in terms of the existential quantifier by taking $A \vee B$ to be $\exists x\,((x=0 \rightarrow A) \wedge (x \neq 0 \rightarrow B))$.)

14.1.3. Prove carefully that $e\,\mathbf{r}\,(A[t/x])$ is the same formula as $(e\,\mathbf{r}\,A)[t/x]$, assuming that e is different from x and is not free in A or t.

14.1.4. Prove the claim right before Theorem 14.1.1: if A has free variables $\vec{x}$ and u is a fresh variable, then $u\,\mathbf{r}\,A$ holds in the standard model with u assigned to e and $\vec{x}$ assigned to $\vec{n}$ if and only if e realizes $A[\vec{n}/\vec{x}]$.

14.1.5. Complete the cases left out of the proof of Proposition 14.1.3.

14.1.6. Complete the cases left out of the proof of Theorem 14.1.2.

14.1.7. Show that HA does not in general prove $A \leftrightarrow \exists e\,(e\,\mathbf{r}\,A)$, but that it does prove that this statement is realizable.

14.1.8. Sketch a proof of Theorem 11.5.2, the recursion theorem, in first-order primitive recursive arithmetic, assuming the general facts about formalizing computability that were enumerated in this section.

14.2 Metamathematical Applications

Suppose HA proves $\forall x\,\exists y\,R(x, y)$, then Theorem 14.1.4 tells us that there is a term realizing $\forall x\,\exists y\,R(x, y)$, provably in HA. Such a realizer is the index of a function that, for every x, returns a pair consisting of a value y and a realizer for $R(x, y)$. Composing this function with the first projection gives rise to an index e such that HA proves $\forall x\,(\varphi_e(x)\downarrow \wedge R(x, \varphi_e(x)))$. Theorem 10.4.6, which says that PA is Π_2 conservative over HA, tells us that the conclusion also holds if $\forall x\,\exists y\,R(x, y)$ is provable in PA instead. But neither of these facts is surprising: HA can prove that $\forall x\,\exists y\,R(x, y)$ implies that the computable function $\varphi_e(x) \simeq \mu y\,R(x, y)$ has this property, and we do not need to invoke realizability at all. In the next section, we will see that a variation on the realizability relation provides us with more information about a class of computable functions that can realize the Π_2 theorems of arithmetic in this way. In this section, however, we will use the realizability relation we already have to uncover some interesting metamathematical properties of HA.

One important application of realizability is to show independence results. For example, we have used tools like Kripke models and the cut elimination theorem to show that the schemas $A \vee \neg A$ and $(A \rightarrow \exists x\,B) \rightarrow \exists x\,(A \rightarrow B)$ are not intuitionistically valid. Although it should seem unlikely, it is a priori conceivable that they are derivable in HA. But Theorem 14.1.2 shows that this is not the case. For example, HA cannot prove $\exists s\,T(e, 0, s) \vee \neg\exists s\,T(e, 0, s)$, because we know this formula is not realizable. Similarly, it cannot prove

$$(\exists s'\,T(e, 0, s') \rightarrow \exists s\,T(e, 0, s)) \rightarrow \exists s\,(\exists s'\,T(e, 0, s') \rightarrow T(e, 0, s)),$$

because otherwise it could prove the conclusion, which is again not realizable. We will establish stronger results than these in this section. But first we will consider some additional axioms that are provably realized in HA.

Let (CT) be the following schema:

$$\forall x\,\exists y\,A(x, y) \rightarrow \exists e\,\forall x\,(\varphi_e(x)\downarrow \wedge A(x, \varphi_e(x))).$$

In words, (CT) says that any true forall-exists statement is witnessed by a computable function. Here the initials *CT* stand for *Church's thesis*. Let ($\mathrm{IP_{an}}$) be the following schema:

$$(A \to \exists x\, B) \to \exists x\, (A \to B),$$

where A is assumed to be almost negative and x is not free in A. Here the initials *IP* stand for *independence of premise*. The formulas in both (CT) and ($\mathrm{IP_{an}}$) can have free variables other than the ones shown. Neither of these is derivable in HA, and, indeed, the principle (CT) is false in the standard model. But the following shows that they can be consistently added to HA.

Proposition 14.2.1. HA *proves that there are explicit realizers for every instance of* (CT) *and* ($\mathrm{IP_{an}}$).

Proof In both cases, for brevity, I will not mention any free variables other than the ones shown, but they should be taken as arguments to all the functions we are about to describe.

For (CT), if e' realizes $\forall x\, \exists y\, A(x, y)$, choose e such that $\varphi_e(x) \simeq (\varphi_{e'}(x))_0$ for every x. The function $\varphi_e(x)$ is total because $\varphi_{e'}(x)$ is. Also, for every x, $(\varphi_{e'}(x))_1$ is defined and realizes $A(x, \varphi_e(x))$. Since the assertion $\varphi_e(x)\downarrow$ is Σ_1, it has a canonical realizer $\varphi_t(x)$ by Proposition 14.1.3, and $\varphi_t(x)$ is total because $\varphi_e(x)$ is total. So (CT) is realized by an index for the computable function that maps e' to the pair (e, e''), where $\varphi_{e''}(x) \simeq (\varphi_t(x), (\varphi_{e'}(x))_1)$.

For ($\mathrm{IP_{an}}$), suppose e realizes $A \to \exists x\, B$, let $\varphi_t()$ be the canonical realizer for A, and suppose $\varphi_{e'}(a) \simeq (\varphi_e(\varphi_t()))_1$ for every a. Then the pair $((\varphi_e(\varphi_t()))_0, e')$ realizes $\exists x\, (A \to B)$. So the realizer for ($\mathrm{IP_{an}}$) is an index for the computable function that maps e to the pair $((\varphi_e(\varphi_t()))_0, e')$. □

(CT) and ($\mathrm{IP_{an}}$) can therefore be added to the set Γ of provably realized axioms in the statement of Theorem 14.1.4. We now show that there is a sense in which these are the strongest such axioms: in their presence, every formula that is provably realized is provable outright.

Theorem 14.2.2. *Over* HA, *the schemas* (CT) *and* ($\mathrm{IP_{an}}$) *together are equivalent to the schema* $A \leftrightarrow \exists e\, (e\, \mathbf{r}\, A)$, *in the following sense:*

- *For every formula* A, HA + (CT) + ($\mathrm{IP_{an}}$) *proves* $A \leftrightarrow \exists e\, (e\, \mathbf{r}\, A)$.
- HA *together with the schema* $A \leftrightarrow \exists e\, (e\, \mathbf{r}\, A)$ *proves every instance of* (CT) *and* ($\mathrm{IP_{an}}$).

Proof The first claim is proved by induction on formulas, and the second claim follows from Proposition 14.2.1. □

This provides a neat characterization of the formulas that are provably realized in HA.

Corollary 14.2.3. *For every formula* A *in the language of arithmetic,* HA *proves* $\exists e\, (e\, \mathbf{r}\, A)$ *if and only if* HA + (CT) + ($\mathrm{IP_{an}}$) *proves* A.

Proof The forward direction follows from the first claim of Theorem 14.2.2, and the reverse direction follows from Theorem 14.1.4 and Proposition 14.2.1. □

The theory HA + (CT) + ($\mathrm{IP_{an}}$), though classically false, has two important constructive properties. We will call the first the *explicit definability property*, though it is really a variation on the property referred to by the same name in Section 7.2.

Theorem 14.2.4. *If* HA + (CT) + ($\mathrm{IP}_{\mathrm{an}}$) *proves a formula* $\exists y\, A(\vec{x}, y)$ *with free variables* $\vec{x}$*, there is a closed term* t *such that it also proves* $\varphi_t(\vec{x}) \downarrow \wedge A(\vec{x}, \varphi_t(\vec{x}))$.

Proof Suppose HA + (CT) + ($\mathrm{IP}_{\mathrm{an}}$) proves $\exists y\, A(x, y)$. Then for some term t', HA proves that $\varphi_{t'}(\vec{x})$ is defined and realizes $\exists y\, A(\vec{x}, y)$. This implies that $(\varphi_{t'}(\vec{x}))_0$ is defined and $(\varphi_{t'}(\vec{x}))_1$ realizes $A(\vec{x}, (\varphi_{t'}(\vec{x}))_0)$. But in HA + (CT) + ($\mathrm{IP}_{\mathrm{an}}$), if a formula is realized, it is true, so we can take t to be an index for the computable function that maps $\vec{x}$ to $(\varphi_{t'}(\vec{x}))_0$. □

It is not hard to show that this implies the disjunction property, since $A \vee B$ is equivalent to $\exists x\, ((x = 0 \to A) \wedge (x \neq 0 \to B))$.

Theorem 14.2.5. *If* HA + (CT) + ($\mathrm{IP}_{\mathrm{an}}$) *proves a sentence* $A \vee B$*, then it proves* A *or it proves* B.

We can strengthen Theorems 14.2.4 and 14.2.5 to eliminate the dependence on (CT) and ($\mathrm{IP}_{\mathrm{an}}$). In the proofs of these theorems, we needed the extra axioms to prove that realizability implies truth. We can achieve that instead by modifying the definition of realizability. Define the relation $e\, \mathbf{rt}\, A$, *realizability with truth*, just as we defined of ordinary realizability, except that the clause for implication is as follows:

- $e\, \mathbf{rt}\, (A \to B) \equiv \forall a\, (a\, \mathbf{rt}\, A \to \varphi_e(a) \downarrow \wedge\, \varphi_e(a)\, \mathbf{rt}\, B) \wedge (A \to B)$.

In other words, we add the conjunct $A \to B$ to the corresponding clause in the definition of ordinary realizability. This clause covers negation also, since $\neg A$ is defined to be $A \to \bot$. Now a straightforward induction on formulas yields the following:

Proposition 14.2.6. *For every formula* A*,* HA *proves* $\forall e\, ((e\, \mathbf{rt}\, A) \to A)$.

This has the following corollary:

Corollary 14.2.7. *For every formula* A*,* HA *proves* $\neg A \to \forall e\, (e\, \mathbf{rt}\, \neg A)$.

Proposition 14.1.3, which asserts that realizability and truth coincide for almost negative formulas, extends to realizability with truth, with the same canonical realizers. We still have the soundness property, and, in fact, the realizers don't change.

Proposition 14.2.8. *If the free variables of* A *are among* $\vec{x}$ *and* HA *proves* A*, there is a closed term* t *such that* HA *proves* $\varphi_t(\vec{x}) \downarrow \wedge\, (\varphi_t(\vec{x})\, \mathbf{r}\, A) \wedge (\varphi_t(\vec{x})\, \mathbf{rt}\, A)$.

In analogy to Theorem 14.1.4, we have:

Theorem 14.2.9. *Suppose* HA + Δ *proves* A*, where the free variables of* A *are among* $\vec{x}$ *and for each sentence* C *in* Δ *there is a closed term* t_C *such that* HA *proves* $C \to \varphi_{t_C}()\, \mathbf{rt}\, C$. *Then there is a closed term* t *such that* HA + Δ *proves* $\varphi_t(\vec{x}) \downarrow$ *and* $\varphi_t(\vec{x})\, \mathbf{rt}\, A$.

Corollary 14.2.10. *With* Δ *as above,* HA + Δ *has the explicit definability property and the disjunction property.*

By Proposition 14.1.3, any almost negative formula meets the criterion for inclusion in Δ. It is not hard to show that (CT) and ($\mathrm{IP}_{\mathrm{an}}$) can be included in Δ as well. By Corollary 14.2.7, we can also add arbitrary negated sentences to Δ. (Since HA proves the law of the excluded middle for atomic formulas, by Exercise 7.2.1 this is equivalent to saying that we can add any sentence that is Harrop.)

We now consider an axiom schema known as *Markov's principle*, denoted by (MP):

$$\forall x\,(A(x) \lor \neg A(x)) \land \neg\neg\exists x\, A(x) \to \exists x\, A(x).$$

As usual A may have free variables other than the one shown. This principle, which was accepted by a community of mathematicians in Russia that aimed to develop analysis in constructive terms, has a natural computational interpretation. If we read the first hypothesis as saying that we can algorithmically determine whether or not $A(x)$ holds for any x, then we can find an x satisfying $\exists x\, A(x)$ simply by searching systematically and trying each instance. The second hypothesis, $\neg\neg\exists x\, A(x)$, can be interpreted as saying that the search must succeed, in that $\forall x\, \neg A(x)$ is contradictory.

The restriction of the principle to the case where A is primitive recursive is called $(\mathrm{MP}_{\mathrm{pr}})$. In that case, since HA proves $\forall x\,(A(x) \lor \neg A(x))$, the principle reduces to $\neg\neg\exists x\, A(x) \to \exists x\, A(x)$, or, equivalently, $\neg\forall x\, A(x) \to \exists x\, \neg A(x)$. In fact, since HA proves that every primitive recursive function is computable, the schema can be replaced by a single instance:

$$\forall e, x\,(\neg\neg\exists s\, T(e, x, s) \to \exists s\, T(e, x, s)).$$

Exercise 14.2.5 asks you to spell out the details.

Because $(\mathrm{MP}_{\mathrm{pr}})$ is almost negative, it can be included in the set Δ in Theorem 14.2.9 and Corollary 14.2.10. Exercise 14.2.6 asks you to show that in HA + (CT) the schema (MP) is equivalent to $(\mathrm{MP}_{\mathrm{pr}})$. With Corollary 14.1.5, this implies that HA + (CT) + $(\mathrm{IP}_{\mathrm{an}})$ + (MP) is conservative over HA + $(\mathrm{MP}_{\mathrm{pr}})$ for the class of formulas Π described there. Since HA + $(\mathrm{MP}_{\mathrm{pr}})$ is contained in PA, it is true of the standard model of arithmetic, so the previous sentence tells us in particular that the theory HA + (CT) + $(\mathrm{IP}_{\mathrm{an}})$ + (MP) is consistent.

We close this section by showing how the results developed in this section give rise to independence results for HA.

Proposition 14.2.11. *1. There is a* Σ_1 *sentence* A *such that* HA *does not prove* $A \lor \neg A$.

2. *There is a* Σ_1 *formula* $A(x)$ *with only the free variable shown such that* HA *does not prove* $\neg\neg\forall x\,(A(x) \lor \neg A(x))$.
3. HA *does not prove the double-negation shift schema,* $\forall x\, \neg\neg A(x) \to \neg\neg\forall x\, A(x)$.
4. HA *does not prove* $(\mathrm{MP}_{\mathrm{pr}})$.
5. *There are a* Π_1 *formula* A *and an atomic formula* B *such that* HA *doesn't prove the instance of* $(\mathrm{IP}_{\mathrm{an}})$ *given by* $(\neg A \to \exists x\, B) \to \exists x\,(\neg A \to B)$.

Proof For claim 1, let B the Gödel sentence, so that $B \leftrightarrow \neg\exists x\, \mathrm{Pr}_{\mathrm{HA}}(x, \ulcorner B \urcorner)$ is provable in HA. Let A be $\exists x\, \mathrm{Pr}_{\mathrm{HA}}(x, \ulcorner B \urcorner)$. Since HA satisfies the disjunction property, it does not prove $A \lor \neg A$. (We also have that B is a Π_1 sentence such that HA doesn't prove $B \lor \neg B$.)

For claim 2, notice that (CT) implies $\neg\forall x\, \exists s\, T(x, 0, s)$ by the undecidability of the halting problem (specifically, Proposition 11.3.5). Since (CT) is realizable, it is consistent with HA, so HA doesn't prove $\neg\neg\forall x\, \exists s\, T(x, 0, s)$.

For claim 3, Exercise 4.6.2 shows that $\neg\neg(A(x) \lor \neg A(x))$ is equivalent to double-negation shift for the formula $A(x) \lor \neg A(x)$.

For claim 4, let $\forall x\, A(x)$ be a Rosser sentence for HA, and suppose HA proves $\neg\forall x\, A(x) \to \exists x\, \neg A(x)$. Then HA proves that it is realized by a closed term, and unfolding the definition of realizability, we get a closed term t such that HA proves $\neg\forall x\, A(x) \to \neg A(t)$. Since $A(x)$ is primitive recursive, either HA proves $A(t)$ or $\neg A(t)$. In the second case, it proves $\neg\forall x\, A(x)$, contradicting the assumption that $\forall x\, A(x)$ is a Rosser sentence. So HA proves $A(t)$, and hence

$\neg\neg\forall x\, A(x)$. But this implies $\forall x\, \neg\neg A(x)$ and hence $\forall x\, A(x)$, again contradicting the fact that $\forall x\, A(x)$ is a Rosser sentence.

Finally, for claim 5, suppose HA proves

$$(\neg\forall s'\, \neg T(x, 0, s') \to \exists s\, T(x, 0, s)) \to \exists s\, (\neg\forall s'\, \neg T(x, 0, s) \to \exists s\, T(x, 0, s)).$$

In HA + (MP_{pr}), the antecedent is equivalent to $\exists s\, T(x, 0, s) \to \exists s'\, T(x, 0, s')$, which is trivially provable. So HA + (MP_{pr}) proves the conclusion, and therefore

$$\forall x\, \exists s\, (\exists s'\, T(x, 0, s') \to T(x, 0, s)).$$

By Theorem 14.1.4, this last sentence is realizable. But this is a contradiction, since any realizer can be used to solve the halting problem. □

Exercises

14.2.1. Let (ECT), or *extended Church's thesis*, be the following schema:

$$\forall x\, (A(x) \to \exists y\, B(x, y)) \to \exists e\, \forall x\, (A(x) \to \varphi_e(x) \downarrow \wedge B(x, \varphi_e(x))).$$

Show that over HA, the schema (ECT) is equivalent to the combination of the schemas (CT) and (IP_{an}).

14.2.2. Prove the first claim in Theorem 14.2.2, namely, that for every formula A, HA + (CT) + (IP_{an}) proves $A \leftrightarrow \exists e\, (e\, \mathbf{r}\, A)$.

14.2.3. Show that, in any theory extending PRA, the disjunction property in Theorem 14.2.5 implies the explicit definability property in Theorem 14.2.4. (For the converse, see the notes at the end of this chapter.)

14.2.4. Prove Proposition 14.2.6 and Corollary 14.2.7.

14.2.5. Show that over first-order PRA, the schema MP_{pr} is equivalent to the single instance $\forall e, x\, (\neg\neg\exists s\, T(e, x, s) \to \exists s\, T(e, x, s))$.

14.2.6. Show that in HA + (CT), the schema (MP) is equivalent to (MP_{pr}).

14.2.7. Some of the metamathematical properties obtained in this section can be alternatively obtained using *slash relations*, essentially a form of realizability that abstracts away the realizers. For example, the *Aczel slash* is defined for sentences inductively, where $\Gamma \vdash A$ means HA $\cup\, \Gamma$ proves A:

- $\Gamma \mid A$ if and only if $\Gamma \vdash A$ for A atomic.
- $\Gamma \mid A \wedge B$ if and only if $\Gamma \mid A$ and $\Gamma \mid B$.
- $\Gamma \mid A \vee B$ if and only if $\Gamma \mid A$ or $\Gamma \mid B$.
- $\Gamma \mid A \to B$ if and only if, first, $\Gamma \vdash A \to B$, and, second, $\Gamma \mid A$ implies $\Gamma \mid B$.
- $\Gamma \mid \forall x\, A(x)$ if and only if $\Gamma \vdash \forall x\, A(x)$ and $\Gamma \mid A(n)$ for every numeral n.
- $\Gamma \mid \exists x\, A(x)$ if and only if $\Gamma \mid A(n)$ for some numeral n.

$\Gamma \mid A$ is read "Γ slashes A."

a. Show by induction on sentences A that if $\Gamma \mid A$ then $\Gamma \vdash A$.
b. Show that if t is a closed term that evaluates to n, then $\Gamma \mid A(t)$ if and only if $\Gamma \mid A(n)$.
c. Show that if $\Gamma \vdash A$ and $\Gamma \mid B$ for every B in Γ, then $\Gamma \mid A$.
d. Use this to show that HA has the disjunction property and the explicit definability property.

For more information on slash relations, see the bibliographical notes at the end of this chapter.

14.3 Modified Realizability

The relation $e \mathbf{r} A$ is a relation between a potential realizer, e, and a formula, A. Theorem 14.1.2 says that if HA proves A, then not only is there realizer for A, but, moreover, HA proves that the realizer meets its specification. The proof shows that the realizer can be straightforwardly read off from the derivation in HA. This provides a general template for interpreting formal theories in computational terms. We can vary virtually every component of the analysis:

- *the class of realizers.* So far, we have used natural numbers, which we interpreted in some cases as representatives of partial computable functions and in some cases as pairs. In general, we can use more restricted classes of computable functions to interpret implication and the universal quantifier, or broader classes of functions, which need not be computable.
- *the class of statements.* We can extend realizability to formulas in the language of second- or higher-order arithmetic, or set theory, or languages with other primitives.
- *the realizability relation.* Even restricting to particular classes of realizers and formulas, we can vary the specification of realizability, thereby varying the type of information we want our realizers to encode.
- *the theory being realized.* Even if we fix a language, we can consider any theory that proves formulas in that language, and we can vary the axioms, as we did in the last section.
- *the verifying theory.* It is often interesting to ask what is needed to prove the correctness of the realizers we extract. We can also omit this component entirely and prove correctness informally, using ordinary mathematics.

These components are interrelated, and varying any one of them generally requires making adjustments to the others. We can think of the realizers as intentional or extensional objects; for example, we can talk about a function realizing an implication, or we can talk about a term denoting that function. In general, if two terms denote the same function, one will realize the implication if and only if the other does, so it is reasonable to think about realizers extensionally. In contrast, it is hard to think of the formulas that are realized in anything but intensional terms. In general, equivalent statements will not have the same realizers. For example, any sentence that is provable in a theory is provably equivalent to $\top$, whose realizers carry no information.

Realizability should remind you of the propositions as types interpretation of intuitionistic logic, described in Section 13.8. Propositions as types can be seen as a special case of realizability in which the statements and the realizability relation are conflated, in that the statement itself is interpreted as a specification of the realizing data. The theory being realized, the realizers, and the verifying theory are also conflated, in that proofs in the realizing theory are interpreted as both the realizers and evidence of their own correctness. The general realizability framework is more flexible in that it enables us to separate the components.

In this section, we will continue to focus on the language of arithmetic as the language that is realized, but we will refine the realizability relation to give us more information about the class of realizers. Specifically, we will use terms denoting primitive recursive functionals of finite type, as defined in Section 13.7. The corresponding realizability relation is known as *modified realizability*. We will use HA^{ω} as the verifying theory, since its language is rich enough to define the relation formally. It is straightforward to generalize the source theory to HA^{ω} as well, and we will later consider additional axioms in that language.

To each formula $A(\vec{x})$ in the language of arithmetic, we associate a finite type τ_A and a formula $t\ \mathbf{mr}\ A$, inductively, as follows:

- If A is atomic, $\tau_A = \text{NAT}$, and $t\ \mathbf{mr}\ A \equiv A$.
- $\tau_{A \wedge B} = \tau_A \times \tau_B$, and $t\ \mathbf{mr}\ (A \wedge B) \equiv (t)_0\ \mathbf{mr}\ A \wedge (t)_1\ \mathbf{mr}\ B$.
- $\tau_{A \vee B} = \text{NAT} \times \tau_A \times \tau_B$, and

$$t\ \mathbf{mr}\ (A \vee B) \equiv ((t)_0 = 0 \to ((t)_1)_0\ \mathbf{mr}\ A) \wedge ((t)_0 \neq 0 \to ((t)_1)_1\ \mathbf{mr}\ B).$$

- $\tau_{A \to B} = \tau_A \to \tau_B$, and $t\ \mathbf{mr}\ (A \to B) \equiv \forall u\ (u\ \mathbf{mr}\ A \to t\, u\ \mathbf{mr}\ B)$.
- $\tau_{\forall x^\alpha A} = \alpha \to \tau_A$, and $t\ \mathbf{mr}\ \forall x^\alpha\ A \equiv \forall x\ (t\, x\ \mathbf{mr}\ A)$.
- $\tau_{\exists x^\alpha A} = \alpha \times \tau_A$, and $t\ \mathbf{mr}\ \exists x^\alpha\ A \equiv (t)_1\ \mathbf{mr}\ A[(t)_0/x]$.

As with ordinary realizability, the realizers in the base case are irrelevant: if an atomic formula A is true, then anything realizes it, and if A is false, nothing does. The type NAT is a reasonable default, but if we include among the finite types a singleton UNIT type, as described in Section 13.8, it would be more natural to use that instead. Another choice is to track the lack of information in the metatheory, by associating to any formula A *either* a type *or* the symbol UNIT, indicating that no realizer is needed. We can then adopt the following conventions:

- $\tau_A \times \text{UNIT}$ is τ_A
- $\text{UNIT} \times \tau_A$ is τ_A
- $\tau_A \to \text{UNIT}$ is UNIT
- $\text{UNIT} \to \tau_A$ is τ_A.

The specification of modified realizability then has to be interpreted accordingly. Yet another option is to omit product types and assign to each formula A a finite *tuple* of types, possibly empty. We will follow that strategy in the next section, when we introduce the Dialectica interpretation. Here, however, for simplicity, we will stick to using product types so that every realizer is a single object.

The type $\text{NAT} \times \tau_A \times \tau_B$ in the third clause is a proxy for the sum type $\tau_A + \tau_B$ that we discussed in Section 13.8. We rely on the fact that for every finite type α there is a constant 0^α defined recursively by $0^{\text{NAT}} = 0$, $0^{\alpha \to \beta} = \lambda x^\alpha.\, 0^\beta$, and $0^{\alpha \times \beta} = (0^\alpha, 0^\beta)$. We can then simulate the inl and inr constructors for $\alpha + \beta$ by defining $\text{inl}\, x = (0, x, 0^\beta)$ and $\text{inr}\, x = (1, 0^\alpha, x)$. Here I am using the triple notation (a, b, c) as a convenient shorthand for $(a, (b, c))$. Given $f : \alpha \to \gamma$ and $g : \beta \to \gamma$, we can define

$$\text{cases}\ u\ f\ g = \begin{cases} f\ (u)_1 & \text{if } (u)_0 = 0 \\ g\ (u)_2 & \text{otherwise.} \end{cases}$$

Here we use primitive recursion to define the conditional, and we adopt the suggestive notation $(u)_0$, $(u)_1$, $(u)_2$ for triples rather than $(u)_0$, $((u)_1)_0$, $((u)_1)_1$.

As in Section 14.1, we can show that $(t\ \mathbf{mr}\ A)[s/x] = t\ \mathbf{mr}\ A[s/x]$ for every variable x and term s of the same type, assuming x is not free in t. In analogy to Proposition 14.1.3, we have the following:

Proposition 14.3.1. *If A is an $\exists$-free formula in the language of* HA^ω, *there is a term t of type τ_A such that the following are provably equivalent in* HA^ω:

1. A
2. $t\,\mathbf{mr}\,A$
3. $\exists x\,(x\,\mathbf{mr}\,A)$.

The proposition is restricted to $\exists$-free formulas rather than almost negative formulas since we can no longer use unbounded search to find a witness to an existential formula. In the case of an $\exists$-free formula, the realizer t carries no useful information at all; you can check that if we introduce UNIT with the conventions above, then the type of an $\exists$-free formula is UNIT. In the setting of HA^ω, we cannot in general prove the law of the excluded middle for formulas of the form $s =_\alpha t$ for arbitrary types α, so not every $\exists$-free formula is provably equivalent to one that is negative.

Theorem 14.3.2. *If* HA^ω *proves a formula* A*, there is a term* t *such that* HA^ω *also proves* $t\,\mathbf{mr}\,A$*.*

The proof is similar to the proof of Theorem 14.1.2, and even more natural given that we have good notation for defining primitive recursive functionals. Specifically, we prove the following by induction on natural deduction derivations: whenever a sequent $A_0, \ldots, A_{k-1} \vdash B$ is derivable, there is a term t such that HA^ω also proves $u_0\,\mathbf{mr}\,A_0, \ldots, u_{k-1}\,\mathbf{mr}\,A_{k-1} \vdash t\,\mathbf{mr}\,B$, where each u_i has type τ_{A_i}, and the free variables of t are among those of $A_0, \ldots, A_{k-1}, B$ together with $u_0, \ldots, u_{k-1}$. For each rule, we inductively assume that each premise satisfies this property. Handling the logical rules is straightforward:

- assumption: if B is the assumption A_i, then u_i realizes B.
- $\wedge$ introduction: if t_0 realizes B and t_1 realizes C, then (t_0, t_1) realizes $B \wedge C$.
- $\wedge$ elimination: if t realizes $B \wedge C$, then $(t)_0$ realizes B and $(t)_1$ realizes C.
- $\vee$ introduction: if t realizes B, then $\mathrm{inl}\,t$ realizes $B \vee C$ and $\mathrm{inr}\,t$ realizes $C \vee B$.
- $\vee$ elimination: if t realizes $B \vee C$, r realizes D on the assumption that the variable v realizes B, s realizes D on that assumption w realizes C, then $\mathrm{cases}\,t\,(\lambda v.\,r)\,(\lambda w.\,s)$ realizes D.
- $\to$ introduction: if t realizes C on the assumption that v realizes B, then $\lambda v.\,t$ realizes $B \to C$.
- $\to$ elimination: if s realizes $B \to C$ and t realizes B, then $s\,t$ realizes C.
- $\forall$ introduction: if t realizes B for every x, then $\lambda x.\,t$ realizes $\forall x\,B$.
- $\forall$ elimination: if t realizes $\forall x\,B$, then for any term s of the same type as x, $t\,s$ realizes $B[s/x]$.
- $\exists$ introduction: if t realizes $B[s/x]$, then (s, t) realizes $\exists x\,B(x)$.
- $\exists$ elimination: if t realizes $\exists x\,B(x)$, and, for very x, $s(x, v)$ realizes C assuming v realizes $B(x)$, then $s((t)_0, (t)_1)$ realizes C.

We leave it as an exercise to show that in each case the claim in question is provable in HA^ω. The quantifier-free axioms of HA^ω are all $\exists$-free, so HA^ω proves that they are trivially realized. That leaves only the induction rule, for which we have to verify the following:

- Suppose s realizes $B(0)$, and for every x, $t(x, v)$ realizes $B(\mathrm{succ}\,x)$ on the assumption that v realizes $B(x)$. Then for every u of type NAT, $R\,s\,(\lambda x, v.\,t)\,u$ realizes $B(u)$.

This can be proved by induction on u in HA^ω. As a corollary, in analogy to Theorem 14.1.4, we have:

Theorem 14.3.3. *Suppose* $\mathrm{HA}^\omega + \Gamma + \Delta$ *proves* A*, where for every sentence* B *in* Γ *there is a term* t_B *such that* HA^ω *proves* $t_B \,\mathbf{mr}\, B$*, and for every sentence* C *in* Δ *there is a term* t_C *such that* HA^ω *proves* $C \to t_C \,\mathbf{mr}\, C$*. Then there is a term* t *such that* $\mathrm{HA}^\omega + \Delta$ *proves* $t \,\mathbf{mr}\, A$.

By Proposition 14.3.1, we can include any $\exists$-free formula in Δ. Since the extensionality axioms (ext) described in Section 13.5 are $\exists$-free, Theorem 14.3.3 therefore holds if we replace HA^ω with the theory E-HA^ω, that is, HA^ω + (ext), in both the hypothesis and conclusion.

Let Π be the smallest set containing the $\exists$-free formulas and closed under $\wedge$, $\vee$, $\forall$, and $\exists$, with the additional property that if A is $\exists$-free and B is in Π, then $A \to B$ is in Π. In analogy to Corollary 14.1.5, we have all the following:

Corollary 14.3.4. *Let* Γ *and* Δ *be as in Theorem 14.3.3. Then* $\mathrm{HA} + \Gamma + \Delta$ *is conservative over* $\mathrm{HA} + \Delta$ *for formulas in* Π.

As we did for realizability, we can characterize the principles needed to show that truth and modified realizability coincide. In the language of HA^ω, the *axiom of choice* is defined to be the schema (AC):

$$\forall x^\alpha \, \exists y^\beta \, A(x, y) \to \exists f^{\alpha \to \beta} \, \forall x^\alpha \, A(x, f\, x).$$

In this schema, A is an arbitrary formula of HA^ω and α and β are arbitrary types. The axiom schema of independence of premise ($\mathrm{IP_{ef}}$) for $\exists$-free formulas is given by

$$(A \to \exists x^\alpha \, B) \to \exists x^\alpha \, (A \to B),$$

where A is $\exists$-free, α is an arbitrary type, and x is not free in A. It is not hard to check that HA^ω proves that these axioms are realized. In analogy to Theorem 14.2.2, Corollary 14.2.3, and Theorems 14.2.4 and 14.2.5, we have:

Theorem 14.3.5. *Over* HA^ω*, the schemas* (AC) *and* ($\mathrm{IP_{ef}}$) *together are equivalent to the schema* $A \leftrightarrow \exists x \, (x \,\mathbf{mr}\, A)$*, in the following sense:*

- *For every formula* A*,* HA^ω + (AC) + ($\mathrm{IP_{ef}}$) *proves* $A \leftrightarrow \exists x \, (x \,\mathbf{mr}\, A)$.
- HA^ω *together with the schema* $A \leftrightarrow \exists x \, (x \,\mathbf{mr}\, A)$ *proves every instance* (AC) *and* ($\mathrm{IP_{ef}}$).

Corollary 14.3.6. *For every formula* A *in the language of arithmetic,* HA^ω *proves* $\exists x \, (x \,\mathbf{mr}\, A)$ *if and only if* HA^ω + (AC) + ($\mathrm{IP_{ef}}$) *proves* A.

Theorem 14.3.7. *The theory* HA^ω + (AC) + ($\mathrm{IP_{ef}}$) *has the explicit definability property and the disjunction property*

As we did with HA, we can show the HA^ω itself has the explicit definability property and the disjunction property using a variation of modified realizability, namely *modified realizability with truth,* $t \,\mathbf{mrt}\, A$. This is defined just like modified realizability except that the clause for implication is replaced by the following:

- $t \,\mathbf{mrt}\, (A \to B) \equiv \forall u \, (t \,\mathbf{mrt}\, A \to t\, u \,\mathbf{mrt}\, B) \wedge (A \to B)$.

With this change, we can prove the following by induction on A:

Proposition 14.3.8. *For every formula* A*,* HA^ω *proves* $u \,\mathbf{mrt}\, A \to A$.

In analogy to Theorem 14.2.9 and Corollary 14.2.10, we have

Theorem 14.3.9. *Suppose* $\mathrm{HA}^\omega + \Delta$ *proves* A*, where the free variables of* A *are among* $\vec{x}$ *and for each sentence* C *in* Δ *there is a closed term* t_C *such that* HA *proves* $C \to t_C \,\mathbf{mrt}\, C$*. Then there is a closed term* t *such that* $\mathrm{HA} + \Delta$ *proves* $t \,\mathbf{mrt}\, A$*.*

Corollary 14.3.10. *With* Δ *as above, if* $\mathrm{HA}^\omega + \Delta$ *proves a formula* $\exists y\, A(y)$*, there is a closed term* t *such that it also proves* $A(t)$*. If it proves a disjunction* $A \vee B$*, either it proves* A *or it proves* B*. With* Δ *as above,* $\mathrm{HA}^\omega + \Delta$ *has the explicit definability property and the disjunction property.*

In particular, (AC) and ($\mathrm{IP_{ef}}$) can be added to Δ, as well as any $\exists$-free formulas, such as the extensionality axioms (ext), and any negated or Harrop formula. In contrast to Theorem 14.2.9 and Corollary 14.2.10, however, we cannot add ($\mathrm{MP_{pr}}$), since there is no term that can possibly realize it. The exercises ask you to show that, since the type 1 terms of Gödel's T denote total computable functions in the standard interpretation, any term realizing

$$\forall e, x\, (\neg\neg\exists s\, T(e, x, s) \to \exists s\, T(e, x, s))$$

would give rise to a universal total computable function, and in Section 11.3 we saw that this is impossible. This shows, incidentally, that no theory satisfying the hypotheses of Theorem 14.3.3 can prove ($\mathrm{MP_{pr}}$).

Modified realizability sheds light on the computational strength of HA^ω and HA, as well as their classical counterparts, PA^ω and PA. For a theory T in a language that includes or interprets the language of arithmetic, say a function $f(\vec{x})$ is a *provably total computable function* of T if the graph of f is defined by a Σ_1 formula $A(\vec{x}, y)$ such that T proves $\forall \vec{x}\, \exists y\, A(\vec{x}, y)$. In the exercises below, you are asked to show that if T includes or interprets PRA, this is equivalent to saying that $f = \varphi_e$ for some index e such that T proves $\forall \vec{x}\, \varphi(\vec{x}) \downarrow$.

By the argument at the beginning of Chapter 12, no computably axiomatized theory can prove every computable function to be total. Thus the set of provably total computable functions provides a measure of a theory's computational strength. For example, Section 10.4 shows that every primitive recursive function is a provably total computable function of $\mathrm{I}\Sigma_1^{\mathrm{i}}$. If $\exists z\, B(\vec{x}, y, z)$ is the graph of a function $f(\vec{x})$ and B is Δ_0, and if $\mathrm{I}\Sigma_1$ proves $\forall \vec{x}\, \exists y, z\, B(\vec{x}, y, z)$, then Theorem 10.6.1 shows that $f(\vec{x})$ is primitive recursive. So the provably total computable functions of $\mathrm{I}\Sigma_1^{\mathrm{i}}$ and $\mathrm{I}\Sigma_1$ are exactly the primitive recursive functions.

In a similar way, the computational interpretation of the primitive recursive functionals in HA given by Proposition 13.7.2 shows that every type 1 primitive recursive functional is a provably total computable function of HA and PA. Conversely, Theorem 14.3.2 and the Π_2 conservativity of PA over HA shows that every provably total computable function of PA and HA is a type 1 primitive recursive functional. Thus, we now have an exact characterization of the provably total computable functions of PA and HA as well.

Exercises

14.3.1. Prove Proposition 14.3.1.

14.3.2. Check that the realizers defined to establish Theorem 14.3.3 meet their specifications.

14.3.3. Show that for every closed instance A of (AC) or ($\mathrm{IP_{ef}}$) there is a closed term t such that HA^ω proves $t \,\mathbf{mr}\, A$.

14.3.4. Suppose T is a theory extending PRA. Show that a function $f(\vec{x})$ is a provably total computable function of T if and only if there is an index e such that f is φ_e and T proves $\forall \vec{x}\, \exists s\, T(e, (\vec{x}), s)$.

14.3.5. Suppose T is a theory extending PRA and every Π_2 theorem $\forall \vec{x}\, \exists y\, R(\vec{x}, y)$ of T, where R is primitive recursive, is witnessed by a function $f(\vec{x})$ in a set of functions Γ. Then every provably total computable function of T can be obtained by composing a function in Γ with the function $z \mapsto (z)_0$, which returns the first element of the pair represented by z.

14.4 Finite-Type Arithmetic

The fact that modified realizability smoothly applies to both HA^ω and its extensional variant, $\mathrm{E\text{-}HA}^\omega$, and the fact that these theories provide expressive means for reasoning about higher-type objects, makes them natural settings to study the computational content of common mathematical arguments. But, as we saw in Sections 13.5 and 13.6, reasoning about higher-type objects and equality between them is sometimes subtle. The goal of this section is to take these subtleties head on, with two goals in mind.

The first is to clarify the relationship between classical theories of finite type and their intuitionistic versions, that is, relationships between variants of PA^ω and HA^ω. The second is to isolate a quantifier-free theory of primitive recursive functionals, analogous to the quantifier-free theory PRA of primitive recursive functions. Such a theory is sometimes denoted PRA^ω but it is more common to use Gödel's original but less informative designation, T.

The Friedman–Dragalin translation, Theorem 10.4.6, shows that PA is conservative over HA for Π_2 formulas. The proof uses the fact that HA proves the law of the excluded middle for atomic formulas. In HA^ω, it is not possible to prove the law of the excluded middle for equalities $s =_\alpha t$ between terms of arbitrary type α, though it can still be done when α is NAT. Thus, if PA^ω proves $\forall \vec{x}\, \exists \vec{y}\, A$, where A is quantifier-free and uses only equality at type NAT, then HA^ω proves it as well. If we add the law of the excluded middle for equality at arbitrary types as axioms to HA^ω, the conclusion holds for arbitrary quantifier-free A.

In addition to the choice of classical or intuitionistic logic, we can also consider the use of the extensionality axioms (ext). Let Π be the class of formulas of the form $\forall \vec{x}\, \exists \vec{y}\, A$ where A is quantifier-free involving only equality of type NAT, and all the variables have type level at most 1. Then using the translation described at the end of Section 13.6, we have the following:

Theorem 14.4.1. PA^ω, HA^ω, $\mathrm{E\text{-}PA}^\omega$, $\mathrm{E\text{-}HA}^\omega$ *all prove the same sentences in* Π.

Remember that, in the presence of ext, we can define higher-type equality in terms of equality at type NAT.

Gödel's theory T is based on combinatory logic with a single basic type NAT, with constants 0, succ, and recursors R_α. The system includes classical quantifier-free logic, the axiom $0 \neq 1$, axioms $R\, f\, g\, 0 = f$ and $R\, f\, g\, (x+1) = g\, x\, (R\, f\, g\, x)$ for each recursor, and an induction rule for quantifier-free formulas:

From $A(0)$ and $A(x) \to A(\mathrm{succ}(x))$ conclude $A(t)$ for any term t.

We can prove in T that the successor function is injective and that zero is not a successor just as in primitive recursive arithmetic. (See Section 9.1 and Exercise 9.1.1.) T is sometimes used instead to refer to an axiomatic theory based on the simply typed lambda calculus instead of combinatory logic, but we will commit to combinatory logic here. It is also possible to consider a version of T based on intuitionistic logic in which it is not possible to prove the law of the excluded middle for equality at types other than NAT, but we will not consider such a theory here. In this formulation, we can view $HA^{\omega} + (LEM)_=$ as the extension of T with quantifiers and full induction, where $(LEM)_=$ denotes the law of the excluded middle for arbitrary atomic formulas in the language, also known as *decidable equality*.

T is designed to be a quantifier-free calculus, whereas the extensionality axioms are formulated using universal quantifiers in the antecedent of an implication. We can, however, approximate these axioms with a rule. If we restrict the language to equality at type NAT, we can interpret a higher-type equation $s =_\alpha t$ as $s\vec{x} =_{\mathrm{NAT}} t\vec{x}$, where the variables $\vec{x}$ are chosen so that $s\vec{x}$ and $t\vec{x}$ have type NAT. In a natural deduction formulation, we can then add the rule

$$\frac{s\vec{x} = t\vec{x}}{rs = rt}$$

where $\vec{x}$ is not free in the conclusion or any hypothesis. In an axiomatic formulation, this is tantamount to adding the rule

$$\frac{A \to s\vec{x} = t\vec{x}}{A \to rs = rt}$$

where A is quantifier-free and none of the variables $\vec{x}$ occur in A. We will refer to this version, where only equality at NAT is taken to be primitive, as WE-T, the *weakly extensional* version of T. In the next section, we will consider the corresponding version of HA^{ω}, which similarly restricts equality to type NAT and adds the formulation of the extensionality rule we have just seen, so that when $A \to s\vec{x} = t\vec{x}$ is provable without hypotheses, we can conclude $A \to rs = rt$. This theory is known as WE-HA^{ω}. The restriction to quantifier-free formulas A in the extensionality rule is somewhat ad hoc, and we will see that, as a result, WE-HA^{ω} does not satisfy the deduction theorem. The theory is nonetheless useful, and, in particular, proves the law of the excluded middle for quantifier-free formulas. The results of the last section still hold with WE-HA^{ω} in place of HA^{ω}.

We can also jettison extensionality and think of the objects of T in intensional terms, say, as programs or simply expressions. This perspective is compatible with the law of the excluded middle for equality between objects of higher type. We can then go a step further and add a constant $\chi_=$ of type $\alpha \to \alpha \to \mathrm{NAT}$ for each type α representing the characteristic function of equality. The theory I-T adds an axiom $\chi_{=_\alpha} xy = 1 \leftrightarrow x =_\alpha y$ for each α, and I-HA^{ω}, and I-PA^{ω} are defined similarly. Once again, the results of the last section carry over to I-HA^{ω}.

Thus we have two intensional versions of intuitionistic arithmetic in finite types, HA^{ω} and I-HA^{ω}, and two extensional versions, WE-HA^{ω} and E-HA^{ω}. We also have the corresponding versions of PA^{ω} and T, except that there is no quantifier-free version of E-T. The double-negation translation interprets each version of PA^{ω} in the corresponding version of HA^{ω}, so each classical theory proves the same negative formulas as its intuitionistic counterpart. Only I-HA^{ω} proves the law of the excluded middle for arbitrary equations, but from an intensional point of view, it also makes sense to add this as an axiom to HA^{ω}.

14.5 The Dialectica Interpretation

We now turn to another interpretation of arithmetic, due to Gödel, which is named the *Dialectica* interpretation after the journal in which it was first published. Like modified realizability, the Dialectica interpretation serves to characterize the provably total computable functions of first-order arithmetic in terms of the primitive recursive functionals of finite type, but some of its features set it apart. For one thing, instead of interpreting HA^ω into itself, it interprets HA^ω into Gödel's quantifier-free theory T. This is a more dramatic reduction, which trades quantifiers for higher types. Moreover, the Dialectica interpretation keeps track of information that is erased by realizability interpretations. For example, the realizer of a formula $\neg A$ never carries useful information, whereas a term extracted by the Dialectica interpretation often does. In particular, when A is quantifier-free, a term witnessing the Dialectica interpretation of $\neg\neg\exists x\, A$ provides a witness to $\exists x\, A$. Also, if A and B are quantifier-free, a realizer for $\forall x\, A \to \forall y\, B$ carries no useful information, whereas a witness to the Dialectica interpretation is a function that takes a counterexample to $\forall y\, B$ to a counterexample to $\forall x\, A$.

The Dialectica interpretation assigns to each formula A in the language of arithmetic a translation A^D of the form

$$A^D = \exists \vec{x}\, \forall \vec{y}\, A_D(\vec{x}, \vec{y}),$$

where A_D is a (quantifier-free) formula in the language of T. The variables of A_D consist of those of A together with tuples of new variables $\vec{x}$ and $\vec{y}$, possibly empty. The translation goes by recursion on formulas, and in the following definition, we assume that A^D is $\exists \vec{x}\, \forall \vec{y}\, A_D$ and B^D is $\exists \vec{u}\, \forall \vec{v}\, B_D$. We require that the variables $\vec{x}, \vec{y}$ are distinct from the variables of B_D and that the variables $\vec{u}, \vec{v}$ are distinct from the variables of A_D, renaming them if necessary. The clauses are as follows:

- $A^D = A_D = A$ when A is atomic
- $(A \wedge B)^D = \exists \vec{x}, \vec{u}\, \forall \vec{y}, \vec{v}\, (A_D \wedge B_D)$
- $(A \vee B)^D = \exists z, \vec{x}, \vec{u}\, \forall \vec{y}, \vec{v}\, ((z = 0 \wedge A_D) \vee (z = 1 \wedge B_D))$
- $(\forall z\, A(z))^D = \exists \vec{X}\, \forall z, \vec{y}\, A_D(\vec{X}\, z, \vec{y}, z)$
- $(\exists z\, A(z))^D = \exists z, \vec{x}\, \forall \vec{y}\, A_D(\vec{x}, \vec{y}, z)$
- $(A \to B)^D = \exists \vec{U}, \vec{Y}\, \forall \vec{x}, \vec{v}\, (A_D(\vec{x}, \vec{Y}\, \vec{x}\, \vec{v}) \to B_D(\vec{U}\, \vec{x}, \vec{v}))$.

In the first clause, we identify atomic formulas in the language of arithmetic with atomic formulas in the language of T by identifying the function symbols of HA with symbols for the corresponding primitive recursive functionals. In the last three clauses, we adopt the following conventions: if $\vec{x}$ is a tuple of variables $x_0, \ldots, x_{n-1}$, then $\vec{X}$ is a tuple of functions of the same length, and $\vec{X}\, z$ abbreviates the tuple $X_0\, z, \ldots, X_{n-1}\, z$. In the last clause $\vec{U}\, \vec{x}$ abbreviates the tuple $U_0\, \vec{x}, \ldots, U_{m-1}\, \vec{x}$ of the same length as $\vec{u}$, and similarly for $\vec{Y}\, \vec{x}\, \vec{v}$.

In the treatment of the universal quantifier, naively putting $\forall z$ in front of A^D would result in the formula $\forall z\, \exists \vec{x}\, \forall \vec{y}\, A_D(\vec{x}, \vec{y}, z)$. The definition of $(\forall z\, A)^D$ is the result of Skolemizing the xs so that they become functions of z. The most interesting clause in the definition is the one for implication. The formula $A^D \to B^D$ is of the following form:

$$\exists \vec{x}\, \forall \vec{y}\, A_D \to \exists \vec{u}\, \forall \vec{v}\, B_D.$$

To motivate the translation, bring the quantifiers to the front in the following order:

$$\forall \vec{x}\, \exists \vec{u}\, \forall \vec{v}\, \exists \vec{y}\, (A_D \to B_D).$$

We can interpret this as saying that from witnesses to the antecedent we obtain witnesses to the conclusion, such that any counterexample to the conclusion yields a counterexample to the antecedent. The Dialectica interpretation is the result of Skolemizing $\vec{u}$ and $\vec{y}$, that is, replacing $\vec{u}$ by $\vec{U}\,\vec{x}$ and replacing $\vec{y}$ by $\vec{Y}\,\vec{x}\,\vec{v}$. Not all of the quantifier transformations just described are intuitionistically valid, and we will clarify the additional assumptions that are needed to justify them later in this section.

The interpretation of $\neg A$ is obtained by replacing B by the formula $\bot$ in the clause for implication, so we have:

- $(\neg A)^D = \exists \vec{Y}\, \forall \vec{x}\, \neg A_D(\vec{x}, \vec{Y}\,\vec{x})$.

We can think of this as saying that $\vec{Y}$ demonstrates the falsity of $\exists \vec{x}\, \forall \vec{y}\, A(\vec{x}, \vec{y})$ by providing, for each $\vec{x}$, a counterexample to $\forall \vec{y}\, A(\vec{x}, \vec{y})$. The formula $\neg\neg A$ is interpreted as $\exists \vec{X}\, \forall \vec{Y}\, A_D(\vec{X}\,\vec{Y}, \vec{Y}\,(\vec{X}\,\vec{Y}))$, which provides a weak way of asserting $\exists \vec{x}\, \forall \vec{y}\, A_D(\vec{x}, \vec{y})$: rather than providing a tuple of witnesses $\vec{x}$, it provides a tuple of functions $\vec{X}$ that foil any putative counterexample $\vec{Y}$. This is an instance of Kreisel's *no counterexample interpretation*, discussed further below.

The fact that the principles needed to show that every formula A is equivalent to A^D are not all intuitionistically valid makes it all the more interesting that we have the following result:

Theorem 14.5.1. *If* HA *proves a formula A, there is a tuple of terms $\vec{t}$ such that* T *proves $A_D(\vec{t}, \vec{y})$.*

As usual, to prove Theorem 14.5.1, we need to pick a proof system and use induction on derivations. In contrast to the situation for realizability and modified realizability, natural deduction is not a good choice: the interpretation of implication goes more smoothly with an axiomatic system. We will use the following axiomatization of first-order logic, due to Gödel. It was designed with the Dialectica interpretation in mind, and it has the virtue of keeping the axioms and rules short and simple.

1. from A and $A \to B$ conclude B
2. from $A \to B$ and $B \to C$ conclude $A \to C$
3. $A \lor A \to A$
4. $A \to A \land A$
5. $A \to A \lor B$
6. $A \land B \to A$
7. $A \lor B \to B \lor A$
8. $A \land B \to B \land A$
9. from $A \to B$ conclude $C \lor A \to C \lor B$
10. from $A \to (B \to C)$ conclude $A \land B \to C$
11. from $A \land B \to C$ conclude $A \to (B \to C)$
12. $\bot \to C$
13. from $A \to B$ conclude $A \to \forall z\, B$, assuming z is not free in A
14. $\forall z\, A \to A[t/z]$, assuming t is free for z in A
15. $A[t/z] \to \exists z\, A$, assuming t is free for z in A
16. from $A \to B$ conclude $\exists z\, A \to B$, assuming z is not free in B.

Showing that the conclusion of Theorem 14.5.1 holds for the axioms and is maintained by the rules is not hard. For example, for 1, suppose we are given a tuple of terms $\vec{r}$ such that T proves $A_D(\vec{r}, \vec{y})$ and terms $\vec{s}$ and $\vec{t}$ such that T proves $A_D(\vec{x}, \vec{s}\,\vec{x}\,\vec{v}) \to B_D(\vec{t}\,\vec{x}, \vec{v})$. We want a tuple $\vec{q}$ such that T proves $B_D(\vec{q}, \vec{v})$. By substituting $\vec{s}\,\vec{r}\,\vec{v}$ for $\vec{y}$ in the first hypothesis and $\vec{r}$ for $\vec{x}$ in the second, we see that taking $\vec{q} = \vec{t}\,\vec{r}$ works.

Verifying the translation of the axiom $A \to A \wedge A$ raises a subtle issue. If the translation of the hypothesis is $\exists \vec{x}\, \forall \vec{y}\, A_D(\vec{x}, \vec{y})$, the translation of the conclusion is of the form

$$\exists \vec{x}', \vec{x}''\, \forall \vec{y}', \vec{y}''\, (A_D(\vec{x}', \vec{y}') \wedge A_D(\vec{x}'', \vec{y}'')).$$

We need to provide tuples of terms $\vec{U}$ and $\vec{U}'$ such that if $\vec{x}$ witnesses the antecedent, then $\vec{U}\,\vec{x}$, $\vec{U}'\,\vec{x}$ witnesses the conclusion. To that end, we can take $\vec{U}$ and $\vec{U}'$ to be identity functions. But then we also need terms $\vec{Y}$ such that given $\vec{y}'$ and $\vec{y}''$ falsifying the conclusion, $\vec{Y}\,\vec{y}'\,\vec{y}''$ is a counterexample to $A_D(\vec{x}, \vec{y})$. We can achieve that for defining, for each i less than the length of $\vec{y}$, functionals Y_i that determine which of $A_D(\vec{x}, \vec{y}')$ or $A_D(\vec{x}, \vec{y}'')$ is false, if either one is, and return y'_i or y''_i accordingly. Using the primitive recursive cases function and the fact that $A_D(\vec{x}, \vec{y})$ is a primitive recursive relation, we can define each Y_i so that

$$Y_i\,\vec{x}\,\vec{y}'\,\vec{y}'' = \begin{cases} y''_i & \text{if } A_D(\vec{x}, \vec{y}') \\ y'_i & \text{otherwise,} \end{cases}$$

and these do what we want.

The quantifier-free axioms describing 0, succ, $+$, and $\cdot$ in HA are provable in T. To interpret induction, suppose that we are given terms $\vec{r}$, $\vec{s}$, and $\vec{t}$ so that T proves $A_D(0, \vec{r}, \vec{y})$ and

$$A_D(u, \vec{x}, \vec{s}\,u\,\vec{x}\,\vec{y}) \to A_D(\text{succ}\, u, \vec{t}\,u\,\vec{x}, \vec{y}).$$

We want terms $\vec{q}$ such that $A_D(u, \vec{q}\,u, \vec{y})$ holds for every u and $\vec{y}$. By recursion, define $\vec{q}$ so that $\vec{q}\,0 = \vec{r}$ and $\vec{q}\,(\text{succ}\,u) = \vec{t}\,u\,(\vec{q}\,u)$. If we write $B(u, \vec{y})$ for $A_D(u, \vec{q}\,u, \vec{y})$, we then have $B(0, \vec{y})$ and $B(u, \vec{s}\,u\,(\vec{q}\,u)\,\vec{y}) \to B(\text{succ}\,u, \vec{y})$. We need to prove $B(u, \vec{y})$. The inference is an instance of quantifier-free induction with parameters, and it can be justified in T in a manner similar to the way it was justified in PRA by Proposition 9.1.3.

Just as modified realizability applies not just to HA but also to HA$^\omega$, the Dialectica interpretation can be extended to HA$^\omega$ as well. However, a complication arises with the treatment of the axiom $A \to A \wedge A$, which required a definition by cases on a quantifier-free formula in the argument above. When we extend the argument to the language of HA$^\omega$, we have new atomic formulas $s =_\alpha t$ at types α other than NAT, and HA$^\omega$ does not provide the means to carry out the requisite definition. One solution is to use I-T, described in the last section, as the target of the translation, in which case we can take I-HA$^\omega$ to be the source of the interpretation.

Theorem 14.5.2. *If* I-HA$^\omega$ *proves a formula A, there is a tuple of terms* $\vec{t}$ *such that* I-T *proves* $A_D(\vec{t}, \vec{y})$.

An alternative solution is to restrict to equality at type NAT in HA$^\omega$. Using the theory WE-HA$^\omega$ allows us to do that, as long as we include the corresponding rule in the target theory.

Theorem 14.5.3. *If* WE-HA$^\omega$ *proves a formula A, there is a tuple of terms $\vec{t}$ such that* WE-T *proves $A_D(\vec{t}, \vec{y})$.*

Yet another means of dealing with the axiom $A \to A \wedge A$, using a variant of the Dialectica interpretation called the *Diller–Nahm interpretation*, is described in the exercises below. Exercise 14.5.9 shows that the Dialectica interpretation of the full extensionality axiom, (ext), is not computationally valid, so there is no hope of extending the interpretation to E-HA$^\omega$.

Since WE-T$^\omega$ is contained in WE-HA$^\omega$ and I-T$^\omega$ is contained in I-HA$^\omega$, the Dialectica translation interprets WE-HA$^\omega$ and I-HA$^\omega$ in themselves. As with modified realizability, we can add additional axioms to the source theory whenever their D-translations can be verified in the target, in the sense of Theorems 14.5.2 and 14.5.3. When B is quantifier-free, the Dialectica translation of the universal sentence $\forall \vec{x}\, B$ is $\forall \vec{x}\, B$ itself, so we can also add such sentences as axioms in the source theory, provided they are added as axioms to the target as well. When the target is one of the theories WE-T or I-T, which are based on quantifier-free logic, this means adding arbitrary substitution instances.

Let us now reflect on the principles that are needed to prove a formula equivalent to its Dialectica interpretation. Skolemization requires the axiom of choice, (AC), discussed in Section 14.3. Bringing the quantifier $\exists u$ to the front of the formula requires the principle of independence of premise for universal formulas, a schema that I will denote by (IP$_\forall$):

$$(\forall y\, A \to \exists u\, B) \to \exists u\, (\forall y\, A \to B).$$

In this schema B is arbitrary but A is assumed to be quantifier-free. Bringing out the last quantifier requires making use of equivalences $(\forall y\, A \to B) \leftrightarrow \exists y\, (A \to B)$ where A and B are quantifier-free. The right-to-left direction is intuitionistically valid, and I leave it as an exercise for you to check that the other direction can be justified by an appeal to an instance of Markov's principle, $\neg \forall y\, A \to \exists y\, \neg A$, where A is assumed to be quantifier-free. I will denote that schema by (MP$_{\mathrm{qf}}$). Note that what counts as quantifier-free depends on the underlying language; the language of WE-HA$^\omega$ restricts to equality at type NAT, whereas the language of I-HA$^\omega$ adds characteristic functions $\chi_=$ at each type α. The statements below should be interpreted accordingly, but the next proposition is not sensitive to the language.

Proposition 14.5.4. *Over intuitionistic logic, the schema $A \leftrightarrow A^D$ is equivalent to* (AC) + (IP$_\forall$) + (MP$_{\mathrm{qf}}$).

Proof By induction on A, it is straightforward to show that the second schema proves the first. The converse direction follows from the fact that the Dialectica interpretation of any instance of (AC), (IP$_\forall$), or (MP$_{\mathrm{qf}}$) is easily witnessed. □

Corollary 14.5.5. *For every formula A,* WE-HA$^\omega$ + (AC) + (IP$_\forall$) + (MP$_{\mathrm{qf}}$) *proves A if and only if there is a tuple of terms $\vec{t}$ such that* WE-T *proves $A_D(\vec{t}, \vec{y})$. The same holds with* I-HA$^\omega$ *and* I-T *replacing* WE-HA$^\omega$ *and* WE-T*, respectively.*

The Dialectica translation extends to the classical theories WE-PA$^\omega$ and I-PA$^\omega$ via the double-negation translation. It is rather pleasant that the Dialectica translation of $\neg\neg\exists \vec{x}\, A$, where A is quantifier-free, is equivalent to $\exists \vec{x}\, A$. Thus if A is quantifier-free and there is a proof of $\forall \vec{x} \, \exists \vec{y} \, A(\vec{x}, \vec{y})$ in I-PA$^\omega$ or WE-PA$^\omega$, there is a tuple of terms $\vec{t}$ such that $A(\vec{x}, \vec{t}\,\vec{x})$ is provable in the corresponding version of T.

What additional principles can we add to PA$^\omega$? There is no point to adding versions of independence of premise or Markov's principle, because these are already derivable in classical logic. The double-negation translation of an instance of the axiom of choice is of the form

$$\forall x \, \neg\neg\exists y \, A(x, y) \rightarrow \neg\neg\exists f \, \forall x \, A(x, f\,x),$$

and the Dialectica interpretation of this is not derivable in any version of T. We will see in Chapter 15 that adding the axiom of choice (AC) to PA$^\omega$ results in a much stronger theory. On the other hand, when A is assumed to be quantifier-free, the Dialectica interpretation of the antecedent is equivalent to $\exists f \, \forall x \, A(x, f\,x)$, which implies the Dialectica interpretation of the conclusion. Thus, we can add the restricted schema (AC$_{\mathrm{qf}}$) for quantifier-free A. Conversely, that schema is enough to show that every formula is equivalent to the Dialectica interpretation of its double-negation translation. Thus we have:

Proposition 14.5.6. *Over classical logic, the schemas* (AC$_{\mathrm{qf}}$) *and* $A \leftrightarrow (A^N)^D$ *are equivalent.*

Corollary 14.5.7. *For every formula A,* WE-PA$^\omega$ + (AC$_{\mathrm{qf}}$) *proves A if and only if there is a tuple of terms $\vec{t}$ such that* WE-T *proves* $(A^N)_D(\vec{t}, \vec{y})$. *The same holds with* I-PA$^\omega$ *and* I-T *replacing* WE-PA$^\omega$ *and* WE-T, *respectively.*

The Dialectica interpretation provides another route to showing that every provably total computable functions of PA and HA is given by a primitive recursive functional of type 1. It also yields a nice computational interpretation of arbitrary prenex consequences of classical arithmetic. Suppose PA (or even PA$^\omega$ + (AC$_{\mathrm{qf}}$)) proves a prenex formula $\forall x_0 \, \exists y_0 \ldots \forall x_{n-1} \, \exists y_{n-1} \, A$, where A is quantifier-free. By Exercise 4.6.8, the double-negation translation of this formulas implies

$$\neg\exists x_0 \, \forall y_0 \ldots \exists x_{n-1} \, \forall y_{n-1} \, \neg A,$$

whose Dialectica translation is the same as that of

$$\forall X_0, \ldots, X_{n-1} \, \exists y_0, \ldots, y_{n-1} \, A(X_0, (X_1\, y_0), \ldots, (X_{n-1}\, y_0 \ldots y_{n-1})).$$

By the Dialectica interpretation we can extract terms $t_0(\vec{X}), \ldots, t_{n-1}(\vec{X})$ to serve as the witnesses for $y_0, \ldots, y_{n-1}$. This is Kreisel's *no counterexample interpretation* for arithmetic, and it can be understood as follows: given functions $X_0, \ldots, X_{n-1}$ providing potential counterexamples to the original formula, $t_0, \ldots, t_{n-1}$ show, effectively, that the counterexamples do not work.

Exercises

14.5.1. It should be clear that Gödel's axiomatization is valid for first-order intuitionistic logic. Conversely, show that it is a complete set of axioms, by showing it can derive all the axioms and rules for the axiomatic system presented in Chapter 4.

14.5.2. Fill in the details of the proof of Theorem 14.5.1.

14.5.3. Suppose A^D is $\exists x\, \forall y\, A_D(x, y)$. Find terms witnessing the validity of $(\neg(A \wedge \neg A))^D$.

14.5.4. Do the same for $(\neg\neg(A \vee \neg A))^D$.

14.5.5. Suppose A^D is as in the previous problems and B^D is $\exists u\, \forall v\, B_D(u, v)$. Given terms witnessing the Dialectica interpretations of $\neg(A \wedge B)$ and A, construct a term witnessing the Dialectica interpretation of $\neg B$.

14.5.6. For any quantifier-free formulas $A(x, y)$ and $B(u, v)$, PA proves

$$\exists x\, \forall y\, \exists x' \leq x\, A(x', y) \wedge \exists u\, \forall v\, \exists u' \leq u\, B(u', v) \to \exists w\, \forall y, v\, (\exists x' \leq w\, A(x', y) \wedge \exists u' \leq w\, B(u', v)).$$

Taking PA to include PRA, you can treat the bounded quantifiers as part of the quantifier-free part. Find terms witnessing the *ND*-translation of that sentence.

14.5.7. The *Diller–Nahm* interpretation is a variant on the Dialectica interpretation that avoids the need to use a case functional to interpret the axiom $A \to A \wedge A$. Heuristically, the idea is to modify the clause for implication so that an implication $\exists x\, \forall y\, A(x, y) \to \exists u\, \forall v\, B(u, v)$ is understood as

$$\forall x\, \exists u\, \forall v\, \exists s\, (\forall y \in s\, A(x, y) \to B(u, v)),$$

where s represents a finite set of elements of the relevant type. The Diller–Nahm translation of the implication then takes the form

$$\exists U, S\, \forall x, v\, (\forall y \in S\, x\, v\, A(x, y) \to B(U\, x, v)).$$

Intuitively, rather than returning a counterexample to the antecedent, S returns a finite set of possible counterexamples. Explain how this solves the problem of interpreting the axiom $A \to A \wedge A$.

14.5.8. Suppose A and B are formulas of HA, A^D is $\exists x\, \forall y\, A_D(x, y)$, and B^D is $\exists u\, \forall v\, B_D(u, v)$. Then the Dialectica interpretation of $\neg\neg A$ is

$$\exists X\, \forall Y\, A_D(X\, Y, Y\, (X\, Y))$$

and the interpretation of $\neg\neg B$ is

$$\exists U\, \forall V\, B_D(U\, V, V\, (U\, V)).$$

Confirm that the interpretation of $\neg\neg(A \wedge B)$ is equivalent in HA^ω to

$$\exists X, U\, \forall Y, V\, (A_D(X\, Y\, V, Y\, (X\, Y\, V)\, (U\, Y\, V)) \wedge B_D(U\, Y\, V, V\, (X\, Y\, V)\, (U\, Y\, V))).$$

Since $\neg\neg A \wedge \neg\neg B \to \neg\neg(A \wedge B)$ is intuitionistically valid, given terms X and U satisfying the interpretations of $\neg\neg A$ and $\neg\neg B$ respectively, we should be able to find terms X' and U' satisfying the interpretation of $\neg\neg(A \wedge B)$. This is surprisingly tricky; see the bibliographical notes.

14.5.9. Consider the following instance of an extensionality axiom, where f and g have type $\mathrm{NAT} \to \mathrm{NAT}$ and F has type $(\mathrm{NAT} \to \mathrm{NAT}) \to \mathrm{NAT}$:

$$\forall F, f, g\, (\forall x\, (f\, x = g\, x) \to F\, f = G\, g).$$

Show that $\mathrm{HA}^\omega + (\mathrm{MP}_{\mathrm{qf}})$ proves that this implies

$$\forall F, f, g\, \exists x\, (f\, x = g\, x \to F\, f = F\, g).$$

There is no primitive recursive functional $X\, F\, f\, g$ witnessing the existential quantifier in this last sentence; see the reference at the end of this chapter. Hence the Dialectica interpretation validates $\mathrm{HA}^\omega + (\mathrm{MP}_{\mathrm{qf}})$ but not the conclusion, which shows that it does not validate extensionality.

14.5.10. Show that if A and B are quantifier-free, then Markov's principle for quantifier-free formulas, $(\mathrm{MP}_{\mathrm{qf}})$, implies $(\forall y\, A \to B) \leftrightarrow \exists y\, (A \to B)$ in I-HA^ω and WE-HA^ω.

14.5.11. Prove Proposition 14.5.4.

Bibliographical Notes

Realizability Troelstra (1998) provides a thorough overview of realizability interpretations. The presentation in Sections 14.1 and 14.2 also draws heavily on Beeson (1985), Troelstra (1973), and Troelstra and van Dalen (1988). The retrospective survey Kleene (1973) provides historical background and motivation. The Aczel slash, described in Exercise 14.2.7, is discussed in Troelstra (1998) as well as Troelstra and van Dalen (1988). The notation used in this chapter for schemas like (MP) and (IP) is similar to that used by other authors, but there are variations in the literature. For example, Church's thesis, which I have denoted as (CT) here, is denoted (CT_0) in Troelstra (1998) and Troelstra and van Dalen (1988).

Metamathematics of intuitionistic arithmetic Troelstra (1973) is a definitive reference for metamathematical properties of intuitionistic arithmetic. See also Troelstra and van Dalen (1988) and Beeson (1985). For more recent developments, see Moschovakis (2018), Fujiwara and Kohlenbach (2018), and the references there.

In Section 14.2, we observed that the explicit definability property implies the disjunction property. For a proof that under general conditions the converse also holds, see Friedman (1975).

Modified realizability and the Dialectica intepretation For more about modified realizability, see Troelstra (1998), Troelstra (1973), Kohlenbach (2008), and Sørensen and Urzyczyn (2006). For the Dialectica interpretation, see Avigad and Feferman (1998), Troelstra (1973), Kohlenbach (2008), and Troelstra's notes to Gödel's 1958 *Dialectica* paper in Gödel (1986–2003). For the Diller–Nahm translation, described in Exercise 14.5.7, see Diller and Nahm (1974) or Troelstra (1973). Oliva (2014) surveys a unified framework for modified realizability, the Dialectica interpretations, and variations.

Kohlenbach (2008) provides an introduction to *proof mining*, which uses methods based on modified realizability and the Dialectica interpretation to extract useful information from nonconstructive proofs in analysis, such as bounds on rates of convergence. That book includes, in particular, a number of interesting variations on the two interpretations.

Using a restriction of the primitive recursive functionals of finite type described in the notes to Chapter 13, we can use modified realizability to characterize the provably total computable functions of $\mathrm{I}\Sigma^{\mathrm{i}}_1$, and once can use the Dialectica interpretation to characterize the provably total computable functions of both $\mathrm{I}\Sigma_1$ and $\mathrm{I}\Sigma^{\mathrm{i}}_1$. See Section 5 of Avigad and Feferman (1998) and the analogous interpretations of systems of bounded arithmetic in Cook and Urquhart (1993).

For a solution to Exercise 14.5.8, see Section 3.1 of Escardó and Oliva (2010). For Dialectica-style interpretations of stronger theories based on the ideas developed there, see Escardó and Oliva (2017). For a proof that there are no primitive recursive functionals witnessing the extensionality axiom along the lines described in Exercise 14.5.9, see the appendix by W. A. Howard in Troelstra (1973).

Finite type arithmetic For more information on theories of finite type arithmetic, see Feferman (1977), Troelstra and van Dalen (1988), Beeson (1985), Troelstra (1973), Kohlenbach (2008), Avigad and Feferman (1998), and Troelstra (1998).

Ordinal analysis Another way of measuring the strength of an axiomatic theory is in terms of the length of the orders that it can prove to be well ordered, or in terms of the types of ordinal-recursive function definitions it can justify, along the lines described in Section 8.5. Mancosu et al. (2021) and Pohlers (2009) provide introductions to this program, known as *ordinal analysis*. The proof-theoretic ordinal of first-order arithmetic, in this sense, is an ordinal known as ε_0.

15

Second-Order Logic and Arithmetic

In the late nineteenth century, mathematicians began to use the concepts of *set*, *function*, and *relation* to support abstract reasoning in algebra, geometry, number theory, and analysis. These notions are so fundamental today that it is hard to imagine mathematics without them.

In Chapter 13, we considered the simply typed lambda calculus as a minimal theory of functions. Adding primitive recursors enables us to define functions on the natural numbers, but there are models of Gödel's T in which every function is computable, so this does not provide us with the means to define noncomputable functions. Since we have generally identified sets and relations with their characteristic functions, this shows that we do not yet have formal means to establish the existence of noncomputable sets and relations either.

In an article in the *Mathematische Annalen* in 1895, Georg Cantor wrote:

> By a *set* we mean any collection M of determinate, distinct objects (called the elements of M) of our intuition or thought into a whole.

This definition begs the question as to what exactly it means to collect objects of our intuition or thought, but it suggests that, at least, any property we can describe in precise terms should give rise to a set. In particular, fixing any formal language for which we have a precise semantics, any formula $A(\vec{x})$ should give rise to the relation $R(\vec{x})$ that holds exactly of the elements satisfying A. This is known as a *comprehension principle*, since the relation R is taken to comprehend the elements satisfying A.

In this chapter, we will study a formal theory, *second-order logic*, based on such a comprehension principle. The logic describes a first-order universe of individual objects together with a second-order universe of relations on those objects. In contrast to first-order logic, the syntactic and semantic notions of consequence come apart dramatically, in the sense that there is no computable deductive system that can account for the entailments that are valid in the full semantic interpretation. But this does not preclude us from using deductive systems for second-order logic to derive *some* of the valid entailments. The deductive limitations are an instance of the incompleteness phenomena we explored in Chapter 12. The fact that the language of second-order logic is rich enough to express strong mathematical statements makes it all the more important to develop suitable means of deductive reasoning.

We will also consider *second-order arithmetic*, a second-order theory of the natural numbers. And we will see that there is no reason to stop at level two; Section 15.7 briefly introduces higher-order logic and arithmetic, which serve as a prelude to *simple type theory*, a formal axiomatic foundation discussed in Chapter 17.

15.1 Second-Order Logic

Second-order logic can be viewed as a many-sorted logic in which we introduce, for each n, a sort ranging over n-place relations on the universe and a function symbol $\mathrm{app}_n(R, \vec{x})$ to denote the result $R(\vec{x})$ of applying the relation R to elements $\vec{x}$ in the first-order universe. We will refer to the new sorts collectively as the *second-order* sorts. By convention, we write $R(\vec{x})$ instead of $\mathrm{app}_n(R, \vec{x})$. The language of second-order logic allows us to quantify over the first- and second-order sorts, enabling us to quantify over predicates and relations as well as over elements of the first-order universe.

For example, recall from Section 5.5 that a *directed graph* is a first-order structure (V, E) where E is a binary relation on V. We think of V as the set of vertices and $E(a, b)$ as saying that there is an edge between a and b. We saw in Section 5.5 that we can define the set of *undirected graphs* with the first-order formula $\forall x, y\, (E(x, y) \to E(y, x))$, which says that the relation is symmetric. An undirected graph is *disconnected* if the vertices can be divided into two nonempty sets such that there is no edge between a vertex in one and a vertex in the other. Using a unary predicate P, we can express this with a second-order sentence,

$$\exists P\, (\exists x\, P(x) \wedge \exists y\, \neg P(y) \wedge \forall u, v\, (E(u, v) \to (P(u) \leftrightarrow P(v)))).$$

The semantics of second-order logic, presented in the next section, provides a precise sense in which this defines disconnectedness on the class of undirected graphs.

If L is any first-order language, we can consider the set of second-order formulas over L in the many-sorted language that includes the function and relation symbols of L on the first-order sort. Variations are possible; we might want to add other first-order sorts, or operations other than app on the second-order sorts. For our purposes, restricting attention to the second-order extension of a one-sorted first-order language will suffice. Notice that the relation symbols of L can be treated instead as constant symbols of the corresponding second-order sort. As usual, we can replace function symbols with functional relations. If L is the empty language with no function or relation symbols at all, the corresponding second-order language is called *pure second-order logic*.

As a formal theory, second-order logic extends many-sorted logic, so we have the usual propositional connectives, quantifiers ranging over the first-order universe, and quantifiers ranging over n-ary predicates for each n. We add the following *comprehension schema*:

$$\exists R\, \forall \vec{x}\, (R(\vec{x}) \leftrightarrow A(\vec{x})).$$

As usual, A may have free variables other than $\vec{x}$, in which case, we leave the universal quantifiers over those variables implicit. We require that R is not a free variable of A; otherwise, we could take $A(\vec{x})$ to be the formula $\neg R(\vec{x})$ and derive a contradiction. We also include the following axiom of extensionality for relations:

$$\forall R, R'\, (\forall \vec{x}\, (R(\vec{x}) \leftrightarrow R'(\vec{x})) \to R = R').$$

The converse implication follows from the usual equality axioms. The axiom tells us that whenever two relations agree on all possible arguments, each can be substituted for the other. Alternatively, we can eliminate the equality symbol on the second-order sorts and take the left-hand side of the implication above as a *definition* of $R = R'$. The exercises ask you to show that, with this definition, second-order logic proves the usual equality axioms, including substitution.

We can even eliminate the equality symbol on the first-order universe. In second-order logic, we can prove

$$\forall x, y\,(x = y \leftrightarrow \forall P\,(P(x) \leftrightarrow P(y))).$$

This says that two objects are equal if and only if they have all the same properties, characterization known as *Leibniz equality* or *identity of indiscernibles*. If we get rid of the equality symbol and take the right-hand side to be the definition of equality on the first-order universe, the usual laws of equality are derivable.

Given any formula A that contains an n-ary relation variable $R(\vec{x})$, we can imagine interpreting R by any formula $B(\vec{x})$. We use the notation $A[\lambda\vec{x}.\,B/R]$ to denote the result of replacing each atomic subformula $R(t_0, \ldots, t_{n-1})$ of A by the formula $B(t_0, \ldots, t_{n-1})$, even though the syntax of second-order logic, as we have just described it, does not include such lambda expressions. In a natural deduction formulation, the following rules can be derived from the comprehension axiom for the formula B:

$$\frac{\Gamma \vdash A[\lambda\vec{x}.\,B/R]}{\Gamma \vdash \exists R\, A} \qquad \frac{\Gamma \vdash \forall R\, A}{\Gamma \vdash A[\lambda\vec{x}.\,B/R]}$$

Conversely, we can derive the comprehension schema from the first of these rules. In natural deduction, therefore, the quantifier rules and comprehension are generally expressed in this way.

The presence of second-order quantification allows us to simplify our syntax. Let PROP denote the sort consisting of 0-ary relations, that is, propositions. For any formula $A(\vec{z})$ with the free variables shown, the comprehension schema gives us $\forall\vec{z}\,\exists P\,(P \leftrightarrow A(\vec{z}))$, which means that every formula is represented by a proposition. With this observation, we can define all the logical connectives in terms of universal quantification and implication as follows:

$$\begin{aligned}
A \wedge B &\equiv \forall P\,((A \to (B \to P)) \to P)\\
A \vee B &\equiv \forall P\,((A \to P) \to ((B \to P) \to P))\\
\bot &\equiv \forall P\, P\\
\exists x\, A &\equiv \forall P\,(\forall x\,(A \to P) \to P).
\end{aligned}$$

The exercises ask you to show that the usual rules for $\wedge$, $\vee$, $\bot$, and $\exists$ are derivable from these definitions. As a result, we can formulate second-order logic with the universal quantifier and implication as the only logical connectives.

In fact, the correspondence between formulas and elements of PROP suggests that we can simplify matters even further by taking formulas to be terms of sort PROP. The propositional connectives are then functions on that sort, and universal quantification becomes a variable-binding term construction. We will not follow this option here, but it will become the basis for our formulation of simple type theory in Section 17.1.

It is reasonable to want to quantify over functions in addition to relations. This can be reduced to quantification over relations by interpreting an n-ary function variable f as an $(n+1)$-ary relation symbol R_f and relativizing to *functional* relations, that is, relations satisfying $\forall\vec{x}\,\exists! y\, R(\vec{x}, y)$. We can then use the method of definite descriptions, presented in Section 4.7, to interpret uses of the function symbol f. Under this interpretation, both classical and intuitionistic versions of second-order logic can prove the following comprehension schema for functions:

$$\forall \vec{x}\, \exists! y\, A(\vec{x}, y) \to \exists f\, \forall \vec{x}\, A(\vec{x}, f(\vec{x})).$$

Here once again $\vec{z}$ denotes the free variables of A other than the ones shown. The stronger schema obtained by dropping the uniqueness requirements in the hypothesis is the *axiom of choice* for elements of the first-order universe, which does not follow from the other axioms of second-order logic but can be included as an additional axiom. We saw in Section 14.3 that adding choice to an intuitionistic theory like HA^{ω} does not add logical strength, but we will see in the next paragraph that, in contrast, adding the comprehension principle for functions to a classical theory like PA^{ω} is tantamount to adding the comprehension schema for relations. We can also consider adding a second-order version of the axiom of choice,

$$\forall \vec{x}\, \exists f\, A(\vec{x}, f) \to \exists F\, \forall \vec{x}\, A(\vec{x}, F_{\vec{x}}),$$

where F is a second-order variable whose arity is the sum of the length of $\vec{x}$ and the arity of f and $F_{\vec{s}}(\vec{t})$ is interpreted as $F(\vec{s}, \vec{t})$.

In classical second-order logic, it is possible interpret relations as functions instead. Suppose we assume that the first-order universe has at least two distinct elements $\top$ and $\bot$. Then, for any formula $A(\vec{x})$, classical logic proves the sentence $\forall \vec{x}\, \exists! y\, (y = \top \leftrightarrow A(\vec{x}))$. If $f(\vec{x})$ meets the specification of y, then $A(\vec{x})$ is equivalent to $f(\vec{x}) = \top$. Thus we can identify relations with their characteristic functions, and if we interpret $R(t_0, \ldots, t_{n-1})$ as an equation $\chi_R(t_0, \ldots, t_{n-1}) = \top$, the function comprehension schema implies the comprehension schema for relations.

In contrast, in intuitionistic logic, two truth values are not sufficient to interpret relations as functions. We have already noted that comprehension gives us $\forall \vec{x}\, \exists P\, (P \leftrightarrow A(\vec{x}))$ where P is of the second-order sort PROP, so PROP can be viewed as a sort of truth values. But any map $\vec{x} \mapsto P$ is a function from the sort of individuals to the sort of propositions, which takes us beyond second-order logic. It is, however, within the scope of *higher-order logic*, which we will discuss in Section 15.7

Exercises

15.1.1. Show that if we formulate second-order logic in a language without an equality symbol on the second-order sort and define $R = R'$ by the formula $\forall \vec{x}\, (R(\vec{x}) \leftrightarrow R'(\vec{x}))$, then the usual axioms for equality on that sort are derivable.

15.1.2. Show that in second-order logic, if we define first-order equality $x = y$ by the formula $\forall P\, (P(x) \leftrightarrow P(y))$, then the axioms for equality on the first-order sort are derivable.

15.1.3. Show that the natural-deduction formulation of second-order logic described above is equivalent to a formulation in which the universal quantifier can only be instantiated by a suitable relation variable in the $\forall$ elimination rule, and similarly for the $\exists$ introduction rule, if we also include the comprehension schema as axioms.

15.1.4. Show that with the definitions of $\wedge$, $\vee$, $\bot$, and $\exists$ given above, we can derive the usual axioms and rules.

15.1.5. Show that with the interpretation of functions in second-order logic as functional relations, second-order logic proves the comprehension schema for functions.

15.2 Semantics of Second-Order Logic

We have seen that second-order logic can be formulated as an axiomatic system or as a system of natural deduction. As usual, we can consider versions based on classical or intuitionistic logic. (Since we can define falsity as $\forall P\, P$, there is not much of a difference between intuitionistic and minimal logic.) We will use the phrase *second-order logic* to refer to the systems based on second-order comprehension, though at times we may wish to add extensionality and/or choice as well.

In this section, we will focus on classical logic, whereas in Section 15.6, we will provide a realizability interpretation that can be generalized to provide a computational semantics for second-order logic. If we view classical second-order logic as a many-sorted first-order logic, we already know what a model is: it is just a many-sorted first-order model satisfying the comprehension axioms, as well as extensionality and choice if we choose to include them. This is what is known as a *Henkin structure* or *Henkin model* for second-order logic. If $\mathfrak{M}$ is such a model, I will use $\mathcal{S}^{\mathfrak{M},n}$ to denote the second-order universe of n-ary relations, and I will continue to use $|\mathfrak{M}|$ to denote the first-order universe. Given an element U in $\mathcal{S}^{\mathfrak{M},n}$ and $a_0, \ldots, a_{n-1}$ in $|\mathfrak{M}|$, I will write $U(\vec{a})$ instead of $\mathrm{app}_n^{\mathfrak{M}}(U, \vec{a})$ to denote the statement that U is true at $\vec{a}$ according to $\mathfrak{M}$.

To make the semantics explicit, an assignment σ is now a function that assigns an element of $|\mathfrak{M}|$ to each first-order variable and an element of $\mathcal{S}^{\mathfrak{M},n}$ to each n-ary relation variable R. The recursive specification of truth values extends the one in Section 5.1 with the following new clauses:

- If R is an n-ary relation variable, then $\mathfrak{M} \models_\sigma R(t_0, \ldots, t_{n-1})$ if and only if $\sigma(R)(t_0^{\mathfrak{M},\sigma}, \ldots, t_{n-1}^{\mathfrak{M},\sigma})$.
- $\mathfrak{M} \models_\sigma \forall R\, A$ if and only if for every $U \in \mathcal{S}^{\mathfrak{M},n}$, $\mathfrak{M} \models_{\sigma[R \mapsto U]} A$.
- $\mathfrak{M} \models_\sigma \exists R\, A$ if and only if for some $U \in \mathcal{S}^{\mathfrak{M},n}$, $\mathfrak{M} \models_{\sigma[R \mapsto U]} A$.

Because this semantics is nothing more than first-order many-sorted semantics in disguise, we have:

Theorem 15.2.1. *Second-order logic with or without extensionality and choice is sound and complete for Henkin semantics with the corresponding classes of models.*

As was the case for models of the simply typed lambda calculus and combinatory logic in Section 13.6, there is a nicer description of the models that satisfy the extensionality axiom. The relation $\mathrm{app}_n^{\mathfrak{M}}(U, a_0, \ldots, a_{n-1})$ associates each element U of $\mathcal{S}^{\mathfrak{M},n}$ with an n-ary relation on the first-order universe, and extensionality says that if U and U' give rise to the same relation, then $U = U'$. As a result, we can simply take $\mathcal{S}^{\mathfrak{M},n}$ to consist of the n-ary relations themselves and dispense with the functions $\mathrm{app}_n(U, a_0, \ldots, a_{n-1})$. In other words, an *extensional Henkin structure* consists of a first-order structure, $\mathfrak{M}$, and for each n, a collection of n-ary relations on $|\mathfrak{M}|$, such that the resulting structure satisfies the comprehension schema.

Now let $\mathfrak{M}$ be an ordinary first-order model of a language, L, and consider the Henkin structure $\mathfrak{M}'$ for L in which for each n the sort of n-ary relations is interpreted a set of *all* n-ary relations on $|\mathfrak{M}|$. This is called the *full* second-order model over $\mathfrak{M}$, and for any second-order formula A and second-order assignment σ, we say $\mathfrak{M} \models_\sigma A$ in the *full* or *standard* second-order semantics if and only if $\mathfrak{M}' \models_\sigma A$ as a Henkin structure. This amounts to interpreting second-order quantifiers by the following clauses:

- $\mathfrak{M} \models_\sigma \forall R\, A$ if and only if for every n-ary relation U on $|\mathfrak{M}|$, $\mathfrak{M} \models_{\sigma[R \mapsto U]} A$.
- $\mathfrak{M} \models_\sigma \exists R\, A$ if and only if for some n-ary relation U on $|\mathfrak{M}|$, $\mathfrak{M} \models_{\sigma[R \mapsto U]} A$.

This model satisfies extensionality and choice, as well as any other principle that is validated by the ordinary set-theoretic understanding of predicates and relations. By default, the phrase *second-order definable* means definable in the full second-order semantics.

In Section 5.5, we used the compactness theorem to show that a number of classes of structures are not first-order definable. Here we will show that those examples are second-order definable, which implies, as a corollary, that compactness fails with respect to full second-order semantics. To start with, recall that Corollary 5.5.4 of Section 5.5 says that the class of finite sets is not definable by any set of first-order sentences.

Proposition 15.2.2. *There is a sentence A of pure second-order logic such that a structure $\mathfrak{M}$ satisfies A if and only if $|\mathfrak{M}|$ is infinite, and hence any structure $\mathfrak{M}$ satisfies $\neg A$ if and only if $|\mathfrak{M}|$ is finite.*

Proof Let A be the following sentence:

$$\exists f\ (\forall x, y\ (f(x) = f(y) \to x = y) \land \exists z\ \forall x\ (f(x) \neq z)).$$

This says that there is an injective function from the universe to itself that is not surjective. Since any injective function from a finite set to itself has to have just as many elements in the image as in the domain, this implies that the universe is infinite. Conversely, by the axiom of choice, if the universe is infinite there is always such a function. Choose a sequence $a_0, a_1, a_2, \ldots$ of distinct elements of the universe; the fact that the universe is infinite means that once $a_0, \ldots, a_n$ have been chosen, there is always a choice of a_{n+1} that is distinct from all of these. Then let f map each a_i to a_{i+1} and set $f(x) = x$ for every x not in the sequence. □

For another illustration of the power of second-order logic, recall Theorem 5.5.6, which says that the set of well-founded relations is not first-order definable.

Proposition 15.2.3. *Let L be a language with a binary relation symbol, R. There is a second-order formula in L that holds if and only if the interpretation of R is well founded.*

Proof Let A be the sentence $\forall P\ (\exists x\, P(x) \to \exists x\ (P(x) \land \forall y\ (P(y) \to \neg R(y, x))))$. Suppose $(X, R) \models A$. Since P can be interpreted by any subset of X, A says that every nonempty subset of X has an R-minimal element. □

Let us now show that second-order logic provides a categorical axiomatization of the standard interpretation $\mathfrak{N} = (\mathbb{N}, 0, \text{succ}, +, \cdot, \leq)$ of the language of arithmetic.

Proposition 15.2.4. *Let L be the language of arithmetic. Then there is a second-order sentence A in L such that any full model $\mathfrak{M}$ of A is isomorphic to $\mathfrak{N}$.*

Proof Consider the following sentences:

- $\forall x, y\ (\text{succ}(x) = \text{succ}(y) \to x = y)$
- $\forall x\ (\text{succ}(x) \neq 0)$
- $\forall P\ (P(0) \land \forall x\ (P(x) \to P(\text{succ}(x))) \to \forall x\, P(x))$.

The first says that succ(x) is injective, the second says that 0 is not in the image of succ, and the third is the second-order axiom of induction. Any structure $\mathfrak{M} = (|\mathfrak{M}|, 0^{\mathfrak{M}}, \text{succ}^{\mathfrak{M}})$

that satisfies these is inductively and freely generated by $0^{\mathfrak{M}}$ and $\text{succ}^{\mathfrak{M}}$, and Theorem 1.5.4 guarantees that it is isomorphic to $(\mathbb{N}, 0, \text{succ})$. Adding the recursive defining equations of $+$ and $\cdot$ and the explicit definition of $\leq$ in terms of $+$, as presented in Section 10.1, guarantees that the isomorphism extends to the interpretation of these symbols as well. Let A be the conjunction of all of these. □

Let $B(a, f)$ be the conjunction of the first three sentences in the proof of Proposition 15.2.4 with 0 and succ replaced by variables a and f respectively. Then the sentence $\exists a, f\, B(a, f)$ holds of a structure $\mathfrak{M}$ if and only if it is possible to pick an element $a^{\mathfrak{M}}$ and a function $f^{\mathfrak{M}}$ such that the structure $(|\mathfrak{M}|, a^{\mathfrak{M}}, f^{\mathfrak{M}})$ is isomorphic to $(\mathbb{N}, 0, \text{succ})$ – in other words, if and only if $\mathfrak{M}$ is countably infinite.

Proposition 15.2.5. *For any fixed language L, the class of structures with countably infinite universe is second-order definable.*

Similarly, let $C(a, f, g, h, R)$ be the axiomatic description of $(\mathbb{N}, 0, \text{succ}, +, \cdot, \leq)$ given in the proof of Proposition 15.2.4 with the symbols 0, succ, $+$, $\times$, and $\leq$ replaced by the variables a, f, g, h, and R. Let A be any second-order sentence in the language of arithmetic. Then A is true in the standard model if and only if the sentence $\forall a, f, g, h, R\, (C(a, f, g, h, R) \rightarrow A')$ is valid, where A' is A with 0, succ, $+$, $\cdot$, and $\leq$ replaced by a, f, g, h, and R. To see this, first notice that if the sentence is valid, it is true in $\mathfrak{N}$ when the variables a, f, g, h and R are given their standard interpretations. Conversely, suppose A is true in $\mathfrak{N}$, $\mathfrak{M}$ is any structure, and the variables a, f, g, h and R are assigned interpretations such that $C(a, f, g, h, R)$ is true. Then $|\mathfrak{M}|$ with these interpretations is isomorphic $\mathfrak{N}$, and the fact that A' is true under these interpretations implies that A is true in $\mathfrak{N}$. Thus we have:

Theorem 15.2.6. *The set of true sentences of second-order arithmetic is many-one reducible to the set of valid sentences of pure second-order logic.*

This implies that the set of sentence A of pure second-order logic that are valid in full second-order semantics is not computably enumerable, and, in fact, not even arithmetically definable. This shows that there is no proof system that is complete for second-order logic, for any computable notion of what it means to be a proof.

We have used second-order logic to say things about relations and functions in two distinct but related ways. If R is a fixed symbol in a language L, a sentence $A(R)$ can be used to define a class of L-structures with such a relation. For example, we can define the class of well-orders. If R is instead a variable, a sentence $A(R)$ can be used to define the set of relations that satisfy R in standard second-order semantics with respect to a fixed structure. For example, we can define the class of well-ordered relations on the universe of that structure. Both ways of using second-order logic are important.

Second-order logic can also be used to define predicates and relations on the first-order part of a structure. For example, over the class of undirected graphs, we can use a second-order formula $A(x, y)$ to say that x and y are in the same connected component. For another important example, for any binary relation $R(x, y)$ on a set A, the *well-founded part* of R is defined to be the smallest set S satisfying the following: for any y, if x is in S whenever $R(x, y)$, then y is in S. Saying that R is well founded is equivalent to saying that every element of A is in the well-founded part, and if R is a strict linear order, the well-founded part of R

is the largest initial segment of R that is a well-order. The exercises below ask you to write down a second-order formula $B(R, y)$ that says that y is in the well-founded part of R.

Various semantics for intuitionistic first-order logic can be extended to second-order logic. Kripke models with sorts for the second-order relations provide a natural Henkin-like semantics, and spaces of *sheaves* provide natural intuitionistic analogues of the full second-order semantics. Algebraic semantics for first-order logic can also be extended to second-order logic. Classical and intuitionistic interpretations can take values in complete Boolean algebras or Heyting algebras, respectively, as well as more general classes of structures. For more information, see the bibliographical notes at the end of this chapter.

Exercises

15.2.1. Present a second-order sentence that defines the class of finite structures such that the cardinality of the universe is a multiple of 3.

15.2.2. Show that the following sentence defines the class of infinite models:

$$\begin{aligned}&\exists R\,(\forall x\,\exists y\,R(x,y)\,\wedge\\&\quad\forall x,x',y\,(R(x,y)\wedge R(x',y)\to x=x')\,\wedge\\&\quad\exists y\,\forall x\,\neg R(x,y)).\end{aligned}$$

15.2.3. Show that the following sentence also defines the class of infinite models:

$$\begin{aligned}&\exists R\,(\forall x\,\exists y\,R(x,y)\,\wedge\\&\quad\forall x\,\neg R(x,x)\,\wedge\\&\quad\forall x,y,z\,(R(x,y)\wedge R(y,z)\to R(x,z))).\end{aligned}$$

Remember that, by definition, the universe of a model is nonempty.

15.2.4. Show that the statement that a relation R is well founded can be formulated equivalently as follows:

$$\forall Q\,(\forall x\,(\forall y\,(R(y,x)\to Q(y))\to Q(x))\to\forall x\,Q(x)).$$

This is a principle of induction for R. From an intuitionistic perspective, it is more natural than the formulation in the proof of Proposition 15.2.3.

15.2.5. Given a variable R ranging over binary relations, write down a second-order formula $B(R, y)$ that says that y is in the well-founded part of R.

15.2.6. Use second-order logic to provide a categorical axiomatization of the real numbers as an ordered field, $(\mathbb{R}, 0, 1, +, \cdot, \leq)$. (Hint: say that it is an ordered field with the property that every nonempty subset has a least upper bound.)

15.2.7. Show that the set of natural numbers in the structure $(\mathbb{R}, 0, 1, +, \cdot, \leq)$ is definable by a second-order formula, and similarly for the integers and the rationals.

15.2.8. Write down a sentence of pure second-order logic that is true in the standard semantics if and only if the continuum hypothesis is true.

15.2.9. Show that the algebraic semantics of Chapter 5 carries over to second-order logic when formulas are evaluated in any complete Boolean algebra or Heyting algebra.

15.2.10. Find a second-order formula that defines the relation "there is a path from x to y" in any undirected graph.

15.2.11. Show that second-order logic is conservative over first-order logic: if $\Gamma \cup \{A\}$ is any set of first-order sentence it is possible to prove A from Γ in second-order logic, then it is possible to prove A from Γ in first-order logic.

15.2.12. After reading the rest of this chapter, show that third-order logic is not conservative over second-order logic in the same sense. (Hint: third-order arithmetic proves the consistency of second-order arithmetic, although both are axiomatized by the same set of second-order axioms.)

15.3 Cut Elimination

We now consider sequent calculi for second-order logic. The extensions needed to incorporate the second-order quantifiers and the comprehension schema are straightforward. To the two-sided classical calculus, we add:

$$\frac{\Gamma, A[\lambda\vec{x}.\, B/X] \vdash \Delta}{\Gamma, \forall X\, A \vdash \Delta}\ \forall^2\mathrm{L} \qquad \frac{\Gamma \vdash A}{\Gamma \vdash \forall X\, A}\ \forall^2\mathrm{R}$$

$$\frac{\Gamma, A \vdash \Delta}{\Gamma, \exists X\, A \vdash \Delta}\ \exists^2\mathrm{L} \qquad \frac{\Gamma \vdash \Delta, A[\lambda\vec{x}.\, B/X]}{\Gamma \vdash \Delta, \exists X\, A}\ \exists^2\mathrm{R}$$

As usual, in the left $\exists$ rule and the right $\forall$ rule we require that X is not free in Γ or Δ, and for intuitionistic second-order logic, we restrict the rules so that only one formula occurs on the right side. For the classical one-sided sequent calculus, we use the following rules:

$$\frac{\Gamma, A}{\Gamma, \forall X\, A}\ \forall^2 \qquad \frac{\Gamma, A[\lambda\vec{x}.\, B/X]}{\Gamma, \exists X\, A}\ \exists^2$$

To include equality on the first-order universe, we add the following to the two-sided calculus, as suggested in Section 6.6:

$$\frac{\Gamma, t = t \vdash \Delta}{\Gamma \vdash \Delta} \qquad \frac{\Gamma, s = t, A(s), A(t) \vdash \Delta}{\Gamma, s = t, A(s) \vdash \Delta}$$

Exercise 6.6.2 showed that in the one-sided calculus, we can implement equality as follows:

$$\frac{\Gamma, t \neq t}{\Gamma} \qquad \frac{\Gamma, s \neq t, {\sim}A(s), {\sim}A(t)}{\Gamma, s \neq t, {\sim}A(s)}$$

We restrict the second rule to atomic formulas A. If we also include equality on the second-order sorts, we need rules to handle both equality and the extensionality axiom. Some care is needed to choose rules that are compatible with cut elimination, and we will return to this issue below.

In this section, we will focus on the classical one-sided calculus. Our goal is to prove the following:

Theorem 15.3.1. *Anything provable in the one-sided classical sequent calculus for second-order logic has a cut-free proof.*

We will start by proving this for second-order logic without equality. After that, we will show how to augment the proof to handle equality on the first-order universe, and then we will handle second-order equality as well.

Since the rules above are clearly sound for second-order logic, we will prove Theorem 15.3.1 by showing that the cut-free calculus is complete, as we did for first-order logic in Section 6.3. A priori, it is also reasonable to ask for an explicit algorithm for eliminating cuts that is similar to the procedures for first-order logic presented in Section 6.4. But in Section 15.4 we will prove the following:

Theorem 15.3.2. *In* PRA, *Theorem 15.3.1 implies the consistency of second-order arithmetic.*

By the second incompleteness theorem, this means that Theorem 15.3.1 is not provable in second-order arithmetic, so the correctness of any algorithm we might be able to describe cannot be established in that system. We will see that second-order arithmetic is quite strong, which suggests that it will be hard to come up with an algorithm that is explicit than a systematic search.

Another thing to notice about the cut elimination theorem for second-order logic is that, from the point of view of proof search, it is not nearly as useful as the cut elimination theorems for propositional and first-order logic. The benefit to using a cut-free calculus is the avoidance of the cut rule, which requires choosing an arbitrary cut formula in a backward search from the goal. In first-order logic, this shifts the challenge to handling the $\exists$ rule, which requires choosing terms to instantiate a existential quantifier. The second-order $\exists$ rule is even worse, since it requires choosing arbitrary formulas to instantiate it, which seems just as bad as having to choose arbitrary cut formulas. Nonetheless, the cut-free calculus provides at least some guidance as to how to carry out a backward search, and perhaps leaves hope that the choices can be deferred until the state of the search suggests promising candidates.

Say a set Π is a *second-order Hintikka set* if it satisfies all the properties of a Hintikka set enumerated in Section 6.3, as well as the following:

- If $\forall X\, B$ is in Π, then so is $B[\lambda \vec{x}.\, C/X]$ for every tuple $\vec{x}$ of the same arity at X and every formula C with at most those variables free.
- If $\exists X\, B$ is in Π, then so is $B[R/X]$ for some relation symbol R of the right arity.

Then just as in the proof of the cut-free completeness theorem for classical first-order logic, Theorem 6.3.4, we can prove the following:

Lemma 15.3.3. *Let Γ be a sequent in the language of second-order logic. If there is no cut-free proof of Γ, then there is a second-order Hintikka set $\Pi \supseteq {\sim}\Gamma$.*

As a result, to prove Theorem 15.3.2, it suffices to prove the following.

Lemma 15.3.4. *Let Π be a second-order Hintikka set in the language of second-order arithmetic. Then there is an extensional Henkin structure $\mathfrak{M}$ for second-order logic such that $\mathfrak{M} \models \Pi$.*

There is no harm in restricting attention to extensional Henkin structures, since, at the moment, we are not including second-order equality as a primitive. We will return to the possibility of constructing non-extensional structures below.

As we did for first-order logic, we will define $|\mathfrak{M}|$ to be the set of closed terms in the language of Π and interpret all first-order relation symbols R in the language by setting $R^{\mathfrak{M}}(t_0, \ldots, t_{n-1})$ true if and only if the atomic formula $R(t_0, \ldots, t_{n-1})$ is in Π. So we only have to determine how to define the relations in the model, that is, the sets $\mathcal{S}^{\mathfrak{M},n}$. One idea that might come to mind is to use the full second-order model, but that won't work: since second-order logic is not complete for the full semantics, there are consistent sets of sentences that cannot be satisfied by any full model. For example, take the axiomatic characterization of the natural numbers given by Proposition 15.2.4 and add the statement that second-order logic together with that characterization is inconsistent. By the second incompleteness theorem, that set of sentences is consistent, although it is false in the standard model. As a result, the first-order part of the model must be nonstandard, so the set of standard numbers does not satisfy the induction axiom. That set cannot be an element of $\mathcal{S}^{\mathfrak{M},1}$. Another idea is to include only the relations that are guaranteed to exist by the comprehension schema, that is, all the relations that are defined by formulas in the language of second-order arithmetic. But now we have a vicious circle: such formulas can include quantifiers over the second-order universes, and it is exactly the contents of those universes that need to be determined. There is a clever solution to the problem, however: we will include all relations that are not ruled out by the Hintikka set Π.

Definition 15.3.5. Let $A(x_0, \ldots, x_{n-1})$ be a formula with at most the free first-order variables shown. Say that an n-ary relation S on the set of closed terms is *compatible with* $A(\vec{x})$ with respect to a fixed Hintikka set Π if the following hold for every n-tuple of closed terms $\vec{t}$:

- If $A(\vec{t})$ is in Π, then $S(\vec{t})$ is true.
- If $\sim A(\vec{t})$ is in Π, then $S(\vec{t})$ is false.

We will show that the following surprisingly simple interpretation of the second-order universe does the trick: for each n, set

$$\mathcal{S}^{\mathfrak{M},n} = \{S \mid S \text{ is compatible with some formula } A(\vec{x})\}.$$

For any formula $A(\vec{x})$, there is always at least one relation compatible with it, namely the relation $S(\vec{t})$ that holds if and only if $A(\vec{t})$ is in Π.

Lemma 15.3.6. *Suppose $A(x_0, \ldots, x_{m-1}, Y_0, \ldots, Y_{n-1})$ is a formula with the free variables shown, where $\vec{x}$ are the first-order variables and $\vec{Y}$ are variables ranging over relations. Let σ be a variable assignment from $\mathfrak{M}$, and for each i, suppose $\sigma(Y_i)$ is compatible with $B_i(\vec{z}^i)$. If*

$$A(\sigma(x_0), \ldots, \sigma(x_{m-1}), \lambda\vec{z}^0 . B_0, \ldots, \lambda\vec{z}^{n-1} . B_{n-1})$$

is in Π, then $\mathfrak{M} \models_\sigma A(\vec{x}, \vec{Y})$.

The novelty compared to the proof of completeness for first-order logic is that, for the second-order variables, we do not know whether the lambda abstractions shown will end up naming the sets Y_i in the final model. At this stage, it is enough to know that Π doesn't rule out that possibility.

Proof Use induction over formulas in negation normal form. There are four base cases:

- $A(\vec{x}, \vec{Y})$ is an atomic formula of the form $R(u_0(\vec{x}), \ldots, u_{k-1}(\vec{x}))$ for some relation symbol R, or the negation of such a formula.
- $A(\vec{x}, \vec{Y})$ is an atomic formula of the form $Y_i(u_0(\vec{x}), \ldots, u_{k-1}(\vec{x}))$ for some relation variable Y_i, or the negation of such a formula.

In the first two cases, the claim follows from the definition of a Hintikka set, and in the second two, it follows from the requirement that each $\sigma(Y_i)$ is compatible with $B_i(\vec{z}^i)$. The inductive steps for $\wedge$, $\vee$, and the first-order quantifiers are just as for first-order logic.

Suppose $A(\vec{x}, \vec{Y})$ is $\forall W\, C(\vec{x}, \vec{Y}, W)$, suppose σ is a variable assignment taking values in the first- and second-order universes of $\mathfrak{M}$, and suppose $B_0, \ldots, B_{n-1}$ are as in the statement of the lemma. Suppose also that

$$\forall W\, C(\sigma(x_0), \ldots, \sigma(x_{n-1}), \lambda\vec{z}^0.\, B_0, \ldots, \lambda\vec{z}^{n-1}.\, B_{n-1}, W)$$

is in Π. We need to show that $\forall W\, C(\vec{x}, \vec{Y}, W)$ is true under assignment σ. Let S be any relation in the second-order universe of $\mathfrak{M}$ that is of the right arity to interpret W and let σ' be $\sigma[W \mapsto S]$. By the definition of $\mathfrak{M}$, $\sigma(W)$ is compatible with some formula $D(\vec{v})$. By the definition of a Hintikka set,

$$C(\sigma(x_0), \ldots, \sigma(x_{n-1}), \lambda\vec{z}^0.\, B_0, \ldots, \lambda\vec{z}^{n-1}.\, B_{n-1}, \lambda\vec{v}.\, D(\vec{v}))$$

is in Π. By the inductive hypothesis, $C(\vec{x}, \vec{Y}, W)$ is true under assignment σ', which is what we want.

Finally, the case for the second-order existential quantifier is straightforward, using the inductive hypothesis and the definition of a Hintikka set. □

Lemma 15.3.6 shows that, in particular, every sentence in Π is true in $\mathfrak{M}$. All that is left to do is to show that the comprehension schema is true in $\mathfrak{M}$.

Lemma 15.3.7. *Suppose for each $i < n$, S_i is a relation that is compatible with the formula $B_i(\vec{z}^i)$, and suppose σ assigns S_i to the variable Z_i. Let U be the relation on closed terms such that $U(\vec{t})$ holds if and only if $\mathfrak{M} \models_{\sigma[\vec{x} \mapsto \vec{t}]} A(\vec{x}, \vec{Z})$. Then U is compatible with $A(\vec{x}, \lambda\vec{z}^0.\, B_0, \ldots, \lambda\vec{z}^{n-1}.\, B_{n-1})$.*

Proof Appealing to Definition 15.3.5 and the definition of U, we need to show:

- If $A(\vec{t}, \lambda\vec{z}^0.\, B_0, \ldots, \lambda\vec{z}^{n-1}.\, B_{n-1})$ is in Π, then $\mathfrak{M} \models_{\sigma[\vec{x} \mapsto \vec{t}]} A(\vec{x}, \vec{Z})$.
- If $\sim A(\vec{x}, \lambda\vec{z}^0.\, B_0, \ldots, \lambda\vec{z}^{n-1}.\, B_{n-1})$ is in Π, then $\mathfrak{M} \models_{\sigma[\vec{x} \mapsto \vec{t}]} \sim A(\vec{t}, \vec{Z})$.

Both of these follow from Lemma 15.3.6. □

Lemma 15.3.8. *$\mathfrak{M}$ satisfies the comprehension schema.*

Proof We need to show that for any formula $A(\vec{x}, \vec{y}, \vec{Z})$ and any assignment σ,

$$\mathfrak{M} \models_\sigma \exists W\, \forall \vec{x}\, (W(\vec{x}) \leftrightarrow A(\vec{x}, \vec{y}, \vec{Z})).$$

For each variable Z_i, suppose $\sigma(Z_i)$ is compatible with B_i. For any tuple of elements $\vec{t}$ in $|\mathfrak{M}|$, let $U(\vec{t})$ hold if and only if $\mathfrak{M} \models_{\sigma[\vec{x} \mapsto \vec{t}]} A(\vec{x}, \vec{y}, \vec{Z})$. By Lemma 15.3.7, U is compatible with the formula

$$A(\vec{x}, \sigma(y_0), \ldots, \sigma(y_{m-1}), \lambda\vec{z}^0.\, B_0, \ldots, \lambda\vec{z}^{n-1}.\, B_{n-1})$$

and hence it is an element of the second-order universe of the model. Then we have

$$\mathfrak{M} \models_{\sigma[W \mapsto U]} \forall \vec{x}\, (W(\vec{x}) \leftrightarrow A(\vec{x}, \vec{y}, \vec{Z})),$$

which is what we need. □

The proof of Theorem 15.3.1 can be extended to include the equality symbol on the first-order universe following the recipe of Exercise 6.6.2 for first-order logic. With the right construction of a Henkin set, we can arrange that $t = t$ is in Π for every closed term t, and that whenever $s = t$ and $A(s)$ are in Π for an atomic formula $A(x)$, then so is $A(t)$. This is enough to guarantee that, in the construction we have described, the equality symbol is interpreted by an equivalence relation that is respected by all the function and relation symbols in the language, including the elementhood relation $t \in X$. Quotienting by that equivalence relation yields a model in which the equality symbol is interpreted by equality on the nose.

Adding a second-order equality symbol is more subtle. If we have the axioms of extensionality on the deductive side, there is no choice as to how to define equality on the second-order sorts of the model. The model we have already built is extensional: relation variables range over relations on the first-order universe, so if the interpretations of two relation variables hold of the same tuples of elements in the first-order universe, the interpretations are the same. We only need to guarantee that the extensional interpretation of equality is consistent with the Hintikka set, so that if $U = V$ is in Π then Π assigns $U(\vec{t})$ and $V(\vec{t})$ the same truth values for every tuple $\vec{t}$, and if $U \neq V$ is in Π then Π assigns $U(\vec{t})$ and $V(\vec{t})$ different truth values. We start with the usual equality rules for the second-order sort:

$$\frac{\Gamma, U \neq U}{\Gamma} \qquad \frac{\Gamma, U \neq V, {\sim}A(U), {\sim}A(V)}{\Gamma, U \neq V, {\sim}A(U)}$$

We can restrict to atomic formulas $A(X)$ in the second rule. These are enough to guarantee that if the sentences $U = V$ and $U(\vec{t})$ are in Π, then so is $V(\vec{t})$, satisfying the first requirement. For the second requirement, we add the following rules to the deductive system:

$$\frac{\Gamma, {\sim}U(\vec{x}), V(\vec{x}) \qquad \Gamma, {\sim}V(\vec{x}), U(\vec{x})}{\Gamma, U = V}$$

Here, the tuple of variables $\vec{x}$ is assumed to not occur in Γ. Exercise 15.3.1 asks you to show that these rules are equivalent to the extensionality schema. If we now construct the model as before and define equality between relations extensionally, we need only check that the base case of Lemma 15.3.6 holds for the new atomic formulas. You are asked to do this in Exercise 15.3.2.

In the absence of the extensionality rules, we can't use the same construction. After all, we might start with a consistent set Γ that contains the negation of an axiom of extensionality. So instead we define the second-order universes to consist of pairs $(A(\vec{x}), R)$ where A is a formula and R a relation is compatible with A. We first build a model $\mathfrak{M}$ where $\mathrm{app}^n_{\mathfrak{M}}((A(\vec{x}), R), \vec{t})$ is interpreted as $R(\vec{t})$, and each $U^{\mathfrak{M}}$ in the expanded language of Π is interpreted as $(U(\vec{x}), S)$ where $S(\vec{t})$ holds if and only if $U(\vec{t})$ is in Π. The equality symbol on relations is interpreted as the equivalence relation $\equiv$, where $(A, S) \equiv (B, R)$ holds if and only if $A(\vec{x})$ is an atomic formula $U(\vec{x})$, $B(\vec{x})$ is an atomic formula $V(\vec{x})$, and $U = V$ is in Π. Exercise 15.3.3 asks you to verify that $\equiv$ is indeed an equivalence relation that is respected by application, and if $U = V$ is in Π, then $\mathfrak{M} \models U \equiv V$. As usual, quotienting by that equivalence relation yields a second-order model where the equality symbol is interpreted as equality on the nose.

Exercises

15.3.1. Show that the rules introduced at the end of this section are equivalent to the second-order schema of extensionality.

15.3.2. Show that the closure conditions of a Hintikka set Π that result from adding the rules for second-order equality yield the relevant base cases for Lemma 15.3.6: if $\sigma(U)$ and $\sigma(V)$ are compatible with U and V respectively, then if $U = V$ is in Π then $\sigma(U) = \sigma(V)$, and if $U \neq V$ is in Π then $\sigma(U) \neq \sigma(V)$.

15.3.3. Fill out the proof of the cut-free completeness theorem for second-order logic with intensional second-order equality symbols, as described at the end of this section.

15.4 Second-Order Arithmetic

By *second-order arithmetic* we mean the second-order theory in the language of arithmetic whose axioms are those of first-order arithmetic, except that the induction schema is replaced by a single second-order induction axiom:

$$\forall P\,(P(0) \wedge \forall x\,(P(x) \to P(\mathrm{succ}(x))) \to \forall x\,P(x)).$$

As usual, we can consider versions based on either classical or intuitionistic logic, which are called *second-order Peano arithmetic* and *second-order Heyting arithmetic* respectively. With comprehension, the induction axiom implies the full schema of induction in which P can be replaced by an arbitrary formula. In the next chapter, we will consider subsystems of second-order arithmetic with set existence principles weaker than the full schema of comprehension, and in such settings it sometimes makes sense to extend the induction axiom to an induction principle for a larger class of formulas.

Second-order arithmetic includes first-order arithmetic and hence also primitive recursive arithmetic, so we can code finite objects as natural numbers as we did in Chapters 8–14. But now we can use variables ranging over the second-order universe can quantify over and reason about infinitary objects as well. Just as primitive recursive arithmetic and first-order arithmetic prove to be robust theories of finitary mathematical objects, so, too, does second-order arithmetic prove to be a good general theory of *countable* mathematical objects and structures.

Because we can code tuples of elements as a single element, there is no need to include relation variables of each arity. Instead, we can interpret an n-ary relation as a unary relation on n-tuples. We will adopt the convention of writing $x \in Y$ instead of $Y(x)$ to express that Y holds of X. In other words, we think of the elements of the second-order universe as being sets of natural numbers rather than predicates on the natural numbers.

As described in Section 11.8, we can represent a pair of sets X and Y by the single set $X \oplus Y = \{2x \mid x \in X\} \cup \{2y+1 \mid y \in Y\}$. Finite tuples of sets can be represented by iterated pairing. We can even represent an infinite sequence of sets indexed by the natural numbers by interpreting $y \in (X)_i$ as $(i, y) \in X$.

With the availability of set variables, we can interpret the primitive recursive functions as second-order objects. If f is an n-ary function represented as a set, $f(\vec{x})$ is defined to be the unique value of y such that $f(\vec{x}) = y$. With comprehension, we can prove the existence of the successor function, for example, as the set $\{(x, y) \mid \mathrm{succ}(x) = y\}$. Moreover, we have:

Proposition 15.4.1. *Second-order arithmetic proves that the universe of functions is closed under composition and primitive recursion.*

Proof We need only translate the first-order descriptions from Section 10.2 to formulas with function variables. If g is a k-ary function and each of $h_0, \ldots, h_{k-1}$ are ℓ-ary functions, then the composition of g with $h_0, \ldots, h_{k-1}$ is the function whose graph is

$$\{(\vec{x}, z) \mid \exists \vec{y}\ (h_0(\vec{x}) = y_0 \wedge \cdots \wedge h_{k-1}(\vec{x}) = y_{k-1} \wedge g(\vec{y}) = z)\}.$$

Similarly, if $g(\vec{z})$ is a k-ary function and $h(x, w, \vec{z})$ is a $(k+2)$-ary function, the function f defined by $f(0, \vec{z}) = g(\vec{z})$ and $f(x+1, \vec{z}) = h(x, f(x, \vec{z}), \vec{z})$ has the graph

$$\{(x, \vec{z}, y) \mid \exists s\ (g(\vec{z}) = (s)_0 \wedge \forall w < x\ ((s)_{w+1} = h(w, (s)_w, \vec{z})) \wedge (s)_x = y)\},$$

and by induction we can prove that the function with this graph satisfies the defining equations for f. □

With the schema of comprehension, we can prove the existence of a lot more than the primitive recursive functions, however. We can use the full language of second-order arithmetic to define relations in terms of their graphs, and we can use the full strength of the axioms to prove that the relations that we define that way are total functions. For example, if $f(x, \vec{z})$ is any function, then the formula $A(\vec{z}, y)$ given by

$$f(y, \vec{z}) = 0 \wedge \forall x < y\ (f(x, \vec{z}) \neq 0)$$

defines the graph of the partial function of $\mu x\ (f(x, \vec{z}) = 0)$, and assuming f satisfies $\forall \vec{z}\ \exists x\ (f(x, \vec{z}) = 0)$, second-order arithmetic can prove that $A(\vec{z}, y)$ defines the graph of a total function, using the least-element principle. The exercises ask you to show that if we represent partial functions as sets, second-order arithmetic proves that the set of partial functions is closed under composition, primitive recursion, and unbounded search. So second-order arithmetic proves the existence of arbitrary computable functions. We can define the graph of a version of the halting problem with the formula $A(x, y)$ given by

$$(\exists s\ T(x, 0, s) \wedge y = 1) \vee (\neg \exists s\ T(x, 0, s) \wedge y = 0),$$

and using classical logic we can prove that the function so defined is total. But that only scratches the surface. The exercises below ask you to show that in second-order arithmetic we can prove the existence of a function that evaluates the truth of formulas in the arithmetic hierarchy, and in the next section we will extend this to formulas with second-order quantifiers.

Another useful fact is that in second-order arithmetic the axiom of choice for first-order variables follows from comprehension: assuming $\forall \vec{x}\ \exists y\ A(\vec{x}, y)$, for every $\vec{x}$ we can define $f(\vec{x})$ to be the least y satisfying $A(\vec{x}, y)$. We can also state the second-order axiom of choice,

$$\forall x\ \exists Y\ A(x, Y) \to \exists Y\ \forall x\ A(x, (Y)_x),$$

and the even stronger *axiom of dependent choice*,

$$\forall x, Y\ \exists Z\ A(x, Y, Z) \to \forall Y\ \exists Z\ ((Z)_0 = Y \wedge \forall x\ A(x, (Z)_x, (Z)_{x+1})).$$

Neither of these schemas follows from comprehension and the other axioms of second-order logic, but they are true in the standard model, which is to say, they are provable in Zermelo–Fraenkel set theory with the axiom of choice.

Just as finitary mathematical objects can be represented by natural numbers, infinitary mathematical objects can be represented by sets of natural numbers as long as they are determined by a countable set of data. For example, a sequence $(a_i)_{i \in \mathbb{N}}$ of rational numbers is said to be *Cauchy* if for every (rational) $\varepsilon > 0$, there is an N such that for every $i, j \geq N$, $|a_i - a_j| < \varepsilon$. We can represent Cauchy sequences as sets of numbers in second-order arithmetic. A common mathematical construction is to take the real numbers to be equivalence classes of Cauchy sequences under the relation $(a_i)_{i \in \mathbb{N}} \equiv (b_i)_{i \in \mathbb{N}}$ which holds when for every $\varepsilon > 0$ there is an N such that for every $i \geq N$, $|a_i - b_i| < \varepsilon$. We would need third-order variables to represent equivalence classes, but we can instead consider the reals to be the Cauchy sequences and use the equivalence relation in place of equality. Similarly, we cannot talk about arbitrary sets of reals or arbitrary functions from reals to reals, but via countable representations we can talk about open and closed sets, and even arbitrary Borel sets, as well as continuous functions.

We can, similarly, represent countable algebraic structures. A countable group consists of a tuple (G, m, i, e) where G is a set, m is a binary function from G to G, i is a unary function from G to G, and e is an element of G, such that G, m, i, and e satisfy the usual group axioms. The element e can be represented by the set $\{e\}$, so the whole structure can be represented as a tuple of sets. We can then consider the extent to which the theorems of algebra can be proved in second-order arithmetic. We can even treat mathematical logic itself in these terms: sets in second-order arithmetic can represent countable models, both classical and intuitionistic, as described in Chapter 5. The exercises at the end of this section encourage you to start thinking about these issues. We will return to the topic of formalization in Section 16.3.

An alternative axiomatization of second-order arithmetic starts with pure second-order logic and adds an axiom of infinity:

$$\exists f, a\, (\forall x, y\, (f(x) = f(y) \to x = y) \land \neg \exists x\, (f(x) = a)).$$

As suggested by the proof of Theorem 15.2.6, we can interpret second-order arithmetic in this theory. More precisely, using comprehension we can prove the existence of a unique set N equal to the intersection of all sets containing a and closed under f:

$$x \in N \leftrightarrow \forall Y\, (a \in Y \land \forall z\, (z \in Y \to f(z) \in Y) \to x \in Y).$$

We can similarly define the graphs of addition and multiplication and show that they uniquely meet their recursive specifications. Hence, whenever second-order arithmetic proves a sentence A, second-order logic with the axiom of infinity proves that, for any f and a satisfying the axiom of infinity, the relativization of A to the set N and operations just defined is also true. In particular, if second-order arithmetic proves $\bot$, then second-order logic with the axiom of infinity proves it as well. Since this argument is entirely syntactic, it can be carried out in PRA.

Proposition 15.4.2. PRA *proves the second-order arithmetic is equiconsistent with second-order logic together with the axiom of infinity.*

In other words, PRA proves that one is consistent if and only if the other is. We can use this fact to fulfill a promise made in Section 15.3.

Proposition 15.4.3. *In* PRA, *the cut elimination theorem for second-order logic implies the consistency of second-order logic with the axiom of infinity.*

Proof Argue in PRA. Suppose second-order logic with the axiom of infinity is inconsistent. Then second-order logic proves the negation of the axiom of infinity,

$$\forall f, a\, (\exists x, y\, (f(x) = f(y) \wedge x \neq y) \vee \exists x\, (f(x) = a)).$$

If therefore proves the formula in which we drop the universal quantifiers in front. By the second-order cut elimination theorem, there is a cut-free proof of this formula, and because there are no second-order quantifiers, it is actually a cut-free proof in first-order logic. By Herbrand's theorem, there is a quantifier-free proof of a finite disjunction of instances of this formula, in which the existential quantifiers are replaced by terms involving only f and a. But any instance is clearly true when a is interpreted as 0 and f is interpreted as the successor function, contradicting the soundness of quantifier-free logic. So second-order logic with the axiom of infinity is consistent. □

Exercises

15.4.1. Show how to represent partial functions from $\mathbb{N}$ to $\mathbb{N}$ in the language of second-order arithmetic, and show that second-order arithmetic proves that the set of such partial functions is closed under composition, primitive recursion, and the unbounded search operator, μ, as defined in Section 11.1.

15.4.2. This exercise asks you to show that classical second-order arithmetic proves the existence of an evaluation function for sentences in the language of first-order arithmetic. We already know, from Section 10.5, that there are primitive recursive functions evaluating closed terms and atomic formulas, the correctness of which can be established in primitive recursive arithmetic.

a. Define the complexity of a formula by counting the number of quantifiers and connectives. Write down a definition of what it means for a function f to provide a correct truth evaluation for sentences of complexity at most n.
b. Show that second-order arithmetic proves that if there is a correct truth evaluation for sentences of complexity at most n, then there is a correct truth evaluation for sentences of complexity at most $n + 1$.
c. Show that if f and g are both correct truth evaluations for sentences of complexity at most n, then they agree in their assignments to those sentences.
d. Let f be the function that assigns a value of *true* to a sentence of complexity at most n if there exists a truth assignment to sentences of complexity at most n that assigns a value of *true*, and similarly for *false*. (If the input doesn't represent a sentence of first-order arithmetic, let f return 0.) Show that second-order arithmetic proves that f is a correct truth evaluation for all sentences of first-order arithmetic.

It might help to consult the proof of Theorem 1.5.3.

15.4.3. Show that a similar argument goes through in intuitionistic second-order arithmetic to prove the existence of the set of all true sentences in the language of first-order arithmetic. (We need to use a set rather than a function because intuitionistic second-order arithmetic cannot prove that every formula is either true or false.)

15.4.4. As a slight variation on the proof described in Exercise 15.4.2, show that classical second-arithmetic proves, by induction on n, that there is a truth evaluation for Σ_n sentences in the language of arithmetic.

15.4.5. Before reading Section 16.3 in the next chapter, think about how you would define basic operations on the real numbers and prove basic facts about the real numbers in second-order arithmetic.

15.4.6. Do the same for operations and theorems that have to do with countable algebraic structures like groups, rings, and fields.

15.5 The Analytic Hierarchy

The arithmetic hierarchy, defined in Section 10.2, can be extended to the *analytic hierarchy* as follows. In the context of the second-order language of arithmetic, we take the arithmetic formulas to be the ones without second-order quantifiers, although they may have second-order variables. Formulas at the nth level of the arithmetic part of the analytic hierarchy are said to be Σ^0_n, Π^0_n, and Δ^0_n accordingly, where the superscripted 0 indicates that we are dealing with the second-order language. These levels can be defined, inductively, as in Section 10.2, the only difference being that now we include expressions $t \in Y$ among the atomic formulas. To obtain the analytic hierarchy, we start with the arithmetic formulas and add second-order quantifiers. We take all of the notation Σ^1_0, Π^1_0, and Δ^1_0 to denote the set of arithmetic formulas, and we define the higher levels inductively as follows:

- If A is Σ^1_n, then $\forall \vec{X}\, A$ is Π^1_{n+1}.
- If A is Π^1_n, then $\exists \vec{X}\, A$ is Σ^1_{n+1}.

More loosely, we often say that A is Σ^1_n or Π^1_n if it is semantically equivalent to a formula of the corresponding syntactic form, and we say that A is Δ^1_n if it is semantically equivalent to one formula that is Σ^1_n and another one that is Π^1_n.

A set S of natural numbers is said to be Σ^1_n-definable if there is a Σ^1_n formula $A(x)$ such that $S = \{n \in \mathbb{N} \mid \mathfrak{N} \models A(n)\}$, where $\mathfrak{N}$ denotes the standard model of arithmetic and the double-turnstile symbol denotes full second-order semantics. The Π^1_n- and Δ^1_n-definable sets of natural numbers are defined analogously. But now the availability of set variables in the language means that we can also talk about the definability of a collection of sets: such a collection $\mathcal{S}$ is said to be definable by a formula $A(X)$ if $\mathcal{S} = \{U \subseteq \mathbb{N} \mid \mathfrak{N} \models A(U)\}$. The terminology extends more generally to relations $R(\vec{x}, \vec{Y}, \vec{f})$ on numbers, sets, and functions from $\mathbb{N}$ to $\mathbb{N}$. Saying that a set variable f represents a total function requires the Π^0_2 formula $\forall x\, \exists! y\, ((x, y) \in f)$, so, if we take functions to be represented by sets, it makes a difference whether we take the statement that f represents a functional relation to be part of the defining formula. When talking about the analytic hierarchy it does not matter, since arbitrary arithmetic formulas are included in the bottom-most level.

As with the arithmetic hierarchy, we say that a formula A is Σ^1_n or Π^1_n in a particular theory if the theory proves that A is equivalent to a Σ^1_n or Π^1_n formula. When we talk about a formula being Σ^1_n or Π^1_n outright, we sometimes mean that the relation it defines is Σ^1_n or Π^1_n in the standard model, which is equivalent to saying that the formula is Σ^1_n or Π^1_n relative to the set of true sentences of second-order arithmetic. In Section 10.2, we showed that levels of the arithmetic hierarchy are closed under bounded quantification. In a similar way, using the second-order axiom of choice, we can show that the levels of the analytic hierarchy are closed under quantification over $\mathbb{N}$: if $R(\vec{x}, \vec{Y}, \vec{f}, w)$ is Σ^1_n, then so are $\forall w\, R(\vec{x}, \vec{Y}, \vec{f}, w)$ and $\exists w\, R(\vec{x}, \vec{Y}, \vec{f}, w)$, and similarly for the Π^1_n-definable relations.

Exercise 15.5.1 asks you to show that, in analogy with Proposition 10.2.2, any Σ^1_n-definable function $f(\vec{x})$ from $\mathbb{N}$ to $\mathbb{N}$ is Δ^1_n definable. Combining this with Exercise 15.4.2, we have the following:

Theorem 15.5.1. *The set of true sentences in the language of arithmetic is a Δ^1_1-definable set, and the set of pairs (A, σ) such that $\mathfrak{N} \models_\sigma A$ is also Δ^1_1 definable.*

Theorems about the arithmetic hierarchy from Sections 10.2 and 10.5 carry over with similar proofs. In particular, in analogy to Theorem 10.5.2 and Proposition 11.4.10, we have the following.

Theorem 15.5.2. *For every $n \geq 1$, let $\mathrm{Tr}_{\Sigma^1_n}(x, y, Z)$ be the relation that holds if and only if x represents a Σ^1_n formula $A(\vec{u}, \vec{W})$, y represents a tuple $\vec{m}$, Z represents a tuple of sets $\vec{S}$, and $\mathfrak{N} \models A(\vec{m}, \vec{S})$. Then $\mathrm{Tr}_{\Sigma^1_n}(x, y, Z)$ is Σ^1_n definable but not Π^1_n definable.*

Theorem 15.5.3. *For any n, the set of true Σ^1_n sentences is a complete Σ^1_n set of natural numbers under many-one reduction.*

Theorem 15.5.4. *For any n, the collection $\mathcal{S}$ of sets U such that U represents a pair $(A(X), V)$, where $A(X)$ is a Σ^1_n formula that is true of V, is a complete Σ^1_n collection of sets of natural numbers in the following sense: if $\mathcal{T}$ is any Σ^1_n-definable collection of sets of natural numbers, there is a computable functional $F(U)$ such that for every U, U is in $\mathcal{T}$ if and only if $F(U)$ is in $\mathcal{S}$.*

In the context of second-order definability, weaker forms of reducibility are also useful. For many purposes, it is often sufficient to know that a set S can be reduced to a set T by an arithmetic or even Δ^1_1-definable function f, in the sense that for every x, x is in S if and only if $f(x)$ is in T. Both these preserve Σ^1_n and Π^1_n definability. The notions carry over to collections of sets as well: a set $\mathcal{S}$ is Δ^1_1 reducible to $\mathcal{T}$ if there is a Δ_1 formula $A(X, Y)$ defining the graph of a function $F(X)$ such that for every X, X is in $\mathcal{S}$ if and only if $F(X)$ is in $\mathcal{T}$. In that case, if $\mathcal{T}$ is Σ^1_n or Π^1_n definable, then so is $\mathcal{S}$.

In addition to defining sets of numbers and collections of sets of numbers with formulas in the language of second-order arithmetic, we can also consider collections of sets of numbers that are definable relative to a set parameter. A collection of sets $\mathcal{S}$ is said to be $\boldsymbol{\Sigma}^1_n$, read *boldface* Σ^1_n, if there is a Σ^1_n formula $A(X, Y)$ and a set V such that $\mathcal{S} = \{U \mid \mathcal{N} \models A(U, V)\}$. The classes $\boldsymbol{\Pi}^1_n$, $\boldsymbol{\Delta}^1_n$, $\boldsymbol{\Sigma}^0_n$, $\boldsymbol{\Pi}^0_n$, and $\boldsymbol{\Delta}^0_n$ are defined analogously. These notions are not useful when it comes to definability of sets of numbers because any set V is defined by the formula $x \in V$ if we are allowed to use V as a parameter. But since there are uncountably many sets of natural numbers, saying that a set $\mathcal{S}$ is in one of the boldface classes tells us something substantial, namely, that the potentially uncountable set can be defined by a single formula making use of a countable set of data.

The study of boldface complexity classes takes us to the realm of *descriptive set theory*. The $\boldsymbol{\Sigma}^1_1$ collections of sets are said to be *analytic*, the $\boldsymbol{\Pi}^1_1$ collections of sets are said to be *co-analytic*, and the $\boldsymbol{\Delta}^1_1$ collections of sets are said to be *Borel*. With the usual representations of real numbers as sets, these notions coincide with notions used in real analysis. We will not pursue the study of boldface complexity classes here, but we note in passing that many ideas carry over from the study of the corresponding lightface classes.

We now introduce two important Π^1_1 sets:

$$\text{Wo} = \{e \in \mathbb{N} \mid e \text{ is the index of a computable well-ordering of } \mathbb{N}\}$$
$$\text{WO} = \{U \in \mathcal{P}(\mathbb{N}) \mid U \text{ is a well-ordering of } \mathbb{N}\}.$$

In the definition of WO, we are relying on the representation of a binary relation as a set U of ordered pairs. To see that Wo is Π^1_1, note that we have $e \in \text{Wo}$ if and only if

> $\varphi^2_e(x, y)$ is the characteristic function of relation $R(x, y)$ such that $R(x, y)$ is a linear order and for every nonempty set S, there is an R-least element of S.

Formalizing the phrase "for every set S" requires a second-order universal quantifier. Saying that $\varphi^2_e(x, y)$ is total is Π_2, and the exercises below ask you to confirm that, modulo that assumption, saying that $\varphi_e(x, y)$ is the characteristic function of a linear-order is Π_1. The verification that WO is a Π^1_1-definable collection of sets is similar.

The notions of many-one reducibility and Turing reducibility can be used to study relationships between definable sets. As with the arithmetic hierarchy, many-one reducibility preserves the Σ and Π classes, although Turing reducibility does not.

Theorem 15.5.5. Wo *is a complete* Π^1_1 *set of numbers in the sense that any* Π^1_1 *set is many-one reducible to it.* WO *is a complete* Π^1_1 *collection of sets of numbers in the sense that for every* Π^1_1 *collection* $\mathcal{S}$, *there is a computable functional* $F(X)$ *such that for every set* X, X *is in* $\mathcal{S}$ *if and only if* $F(X)$ *is in* WO.

We will prove this theorem using two lemmas. In the next one, the word "equivalent" means "equivalent in the standard model," though in Section 16.5 we will make use of the fact that the equivalence is provable in a subsystem of second-order arithmetic known as ACA_0.

Lemma 15.5.6. *Let* $A(\vec{x}, \vec{Y})$ *be a* Π^1_1 *formula. Then* A *is equivalent to a formula of the form* $\forall f\, \exists \sigma \subset f\, B(\vec{x}, \vec{Y}, \sigma)$, *where* B *is* Δ^0_0 *and for every* $\vec{x}$ *and* $\vec{Y}$ *the set* $\{\sigma \mid \neg B(\vec{x}, \vec{Y}, \sigma)\}$ *is a tree.*

Proof Suppose $A(\vec{x}, \vec{Y})$ is the formula

$$\forall W\, \exists u_0\, \forall v_0 \ldots \exists u_{k-1}\, \forall v_{k-1}\, C(\vec{x}, \vec{Y}, u_0, v_0, \ldots, u_{k-1}, v_{k-1}, W).$$

By Skolemizing the negation we have that $A(\vec{x}, \vec{Y})$ is equivalent to

$$\forall W, g_0, \ldots, g_{k-1}\, \exists u_0, \ldots, u_{k-1}\, C(\vec{x}, \vec{Y}, u_0, g_0(u_0), \ldots, u_{k-1}, g_{k-1}(u_0, \ldots, u_{k-1}), W).$$

Combining $W, g_0, \ldots, g_{k-1}$ into a single function, this can be expressed in the form

$$\forall h\, \exists \vec{u}\, D(\vec{x}, \vec{Y}, \vec{u}, h),$$

where D is Σ^0_1. Theorem 11.8.7 then yields the desired conclusion. □

Let T be a tree on $\mathbb{N}$. Define the *Kleene–Brouwer order* on the nodes of T as follows: say $\sigma \preceq \tau$ if and only if $\sigma \supseteq \tau$ or for some i, $(\sigma)_i \neq (\tau)_i$ and $(\sigma)_i < (\tau)_i$ for the least such i. In other words, $\sigma \preceq \tau$ if and only if σ extends τ or is to the left of τ in the tree.

Lemma 15.5.7. *For any tree* T *on* $\mathbb{N}$, *the Kleene–Brouwer order on* T *is a linear order, and* T *is well founded if and only if the Kleene–Brouwer order on* T *is a well-order.*

Proof Showing that the Kleene–Brouwer order is a linear order is tedious but straightforward. If T is not well founded, there is an infinite path $\{\sigma_0, \sigma_1, \sigma_2, \ldots\}$ through T with $\sigma_0 \subseteq \sigma_1 \subseteq \sigma_2 \ldots$, in which case $\sigma_0, \sigma_1, \sigma_2, \ldots$ is an infinite descending sequence in the Kleene–Brouwer order. Exercise 1.4.4 suggests a few ways of proving the converse; for completeness, I will spell out one of them here.

Suppose the Kleene–Brouwer order on T is not a well-order, and let $\sigma_0, \sigma_1, \sigma_2, \ldots$ be an infinite descending sequence. Define a sequence of nodes $\tau_0, \tau_1, \tau_2, \ldots$ recursively as follows: $\tau_0 = ()$, and assuming τ_i has been defined, let $\tau_{i+1} = \tau_i \frown x$, where x is the least element of $\mathbb{N}$ such that one of the σ_is extends $\sigma_i \frown x$. To justify this definition, we show inductively that for each i, there is some σ_j extending τ_i and that there are no elements σ_k to the left of τ_i. This is clearly true for $\tau_0 = ()$. Suppose it is true for τ_i. The choice of x above ensures that there are no elements σ_k to the left of $\tau_{i+1} = \tau_i \frown x$, and if $\sigma_j \preceq \tau_{i+1}$, then all of $\sigma_{j+1}, \sigma_{j+2}, \ldots$ are strictly less than τ_{i+1} in the Kleene–Brouwer order, and hence extend it. □

Proof of Theorem 15.5.5 Suppose $A(x)$ is a Π^1_1 formula, let $B(x, \sigma)$ be the formula given by Lemma 15.5.6, and let $T_x = \{\sigma \mid \neg B(x, \sigma)\}$. Then by Lemmas 15.5.6 and 15.5.7, $A(x)$ holds if and only if the Kleene–Brouwer order on T_x is a well-order. For every x, let $f(x)$ return an index for the relation that agrees with the Kleene–Brouwer order on T_x, agrees with the usual order on $\mathbb{N}$ for the natural numbers not in T_x, and put the elements of T_x before the non-elements. Then $f(x)$ is a many-one reduction of $\{x \mid A(x)\}$ to Wo. The argument for the second-claim is similar: given $B(X, \sigma)$ as in Lemma 15.5.6, let $T_X = \{\sigma \mid \neg B(X, \sigma)\}$, and let $F(X)$ return the same order on T_X. □

In Chapters 11 and 12, we made use of the fact that if $R(x, y)$ is a Σ_1-definable relation, there is a Σ_1-definable relation $R' \subseteq R$ such that for every x there is a unique y such that $R'(x, y)$ holds, and hence there is a computable function $f(x)$ that, for every x, satisfies $R(x, f(x))$. This is known as the *uniformization property* of Σ_1 relations. We can define $R'(x, y)$ by saying that y is the first component of the least pair consisting of a y and a witness to the truth of $R(x, y)$. Interestingly, the fact that we can associate well-orders to Π^1_1 formulas makes them analogous to Σ_1 formulas.

Theorem 15.5.8. *The Π^1_1-definable relations $R(x, y)$ on $\mathbb{N}$ have the uniformization property.*

The proof requires a fundamentally important fact, namely, that any two well-orders can be compared to one another. Given an order $(A, \prec)$, an *initial segment* of A is a set I such that if $a \in I$ and $b \prec a$ then $b \in I$. An initial segment of A is *proper* if it is not all of A.

Proposition 15.5.9. *Let $(A, \prec)$ and $(B, \prec')$ be two well-orders. Then exactly one of the following statements holds: either there is an isomorphism between A and B, or A is isomorphic to a proper initial segment of B, or B is isomorphic to a proper initial segment of A.*

Proof If a is any element of A, we let $A\restriction_a$ denote the initial segment $\{b \mid b \prec a\}$. It isn't hard to show that if f and g are any two isomorphisms between initial segments of A and initial segments of B, then f and g agree on their common domain. (Otherwise, consider the least element where they differ.)

If, for every $a \in A$, $A\restriction_a$ is isomorphic to an initial segment of B, the union of all of these is an isomorphism between A and an initial segment of B – that is, all of B or a proper initial segment of B – and we are done. So suppose, for some a, $A\restriction_a$ is not isomorphic to an initial segment of B. Consider the least such a. Then for every $a' \prec a$, $A\restriction_{a'}$ is isomorphic to an initial segment of B, and hence so is their union. If there is no largest element $a' \prec a$, then this union is all of $A\restriction_a$, a contradiction. So there is a largest $a' \prec a$ such that $A\restriction_{a'}$ is isomorphic to an initial segment of B. The image of this isomorphism must be all of B; otherwise we could map a' to the least b' not in the image, thereby extending it to an isomorphism of $A\restriction_a$ with an initial segment of B. □

Proof of Theorem 15.5.8 Let $R(x, y)$ be a Π^1_1 definable relation. Then by Theorem 15.5.5, there is a computable function $f(x, y)$ such that, for every x and y, $R(x, y)$ holds if and only if $f(x, y)$ is the index for a well-order. Define $R'(x, y)$ by the following condition:

> $f(x, y)$ is the index of a well-order, and for every y', either $f(x, y')$ is not the index of a linear order or it is not the case that $y' < y$ and there is an order preserving mapping of $f(x, y')$ into $f(x, y)$.

In other words, $R'(x, y)$ chooses y such that $f(x, y)$ has the smallest order type among all ys satisfying $R(x, y)$, and among those ys, it chooses the smallest one. It is not hard to check that this is Π^1_1. □

Similar methods can be used to show that the Π^1_1 relations $R(X, Y)$ on sets of numbers also have the uniformization property, but this is trickier; see the bibliographical notes.

Exercises

15.5.1. Show that any Σ^1_n-definable function $f(\vec{x})$ from $\mathbb{N}$ to $\mathbb{N}$ is Δ^1_n definable.

15.5.2. Prove Theorems 15.5.2, 15.5.3, and 15.5.4.

15.5.3. Spell out the details of showing that the set Wo is a Π^1_1-definable set of natural numbers and that the set WO is a Π^1_1-definable collection of sets of natural numbers.

15.5.4. In the terminology introduced in this section, show that there is a boldface Σ^1_1 set that is not boldface Π^1_1. In other words, not every analytic set is coanalytic. (Hint: any complete lightface Σ^1_1 set will do).)

15.5.5. Show that Π^1_1 collections of sets of numbers have the reduction property and Σ^1_1 collections of sets have the separation property. (See Exercises 10.2.3 and 10.2.4.)

15.6 The Second-Order Typed Lambda Calculus

In Section 14.3 we learned that the provably total computable functions of PA and HA can be characterized as the type 1 primitive recursive functionals of finite type. In this section, we will obtain a similar characterization of the provably total computable functions of second-order PA and HA.

We start by generalizing the primitive recursive functions of finite type. We extend the simply typed lambda calculus over $\mathbb{N}$ to allow *second-order polymorphism* as follows. Start with a set of variable $X, Y, Z, \ldots$ ranging over types. The set of polymorphic types is defined inductively as follows:

- NAT is a type.
- Each type variable $X, Y, Z, \ldots$ is a type.

- If α and β are types, so are $\alpha \to \beta$ and $\alpha \times \beta$.
- If X is a type variable and β is a type, $\Pi X.\, \beta$ is a type.

In the last clause, the variable X is bound in the expression $\Pi X.\, \beta$, and, as usual, we identify expressions that are the same up to renaming of bound variables. The idea is that a term t of type $\Pi X.\, \beta$ is an expression that is polymorphic over types: when applied a to a specific type α, t becomes a term of type $\beta[\alpha/X]$. Substitution $\beta[\alpha/X]$ for a type variable X, as well as simultaneous substitution $\beta[\vec{\alpha}/\vec{X}]$, is defined as for first-order formulas.

In the second-order lambda calculus, not only can terms depend on variables $x, y, z, \ldots$ of various types, but both terms and their types can depend on variables $X, Y, Z, \ldots$ ranging over types. Terms are built up starting from variables x^β, where β may have free type variables. Consider a term like $\lambda x^X.\, x$, which has type $X \to X$, where X is a type variable. If we substitute any type α for X, we get the identity function for α. It therefore makes sense to abstract over this to obtain a new expression $\Lambda X.\, \lambda x^X.\, x$ of type $\Pi X.\, X \to X$. We also allow the application of any term of type t of type $\Pi X.\, \beta$ to an arbitrary type α to obtain a term $t\,\alpha$ of type $\beta[\alpha/X]$. So applying the polymorphic identity function $\Lambda X.\, \lambda x^X.\, x$ to a type α yields the identity function of type $\alpha \to \alpha$.

We extend the syntax of the simply typed lambda calculus with recursors as follows:

- For each type α, there are infinitely many variables $x, y, z, \ldots$ of type α.
- Each constant $c \in \mathcal{C}$ is a term of the associated type.
- If t is a term of type β and x is a variable of type α, then $(\lambda x.\, t)$ is a term of type $\alpha \to \beta$.
- If t is a term of type $\alpha \to \beta$ and s is a term of type α, then $(t\, s)$ is a term of type β.
- If s and t are terms of type α and β respectively, then (s, t) is a term of type $\alpha \times \beta$.
- If t is a term of type $\alpha \times \beta$, then $(t)_0$ is a term of type α and $(t)_1$ is a term of type β.
- 0 is a constant of type NAT
- succ is a constant of type NAT $\to$ NAT.
- For every type α and terms $f : \alpha$, $g : \text{NAT} \to \alpha \to \alpha$, and $t : \text{NAT}$, $(R\, f\, g\, t)$ is a term of type α.
- If t is a term of type β, X is a type variable, and X is not free in the type of any variable of t, then $\Lambda X.\, t$ is a term of type $\Pi X.\, \beta$.
- If t is a term of type $\Pi X.\, \beta$ and α is a type, $t\,\alpha$ has type $\beta[\alpha/X]$.

Only the last two clauses are new. To understand the restriction in the second-to-last clause, suppose a term t has a free variable y of type α, where X is free in α. Then y should still be a free variable of $\Lambda X.\, t$, but now what type could it possibly have? The fact that X occurs in α means that α *depends* on X. We will learn more about dependent types in Section 17.5.

The notion of substitution $t[\alpha/X]$ for a type variable in a term is defined as one would expect. We add the following reduction rule to the ones already introduced in Chapter 13:

$$(\Lambda X.\, t)\, \alpha \rhd t[\alpha/X].$$

This gives rise to notions of one-step reduction $\to_1$ and reduction $\twoheadrightarrow$ as described in Chapter 13. Below we will show that the reduction relation is strongly normalizing.

This system is conventionally known as *system* F or the *Girard–Reynolds polymorphic second-order lambda calculus*. Strictly speaking, system F is usually taken to describe a system without products and recursors, since products can be defined in a manner similar to the way we defined conjunction in Section 15.1, and recursors can be defined in a manner similar to the way we we interpreted the natural numbers in second-order logic with an axiom of infinity in Section 15.4. For the realizability interpretation we will carry out later in this section, it is convenient to include them.

To prove strong normalization, we would like to extend the method we used to prove strong normalization for the simply typed lambda calculus, which assigns a set of strongly computable terms to each type. As with the second-order cut elimination theorem, the circularity inherent in second-order abstraction raises a problem: the assignment of a notion of strong computability to a type of the form $\Pi X.\, \alpha$ ought to depend on the notion of strong computability for any type that can be substituted for X, but that includes *all* the types, including $\Pi X.\, \alpha$ itself. The solution is similar to that used in Section 15.3: we will quantify over a collection of sets, the *saturated* sets of terms of any given type, that includes all possible interpretations of strong computability.

In Section 13.2, we showed inductively that for every type α, the set of strongly computable terms of type α has the following three properties:

1. If t is strongly computable then t is strongly normalizing.
2. If t is strongly computable and $t \to t'$, then t' is strongly computable.
3. If t is neutral and t' is strongly computable for every t' such that $t \to_1 t'$ then t is strongly computable.

We now extend the definition of the neutral terms to be the terms that are not of the form $\lambda x.\, t$, (s, t), or $\Lambda X.\, t$, and if α is any type, a *reducibility candidate* for α is any set of terms of type α satisfying these three properties. Suppose $\mathcal{S}$ is a reducibility candidate for a type α and $\mathcal{T}$ is a reducibility candidate for another type β. Define $\mathcal{S} \Rightarrow \mathcal{T}$ to be

$$\{t \colon \alpha \to \beta \mid \forall s \in \mathcal{S}\, (t\, s \in \mathcal{T})\}$$

and define $\mathcal{S} \otimes \mathcal{T}$ to be

$$\{p \colon \alpha \times \beta \mid (p)_0 \in \mathcal{S} \text{ and } (p)_1 \in \mathcal{T}\}.$$

The proofs in Section 13.2 show that $\mathcal{S} \Rightarrow \mathcal{T}$ is a reducibility candidate for $\alpha \to \beta$ and $\mathcal{S} \otimes \mathcal{T}$ is a reducibility candidate for type $\alpha \times \beta$.

Suppose α is a type. Let $X_0, \ldots, X_{n-1}$ be a tuple of variables, let μ be an assignment of types to the variables $\vec{X}$, and for each $i < n$, let $\nu(X_i)$ be a reducibility candidate for $\mu(X_i)$. It will be convenient to write $\alpha[\mu/\vec{X}]$ for the simultaneous substitution $\alpha[\mu(X_0)/X_0, \ldots \mu(X_{n-1})/X_{n-1}]$. We are about to define, inductively, what it means for a term of type $\alpha[\mu/\vec{X}]$ to be strongly computable modulo the assumption that at each place where X_i is replaced by $\mu(X_i)$, strong computability for terms of type $\mu(X_i)$ is interpreted as membership in $\nu(X_i)$. This avoids the circularity alluded to above: we define the notion of strong computability for a type $\Pi Y.\, \alpha$ by making reference to all the types β that can instantiate Y and all *possible* interpretations of strong computability at type β. The *actual* definition of strong computability for type expressions without variables, corresponding to the case where $n = 0$, is determined only at the end.

Formally, we assign to each such α, μ, and ν a set of terms $\mathrm{SC}_{\alpha,\mu,\nu}$ of type $\alpha[\mu/\vec{X}]$ as follows:

- $\mathrm{SC}_{X_i,\mu,\nu} = \nu(X_i)$ for each $i < n$.

- If α is NAT or a type variable other than an X_i, $\mathrm{SC}_{\alpha,\mu,\nu}$ is the set of strongly normalizing terms of type $\alpha[\mu/\vec{X}]$.
- $\mathrm{SC}_{\alpha\to\beta,\mu,\nu}$ is $\mathrm{SC}_{\alpha,\mu,\nu} \Rightarrow \mathrm{SC}_{\beta,\mu,\nu}$.
- $\mathrm{SC}_{\alpha\times\beta,\mu,\nu}$ is $\mathrm{SC}_{\alpha,\mu,\nu} \otimes \mathrm{SC}_{\beta,\mu,\nu}$.
- A term t is in $\mathrm{SC}_{\Pi Y.\,\alpha,\mu,\nu}$ if and only if for every type β and reducibility candidate $\mathcal{R}$ for β, $t\,\beta$ is in $\mathrm{SC}_{\alpha,\mu[Y\mapsto\beta],\nu[Y\mapsto\mathcal{R}]}$.

In words, the last condition says that a term t of type $(\Pi Y.\,\alpha)[\mu/\vec{X}]$ is strongly computable modulo the interpretations given by μ and ν if and only if for every type β and reducibility candidate $\mathcal{R}$ for β, $t\,\beta$ is strongly computable if we also interpret strong computability for β as membership in $\mathcal{R}$.

We can now cast the proof of strong normalization as an extension of the proof of strong normalization for the simply typed lambda calculus. First, we prove the following by induction on types:

Lemma 15.6.1. *For every α, μ, and ν, the set $\mathrm{SC}_{\alpha,\mu,\nu}$ is a reducibility candidate.*

Proof The proof goes by induction on types and extends the proof of Lemma 13.2.7. The only new case is where α is the form $\Pi Y.\,\beta$. For condition 1, suppose t is in $\mathrm{SC}_{\Pi Y.\,\beta,\mu,\nu}$. Let Z be any type variable and let $\mathcal{R}$ be the set of strongly normalizing terms of type Z. Then $\mathcal{R}$ is a reducibility candidate, and by definition $t\,Z$ is in $\mathrm{SC}_{\beta,\mu[Y\mapsto Z],\nu[Y\mapsto\mathcal{R}]}$. By the inductive hypothesis, $t\,Z$ is strongly normalizing. But an infinite sequence of reductions for t would give rise to an infinite sequence of reductions for $t\,Z$, so t is strongly normalizing as well.

For claim 2, suppose t is in $\mathrm{SC}_{\Pi Y.\,\beta,\mu,\nu}$ and $t \to t'$. Let γ be any type and let $\mathcal{R}$ be a reducibility candidate for γ. Then, by definition, $t'\,\gamma$ is in $\mathrm{SC}_{\beta,\mu[Y\mapsto\gamma],\nu[Y\mapsto\mathcal{R}]}$. Since $t\,\gamma \to t'\,\gamma$, we have that $t'\,\gamma$ is in $\mathrm{SC}_{\beta,\mu[Y\mapsto\gamma],\mu[Y\mapsto\mathcal{R}]}$ by the inductive hypothesis. Since this holds for every γ and $\mathcal{R}$, we have that t' is in $\mathrm{SC}_{\Pi Y.\,\beta,\mu,\nu}$.

Finally, suppose t of type $(\Pi Y.\,\beta)[\mu/\vec{X}]$ is neutral, and suppose that whenever $t \to_1 t'$, t' is in $\mathrm{SC}_{\Pi Y.\,\beta,\mu,\nu}$. Let γ be any type and $\mathcal{R}$ be any reducibility candidate for γ. Then every one-step reduction of $t\,\gamma$ yields a term of type $t'\,\gamma$, where $t \to_1 t'$. By the definition of $\mathrm{SC}_{\Pi Y.\,\beta,\mu,\nu}$, the term $t'\,\gamma$ is in $\mathrm{SC}_{\beta,\mu[Y\mapsto\gamma],\nu[Y\mapsto\mathcal{R}]}$, and so, by the inductive hypothesis, $t\,\gamma$ is in $\mathrm{SC}_{\beta,\mu[Y\mapsto\gamma],\nu[Y\mapsto\mathcal{R}]}$. Once again by the definition of $\mathrm{SC}_{\Pi Y.\,\beta,\mu,\nu}$, the term t is in that set. □

We now need to show that every term of the simply typed lambda calculus is strongly computable. The following lemma is analogous to Lemma 13.2.8 of Section 13.2.

Lemma 15.6.2. *Suppose β is a type, μ is an assignment of types to $X_0, \ldots, X_{n-1}$, $\nu(X_i)$ is a reducibility candidate for $\mu(X_i)$ for each $i < n$, and t is a term of type $\beta[\mu/\vec{X}]$. If $t[\gamma/Y]$ is in $\mathrm{SC}_{\beta,\mu[Y\mapsto\gamma],\nu[Y\mapsto\mathcal{R}]}$ for every type γ and reducibility candidate $\mathcal{R}$ for γ, then $\Lambda Y.\,t$ is in $\mathrm{SC}_{\Pi Y.\,\beta,\mu,\nu}$.*

Proof By the definition of $\mathrm{SC}_{\Pi Y.\,\beta,\mu,\nu}$, our task is to show that $(\Lambda Y.\,t)\,\gamma$ is in $\mathrm{SC}_{\beta,\mu[Y\mapsto\gamma],\nu[Y\mapsto\mathcal{R}]}$ for every γ and reducibility candidate $\mathcal{R}$ for γ. Taking γ to be Y and $\mathcal{R}$ to be the strongly normalizing terms of type Y in the hypothesis of the lemma, we have that t is in $\mathrm{SC}_{\beta,\mu[Y\mapsto\gamma],\nu[Y\mapsto\mathcal{R}]}$, and so by Lemma 15.6.1, we have that t is strongly normalizing. We can therefore use induction on the maximum number of steps $h(t)$ in a reduction sequence starting from t. Again by Lemma 15.6.1, it suffices to show that every one-step reduction of $(\Lambda Y.\,t)\,\gamma$ is in $\mathrm{SC}_{\beta,\mu[Y\mapsto\gamma],\nu[Y\mapsto\mathcal{R}]}$. But in one step, $(\Lambda Y.\,t)\,\gamma$ reduces to

- $(\Lambda Y.\, t')\, \gamma$ for some t' such that $t \rightarrow_1 t'$, or
- $t[\gamma/Y]$.

In the first case, the result follows from the inductive hypothesis, given that $h(t') < h(t)$. In the second case, the desired conclusion follows from the hypothesis of the lemma. □

Lemma 15.6.2 gives us the means to show that a second-order abstraction $\Lambda Y.\, t$ is strongly computable. The next lemma does the same for a type application $t\, \gamma$.

Lemma 15.6.3. *Suppose β is a type, μ is an assignment of types to $X_0, \ldots, X_{n-1}$, $\nu(X_i)$ is a reducibility candidate for $\mu(X_i)$ for each $i < n$, and t is a term of type $\Pi Y.\, \beta[\mu/\vec{X}]$ in $\mathrm{SC}_{\Pi Y.\, \beta, \mu, \nu}$. Then for any type γ, $t\, \gamma$ is in $\mathrm{SC}_{\beta[\gamma/Y], \mu, \nu}$.*

Proof By the definition of $\mathrm{SC}_{\Pi Y.\, \beta, \mu, \nu}$, for any type γ and reducibility candidate $\mathcal{R}$ for γ, $t\, \gamma$ is in $\mathrm{SC}_{\beta, \mu[Y \mapsto \gamma], \nu[Y \mapsto \mathcal{R}]}$. We only need to check that if we take $\mathcal{R}$ to be $\mathrm{SC}_{\gamma, \mu, \nu}$, this last expression is equal to $\mathrm{SC}_{\beta[\gamma/Y], \mu, \nu}$. In other words, we need the identity

$$\mathrm{SC}_{\beta[\gamma/Y], \mu, \nu} = \mathrm{SC}_{\beta, \mu[Y \mapsto \gamma], \nu[Y \mapsto \mathrm{SC}_{\gamma, \mu, \nu}]}.$$

This is easily verified by induction on β. □

We are finally ready to use an induction on terms to show that every term is strongly computable. As in the proof of Lemma 13.2.10, we prove a more general statement.

Lemma 15.6.4. *Suppose t is a term of type α, $x_0, \ldots, x_{n-1}$ are variables, and $s_0, \ldots, s_{k-1}$ are terms such that each s_i has the same type as x_i. Suppose also that $X_0, \ldots, X_{n-1}$ is a sequence of type variables, μ is an assignment of types to $\vec{X}$, and $\nu(X_i)$ is a reducibility candidate for $\mu(X_i)$ for each $i < n$. Then $t[\vec{s}/\vec{x}][\mu/\vec{X}]$ is in $\mathrm{SC}_{\alpha, \mu, \nu}$.*

Proof The proof is similar to the proof of Lemma 13.2.10. The case where t is of the form $\Lambda X_n.\, q$ is handled by Lemma 15.6.2, and the case where t is a second-order application $q\, \gamma$ is handled by Lemma 15.6.3. □

In the case where $n = 0$ and $k = 0$, we have that t is strongly computable, and hence strongly normalizing.

Theorem 15.6.5. *System* F *is strongly normalizing.*

As for the ordinary simply typed lambda calculus, the only closed terms of type NAT in the second-order simply typed lambda calculus are numerals. Strong normalization and confluence imply that for every term t of type $\mathrm{NAT} \rightarrow \cdots \rightarrow \mathrm{NAT} \rightarrow \mathrm{NAT}$ in system F and any tuple $n_0, \ldots, n_{k-1}$ of natural numbers of the right length, the term $t\, n_0\, \cdots\, n_{k-1}$ reduces to a unique numeral m. The term t is said to *represent* the corresponding k-ary function f from $\mathbb{N}$ to $\mathbb{N}$. The rest of this section is devoted to the following:

Theorem 15.6.6. *The provably total computable functions of classical or intuitionistic second-order arithmetic are exactly the functions that are represented by terms of system* F.

One direction follows from the fact that for each fixed term t of system F, we can formalize the proof above to show, in second-order arithmetic, that t is strongly computable. Hence second-order arithmetic can prove

$$\forall n_0, \ldots, n_{k-1}\, \exists m\, (t\, n_0\, \cdots\, n_{k-1} \twoheadrightarrow m).$$

Since the expression in parentheses can be expressed as a Σ_1 formula, the function represented by t is a provably total computable function of second-order arithmetic.

I will only sketch the details of that argument here. First, notice that there is an arithmetic formula that expresses the fact that a set of terms $\mathcal{R}$ is a reducibility candidate. The key observation is then that for each type α and tuple of terms $\beta_0, \ldots, \beta_{n-1}$, there is a second-order formula $A(t, \mathcal{R}_0, \ldots, \mathcal{R}_{n-1})$ that expresses that t is an element of $\mathrm{SC}_{\alpha, X_i \mapsto \beta_i, X_i \mapsto \mathcal{R}_i}$. It is straightforward to read off such a formula from the inductive definition of $\mathrm{SC}_{\alpha,\mu,\nu}$. Since the clause for $\mathrm{SC}_{\Pi Y.\, \beta,\mu,\mu}$ requires a universal quantifier over reducibility candidates, as the type α becomes more complex, the formula A rises higher in the analytic hierarchy. With the ability to talk about the sets $\mathrm{SC}_{\alpha,\mu,\nu}$ for each fixed α, the proofs above can be formalized straightforwardly in intuitionistic second-order arithmetic. A subtle use of the comprehension schema occurs in the proof of Lemma 15.6.3: to carry out the substitution described there, we need to know that the predicate $t \in \mathrm{SC}_{\gamma,\mu,\nu}$ gives rise to a set of terms $\mathcal{R}$, which can be used to instantiate a variable ranging over reducibility candidates.

We will use a realizability argument to prove the other direction of Theorem 15.6.6, that is, to show that every provably total computable function of second-order arithmetic is represented by a term of system F. First we will show that it suffices to prove the claim for second-order Heyting arithmetic.

Proposition 15.6.7. *Classical second-order arithmetic is Π_2 conservative over intuitionistic second-order arithmetic.*

Proof The Friedman–Dragalin translation, which was described in the proof of Theorem 10.4.6, extends straightforwardly to second-order arithmetic. The comprehension schema implies its double-negation translation, and it is preserved under replacement of $\bot$ by any other formula. □

So it suffices to show that whenever second-order Heyting arithmetic proves a Π_2 formula $\forall \vec{x}\, \exists y\, A(\vec{x}, y)$, there is a term t of system F representing a function $f(\vec{x})$ such that for every $\vec{n}$, $A(\vec{n}, f(\vec{n}))$ is true.

We will use a variant of the modified realizability relation that was described in Section 14.3, but we will define the realizability in ordinary mathematical terms rather than in a formal axiomatic theory. To each formula $A(\vec{Y})$ of second-order arithmetic with free set variables $\vec{Y}$, we will assign a type and a set of realizers of that type, modulo a similar assignment to the property of membership in each of the elements of $\vec{Y}$. (We don't have to worry about first-order variables because we can instantiate them with numerals.) The arguments in Section 14.3 depend on the fact that anything that reduces to a realizer is itself a realizer; for example, if s realizes a formula A and t realizes a formula B then (s, t) realizes $A \wedge B$, since $(s, t)_0$ reduces to s and $(s, t)_1$ reduces to t. Our definition will straightforwardly ensure that the set of realizers R of a formula is closed under β-expansion, meaning that if s reduces to t and t is in R, then s is in R. We will interpret second-order variables Y by a type α and function $F(n)$ which, for each natural number n, returns a set of terms of type α that is closed under β-expansion. The idea is that for each n, $F(n)$ is the set of realizers for the formula $n \in Y$.

It helps to choose a convenient formulation of second-order HA. We can take the first-order part to include arbitrary primitive recursive functions and predicates, as we do in Section 14.1. It will save us effort if we only allow the universal quantifier on the second-order part. This is not a loss, since we can define the second-order existential quantifier in terms of the second-order universal quantifier as described in Section 15.1. In a natural deduction formulation, in addition to the universal axioms of arithmetic, we have the following:

- the introduction and elimination rules for universal quantification over sets:

$$\frac{A}{\forall X\, A} \qquad \frac{\forall X\, A}{A[\lambda x.\, B/X]}$$

- the schema of induction for second-order formulas.

In the elimination rule for the universal quantifier, we interpret an atomic formula $t \in \lambda x.\, B$ as $B[t/x]$.

To each formula $A(\vec{Y})$ in the language of second-order arithmetic whose only free variables are the ones shown, we assign a type τ_A of system F, where τ_A depends on a tuple of type variables $\vec{U}$ of the same length as $\vec{Y}$. Given an assignment $\mu(U_j)$ of a type to each U_j, we will also say what it means for a term t of type $\tau_A[\mu/\vec{U}]$ to realize $A(\vec{Y})$ relative to an interpretation $\rho(Y_j)$ of Y_j, where $\rho(Y_j)$ is a function that takes each element of $\mathbb{N}$ to a set of terms of type $\mu(U_j)$.

- If A is the formula $u \in Y$ for some term u, then τ_A is the variable U corresponding to Y, and a term t realizes A relative to ρ if and only if t is in $\rho(Y)(n)$, where n is the result of evaluating u.
- If A is any other atomic formula, it has no free variables. We assign $\tau_A = \mathrm{NAT}$, and say t realizes A relative to ρ if and only if A is true (in which case ρ and the realizer are irrelevant).
- $\tau_{A \wedge B} = \tau_A \times \tau_B$, and t realizes $A \wedge B$ relative to ρ if and only if $(t)_0$ realizes A and $(t)_1$ realizes B relative to ρ.
- $\tau_{A \vee B} = \mathrm{NAT} \times \tau_A \times \tau_B$, and t realizes $A \vee B$ relative ρ if and only if $(t)_0$ normalizes to 0 and $(t)_1$ realizes A relative to ρ or $(t)_0$ normalizes to 1 and $(t)_2$ realizes B relative to ρ.
- $\tau_{A \to B} = \tau_A \to \tau_B$, and t realizes $A \to B$ relative to ρ if and only if whenever u realizes A relative to ρ, $t\,u$ realizes B relative to ρ.
- $\tau_{\forall x\, A} = \mathrm{NAT} \to \tau_A$, and t realizes $\forall x\, A$ relative to ρ if and only if for every numeral n, $t\,n$ realizes $A[n/x]$ relative to ρ.
- $\tau_{\exists x\, A} = \mathrm{NAT} \times \tau_A$, and t realizes $\exists x\, A$ relative to ρ if and only if $(t)_1$ realizes $A[n/x]$ relative to ρ, where $(t)_0$ reduces to n.
- $\tau_{\forall X\, A} = \Pi_X \tau_A$, and t realizes $\forall X\, A$ relative to ρ if and only if for every type α and every function F that assigns to every natural number n a set $F(n)$ of terms of type α closed under β expansion, $t\,\alpha$ realizes A relative to $\rho[X \mapsto F]$.

Only the first clause and the last clause are genuinely new. The others are transported from the proof of Theorem 14.3.2.

Proposition 15.6.8. *Suppose a formula $A(\vec{x}, \vec{Y})$ with at most the variables shown is provable in second-order arithmetic. Then there is a term t of system* F *with at most the variables $\vec{x}$, $\vec{U}$ such that for every assignment μ to $\vec{U}$ and ρ to $\vec{Y}$ as above, and every tuple of numerals $\vec{n}$, $t[\vec{n}/\vec{x}, \mu/\vec{U}]$ realizes $A(\vec{n}, \vec{Y})$ relative to ρ.*

Proof We use the system of natural deduction described above. The proof of Theorem 14.3.2 carries over essentially unchanged, and we only have to deal with the rules for the second-order universal quantifier. We continue with the conventions established in the proof of Theorem 14.3.2: we assume that we are dealing with a sequent $A_0, \ldots, A_{k-1} \vdash B$, having obtained, inductively, a term that realizes the hypothesis to the last premise, depending on variables that are assumed to realize the formulas in the context.

To handle second-order $\forall$ introduction, suppose the last inference of the proof results in a sequent with $\forall Z\, B(\vec{x}, \vec{Y}, Z)$ in the conclusion. Let μ be an assignment of types to the variables $\vec{U}$ corresponding to $\vec{Y}$ and let ρ be an assignment of interpretations to $\vec{Y}$ as described above. Then, by the inductive hypothesis, there is a term t with at most the variables $\vec{x}$, $\vec{U}$, V such that for any type α, every function $F(j)$ assigning a set of terms of type α closed under β expansion, and every tuple $\vec{n}$, $t[\mu/U, \alpha/V, \vec{n}/\vec{x}]$ realizes $B(\vec{n}, \vec{Y}, Z)$ relative to $\rho[Z \mapsto F]$. By definition, $\Pi_V t$ realizes $\forall Z\, B(\vec{n}, \vec{Y}, Z)$ relative to ρ.

To handle second-order $\forall$ elimination, suppose t satisfies the inductive hypothesis for $\forall Z\, B(\vec{x}, \vec{Y}, Z)$. Let $C(w, \vec{x}, \vec{Y})$ be any formula. Let α be the type of realizers of $C(j, \vec{n}, \vec{Y})$ for every j and $\vec{n}$. (It is easy to check that for any formula $A(\vec{x}, \vec{Y})$, the type of realizers $\tau_{A(\vec{n}, \vec{Y})}$ does not depend on $\vec{n}$.) For every natural number j, let $F(j)$ be the set of terms of type α that realize $C(j, \vec{n}, \vec{Y})$. By hypothesis, $t[\vec{n}/\vec{x}, \mu/\vec{U}]$ realizes $\forall Z\, B(\vec{n}, \vec{Y}, Z)$ relative to ρ. By definition this means that $t[\vec{n}/\vec{x}, \mu/\vec{U}]\, \alpha$ realizes $B(\vec{n}, \vec{Y}, Z)$ relative to $\rho[Z \mapsto F]$.

Since $t[\vec{n}/\vec{x}, \mu/\vec{U}]\, \alpha$ is equal to $(t\, \alpha)[\vec{n}/\vec{x}, \mu/\vec{U}]$, we would be done if we knew that realizing the formula $B(\vec{n}, \vec{Y}, Z)$ relative to $\rho[Z \mapsto F]$ is the same as realizing $B(\vec{n}, \vec{Y}, \lambda w.\; C(w, \vec{n}, \vec{Y}))$ relative to ρ, given that for every j, $F(j)$ is the set of realizers for $C(j, \vec{n}, \vec{Y})$ relative to ρ. This can be established by induction on B. □

Now it is easy to check that if $R(\vec{x}, y)$ is any primitive recursive relation and t realizes $\forall \vec{x}\, \exists y\, R(\vec{x}, y)$ then the function $f(\vec{x})$ represented by $\lambda \vec{x}.\, (t\, \vec{x})_0$ satisfies $R(\vec{x}, f(\vec{x}))$ for every $\vec{x}$. So every provably total computable function of second-order arithmetic is represented by a term in system F, and we have completed the proof of Theorem 15.6.6.

Exercise

15.6.1. Show that system F is confluent.

15.7 Higher-Order Logic and Arithmetic

Having opened the door to quantifying over predicates and relations on a first-order universe, there is no reason to stop there. One way to describe higher-order logic is to start with a collection $\mathcal{B}$ of basic sorts and define a set of higher-order sorts inductively as follows:

- Every basic sort is a sort.
- If $\alpha_0, \ldots, \alpha_{n-1}$ are sorts, so is $[\alpha_0, \ldots, \alpha_{n-1}]$.

Here, $[\alpha_0, \ldots, \alpha_{n-1}]$ is intended to denote the sort of relations $R(x_0, \ldots, x_{n-1})$ where each x_i is a variable of set α_i. If A is any formula and the variables $\vec{x}$ range over the sorts $\vec{\alpha}$, we have the comprehension axiom $\exists R\, \forall \vec{x}\, (R(\vec{x}) \leftrightarrow A(\vec{x}))$, where R has sort $[\vec{\alpha}]$. As in the case of second-order logic, the comprehension axioms can be formulated instead in terms of the $\exists$ introduction rule.

In contrast to the second-order case, here we cannot easily eliminate the need for extensionality axioms. For example, combining the axiom of extensionality for unary second-order predicates with the substitution rule for equality yields the following principle for any third-order relation $\mathcal{R}$:

$$\forall x\,(P(x) \leftrightarrow Q(x)) \rightarrow (\mathcal{R}(P) \leftrightarrow \mathcal{R}(Q)).$$

In the second-order setting, we could define $P = Q$ to be $\forall x\,(P(x) \leftrightarrow Q(x))$, in which case the extensionality axiom holds by definition. But when we take equality as a primitive in the higher-order setting, we have $P = Q \rightarrow \mathcal{R}(P) = \mathcal{R}(Q)$ as an instance of the usual substitution axiom for equality. If, instead, we define $P = Q$ to be $\forall x\,(P(x) \leftrightarrow Q(x))$, we have to add implications like the one above as additional axioms. In other words, there are two interpretations of an equation $P = Q$ to contend with: Leibniz equality, which says that P and Q can be exchanged in any formula, and extensional equality, which says that P and Q hold of the same elements. The first always implies the second, and extensionality is equivalent to saying that the second implies the first.

If we add a basic type of natural numbers or an axiom of infinity, we have both classical and intuitionistic versions of *higher-order arithmetic*. We will see in Chapter 17 that these theories are expressive enough to serve as a foundation for a fair amount of everyday mathematics. In that chapter, we will consider a formulation of higher-order logic based on the simply typed lambda calculus that takes functions as basic and interprets relations as functions that take values in a type PROP of propositions.

Whether we formulate higher-order logic with relations or functions, it is possible to interpret higher-order logic with extensionality in higher-order logic without it, by relativizing quantifiers and variables to the hereditarily extensional objects as described in Section 13.6. This shows that, in a theoretical sense, the principle of extensionality does not add logical strength. But the extensional view of functions is fundamental to classical mathematics, and most implementations of classical higher-order logic include extensionality.

The classical semantics of second-order logic generalizes straightforwardly to higher-order logic, and once again we can consider both Henkin models and full models over a first-order universe. (When it comes to formulations of higher-order logic based on the simply typed lambda calculus, as explained in Section 13.6, it takes more work to describe a semantics in which extensionality does not hold but equality is nonetheless preserved by lambda abstraction.)

Exercises

15.7.1. Use the double-negation translation and the Friedman–Dragalin translation to show that classical higher-order arithmetic is Π_2 conservative over intuitionistic higher-order arithmetic.

15.7.2. Verify that the definition of the hereditarily extensional relations, along the lines described in Section 13.6, interprets extensional higher-order logic in intensional higher-order logic.

Bibliographical Notes

Second-order logic Väänänen (2020) is a good reference for second-order logic. The relationships between various formulations of the axiom of choice and the axiom of infinity in second-order logic are subtle, and more information is provided there. Some methods of understanding the relationships between them carry over from the study of the axiom of choice in set theory, as described by Jech (1973).

Cut elimination The proof of the cut elimination theorem for second-order logic is adapted from the version for higher-order logic in Takeuti (1987). Takeuti's book is a definitive source for that and related results. For algebraic approaches that work for intuitionistic logic as well, see Okada (2002) and the references there.

Second-order arithmetic For more on second-order arithmetic, see Chapter 16 and Simpson (2009). The idea of expressing the fact that the universe is infinite by saying that there is an injective function that is not surjective can be traced to Dedekind, and so is said to express the fact that the universe is *Dedekind infinite.*

It is folklore that one can show that the axiom of choice is equiconsistent with second-order arithmetic using the method that Gödel used to show that the axiom of choice is equiconsistent with set theory. The schema of dependent choice of Σ^1_2 formulas is provable, but already the Σ^1_3 axiom of choice is independent of second-order arithmetic. Section VII.3 of Simpson (2009) and the notes at the end of Chapter 17 provide more information as to how to interpret the language of set theory in the language of second-order arithmetic, and Section VII.6 of Simpson (2009) provides more information about the provability and independence of various choice principles.

Descriptive set theory Kechris (1995) and Moschovakis (2009) are good references for the boldface analytic hierarchy, including proofs of the uniformization property for Π^1_1 relations, the reduction property for Π^1_1 sets, and the separation property for Σ^1_1 sets. Mansfield and Weitkamp (1985) is a good reference for *effective descriptive set theory*, that is, the study of the lightface analytic hierarchy.

The second-order typed lambda calculus The proof of strong normalization here follows Girard et al. (1989), but there are also good presentations in Troelstra and Schwichtenberg (2000) and Sørensen and Urzyczyn (2006). Gallier (1989) surveys various approaches to proving the result.

Higher-order logic For more on higher-order logic, see Takeuti (1987), Andrews (2002), Lambek and Scott (1988), and Section 17.1.

Intuitionistic higher-order logic has important algebraic and category-theoretic interpretations. It can be seen as the *internal logic* of a *topos*, and versions of intuitionistic logic are sound and complete with respect to semantics in an elementary topos. The latter is a generalization of the notion of a topos of sheaves on a Grothendieck site, and that, in turn, is a far-reaching generalization of the notion of a topos of sheaves on a topological space, first developed by Grothendieck for its applications to algebraic geometry. For more information, see Troelstra and van Dalen (1988), Lambek and Scott (1988), or Mac Lane and Moerdijk (1992).

16

Subsystems of Second-Order Arithmetic

In Chapters 9 and 10, we saw that primitive recursive arithmetic and first-order arithmetic can be viewed as foundations for reasoning about finite objects. In the last chapter, we saw that, in a similar way, second-order arithmetic provides a foundation for reasoning about mathematical objects and structures that can be represented as sets of natural numbers. That gives us a lot to work with. Any real number can be represented by a sequence of rational numbers, which can, in turn, be represented by a set of natural numbers. We can't represent arbitrary sets of real numbers or arbitrary functions from the reals to the reals, but we can represent sufficiently nice sets and functions. For that reason, logicians sometimes refer to second-order arithmetic as *analysis*.

It turns out that most mathematical arguments do not require anything close to the full strength of second-order arithmetic. In this chapter, we will study subsystems of second-order arithmetic in which the full schema of comprehension is replaced by weaker set existence principles. We will consider one subsystem, ACA_0, that is a conservative extension of Peano arithmetic. We will consider two weaker theories, RCA_0 and WKL_0, that are conservative extensions of $\mathrm{I}\Sigma_1$. We will also consider two theories, ATR_0 and Π^1_1-CA_0, that are strictly stronger than ACA_0. The last of these is just strong enough to establish an abstract principle of inductive definition like the ones introduced in Section 1.2.

The subsystems of second-order arithmetic that we discuss should not be viewed as full-blown foundations for infinitary mathematics. They are best understood as a basis for exploring the extent to which infinite objects of mathematics can be represented as sets of natural numbers and the axiomatic principles that are implicit in conventional mathematical reasoning about them. The study of these subsystems ties into other important branches of logic. They provide axiomatic bases for *computable analysis*, that is, the study of the extent to which the concepts of analysis can be defined and studied in computable terms, and *descriptive set theory*, which studies such concepts in terms of their complexity in the analytic hierarchy. They also provide a starting point for *proof mining*, which tries to extract useful information from proofs in restricted mathematical theories, and *reverse mathematics*, which aims to calibrate the strength of mathematical theorems in terms of the axioms needed to prove them.

Although we focus on classical subsystems of second-order arithmetic here, the bibliographical notes at the end of this chapter show that intuitionistic subsystems of second-order arithmetic are of interest as well.

16.1 Arithmetic Comprehension

Remember that formula A in the language of second-order arithmetic is said to be *arithmetic* if it has no second-order quantifiers, although it may have set variables. The theory ACA_0 is the subsystem of second-order arithmetic in which the comprehension axiom is restricted to such formulas. In other words, we replace the full comprehension principle by the *arithmetic comprehension axiom* schema, denoted (ACA):

$$\exists Y\, \forall x\, (x \in Y \leftrightarrow A(x)).$$

Throughout this chapter, when we write an axiom schema like this, it should be understood that $A(x)$ may have free variables other than x. Universal quantifiers over those variables are left implicit. As described in Section 15.1, we omit the second-order equality symbol and interpret $X = Y$ as the formula $\forall z\, (z \in X \leftrightarrow z \in Y)$. Aside from the basic axioms of Peano arithmetic, the only other non-logical axiom is the principle of induction,

$$\forall X\, (0 \in X \wedge \forall y\, (y \in X \to y+1 \in X) \to \forall y\, (y \in X)).$$

Combined with the arithmetic comprehension axiom schema, this yields the principle of induction for any arithmetic formula. The subscripted 0 indicates that, in ACA_0, we do not have the full schema of induction. The exercises below ask you to show that the stronger theory named ACA, which adds induction for all formulas in the language, is strictly stronger than ACA_0.

A Henkin model $\mathfrak{M}$ of ACA_0 consists of a first-order universe, $|\mathfrak{M}|$, interpretations of the symbols in the language of arithmetic, and a collection of sets $\mathcal{S}^{\mathfrak{M}}$ to serve as the second-order universe. The structure is moreover required to satisfy the axioms of the theory. When studying subsystems of second-order arithmetic, it is often illuminating to consider models in which the first-order part is the standard interpretation $\mathfrak{N} = (\mathbb{N}, 0, \mathrm{succ}, +, \cdot, \le)$. Such a model is called an *ω-model* and can be identified with the collection $\mathcal{S}^{\mathfrak{M}}$ of subsets of $\mathbb{N}$. An ω-model always satisfies the basic axioms of arithmetic as well as the full schema of induction.

Proposition 16.1.1. *An ω-model $\mathfrak{M}$ is a model of* (ACA) *if and only if $\mathcal{S}^{\mathfrak{M}}$ is closed under relative arithmetic definability, meaning that whenever $A(x, \vec{Z})$ is an arithmetic formula with the free variables shown and $\vec{S}$ is a tuple of sets in $\mathcal{S}^{\mathfrak{M}}$ then the set $\{n \mid \mathfrak{N} \models A(n, \vec{S})\}$ is in $\mathcal{S}^{\mathfrak{M}}$.*

Proof The comprehension axiom is equivalent to saying that for every formula $A(x, \vec{z}, \vec{Z})$, every tuple of elements $\vec{m}$ in $|\mathfrak{M}|$, and every tuples of sets $\vec{S}$ in $\mathcal{S}^{\mathfrak{M}}$, the set $\{n \mid \mathfrak{M} \models A(n, \vec{m}, \vec{S})\}$ is in $\mathcal{S}^{\mathfrak{M}}$. In the case of an ω-model, we can replace $\vec{m}$ with numerals, and $\mathfrak{M} \models A(n, \vec{m}, \vec{S})$ is equivalent to $\mathfrak{N} \models A(n, \vec{S})$. □

In particular, the smallest ω-model of (ACA) consists of the arithmetically definable sets. The following proposition provides an alternative characterization.

Proposition 16.1.2. *$\mathcal{S}^{\mathfrak{M}}$ is an ω-model of* (ACA) *if and only if it is closed under the pairing operation $U, V \mapsto U \oplus V$, the Turing jump $U \mapsto U'$, and relative computability.*

The last condition means that whenever U is in $\mathcal{S}^{\mathfrak{M}}$ and $U \le_T V$ then V is in $\mathcal{S}^{\mathfrak{M}}$. Proposition 16.1.2 is a straightforward consequence of Proposition 16.1.1 and Theorem 11.8.8, but we can also extract a direct proof from the proof of the following.

Proposition 16.1.3. ACA_0 *is finitely axiomatizable.*

Proof We will reduce ACA_0 to a finite set of axioms in stages. First, we show that we can restrict the arithmetic comprehension schema to Σ^0_1 formulas. To that end, it suffices to show by induction that the restricted principle implies Σ^0_n comprehension for every n. Suppose the claim is true for n, and let $\exists y\, B(x, y, \vec{z}, \vec{W})$ be a Σ^0_{n+1} formula with the parameters shown, where B is Π^0_n. Using Σ^0_n comprehension and a pairing operation on the natural numbers, we obtain the set $Y = \{u \mid \neg B((u)_0, (u)_1, \vec{z}, \vec{W})\}$. Using Σ^0_1 comprehension we obtain the set $Z = \{x \mid \exists y\, ((x, y) \notin U)\}$. But then we have that for every x, x is in Z if and only if $\exists y\, B(x, y, \vec{z}, \vec{W})$, as required.

Next, observe that we can restrict the Σ^0_1 comprehension schema to a single set parameter W if add a pairing axiom, $\forall U, V\, \exists Z\, (Z = U \oplus V)$. With this axiom, any tuple $W_0, \ldots, W_{k-1}$ can be represented by a single set W, and with such a set W, for any Σ^0_1 formula $A(x, \vec{z}, \vec{W})$ with the set parameters shown we can find an equivalent Σ^0_1 formula $A'(x, \vec{z}, W)$.

Finally, we replace the Σ^0_1 comprehension schema with one comprehension axiom for a universal Σ^0_1 formula $A(x, z, W)$ that says that the Σ^0_1 formula coded by z is true of x and the parameter W. As in the proof of Theorem 10.5.6, finitely many additional axioms, all consequences of ACA_0, suffice to prove that every Σ^0_1 formula is equivalent to a particular instance of the universal one. □

Clearly ACA_0 extends classical first-order arithmetic, PA. The following theorem shows that the two theories prove the same formulas in their common language.

Theorem 16.1.4. ACA_0 *is a conservative extension of* PA.

We will consider two proofs of this fact. The first, a model-theoretic proof, is quick.

Proof Suppose A is a sentence in the language of arithmetic, and suppose PA does not prove A. Let $\mathfrak{M}$ be a first-order model of $\mathrm{PA} \cup \{\neg A\}$. Extend $\mathfrak{M}$ to a second-order Henkin structure $\mathfrak{M}'$ by taking $S^{\mathfrak{M}'}$ to be the collection of all subsets of $|\mathfrak{M}|$ that are definable by a first-order formula with parameters from $|\mathfrak{M}|$.

Clearly $\mathfrak{M}' \models \neg A$, since the interpretation of any formula in the language of first-order arithmetic is the same in $\mathfrak{M}$ and $\mathfrak{M}'$. The induction axiom in (ACA_0) follows from induction schema in $\mathfrak{M}$, so it suffices to show that $\mathfrak{M}'$ is a model of (ACA). But it is easy to check that if $A(x, \vec{y}, \vec{Z})$ is any arithmetic formula, $\vec{b}$ is any tuple of elements of $|\mathfrak{M}|$, and $\vec{S}$ is any tuple of subsets of $|\mathfrak{M}|$ first-order definable from parameters in $\mathfrak{M}$, then the set $\{a \mid \mathfrak{M} \models A(a, \vec{b}, \vec{S})\}$ is again first-order definable from parameters in $\mathfrak{M}$: just replace the sets $\vec{S}$ by their definitions. □

The second proof, based on the cut elimination theorem, is less straightforward, but it still draws on the intuition that in ACA_0 we can think of the second-order quantifiers as ranging over sets defined by first-order formulas.

Proof Let us extend the language of second-order arithmetic to include a function symbol $S_{A(x,\vec{y},\vec{Z})}$ for every arithmetic formula $A(x, \vec{y}, \vec{Z})$, where $S_{A(x,\vec{y},\vec{Z})}(\vec{y}, \vec{Z})$ denotes an element of the second-order sort. Let T be the two-sorted theory in that language consisting of the axioms of PA, the induction axiom, and axioms

$$\forall x, \vec{y}, \vec{Z}\, (x \in S_{A(x,\vec{y},\vec{Z})}(\vec{y}, \vec{Z}) \leftrightarrow A(x, \vec{y}, \vec{Z})).$$

In other words, we assert that $S_{A(x,\vec{y},\vec{Z})}$ is the set asserted to exist by the comprehension axiom. The theory T clearly includes ACA_0, so it suffices to show that T is conservative over PA.

Represent T with a one-sided sequent calculus, with induction and quantifier-free axioms of PA as described in Section 10.6. We add rules governing the new function symbols analogous to the treatment of equality in Theorem 6.6.5 and the treatment of indefinite descriptions in the syntactic proof of Theorem 7.4.1:

$$\frac{\Gamma, A(t, \vec{u}, \vec{V}), t \in S_{A(x,\vec{y},\vec{Z})}(\vec{u}, \vec{V})}{\Gamma, A(t, \vec{u}, \vec{V})} \qquad \frac{\Gamma, t \in S_{A(x,\vec{y},\vec{Z})}(\vec{u}, \vec{V}), A(t, \vec{u}, \vec{V})}{\Gamma, t \in S_{A(x,\vec{y},\vec{Z})}(\vec{u}, \vec{V})}$$

The intuition is that since $t \in S_{A(x,\vec{y},\vec{Z})}(\vec{u}, \vec{V})$ and $A(t, \vec{u}, \vec{V})$ say the same thing, having both in a sequent is redundant, and so either can be dropped. We can use these to prove the previous axioms as follows, where S is $S_{A(x,\vec{y},\vec{Z})}$:

$$\dfrac{\dfrac{\dfrac{x \notin S(\vec{y}, \vec{Z}), x \in S(\vec{y}, \vec{z}), A(x, \vec{y}, \vec{Z})}{x \notin S(\vec{y}, \vec{Z}), A(t, \vec{y}, \vec{Z})}}{x \notin S(\vec{y}, \vec{Z}) \vee A(x, \vec{y}, \vec{Z})} \qquad \dfrac{\dfrac{\neg A(x, \vec{y}, \vec{Z}), A(x, \vec{y}, \vec{Z}), x \in S(\vec{y}, \vec{z})}{\neg A(x, \vec{y}, \vec{Z}), x \in S(\vec{y}, \vec{Z})}}{\neg A(x, \vec{y}, \vec{Z}) \vee x \in S(\vec{y}, \vec{Z})}}{\dfrac{x \in S(\vec{y}, \vec{Z}) \leftrightarrow A(x, \vec{y}, \vec{Z})}{\forall \vec{y}, \vec{Z}\, (x \in S(\vec{y}, \vec{Z}) \leftrightarrow A(x, \vec{y}, \vec{Z}))}}$$

The new rules violate the subformula property: moving up the proof tree, the second-order existential quantifier in a formula $\exists X\, B(X)$ may be replaced by an arbitrary term involving the new functions S, which can, in turn, be expanded into other formulas. But because the new rules are closed under substitution, the cut elimination theorem still holds, as discussed in Section 6.7. And even though the new rules violate the subformula property, they do not introduce any second-order quantifiers or second-order variables as we traverse them upward. (In an instance of the second-order $\exists$ rule that derives $\Gamma, \exists X\, C(X)$ from $\Gamma, C(T)$, T may have set variables that are not free in the conclusion. But in that case the proof is still valid if we substitute any closed set term for those variables.)

We need to show that any finite set of formulas in the language of first-order arithmetic that is provable in the one-sided calculus we have just designed is also provable in PA. Let $B \mapsto B^*$ denote the obvious translation of formulas in T without second-order quantifiers and second-order variables to formulas in the language of PA that iteratively expands the new S-terms to their definitions. In other words, $(t \in S_{A(x,\vec{y},\vec{Z})}(\vec{u}, \vec{V}))^*$ is defined to be $(A(t, \vec{u}, \vec{V}))^*$, and otherwise the translation commutes with the logical connectives and quantifiers. It suffices to show that if Γ is any finite set of formulas in the language of T without second-order quantifiers and second-order variables and Γ has a cut-free proof in the one-sided sequent calculus for T, then Γ^* is provable in PA. But in such a cut-free proof, there are no instances of the second-order quantifier rules. In a translation of any of the new rules, the premise is the same as the conclusion, so that step can be deleted. The logical rules, the induction rules, and basic axioms of arithmetic are handled by their counterparts in PA. □

Theorem 16.1.4 can be strengthened to show that ACA_0 is Π^1_1 conservative over PA, in the following sense. Let $\mathrm{PA}(X)$ be Peano arithmetic extended with a free predicate $X(y)$ that can appear in the induction formulas. Both the model-theoretic and proof-theoretic proofs can easily be adapted to show that if ACA_0 proves $\forall X\, A(X)$, then $\mathrm{PA}(X)$ proves $A(X)$, where in the latter theory formulas $t \in X$ occurring in $A(X)$ are understood as $X(t)$.

In fact, the conservation result just proved is analogous to Theorem 9.4.1, which shows that PRA with quantifiers is conservative over the quantifier-free version. Just as that theorem can be extended to Π_2 formulas in an appropriate sense, so, too, can Theorem 16.1.4 be extended to Π^1_2 formulas, in an appropriate sense. Other conservations results for second-order theories over first-order theories stated in this chapter can be generalized in similar ways.

Theorem 16.1.4 can be strengthened in a more interesting way. The theory Σ^1_1-AC_0 is obtained by adding to ACA_0 the following choice principle, (Σ^1_1-AC), for Σ^1_1 formulas $A(x, Y)$:

$$\forall x\, \exists Y\, A(x, Y) \rightarrow \exists Y\, \forall x\, A(x, (Y)_x).$$

In the conclusion, $(Y)_x$ refers to the representation of sequences of sets described in Section 15.4, whereby $u \in (Y)_v$ is interpreted as $(v, u) \in Y$.

Theorem 16.1.5. *Σ^1_1-AC_0 is a conservative extension of* PA.

The exercises suggest two approaches to proving this theorem, one proof theoretic and one model theoretic.

The proof-theoretic proof of Theorem 16.1.4 provides an explicit translation of proofs in ACA_0 to proofs in PA, but the fact that we rely on cut elimination allows for the possibility that the translated proof is much longer. We now show that such an increase cannot be avoided. Recall that the iterated exponential function 2^y_n is defined recursively by $2^y_0 = y$ and $2^y_{n+1} = 2^{2^y_n}$.

Theorem 16.1.6. *There is a sequence of sentences $A_0, A_1, A_2, \ldots$ and a polynomial $p(n)$ with the following property. For every n, there is a proof of A_n in* ACA_0 *with length at most $p(n)$, whereas the shortest proof of A_n in* PA *has length at least 2^0_n.*

I will be somewhat cavalier about the precise measure of the length of a proof that is used here, and I will only sketch a proof of Theorem 16.1.6. The bibliographical notes suggest more rigorous accounts, but the argument presented here conveys the main ideas. Note that by interpreting primitive recursive arithmetic in ACA_0, we can use primitive recursive functions to represent the functions 2^x and 2^0_x in an efficient way. The following theorem introduces the central method.

Theorem 16.1.7. *Let T be a theory that contains primitive recursive arithmetic and let $C(x)$ be a formula such that T proves $C(0)$ and $C(x) \rightarrow C(x+1)$. Then for every n, there is a proof of $C(2^0_n)$ whose length is bounded by a polynomial in n.*

Proof Say that a formula $C(x)$ satisfying the hypothesis of the theorem is a *cut* in T. We first show that if $C(x)$ is any cut in T and $C'(x)$ is the formula $\forall y\, (C(y) \rightarrow C(y+2^x))$ then the following hold:

- $C'(x)$ is a cut in T.
- T proves $\forall x\, (C'(x) \rightarrow C(2^x))$.

The fact that T proves $C'(0)$ follows from the fact that it proves $\forall y\, (C(y) \rightarrow C(y+1))$. Also, if T proves $\forall y\, (C(y) \rightarrow C(y+2^x))$, then instantiating the universal quantifier consecutively at y and $y+2^x$ yields $C(y) \rightarrow C(y+2^{x+1})$. The fact that $C'(x)$ implies $C(2^x)$ is obtained by taking y to be 0.

Now, starting with $C(x)$, we want to construct a sequence of cuts

$$C(x) = C_0(x),\ C_1(x),\ \ldots,\ C_{n+1}(x)$$

such for each $i < n+1$, T proves $C_{i+1}(x) \rightarrow C_i(2^x)$. In particular, this implies that T proves $C_n(0)$ and hence $C_0(2^0_{n+1})$.

Unfortunately, defining each $C_{i+1}(x)$ to be $\forall y\, (C_i(y) \rightarrow C_i(y+2^x))$ results in an exponential growth in the length of the formulas, since each C_i is used twice in the definition of C_{i+1}. With a trick the exponential growth can be avoided. The idea is to define a sequence of formulas $D_i(x, z)$ with the property that for each i, $D_i(x, z)$ is equivalent to $(C_i(x) \wedge z = 1) \vee (\neg C_i(x) \wedge z = 0)$, but the length of $D_i(x, z)$ is bounded by a polynomial in i. To do that, first define $D_0(x)$ to $(C(x) \wedge z = 1) \vee (\neg C(x) \wedge z = 0)$. Then, given i, define $E_i(x, y, u_0, u_1)$ to be

$$\forall w, z\, (D_i(w, z) \rightarrow ((w = y \rightarrow u_0 = z) \wedge (w = y + 2^x \rightarrow u_1 = z))).$$

This is just a clever way of expressing

$$(D_i(y, z) \rightarrow u_0 = z) \wedge (D_i(y + 2^x, z) \rightarrow u_1 = z)$$

without having to duplicate D_i. We can then define $D_{i+1}(x, z)$ as follows:

$$\forall y, u_0, u_1\, (E_i(x, y, u_0, u_1) \rightarrow ((u_0 = 1 \rightarrow u_1 = 1) \rightarrow z = 1) \wedge (\neg(u_0 = 1 \rightarrow u_1 = 1) \rightarrow z = 0)).$$

If we now redefine $C_i(x)$ to be $D_i(x, 1)$, the sequence of formulas $C_i(x)$ has the right properties. □

This method is known as *shortening of cuts*, since each cut $C_{i+1}(x)$ is contained in the previous one.

Proof of Theorem 16.1.6 For each n, take A_n to be a formalization of the statement that $\mathrm{I}\Sigma_{2^0_n}$ is consistent. By the incompleteness theorem, $\mathrm{I}\Sigma_{2^0_n}$ cannot prove A_n. So any proof in PA must use an instance of induction with a formula that is not $\Sigma_{2^0_n}$. Even writing down such a formula requires at least 2^0_n symbols.

All that is left to do is to describe a series of short proofs of A_n in ACA_0. Let $B(x)$ be a formula that says

> there exists a set Y that is a truth predicate for Σ_x sentences.

Then ACA_0 proves $B(0)$ and, using arithmetic comprehension, it also proves $\forall x\, (B(x) \rightarrow B(x+1))$. By Theorem 16.1.7, ACA_0 has a short proof of $B(2^0_n)$ for every n, and ACA_0 proves that $B(2^0_n)$ implies A_n. □

This argument assumes that we are measuring the length of a proof in terms of the sizes of the formulas it contains. With a more work, we can also obtain a lower bound on the number of steps in the proof. The idea is that if a proof that uses an instance of induction on a complex formula $A(x)$ is not sufficiently long, then not every subformula of $A(x)$ is used, and $A(x)$ can be replaced by a simpler formula without breaking the proof.

Exercises

16.1.1. Show that the theory ACA, which extends ACA_0 with the full induction schema, proves the consistency of ACA_0. Show that, for that purpose, we only need induction on Δ^1_1 formulas.

16.1.2. Prove Proposition 16.1.2.

16.1.3. Show that over Δ^0_0 comprehension and the pairing axiom for sets, (ACA) is equivalent to the statement that for every function f, the range of f exists:

$$\forall f\, \exists Z\, \forall y\, (y \in Z \leftrightarrow \exists x\, (f(x) = y)).$$

16.1.4. Verify the key claim in the model-theoretic proof of Theorem 16.1.4: if $A(x, \vec{y}, \vec{Z})$ is a formula in the language of arithmetic, $\vec{b}$ is a tuple of elements of $|\mathfrak{M}|$, and $\vec{S}$ is a tuple of sets arithmetically definable from parameters in $\mathfrak{M}$, then $\{a \mid \mathfrak{M} \models A(a, \vec{b}, \vec{S})\}$ is arithmetically definable with parameters in $\mathfrak{M}$.

16.1.5. Show that intuitionistic ACA_0 is conservative over HA, using either a proof-theoretic or model-theoretic argument.

16.1.6. Following the discussion after the two proofs of Theorem 16.1.4, show that if ACA_0 proves $\forall X\, \exists Y\, A(X, Y)$ for some arithmetic formula A, then there is a formula $B(z, X)$ in the language of PA(X) such that PA(X) proves $A(X, \lambda z.\, B(z, X))$. In this last formula, we interpret $t \in \lambda z.\, B(z, X)$ as $B(t, X)$.

16.1.7. Use cut elimination to prove Theorem 16.1.5. (Hint: First show that, using pairing, we can restrict to arithmetic formulas in the choice principle. In a one-sided calculus, use the following rule:

$$\frac{\Gamma, \exists Y\, A(z, Y)}{\Gamma, \exists Y\, \forall x\, A(x, (Y)_x)}$$

Here we assume that z is not free in Γ. Show that given any cut-free proof of a sequent consisting of arithmetic formulas and formulas of the form $\exists Z\, B$ where B is arithmetic, we can extract explicit witnesses for the second-order existential quantifiers. It may be helpful to consult the proof of Lemma 10.6.6.)

16.1.8. Use a model-theoretic argument to prove Theorem 16.1.5. (Hint: Suppose PA does not prove a formula A. Let $\mathfrak{M}$ be a countably saturated model of PA $\cup \{\neg A\}$ (cf. Section 5.6), and let $\mathfrak{M}'$ be the second-order Henkin structure in which the second-order universe consists of the arithmetically definable subsets of $|\mathfrak{M}|$. It suffices to show $\mathfrak{M}'$ is a model of (Σ^1_1-AC). To see this, suppose $\mathfrak{M}' \models_\sigma \forall x\, \exists Y\, A(x, Y)$. Then the type $\{\neg A(x, (\lambda z.\, B(z))_x)\}$ is not realized, where $B(z)$ ranges over arithmetic formulas. Use countable saturation to find an arithmetically definable set Y satisfying $\forall x\, A(x, (Y)_x)$.)

16.1.9. Show that with the efficient definition of the sequence of formulas C_i in the proof of Theorem 16.1.7, each C_i is a cut in T, and for each i, T proves $\forall x\, (C_{i+1}(x) \to C_i(2^x))$.

16.1.10. Use shortening of cuts to show that there is an iterated exponential increase in proof length associated with Theorem 10.6.4, the conservation of $I\Sigma_1$ over PRA.

16.2 Recursive Comprehension

We have seen that the smallest ω-model of ACA_0 is the collection of arithmetically definable sets. We now consider a weaker theory, RCA_0, whose smallest ω-model is the collection of computable sets. Recall that a set Y of natural numbers is said to be Δ^0_1-definable from some set parameters if and only if it has both a Σ_1^0 and a Π_1^0 definition from those parameters. Define the *recursive comprehension axiom* schema (RCA) as follows:

$$\forall x\,(A(x) \leftrightarrow B(x)) \to \exists Y\,\forall x\,(x \in Y \leftrightarrow A(x)).$$

Here A and B are Σ_1^0 and Π_1^0, respectively, and Y does not occur in either formula. In words, (RCA) says that if A and B are equivalent descriptions of a set computable from their parameters, then that set exists. In this context, it is helpful and natural to replace the induction axiom with the stronger schema of induction for Σ_1^0 formulas. We therefore take the theory RCA_0 to be the subsystem of second-order arithmetic with the following axioms:

- the usual quantifier-free axioms of arithmetic
- the induction schema restricted to Σ_1^0 formulas, (Σ_1^0-IND)
- the recursive comprehension schema, (RCA).

RCA_0 proves the existence of a set coding the pair $X \oplus Y$. In analogy to Propositions 16.1.1, 16.1.2, and 16.1.3, we have the following two propositions:

Proposition 16.2.1. *A collection of sets is an ω-model of* RCA_0 *if and only if it is closed under Δ_1^0 definability, or equivalently, if and only if it is nonempty and closed under pairing and relative computability.*

Proposition 16.2.2. RCA_0 *is finitely axiomatizable.*

The fact that any theorem provable in RCA_0 is true when the second-order variables are interpreted as ranging over computable sets and functions makes RCA_0 a natural axiomatic theory with which to formalize the theory of computability and computable mathematics. Remember that we take function variables f in the language of second-order arithmetic to be set variables satisfying $\forall x\,\exists! y\,((x, y) \in f)$. In this last statement, we can use the Δ_0-definable pairing function $J(x, y)$, and Section 11.2 shows that Δ_0^0 formulas in which f is treated as a function symbol can be interpreted as Δ_1^0 formulas in the corresponding set variable. Along the lines of Corollary 11.2.2, which asserts that a function is computable if and only if it has a Σ_1-definable graph, we can prove the following in RCA_0 for every Σ_1^0 formula A:

$$\forall \vec{x}\,\exists! y\,A(\vec{x}, y) \to \exists f\,\forall \vec{x}\,A(\vec{x}, f(\vec{x})).$$

The following proposition strengthens this by dropping the uniqueness assumption in the hypothesis.

Proposition 16.2.3. RCA_0 *proves*

$$\forall \vec{x}\,\exists y\,A(\vec{x}, y) \to \exists f\,\forall \vec{x}\,A(\vec{x}, f(\vec{x}))$$

for every Σ_1^0 formula A.

Proof Without loss of generality, using J, we can assume that there is only a single universally quantified variable x. We can also absorb any unbounded existential quantifiers of A: if $A(x, y)$ is the formula $\exists z\,B(x, y, z)$, then from the hypothesis we have $\forall x\,\exists y\,B(x, (y)_0, (y)_1)$, and given a function g satisfying $\forall x\,B(x, (g(x))_0, (g(x))_1)$, we can let $f(x) = (g(x))_0$. The function g is Δ_1^0 definable from f since $(x, z) \in f$ is equivalent to $\exists y\,((x, y) \in g \wedge (y)_0 = z)$

as well as $\forall y\,((x, y) \in g \rightarrow (y)_0 = z)$, so the existence of g is provable in RCA_0 from the existence of f.

So, without loss of generality, assume $A(x, y)$ is Δ^0_0. Define $A'(x, y)$ to be the formula $A(x, y) \wedge (\forall z < y\, \neg A(x, z))$, which asserts that y is a least witness to $\exists x\, A(x, y)$. $A'(x, y)$ is a Δ^0_0 formula, and so RCA_0 proves the existence of a set f such that for every x and y, $(x, y) \in f \leftrightarrow A'(x, y)$. The hypothesis $\forall x\, \exists y\, A(x, y)$ and the least-element principle for Δ^0_0 formulas imply that f is a function satisfying $\forall x\, A(x, f(x))$. □

Since RCA_0 includes $\mathrm{I}\Sigma_1$, all the primitive primitive recursive functions can be introduced in a definitional extension. But now that functions are bona fide objects, we can make the stronger claim that each primitive recursive function is an element of the second-order universe. The proof is left as an exercise.

Proposition 16.2.4. RCA_0 *proves that the universe of functions is closed under composition and primitive recursion.*

In the next section, we will see that we can formalize portions of analysis in RCA_0. It is therefore a striking fact that the theory is no stronger than $\mathrm{I}\Sigma_1$, and hence no stronger than primitive recursive arithmetic. As we did with ACA_0, we will consider both model-theoretic and proof-theoretic proofs of this fact.

Theorem 16.2.5. RCA_0 *is a conservative extension of* $\mathrm{I}\Sigma_1$.

Proof Suppose $\mathrm{I}\Sigma_1$ does not prove A, and let $\mathfrak{M}$ be a model of $\mathrm{I}\Sigma_1$ in which A is false. Turn $\mathfrak{M}$ into a Henkin-model $\mathfrak{M}'$ for the language of second-order arithmetic by taking $S^{\mathfrak{M}'}$ to be the collection of Δ_1-definable subsets of $|\mathfrak{M}|$. As in the model-theoretic proof of Theorem 16.1.4, it suffices to show that any subset of $|\mathfrak{M}|$ that is Δ^0_1 definable from parameters in $\mathfrak{M}'$ is Δ_1-definable in $\mathfrak{M}$, and that Σ^0_1 induction in $\mathfrak{M}'$ follows from Σ_1 induction in $\mathfrak{M}$.

Both are consequences of the following claim: if $A(\vec{x}, \vec{Z})$ is any Σ^0_1 formula in the language of second-order arithmetic and $\vec{S}$ are elements of $S^{\mathfrak{M}'}$, then there are a Σ_1 formula $A'(\vec{x}, \vec{y})$ and elements $\vec{b}$ in $|\mathfrak{M}|$ such that for every tuple of elements $\vec{a}$ in $|\mathfrak{M}|$, $\mathfrak{M}' \models A(\vec{a}, \vec{S})$ if and only if $\mathfrak{M} \models A'(\vec{a}, \vec{b})$. This tells us that Σ^0_1, Π^0_1, and Δ^0_1 formulas with set parameters from $S^{\mathfrak{M}'}$ have Σ_1, Π_1, and Δ_1 counterparts using only first-order parameters from $\mathfrak{M}$. To prove the claim, intuitively, we only need to replace the sets $\vec{S}$ in $A(\vec{x}, \vec{S})$ by their Δ_1 definitions in $\mathfrak{M}$; more formally, use induction on A, as in the proof of Theorem 10.3.8. □

Once again, the proof-theoretic argument is similar in spirit to the model-theoretic one. But in contrast to the proof of conservation of ACA_0 over PA, this one provides a direct interpretation of RCA_0 in $\mathrm{I}\Sigma_1$ that does not yield a substantial increase in the length of proof.

Proof We will show how to translate a proof of a first-order statement in RCA_0 to a proof of the same first-order statement in $\mathrm{I}\Sigma_1$. The idea is to describe the minimal ω-model of RCA_0 within the language $\mathrm{I}\Sigma_1$, and interpret references to sets in RCA_0 by references to appropriate numeric representations of Δ_1 sets in $\mathrm{I}\Sigma_1$.

Using Σ_1 and Π_1 truth predicates in the language of arithmetic, let $\mathrm{Set}(u)$ be a formula that expresses the following:

> u represents a pair $(A(x), B(x))$ where $A(x)$ is a Σ_1 formula, $B(x)$ is a Π_1 formula, and for every x, $A(\bar{x})$ is true if and only if $B(\bar{x})$ is true.

In other words, $\mathrm{Set}(u)$ says that u represents a Δ^0_1 definable set. To each set variable U in the language of RCA_0, assign a number variable u in the language of $\mathrm{I}\Sigma_1$, and keep these separate from the variables that are used to range over numbers. Then translate any formula $A(\vec{U})$ in the language of RCA_0 with the set variables shown to a formula $A'(\vec{u})$ in the language of $\mathrm{I}\Sigma_1$ as follows: $A'(u_0, \ldots, u_{k-1})$ is given by

$$\mathrm{Set}(u_0) \wedge \cdots \wedge \mathrm{Set}(u_{k-1}) \rightarrow \hat{A}(u_0, \ldots, u_{k-1}),$$

where $\hat{A}(u_0, \ldots, u_{k-1})$ is obtained by

- translating each atomic formula of the form $t \in U$ in A to the statement that the Σ_1 formula represented by $(u)_0$ holds of t, and
- translating set quantifiers $\exists V$ and $\forall V$ to bounded quantifiers $\exists v\, (\mathrm{Set}(v) \wedge \cdots)$ and $\forall v\, (\mathrm{Set}(v) \rightarrow \cdots)$.

We can show inductively that on the assumption that $\mathrm{Set}(u_0) \wedge \cdots \wedge \mathrm{Set}(u_{k-1})$ holds, if A is Σ_1 (or, respectively Π_1 or Δ_1), then so is A'. With that it is not hard to show that the translations of axioms of RCA_0 are provable in $\mathrm{I}\Sigma_1$, so if a formula A is provable in RCA_0, the translation A' is provable in $\mathrm{I}\Sigma_1$. The result follows from the fact that formulas without second-order quantifiers or variables are unchanged by the translation. □

Exercises

16.2.1. Let (Δ^0_0-CA) be the comprehension principle for Δ^0_0 formulas. Show that in the presence of (Δ^0_0-CA), the schema (RCA) is equivalent to each of the following:

a. the principle $\forall x\, \exists y\, A(x, y) \rightarrow \exists f\, \forall x\, A(x, f(x))$ for Δ^0_0 formulas A
b. the principle that for every pair of functions f and g from $\mathbb{N}$ to $\mathbb{N}$, the composition of f and g exists
c. the principle that for every function $g(x)$, the function $f(x) = L(g(x))$ exists, where the L function is the Δ^0_0 projection function for pairs discussed in Section 11.2.

(Hint: see the proof of Proposition 16.2.3.)

16.2.2. Prove Proposition 16.2.4.

16.3 Formalizing Analysis

We will now consider ways that infinitary mathematics can be carried out in restricted subsystems of second-order arithmetic, focusing on the branch of mathematics known as *analysis*. In the late seventeenth century, the fundamental notions of calculus were seen as being grounded in a geometric understanding of the *continuum*, a fancy name for the real number line. In the nineteenth century, the goal of obtaining a more rigorous foundation for reasoning about limits and infinitesimals led mathematicians to the arithmetization of analysis, which is to say, reducing concepts of analysis to concepts of arithmetic over the natural numbers. This resulted in definitions of the real numbers and real-valued functions using incipient set-theoretic language and methods, otherwise taking only the natural numbers for granted.

We have seen that integers and rational numbers can straightforwardly be represented as natural numbers. The second half of the nineteenth century brought various means of constructing the real numbers. Dedekind used what we now know of as *Dedekind cuts* while Cantor used what are now called *Cauchy sequences*. Since both of these can ultimately be represented as sets of numbers, it is historically and mathematically interesting to ask as to the axioms that are needed to prove theorems about the real numbers when they are expressed in terms of these representations.

It is desirable to choose representations such that common operations on the natural numbers like addition and multiplication are computable, in the sense of type 2 computability. It is also desirable to choose representations in such a way that fundamental properties of the real numbers can be established in RCA_0. The first desideratum gives us means to measure the extent to which the fundamental operations of calculus, like limits, integrals, and derivatives, are computable, a task that falls under the domain of computable analysis and effective descriptive set theory. The second desideratum gives us a means of calibrating the strength of the axioms needed to establish fundamental theorems such as the intermediate value theorem or the extreme value theorem, which posit the existence of second-order objects. One way to show that a set existence principle like arithmetic comprehension is required to prove such a theorem is to show that, over a weak base theory like RCA_0, the theorem implies the axiom. This is known as *reverse mathematics*.

A thorough exploration of computable analysis, descriptive set theory, and reverse mathematics are beyond the scope of this book. The goal here is to offer only a few illustrative examples of the way that notions of analysis can be represented in the language of second-order arithmetic, opening the door to questions having to do with computability, definability, and provability in analysis.

The following definition should be viewed as taking place in the language of second-order arithmetic.

Definition 16.3.1. A *real number* x is a sequence $(x_i)_{i \in \mathbb{N}}$ such that for every i and $j \geq i$, $|x_i - x_j| \leq 1/2^i$. Two real numbers are *equal*, written $x \equiv y$, if $|x_i - y_i| \leq 1/2^{i-1}$ for every $i \in \mathbb{N}$.

In other words, we take real numbers to be Cauchy sequences $(x_i)_{i \in \mathbb{N}}$ with a fixed rate of convergence. We will see soon that there are computable Cauchy sequences of rational numbers with no computable rate of convergence, so equipping our Cauchy sequences with such information is a substantive choice. Since equality between real numbers, as we have defined it, is not the same as equality of the underlying sets, it may be better to think of sequences $(x_i)_{i \in \mathbb{N}}$ as representations of real numbers, which can in turn be identified with the equivalence classes induced by $\equiv$. In this section, we will use the letters $i, j, m, n, k, \ldots$ for natural numbers. We will use $p, q, r, \ldots$ for rational numbers, which are represented by natural numbers, and $x, y, z, \ldots$ for real numbers, which are represented by sets. We can view any rational number p as a real number represented by the sequence $p, p, p, \ldots$.

For any rational number p, the *absolute value* of p, denoted $|p|$, is equal to p if $p \geq 0$ and $-p$ if $p < 0$. The following are easy to verify formally in PRA:

1. $-|p| \leq p \leq |p|$.
2. If $p \leq q$ and $-p \leq q$ then $|p| \leq q$.
3. The *triangle inequality*: $|p + q| \leq |p| + |q|$.

To prove 3, by 2 it suffices to show $p + q \leq |p| + |q|$ and $-(p + q) \leq |p| + |q|$, and both these follow easily from 1. (In fact, these properties hold in any ordered group.) Substituting $p - q$

for p and $q-r$ for q in 3, we obtain the following equivalent formulation of the triangle inequality: $|p-r| \leq |p-q|+|q-r|$ for any p, q, and r. This explains the name, since it asserts that, on the real number line, the distance from p to r is less than or equal to sum of the distances from p to q and from q to r. This is used in the next proof.

Proposition 16.3.2. RCA_0 *proves that the relation* $\equiv$ *is an equivalence relation.*

Proof Reflexivity and transitivity are easily verified. For transitivity, suppose $x \equiv y$ and $y \equiv z$. For any i and $j \geq i$, we have

$$\begin{aligned} |x_i - z_i| &\leq |x_i - x_j| + |x_j - y_j| + |y_j - z_j| + |z_j - z_i| \\ &\leq 1/2^i + 1/2^{j-1} + 1/2^{j-1} + 1/2^i \\ &= 1/2^{i-1} + 1/2^{j-2}. \end{aligned}$$

Since j is arbitrary, we have $|x_i - z_i| \leq 1/2^{i-1}$, and hence $x \equiv z$. □

Given x and y, define $x+y$ to be the sequence $(x_{i+1}+y_{i+1})_{i\in\mathbb{N}}$. To see that this is Cauchy with the required rate of convergence, note that for every $j \geq i$ we have

$$|(x_{i+1}+y_{i+1})-(x_{j+1}+y_{j+1})| \leq |x_{i+1}-x_{j+1}|+|y_{i+1}+y_{j+1}| < 1/2^{i+1}+1/2^{i+1} = 1/2^i,$$

as needed. It is not hard to show that addition, defined this way, is commutative. To see that it respects the equivalence relation, note that if $y \equiv z$, then $x+y \equiv x+z$, since $|(x_i+y_i)-(x_i+z_i)| = |y_i - z_i| \leq 1/2^{i-1}$ for every i. Hence if $x \equiv x'$ and $y \equiv y'$ we have $x+y \equiv x+y' \equiv y'+x \equiv y'+x' \equiv x'+y'$.

If we define $-x$ to be $(-x_i)_{i\in\mathbb{N}}$, it is easy to show that negation respects the equivalence relation as well. We similarly define the absolute value of x pointwise by setting $|x| = (|x_i|)_{i\in\mathbb{N}}$. Defining multiplication is trickier. Notice that if x is a Cauchy sequence, then for every i, $|x_i| \leq |x_i - x_0| + |x_0| \leq 1 + |x_0|$. Given x and y, let N be the least natural number such that 2^N is greater than or equal to $|x_0|+|y_0|+2$, and define $x \cdot y$ to be the sequence $(x_{N+i}\cdot y_{N+i})_{i\in\mathbb{N}}$. Then for every i and $j \geq i$, we have

$$\begin{aligned} |x_{N+i}\cdot y_{N+i} - x_{N+j}\cdot y_{N+j}| &\leq |x_{N+i}\cdot y_{N+i} - x_{N+i}\cdot y_{N+j}| + |x_{N+i}\cdot y_{N+j} - x_{N+j}\cdot y_{N+j}| \\ &= |x_{N+i}||y_{N+i}-y_{N+j}| + |y_{N+j}||x_{N+i}-x_{N+j}| \\ &\leq 1/2^{N+i}\cdot(|x_{N+i}|+|y_{N+j}|) \\ &\leq 1/2^i \cdot 1/2^N \cdot (|x_0|+|y_0|+2) \\ &\leq 1/2^i. \end{aligned}$$

This shows that x is a real number. We leave it as an exercise to show that multiplication respects the equivalence relation.

We define $x \leq y$ by the formula $\forall i\,(x_i \leq y_i + 1/2^{i-1})$, and we sometimes write $y \geq x$ instead of $x \leq y$. It is immediate that $x \leq y$ and $y \leq x$ implies $x \equiv y$. Notice that $x \leq y$ is a Π^0_1 formula in x and y. We define $x < y$ to mean $\neg y \leq x$, which is equivalent to the Σ^0_1 formula $\exists i\,(x_i + 1/2^{i-1} < y_i)$. This provides a computational interpretation of the difference between strict and weak inequality: if x is strictly less than y, then this fact is made apparent by some pair of approximations x_i and y_i, whereas the fact that x is less than or equal to y is only verified in the limit. Assuming that x and y are real numbers, if $x_i + 1/2^{i-1} < y_i$, then for every $j \geq i$ we have

$$x_j + 1/2^i \leq (x_i + 1/2^i) + 1/2^i \leq x_i + 1/2^{i-1} < y_j,$$

which implies $x + 1/2^i \leq y$.

Expressing $x < y$ as a Σ_1^0 formula requires classical reasoning. There is no computable function $F(x)$ that decides whether x is equal to zero: given a sequence of approximations $0, 0, 0, \ldots$ to x, an algorithmic procedure cannot rule out, on the basis of a finite number of queries, the possibility that a later approximation will render it positive or negative. We saw in Chapter 14 that theorems provable in constructive theories are generally witnessed by computational data, which shows that we should not expect properties like $x = 0 \vee x \neq 0$ and $x \leq 0 \vee x \geq 0$ to be provable constructively. In constructive mathematics, therefore, $x < y$ is usually defined in terms of the positive Σ_1^0 characterization just given. The exercises below ask you to show that with only intuitionistic logic, RCA_0 proves the following constructive proxy for the law of the excluded middle: for every $\varepsilon > 0$, either $x < \varepsilon$ or $x > 0$. Computationally speaking, an algorithm can always compute a sufficiently good approximation to confidently assert one or the other, allowing for the fact that there is a grey area in which both disjuncts hold.

Constructive analysis considers the extent to which analysis can be developed on the basis of constructive principles and logic, and computable analysis considers the extent to which the operations on real numbers and countably presented data can be carried out computationally. Provability in RCA_0 is a sort of compromise between the two: set and function variables can be interpreted as ranging over computable sets and functions, but the theory allows us to reason about them classically.

We leave it as an exercise to verify that RCA_0 proves all the following, which amount to the fact that the real numbers are an instance of an ordered field.

Proposition 16.3.3. RCA_0 *proves that the following hold for all real numbers x,y, and z:*

1. $x + 0 \equiv x$
2. $(x + y) + z \equiv x + (y + z)$
3. $x + y \equiv y + x$
4. $x + -x \equiv 0$
5. $x \cdot 1 \equiv x$
6. $(x \cdot y) \cdot z \equiv x \cdot (y \cdot z)$
7. $x \cdot y \equiv y \cdot x$
8. $x \cdot (y + z) \equiv x \cdot y + x \cdot z$
9. $x \not\equiv 0 \to \exists y\, (x \cdot y \equiv 1)$
10. $x \leq x$
11. $x \leq y \wedge y \leq z \to x \leq z$
12. $x \leq y \wedge y \leq x \to x \equiv y$
13. $x \leq y \to x + z \leq y + z$
14. $x \geq 0 \wedge y \geq 0 \to x \cdot y \geq 0.$

All of these can be proved intuitionistically except for 9; constructively, one needs a bound on the distance from zero to define an inverse.

A sequence $(x_i)_{i \in \mathbb{N}}$ of real numbers *converges to* a real number y if for every $\varepsilon > 0$ there is an N such that for every $i \geq N$, $|x_i - y| \leq \varepsilon$. A sequence $(x_i)_{i \in \mathbb{N}}$ *converges* if it converges to some y. The following proposition shows why it matters whether we specify a fixed rate of convergence in the definition of a real number.

Theorem 16.3.4. *in* RCA$_0$, *the following statements are all equivalent to* (ACA)*:*

1. *Every bounded, nondecreasing sequence* $(p_i)_{i\in\mathbb{N}}$ *of rational numbers converges.*
2. *Every Cauchy sequence of rational numbers converges.*
3. *Every Cauchy sequence of real numbers converges.*

Proof We argue in RCA$_0$. Clearly 3 implies 2. To show that 2 implies 1, it suffices to show that every bounded, nondecreasing sequence of rationals is Cauchy. To that end, suppose $(p_i)_{i\in\mathbb{N}}$ is nondecreasing but not Cauchy. Then for some $\varepsilon > 0$, for every i, there is a $j > i$ such that $p_j > p_i + \varepsilon$. By induction, therefore, for every n there is an i such that $p_i > p_0 + n\varepsilon$. This implies that the sequence p_i is not bounded.

So we only need to show that (ACA) implies 3 and that 1 implies (ACA). For the first, suppose we are given a Cauchy sequence $(x_i)_{i\in\mathbb{N}}$. By arithmetic comprehension, define the function f so that, for every n, $f(n)$ is the least i such that for every $j \geq i$, $|x_j - x_i| \leq 1/2^n$. Then $x_{f(0)}, x_{f(1)}, x_{f(2)}, \ldots$ is a Cauchy sequence of real numbers such that for such that for every $j \geq i$, $|x_{f(j)} - x_{f(i)}| \leq 1/2^i$. To turn this into a real number y, we pick sufficiently good rational approximations to each term by setting y_i to $(x_{f(i+1)})_{i+2}$ for every i. Then for every i and $j \geq i$, we have

$$\begin{aligned}
|y_j - y_i| &= |(x_{f(j+1)})_{j+2} - (x_{f(i+1)})_{i+2}| \\
&\leq |(x_{f(j+1)})_{j+2} - x_{f(j+1)}| + |x_{f(j+1)} - x_{f(i+1)}| + |x_{f(i+1)} - (x_{f(i+1)})_{i+2}| \\
&\leq 1/2^{j+2} + 1/2^{i+1} + 1/2^{i+2} \\
&\leq 1/2^{i+2} + 1/2^{i+1} + 1/2^{i+2} \\
&= 1/2^i.
\end{aligned}$$

So $y = (y_i)_{i\in\mathbb{N}}$ is a real number, and a similar calculation confirms that $(x_i)_{i\in\mathbb{N}}$ converges to y.

To show that 1 implies (ACA), let $A(n, m)$ be any Δ^0_0 formula, possibly with set and number variables other than the ones shown. As in the proof of Proposition 16.1.3, it suffices to show that the set $\{m \mid \exists n\, A(n, m)\}$ exists. By (RCA), it suffices to show that $\exists n\, A(n, m)$ is equivalent to a Π^0_1 formula.

For each s, let U_s be the finite set $\{m < s \mid \exists n < s\, A(n, m)\}$, and define the rational number p_s to be $\sum_{m\in U_s} 1/2^m$. When p_s is expressed as a binary decimal, the digits that are 1 correspond to the numbers $m < s$ that are seen to satisfy $\exists n\, A(n, m)$ with witnesses less than s. Clearly the sequence $(p_s)_{s\in\mathbb{N}}$ is increasing and bounded by 1. By the hypothesis that clause 1 holds, the sequence converges to a real number x.

Having x in hand allows us to determine whether $\exists n\, A(m, n)$ holds as follows. Because $(p_s)_{s\in\mathbb{N}}$ is increasing, if we find an $s > m$ such that $x - p_s < 1/2^m$, we know that for every $t \geq s$ we have $p_t - p_s \leq x - p_s < 1/2^m$. This means that if m is not in U_s, it is not in U_t for any t, and so there is no n such that $A(m, n)$ holds. So $\neg\exists n\, A(m, n)$ is equivalent to the Σ^0_1 formula

$$\exists s\, (m < s \wedge x - p_s < 1/2^m \wedge \forall n < s\, \neg A(n, m)).$$

The negation of this formula is the Π^0_1 equivalent of $\exists n\, A(m, n)$ that we are after. □

The real numbers can be characterized uniquely up to isomorphism as a complete, archimedean ordered field. The completeness property can be expressed by saying that every Cauchy sequence converges, and the archimedean property can be expressed by saying that for every real number, there is a natural number n such that $|x| < n$. Theorem 16.3.4 tells us not only that the completeness principle is provable in ACA_0, but moreover that proving it requires arithmetic comprehension.

In Chapters 8 and 9, we saw that there are various ways of representing finite objects as natural numbers. We also that, when working with a theory that includes primitive recursive arithmetic, the choice is not important as long as PRA can establish their key properties. The same is true when it comes to representing countable mathematical objects as sets of natural numbers: differences between choices tend to vanish in full second-order arithmetic, which is generally strong enough to establish the equivalence of any two reasonable choices. But Theorem 16.3.4 shows that in weak subsystems of second-order arithmetic, the differences do matter. If we define a real number to be an arbitrary Cauchy sequence of rationals, we cannot make use of its rate of convergence in other constructions. This shows that we can also view reverse mathematics as an explicit study of representations and the principles needed to establish equivalences between them.

Section 16.4 will describe a relationship between the unit interval $[0, 1]$ in the real numbers and the space $\{0, 1\}^{\mathbb{N}}$ with its standard topology (see Appendix A.4). Similarly, there are formal relationships between the set of all real numbers and the space $\mathbb{N}^{\mathbb{N}}$. If σ is any finite binary sequence, write $[\sigma]$ for the set $\{f \in \{0, 1\}^{\mathbb{N}} \mid f \supset \sigma\}$. Then the collection of sets $[\sigma]$ is a basis for the topology on $\{0, 1\}^{\mathbb{N}}$, and any open set U can be written as a union of such sets. Given the one-to-one correspondence between subsets of $\mathbb{N}$ and elements of $\{0, 1\}^{\mathbb{N}}$ that identifies each subset S with its characteristic function χ_S, Theorem 11.8.5 provides information about subsets of $\{0, 1\}^{\mathbb{N}}$.

Proposition 16.3.5. *Let U be a subset of $\{0, 1\}^{\mathbb{N}}$. Then the following are equivalent:*

1. *U is open.*
2. *The complement of U is the set of paths through some tree T on $\{0, 1\}$.*
3. *There are a Σ^0_1 formula $A(X, Y)$ and $S \subseteq \mathbb{N}$ such that $U = \{X \mid \mathfrak{N} \models A(X, S)\}$.*

Proof We have 2 implies 1 because if T is a tree and U is the complement of the set of paths through T, then $U = \bigcup_{\sigma \notin T} [\sigma]$. We have 1 implies 3 because if $U = \bigcup_{\sigma \in S} [\sigma]$, then U is defined by the formula $\exists \sigma\, (\sigma \in S \wedge \sigma \subset X)$, which is Σ^0_1. Finally, the fact that 3 implies 2 is given by Theorem 11.8.5. □

Analogously, we have the following:

Proposition 16.3.6. *Let U be a subset of $\mathbb{N}^{\mathbb{N}}$. Then the following are equivalent:*

1. *U is open.*
2. *The complement of U is the set of paths through some tree T on $\mathbb{N}$.*
3. *There are a Σ^0_1 formula $A(f, Y)$ and $S \subseteq \mathbb{N}$ such that $U = \{f \mid \mathfrak{N} \models A(f, S)\}$.*

Using the boldface notation introduced in Section 15.5, the equivalence of 1 and 3 in each of these propositions says that U is open if and only if it is $\boldsymbol{\Sigma}^0_1$ definable.

Formally, in RCA_0, we can define an open subset U of $\{0, 1\}^{\mathbb{N}}$ to be a set of binary sequences, and define $X \in U$ to mean $\exists \sigma \in U\, (X \supset \sigma)$. Then for each Σ^0_1 formula $A(X)$, possibly with other number and set variables, RCA_0 can prove that there is an open set U satisfying $\forall X\,(A(X) \leftrightarrow X \in U)$. Open subsets of $\mathbb{N}^{\mathbb{N}}$ are defined similarly and give rise to the analogous correspondence for Σ^0_1 formulas $A(f)$. Provability in RCA_0 follows from the fact that the arguments for Theorems 11.8.5 and 11.8.7 can be formalized straightforwardly to justify the following:

Proposition 16.3.7. *Let $A(X)$ be a Σ^0_1 formula. Then* RCA_0 *proves that there is a tree on $\{0, 1\}$ such that for every set X, $A(X)$ holds if and only if there is a σ in the complement of T such that $\sigma \subset X$. Similarly, for a Σ^0_1 formula $A(f)$, there is a tree T on $\mathbb{N}$ such that for every function f, $A(f)$ holds if and only if there is a σ in the complement of T such that $\sigma \subset f$.*

It would be nice to have a similar correspondence between Σ^0_1 formulas and open subsets of $\mathbb{R}$. A *rational open ball* $B(p, \varepsilon)$ is given by a pair (p, ε) of rational numbers with $\varepsilon > 0$, intended to represent the open interval $(p - \varepsilon, p + \varepsilon)$. Accordingly, in RCA_0, we define $x \in B(p, \varepsilon)$ to mean $|x - p| < \varepsilon$. Mathematically speaking, any open subset of $\mathbb{R}$ can be described as a union of rational balls. In RCA_0, it is more natural and convenient to represent an open subset of $\mathbb{R}$ as an enumerable set of open balls, which is to say, we define an open set U to be a set of triples (n, p, ε) where n is a natural number, p and ε are rational numbers, and $\varepsilon > 0$. Given such a set U, we define $x \in U$ to mean that there is a triple $(n, p, \varepsilon) \in U$ such that $x \in B(p, \varepsilon)$. Notice that $x \in U$ is a Σ^0_1 formula. Conversely, we have the following.

Theorem 16.3.8. *Suppose $A(x)$ is a Σ^0_1 formula. Then* RCA_0 *proves the following. Suppose $A(x)$ respects equality on the real numbers in the sense that for any two real numbers x and y, if $A(x)$ and $x \equiv y$ then $A(y)$. Then there is an open set U such that for every x, x is in U if and only if $A(x)$ holds.*

Proof Viewing x as a function from $\mathbb{N}$ to $\mathbb{N}$, by Proposition 16.3.7 we can assume that $A(x)$ is of the form $\exists \sigma \subset x\, C(\sigma)$, where C is Δ_0.

Suppose $A(x)$ holds, where $x = (x_i)_{i \in \mathbb{N}}$. Let $x' = (x_{i+1})_{i \in \mathbb{N}}$, a variant of x with a slightly faster rate of convergence. It is easy to check that x is equivalent to x', so we have $A(x')$. Thus there is some n such that $C((x_1, \ldots, x_n))$ holds, which implies $A(y)$ holds for any y extending $(x_1, \ldots, x_n)$. I claim that any z in $B(x_n, 1/2^n)$ is equivalent to such a y. To see this, suppose z is in $B(x_n, 1/2^n)$. Then $|z - x_n| < 1/2^n$ and we can find a $j \geq n$ large enough so that $|z - x_n| \leq 1/2^n - 1/2^j$. This implies that whenever $k \geq j$, we have

$$|z_k - x_n| \leq |z_k - z| + |z - x_n| \leq 1/2^k + 1/2^n - 1/2^j \leq 1/2^n.$$

Let y be given by the sequence

$$x_1, \ldots, x_n, z_j, z_{j+1}, \ldots.$$

It is straightforward to show that y is equivalent to z, so we only need to check that y is Cauchy with the required rate of convergence. To that end, the only hard part is to check that for every $i < n$ and $k \geq j$, $|x_i - z_k| \leq 1/2^{i-1}$, keeping in mind that x_i is $(y)_{i-1}$. But this follows from

$$|x_i - z_k| \leq |x_i - x_n| + |x_n - z_k| \leq 1/2^i + 1/2^n \leq 1/2^{i-1}.$$

We would like U to represent the union of all such balls $B(x_n, 1/2^n)$, so let U be the set of triples $(k, p, 1/2^n)$ such that k codes a tuple $x_0, \ldots, x_{n-1}$ with $|x_i - x_j| \leq 1/2^{i+1}$ for every

$i \le j < n$ and $x_{n-1} = p$. The argument in the previous paragraph shows that if x is in U, then $A(x)$ holds.

Conversely, suppose $A(x)$ holds. We need to show that x is in U. As above, we have $C((x_1, \ldots, x_n))$ for some n, but we were conservative in our choice of a ball around x_n, so it is not necessarily the case that x is in $B(x_n, 1/2^n)$. Instead we use the fact that for some n, $C((x_2, \ldots, x_{n+1}))$ holds, where now x_{n+1} is an even better approximation to x. Indeed, we have $|x - x_{n+1}| \le 1/2^{n+1} < 1/2^n$, so x is in $B(x_{n+1}, 1/2^n)$, as required. □

We thus have a foundation for reasoning about open sets of real numbers. What about functions from the real numbers to the real numbers? Specifying an arbitrary function $f : \mathbb{R} \to \mathbb{R}$ requires specifying a real value $f(x)$ for every real number x, which is more information than can be encoded in a single set of natural numbers. (One way to see this is to note that the cardinality of the set of functions from $\mathbb{R}$ to $\mathbb{R}$ is greater than the cardinality of $\mathcal{P}(\mathbb{N})$.) But we can use sets of natural numbers to represent sufficiently nice functions from $\mathbb{R}$ to $\mathbb{R}$. A function $f : \mathbb{R} \to \mathbb{R}$ is *continuous* if for every open set U, the preimage $f^{-1}(U) = \{x \mid f(x) \in U\}$ is open. Moreover, these preimages fully determine f, and in fact it is sufficient to know the preimage of every rational open ball. Since there are only countably many rational open balls, this amounts to a countable sequence of open sets, something which can be represented by a set of natural numbers. Developing the theory of continuous functions in weak subsystems of second-order arithmetic would take us too far afield, but see the bibliographical notes.

Exercises

16.3.1. Show that RCA_0 proves that there are uncountably many real numbers, in the sense that for any sequence $(x_i)_{i \in \mathbb{N}}$ of real numbers, there exists a real number that is not equal to any of the ones in the sequence.

16.3.2. Show that RCA_0 proves that multiplication, as defined here, respects the definition of equality of the real numbers.

16.3.3. Show that $\mathrm{RCA}_0^{\mathrm{i}}$, the intuitionistic version of RCA_0, proves that for every rational number $\varepsilon > 0$ and every real number x, either $x > 0$ or $x < \varepsilon$.

16.3.4. Show that over RCA_0, (ACA) is equivalent to the principle that every bounded sequence of real numbers has a least upper bound.

16.3.5. Show that over RCA_0, (ACA) is equivalent to the Bolzano–Weierstraß theorem, that is, the statement that every bounded sequence of real numbers has a convergent subsequence. (Hint: One approach to proving the Bolzano–Weierstraß theorem is to first show that every sequence of real numbers has either a nondecreasing subsequence or a nonincreasing subsequences. Another approach is to start with a closed interval containing the sequence, and iteratively divide the interval in two and choose the half that has infinitely many elements of the sequence.)

16.3.6. If $(x_i)_{i \in \mathbb{N}}$ is a sequence of real numbers, a *rate of convergence* for the sequence is a function $f : \mathbb{N} \to \mathbb{N}$ such that for every i and $j \ge f(i)$, $|x_j - x_{f(i)}| < 1/2^i$. Show that over RCA_0, (ACA) is equivalent to the statement that every convergent sequence of real numbers has a rate of convergence.

16.3.7. Show that RCA_0 proves that a finite intersection of open sets of real numbers is open.

16.4 Weak Kőnig's Lemma

Kőnig's lemma is the statement that every infinite, finitely branching tree on a set Σ has an infinite path. *Weak Kőnig's lemma*, or (WKL), formalizes the case in which Σ is $\{0, 1\}$. Let $\text{Tree}(T)$ be a formula in the language of second-order arithmetic that says that T is a set of binary sequences closed under initial segments, and let $\text{Path}(P, T) \equiv \forall \sigma \subset P\, (\sigma \in T)$ express the statement that the characteristic function of P is a path through T. Then (WKL) is the sentence

$$\forall T\, (\text{Tree}(T) \wedge \forall y\, \exists \sigma\, (\sigma \in T \wedge \text{length}(\sigma) = y) \rightarrow \exists P\, \text{Path}(P, T)).$$

The theory WKL_0 is RCA_0 together with this additional axiom.

We saw in Section 11.9 that there are infinite computable trees on $\{0, 1\}$ with no computable paths, so WKL_0 is strictly stronger than RCA_0. It is a remarkable fact, however, that WKL_0 is a Π^1_1 conservative extension of RCA_0. So, like RCA_0, WKL_0 proves the same arithmetic statements as $\text{I}\Sigma_1$ and the same Π_2 statements as primitive recursive arithmetic. Our proof of this conservation result will make use of the next proposition, which formalizes Lemma 11.9.6 and part of the proof of Theorem 11.9.1.

Proposition 16.4.1. *The theory* WKL_0 *can be axiomatized by the basic defining equations of arithmetic,* Σ^0_1 *induction, and the* Σ^0_1 *separation principle:*

$$\forall x\, \neg(A(x) \wedge B(x)) \rightarrow \exists Y\, \forall x\, ((A(x) \rightarrow x \in Y) \wedge (B(x) \rightarrow x \notin Y)).$$

Here $A(x)$ *and* $B(x)$ *are* Σ^0_1 *formulas, possibly with free variables other than* x.

Proof To show that WKL_0 proves the Σ^0_1 separation principle, suppose $A(x)$ is the formula $\exists y\, C(x, y)$ and $B(x)$ is the formula $\exists y\, D(x, y)$, where C and D are both Δ^0_0. Argue in WKL_0. Let T to be the tree of sequences σ such that

$$\forall x < \text{length}(\sigma)\, ((\exists y < \text{length}(\sigma)\, C(x, y) \rightarrow (\sigma)_x = 1) \wedge$$
$$(\exists y < \text{length}(\sigma)\, D(x, y) \rightarrow (\sigma_x) = 0)).$$

Using Σ^0_1 induction (in fact, induction on a primitive recursive predicate) and the hypothesis that A and B define disjoint sets, we have that for every z there is a σ of length z in T. By Weak Kőnig's lemma, there is an infinite path P through T. The definition of T implies that P is a separation of $\{x \mid A(x)\}$ and $\{x \mid B(x)\}$.

For the converse direction, notice that the axiom (RCA) is a special case of Σ^0_1 separation: if $A(x)$ is a Σ^0_1 formula and $B(x)$ is an equivalent Π^0_1 formula then $A(x)$ and $\neg B(x)$ satisfy the hypotheses of Σ^0_1 separation, and the set Y in the conclusion satisfies $x \in Y \leftrightarrow A(x)$ for every x. So we only need to show that RCA_0 together with Σ^0_1 separation implies (WKL).

To that end, we use the construction of Lemma 11.9.6. Given T, let $A(\sigma)$ say that there is an extension τ of $\sigma^\frown 0$ in T such that there are no extensions of $\sigma^\frown 1$ of the same length in T, and let $B(\sigma)$ say that there is an extension τ of $\sigma^\frown 1$ in T such that there are no extensions of $\sigma^\frown 0$ of the same length in T. By the Σ^0_1 separation principle, there is a set Y separating A from B. Formalizing the argument of Lemma 11.9.6 in RCA_0 shows the desired path P is Δ^0_1-definable from Y. □

WKL_0 is finitely axiomatized, because it is obtained from RCA_0 by adding a single axiom. The discussion before Theorem 11.9.9 provides two characterizations of its ω-models, one of which we will make use of now.

Theorem 16.4.2. WKL_0 *is* Π^1_1 *conservative over* RCA_0.

As with the other conservation results in this chapter, there are both model-theoretic and proof-theoretic proofs of this result. For a model-theoretic approach, one starts by assuming that RCA_0 does not prove some Π^1_1 sentence $\forall X\, A(X)$, where A is arithmetic. This implies that there is a second-order Henkin structure $\mathfrak{M}$ and an element S of the second-order universe such that $\mathfrak{M} \models \neg A(S)$. It therefore suffices to show that $\mathfrak{M}$ can be expanded to a model $\mathfrak{M}'$ of WKL_0 by adding additional sets to the second-order part, since that does not change the evaluation of $A(S)$. One approach to doing so uses a version of the set-theoretic method known as *forcing*. The challenge is to add paths through all the finite binary trees in the second-order universe of $\mathfrak{M}$, and then paths through all the trees Δ^0_1 definable from those paths, and so on – all while preserving Σ^0_1 induction with parameters from the expanded structure.

Here we will consider a proof-theoretic argument, for two reasons. The first is that we have already laid the groundwork in Section 11.9. The second is that, like the proof-theoretic proof of the conservation of RCA_0 over $\mathrm{I}\Sigma_1$, it provides additional information: it shows that proofs in WKL_0 can be interpreted in RCA_0 without a dramatic increase in length. In fact, the interpretation targets a fragment $\mathrm{I}\Sigma_1(X)$ of RCA_0, that is, $\mathrm{I}\Sigma_1$ with a free set variable X.

Lemma 16.4.3. *Let* $T(x)$ *be a* $\Delta_1(X)$ *formula such that* $\mathrm{I}\Sigma_1(X)$ *proves that* $T(x)$ *is an infinite tree on* $\{0, 1\}$. *Then there is a formula* $P(x)$ *such that the following holds:*

1. *For every formula* $A(x)$ *that is* Σ_1 *relative to* P, $\mathrm{I}\Sigma_1(X)$ *proves the finite separation principle for* $A(x)$*:*

$$\forall y\, \exists s\, \forall x\, (x \in s \leftrightarrow x < y \wedge A(x)).$$

2. $\mathrm{I}\Sigma_1(X)$ *proves that* $P(x)$ *is a path through* $T(x)$*, that is,* $\forall \sigma \subset P\, (T(\sigma))$.

In the statement of this lemma, s ranges over finite sets, σ ranges over finite binary sequences, and $\sigma \subset P$ abbreviates the formula $\forall x < \mathrm{length}(\sigma)\, ((\sigma)_x = 1 \leftrightarrow P(x))$. The first clause implies that $\mathrm{I}\Sigma_1(X)$ proves the finite separation principle for $P(x)$ itself, and hence the fact that for every n, there exists a $\sigma \subset P$ such that $\mathrm{length}(\sigma) = n$.

Proof The discussion after our proof of the low basis theorem, Theorem 11.9.5, shows that the construction at each stage can be cast in terms of a certain pair of finite sets $I_{1,x}$ and $I_{2,x}$. A relativization of Theorem 10.4.8 shows that $\mathrm{I}\Sigma_1(X)$ proves a finite separation property for $\Sigma_1(X)$ formulas, and hence that these sets exist for every x.

Let $P(x)$ be a $\Sigma_1(X)$ formula that says that there exist finite sets $I_{1,x}$ and $I_{2,x}$ meeting their specifications and that for the pair $\sigma_{x+1}, \alpha_{x+1}$ computed from these, $(\sigma_{x+1})_x = 1$. Similarly, let $P'(x)$ be a formula that says that, in these same circumstances, $(\alpha_{x+1})_x = 1$. Then $\mathrm{I}\Sigma_1(X)$ proves the finite separation principle for $P(x)$ and $P'(x)$. From the construction, it also proves that $P(x)$ is a path through $T(x)$ and that $P'(x)$ defines the Turing jump of $P(x)$.

Let $A(x)$ be any formula that is Σ_1 relative to $P(x)$. Then $\mathrm{I}\Sigma_1(X)$ proves that there is a computable function $f(x)$ such that for every x, $A(x)$ holds if and only if $P'(f(x))$. Given y, let $z = \max_{x<y} f(x)$. Using the finite separation principle, let t be $\{w \leq z \mid P'(w)\}$. Then we have $\forall x < y\, (A(x) \leftrightarrow f(x) \in t)$, so we can let s be $\{x < y \mid f(x) \in t\}$. □

Proof of Theorem 16.4.2 Suppose WKL_0 proves a Π^1_1 sentence $\forall Y\, A(Y)$ where A is arithmetic. Given an arbitrary set X, we need to prove $A(X)$ in $\mathrm{I}\Sigma_1(X)$. As we did for RCA_0, we will use the language of $I\Sigma_1(X)$ to describe an ω-model of WKL_0 containing X, and we will use that to interpret the proof in order to conclude that $A(X)$ is true.

Theorem 11.9.9 tells us how to construct an infinite binary tree T, Δ_1-definable from X, such that any path through T yields an ω-model containing x. Let $P(x)$ be the path through T given by Lemma 16.4.3. Interpret WKL_0 in $I\Sigma_1(X)$ by interpreting the set variables Y of WKL_0 as natural numbers and interpreting atomic formulas $t \in Y$ by $P((Y, t))$. Then we have the following:

- Arithmetic formulas are unchanged, so the translation of the basic axioms of arithmetic are provable in $I\Sigma_1(X)$.
- Any Σ^0_1 formula $B(x)$ in the language of WKL_0 translates to a formula in the language of $I\Sigma_1(X)$ that is Σ_1 relative to $P(x)$. By Lemma 16.4.3, $I\Sigma_1(X)$ proves the translation of the induction principle for $B(x)$.
- By the construction of Theorem 11.9.9, $I\Sigma_1(X)$ proves that the collection of sets $\{u \mid P((Y, u))\}_{Y\in\mathbb{N}}$ is closed under Σ_1 separation. So $I\Sigma_1(X)$ proves the translation of the Σ^0_1 separation principle.

Since WKL_0 proves $\forall Y\, A(Y)$, $I\Sigma_1(X)$ proves its translation. Since A is arithmetic and the ω-model contains X, $I\Sigma_1(X)$ proves $A(X)$, as required. □

In the last section, we saw that the statement that every bounded increasing sequence of real numbers has a convergent subsequence is equivalent to (ACA) over RCA_0. The exercises asked you to show that (ACA) is also equivalent to the statement that every bounded sequence of real numbers has a convergent subsequence, which expresses the fact that every closed, bounded interval in $\mathbb{R}$ is *sequentially compact*. A set C in a topological space is said to be *compact* if every every open cover of C has a finite subcover; in other words, given a collection of open sets $\mathcal{S}$ such that $C \subseteq \bigcup \mathcal{S}$, there is a finite subset $\mathcal{S}' \subseteq \mathcal{S}$ such that $C \subseteq \bigcup \mathcal{S}'$. The Heine–Borel principle states that every closed, bounded interval $[a, b]$ in $\mathbb{R}$ is compact. We can express this in the language of second-order arithmetic by saying that if $(U_i)_{i\in\mathbb{N}}$ is any sequence of open sets and $[a, b] \subseteq \bigcup_{i\in\mathbb{N}} U_i$, there is an n such that $[a, b] \subseteq \bigcup_{i<n} U_i$. Since we can apply a linear scaling operation, without loss of generality we can restrict attention to the interval $[0, 1]$.

Theorem 16.4.4. *Over* RCA_0, (WKL) *is equivalent to the Heine–Borel principle.*

Theorem 16.4.4 should not be surprising. Exercise 1.4.2 points out that weak Kőnig's lemma can be viewed as an expression of the compactness of $\{0, 1\}^{\mathbb{N}}$. This space is isomorphic to the *Cantor set*, a closed subset of $[0, 1]$, constructed as follows. Start with $C_0 = [0, 1]$ and remove the middle third, leaving the endpoints, to obtain the set $C_1 = [0, 1/3] \cup [2/3, 1]$. Remove the middle third from each of those sets, yielding the set

$$C_2 = [0, 1/9] \cup [2/9, 1/3] \cup [2/3, 7/9] \cup [8/9, 1].$$

Continuing in this way, we obtain that Cantor set $C = \bigcap_{i\in\mathbb{N}} C_i$. If we label each interval in C_i the with binary sequences of length i corresponding to the sequence of left/right choices that defines it, then any infinite binary sequence corresponds to a nested sequence of closed intervals whose intersection is a single point of C. We can show that this gives rise to a topological isomorphism between $\{0, 1\}^{\mathbb{N}}$ and the Cantor set, both of which are often called *Cantor space* for that reason. Since a closed subset of a compact set is again compact, the compactness of $\mathbb{R}$ implies the compactness of C. It is easier to prove the compactness of $\{0, 1\}^{\mathbb{N}}$ directly and the isomorphism just described is only implicit in the proof below, but the correspondence helps explain why we should think of weak König's lemma as a compactness principle.

It isn't possible to view $[0, 1]$ as embedded in Cantor space, since, for example, $[0, 1]$ is connected while Cantor space is totally disconnected. But thinking of real numbers in $[0, 1]$ in terms of their binary expansions gives a sense of how the proof should go, despite the fact that a number like $1/2$ can be represented in two ways, as $0.100000\ldots$ or $0.011111\ldots$.

Proof To show that the Heine–Borel principle implies (WKL), let T be any infinite binary tree. Associate to each σ in T the closed interval

$$I_\sigma = \left[\sum_{i<\text{length}(\sigma)} \frac{2}{3^{i+1}}\sigma_i, \sum_{i<\text{length}(\sigma)} \frac{2}{3^{i+1}}\sigma_i + \frac{1}{3^{\text{length}(\sigma)}}\right].$$

To each level n of T associate the closed set $C_i = \bigcup_{\sigma\in T,\,\text{length}(\sigma)=1} I_\sigma$. For each i, let U_i be the complement of C_i, an open set. Let $U = \bigcup_{i\in\mathbb{N}} U_i$. Then the sequence $(U_i)_{i\in\mathbb{N}}$ is an increasing sequence of open sets. The fact that T is infinite implies that C_n is nonempty for every n, and so $U_n = \bigcup_{i\le n} U_i$ does not cover $[0, 1]$. By the Heine–Borel principle, $\bigcup_{i\in\mathbb{N}} U_i$ does not cover $[0, 1]$, so there is a real number x in $[0, 1]$ such that for every i, x is in C_i. From x, we can compute a path through T as follows: for each i, since the intervals of C_i are separated by $1/3^i$, we can use a rational approximation of x to within $1/2\cdot 3^i$ to determine the finite sequence σ_i such that x is in I_{σ_i}. The sequence $\sigma_0, \sigma_1, \sigma_2, \ldots$ is a path through T.

To show that (WKL) implies the Heine–Borel principle, suppose we are given a sequence $(U_i)_{i\in\mathbb{N}}$ that covers $[0, 1]$. Since each U_i can be written as a union of finite open intervals with rational endpoints, by reindexing the sequence we can assume without loss of generality that each U_i is of the form (p_i, q_i). This time, to each sequence σ, associate the closed interval

$$I_\sigma = \left[\sum_{i<\text{length}(\sigma)} \frac{1}{2^{i+1}}\sigma_i, \sum_{i<\text{length}(\sigma)} \frac{1}{2^{i+1}}\sigma_i + \frac{1}{2^{\text{length}(\sigma)}}\right].$$

For every n, the intervals $\{I_\sigma \mid \text{length}(\sigma) = n\}$ divide the unit interval into 2^n subintervals of size $1/2^n$ with overlapping endpoints. Let T be the set of sequences σ such that I_σ is not a subset of $\bigcup_{i<\text{length}(\sigma)} U_i$. Think of the elements σ in T that have length n as specifying the intervals I_σ that haven't yet been eliminated by stage n.

By (WKL), it suffices to show that there is no infinite path through T, since that implies that, for some n, there are no intervals of length n in T, which in turn implies $[0, 1] \subseteq \bigcup_{i\le n} U_i$. Suppose P is any set, and let $x = \sum_{i\in\mathbb{N}} \chi_P(i)/2^i$. Then x is in $[0, 1]$, and since we are assuming the sequence $(U_i)_{i\in\mathbb{N}}$ covers $[0, 1]$, we have $p_i < x < q_i$ for some i. But then for sufficiently large n we have

$$p_i < \sum_{i<n} \chi_P(i)/2^i \le x \le \sum_{i<n} \chi_P(i)/2^i + 1/2^n < q_i,$$

and by the definition of T this means that the initial segment of P of length n is not in T. □

With a suitable formalization of continuous functions, WKL_0 proves that every continuous function $f(x)$ on the unit interval has a *modulus of uniform continuity*, that is, a function $g\colon \mathbb{Q}^{>0} \to \mathbb{Q}^{>0}$ such that for every rational $\varepsilon > 0$, $|f(x) - f(y)| < \varepsilon$ whenever

$|x - y| < g(\varepsilon)$. Using this, WKL_0 can prove that every continuous function on the unit interval has a maximum value and attains it. Moreover, each of these principles is equivalent to (WKL) over RCA_0. In a similar way, (WKL) can be shown to be equivalent to formal versions of the completeness and compactness theorems for first order logic.

Exercises

16.4.1. Show that RCA_0 proves that a closed subset of a compact set is compact.

16.4.2. Show that, as a topological space, the Cantor set is isomorphic to $\{0, 1\}^{\mathbb{N}}$.

16.4.3. Show that WKL_0 proves the following generalization of (WKL). Let T be a tree on $\mathbb{N}$ and $f(x)$ a function such that for every σ in T and $i < \mathrm{length}(x)$, $(\sigma)_i < f(i)$. In other words, T is a finitely branching tree, and the labels of the branches on the children of a node of length n are bounded by $f(n+1)$. If T is infinite, then there is a path through T.

16.4.4. This exercise shows that the bound f in the previous exercise is needed. Show that there is an infinite tree T on $\mathbb{N}$ such that every node has at most two children, but the halting problem is computable from any path through T. (Hint: arrange it so that for every σ in T, $(\sigma)_x$ is either 0 or one plus a halting sequence for Turing machine x on input 0, and that if Turing machine x halts on input 0, $(\sigma)_x \neq 0$ when σ is sufficiently long.)

16.4.5. Modify and formalize the previous exercise to show that over RCA_0, the corresponding principle implies (ACA).

16.4.6. If you have sufficient background in real analysis, think about formalizing the definition of a continuous function from $\mathbb{R}$ to $\mathbb{R}$ and establishing the claims at the end of this section.

16.4.7. Show that over RCA_0, (WKL) is equivalent to each of the following statements:
a. The compactness theorem for propositional logic: if Γ is a set of propositional formulas and every finite subset is satisfiable, then Γ is satisfiable.
b. If Γ is a consistent set of sentences then there is a maximally consistent set containing Γ.
c. If Γ is a consistent set of sentences then there is a model $\mathfrak{M}$ and a satisfaction function for this model such that every sentence in Γ is true.

16.5 Π^1_1 Comprehension and Inductive Definitions

We have seen that RCA_0 and WKL_0 are conservative over primitive recursive arithmetic and that ACA_0 is conservative over first-order arithmetic. The fact that a substantial amount of infinitary mathematics can be carried out in these systems is therefore rather striking. This section deals with a mathematical construction, one that we encountered in Chapter 1, that goes beyond the strength of ACA_0.

Let U be a set and let Γ be a monotone function from $\mathcal{P}(U)$ to $\mathcal{P}(U)$. Saying that Γ is monotone means that for any two subsets A and B of U, if $A \subseteq B$, then $\Gamma(A) \subseteq \Gamma(B)$. Exercise 1.2.2 shows that every such operator Γ has a *least fixed point*, F. Saying that F is a fixed point means that $\Gamma(F) = F$, and saying that it is a least fixed point means that it is a subset of any other fixed point. A solution to Exercise 1.2.2 is included in the proof of Theorem 16.5.1 below, which shows that F can be characterized exactly as the intersection of the collection $\{A \subseteq U \mid \Gamma(A) \subseteq A\}$.

In the context of subsystems of second-order arithmetic, we are interested in the case where U is $\mathbb{N}$. In other words, we are dealing with inductively defined sets of numbers. We will focus on the case where Γ is defined by an arithmetic formula $A(x, Y)$ in which Y occurs only positively. Recall from the discussion around Definition 7.3.4 that this is equivalent to saying that when $A(x, Y)$ is expressed in negation normal form, the atomic formulas $t \in Y$ are not negated. In that case, on the basis of pure logic, we can prove $Y \subseteq Z \to (A(x, Y) \to A(x, Z))$. As a result, the function $\Gamma_A(Y) = \{x \mid \mathfrak{N} \models A(x, Y)\}$ defines a monotone function on $\mathcal{P}(\mathbb{N})$. Such a function Γ_A is said to be a *positive arithmetic operator*.

We will consider the following axiom schema:

$$\exists F\ (\forall x\ (A(x, F) \to x \in F) \wedge \forall Y\ (\forall x\ (A(x, Y) \to x \in Y) \to F \subseteq Y)).$$

Here A is an arithmetic formula, possibly free variables other than the ones shown. The schema can be expressed more compactly in terms of Γ_A:

$$\exists F\ (\Gamma_A(F) \subseteq F \wedge \forall Y\ (\Gamma_A(Y) \subseteq Y \to F \subseteq Y)).$$

The first conjunct says that F is closed under Γ_A and the second conjunct says that it is the least such set. By monotonicity, the first fact implies $\Gamma_A(\Gamma_A(F)) \subseteq \Gamma_A(F)$, and then taking Y to be $\Gamma_A(F)$ in the second conjunct implies $\Gamma_A(F) \subseteq F$. So the conjunction implies $\Gamma_A(F) = F$, or, equivalently, $\forall x\ (A(x, F) \leftrightarrow x \in F)$. In short, the schema expresses the fact that every positive arithmetic operator has a least fixed point. As we saw in Chapter 1, the leastness property can be understood as a principle of induction: to show that every element of F has some property, it suffices to show that the set of things with that property is closed under Γ_A.

We use (Π^1_1-CA) to denote the schema of Π^1_1 comprehension and Π^1_1-CA$_0$ to denote the corresponding strengthening of ACA$_0$. Using a universal Π^1_1 formula (see Section 15.5), we can axiomatize Π^1_1-CA$_0$ with a single axiom on top of ACA$_0$. Together with Proposition 16.1.3, this shows that Π^1_1-CA is finitely axiomatizable. The next theorem provides an informative characterization of the strength of Π^1_1 comprehension.

Theorem 16.5.1. *Over the basic axioms of arithmetic and the induction axiom, the schema* (Π^1_1-CA) *is equivalent to the schema asserting the least fixed point of any positive arithmetic operator.*

The proof will make use of two lemmas.

Lemma 16.5.2. *For every Π^1_1 formula $A(x)$,* ACA$_0$ *proves that there is a sequence of trees $(T_x)_{x \in \mathbb{N}}$ such that for every x, $A(x)$ holds if and only if there is no path through T_x.*

Proof This is a straightforward formalization of Lemma 15.5.6. Given $A(x)$, let $B(x, \sigma)$ be the Δ^0_0 formula described there, and let $T = \{(x, \sigma) \mid \neg B(x, \sigma)\}$. □

Let T be a tree on $\mathbb{N}$ and let $A(\sigma, X)$ be the formula $\sigma \in T \wedge \forall y\ (\sigma^\frown y \in X)$. Then $A(\sigma, X)$ defines a positive arithmetic operator $\Gamma(X)$ whose least fixed point, W, is called the *well-founded part of T*. The fact that W is closed under Γ means that whenever σ is a node of the tree and all of the children of σ are in W, then so is σ. In particular, any element σ of T with no children at all is in W. The fact that W is the smallest set closed under Γ means that we have a principle of induction on W. The following justifies calling it the well-founded part of T.

Lemma 16.5.3. ACA$_0$ *proves that if the well-founded part of T exists, then it is equal to the set of elements σ in T such that there is no infinite path through T that extends σ.*

Proof To show that every element of W has the requisite property, it suffices to show that the set of elements σ such that there is no path through T extending σ is closed under Γ. But clearly if there is no path through T that extends any of the elements $\sigma^\frown y$ in T then there is no path through T that extends σ, since any such path has to pass through some $\sigma^\frown y$.

In the other direction, the fact W is closed under Γ means that if σ is in $T \setminus W$, then some child $\sigma^\frown y$ of σ is also in $T \setminus W$. So if σ is in T but not W, we can find a path through T that extends σ by starting with σ and iteratively choosing a child with the same property. □

Proof of Theorem 16.5.1 For the forward direction, we need to show that Π^1_1-CA$_0$ proves the existence of least fixed points of positive arithmetic operators. Given $A(x, Y)$, use (Π^1_1-CA) to obtain F such that for every x,

$$x \in F \leftrightarrow \forall Y\, (\forall z\, (A(z, Y) \to z \in Y) \to x \in Y).$$

Writing Γ for Γ_A, this says that F is equal to $\bigcap\{Y \mid \Gamma(Y) \subseteq Y\}$, the intersection of all sets closed under Γ. We need to show that F is closed under Γ and that it is a subset of any set Y that is closed under Γ.

Arguing in Π^1_1-CA$_0$, we prove the second claim first: for any Y, if $\Gamma(Y) \subseteq Y$, then the fact that F is the intersection of all such sets implies $F \subseteq Y$. To show the first claim, namely $\Gamma(F) \subseteq F$, let Y be an arbitrary set satisfying $\Gamma(Y) \subseteq Y$. By the second claim, we have $F \subseteq Y$, and then by monotonicity we have $\Gamma(F) \subseteq \Gamma(Y) \subseteq Y$. In other words, we have $\Gamma(F) \subseteq Y$ for every Y such that $\Gamma(Y) \subseteq Y$. This implies $\Gamma(F) \subseteq \bigcap\{Y \mid \Gamma(Y) \subseteq Y\} = F$.

For the reverse direction, it suffices to show that the least fixed point schema implies (Π^1_1-CA). We have already noted that the schema implies $\exists F\, (x \in F \leftrightarrow A(x, F))$ for any arithmetic formula in which F occurs positively; this yields arbitrary instances of arithmetic comprehension when F does not occur in A at all. Thus we can freely use arithmetic comprehension toward showing that the existence of least fixed points implies (Π^1_1-CA).

Let $A(x)$ be an arbitrary Π^1_1 formula and argue in ACA$_0$. By Lemma 16.5.2, there is a sequence of trees $(T_x)_{x \in \mathbb{N}}$ such that for every x, $A(x)$ holds if and only if T_x is well founded. Combine these into a single tree T such that for every x, $T\restriction_{(x)}$ is T_x. Using the least-fixed-point principle, the well-founded part W of T exists. By Lemma 16.5.3, $T\restriction_{(x)}$ is well founded if and only if the sequence (x) is in W. So the set $\{x \mid A(x)\}$ is equal to the set $\{x \mid (x) \in W\}$, which exists by the arithmetic comprehension principle. □

Just as ACA$_0$ is a conservative extension of PA, we can view Π^1_1-CA$_0$ as a conservative extension of a first-order theory. Specifically, start with the language of first-order arithmetic, and for every first order formula $A(x, Y)$ with a new predicate symbol Y, add a predicate symbol $F_A(x)$ and the axiom

$$\forall x\, (A(x, F_A) \to F_A(x)) \wedge \forall Y\, (\forall x\, (A(x, Y) \to Y(x)) \to \forall x\, (F_A(x) \to Y(x))).$$

The resulting theory is known as ID$_1$. The theory ID$_2$ adds symbols and axioms for formula $A(x, Y)$ in the language of ID$_1$, and so on, and the theory ID$_{<\omega}$ denotes the union of ID$_1$, ID$_2$, ID$_3$, ... Exercise 16.5.3 asks you to show:

Proposition 16.5.4. Π^1_1-CA$_0$ *is conservative over* ID$_{<\omega}$ *for formulas in the language of first-order arithmetic.*

For any arithmetic formula $A(X)$, an element y is in the set $\bigcap\{X \mid A(X)\}$ if and only if $\forall X\, (A(X) \to y \in X)$. The theory Π^1_1-CA$_0$ therefore proves the existence of $\bigcap\{X \mid A(X)\}$.

This principle can be used to formalize a number of theorems in algebra and analysis, and one often obtains interesting reversals by showing that such theorems imply (Π^1_1-CA) over a weaker base theory like RCA_0 or ACA_0. Since saying that a tree or relation is well founded is a Π^1_1 statement, Π^1_1-CA_0 can also support transfinite constructions, that is, constructions that use induction and recursion on well-founded trees and relations. But for that purpose, we can often get by with a theory that is strictly weaker, which we turn to next.

Exercises

16.5.1. For every set Y, let Wo^Y be the set of indices of well-orders that are computable relative to Y. Show that ACA_0 proves that for every Y, Wo^Y is a complete Π^1_1 set relative to Y, in the following sense: for every Π^1_1 formula $A(x, Y)$ with the free variables shown, there is a primitive recursive function f such that ACA_0 proves that for every x, $A(x, Y)$ holds if and only if $f(x)$ is in Wo^Y.

16.5.2. Show that over RCA_0, (Π^1_1-CA) is equivalent to the statement that for every Y, Wo^Y exists.

16.5.3. Prove Proposition 16.5.4.

16.6 Arithmetic Transfinite Recursion

We have considered four of the "big five" subsystems of second-order arithmetic that play an important role in the reverse mathematics program. This section briefly introduces the fifth, ATR_0, which contains ACA_0 and is contained in Π^1_1-CA_0. Thus we have the following inclusions:

$$RCA_0 \subseteq WKL_0 \subseteq ACA_0 \subseteq ATR_0 \subseteq \Pi^1_1\text{-}CA_0.$$

ATR_0 is based on a principle of *arithmetic transfinite recursion* that allows us to iterate arithmetic comprehension along any well-order. To state the principle, we need some definitions. Fix a second-order variable, $\prec$, which is intended to represent a well-ordering of the natural numbers. If H represents a sequence of sets $(H_a)_{a\in\mathbb{N}}$, we use $H_{\prec a}$ to denote $\{(b, x) \mid b \prec a \wedge x \in H_b\}$. The set $H_{\prec a}$ represents the sequence $(H_b)_{b\prec a}$ of sets coming before a in the sequence. We use $WO(\prec)$ to denote the Π^1_1 formula that says that $\prec$ is a well-order. With that notation in place, ATR_0 is the following schema, for arithmetic formulas $A(x, Y)$:

$$WO(\prec) \to \exists H\, \forall a, x\, (x \in H_a \leftrightarrow A(x, H_{\prec a})).$$

In words, the principle says that there exists a hierarchy of sets (H_a) along $\prec$ such that each H_a is defined by A in terms of the sequence of previous sets in the hierarchy. It is sometimes convenient to let $\prec$ be a well-ordering of an arbitrary set of natural numbers rather than all of $\mathbb{N}$, but there is generally no harm in adding the remaining elements of $\mathbb{N}$ to the end of the order, and for our purposes it will be convenient to take the field of $\prec$ to be all of $\mathbb{N}$.

The theory ATR_0 is given by the axioms of RCA_0 together with (ATR). Over RCA_0, the schema (ATR) trivially implies arithmetic comprehension, since if $\prec$ is the usual order on the natural numbers and $A(x)$ doesn't mention the hierarchy then we have $\forall x\, (x \in H_0 \leftrightarrow A(x))$. As an example of a more interesting use of (ATR), we show that ATR_0 proves a version of Proposition 15.5.9, which says that for any two well-orders there is a comparison map between them. The formal statement in the next proposition is more easily seen to be equivalent to Proposition 15.5.9 if we replace the second ordering by its well-founded part, but as it stands the statement is more general because ATR_0 cannot always prove that the well-founded part of an order exists.

Proposition 16.6.1. ATR_0 *proves the following. Let* $\prec$ *be a well-ordering of the natural numbers and let* $\prec'$ *be a linear order. Then there is a partial function* f *such that either* f *is an isomorphism between* $\prec$ *and an initial segment of* $\prec'$*, or the well-ordered part of* $\prec'$ *exists and* f *is an order-preserving surjection from a proper initial segment of* $\prec$ *to that well-ordered part.*

Proof We use (ATR) to recursively define a partial function with the following property: for every a, if $f(b)$ is defined for every $b \prec a$ and there is a $\prec'$-least element c not in the range of $f\restriction_{\prec a}$, then $f(a) = c$. Formally, we do this by defining a hierarchy $(H_a)_{a \in \mathbb{N}}$ such that each H_a represents $f\restriction_{\prec a}$, that is, each H_a is a partial function on $\{b \mid b \prec a\}$ meeting the specification above. In (ATR), we take $A(x, Y)$ to be a formula that says all of the following:

- x is of the form (a, c).
- For every $b \prec a$ there is an element d such that (b, d) is in Y.
- For every (b, d) in Y with $b \prec a$ we have $d \prec' c$.
- c is the least value such that the previous property holds, that is, there is no $c' \prec c$ with the same property.

Let H be the set that is asserted to exist by (ATR). By induction along $\prec$, we can prove that each H_a represents a partial function and that the domains agree, so it makes sense to let f be equal to the union of these. We can also prove that the range of each H_a, and hence of f, is an initial segment of $\prec'$. If f is total, it is an isomorphism between $\prec$ and an initial segment of $\prec'$. Otherwise, the proof follows that of Proposition 15.5.9: let a be the $\prec$-least element such that f is not defined at a, argue that a has a predecessor, a', and show that f is an isomorphism between $\{b \mid b \preceq a'\}$ and its image. By arithmetic comprehension, this image, W, exists. Since it is isomorphic to an initial segment of $\prec$, it is contained in the well-ordered part of $\prec'$. Since f cannot be extended to a', there is no $\prec'$-least element of the complement of W. So W is exactly the well-founded part of $\prec$. □

Exercise 10.2.4 shows that any two disjoint Π^0_1-definable sets of numbers can be separated by a set that is Δ^0_1. In Section 15.5, we saw that there is an analogy between Σ^0_1 sets and Π^1_1 sets, and hence one between Π^0_1 sets and Σ^1_1 sets. The analogy might lead us to expect that any two disjoint Σ^1_1-definable sets of numbers can be separated by a Δ^1_1 set. A proof of this fact is implicit in the next result.

Theorem 16.6.2. *Over* RCA_0*, (ATR) is equivalent to the principle of* Σ^1_1 *separation, where* $A(x)$ *and* $B(x)$ *are any two* Σ^1_1 *formulas:*

$$\neg\exists x\,(A(x) \wedge B(x)) \to \exists Y\,\forall x\,((A(x) \to x \in C) \wedge (B(x) \to x \notin C)).$$

Proof First, let us show that ATR_0 proves the principle of Σ^1_1 separation. By Lemma 16.5.2, let $(T^A_x)_{x \in \mathbb{N}}$ be a sequence of trees such that, for every x, $A(x)$ holds if and only if there is a path through T^A_x, and let T^B_x be the corresponding sequence for $B(x)$. We combine all this information into a single tree T as follows, where we denote the subtree $T\restriction_{(x)}$ by T_x. For each x and sequence ρ, we put ρ into the subtree T_x of and only if ρ represents a sequence of pairs $(a_0, b_0), \ldots, (a_{n-1}, b_{n-1})$ where $(a_0, \ldots, a_{n-1})$ is in T^A_x and $(b_0, \ldots, b_{n-1})$ is in T^B_x. It will be convenient to write ρ as $[\sigma, \tau]$ when $\sigma = (a_0, \ldots, a_{n-1})$ and $\tau = (b_0, \ldots, b_{n-1})$. So we can think of T_x as the tree whose nodes are pairs $[\sigma, \tau]$ of sequences of the same length such that $\sigma \in T_x^A$ and $\tau \in T_x^B$.

Let KB(T) be the Kleene–Brouwer order on T, which can be defined by recursive comprehension. From the assumption $\neg\exists x\,(A(x) \wedge B(x))$ we have that each T_x is well founded. This implies that T is well founded, so by Lemma 15.5.7, KB(T) is a well-order. The idea behind the proof is to use recursion along KB(T) to assign values $f_x(\rho)$ to the nodes ρ of all the trees T_x simultaneously so that the value $f_x(())$ assigned to the root of T_x tells us whether to put x in C or not. Since the children of any node of T are less than that node in the Kleene–Brouwer order, we can define the value at each node of T_x in terms of the values assigned to its children.

Suppose $[\sigma, \tau]$ is a terminal node of T_x. Then σ and τ cannot both be extended to longer nodes in T_x^A and T_x^B respectively. If τ cannot be extended to a node in T_x^B, then there is no path through T_x^B with initial segment τ; we take that as evidence that, as far as extensions of $[\sigma, \tau]$ are concerned, it is safe to put x in C, and so we assign $f([\sigma, \tau]) = 1$. Otherwise, we have that σ cannot be extended to a node in T_x^A, and we set $f([\sigma, \tau]) = 0$.

More generally, assuming $[\sigma, \tau]$ is in T_x and the value of f_x has been assigned to all its children, we assign $f_x([\sigma, \tau]) = 1$ if and only if

$$\exists i\,\forall j\,(\tau^\frown j \in T_x^B \to \sigma^\frown i \in T_x^A \wedge f([\sigma^\frown i, \tau^\frown j]) = 1).$$

Otherwise, we assign $f_x([\sigma, \tau]) = 0$. By induction along KB(T), we have that f_x assigns a value of 0 or 1 to every node of T_x, and $f_x([\sigma, \tau]) = 0$ if and only if

$$\forall i\,\exists j\,(\tau^\frown j \in T_x^B \wedge (\sigma^\frown i \in T_x^A \to f([\sigma^\frown i, \tau^\frown j]) = 0)).$$

We will show that for every x we have that if $f_x(()) = 1$ there is no path g through T_x^B and if $f_x(()) = 0$ there is no path h through T_x^A. This suffices to prove the forward direction of Theorem 16.6.2, since if we then use recursive comprehension to define $C = \{x \mid f_x(()) = 1\}$, then C is the desired separation. For the first claim, suppose $f_x(()) = 1$ and g is a path through $T^B(x)$. Then, by definition, there is an a_0 such that $(a_0) \in T_x^A$ and $f_x[(a_0), (g(0))] = 1$. Again by definition, there is an a_1 such that $(a_0, a_1) \in T_x^A$ and $f_x[(a_0, a_1), (g(0), g(1))] = 1$. Continuing this way, we obtain an infinite path through T_x, contradicting the fact that T_x is well founded. The second claim is proved similarly.

To prove the other direction of Theorem 16.6.2, we need to use the Σ_1^1 separation principle to prove (ATR). Notice that if $C(x)$ is any arithmetic formula, then $C(x)$ and $\neg C(x)$ are disjoint Σ_1^1 formulas, so Σ_1^1 implies arithmetic comprehension. So we fix an arithmetic formula $A(x, Y)$ and argue in ACA$_0$. Let $\prec$ be a well-order and let $B(a, Z)$ be the formula $\forall b \prec a\,\forall x\,(x \in Z_b \leftrightarrow A(x, Z_{\prec b}))$, which says that Z is a hierarchy up to a. By induction along $\prec$, we can show that, for any a, if $B(a, Z)$ and $B(a, Z')$ both hold, then $Z_{\prec a} = Z'_{\prec a}$. Let $D(x, a)$ be the formula

$$\exists Z\,(B(a, Z) \wedge A(x, Z_{\prec a}))$$

and let $E(x, a)$ be the formula

$$\exists Z\,(B(a, Z) \wedge \neg A(x, Z_{\prec a})).$$

For any x and a, $D(x, a)$ and $E(x, a)$ cannot both be true, so using a pairing operation and Σ^1_1 separation we obtain a set Y such that for every x and a, if $D(x, a)$ then $(x, a) \in Y$ and if $E(x, a)$ then $(x, a) \notin Y$.

To show that Y satisfies the conclusion of (ATR), we need to show that for every a and x, we have $x \in Y_a \leftrightarrow A(x, Y_{\prec a})$. We can prove this by induction along $\prec$. Suppose it holds for every $b \prec a$, so that we have $B(a, Y)$. Then if $A(x, Y_{\prec a})$ holds we have $D(x, a)$ and hence $(x, a) \in Y$, which is by definition $x \in Y_a$. Similarly, if $\neg A(x, Y_{\prec a})$ holds we have $E(x, a)$ and hence $x \notin Y_a$. In either case, we have $x \in Y_a \leftrightarrow A(x, Y_{\prec a})$, as required. □

When $A(x)$ and $B(x)$ are equivalent Σ^1_1 and Π^1_1 formulas, applying Σ^1_1 separation to $A(x)$ and $\neg B(x)$ yields the corresponding instance of Δ^1_1 comprehension. So Theorem 16.6.2 shows that ATR_0 also proves Δ^1_1 comprehension. Using Δ^1_1 comprehension or (ATR) directly, it is not hard to show that ATR_0 proves the consistency of ACA_0. It requires more effort to show that Π^1_1-CA_0 proves the consistency of ATR_0; see the bibliographical notes.

We can also extract from the first part of the proof we have just seen the fact that any two disjoint Σ^1_1 sets can be separated by a set that is Δ^1_1. The tree T defined there is arithmetically definable in the parameters to $A(x)$ and $B(x)$. The formula $D(f, x)$ that describes the assignment f_x to the nodes of T is also arithmetic, and the separation C can be described equivalently as the set of x such that $\exists f\, (D(f, x) \wedge f(()) = 1)$ and the set of x such that $\forall x\, (D(f, x) \rightarrow f(()) = 0)$. The statement that any disjoint Σ^1_1-definable sets of numbers can be separated by a Δ^1_1-definable set extends to lightface and boldface definable collections of sets as well. The proof is similar, though slightly more involved.

We close with a characterization of ATR_0 that provides a nice counterpart to Theorem 16.5.1: whereas (Π^1_1-CA) is equivalent to the principle that every positive arithmetic formula has a least fixed point, (ATR) is equivalent to the principle that every positive arithmetic formula has *some* fixed point.

Theorem 16.6.3. *Over* RCA_0, (ATR) *is equivalent to the following principle, for any arithmetic formula $A(x, Y)$ in which Y occurs positively:*

$$\exists Y\, \forall x\, (x \in Y \leftrightarrow A(x, Y)).$$

The proof will take up the remainder of this section. First, we show that RCA_0 together with the fixed-point principle implies (ATR). To start with, the fixed-point principle yields instances of arithmetic comprehension when Y does not occur in A, so we can work in ACA_0. Let $A(x, Y)$ be an arithmetic formula in which Y occurs positively, and, within ACA_0, suppose $\prec$ is a well-order. We need to show that there exists a set Y such that for every a, we have $x \in Y_a$ if and only if $A(x, Y_{\prec a})$.

We will use the fixed-point principle to define a set Z representing the characteristic function of Y. In other words, Z will have the property that for every a and x, $((a, x), 1)$ is in Z if and only if $((a, x), 0)$ is not in Z, and if we define Y to be the set of pairs (a, x) such that $((a, x), 1)$ is in Z, then Y satisfies the conclusion of (ATR). The idea is to write down an arithmetic formula $B(w, Z)$ describing the inductive buildup of Z and then take a fixed point, although some care has to be taken to ensure that Z occurs only positively.

To obtain $B(w, Z)$, first put $A(x, Y)$ in negation-normal form, so that all negations appear in front of atomic formulas. Let $C(a, x, Z)$ be the result of replacing each formula of the form $t \in X$ by $(t)_0 \prec a \wedge (t, 1) \in Z$ and each formula of the form $t \notin X$ by $(t)_0 \not\prec b \vee (t, 0) \in Z$. The intuition is that $C(a, x, Z)$ says that $A(x, Y_{\prec a})$ holds, assuming the restriction of Z to elements of the form (x, a) represents the characteristic function of $Y_{\prec a}$.

Let $B(w, Z)$ say that w is of the form $((a, x), y)$ and that these two formulas hold:

- $\forall b \prec a \, \forall u \, (((b, u), 0) \in Z \vee ((b, u), 1) \in Z)$
- $(y = 0 \wedge \neg C(a, x, Z)) \vee (y = 1 \wedge C(a, x, Z))$.

Since Z occurs only positively, by the fixed-point principle there is a set Z such that any triple $((a, x), y)$ is in Z if and only if these two formulas hold. The first says that we can determine whether or not to put an element x into Y_a only after all the elements of $Y_{\prec a}$ have been decided, at which point, the second says that we put x into Y_a if and only if $A(x, Y_{\prec a})$. Formally, in ACA_0, we show by induction along $\prec$ that for every a and x we have that either $((x, a), 0)$ or $((x, a), 1)$ is in Z, but not both. If we then define Y to be the set of pairs (x, a) such that $((x, a), 1)$ is in Z, we have that $C(a, x, Z)$ is equivalent to $A(x, Y_{\prec a})$, and so we have $x \in Y_a$ if and only if $A(x, Y_{\prec a})$, as required.

To show that ATR_0 proves the fixed-point principle, we use the method of *pseudohierarchies*. Suppose we are given a positive arithmetic formula $A(x, Y)$, and let $\Gamma(Y)$ be the associated operator. One way of finding the least fixed point of Γ (see Exercise 1.2.2) is to iteratively apply Γ, starting at empty set and taking unions at limits, to obtain the following sequence:

$$\emptyset \subseteq \Gamma(\emptyset) \subseteq \Gamma(\Gamma(\emptyset)) \subseteq \cdots \subseteq \Gamma^{\alpha}(\emptyset) \subseteq \cdots .$$

By cardinality considerations, eventually the sequence has to stabilize, at which point it has reached the least-fixed point. In ATR_0, we cannot in general prove the existence of a well-order that is long enough to ensure this. But using the next lemma, we will see that there is a way of building a hierarchy along a sequence that is *not* a well-order, and using that to find *some* fixed point, though not necessarily the least one.

Lemma 16.6.4. *For any Σ^1_1 formula $B(\prec)$,* ACA_0 *proves* $\neg \forall X \, (B(X) \leftrightarrow \mathrm{WO}(X))$.

The lemma follows from the fact that in ACA_0 we can formalize the proof of Theorem 15.5.5, which says that WO is a complete Π^1_1 set, and then carry out a diagonalization argument to show that there is at least one Π^1_1 formula that is not Σ^1_1, as in the proof of Theorem 10.5.2. This implies that WO is not Σ^1_1, because otherwise every Π^1_1 formula would be Σ^1_1. Exercise 16.6.1 asks you to fill in the details, and Exercise 16.6.2 outlines a more direct argument.

Let $B(\prec)$ say that $\prec$ is a linear order such that every a, other than possibly the $\prec$-greatest element if it exists, has a successor element $a + 1$, and that there exists a set Y satisfying the following:

1. $Y_0 = \emptyset$.
2. For every a and x, if a has a successor, then $x \in Y_{a+1}$ if and only if $A(x, Y_a)$.
3. For every a and x, if a is a limit in the order (that is, it has no immediate predecessor), $Y_a = \bigcup_{b \prec a} Y_b$.
4. For every a and b, if $a \prec b$, then $Y_a \subseteq Y_b$.
5. For every a and x, if x is in Y_a, then there is some b such that x is not in Y_b but it is in Y_{b+1}.

In other words, $B(\prec)$ says that there is a hierarchy along $\prec$ starting with the empty set, applying Γ at successor stages, and taking unions at limit stages, with the additional properties that the hierarchy is increasing (clause 4) and that any number to enter the hierarchy enters at some successor stage (clause 5).

Using arithmetic transfinite recursion, ATR_0 proves $WO(\prec) \to B(\prec)$. By Lemma 16.6.4, we can conclude, in ATR_0, that there is some set $\prec$ such that $B(\prec)$ holds but $\prec$ is not a well-order. Then $\prec$ is a linear order and there is set Y satisfying the conditions set down by B. Let W be a set with no $\prec$-least element; without loss of generality, we can assume W is closed upward. Let $W' = \{b \mid \forall a \in W\,(b \prec a)\}$. So W is an ill-founded part of our linear order, and W' contains the elements beneath W. Note that W' has no greatest element, because if $a \in W'$ is less than every element of W, then so is $a+1$, since otherwise it would be the least element of W. By arithmetic comprehension let $Z = \bigcap_{a \in W} Y_a$, and let $Z' = \bigcup_{b \in W'} Y_b$. In other words, Z is the intersection of the top part of the pseudohierarchy and Z' is the union of the bottom part. We claim that $Z = Z'$ and that this is the fixed point we're looking for.

The fact that $Z' \subseteq Z$ follows from clause 4, since every set in the bottom part is contained in every set in the top part. Conversely, the fact that $Z \subseteq Z'$ follows from clause 5. To see this, suppose $x \in Y_a$ for some a in W. By clause 5 there is a b such that x is not in Y_b but it is in Y_{b+1}. If b is in W' then x is in both Z and Z'; if b is in W then x is in neither.

For every a in W', we have $Y_a \subseteq Y_{a+1} = \Gamma(Y_a) \subseteq \Gamma(Z')$, since $Y_a \subseteq Z'$. So we have $Z' = \bigcup_{a \in W'} Y_a \subseteq \Gamma(Z')$. Similarly, for every b in W, there is some $b' \prec b$ in W, and we have $\Gamma(Z) \subseteq \Gamma(Y_{b'}) = Y_{b'+1} \subseteq Y_b$, and so $\Gamma(Z) \subseteq Z$. Since $Z = Z'$ we have $Z = \Gamma(Z)$, and so Z is the desired fixed point.

Exercises

16.6.1. Prove Lemma 16.6.4 as follows:

a. Show that for every Π^1_1 formula $A(X)$, ACA_0 proves that there is a computable functional $F(X)$ such that $A(X)$ holds if and only if $WO(F(X))$.

b. Show, as in the proof of Theorem 10.5.2, that there is a Π^1_1 formula $A(X)$ that is not Σ^1_1.

16.6.2. Prove Lemma 16.6.4 more directly as follows:

a. If T is a tree on $\mathbb{N} \times \{0, 1\}$ and X is a set, let $X \upharpoonright m$ denote the initial segment of the characteristic function of X of length m, let T^X be the tree on ω given by

$$\sigma \in T^X \leftrightarrow [\sigma, X \upharpoonright \text{length}(\sigma)] \in T,$$

using the notation $[\sigma, \tau]$ from the proof of Theorem 16.6.2 to interpret a pair of sequences as a sequence of pairs. Let $KB(T^X)$ be the Kleene–Brouwer order on T^X. Show that for each Π^1_1 formula $C(X)$, ACA_0 proves the existence of a tree T on $\mathbb{N} \times \{0, 1\}$ such that for any set X, $C(X)$ holds if and only if T^X is well founded. Hence, for that tree T and any set X, $C(X)$ holds if and only if $WO(KB(T^X))$.

b. Given a Σ^1_1 formula $B(X)$, diagonalize by letting T be the tree corresponding to the Π^1_1 formula $C(X)$ that says that X is a tree on $\mathbb{N} \times \{0, 1\}$ and $\neg B(KB(X^X))$. Show that $WO(KB(T^T))$ holds if and only if $\neg B(KB(T^T))$ holds.

So $KB(T^T)$ witnesses the fact that $B(X)$ is not equivalent to $WO(X)$.

16.6.3. Give an alternative proof of the forward direction of Theorem 16.6.3 by using the fixed-point principle to prove the Σ^1_1 separation principle.

16.6.4. It is not hard to show that a collection of sets $\mathcal{C}$ is an ω-model of (Σ^1_1-AC) if and only if it is nonempty, closed under the operation $X, Y \mapsto X \oplus Y$, and has the following property: for every formula $B(x, Y)$ with the free variables shown, if for every x there is a set Y in $\mathcal{C}$ satisfying $B(x, Y)$, then there is a set Y in $\mathcal{C}$ such that for every x, $B(x, (Y)_x)$ holds. Show that over RCA_0, (ATR) is equivalent to the principle that for every set Y, there is a set Z, such that $\mathcal{C} = \{(Z)_i \mid i \in \mathbb{N}\}$ is an ω-model of (Σ^1_1-AC) containing Y.

(Hints: To show that the principle implies (ATR), show that if $\prec$ is a well-order and the parameters of $A(x, Y)$ are in $\mathcal{C}$, then the desired hierarchy is also in $\mathcal{C}$. For the converse direction, use the method of pseudohierarchies to obtain a collection of sets closed under arithmetic comprehension and the choice principle.)

Bibliographical Notes

Subsystems of second-order arithmetic Simpson (2009) offers a thorough treatment of subsystems of second-order arithmetic, including all the subsystems that appear in this chapter. It is also an excellent introduction to reverse mathematics, which calibrates the strength of fundamental theorems in a number of branches of mathematics, including algebra, analysis, logic, and infinitary combinatorics.

For reverse mathematics and combinatorics, see Hirschfeldt (2015). For examples of intuitionistic reverse mathematics, see Ishihara (2005) and Diener (2018).

Conservation of Σ^1_1-AC_0 *over* PA For the conservation of Σ^1_1-AC_0 over PA, see Kaye (1991) or Feferman (1977).

Proof complexity We have already encountered speedup results like Theorem 16.1.6 in Chapters 6 and 10; see the notes at the end of those chapters and Pudlák (1998). The phenomena recurs with other second-order extensions of first-order theories along the same lines. A good example is GBN, a version of set theory due to Gödel, Bernays, and von Neumann, which has two sorts of objects, sets and classes. The relationship between GBN and Zermelo-Fraenkel set theory, ZF, is analogous to the relationship between ACA_0 and PA. GBN is finitely axiomatizable, whereas ZF is not (it proves the consistency of any finitely axiomatized fragment), and GBN is conservative over ZF, but there is a speedup involved. The same sort of argument can be used to show that $\mathrm{I}\Sigma_1$ has speedup over PRA, as indicated at the notes at the end of Chapter 10.

The real numbers and foundations of mathematics For the historical importance of foundational reductions of the real numbers, see Boyer (1959) or Ewald (1996) on the work of Bolzano, Dedekind, Cantor, Hilbert, Brouwer, and others. The formalization in Section 16.3 follows Simpson (2009).

Weak Kőnig's lemma For other approaches to the conservation of Weak Kőnig's lemma over RCA_0, see Simpson (2009), Avigad (2004), Wong (2016), and the references to those. The elimination of compactness also plays a central role in proof mining; see Kohlenbach (2008). Weak Kőnig's lemma is classically equivalent to a semi-constructive principle, the *fan principle*, which is important in constructive mathematics; see Troelstra and van Dalen (1988).

The theories Π^1_1-CA_0 *and* ATR_0 Simpson (2009) provides a wealth of information about these theories and their models. The fact that Π^1_1-CA_0 proves the consistency of ATR_0 is a consequence of Theorems VII.2.7 and VII.2.10 there.

Descriptive set theory The statement that any two Σ^1_1-definable subsets of $\mathcal{P}(\mathbb{N})$ can be separated by a Δ^1_1-definable set falls under the realm of descriptive set theory. Proofs of this fact can be found in Kechris (1995) and Moschovakis (2009). Theorem V.3.9 of Simpson (2009) shows that it is provable in ATR_0.

17

Foundations

This book has been about developing formal models of the fundamental concepts and methods of mathematical reasoning. We have explored the basic vocabulary of mathematics, including logical connectives, quantifiers, and the equality relation, and ways of talking about the natural numbers, sets, functions, and relations. We have also explored the extent to which mathematical definitions give rise to functions and predicates we can compute, and the extent to which we can compute with formal representations.

In this last chapter, we will discuss formal axiomatic foundations for mathematics, which are formal systems designed to model *all* of mathematical reasoning, or, at least, large portions of it. We will consider three general frameworks, namely, simple type theory, set theory, and dependent type theory. These offer complementary perspectives on informal mathematical practice, and they provide bases for the design of interactive proof assistants, which support formalization in practice.

There are no exercises in this chapter, which is meant to serve as a brief introduction to the study of formal foundations and to indicate some of the ways that the topics we have considered in this book converge on that study. We adopt the computer scientists' convention of giving quantifiers the widest scope possible in Sections 17.1 and 17.5, since that is more common in the literature on simple and dependent type theory. We retain the mathematical logicians' notation, however, when we discuss set theory in Section 17.3.

17.1 Simple Type Theory

As a foundation for mathematics, higher-order logic, described in Section 15.7, fares pretty well. With an axiom of infinity, we can construct the natural numbers, and from those, the integers and rationals. In higher-order logic, we can define the real numbers as equivalence classes of Cauchy sequences of rationals. We can then reason about arbitrary functions from the real numbers to the real numbers, sets of functions, function spaces, functionals on function spaces, and so on. Higher-order comprehension ensures that any property we can describe in the language gives rise to a set.

The goal of this section is to describe a particular formulation of higher-order logic, *simple type theory*, that combines the advantages of higher-order logic with the expressiveness of the simply typed lambda calculus. We can view the system as the simply typed lambda calculus augmented with a type of propositions and a comprehension principle, or, alternatively, as higher-order logic with lambda abstraction built in. There are natural classical and intuitionistic variants.

We start with the simply typed lambda calculus with a basic type IND of individuals, as well as a basic type PROP of propositions. In a natural deduction formulation, in addition to reflexivity, symmetry, and transitivity of equality, we add rules to express that equality is preserved by lambda abstraction and application, as described in Section 13.5. We take formulas to be terms of type PROP, in which case, we no longer need to add comprehension axioms, since we now have a term $\lambda\vec{x}.\, A$ to denote the relation $R(\vec{x})$ defined by A. The axiom for β-reduction gives $(\lambda\vec{x}.\, A)\,\vec{t} = A[\vec{t}/\vec{x}]$.

We next introduce a constant $\mathrm{Forall}_\alpha \colon (\alpha \to \mathrm{PROP}) \to \mathrm{PROP}$ for each type α, and define $\forall x \colon \alpha.\, A$ to be $\mathrm{Forall}_\alpha\, (\lambda x.\, A)$. Defining the universal quantifier this way has the advantage that there is only one term-binding operation to worry about, namely, lambda abstraction. A formula is simply a term of type PROP. We introduce another constant $\mathrm{Imp} \colon \mathrm{PROP} \to \mathrm{PROP} \to \mathrm{PROP}$ for implication, we define the existential quantifier and the other propositional connectives as described in Section 15.1, and we define $x =_\alpha y$ at each type α by the formula $\forall P \colon \alpha \to \mathrm{PROP}.\, P\,x \leftrightarrow P\,y$.

In this formulation, there are two types of extensionality axioms to consider:

- Propositional extensionality: $\forall P, Q \colon \mathrm{PROP}.\, (P \leftrightarrow Q) \to P = Q$.
- Function extensionality: $\forall f, g \colon \alpha \to \beta.\, (\forall x.\, f\,x = g\,x) \to f = g$.

As noted in Section 13.5, in the presence of the other equality axioms, function extensionality is equivalent to having an axiom $\forall f.\, (\lambda x.\, f\,x) = f$ for each function type. As described in Section 13.6, we can interpret simple type theory with these extensionality axioms in simple type theory without them, by relativizing quantifiers and variables to the *hereditarily extensional* objects. This shows that, in a theoretical sense, the principle of extensionality does not add logical strength. But the extensional view of functions is fundamental to classical mathematics, and so it is natural to include them in classical theories. From a computational standpoint, it is often natural to adopt an intensional point of view, so intuitionistic theories often omit them.

We adopt a natural deduction formulation, so the proof system establishes sequents of the form $\Gamma \vdash A$, where the elements of Γ as well as A are terms of type PROP. The core axioms and rules of the proof system are as follows. First, we include the natural deduction rules for assumption, implication, and the universal quantifier:

$$\frac{}{\Gamma, A \vdash A}$$

$$\frac{\Gamma, A \vdash B}{\Gamma \vdash A \to B} \qquad \frac{\Gamma \vdash A \to B \qquad \Gamma \vdash A}{\Gamma \vdash B}$$

$$\frac{\Gamma \vdash A}{\Gamma \vdash \forall x\, A} \qquad \frac{\Gamma \vdash \forall x.\, A}{\Gamma \vdash A[t/x]}$$

In the last rule, t must have the same type as x. We add rules expressing that equality is reflexive and transitive, and that it respects application and lambda abstraction, together with an axiom for β reduction:

$$\frac{}{\Gamma \vdash t = t} \qquad \frac{\Gamma \vdash r = s \qquad \Gamma \vdash s = t}{\Gamma \vdash r = t}$$

$$\frac{\Gamma \vdash s = t \qquad \Gamma \vdash u = v}{\Gamma \vdash s\,u = t\,v} \qquad \frac{\Gamma \vdash s = t}{\Gamma \vdash \lambda x.\, s = \lambda x.\, s}$$

$$\frac{}{\Gamma \vdash (\lambda x.\, t)\, s = t[s/x]}$$

In the second-to-last rule, x cannot be free in Γ. This yields an intuitionistic logic. We can add propositional and function extensionality as the axioms above, or as rules.

The axioms and rules presented so far have a (classical) model in which IND consists of a single element. To obtain a stronger system, we add an *axiom of infinity*:

$$\exists f : \mathrm{IND} \to \mathrm{IND}.\, (\forall x, x'.\, f\,x = f\,x' \to x = x') \land \exists y.\, \forall x.\, f\,x \neq y.$$

This says that there is a function from IND to IND that is injective but not surjective.

To obtain a classical system, we can add the law of the excluded middle or the principle of proof by contradiction. In classical systems, it is also natural to add a choice operator $\varepsilon_\alpha : (\alpha \to \mathrm{PROP}) \to \alpha$ with the following axiom, in which P has type $\alpha \to \mathrm{PROP}$ and x has type α:

$$\forall P, x.\, P\,x \to P\,(\varepsilon\, P).$$

This axiom says that if there is anything that satisfies the predicate P, $\varepsilon\, P$ chooses one such value. The operation is often called the *Hilbert epsilon function*, since it was used in the context of first-order arithmetic by David Hilbert and his students in the early 1920s.

A trick due to Radu Diaconescu can be used to derive the law of the excluded middle, $\forall P\,(P \lor \neg P)$, from extensionality and choice. For a sketch of the argument, let P be an arbitrary element of PROP, and define the following two predicates:

$$U \equiv \lambda x.\, ((x = \top) \lor P) \qquad V \equiv \lambda x.\, ((x = \bot) \lor P).$$

From extensionality for propositions and functions, we have $P \to U = V$, and hence $P \to \varepsilon\, U = \varepsilon\, V$. From the choice axiom, we have $(\varepsilon\, U = \top) \lor P$ and $(\varepsilon\, V = \bot) \lor P$. These two disjunctions give rise to four cases. In three of them, P holds, and in the fourth we have $\varepsilon\, U \neq \varepsilon\, V$, and hence $\neg P$.

By providing terms that denote elements satisfying indefinite descriptions, the ε operator, sometimes also known as the *some* operator, also provides a way of denoting the results of definite descriptions. We may choose to omit the choice operator but add a definite description operator, sometimes known as *Russell's iota* or the *the* operator, and denoted by ι. In that case, to introduce the natural numbers, we should then add constants to name the objects asserted to exist by the axiom of infinity, and for classical logic one has to add the law of the excluded middle as a separate axiom.

Implementations of simple type theory generally include variables α, β, γ, ... that range over types. We cannot quantify over these variables, but we can prove theorems generically for expressions involving type variables and then instantiate them to specific types. Implementations of simple type theory also generally include definitional mechanisms for introducing new types that are represented by nonempty predicates on other types. For example, fixing the injective function f asserted to exist by the axiom of infinity and an element c that is not in its range, we can define the predicate on IND that holds on the intersection of all subsets of type of IND containing c and closed under f. We then introduce a new type, NAT, of natural numbers, in one-to-one correspondence with the elements of IND that satisfy this predicate. We can also define types, such as LIST α, that depend on the parameter α.

Given a type α, subsets of α can be represented as predicates $P : \alpha \to \mathrm{PROP}$. In other words, we can define new types SET α as abstractions of $\alpha \to \mathrm{PROP}$, or we can simply take

$x \in P$ to be notation for $P\,x$. We then obtain the expected properties of sets, with the restriction that when we talk about a set, it is always a subset of some underlying type. With this identification of predicates with sets, the principle for defining new types allows us to declare a new type in bijective correspondence with any nonempty subset of an existing type.

17.2 Mathematics in Simple Type Theory

Up to a point, the development of mathematics in simple type theory mirrors the informal set-theoretic approach described in Appendix A.1. Cartesian products $\alpha \times \beta$ can be represented by the set of relations $R\colon \alpha \to \beta \to \text{PROP}$ with the property that for some $a\colon \alpha$ and $b\colon \beta$, we have $\forall x, y.\, R\,a\,b \leftrightarrow x = a \wedge y = b$. In other words, we represent pairs as binary relations on α and β that hold of exactly one pair of elements. Iterating cartesian products, we get arbitrary tuples of data, sometimes called *records*. Quotients can also be defined as usual: if α is a type and R is an equivalence relation on α, we can define a new type α/R represented by the set of all nonempty equivalence classes modulo R. The integers, rationals, and real numbers can be defined in this way.

Structures are generally parameterized by types. For example, a group structure on a type α is an element of $(\alpha \to \alpha \to \alpha) \times \alpha \times (\alpha \to \alpha)$ consisting of the group operation, the identity, and the inverse function. An element of this type is a group if it satisfies a predicate that says that the multiplication is associative and the identity and inverse have the required properties. Proving theorems about arbitrary groups amounts to proving theorems about group structures that satisfy the predicate, where α is left as a variable ranging over the underlying types. This allows us to prove theorems about arbitrary groups, rings, and fields, and then instantiate them to particular instances, like the integers, rationals, and reals.

Simple type theory has its downsides, many of which stem from the fact that the distinction between types and objects is often artificial. For example, it is impossible to define a function f that, for every natural number n, return a type α_n. If we take a group to be a type together with a structure on that type, this means that it is impossible to define a function f that, for every n, returns a group whose carrier type depends on n. In other words, groups can't depend on parameters. We can talk about the additive group of integers, but we can't talk about $\mathbb{Z}/n\mathbb{Z}$, the additive group of integers modulo n. Nor can we quantify over types, beyond the implicit universal quantification we get by proving a theorem with a free type variable.

If instead we want a formal system in which our data types are treated as objects, we have two options. The first is to erase the distinction between types and objects by eliminating types entirely. Axiomatic set theory, which we consider in the next section, does just that. The second is to develop a more elaborate type theory in which types are on par with other expressions, like those denoting numbers and functions. We will consider that option in Section 17.5.

17.3 Set Theory

Set theory was designed in the early twentieth century to provide a uniform foundation for the methods of modern mathematics. It is meant to support the various kinds of algebraic abstraction and nonconstructive methods of reasoning about infinite domains that were introduced in the late nineteenth century. It has the advantages that the underlying logic, first-order logic, is simple and the axioms are intuitive and plausible. It has been so successful at describing modern methods that many mathematicians consider it to be the gold standard for mathematical proof.

The axioms of *Zermelo–Fraenkel set theory* are expressed in classical first-order logic in a language with a single binary relation symbol, $\in$. We think of the entire mathematical universe as consisting of nothing but sets, and if x and y are sets, $x \in y$ says that x *is an element of* y.

The first axiom says that two sets are equal if they have the same elements:

- Extensionality: $\forall x, y\, (\forall z\, (z \in x \leftrightarrow z \in y) \to x = y)$.

Alternatively, we can formulate set theory in first-order logic without equality and take the antecedent of the implication as a definition of $x = y$. In that case, we need the following axiom in order to prove the substitution property of equality:

- $\forall x, y\, (x = y \to \forall z\, (x \in z \leftrightarrow y \in z))$.

The next four axioms postulate the existence of the empty set, denoted $\emptyset$, with no elements; the unordered pair, $\{x, y\}$, whose elements are x and y; the union, $\bigcup x$, whose elements z are exactly the members of some set w in x; and the power set, $\mathcal{P}(x)$, whose members are exactly the subsets of x. In the statement of the power set axiom, $z \subseteq x$ abbreviates $\forall w \in z\, (w \in x)$.

- Empty set: $\exists x\, \forall y\, (y \notin x)$.
- Pairing: $\forall x, y\, \exists z\, \forall w\, (w \in z \leftrightarrow w = x \vee w = y)$.
- Union: $\forall x\, \exists y\, \forall z\, (z \in y \leftrightarrow \exists w\, (w \in x \wedge z \in w))$.
- Power set: $\forall x\, \exists y\, \forall z\, (z \in y \leftrightarrow z \subseteq x)$.

For each formula $A(w, \vec{x})$, the next schema says that for any choice of parameters $\vec{x}$ and any set y, there is a set $\{w \in y \mid A(w, \vec{x})\}$ whose elements are exactly the elements w of y that satisfy A.

- Separation: $\forall \vec{x}, y\, \exists z\, \forall w\, (w \in z \leftrightarrow w \in y \wedge A(w, \vec{x}))$.

The axiom is restricted to separating elements from another set y to avoid Russell's paradox, since an unrestricted comprehension axiom would allow us to build the set $\{w \mid w \notin w\}$. The variable z is not allowed to appear in A, since otherwise we could take $A(w)$ to be $w \notin z$, and taking y to be any nonempty set would then lead to a contradiction.

The next axiom asserts the existence of an infinite set:

- Infinity: $\exists x\, (\emptyset \in x \wedge \forall y\, (y \in x \to y \cup \{y\} \in x))$.

In the next section, we will see how this axiom allows us to define the natural numbers.

Nothing so far rules out the possibility that a set can be an element of itself. The axiom of foundation does so. In words, it says that every nonempty set x has an $\in$-least element y, that is, an element y such that there is no z in x such that z is in y.

- Foundation: $\forall x\, (\exists y\, (y \in x) \to \exists y \in x\, \forall z \in x\, (z \notin y))$.

This axiom does not add logical strength. In set theory without that axiom, we can define what it means for a set to be *well founded* and show that the well-founded sets again satisfy the axioms of set theory, as well as the axiom of foundation.

Once we have defined the set $\mathbb{N}$ of natural numbers, we can use the power set axiom repeatedly to define sets $\mathcal{P}(\mathbb{N})$, $\mathcal{P}(\mathcal{P}(\mathbb{N}))$, $\mathcal{P}(\mathcal{P}(\mathcal{P}(\mathbb{N})))$, ... But it is consistent with the axioms we have seen so far that there is no single set that contains all of these. The next axiom schema implies that such a set exists.

- Replacement:

$$\forall x, \vec{y}\,(\forall z \in x\, \exists! w\, A(z, w, \vec{y}) \rightarrow \exists u\, \forall w\,(w \in u \leftrightarrow \exists z \in x\, A(z, w, \vec{y})).$$

Here, A is any formula that does not contain u. The schema says that if A describes a function on x, then the image of x under that function is again a set.

This constitutes the core axioms of Zermelo–Fraenkel set theory, sometimes abbreviated ZF. *Zermelo–Fraenkel set theory with choice*, abbreviated ZFC, adds the following principle:

- Choice: $\forall x\,(\emptyset \notin x \rightarrow \exists f \in (x \rightarrow \bigcup x)\, \forall y \in x\,(f(y) \in y))$.

In this statement $x \rightarrow \bigcup x$ denotes the set of functions from x to $\bigcup x$, relying on the representation of a function as a set of ordered pairs as described below. In words, it says that if x is a collection of nonempty sets, there is a function f which selects an element $f(y) \in y$ from each element y of x.

The notation $\mathcal{P}(x)$ for power sets and $\bigcup x$ for unions can be viewed as definitional extensions. But now we have to contend with the slightly confusing situation that there are function symbols and relation symbols in the metatheory, which correspond to functions and relations on the universe of any model, as well as functions and relations as set-theoretic objects *within* the theory, which are elements of the universe of any such model. It might help to think of the former as *global* functions and relations associated with their defining formulas.

In particular, if f is a function within the theory and x is a set, the notation $f(x)$ refers to the unique element y such that (x, y) is in f. It should therefore be viewed as shorthand for a term of the form $\mathrm{app}(f, x)$, where app is a metatheoretic function symbol introduced to describe that value. If x is not in the domain of f, we assign a default value to $\mathrm{app}(f, x)$, like the empty set.

17.4 Mathematics in Set Theory

The axioms of set theory provide recipes for constructing sets. If we define the *ordered pair* (x, y) to be the set $\{\{x\}, \{x, y\}\}$, we can prove that $(x, y) = (x', y')$ holds if and only if $x = x'$ and $y = y'$. We can then define the cartesian product $A \times B$ to be the set $\{z \in \mathcal{P}(\mathcal{P}(A \cup B)) \mid \exists x \in A, y \in B\,(z = (x, y))\}$. Relations, functions, and partial functions are defined as in Appendix A.1. We still need to describe a set-theoretic construction of the natural numbers, but modulo that, we can define the integers, $\mathbb{Z}$, the rationals, $\mathbb{Q}$, and the real numbers, $\mathbb{R}$, as suitable quotients.

We learned in Chapter 1 that the natural numbers can be characterized abstractly as a set, $\mathbb{N}$, with a constant, 0, and a function $\mathrm{succ} : \mathbb{N} \rightarrow \mathbb{N}$ satisfying the following:

- 0 is not equal to $\mathrm{succ}(x)$ for any x in $\mathbb{N}$.
- The function $\mathrm{succ}(x)$ is injective.
- If $y \subseteq \mathbb{N}$ contains 0 and is closed under $\mathrm{succ}(x)$, then $y = \mathbb{N}$.

Since these characterize $\mathbb{N}$ uniquely up to isomorphism, we only need to show that the axioms of set theory imply the existence of one such set. Take 0 to be the empty set, $\emptyset$, and take $\text{succ}(x)$ to be $x \cup \{x\}$. The axiom of infinity says that there exists a set y that contains 0 and is closed under succ. As described in Section 1.2, if we define $\mathbb{N}$ to be the intersection of all such subsets of y, $\mathbb{N}$ will satisfy the third property above, the principle of induction. It is clear from the definition that 0 is not equal to $\text{succ}(x)$ for any x, since the empty set does not have any elements and $\text{succ}(x)$ has at least one. With some effort, we can show that ZF proves that $\text{succ}(x)$ is injective on $\mathbb{N}$. We therefore have a suitable definition of the natural numbers. As in Section 1.5, we can justify the principle of recursion on $\mathbb{N}$.

Our construction of the natural numbers is a classic example of an *impredicative* construction: the set of natural numbers is defined with respect to a collection of sets – the collection of inductive subsets of y – that includes the very set, $\mathbb{N}$, that is being defined. In the wake of the set-theoretic paradoxes around the turn of the twentieth century, this circularity raised concerns. The definition is problematic if you think of sets as somehow being created by their definitions. The problem goes away if you think of the definition as merely picking out a particular set from among a well-defined totality, namely, the collection of all subsets of y.

Finally, algebraic structures are represented as tuples of data. A group is a tuple $(G, \circ, e, \cdot^{-1})$ satisfying the group axioms, a metric space (M, d) consists of a set M and a function $d \colon M \times M \to \mathbb{R}$ satisfying the axioms for a metric space, and a topological space $(X, \mathcal{O})$ consists of a set X together with its *topology*, that is, a collection $\mathcal{O}$ of *open* subsets of X closed under finite intersections and arbitrary unions.

17.5 Dependent Type Theory

Although in some ways simple type theory provides a natural framework for carrying out mathematical constructions, the sharp distinction between objects and types belies the fact that mathematical data types can arise as the result of arbitrary mathematical constructions. Set theory's response is to erase the distinction entirely: everything is a set, and we keep track of what type of objects we are dealing with by saying things about them in the language of set theory. Saying that an object x is a natural number is on par with saying that it is greater than 7.

In implementations of formal systems, however, as in programming languages, it is often helpful to associate objects with their data types more directly in the syntax. To start with, type information allows a system to infer information that users can therefore leave implicit. In an expression like $x + y$, if x and y are known to be integers, the system can infer that the $+$ symbol denotes addition on the integers. Type information also helps catch mistakes on the level of syntax; a proof system can flag an error if a function is applied to the wrong number of arguments or to arguments of the wrong type. Finally, type checking provides a kind of automated support for theorem proving: in the example above, syntactic rules dictate that $x + y$ is an integer, a fact that would require proof in a set-theoretic framework.

So, whereas set theory gains added flexibility by eliminating types, dependent type theory takes the opposite approach by making the type system more elaborate. Just as sets are first-class citizens in set theory, types are first-class citizens in dependent type theory. Functions can take types as arguments and return other types, types can be elements of tuples, and so on. For better or for worse, type theory also dictates that every object – even a type – has to be an object *of* a type.

There are many systems of dependent type theory on offer. The one we will consider here is a version of a system known as the *calculus of inductive constructions*. Historically, it was developed as constructive system of logic, one in which every term bears a computational interpretation. But, like simple type theory, we can add classical axioms and thereby obtain a powerful and expressive system for classical mathematics as well.

The syntax of dependent type theory is more subtle than that of set theory or simple type theory. In set theory, the sets of terms, formulas, and proofs are defined by successive inductive definitions, each independent of the next. In simple type theory, there are, similarly, three kinds of syntactic objects: types, terms, and proofs. In contrast, dependent type theory can be presented in such a way that there is only one syntactic category, that of *expressions*. Every expression has a type. Some expressions *are* types, some expressions are propositions, some expressions are proofs, and it is the type of the expression that determines which is the case. The tradeoff is that the definition of the set of well-formed expressions and the relation $t\colon \alpha$ that expresses that t has type α are more complicated than the syntactic definitions that comprise first-order logic and simple type theory.

In dependent type theory, a term t with variables $\vec{x}$ is judged to be well formed with respect to a *context* Γ of the form $x_0 : \alpha_0, \ \ldots, x_{n-1} : \alpha_{n-1}$ that specifies the types of the variables. As with natural deduction and simple type theory, we use the notation $\Gamma, x\colon \alpha$ to denote the extension of Γ with the additional type hypothesis $x\colon \alpha$, and we use Γ, Δ for the union of two contexts. But now there is an important difference in that the order of the hypotheses matters. In the context $x_0 : \alpha_0, \ \ldots, x_{n-1} : \alpha_{n-1}$, a type expression α_j may depend on a variable x_i for some $i < j$, and so exchanging x_i and x_j does not result in a well-formed context. The notation Γ, Δ should therefore be interpreted as a combination of the two contexts that preserves the dependency structure. Theoretical presentations of dependent type theory generally take a context to be a sequence, and they generally include explicit *structural rules* that allow for weakening a context, contracting duplicate elements, and reordering elements subject to the dependency rules. In contrast, implementations of dependent type theory generally use efficient data structures and algorithms to manage and merge contexts.

The following three judgments are usually defined simultaneously with a mutual inductive definition:

- the judgment that Γ is a valid context;
- the judgment $\Gamma \vdash t\colon \alpha$, which says that expression t has type α in context Γ; and
- the judgment $s \equiv t$, which says that expressions s and t are *definitionally equal*.

Definitional equality is also called *judgmental equality*, and sometimes *computational equality*, because it can usually be determined in principle by reducing terms to a normal form. The notion of β-equivalence for terms in the simply typed lambda calculus is a prototypical example of a notion of definitional equality. Saying that two expressions t and t' are definitionally equal says that the logical foundation treats them as being essentially the same. This stands in contrast to *propositional equality*, namely, judgments of the form $\Gamma \vdash p\colon s =_\alpha t$, where the type $s =_\alpha t$ represents the *proposition* that the terms s and t of type α are equal. Expressions that are definitionally equal are propositionally equal, by the reflexivity of propositional equality. But the converse does not hold in general: since all the resources of mathematics can be brought to bear to prove that two expressions are equal, one cannot identify propositional equality with definitional equality without making type-checking undecidable.

Within this general framework, we are now ready to describe the expressions of dependent type theory and their types. We start with a sequence of constants

$$\text{TYPE}_0, \text{TYPE}_1, \text{TYPE}_2, \ldots$$

each of which denotes a *universe* of types. If an expression α has one of these types, we say that α is a *type*. In other words, data types in dependent type theory are expressions of type TYPE_i for some i. Each of these constants must have a type, and we declare that TYPE_i has type TYPE_{i+1}.

We also declare a constant PROP to denote the type of propositions. If an expression α has type PROP, we say that α is a *proposition*. Since PROP also needs to have a type, we declare that it has type TYPE_0. Taking propositions to be expressions of type PROP is not novel; we have already done so in the context of simple type theory. What is more novel is that we also use the expressions of dependent type theory to represent proofs. If α is a proposition, an expression $t\colon \alpha$ represents the assertion that t is proof of α. This is an instance of the *propositions as types* interpretation, discussed in Section 13.8. With this formulation, the fact that t is a proof of a proposition α is checked using the same mechanisms that are used to check that another expression s denotes an element of type NAT.

In general, an expression may depend on variables of various types, and the judgment that an expression has type TYPE_i or PROP has to be taken relative to the context Γ that specifies the types of its variables. We therefore have the following rules for generating contexts:

- The empty context is a context.
- If Γ is a context, α is a type in context Γ, and x is a fresh variable, then $\Gamma, x\colon \alpha$ is a context.

In analogy to the assumption rule in natural deduction, we have:

- If $x\colon \alpha$ is in Γ, then $\Gamma \vdash x\colon \alpha$.

We also specify that definitional equality is reflexive, symmetric, and transitive, and that it respects the typing judgments: if $\Gamma \vdash t\colon \alpha$ and we have $\alpha \equiv \alpha'$, then $\Gamma \vdash t\colon \alpha'$. In Section 17.6, we will see why such a rule is necessary.

Most rules in dependent type theory are associated with a single type construction and fall into one of the following categories:

- *type formation rules*, resulting in a judgment of the form $\Gamma \vdash \alpha\colon \text{TYPE}_i$ or $\Gamma \vdash \alpha\colon \text{PROP}$, where α is an instance of the new type construction
- *introduction rules*, which show how to derive a judgment $\Gamma \vdash t\colon \alpha$
- *elimination rules*, which show how to a judgment $\Gamma \vdash t\colon \alpha$ to derive other judgments
- *conversion rules*, which declare definitional identities, for example asserting that an introduction followed by an elimination can be simplified.

One attractive feature of the calculus of inductive constructions is that all the axiomatic type constructions can be reduced to two general kinds:

- dependent function types
- inductive types.

We will consider dependent function types here and inductive types in the next section.

In dependent type theory, the function types $\alpha \to \beta$ are generalized by the *Pi types*, $\Pi x\colon \alpha.\, \beta$, in which the variable x is bound. An element of this type should be viewed as a function f that maps any element $x\colon \alpha$ to an element of type β, where β can depend on x. Such a function is therefore called a *dependent function*. The formation, introduction, and elimination rules are as follows:

$$\frac{\Gamma \vdash \alpha\colon \text{TYPE} \qquad \Gamma,\, x\colon \alpha \vdash \beta\colon \text{TYPE}}{\Gamma \vdash (\Pi x\colon \alpha.\, \beta)\colon \text{TYPE}}$$

$$\frac{\Gamma,\, x\colon \alpha \vdash t\colon \beta}{\Gamma \vdash (\lambda x\colon \alpha.\, t)\colon (\Pi x\colon \alpha.\, \beta)} \qquad \frac{\Gamma \vdash t\colon (\Pi x\colon \alpha.\, \beta) \qquad \Gamma \vdash a\colon \alpha}{\Gamma \vdash t\, a\colon \beta[a/x]}$$

The conversion rule is $(\lambda x.\, t)\, s \equiv t[s/x]$. The usual construction $\alpha \to \beta$ represents the special case in which β does not depend on x.

In the hypotheses of the first rule, each occurrence of TYPE stands for either TYPE_i for some i or PROP. If the first is TYPE_i and the second is TYPE_j, then the TYPE in the conclusion is TYPE_k, where k is the maximum of i and j. If the second is instead PROP, however, the conclusion is in PROP as well. Below I will resort to the notation $\forall x\colon \alpha.\, \beta\, x$ instead of $\Pi x\colon \alpha.\, \beta\, x$ when $\beta\, x$ is a proposition, but you should keep in mind that the two expressions mean the same thing. Logical implication $\alpha \to \beta$ corresponds to the case where α and β are both propositions and β does not depend on x. Universal quantification $\forall x\colon \alpha.\, \beta$ corresponds to the case where α is some data type and β is in PROP. This explains the rule for calculating universes: quantifying over an arbitrary data type still yields a proposition.

17.6 Inductive Types

In Section 1.2, we developed general means to define a set inductively, namely, as the smallest set closed under some closure operations. In Section 1.5, we saw that if we take measures to ensure that the set so defined is freely generated by the closure operations, we can define functions on that set by recursion. We have used this method to construct the natural numbers in second-order logic, simple type theory, and set theory, in each case starting from a suitable infinite set. We can define other inductive structures, like the set of lists of elements of a set A or the set of well-founded trees on a set Σ, assuming we start with an ambient big enough to generate the structure freely. The primitives of simple type theory and set theory give us the means to construct such universes.

The calculus of inductive constructions inverts this approach: given a signature of constructors, the axiomatic framework asserts the existence of freely generated inductive structures of the type described. We can interpret these as set-theoretic constructions, but here the type theory treats them as axiomatically given primitives. This is a natural thing to do in the sense that we generally don't care what the basic elements of the inductive data type are made of; we only care that they support principles of induction and recursion. In dependent type theory, it is these principles that are fundamental.

We will see that such declarations can be used not only to define data types like the natural numbers and lists, but even to define basic logical operations, like conjunction, disjunction, and existential quantification. Describing the principle in enough generality to do all this work is technically involved, so we will first list a number of important instances that illustrate both way the principle is used and the things it can do.

The cartesian product $\alpha \times \beta$ of two types has one constructor, which forms the pair $(a, b)\colon \alpha \times \beta$ from $a\colon \alpha$ and $b\colon \beta$. A special case of the elimination principle is given by a term

$$\text{rec}\colon (\alpha \to \beta \to \gamma) \to (\alpha \times \beta \to \gamma).$$

For any $f\colon \alpha \to \beta \to \gamma$, rec f is the function from $\alpha \times \beta \to \gamma$ defined by

$$\text{rec}\ f\ (a, b) = f\ a\ b.$$

More generally, we allow γ to be a term that can depend on the pair (a, b), and all the constructions are relativized to a context in which α and β are defined. The full set of rules for the product type is as follows:

$$\frac{\Gamma \vdash \alpha\colon \text{TYPE} \qquad \Gamma \vdash \beta\colon \text{TYPE}}{\Gamma \vdash \alpha \times \beta\colon \text{TYPE}} \qquad \frac{\Gamma \vdash a\colon \alpha \qquad \Gamma \vdash b\colon \beta}{\Gamma \vdash (a, b)\colon \alpha \times \beta}$$

$$\frac{\Gamma,\ z\colon \alpha \times \beta \vdash \gamma\colon \text{TYPE} \qquad \Gamma,\ x\colon \alpha,\ y\colon \beta \vdash t\colon \gamma[(x, y)/z] \qquad \Gamma \vdash p\colon \alpha \times \beta}{\Gamma \vdash \text{prod-rec}\ (\lambda x, y.\, t)\, p\colon \gamma[p/z]}$$

$$\text{prod-rec}\ (\lambda x, y.\, t)\ (a, b)\ \equiv\ t[a/x, b/y]$$

Here each TYPE is TYPE_i for some i, and the index of the universe in the conclusion of the introduction rule is the maximum of the indices in the two premises. The last rule is the conversion rule.

With the availability of Pi types and application, the pairing operation can be presented equivalently as a constant:

$$\text{pair}\colon \Pi\alpha, \beta\colon \text{TYPE}.\, \alpha \to \beta \to \alpha \times \beta.$$

Using constants and application rather than term constructions is analogous to using axioms and modus ponens rather than rules in the presentation of a logical system. This approach is generally favored in practice because it is easy to add new constants once Pi types and application have been implemented correctly. With this manner of presentation, the type of the prod-rec is as follows:

$$\begin{aligned}\text{prod-rec}\colon \Pi\alpha\colon \text{Type}_i,\ \beta\colon \text{Type}_j,\ \gamma\colon \alpha \times \beta \to \text{Type}_k.& \\ (\Pi a\colon \alpha,\ b\colon \beta.\, \gamma\ (a, b)) \to \Pi p\colon \alpha \times \beta.\, \gamma\, p&\end{aligned}$$

Notice that the type arguments α and β are generally left implicit in the expression (a, b), and α, β, and γ are left implicit in the expression prod-rec $(\lambda x, y.\, t)\, p$. Implementations of dependent type theory generally support this using the fact that these arguments can be inferred from the types of arguments that come later. In any case, the full syntax of type theory is quite verbose. For the sake of brevity and readability, I will resort to a shorter, less formal style of presentation from now on.

The sum type $\alpha + \beta$ has two constructors:

$$\text{inl}\colon \alpha \to \alpha + \beta \qquad \text{inr}\colon \beta \to \alpha + \beta.$$

The expression inl a introduces a to $\alpha + \beta$ tagged as an element of α, and inr b introduces b to $\alpha + \beta$ tagged as an element of β. The recursor, sum-rec, allows us to construct a function $f\colon (\Pi z\colon \alpha + \beta.\, \gamma)$ from functions $g\colon (\Pi x\colon \alpha.\, \gamma[\text{inl}\, x/z])$ and $h\colon (\Pi y\colon \alpha.\, \gamma[\text{inr}\, y/z])$. The conversion rules are as follows:

$$\text{sum-rec}\, g\, h\, (\text{inl}\, a) \equiv g\, a \qquad \text{sum-rec}\, g\, h\, (\text{inr}\, b) \equiv h\, a.$$

It is instructive to compare the previous two examples. The product type has one constructor with two arguments, giving rise to conjunctive behavior: constructing an element of $\alpha \times \beta$ requires two pieces of data, an element of α and an element of β. The sum type has two constructors, giving rise to disjunctive behavior: an element of $\alpha + \beta$ is of the form inl a or of the form inr β. In each case, the recursion principle says, in essence, that all of the elements of the type can be viewed as arising from the canonical constructions, in the sense that to define a map out of the type it is sufficient to describe the behavior of the map on the canonically constructed elements.

The *Sigma types* are a generalization of the product types. We start with a type α and another type $\beta\colon \alpha \to \text{Type}$, and think of β as a family of types indexed by elements of α. The type $\Sigma x\colon \alpha.\, \beta\, x$ consists of pairs (a, b) where a has type α and b has type $\beta\, a$. So $\Sigma x\colon \alpha.\, \beta\, x$ has the constructor

$$\text{dpair}\colon \Pi\alpha\colon \text{TYPE},\ \beta\colon \alpha \to \text{TYPE},\ x\colon \alpha.\, \beta\, x \to \Sigma x\colon \alpha.\, \beta\, x.$$

The ordinary product $\alpha \times \beta$ can be seen as the special case of $\Sigma x\colon \alpha.\, \beta\, x$ in which $\beta\, x$ is a constant. Using the recursor for Sigma types, we can define the projections:

$$\text{fst}\colon (\Sigma x\colon \alpha.\, \beta) \to \alpha \qquad \text{snd}\colon \Pi z\colon (\Sigma x\colon \alpha.\, \beta).\, \beta[\text{fst}\, z/x].$$

Given $a\colon \alpha$ and $b\colon \beta[a/x]$, these satisfy fst $(a, b) \equiv a$ and snd $(a, b) \equiv b$. The second of these identities illustrates the necessity of allowing definitional equality in type checking, since the left-hand side has type $\beta[\text{fst}\, (a, b)/x]$ and the right-hand side has type $\beta[a/x]$. We have to reduce fst (a, b) to a in order to recognize that these types are the same.

We can go on to define the empty type, EMPTY, with no constructors, a type UNIT, with one constructor, and the type BOOL with two constructors, tt and ff. What makes inductive types logically powerful is that they support recursive constructions as well. Constructors can take elements of the very inductive type under construction. The type Nat, with constructors $0\colon$ NAT and succ$\colon$ NAT $\to$ NAT, is the most important example. It is worthwhile to consider the type of the dependent form of the recursor:

$$\Pi\gamma\colon \mathbb{N} \to \text{TYPE}_i.\, \gamma\, 0 \to (\Pi x\colon \mathbb{N}.\, \gamma\, x \to \gamma\, (\text{succ}\, x)) \to \Pi x\colon \mathbb{N}.\, \gamma\, x.$$

If we replace TYPE_i by PROP, the type is written more naturally as follows:

$$\forall\gamma\colon \mathbb{N} \to \text{PROP}.\, \gamma\, 0 \to (\forall x\colon \mathbb{N}.\, \gamma\, x \to \gamma\, (\text{succ}\, x)) \to \forall x\colon \mathbb{N}.\, \gamma\, x.$$

In other words, the principle of induction structurally the same as the principle of recursion. We can unify the two principles by allowing i to take on the value -1 in the description of the recursor and interpreting TYPE_{-1} as PROP. The same understanding allows us to prove that a property $\gamma\, z$ holds for an arbitrary member $z\colon \alpha \times \beta$ by proving $\gamma\, (a, b)$ for every a and b, and to prove $\gamma\, z$ for an arbitrary element z of $\alpha + \beta$ by proving γ (inl a) for an arbitrary $a\colon \alpha$ and γ (inr b) for an arbitrary $b\colon \beta$.

The type $\mathrm{WF}\,\alpha$ of well-founded trees on a type α, defined informally in Section 1.4, can be defined inductively with the following constructors:

- empty: $\mathrm{WF}\,\alpha$
- sup: $(\alpha \to \mathrm{WF}\,\alpha) \to \mathrm{WF}\,\alpha$.

Here empty denotes the empty tree, and if $f : \alpha \to \mathrm{WF}\,\alpha$ is any α-indexed sequence of well-founded trees, sup f denotes the tree T such that for each $a : \alpha$, $T\restriction_{(a)}$ is equal to $f\,a$. Similarly, the type $\mathrm{LIST}\,\alpha$ is defined inductively with the following constructors:

- nil: $\mathrm{LIST}\,\alpha$
- cons: $\alpha \to \mathrm{LIST}\,\alpha \to \mathrm{LIST}\,\alpha$.

Here nil is the empty list and cons $a\,\ell$ is the list consisting of a followed by the elements of ℓ. More precisely, all of empty, sup, nil, and cons should be written as dependent functions whose types are of the form $\Pi\alpha : \mathrm{TYPE}.\ \ldots$, so that nil is really nil α and cons $a\,\ell$ is really cons $\alpha\,a\,\ell$. In practice, the type arguments are generally left implicit.

It is a remarkable and useful fact that even logical operations can be defined as inductive definitions that land in PROP. For example, conjunction can be defined inductively with one constructor:

- and-intro: $\Pi\alpha, \beta : \mathrm{PROP}.\ \alpha \to \beta \to \alpha \wedge \beta$.

Disjunction can be defined similarly:

- or-inl: $\Pi\alpha, \beta : \mathrm{PROP}.\ \alpha \to \alpha \vee \beta$
- or-inr: $\Pi\alpha, \beta : \mathrm{PROP}.\ \beta \to \alpha \vee \beta$.

These are just PROP-valued variants of the cartesian product and sum types, respectively. Falsity, $\bot$, is analogous to EMPTY, and truth, $\top$, is analogous to UNIT. We already have implication and the universal quantifier. The existential quantifier $\exists x : \alpha.\ \beta\,x$ is just a PROP-valued version of the Sigma type, with constructor

$$\text{exists-intro}: \Pi\alpha : \mathrm{TYPE},\ \beta : \alpha \to \mathrm{PROP},\ x : \alpha.\ \beta\,x \to \exists x : \alpha.\ \beta\,x.$$

The *subtype* construction lies in between the Σ type construction and the existential quantifier, yielding a type we can denote $\{x : \alpha \mid \beta\,x\}$. The constructor looks the same as that of the existential quantifier; the only difference is that $\{x : \alpha \mid \beta\,x\}$ is taken to be an element of a suitable TYPE rather than PROP. In general, the rules governing the universes in dependent type theory do not allow us to define a function from α to β when α has type PROP and β has type TYPE_i for some i. A proposition is supposed to tell us that something is true, but it does not carry information beyond that. In contrast, given an element (a, p) of the subtype $\{x : \alpha \mid \beta\,x\}$, we can extract the element a, as well as the proof $p : \beta\,a$.

Even equality can be defined as an inductively defined proposition, with a single constructor refl: $\forall\alpha : \mathrm{TYPE}.\ \forall x : \alpha.\ x = x$. The only canonical way to prove an equality is to use reflexivity, and the elimination principle says that given $h : a = b$ as a hypothesis, to prove a proposition $p\,a\,b$, it is enough to prove $p\,a\,a$.

We can now describe the inductive types in full generality. An inductive type I is specified by giving a list of constructors, each taking a list of arguments whose types can depend on the parameters to the inductive type and previous arguments. Arguments to the constructors can be *recursive*, which is to say, they can have types of the form $\Pi\vec{x} : \vec{\alpha}.\ I$, where each type in the tuple $\vec{\alpha}$ can depend on the previous elements of the tuple $\vec{x}$ but not on I. As we have explained, the intuition is that the new type I is freely generated by its constructors.

The universe of the inductive type construction is the maximum of the universes of the types of the nonrecursive arguments to the constructors as well as the universes of the types $\vec{\alpha}$ occurring in the types of the recursive arguments. An exception is that any inductive type can be declared to be of type PROP. This specifies that it does not carry data, a decision that has bearing on the elimination principle. Roughly speaking, the elimination principle for an inductive data type provides a principle of recursion by specifying output values for each value of the constructor. Generally speaking, however, an inductive proposition can only eliminate to PROP. For example, we can use a hypotheses $\exists x\colon \alpha.\ \beta\, x$ to prove another proposition, but, constructively, we cannot define a function from $\exists x\colon \alpha.\ \beta\, x$ to a data type.

Because inductive types are so general, specifying the rules precisely is delicate. For example, there are conditions under which, even constructively, it makes sense to define a function out of a proposition – roughly, when then function uses nothing more than the fact that the proposition is true to determine the output value. Moreover, there are a number of useful generalizations, including nested and mutually-defined inductive types, and simultaneously defined families of inductive types. You can learn more about these from the bibliographical notes at the end of this chapter.

17.7 Mathematics in Dependent Type Theory

With inductive types in place, we are off and running. An algebraic structure like GROUP can be defined as a Σ type:

$$\Sigma G\colon \text{TYPE},\ \text{mul}\colon G \to G \to G,\ \text{id}\colon G,\ \text{inv}\colon G \to G.$$
$$(\forall g, h, k\colon G.\ (\text{mul}\ (\text{mul}\, g\, h)\, k = \text{mul}\, g\ (\text{mul}\, h\, k))) \wedge \cdots$$

A group is just an element of this type, which bundles up the carrier, the multiplication, the identity element, the inverse, and the fact that these pieces of data satisfy the relevant axioms. A more elegant representation is to take the type GROUP to be an inductive type with one constructor that packages up the elements all at once. The recursion principle says that we can define a function with an argument that is a group in terms of the group's components, and the induction principle says that we can reason about any group in terms of its components. For that reason, inductive types with one nonrecursive constructor are sometimes called *structures* or *records*.

An attractive feature of dependent type theory is that the pure system, as we have defined it, has a computational interpretation. Any function $f\colon \text{NAT} \to \text{NAT}$ that is defined in the system, even if the definition makes use of complicated algebraic structures and constructions, is a computable function that can be evaluated on any input. The underlying formal system specifies what it means to evaluate f correctly.

As with simple type theory, we obtain a fully classical system by adding propositional extensionality, function extensionality, and choice. An elegant way to incorporate choice is to add the following term:

$$\text{choice}\colon \Pi\alpha\colon \text{TYPE}.\ (\exists x\colon \alpha.\ \top) \to \alpha.$$

This enables us to extract an element from any nonempty type α. In the case where α is the subtype $\{x : \beta \mid P\,x\}$, the choice principle yields the more intuitive formulation that says that if we know there exists an element x of β with property P, then we can choose such an element. As with simple type theory, the law of the excluded middle follows from extensionality and choice.

The use of choice breaks the computational interpretation entirely. In contrast, propositional extensionality and function extensionality, and even principles like the law of the excluded middle, only add new propositions. The very nature of PROP is that its elements are not intended to carry any data, and methods for extracting executable code from expressions in dependent type theory generally ignore subterms of type PROP. So we can still compute with any expression that does not use the choice operator to produce data.

This means that dependent type theory does not force us to make a sharp choice between constructive and classical methods. When we care about computation, simply avoiding the use of choice endows our definitions with computational meaning. We are also free to reason classically whenever we want to. We can, moreover, use all the resources of classical mathematics to reason about computable functions. So dependent type theory has something to offer both mathematicians and computer scientists, and we can even consider ourselves to be both at the same time.

Bibliographical Notes

Proof assistants In computer science, *formal methods* and *formal verification* refer to the use of computational, logic-based methods for verifying claims about the behavior of hardware and software systems. *Interactive theorem proving* refers to the use of computational proof assistants in particular. Contemporary proof assistants are used to verify mathematical theorems as well as claims about hardware and software. Blanchette and Mahboubi (in press) provides an overview, and the second chapter, Avigad (in press), shares some content with this one.

General references For more on higher-order logic and simple type theory, see Andrews (2002), Lambek and Scott (1988), Mac Lane and Moerdijk (1992), and Takeuti (1987). For set theory see Devlin (1993), Enderton (1977), Kunen (2011), Schindler (2014), and Jech (2003). For more on dependent type theory, see Martin-Löf (1984), Nordström et al. (1990), or Nederpelt and Geuvers (2014).

Constructive mathematics See Bishop and Bridges (1985), Mines et al. (1988), Troelstra and van Dalen (1988), and Beeson (1985).

Logical strength The statement that two theories "have the same logical strength" is vague. It generally means, at the very least, that the two theories prove the same Π^0_1 statements. When comparing classical theories, it often means that they prove the same arithmetic statements, or Π^1_1 statements, or more. When comparing classical and constructive theories, it often means that they prove the same negative arithmetic statements and the same Π^0_2 statements. In the name of brevity, I will omit precise comparisons here and fall back on the minimal interpretation.

By the double-negation interpretation, classical and constructive versions of simple type theory have the same logical strength. The classical version is easily interpreted in Zermelo set theory, which is to say, Zermelo–Fraenkel set theory without the replacement axiom, by interpreting types as straightforwardly sets. Since, in type theory, we can only quantify over the elements of a single type, the interpretation only requires the comprehension schema for *bounded* formulas of set theory, in which every quantifier is relativized to a set. In terms of logical strength, the correspondence is sharp, but proving the other direction is more difficult, requiring either cut elimination or a model theoretic argument. See Mathias (2001) for more information.

As far as the logical strength of dependent type theory is concerned, an important factor is whether the type theory includes an impredicative type PROP like the one presented here. (Here the word "impredicative" means that the proposition $\forall x \colon \alpha.\, \beta$ is in PROP no matter what type universe α lives in.) *Martin–Löf type theory* refers to a family of predicative systems, and they are generally substantially weaker than their impredicative counterparts. For example, a basic system of Martin–Löf type theory with a type of natural numbers and a sequence of universes indexed by the natural numbers but no inductive types, has the same strength as the theory ATR_0 discussed in Chapter 16. For more information, see Griffor and Rathjen (1994).

The classical version of variant of the calculus of inductive constructions described here has the same strength as Zermelo–Fraenkel set theory with countably many universes; see Werner (1997) and Carneiro (2019). The constructive version is substantially weaker, though still fairly strong; see Rathjen (2012).

Appendix A

Background

A.1 Naive Set Theory

From the point of view of Zermelo–Fraenkel set theory, everything is a set, and other mathematical objects are introduced via suitable definitions. The following brief review of informal set-theoretic notions will help establish terminology and notation. A formal axiomatization of set theory is presented in Section 17.3.

The notion of a *set* of objects is assumed to be basic, and if x is an object and A is a set, $x \in A$ denotes the fact that *x is an element of A*. If A and B are sets, then *A is a subset of B*, written $A \subseteq B$, if every element of A is an element of B. Sets are determined by their elements, so $A = B$ if and only if every element of A is an element of B and vice versa. $\mathbb{N}$, $\mathbb{Q}$, and $\mathbb{R}$ denote the sets of natural numbers, rational numbers, and real numbers, respectively. The empty set, i.e. the set with no elements, is denoted by $\emptyset$.

Given a set A and a property P, the notation $\{x \in A \mid P(x)\}$ denotes the subset of A consisting of those elements that satisfy P. This operation is called *separation* because we are separating the elements from a previously defined set, A. A restriction of this sort is needed to avoid Russell's paradox: if we can form the set $S = \{x \mid x \notin x\}$, we have the paradoxical conclusion that S is an element of itself if and only if it isn't. The example also shows that the separation axiom implies that there is no set of all sets: if A is such a set, then $\{x \in A \mid x \notin x\}$ is just $\{x \mid x \notin x\}$.

If A and B are sets, $A \cup B$ denotes their *union*, i.e. the set of elements in either, $A \cap B$ denotes their *intersection*, i.e. the set of elements in both, and $A \setminus B$ denotes the set of element of *set difference* of A and B, i.e. the set of elements of A that are not in B. A and B are said to be *disjoint* if their intersection is empty, i.e. $A \cap B = \emptyset$. When a set B is understood to be a subset of some ambient universe U, then $\overline{B}$, the *complement* of B, is defined to be $U \setminus B$. For example, if B is introduced as a set of natural numbers, $\overline{B}$ is $\mathbb{N} \setminus B$.

If $\mathcal{A}$ is a collection of sets, $\bigcup \mathcal{A}$ and $\bigcap \mathcal{A}$ denote the union and intersection, respectively, of all the sets in $\mathcal{A}$. Similarly, if $\{A_i\}$ is a collection of sets indexed by elements of a set I, $\bigcup_{i \in I} A_i$ and $\bigcap_{i \in I} A_i$ denote their union and intersection, respectively. While $\bigcup \emptyset$ is clearly the empty set, defining $\bigcap \emptyset$ is problematic. When context makes it clear that we are dealing with subsets of an ambient universe, U, $\bigcap \emptyset$ is equal to U.

If A is any set, $\mathcal{P}(A)$, the *power set* of A, denotes the set of all subsets of A. If A and B are sets, $A \times B$, the *cartesian product* of A and B, is the set of all ordered pairs (a, b) consisting of an element $a \in A$ and an element $b \in B$. If c is an ordered pair, $(c)_0$ denotes the first element of c and $(c)_1$ denotes the second element. Iterated pairing leads to notions of ordered triples, quadruples, and so on. By convention we take the products in $A \times B \times C$ to associate to the right. If c is an element of this product and the context is understood, it is also conventional to write $(c)_0$, $(c)_1$, and $(c)_2$ for the elements of c. The n-fold cartesian product of A with itself, that is, the set of all n-tuples from A, is written A^n.

If A and B are sets, a binary relation on A and B is simply a subset $R \subseteq A \times B$, though we usually write $R(x, y)$ or $x \, R \, y$ instead of $(x, y) \in R$. Similar conventions hold for ternary relations and so on. A *function* from A to B is a binary relation f on A and B such that for every x in A there is exactly one y in B such that $x f y$. Of course, we write $f(x) = y$ instead. I will write $f : A \to B$ to indicate that f is a function from A to B, in which case A is called the *domain* of f and B is called the *codomain*. The *image* of f is the set of elements y in B such that for some x in A, $f(x) = y$. (The word *range* is sometimes used to refer to the codomain of a function and sometimes used to refer to its image, so we will avoid it here.) If f and g are two functions from A to B, then $f = g$ if and only if $f(x) = g(x)$ for every x in A. A function f is said to be *surjective*, or *onto*, if its image is its codomain; in other words, for every y in B, there is some x in A such that $f(x) = y$. f is said to be *injective*, or *one to one*, if no two elements get mapped to the same element, i.e. if $f(x_0) = f(x_1)$ implies $x_0 = x_1$ for every x_0 and x_1 in A. f is said to be *bijective*, or *a one-to-one correspondence*, if it is both injective and surjective.

The set of functions from A to B is denoted B^A. A function $f(x, y, z)$ of multiple arguments can be understood as a function $f : A \times B \times C \to D$, where x, y, and z range over A, B, and C respectively. When we view f as a function of multiple arguments, the number of arguments is called its *arity*.

If f is a function from A to B and g is a function from B to C, then the *composition* $g \circ f$ is the function from A to C defined by $(g \circ f)(x) = g(f(x))$. If A is any set, the function id_A mapping each element x of A to itself is called the *identity* on A. If $f : A \to B$ and $g : B \to A$ are functions and $g \circ f = \mathrm{id}_{\mathrm{A}}$, then g is said to be a *left inverse* to f and f is said to be a *right inverse* to g. If f is both a left and right inverse to g, then f is said to be an *inverse* to g.

We may also consider *partial functions* from a set A to a set B. By definition, these are binary relations f on A and B such that for every x in A there is at most one y in B such that $x f y$. In other words, f is just like a function, except that for some values of x there is no corresponding value of y. This is equivalent to saying that f is an ordinary function from some subset of A to B. If f is a partial function from A to B and x is in A, then $f(x)\downarrow$ means that x is in the domain of f. In this case, we say that $f(x)$ is *defined*. Other conventions for dealing with partial functions are discussed in Section 4.8. Sometimes the word *total* is used to distinguish ordinary functions from those that are only partially defined on their domain, but, by default, the word "function" always means "total function."

There is a problem with the definitions above, in that the codomain of a function, and both the domain and codomain of a partial function, are not determined unambiguously. For example, if $f = \{(x, y) \in \mathbb{N} \times \mathbb{N} \mid y = 2x\}$, then f can be considered a function from $\mathbb{N} \to \mathbb{N}$, a function from $\mathbb{N} \to \mathbb{R}$, a function from $\mathbb{N}$ to the set of even numbers, or a partial function from $\mathbb{R} \to \mathbb{R}$. Thus it is only meaningful to talk about the totality or surjectivity of a partial function relative to a particular domain and codomain. In such cases, the phrase "let f be a partial function from A to B" or the phrase "let $g : A \to B$ be a function" can be taken to fix the choice of domain and codomain.

The set of all *finite sequences* of elements from a set A is denoted by $A^{<\omega}$. This can be defined formally in a number of ways. What is important is that each such sequence s has a *length*, denoted length(s), and an *ith element*, denoted $(s)_i$, for each $i < \text{length}(s)$. Notice that we are reusing notation introduced in the context of ordered pairs and tuples, and, in fact, there is a natural correspondence between the set of finite sequences from A of length n and the n-fold product, A^n. I will use $(a_0, \ldots, a_{n-1})$ to denote the finite sequence of length n whose ith element is a_i, and I wil use () to denote the empty sequence, whose length is 0. If s and t are finite sequences, $s^\frown t$ denotes their concatenation and $s \subseteq t$ means that s is an *initial segment* of t. So $s \subseteq t$ holds if and only if there is an r, possibly empty, such that $s^\frown r = t$. I will often drop brackets in the metatheory and speak informally of a sequence of elements $a_0, \ldots, a_{n-1}$, and I will often use $\vec{a}$ to denote such a sequence.

A.2 Orders and Equivalence Relations

A binary relation R on a set A is said to be

- *reflexive* if $R(x, x)$ holds for every x in A;
- *symmetric* if $R(x, y)$ implies $R(y, x)$ for every x and y in A; and
- *transitive* if $R(x, y)$ and $R(y, z)$ imply $R(x, z)$ for every x, y, z in A.

A *partial order* (or *partial ordering*) is a relation R on a set A that is reflexive, *antisymmetric*, and transitive, where antisymmetry is the property that if $R(x, y)$ and $R(y, x)$, then $x = y$. A *strict partial order* is a relation that is transitive and *asymmetric*, which means that $R(x, y)$ implies not $R(y, x)$. In particular, if R is asymmetric, then $R(x, x)$ does not hold. It is common to use $\leq$ and $\preceq$ to denote partial orders and $<$ and $\prec$ to denote strict partial orders. Partial orders and strict partial orders come in pairs: if $\leq$ is a partial order then the relation $x < y$ which holds when $x \leq y$ and $x \neq y$ is a strict order, and if $<$ is a strict order, the relation $x \leq y$ which holds when $x < y$ or $x = y$ is a partial order. Moreover, these two transformations are inverse to each other.

A partial order $\leq$ on a set A is said to be a *total order*, or a *linear order*, if R additionally satisfies $x \leq y$ or $y \leq x$ for every x and y in A; in other words, every two elements are comparable. The corresponding property for a strict partial order R on a set A is that for every x and y, either $x < y$ or $x = y$ or $y < x$. It is not hard to show that these alternatives are mutually exclusive. As examples, the subset relation $\subseteq$ is a partial order on the set of subsets of any set B but is not a linear order unless B has at most one element. The usual orders $\leq$ on $\mathbb{N}$, $\mathbb{Q}$, $\mathbb{Z}$, or $\mathbb{R}$ are, of course, total orders.

A binary relation $\prec$ on a set A is said to be *well founded* if every nonempty subset $S \subseteq A$ has a $\prec$-minimal element; in other words, there is an x in S such that for no z in S do we have $z \prec x$. Here are two equivalent formulations of being well founded:

1. Suppose $S \subseteq A$ is *inductive*, meaning that whenever y is in S for every $y \prec x$, then x is in S. Then $S = A$.
2. There is no infinite $\prec$-decreasing sequence $x_0 \succ x_1 \succ x_2 \succ \cdots$.

The notion is often used in the case where $\prec$ is a strict partial order. If $\prec$ is a strict total order, then being well founded is equivalent to saying that every nonempty subset S of A has a $\prec$-least element, that is, an element x in A such that $x \preceq y$ for every y in A. In that case, $\prec$ (or the associated partial order, $\preceq$) is said to be a *well-order*.

A relation on a set A that is reflexive, symmetric, and transitive is an *equivalence relation*. Equality is one instance, and for every integer m, so is the relation $x \equiv y \pmod{m}$, which holds if and only if m divides $x - y$. If $\sim$ is an equivalence relation on A, then for each x in A the set $[x] = \{y \in A \mid y \sim x\}$ is called *the equivalence class of* x. It is not hard to show that $x \sim y$ holds if and only if $[x] = [y]$. The set of equivalence classes $\{[x] \mid x \in A\}$ is written $A/\!\sim$ and called the *quotient* of A by $\sim$, or the result of starting with A and *modding out* by $\sim$. For example, the quotient of the integers by equivalence modulo 5 is the set $\{[0], [1], [2], [3], [4]\}$.

A function $f(x_0, \ldots, x_{k-1})$ from a set A to another set B is said to *respect* an equivalence relation $\sim$ on A, or to be a *congruence with respect to* $\sim$, if whenever $x_0 \sim y_0, \ldots, x_{k-1} \sim y_{k-1}$, we have $f(x_0, \ldots, x_{k-1}) = f(y_0, \ldots, y_{k-1})$. Mathematicians say that such a function f *descends* to a function $\hat{f}$ from $A/\!\sim$ to B defined by

$$\hat{f}([x_0], \ldots, [x_{k-1}]) = f(x_0, \ldots, x_{k-1}).$$

Computer scientists say instead that f *lifts* to $\hat{f}$. The fact that f is a congruence is needed to ensure that this expression really defines a function, in the sense that the right hand side does not depend on which representatives of $[x_0], \ldots, [x_{k-1}]$ we choose. A relation $R(x_0, \ldots, x_{k-1})$ is said to respect the equivalence relation if whenever $x_0 \sim y_0, \ldots, x_{k-1} \sim y_{k-1}$, then $R(x_0, \ldots, x_{k-1})$ holds if and only if $R(y_0, \ldots, y_{k-1})$ does. Such a relation R is also called a *congruence* with respect to $\sim$, and it descends to a relation on $A/\!\sim$ via the equivalence

$$\hat{R}([x_0], \ldots, [x_{k-1}]) \Leftrightarrow R(x_0, \ldots, x_{k-1}).$$

Any relation can be extended to a transitive one in a conservative way: if $R(x, y)$ is a binary relation on a set A, let S be the intersection of all transitive relations on A that include R, where we view each binary relation on A as a set of ordered pairs. This is called the *transitive closure of* R. We can show that the transitive closure of R is in fact transitive, and that it is the smallest transitive relation R, in the sense that it is included in every transitive relation extending R. We can also extend any binary relation to the smallest reflexive relation that contains it, or the smallest equivalence relation that contains it, in similar ways.

If $\leq$ is an order on a set A, we obtain an order on the set of finite sequences of elements of A by defining $\sigma \leq \tau$ to mean that either σ is an initial segment of τ or we have $(\sigma)_i \neq (\tau)_i$ for some i and $(\sigma)_i < (\tau)_i$ for the least such i. This is known as the *lexicographic* order.

A.3 Cardinality and Zorn's Lemma

Two sets A and B are said to have the *same cardinality* if there is a bijection between them. This is clearly an equivalence relation. In classical set theory with the axiom of choice, it is possible to define a unique representative of each equivalence class, and $|A|$ is used to denote the one with the same cardinality as A. Thus A and B have the same cardinality if and only if $|A| = |B|$. Such representatives are known as *cardinals* or *cardinal numbers*, and they are usually denoted by letters like κ and λ.

A set A is said to be *finite* if it has the same cardinality as $\{0, \ldots, n-1\}$ for some natural number n. In fact, in set theory, the natural numbers are usually defined so that n is the same object as $\{0, \ldots, n-1\}$, which is, in turn, a cardinal. So it makes sense to say that A has cardinality n. A set is *countably infinite* if it has the same cardinality as the natural numbers, $\mathbb{N}$, and a set is *countable* if it is finite or countably infinite.

Proposition A.3.1. *Let A be any set. The following are equivalent:*

1. *A is countable.*
2. *A is empty or there is a surjective function* $f : \mathbb{N} \to A$.
3. *There is an injective function* $f : A \to \mathbb{N}$.

Proposition A.3.2. *The class of countable sets enjoys the following closure properties:*

- *Any subset of a countable set is countable.*
- *If A and B are countable, then so is* $A \times B$.
- *If I is countable and for each i,* A_i *is countable, then* $\bigcup_i A_i$ *is countable.*

The second fact in Proposition A.3.2 follows from the fact that $\mathbb{N} \times \mathbb{N}$ is countable. Cantor's pairing function, discussed in Section 8.3, provides an explicit bijection. If, for each $i \in \mathbb{N}$, f_i is a surjection from $\mathbb{N}$ to a set A_i, then the map $(i, j) \mapsto f_i(j)$ is a surjection from $\mathbb{N} \times \mathbb{N}$ to $\bigcup_i A_i$. The third fact follows from this.

These closure properties imply that for every countable set A and natural number n, the set A^n of n-tuples of elements of A is countable, and hence the set $\bigcup_i A^i$ of all finite tuples of elements of A is countable. The same properties can be used to show that lots of other mathematical sets are countable. For example, we can represent any integer by a pair of natural numbers (i, j), where taking $i = 0$ represents the integer $+j$ and $i = 1$ represents the integer $-j$. This provides an injection of $\mathbb{Z}$ into $\mathbb{N} \times \mathbb{N}$, so $\mathbb{Z}$ is countable. Similarly, the map from $\mathbb{Z} \times \mathbb{Z}$ that takes (x, y) to x/y if $y \neq 0$ and 0 otherwise is a surjection from $\mathbb{Z} \times \mathbb{Z}$ to $\mathbb{Q}$, so $\mathbb{Q}$ is also countable. Indeed, the various representation schemes described in Chapters 8 and 10 can be seen as computationally explicit ways of showing that, in each case, the set of objects being represented is a countable set.

In particular, if L is a language with countably many constant, function, and relation symbols, there are only countably many terms of L, formulas of L, proofs in L, and so on. The construction behind the proof of the completeness theorem in Section 5.1 shows that any consistent set of sentences in L has a model with a countable universe. When approaching logic from a syntactic perspective, there is usually no harm in restricting attention to countable languages and sets. This explains the focus on countable languages in this book.

Nonetheless, many common mathematical structures like the real numbers are uncountable, and, from a semantic perspective, it is important to know how cardinality considerations bear on questions involving definability, or involving the existence of structures with certain properties. For that purpose, it is often useful to consider first-order languages with arbitrary sets of symbols, and to consider generalizations of first-order logic, say, with infinite conjunctions and disjunctions.

Many constructions described in this book take advantage of countability to carry out a construction in finite stages $s_0, s_1, s_2, \ldots$. In such arguments, countability ensures that everything that needs to happen in the construction happens at one of these stages. Zorn's lemma provides a powerful means of generalizing such constructions to higher cardinalities. Let $(P, \leq)$ partial order. A *chain* in P is a subset $C \subseteq P$ that is totally ordered. An element x in P is an *upper bound* for C if $y \leq x$ for every $y \in C$. Finally, a *maximal element* of P is an element x in P such that there is nothing strictly greater than it; in other words, if $y \geq x$, then $y = x$.

Theorem A.3.3 (Zorn's lemma). *Let* $(P, \leq)$ *be partial order, and suppose every chain in P has an upper bound. Then P has a maximal element.*

Zorn's lemma is a consequence of the axiom of choice, but, intuitively, the proof goes something like this. Let a_0 be any element of P. If a_0 is not maximal, pick $a_1 > a_0$. If that is not maximal, pick $a_2 > a_1$. This process may not terminate at any finite stage, and so yield a chain $a_0, a_1, a_2, \ldots$. The hypothesis says we can keep going and find an element a_ω bounding those. If that is not maximal, we continue to choose $a_{\omega+1} > a_\omega$, and so on. Eventually, we have to obtain a maximal element, because ultimately we will run out of elements of P.

Zorn's lemma can be applied in situations where we can describe a construction such that

- at each stage, if we haven't finished yet, there is something more we can do, and
- given any unending sequence of construction steps, there is some way of putting them together and continuing on.

When such a construction does not stop at any finite stage, it is known as a *transfinite* construction.

Cantor's *diagonal argument* provides a flexible method of showing the existence of uncountable sets.

Theorem A.3.4. *For any set A, $\mathcal{P}(A)$ does not have the same cardinality as A.*

Proof Let f be any function from A to $\mathcal{P}(A)$. It suffices to show that f is not surjective. Define $B \subseteq A$ by $B = \{x \in A \mid x \notin f(x)\}$. B is not in the image of f: if $B = f(a)$ for some a in A, then, by the definition of B, we have that a is in $f(a)$ if and only if a is not in $f(a)$, a contradiction. □

Imagine a grid with rows and columns indexed by A. Given f from A to $\mathcal{P}(A)$, put a tick mark at row x and column y if y is in $f(x)$. Then the diagonal of the grid contains a tick mark for every x such that $x \in f(x)$, and B is the complement of that diagonal. We can think of the construction as follows: to obtain a set B that is not in the image of f, we have a set of A-many conditions to satisfy, namely, $B \neq f(x)$ for each $x \in A$. We satisfy the condition for x by putting x in B if and only if it is not in $f(x)$. The diagonalization arguments in Sections 8.6 and 10.5, as well as the proofs of Gödel's incompleteness theorem in Chapter 12, can be seen as variations on this idea.

Note that the function that maps x to $\{x\}$ is an injective function from A to $\mathcal{P}(A)$, so in a sense, A is smaller than $\mathcal{P}(A)$. The Cantor–Schröder–Bernstein theorem says that if there is an injective function from A to B and also an injective function from B to A then there is a bijection between them. It can be proved without the axiom of choice.

A.4 Topology

Let X be any set. A *topology on X* is a collection $\mathcal{T}$ of subsets of X, called *open sets*, satisfying the following:

- The empty set, $\emptyset$, and X itself are elements of $\mathcal{T}$.
- An arbitrary union of open sets is open. In other words, if $\mathcal{A} \subseteq \mathcal{T}$ then $\bigcup \mathcal{A} \in \mathcal{T}$.
- The intersection of any two (and hence any finite number of) open sets is open. In other words, if A and B are elements of $\mathcal{T}$, then so is $A \cap B$.

A *topological space* is a pair $(X, \mathcal{T})$ consisting of a set X and a topology $\mathcal{T}$ on X. Intuitively, if an element $x \in X$ is in an open set U, then x is comfortably inside of U, in the sense that some *neighborhood* N of x is a subset of U. This can be made precise by defining a neighborhood N of x to be any set that includes an open set that contains x.

Here are two important examples of topological spaces:

- Let $(X, <)$ by any strict linear order, and say a set U is open if every x in U is contained in an open subinterval $(a, b) \subseteq U$. This is called the *order topology* on $(X, <)$.
- Let (X, d) be a metric space, where $d(x, y)$ denotes the distance from x to y. For any $\varepsilon > 0$, define the *open ball* $B(x, \varepsilon)$ around any point x to be the set $\{y \in X \mid d(x, y) < \varepsilon\}$. Say a set U is open if for every $x \in U$, there is a ball $B(x, \varepsilon) \subseteq U$. This is called the *metric space topology* on (X, d).

The standard topology on the real numbers is an instance of order topology with the usual order on the real numbers. It is also an instances of the metric space topology with the distance function $d(x, y) = |x - y|$.

Many important notions of analysis can be defined in topological terms. A function $f: X \to Y$ between two topological spaces X and Y is said to be *continuous at* x if the preimage $f^{-1}(N)$ of any neighborhood of N of $f(x)$ is a neighborhood of X. This is equivalent to saying that for every open set V containing $f(x)$, there is an open set U containing x such that $f(U) \subseteq V$. For the topology on the real numbers, this is equivalent to the familiar definition: for every $\varepsilon > 0$, there is a $\delta > 0$ such that for every y, if $|x - y| < \delta$, then $|f(x) - f(y)| < \varepsilon$.

A *basis* $\mathcal{B}$ for a topology on X is defined to be a collection of subsets of X such that $\bigcup \mathcal{B} = X$ and for every B_1 and B_2 in $\mathcal{B}$, there is a set $B_3 \subseteq B_1 \cap B_2$ such that B_3 is in $\mathcal{B}$. Such a basis *generates* the topology $\mathcal{T}$ defined by taking the open sets U to be arbitrary unions of basis elements. Equivalently, U is open in the topology generated by $\mathcal{B}$ if for every x in U there is a basis element B such that $x \in B \subseteq U$. The set of open intervals in any linear order form a basis, as well as the set of open balls in any metric space, giving rise to the examples above.

In addition to the usual topology on the reals, the topologies on $\{0, 1\}^{\mathbb{N}}$ and $\mathbb{N}^{\mathbb{N}}$ play a prominent role in this book. If Σ is any set, then $\Sigma^{\mathbb{N}}$ denotes the set of sequences $(x_i)_{i \in \mathbb{N}}$ of elements of Σ. For any finite sequence σ of elements of Σ, let $[\sigma] = \{f \in \Sigma^{\omega} \mid f \supset \sigma\}$ be the collection of infinite sequences that have σ as an initial segment. The collection of sets $[\sigma]$ is the basis for a topology on $\Sigma^{\mathbb{N}}$. The space $\{0, 1\}^{\mathbb{N}}$ with this topology is known as *Cantor space* and is often also denoted by 2^{ω}. The space $\mathbb{N}^{\mathbb{N}}$ with this topology is known as *Baire space* and is often denoted by ω^{ω}.

If $(X, \mathcal{T})$ is a topological space and S is a subset of X, then S has the *subspace topology* or *induced topology* in which a subset U of S is deemed to be open and only if there is a subset U' of X, open in the topology on X, such that $U = S \cap U'$. For example, $[0, 1/2)$ is an open subset of $[0, 1]$ in the topology induced by the usual topology on the real numbers because it is equal to $[0, 1] \cap (-1, 1/2)$.

In many applications in analysis, the *compact* sets have properties that are analogous to finite sets in discrete mathematics. Given a topological space $(X, \mathcal{T})$, a set $C \subseteq X$ is said to be compact if every covering of C by open sets has a finite subcover. In other words, C is compact if whenever $\mathcal{A}$ is a collection of open sets and $C \subseteq \bigcup \mathcal{A}$, there is a finite subset $\mathcal{B} \subseteq \mathcal{A}$ such that $C \subseteq \bigcup \mathcal{B}$.

The topics surveyed here belong to *point set topology*. Munkres (2000) provides a nice introduction. Cantor space and Baire space are studied in *descriptive set theory*, which is presented in Kechris (1995) and Moschovakis (2009).

A.5 Category Theory

Consider the collection of all groups and homomorphisms between them. If f is a homomorphism from G to H and g is a homomorphism from H to K, then the composition $g \circ f$ is a homomorphism from G to K. Composition is associative, and for any group G there is an identity homomorphism id from G to G, which satisfies $f \circ \mathrm{id} = f$ for any homomorphism f with domain G and $\mathrm{id} \circ g = g$ for any homomorphism g with codomain G.

The notion of a *category* is an abstraction of this structure. A category consists of:

- a set Obj of *objects*,
- for any two objects a and b a set $\mathrm{Hom}(a, b)$ of *morphisms* or *arrows* between them,
- an identity morphism 1_a for any object a, and
- a composition operation taking an element of $\mathrm{Hom}(a, b)$ and an element of $\mathrm{Hom}(b, c)$ to an element of $\mathrm{Hom}(a, c)$, for any objects a, b, and c.

We write the composition of $f \in \mathrm{Hom}(a, b)$ and $g \in \mathrm{Hom}(b, c)$ as $g \circ f$, and we require that composition is associative and that we have $f \circ 1_a = 1_b \circ f = f$ for any $f \in \mathrm{Hom}(a, b)$.

In set theory, the collection of all groups cannot be the set of objects of a category, because it is too big to be a set. But the set of all groups whose carriers inhabit some large, fixed universe of sets forms a category, and using considerations like this, foundational concerns can be managed appropriately.

Many important mathematical constructions can be characterized in category-theoretic terms. For example, products $A \times B$ in the category of groups (or sets) are characterized uniquely up to isomorphism by the property that there are projections $\pi_0 \colon A \times B \to A$ and $\pi_1 \colon A \times B \to B$ such that for any pair of morphisms $f \colon C \to A$ and $g \colon C \to B$ there is a unique morphism $(f, g) \colon C \to A \times B$ such that $f = \pi_0 \circ (f, g)$ and $g = \pi_1 \circ (f, g)$. This is depicted in the diagram on the left in Figure A.1. Dually, the *coproduct* $A \oplus B$ of two groups or sets is characterized by the fact that there are insertion maps $\iota_0 \colon A \to A \oplus B$ and $\iota_1 \colon B \to A \oplus B$ such that any pair of morphisms $f \colon A \to C$ and $g \colon B \to C$ gives rise to a unique morphism $f \oplus g$ such that $f = (f \oplus g) \circ \iota_0$ and $g = (f \oplus g) \circ \iota_1$.

Although it is not used explicitly in this book, category theory offers valuable algebraic perspectives on a number of the central themes. For example, systems like the simply typed lambda calculus have natural category-theoretic interpretations whereby types are interpreted as objects in a suitable category and terms are interpreted as morphisms. Similarly, intuitionistic theories often have natural category-theoretic interpretations. A *functor* F between two categories $\mathcal{C}$ and $\mathcal{D}$ consists of pair of maps, one that takes any object a in $\mathcal{C}$ to an object $F(a)$ in $\mathcal{D}$, and another that takes a morphism $f \colon a \to b$ to a morphism

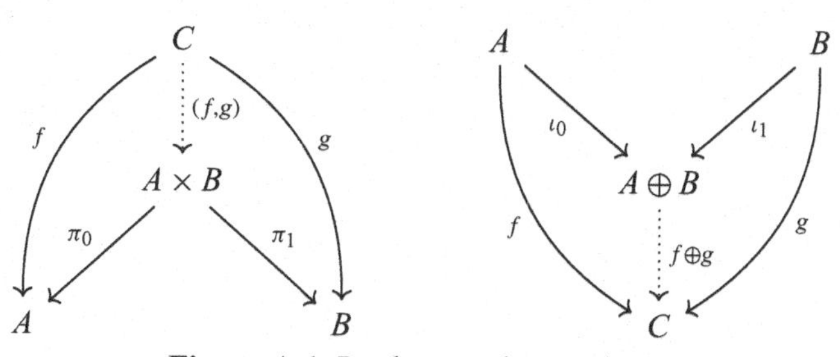

Figure A.1 Products and coproducts.

$F(f)\colon F(a) \to F(b)$, in a manner that preserves the identities and respects composition. A partial order $(A, \leq)$ can be viewed as a category in which there is at most one morphism between two elements a and b, whereby the existence of such a morphism indicates $a \leq b$. The assignment which takes each node of a first-order Kripke structure to the interpretation of the universe at that node is an instance of a functor, and the interpretation of functions and relation can also be expressed in category-theoretic terms. The notion of a *cover* introduced in Section 3.5 is a special case of a basis for a Grothendieck topology, an important notion in category theory. The natural deduction rules for $\wedge$ and $\to$ reflect the fact that products and exponents can be viewed as *adjoint* operations in a categorical interpretation, and the quantifier rules can also be interpreted in those terms.

In short, many of the semantic notions considered here have category-theoretic generalizations, and these provide an algebraic perspective that is midway between the syntactic and model-theoretic perspectives offered in this book. For more information, see Awodey (2010), Lambek and Scott (1988), Mac Lane and Moerdijk (1992) and the bibliographical notes to Chapters 3, 5, 13, and 15.

References

Aczel, Peter. An introduction to inductive definitions. In Barwise, Jon, editor, *The Handbook of Mathematical Logic*, pages 739–782. North-Holland Publishing Co., Amsterdam, 1977.

Andrews, Peter B. *An Introduction to Mathematical Logic and Type Theory: To Truth through Proof.* Kluwer Academic Publishers, Dordrecht, second edition, 2002.

Aschieri, Federico. On natural deduction for Herbrand constructive logics III: The strange case of the intuitionistic logic of constant domains. In Berardi, Stefano and Miquel, Alexandre, editors, *Classical Logic and Computation*, pages 1–9. Electronic Proceedings in Theoretical Compututer Science (EPTCS), Oxford, UK, 2018.

Avigad, Jeremy. Forcing in proof theory. *The Bulletin of Symbolic Logic*, 10(3):305–333, 2004.

Avigad, Jeremy. Foundations. In Blanchette, Jasmin and Mahboubi, Assia, editors, *Proof Assistants and Their Use in Mathematics and Computer Science*. Springer-Verlag, in press.

Avigad, Jeremy and Brattka, Vasco. Computability and analysis: The legacy of Alan Turing. In Downey, Rod, editor, *Turing's Legacy: Developments from Turing's Ideas in Logic*, pages 1–47. Association for Symbolic Logic, La Jolla, CA; Cambridge University Press, Cambridge, 2014.

Avigad, Jeremy and Feferman, Solomon. Gödel's functional ("Dialectica") interpretation. In Buss, Samuel R., editor, *The Handbook of Proof Theory*, pages 337–405. North-Holland Publishing Co., Amsterdam, 1998.

Avigad, Jeremy and Zach, Richard. The epsilon calculus. In Zalta, Edward N., editor, *The Stanford Encyclopedia of Philosophy*. Metaphysics Research Lab, Stanford University, 2002.

Awodey, Steve. *Category Theory*. Oxford University Press, Oxford, second edition, 2010.

Baader, Franz and Nipkow, Tobias. *Term Rewriting and All That.* Cambridge University Press, Cambridge, 1998.

Baaz, Matthias, Egly, Uwe, and Leitsch, Alexander. Normal form transformations. In Robinson, John Alan and Voronkov, Andrei, editors, *Handbook of Automated Reasoning*, pages 273–333. Elsevier and MIT Press, Amsterdam, 2001.

Baaz, Matthias, Hetzl, Stefan, and Weller, Daniel. On the complexity of proof deskolemization. *The Journal of Symbolic Logic*, 77(2):669–686, 2012.

Bachmair, Leo and Ganzinger, Harald. Resolution theorem proving. In Robinson, John Alan and Voronkov, Andrei, editors, *Handbook of Automated Reasoning*, pages 19–99. Elsevier and MIT Press, Amsterdam, 2001.

Balbes, Raymond and Dwinger, Philip. *Distributive Lattices*. University of Missouri Press, Columbia, MO, 1974.

Barendregt, Henk. *The Lambda Calculus: Its Syntax and Semantics*. North-Holland Publishing Co., Amsterdam and New York, 1981.

Barendregt, Henk, Dekkers, Wil, and Statman, Richard. *Lambda Calculus with Types*. Association for Symbolic Logic, Ithaca, NY; Cambridge University Press, Cambridge, 2013.

Barwise, Jon, editor. *The Handbook of Mathematical Logic*. North-Holland Publishing Co., Amsterdam, 1977.

Basu, Saugata, Pollack, Richard, and Roy, Marie-Françoise. *Algorithms in Real Algebraic Geometry*. Springer-Verlag, Berlin, second edition, 2006.

Beeson, Michael J. *Foundations of Constructive Mathematics*. Springer-Verlag, Berlin, 1985.

Bell, John Lane and Slomson, Alan B. *Models and Ultraproducts: An Introduction*. North-Holland Publishing Co., Amsterdam and London, 1969.

Bergman, George M. *An Invitation to General Algebra and Universal Constructions*. Springer-Verlag, Cham, second edition, 2015.

Biere, Armin, Heule, Marijn, van Maaren, Hans, and Walsh, Toby, editors. *Handbook of Satisfiability*. IOS Press, Amsterdam, second edition, 2021.

Bishop, Errett and Bridges, Douglas. *Constructive Analysis*. Springer-Verlag, Berlin, 1985.

Blanchette, Jasmin and Mahboubi, Assia, editors. *Proof Assistants and Their Use in Mathematics and Computer Science*. Springer-Verlag, in press.

Blekherman, Grigoriy, Parrilo, Pablo A., and Thomas, Rekha R., editors. *Semidefinite Optimization and Convex Algebraic Geometry*. Society for Industrial and Applied Mathematics (SIAM), Philadelphia, PA, 2013.

Bochnak, Jacek, Coste, Michel, and Roy, Marie-Françoise. *Real Algebraic Geometry*. Springer-Verlag, Berlin, 1998.

Boolos, George. *Logic, Logic, and Logic*. Harvard University Press, Cambridge, MA, 1998.

Börger, Egon, Grädel, Erich, and Gurevich, Yuri. *The Classical Decision Problem*. Springer-Verlag, Berlin, 2001.

Boyer, Carl B. *The History of the Calculus and Its Conceptual Development*. Dover Publications, Inc., New York, 1959.

Bradley, Aaron R. and Manna, Zohar. *The Calculus of Computation: Decision Procedures with Applications to Verification*. Springer-Verlag, Berlin, 2007.

Burr, Wolfgang. Fragments of Heyting arithmetic. *The Journal of Symbolic Logic*, 65(3):1223–1240, 2000.

Buss, Samuel R., editor. *The Handbook of Proof Theory*. North-Holland Publishing Co., Amsterdam, 1998a.

Buss, Samuel R. First-order proof theory of arithmetic. In Buss, Samuel R., editor, *The Handbook of Proof Theory*, pages 79–147. North-Holland Publishing Co., Amsterdam, 1998b.

Buss, Samuel R. An introduction to proof theory. In Buss, Samuel R., editor, *The Handbook of Proof Theory*, pages 1–78. North-Holland Publishing Co., Amsterdam, 1998c.

Butz, Carsten and Johnstone, Peter. Classifying toposes for first-order theories. *Annals of Pure and Applied Logic*, 91(1):33–58, 1998.

Carneiro, Mario. The Type Theory of Lean. Master's thesis, Carnegie Mellon University, 2019.

Chang, Chen Chung and Keisler, H. Jerome. *Model Theory*. North-Holland Publishing Co., Amsterdam, third edition, 1990.

Charguéraud, Arthur. The locally nameless representation. *Journal of Automated Reasoning*, 49(3): 363–408, 2012.

Cook, Stephen and Urquhart, Alasdair. Functional interpretations of feasibly constructive arithmetic. *Annals of Pure and Applied Logic*, 63(2):103–200, 1993.

Cooper, S. Barry. *Computability Theory*. Chapman & Hall/CRC Press, Boca Raton, FL, 2004.

Coquand, Thierry and Hofmann, Martin. A new method for establishing conservativity of classical systems over their intuitionistic version. *Mathematical Structures in Computer Science*, 9(4):323–333, 1999.

Cutland, Nigel. *Computability: An Introduction to Recursive Function Theory*. Cambridge University Press, Cambridge and New York, 1980.

Davey, Brian A. and Priestley, Hilary A. *Introduction to Lattices and Order*. Cambridge University Press, New York, second edition, 2002.

Davis, Martin. *Computability and Unsolvability*. McGraw-Hill, New York, Toronto, and London, 1958.

Davis, Martin. *The Undecidable: Basic Papers on Undecidable Propositions, Unsolvable Problems and Computable Functions*. Dover Publications, Inc. New york, 2004.

Devlin, Keith. *The Joy of Sets: Fundamentals of Contemporary Set Theory*. Springer-Verlag, New York, second edition, 1993.

Diamondstone, David E., Dzhafarov, Damir D., and Soare, Robert I. Π^0_1 classes, Peano arithmetic, randomness, and computable domination. *Notre Dame Journal of Formal Logic*, 51(1):127–159, 2010.

Diener, Hannes. *Constructive Reverse Mathematics*. Habilitation, Universität Siegen, 2018.

Diller, Justus and Nahm, Werner. Eine Variante zur Dialectica-Interpretation der Heyting-Arithmetik endlicher Typen. *Archiv für Mathematische Logik und Grundlagenforschung*, 16(1):49–66, 1974.

Downey, Rodney G. and Hirschfeldt, Denis R. *Algorithmic Randomness and Complexity*. Springer-Verlag, New York, 2010.

Dyckhoff, Roy. Contraction-free sequent calculi for intuitionistic logic. *The Journal of Symbolic Logic*, 57(3):795–807, 1992.

Dyckhoff, Roy. Contraction-free sequent calculi for intuitionistic logic: A correction. *The Journal of Symbolic Logic*, 83(4):1680–1682, 2018.

Ebbinghaus, Heinz-Dieter and Flum, Jörg. *Finite Model Theory*. Springer-Verlag, Berlin, enlarged edition, 2006.

Enderton, Herbert B. *Elements of Set Theory*. Academic Press [Harcourt Brace Jovanovich, Publishers], New York and London, 1977.

Escardó, Martín and Oliva, Paulo. Selection functions, bar recursion and backward induction. *Mathematical Structures in Computer Science*, 20(2):127–168, 2010.

Escardó, Martín and Oliva, Paulo. The Herbrand functional interpretation of the double negation shift. *The Journal of Symbolic Logic*, 82(2):590–607, 2017.

Ewald, William. *From Kant to Hilbert: A Source Book in the Foundations of Mathematics*. Clarendon Press, Oxford, 1996.

Feferman, Solomon. Theories of finite type related to mathematical practice. In Barwise, Jon, editor, *The Handbook of Mathematical Logic*, pages 913–971. North-Holland Publishing Co., Amsterdam, 1977.

Feferman, Solomon. Definedness. *Erkenntnis*, 43(3):295–320, 1995.

Feferman, Solomon. Harmonious logic: Craig's interpolation theorem and its descendants. *Synthese*, 164(3):341–357, 2008.

Feferman, Solomon, Parsons, Charles, and Simpson, Steven G., editors. *Kurt Gödel: Essays for His Centennial*. Association for Symbolic Logic, La Jolla, CA; Cambridge University Press, Cambridge, 2010.

Ferreira, Gilda and Oliva, Paulo. On the relation between various negative translations. In Berger, Ulrich and Schwichtenberg, Helmut, editors, *Logic, Construction, Computation*, pages 227–258. Ontos Verlag, Heusenstamm, 2012.

Fitting, Melvin Chris. *Intuitionistic Logic, Model Theory and Forcing*. North-Holland Publishing Co., Amsterdam and London, 1969.

Franzén, Torkel. *Gödel's Theorem: An Incomplete Guide to Its Use and Abuse*. A K Peters, Ltd., Wellesley, MA, 2005.

Friedman, Harvey. The disjunction property implies the numerical existence property. *Proceedings of the National Academy of Sciences*, 72(8):2877–2878, 1975.

Fujiwara, Makoto and Kohlenbach, Ulrich. Interrelation between weak fragments of double negation shift and related principles. *The Journal of Symbolic Logic*, 83(3):991–1012, 2018.

Gallier, Jean H. On Girard's "candidats de reducibilité." Technical report, University of Pennsylvania, 1989.

Gentzen, Gerhard. *The Collected Papers of Gerhard Gentzen*. North-Holland Publishing Co., Amsterdam and London, 1969. Edited by M. E. Szabo.

Girard, Jean-Yves, Taylor, Paul, and Lafont, Yves. *Proofs and Types*. Cambridge University Press, Cambridge, 1989.

Gödel, Kurt. *Collected Works*. Oxford University Press, New York, 1986–2003. Solomon Feferman et al. eds. Volumes I–V.

Goodstein, Reuben L. *Recursive Number Theory: A Development of Recursive Arithmetic in a Logic-Free Equation Calculus*. North-Holland Publishing Co., Amsterdam, 1957.

Goodstein, Reuben L. *Recursive Analysis*. North-Holland Publishing Co., Amsterdam, 1961.

Görnemann, Sabine. A logic stronger than intuitionism. *The Journal of Symbolic Logic*, 36(2):249–261, 1971.

Griffor, Edward and Rathjen, Michael. The strength of some Martin–Löf type theories. *Archive for Mathematical Logic*, 33(5):347–385, 1994.

Griffor, Edward R., editor. *Handbook of Computability Theory*. North-Holland Publishing Co., Amsterdam, 1999.

Hähnle, Reiner. Tableaux and related methods. In Robinson, John Alan and Voronkov, Andrei, editors, *Handbook of Automated Reasoning*, pages 100–178. Elsevier and MIT Press, Amsterdam, 2001.

Hájek, Petr and Pudlák, Pavel. *Metamathematics of First-Order Arithmetic*. Springer-Verlag, Berlin, 1998. Reprint.

Harper, Robert. *Practical Foundations for Programming Languages.* Cambridge University Press, Cambridge, second edition, 2016.

Harrison, John. *Handbook of Practical Logic and Automated Reasoning.* Cambridge University Press, Cambridge, 2009.

Hindley, J. Roger and Seldin, Jonathan P. *Lambda-Calculus and Combinators: An Introduction.* Cambridge University Press, Cambridge, 2008.

Hirschfeldt, Denis R. *Slicing the Truth: On the Computable and Reverse Mathematics of Combinatorial Principles.* World Scientific, Hackensack, NJ, 2015.

Hodges, Wilfrid. *Model Theory.* Cambridge University Press, Cambridge, 1993.

Ishihara, Hajime. Constructive reverse mathematics: Compactness properties. In Crosilla, Laura and Schuster, Peter, editors, *From Sets and Types to Topology and Analysis*, pages 245–267. Oxford University Press, Oxford, 2005.

Jech, Thomas. *Set theory.* Springer-Verlag, Berlin, 2003. The third millennium edition, revised and expanded.

Jech, Thomas J. *The Axiom of Choice.* North-Holland Publishing Co., Amsterdam and London, 1973.

Joachimski, Felix and Matthes, Ralph. Short proofs of normalization for the simply typed λ-calculus, permutative conversions and Gödel's *T*. *Archive for Mathematical Logic*, 42(1):59–87, 2003.

Johnstone, Peter T. *Stone Spaces.* Cambridge University Press, Cambridge, 1986. Reprint of the 1982 edition.

Kanamori, Akihiro. *The Higher Infinite.* Springer-Verlag, Berlin, second edition, 2003.

Kaye, Richard. *Models of Peano Arithmetic.* The Clarendon Press, Oxford University Press, New York, 1991.

Kechris, Alexander S. *Classical Descriptive Set Theory.* Springer-Verlag, New York, 1995.

Kleene, Stephen Cole. *Introduction to Metamathematics.* D. Van Nostrand Co., New York, 1952.

Kleene, Stephen Cole. Realizability: A retrospective survey. In Mathias, A. R. D and Rogers, H., editors, *Cambridge Summer School in Mathematical Logic*, pages 95–112. Springer-Verlag, Berlin, 1973.

Kneale, William and Kneale, Martha. *The Development of Logic.* Clarendon Press, Oxford, 1962.

Kohlenbach, Ulrich. *Applied Proof Theory: Proof Interpretations and Their Use in Mathematics.* Springer-Verlag, Berlin, 2008.

Krajíček, Jan. *Bounded Arithmetic, Propositional Logic, and Complexity Theory.* Cambridge University Press, Cambridge, 1995.

Kreisel, Georg and Krivine, Jean-Louis. *Elements of Mathematical Logic. Model Theory.* North-Holland Publishing Co., Amsterdam, 1967.

Kroening, Daniel and Strichman, Ofer. *Decision Procedures: An Algorithmic Point of View.* Springer-Verlag, Berlin, second edition, 2016.

Kunen, Kenneth. *Set Theory.* College Publications, London, 2011.

Lambek, Joachim and Scott, Philip J. *Introduction to Higher-Order Categorical Logic.* Cambridge University Press, Cambridge, 1988.

Libkin, Leonid. *Elements of Finite Model Theory.* Springer-Verlag, Berlin, 2004.

Mac Lane, Saunders and Moerdijk, Ieke. *Sheaves in Geometry and Logic.* Springer-Verlag, New York, 1992.

Mancosu, Paolo. *The Philosophy of Mathematical Practice.* Oxford University Press, Oxford, 2008.

Mancosu, Paolo, Zach, Richard, and Badesa, Calixto. The development of mathematical logic from Russell to Tarski, 1900–1935. In Haaparanta, Leila, editor, *The Development of Modern Logic*, pages 318–470. Oxford University Press, Oxford, 2009.

Mancosu, Paolo, Galvan, Sergio, and Zach, Richard. *An Introduction to Proof Theory: Normalization, Cut-Elimination, and Consistency Proofs.* Oxford University Press, Oxford, 2021.

Mansfield, Richard and Weitkamp, Galen. *Recursive Aspects of Descriptive Set Theory.* The Clarendon Press, Oxford University Press, New York, 1985.

Marker, David. *Model Theory: An Introduction.* Springer-Verlag, New York, 2002.

Martin-Löf, Per. *Intuitionistic Type Theory.* Bibliopolis, Naples, 1984.

Mathias, Adrian R. D. The strength of Mac Lane set theory. *Annals of Pure and Applied Logic*, 110(1–3):107–234, 2001.

Matiyasevich, Yuri V. *Hilbert's Tenth Problem*. MIT Press, Cambridge, MA, 1993. Translated from the 1993 Russian original by the author.

McCune, William, Veroff, Robert, Fitelson, Branden, et al. Short single axioms for Boolean algebra. *Journal of Automated Reasoning*, 29(1):1–16, 2002.

Mendelson, Elliott. *Introduction to Mathematical Logic*. CRC Press, Boca Raton, FL, sixth edition, 2015.

Mines, Ray, Richman, Fred, and Ruitenburg, Wim. *A Course in Constructive Algebra*. Springer-Verlag, New York, 1988.

Mints, Grigori. Axiomatization of a Skolem function in intuitionistic logic. In Faller, Martina, Kaufmann, Stefan, and Pauly, Marc, editors, *Formalizing the Dynamics of Information*, pages 105–114. Center for the Study of Language and Information, Stanford, 2000a.

Mints, Grigori. *A Short Introduction to Intuitionistic Logic*. Kluwer Academic/Plenum Publishers, New York, 2000b.

Moschovakis, Joan. Intuitionistic logic. In Zalta, Edward N., editor, *The Stanford Encyclopedia of Philosophy*. Metaphysics Research Lab, Stanford University, 2018.

Moschovakis, Yiannis N. *Descriptive Set Theory*. American Mathematical Society, Providence, RI, second edition, 2009.

Moschovakis, Yiannis N. Kleene's amazing second recursion theorem. *The Bulletin of Symbolic Logic*, 16(2):189–239, 2010.

Munkres, James R. *Topology*. Prentice Hall, Inc., Upper Saddle River, NJ, second edition, 2000.

Nederpelt, Rob and Geuvers, Herman. *Type Theory and Formal Proof*. Cambridge University Press, Cambridge, 2014.

Negri, Sara and von Plato, Jan. *Structural Proof Theory*. Cambridge University Press, Cambridge, 2001.

Negri, Sara and von Plato, Jan. *Proof Analysis*. Cambridge University Press, Cambridge, 2011.

Nies, André. *Computability and Randomness*. Oxford University Press, Oxford, 2009.

Nordström, Bengt, Petersson, Kent, and Smith, Jan M. *Programming in Martin-Löf's Type Theory: An Introduction*. Clarendon Press, New York, 1990.

Odifreddi, Piergiorgio. *Classical Recursion Theory*. North-Holland Publishing Co., Amsterdam, 1989.

Odifreddi, Piergiorgio. *Classical Recursion Theory*. Vol. II. North-Holland Publishing Co., Amsterdam, 1999.

Okada, Mitsuhiro. A uniform semantic proof for cut-elimination and completeness of various first and higher order logics. *Theoretical Computer Science*, 281(1–2):471–498, 2002.

Oliva, Paulo. Unifying functional interpretations: Past and future. In Schroeder-Heister, Peter, Heinzmann, Hodges, Wilfrid, and Bour, Pierre Edouard, editors, *Logic, Methodology and Philosophy of Science*, pages 97–122. College Publications, London, 2014.

Palmgren, Erik. Constructive sheaf semantics. *Mathematical Logic Quarterly*, 43(3):321–327, 1997.

Pierce, Benjamin C. *Types and Programming Languages*. MIT Press, Cambridge, MA, 2002.

Pohlers, Wolfram. *Proof Theory: The First Step into Impredicativity*. Springer-Verlag, Berlin, 2009.

Poonen, Bjorn. Characterizing integers among rational numbers with a universal-existential formula. *American Journal of Mathematics*, 131(3):675–682, 2009.

Poonen, Bjorn. Undecidable problems: A sampler. In Kennedy, Juliette, editor, *Interpreting Gödel*, pages 211–241. Cambridge University Press, Cambridge, 2014.

Pour-El, Marian B. and Richards, J. Ian. *Computability in Analysis and Physics*. Springer-Verlag, Berlin, 1989.

Pudlák, Pavel. The lengths of proofs. In Buss, Samuel R., editor, *The Handbook of Proof Theory*, pages 547–637. North-Holland Publishing Co., Amsterdam, 1998.

Quine, Willard Van Orman. *The Ways of Paradox and Other Essays*. Harvard University Press, Cambridge, MA, 1997.

Rathjen, Michael. Constructive Zermelo-Fraenkel set theory, power set, and the calculus of constructions. In Dybjer, Peter, Lindstrom, Sten, Palmgren, Erik, and Sundholm, Göran, editors, *Epistemology versus Ontology*, pages 313–349. Springer, Dordrecht, 2012.

Robinson, John Alan and Voronkov, Andrei, editors. *Handbook of Automated Reasoning*. Elsevier and MIT Press, Amsterdam, 2001.

Robinson, Julia. Definability and decision problems in arithmetic. *The Journal of Symbolic Logic*, 14(2):98–114, 1949.

Rose, Harvey E. *Subrecursion: Functions and Hierarchies*. The Clarendon Press, Oxford University Press, New York, 1984.

Rotman, Joseph J. *An Introduction to the Theory of Groups*. Springer-Verlag, New York, fourth edition, 1995.

Russell, Bertrand. On denoting. *Mind*, 14(56):479–493, 1905.

Schindler, Ralf. *Set Theory: Exploring Independence and Truth*. Springer-Verlag, Cham, 2014.

Schwichtenberg, Helmut. Proof theory: Some aspects of cut-elimination. In Barwise, Jon, editor, *The Handbook of Mathematical Logic*, pages 867–895. North-Holland Publishing Co., Amsterdam, 1977.

Shapiro, Stewart. Incompleteness, mechanism, and optimism. *The Bulletin of Symbolic Logic*, 4(3): 273–302, 1998.

Shapiro, Stewart, editor. *The Oxford Handbook of Philosophy of Mathematics and Logic*. Oxford University Press, Oxford, 2005.

Shoenfield, Joseph R. *Mathematical Logic*. Association for Symbolic Logic, Urbana, IL, 2001. Reprint of the 1973 second printing.

Sieg, Wilfried. *Hilbert's Programs and Beyond*. Oxford University Press, Oxford, 2013.

Simpson, Stephen G. *Subsystems of Second Order Arithmetic*. Cambridge University Press, Cambridge; Association for Symbolic Logic, Poughkeepsie, NY, second edition, 2009.

Smith, Peter. *An Introduction to Gödel's Theorems*. Cambridge University Press, Cambridge, second edition, 2013.

Smoryński, Craig. Applications of Kripke models. In Troelstra, Anne S., editor, *Metamathematical Investigation of Intuitionistic Arithmetic and Analysis*, pages 324–391. Springer-Verlag, Berlin and New York, 1973.

Smorynski, Craig. The incompleteness theorems. In Barwise, Jon, editor, *The Handbook of Mathematical Logic*, pages 821–865. North-Holland Publishing Co., Amsterdam, 1977.

Smoryński, Craig. The axiomatization problem for fragments. *Annals of Mathematical Logic*, 14(2): 193–221, 1978.

Smullyan, Raymond M. *Gödel's Incompleteness Theorems*. The Clarendon Press, Oxford University Press, New York, 1992.

Smullyan, Raymond M. *First-Order Logic*. Dover Publications, Inc., New York, 1995. Corrected reprint of the 1968 original.

Soare, Robert I. *Recursively Enumerable Sets and Degrees*. Springer-Verlag, Berlin, 1987.

Soare, Robert I. *Turing Computability*. Springer-Verlag, Berlin, 2016.

Sørensen, Morten Heine and Urzyczyn, Pavel. *Lectures on the Curry–Howard Isomorphism*. Elsevier, Amsterdam, 2006.

Takeuti, Gaisi. *Proof Theory*. North-Holland Publishing Co., Amsterdam, second edition, 1987.

Tarski, Alfred. *Undecidable Theories*. North-Holland Publishing Co., Amsterdam, 1968. In collaboration with Andrzej Mostowski and Raphael M. Robinson.

Troelstra, Anne S., editor. *Metamathematical Investigation of Intuitionistic Arithmetic and Analysis*. Springer-Verlag, Berlin and New York, 1973.

Troelstra, Anne S. Realizability. In Buss, Samuel R., editor, *The Handbook of Proof Theory*, pages 407–473. North-Holland Publishing Co., Amsterdam, 1998.

Troelstra, Anne S. and Schwichtenberg, Helmut. *Basic Proof Theory*. Cambridge University Press, Cambridge, second edition, 2000.

Troelstra, Anne S. and van Dalen, Dirk. *Constructivism in Mathematics: An Introduction*. North-Holland Publishing Co., Amsterdam, 1988. Volumes 1 and 2.

Urquhart, Alasdair. The complexity of propositional proofs. *Bulletin of Symbolic Logic*, 1(4):425–467, 1995.

Väänänen, Jouko. Second-order and higher-order logic. In Zalta, Edward N., editor, *The Stanford Encyclopedia of Philosophy*. Metaphysics Research Lab, Stanford University, fall 2020 edition, 2020.

van den Dries, Lou. Alfred Tarski's elimination theory for real closed fields. *The Journal of Symbolic Logic*, 53(1):7–19, 1988.

van Heijenoort, Jean. *From Frege to Gödel: A Sourcebook in Mathematical Logic, 1879–1931*. Harvard University Press, Cambridge, MA, 1967.

Weber, Rebecca. *Computability Theory*. American Mathematical Society, Providence, RI, 2012.

Weihrauch, Klaus. *Computable Analysis: An Introduction*. Springer-Verlag, Berlin, 2000.

Werner, Benjamin. Sets in types, types in sets. In Abdali, Martin and Ito, Takayasu, editors *Theoretical Aspects of Computer Software*, pages 530–546. Springer-Verlag, Berlin, 1997.

Wolfram, Stephen. *Combinators: A Centennial View*. Wolfram Media Inc., Champaign, 2021.

Wong, Tin Lok. Interpreting weak König's lemma using the arithmetized completeness theorem. *Proceedings of the American Mathematical Society*, 144(9):4021–4024, 2016.

Zach, Richard. The practice of finitism: Epsilon calculus and consistency proofs in Hilbert's program. *Synthese*, 137(1–2):211–259, 2003.

Notation

See also the general mathematical conventions and notation described in the Appendix.

Typographical conventions

$\sigma, \tau, \rho, \ldots$	sequences
$L, L_1, L_2, \ldots$	first-order languages
$a, b, c, \ldots$	constants and constant symbols
$f, g, h, \ldots$	functions and function symbols
$\mathrm{length}(\sigma), \ldots$	specific functions
$R, S, T, \ldots$	relations and relation symbols
$x, y, z, \ldots$	variables
$r, s, t, \ldots$	terms
$A, B, C, \ldots$	formulas
$X, Y, Z, \ldots$	second-order variables
$\Gamma, \Delta, \Pi, \ldots$	sets of formulas (finite or infinite)
$\mathfrak{M}, \mathfrak{N}, \mathfrak{A}, \mathfrak{B}, \ldots$	models (also known as structures)
$\alpha, \beta, \gamma, \ldots$	types
NAT, BOOL, ...	specific types
$\Gamma, \Delta, \Pi, \ldots$	contexts

Symbols

(a, b)	an ordered pair
$(a_0, \ldots, a_{n-1})$	an n-tuple or sequence of length n
$s^\frown t$	concatenation of sequences s and t
$s^\frown a$	concatenation of a single element, i.e. $s^\frown(a)$
$T\restriction_\sigma$	the subtree of T rooted a σ
$\sup_{a\in\Sigma} T_a$	the tree whose immediate subtrees are $(T_a)_{a\in\Sigma}$
$\vec{x}$	a tuple of variables $x_0, \ldots, x_{n-1}$
$\vec{t}$	a tuple of terms $t_0, \ldots, t_{n-1}$
$A \wedge B$	conjunction: "A and B"
$A \vee B$	disjunction: "A or B"
$A \to B$	implication: "A implies B"
$\neg A$	negation: "not A"
$\forall x\, A$	the universal quantifier: "for every x, A"

$\exists x\, A$	the existential quantifier: "for some x, A"
$\forall x.\, A$	the universal quantifier, computer science convention
$\exists x.\, A$	the existential quantifier, computer science convention
$\sim A$	negation for formulas in negation normal form
$\vdash A$	A is provable
$\Gamma \vdash A$	Γ proves A
$\Gamma \vdash_C A$	Γ proves A in classical logic
$\Gamma \vdash_I A$	Γ proves A in intuitionistic logic
$\Gamma \vdash_M A$	Γ proves A in minimal logic
Γ, A	$\Gamma \cup \{A\}$
Γ, Δ	$\Gamma \cup \Delta$
$\Gamma \vdash A$	a sequent in natural deduction or the sequent calculus
$\Gamma \vdash \Delta$	a sequent in the classical sequent calculus
A^N	the double-negation translation of A
$[\![A]\!]^v$	the truth value of A under assignment v
$\lvert\mathfrak{A}\rvert$	the universe of the model $\mathfrak{A}$
$f^{\mathfrak{M}}, R^{\mathfrak{M}}$	the interpretations of f and R in $\mathfrak{M}$
$[\![t]\!]^{\mathfrak{A},\sigma}$	the value of t in $\mathfrak{A}$ under assignment σ
$\sigma[x \mapsto a]$	the modification of σ that assigns a to x
$\mathfrak{M} \models A$	$\mathfrak{M}$ satisfies A, i.e. A is true in $\mathfrak{M}$
$\mathfrak{M} \models_\sigma A$	$\mathfrak{M}$ satisfies A under assignment σ
$\models A$	A is valid
$\Gamma \models A$	Γ semantically entails A
$\alpha \Vdash A$	A is true at world α in a Kripke model
$L(\mathfrak{M})$	the extension of language L with constants for $\lvert\mathfrak{M}\rvert$
$\mathrm{Th}(\mathfrak{M})$	the theory of $\mathfrak{M}$
$\mathrm{Mod}(\Gamma)$	the class of models of Γ
$t\downarrow$	t is defined, or t denotes
$t\uparrow$	t is undefined
$s \simeq t$	equality for partial terms, known as *Kleene equality*
$\Gamma \vdash t : \alpha$	t is a term of type α in context Γ
χ_R	the characteristic function of R
$\mathrm{succ}(x)$	the successor function
$p^n_i(x_0, \ldots, x_{n-1})$	the ith projection function
$\mathrm{pred}(x)$	the predecessor function
$x \dot{-} y$	truncated subtraction
$\mathrm{cond}(x, y, z)$	the conditional function
$\exists x < y\, A$	bounded existential quantification
$\forall x < y\, A$	bounded universal quantification
$\min_{x<y} R(x, \vec{z})$	bounded search
$\Sigma_n, \Pi_n, \Delta_n$	classes in the arithmetic hierarchy
$\Sigma_n(S), \Pi_n(S), \Delta_n(S)$	relative definability
PA	Peano arithmetic
HA	Heyting arithmetic
$(\mathrm{I}\Gamma)$	induction for formulas in Γ
$(\mathrm{B}\Gamma)$	collection for formulas in Γ

$I\Delta_0, I\Sigma_n, \ldots$	corresponding subsystems of first-order arithmetic
$\ulcorner A \urcorner$	corner quotes, the numeral representing A
Q	Robinson's theory of arithmetic
$\varphi_0, \varphi_1, \varphi_2, \ldots$	an enumeration of the partial computable functions
$\varphi_e(x) \downarrow = y$	$\varphi_e(x)$ is defined and equal to y
$\mathrm{Univ}(e, x)$	a universal partial computable function
s^m_n	the functions in the s-m-n theorem
$T(e, x, s)$, $U(x)$	Kleene's T and U
$\lambda x.\, t$	lambda abstraction
$t\, s$	application of t to s
$A \oplus B$	the set $\{2n \mid n \in A\} \cup \{2n+1 \mid n \in B\}$
φ^X_e	the eth partial computable function relative to X
$\sigma \subset X$	σ is an initial segment of the set X
$\sigma \subset f$	σ is an initial segment of the function f
$\varphi^\sigma_e(y) \downarrow = z$	$\varphi^X_e(y) \downarrow = z$ witnessed by $\sigma \subset X$
x^α or $x \colon \alpha$	x is a variable of type α
t^α or $t \colon \alpha$	t is a term of type α
$\lambda x.\, t$ or $\lambda x, y, z.\, t$	lambda abstraction
$s\, t$	application
$s \rhd_\beta t$	s β-contracts to t
$s \to_{\beta,1} t$	s β-reduces to t in one step
$s \to_\beta t$	s β-reduces to t
$s \equiv_\beta t$	s is β-equivalent to t
$s \to_{\beta\eta} t$	s $\beta\eta$-reduces to t
$s \to_w t$	s weakly reduces to t
$s \to t$	s reduces to t in a rewrite system
$\lambda\beta$	the equational theory of the lambda calculus
$\lambda\beta\eta$	the equational theory with η
CL	the equational theory of combinatory logic
(ext)	the axiom of extensionality
HRO	the hereditarily recursive operations
HEO	the hereditarily effective operations
T or PRA^ω	Gödel's theory T
HA^ω	finite type Heyting arithmetic
PA^ω	finite type Peano arithmetic
$e\, \mathbf{r}\, A$	e realizes A
$e\, \mathbf{rt}\, A$	realizability with truth
$e\, \mathbf{mr}\, A$	modified realizability
$e\, \mathbf{mrt}\, A$	modified realizability with truth
(CT)	Church's thesis
(IP)	independence of premise
(MP)	Markov's principle
(AC)	the axiom of choice
PROP	the type of propositions
TYPE	the type of all (small) types
Σ^0_n, Π^0_n, Δ^0_n	the arithmetic part of the analytic hierarchy

Σ^1_n, Π^1_n, Δ^1_n classes in the analytic hierarchy

Σ^i_n, Π^i_n, Δ^i_n boldface definability classes

Wo set of computable well-orders of $\mathbb{N}$

WO set of well-orders of $\mathbb{N}$

$\Lambda X.\, t$ second-order lambda abstraction

F polymorphic second-order lambda calculus

(RCA) recursive comprehension axiom

(WKL) weak König's lemma

(ACA) arithmetic comprehension axiom

(ATR) arithmetic transfinite recursion

(Π^1_1-CA) Π^1_1 comprehension axiom

RCA_0, etc. subsystems of second-order arithmetic

εP the Hilbert epsilon function

ιP Russell's iota

$\Pi x \colon \alpha.\, \beta$ a Pi type

$\Sigma x \colon \alpha.\, \beta$ a Sigma type

Index

Printed in the United States
by Baker & Taylor Publisher Services